Advances in

BOTANICAL RESEARCH

VOLUME 52

Advances in

BOTANICAL RESEARCH

Series Editors

JEAN-CLAUDE KADER — *Laboratoire Physiologie Cellulaire et Moléculaire des Plantes, CNRS, Université de Paris, Paris, France*

MICHEL DELSENY — *Laboratoire Génome et Développement des Plantes, CNRS IRD UP, Université de Perpignan, Perpignan, France*

Advances in

BOTANICAL RESEARCH

Oxidative stress and redox regulation in plants

Editor

JEAN-PIERRE JACQUOT

Nancy University France,
UMR IAM Faculte des Sciences,
Vandoeuvre Cedex, France

VOLUME 52

ELSEVIER

AMSTERDAM • BOSTON • HEIDELBERG • LONDON
NEW YORK • OXFORD • PARIS • SAN DIEGO
SAN FRANCISCO • SINGAPORE • SYDNEY • TOKYO
Academic Press is an imprint of Elsevier

Academic Press is an imprint of Elsevier

32 Jamestown Road, LondonNW17BY, UK
30 Corporate Drive, Suite 400, Burlington, MA 01803, USA
525 B Street, Suite 1900, San Diego, CA 92101-4495, USA
Radarweg 29, PO Box 211, 1000 AE Amsterdam, The Netherlands

First edition 2009

ISBN: 978-0-12-378622-7
ISSN: 0065-2296

Printed and bound by CPI Group (UK) Ltd, Croydon, CR0 4YY

Transferred to Digital Print 2011

CONTENTS

Roles of γ-Glutamyl Transpeptidase and γ-Glutamyl Cyclotransferase in Glutathione and Glutathione-Conjugate Metabolism in Plants

NAOKO OHKAMA-OHTSU, KEIICHI FUKUYAMA AND DAVID J. OLIVER

The Redox State, a Referee of the Legume–*Rhizobia* Symbiotic Game

DANIEL MARINO, CHIARA PUCCIARIELLO, ALAIN PUPPO AND PIERRE FRENDO

Reactive Oxygen Species in *Phanerochaete chrysosporium*: Relationship Between Extracellular Oxidative and Intracellular Antioxidant Systems

MÉLANIE MOREL, ANDREW A. NGADIN, JEAN-PIERRE JACQUOT AND ERIC GELHAYE

Physiological Impact of Thioredoxin- and Glutaredoxin-Mediated Redox Regulation in Cyanobacteria

YOSHITAKA NISHIYAMA AND TORU HISABORI

Use of Transgenic Plants to Uncover Strategies for Maintenance of Redox Homeostasis During Photosynthesis

GUY T. HANKE, SIMONE HOLTGREFE, NICOLAS KÖNIG, INGA STRODTKÖTTER, INGO VOSS AND RENATE SCHEIBE

Redundancy and Crosstalk Within the Thioredoxin and Glutathione Pathways: A New Development in Plants

J.-P. REICHHELD, T. BASHANDY, W. SIALA, C. RIONDET, V. DELORME, A. MEYER AND Y. MEYER

Protein Import in Chloroplasts: An Emerging Regulatory Role for Redox

MÓNICA BALSERA, JÜRGEN SOLL AND BOB B. BUCHANAN

Glutaredoxins in Development and Stress Responses of Plants

SHUTIAN LI AND SABINE ZACHGO

Glutathionylation in Photosynthetic Organisms

XING-HUANG GAO, MARIETTE BEDHOMME, LAURE MICHELET, MIRKO ZAFFAGNINI AND STÉPHANE D. LEMAIRE

Glutaredoxin: The Missing Link Between Thiol-Disulfide Oxidoreductases and Iron Sulfur Enzymes

BENJAMIN SELLES, NICOLAS ROUHIER, KAMEL CHIBANI, JEREMY COUTURIER, FILIPE GAMA AND JEAN-PIERRE JACQUOT

Oxidative Stress and Thiol-Based Antioxidants in Cereal Seeds

PABLO PULIDO, FERNANDO DOMÍNGUEZ AND FRANCISCO JAVIER CEJUDO

Molecular Recognition in NADPH-Dependent Plant Thioredoxin Systems—Catalytic Mechanisms, Structural Snapshots and Target Identifications

PER HÄGGLUND, KRISTINE GROTH KIRKENSGAARD, KENJI MAEDA, CHRISTINE FINNIE, ANETTE HENRIKSEN AND BIRTE SVENSSON

CONTRIBUTORS TO VOLUME 52

MADHOOLIKA AGRAWAL *Laboratory of Air Pollution and Global Climate Change, Ecology Research Circle, Department of Botany, Banaras Hindu University, Varanasi 221005, India*

S. B. AGRAWAL *Laboratory of Air Pollution and Global Climate Change, Ecology Research Circle, Department of Botany, Banaras Hindu University, Varanasi 221005, India*

MÓNICA BALSERA *Departamento de Estrés Abiótico, Instituto de Recursos Naturales y Agrobiología de Salamanca (IRNASA-CSIC), Salamanca, Spain, and Department of Structural Biology, Paul Scherrer Institute, Villigen, Switzerland*

T. BASHANDY *Genome et Developement des Plantes, Universite de Perpignan, CNRS-UP-IRD UMR 5096, Av P. Alduy, F 66860 Perpignan Cedex, France*

MARIETTE BEDHOMME *Institut de Biotechnologie des Plantes, UMR 8618, CNRS, Univ Paris-Sud, Bâtiment 630, 91405 Orsay Cedex, France*

BOB B. BUCHANAN *Department of Plant and Microbial Biology, University of California, Berkeley, USA*

FRANCISCO JAVIER CEJUDO *Instituto de Bioquímica Vegetal y Fotosíntesis, Universidad de Sevilla and CSIC, Avda Américo Vespucio 49, 41092-Sevilla, Spain*

KAMEL CHIBANI *Unité Mixte de Recherches 1136 INRA UHP Interactions Arbre-Microorganismes, IFR 110 EFABA, Ecosystèmes Forestiers, Agroressources, Bioprocédés et Alimentation, Faculté des Sciences, Nancy Université, BP 239, 54506 Vandoeuvre Cedex, France*

JEREMY COUTURIER *Unité Mixte de Recherches 1136 INRA UHP Interactions Arbre-Microorganismes, IFR 110 EFABA, Ecosystèmes Forestiers, Agroressources, Bioprocédés et Alimentation, Faculté des Sciences, Nancy Université, BP 239, 54506 Vandoeuvre Cedex, France*

V. DELORME *Genome et Developement des Plantes, Universite de Perpignan, CNRS-UP-IRD UMR 5096, Av P. Alduy, F 66860 Perpignan Cedex, France*

FERNANDO DOMÍNGUEZ *Instituto de Bioquímica Vegetal y Fotosíntesis, Universidad de Sevilla and CSIC, Avda Américo Vespucio 49, 41092-Sevilla, Spain*

CHRISTINE FINNIE *Enzyme and Protein Chemistry, Department of Systems Biology, Technical University of Denmark, DK-2800 Kgs. Lyngby, Denmark*

PIERRE FRENDO *Interactions Biotiques et Santé Végétale UMR INRA 1301-CNRS 6243-Université de Nice-Sophia Antipolis, 400 Route des Chappes, BP167, 06903 Sophia Antipolis Cedex, France*

KEIICHI FUKUYAMA *Department of Biological Sciences, Graduate School of Science, Osaka University, Toyonaka, Osaka 560-0043, Japan*

FILIPE GAMA *Unité Mixte de Recherches 1136 INRA UHP Interactions Arbre-Microorganismes, IFR 110 EFABA, Ecosystèmes Forestiers, Agroressources, Bioprocédés et Alimentation, Faculté des Sciences, Nancy Université, BP 239, 54506 Vandoeuvre Cedex, France*

XING-HUANG GAO *Institut de Biotechnologie des Plantes, UMR 8618, CNRS, Univ Paris-Sud, Bâtiment 630, 91405 Orsay Cedex, France*

ERIC GELHAYE *IFR 110 Ecosystèmes Forestiers, Agroressources, Bioprocédés et Alimentation, Unité Mixte de Recherches INRA UHP 1136 Interaction Arbres Microorganismes, Université Nancy I BP 239, 54506 Vandoeuvre-lès-Nancy Cedex, France*

PER HÄGGLUND *Enzyme and Protein Chemistry, Department of Systems Biology, Technical University of Denmark, DK-2800 Kgs. Lyngby, Denmark*

KYU-WOONG HAHN *Department of Biological Sciences, Hannam University, Daejeon 306-791, Korea*

GUY T. HANKE *Department of Plant Physiology, University of Osnabrueck, D-49069 Osnabrueck, Germany*

ANETTE HENRIKSEN *Protein Chemistry Group, Carlsberg Laboratory, Gamle Carlsberg Vej 10, DK-2500 Valby, Denmark*

TORU HISABORI *Chemical Resource Laboratory, Tokyo Institute of Technology, 4259 Nagatsuta, Midori-Ku, Yokohama 226-8503, Japan*

SIMONE HOLTGREFE *Department of Plant Physiology, University of Osnabrueck, D-49069 Osnabrueck, Germany*

JEAN-PIERRE JACQUOT *IFR 110 Ecosystèmes Forestiers, Agroressources, Bioprocédés et Alimentation, Unité Mixte de Recherches INRA UHP 1136 Interaction Arbres Microorganismes, Université Nancy I BP 239, 54506 Vandoeuvre-lès-Nancy Cedex, France*

JAE-HEUNG JEON *Plant Systems Engineering Center, Korea Research Institute of Bioscience and Biotechnology (KRIBB), Daejon 305-333, South Korea*

HYOUK JOUNG *Plant Systems Engineering Center, Korea Research Institute of Bioscience and Biotechnology (KRIBB), Daejon 305-333, South Korea*

NICOLAS KÖNIG *Department of Plant Physiology, University of Osnabrueck, D-49069 Osnabrueck, Germany*

HYUN-SOON KIM *Plant Systems Engineering Center, Korea Research Institute of Bioscience and Biotechnology (KRIBB), Daejon 305-333, South Korea*

YOON-SIK KIM *Department of Biological Sciences, Hannam University, Daejeon 306-791, Korea, and Plant Systems Engineering Center, Korea Research Institute of Bioscience and Biotechnology (KRIBB), Daejon 305-333, South Korea*

KRISTINE GROTH KIRKENSGAARD *Enzyme and Protein Chemistry, Department of Systems Biology, Technical University of Denmark, DK-2800 Kgs. Lyngby, Denmark*

STÉPHANE D. LEMAIRE *Institut de Biotechnologie des Plantes, UMR 8618, CNRS, Univ Paris-Sud, Bâtiment 630, 91405 Orsay Cedex, France*

SHUTIAN LI *Department of Botany, Osnabrück University, 49076 Osnabrück, Germany*

KENJI MAEDA *Enzyme and Protein Chemistry, Department of Systems Biology, Technical University of Denmark, DK-2800 Kgs. Lyngby, Denmark*

DANIEL MARINO *Interactions Biotiques et Santé Végétale UMR INRA 1301-CNRS 6243-Université de Nice-Sophia Antipolis, 400 Route des Chappes, BP167, 06903 Sophia Antipolis Cedex, France*

A. MEYER *Heidelberg Institute for Plant Science, Heidelberg University, Im Neuenheimer Feld 360, 69120 Heidelberg, Germany*

Y. MEYER *Genome et Developement des Plantes, Universite de Perpignan, CNRS-UP-IRD UMR 5096, Av P. Alduy, F 66860 Perpignan Cedex, France*

LAURE MICHELET *Institut de Biotechnologie des Plantes, UMR 8618, CNRS, Univ Paris-Sud, Bâtiment 630, 91405 Orsay Cedex, France*

IAN M. MØLLER *Department of Genetics and Biotechnology, Faculty of Agricultural Sciences, Aarhus University, Forsøgsvej 1, DK-4200 Slagelse, Denmark*

MÉLANIE MOREL *IFR 110 Ecosystèmes Forestiers, Agroressources, Bioprocédés et Alimentation, Unité Mixte de Recherches INRA UHP 1136 Interaction Arbres Microorganismes, Université Nancy I BP 239, 54506 Vandoeuvre-lès-Nancy Cedex, France*

ANDREW A. NGADIN *IFR 110 Ecosystèmes Forestiers, Agroressources, Bioprocédés et Alimentation, Unité Mixte de Recherches INRA UHP 1136 Interaction Arbres Microorganismes, Université Nancy I BP 239, 54506 Vandoeuvre-lès-Nancy Cedex, France*

YOSHITAKA NISHIYAMA *Department of Biochemistry and Molecular Biology, Graduate School of Science and Engineering, Saitama University, 255 Shimo-Okubo, Sakura-ku, Saitama 338-8570, Japan*

NAOKO OHKAMA-OHTSU *RIKEN Plant Science Center, 1-7-22 Suehiro-cho, Tsurumi-ku, Yokohama-City, Kanagawa 230-0045, Japan*

DAVID J. OLIVER *Department of Genetics, Development and Cell Biology, Iowa State University, Ames, IA 50011, USA*

CHIARA PUCCIARIELLO *Plant & Crop Physiology Laboratory, Scuola Superiore Santa Anna, Via Mariscoglio 34, 56124 Pisa, Italy*

PABLO PULIDO *Instituto de Bioquímica Vegetal y Fotosíntesis, Universidad de Sevilla and CSIC, Avda Américo Vespucio 49, 41092-Sevilla, Spain*

ALAIN PUPPO *Interactions Biotiques et Santé Végétale UMR INRA 1301-CNRS 6243-Université de Nice-Sophia Antipolis, 400 Route des Chappes, BP167, 06903 Sophia Antipolis Cedex, France*

J.-P. REICHHELD *Genome et Developement des Plantes, Universite de Perpignan, CNRS-UP-IRD UMR 5096, Av P. Alduy, F 66860 Perpignan Cedex, France*

C. RIONDET *Genome et Developement des Plantes, Universite de Perpignan, CNRS-UP-IRD UMR 5096, Av P. Alduy, F 66860 Perpignan Cedex, France*

NICOLAS ROUHIER *Unité Mixte de Recherches 1136 INRA UHP Interactions Arbre-Microorganismes, IFR 110 EFABA, Ecosystèmes Forestiers, Agroressources, Bioprocédés et Alimentation, Faculté des Sciences, Nancy Université, BP 239, 54506 Vandoeuvre Cedex, France*

RENATE SCHEIBE *Department of Plant Physiology, University of Osnabrueck, D-49069 Osnabrueck, Germany*

BENJAMIN SELLES *Unité Mixte de Recherches 1136 INRA UHP Interactions Arbre-Microorganismes, IFR 110 EFABA, Ecosystèmes Forestiers, Agroressources, Bioprocédés et Alimentation, Faculté des Sciences, Nancy Université, BP 239, 54506 Vandoeuvre Cedex, France*

W. SIALA *Genome et Developement des Plantes, Universite de Perpignan, CNRS-UP-IRD UMR 5096, Av P. Alduy, F 66860 Perpignan Cedex, France*

SURUCHI SINGH *Laboratory of Air Pollution and Global Climate Change, Ecology Research Circle, Department of Botany, Banaras Hindu University, Varanasi 221005, India*

JÜRGEN SOLL *Department of Biology I, Botany, Ludwig Maximilians University, Martinsried, Germany*

INGA STRODTKÖTTER *Department of Plant Physiology, University of Osnabrueck, D-49069 Osnabrueck, Germany*

BIRTE SVENSSON *Enzyme and Protein Chemistry, Department of Systems Biology, Technical University of Denmark, DK-2800 Kgs. Lyngby, Denmark*

LEE J. SWEETLOVE *Department of Plant Sciences, University of Oxford, South Parks Road, Oxford OX1 3RB, United Kingdom*

INGO VOSS *Department of Plant Physiology, University of Osnabrueck, D-49069 Osnabrueck, Germany*

SABINE ZACHGO *Department of Botany, Osnabrück University, 49076 Osnabrück, Germany*

MIRKO ZAFFAGNINI *Institut de Biotechnologie des Plantes, UMR 8618, CNRS, Univ Paris-Sud, Bâtiment 630, 91405 Orsay Cedex, France*

PREFACE

Since 1963, *Advances in Botanical Research* has published more than 50 volumes dedicated to plant research. The present volume (#52) is named "Oxidative Stress and Redox Regulation in Plants". It is divided into 15 chapters written by expert scientists from 10 different countries.

The leading articles describe the molecular effects of reactive oxygen species (ROS) and their importance in signalling and control of transcriptional and translational events. The opening chapter by Lee Sweetlove and Ian Max Møller describes the "Oxidation of Proteins in Plants – Mechanisms and Consequences" and shows how ROS affect several macromolecules, and, in particular, proteins via the oxidation of cysteines but also through the carbonylation of many other amino acid residues. The contribution of the Jae-Heung Jeon group in South Korea (Kim *et al.*, Reactive Oxygen Species: Regulation of Plant Growth and Development) shows how ROS can be toxic to the cell but also necessary for signalling. It also details the interconnections between hormones and ROS, a timely topic. The adaptation of plants to UV light is also discussed in relation with the oxidative stress linked to these conditions by S. B. Agrawal, S. Singh and M. Agrawal (Agrawal *et al.*, Ultraviolet-B Induced Changes in Gene Expression and Antioxidants in Plants). This chapter also demonstrates that UV light promotes ROS formation, which in turn induces a modification of the gene transcription pattern in plants.

It is well known that the most important redox players in the cell are glutathione, often linked to glutaredoxins and thioredoxins. The metabolism of glutathione is described in a chapter by Naoko Ohkama-Ohtsu and colleagues (Okhama-Ohtsu, Fukuyama and Oliver, "Roles of γ-Glutamyl Transpeptidase and γ-Glutamyl Cyclotransferase in Glutathione and Glutathione-Conjugate Metabolism in Plants"). While the synthesis of glutathione is a multi-step process that occurs essentially in chloroplast, its degradation takes place outside this organelle in the apoplast and in the cytosol and it requires either γ-glutamyl transpeptidase or γ-glutamyl cyclotransferase and 5-oxoprolinase which are the focus of this chapter. The importance of glutathione as a signal molecule in the *Rhizobium* legume symbiosis is described in a chapter by the group of Pierre Frendo (Marino *et al.*, The Redox State, a Referee of the Legume-*Rhizobia* Symbiotic Game). Abundant evidence, including genetic knockouts, indicates that the redox state is absolutely essential for controlling this type of symbiotic interaction.

The following chapters deal mostly with thioredoxins and glutaredoxins and their importance in plants and fungi. The only one that describes these systems in a non-photosynthetic organism is a contribution from the group of Eric Gelhaye (Morel *et al.*, Reactive Oxygen Species in *Phanerochaete chrysosporium*: Relationship Between Extracellular Oxidative and Intracellular Antioxidant Systems). This chapter describes the ligninolytic enzymes of this wood-degrading fungus and the many systems necessary to control ROS levels in this highly oxidizing extracellular environment. The next chapter by Yoshitaka Nishiyama and Toru Hisabori "Physiological Impact of Thioredoxin- and Glutaredoxin-Mediated Redox Regulation in Cyanobacteria" describes the thioredoxin and glutaredoxin systems in cyanobacteria, the influence of the redox poise on gene transcription and protein synthesis, plus an interesting complement that shows that the protein targets of these systems in cyanobacteria are probably very different from those in land plants.

Moving to land plants, the group of Renate Scheibe (Hanke *et al.*, Use of Transgenic Plants to Uncover Strategies for Maintenance of Redox-Homeostasis During Photosynthesis) offers a very comprehensive analysis of the literature concerning the analysis of redox mechanisms via the generation of transgenic plants. They demonstrate how difficult it is to dissect those mutants, due to the gene redundancy. The generation of multiple mutants is sometimes required, and besides, the fine-tuning of subtle physiological conditions may also be necessary to observe a phenotype. Similar genetic approaches in the laboratory of Yves Meyer (Reichheld *et al.*, Redundancy and Crosstalk Within the Thioredoxin and Glutathione Pathways: A New Development in Plants) have led to the conclusion that the glutaredoxins and thioredoxin systems are interconnected in plants. This was initially suggested by biochemically based experiments and then comforted by analyses of double mutants of the glutathione and thioredoxin pathways and Reichheld *et al.* describe these recent developments. Very recent progress on the protein import mechanisms into organelles is also described by Mónica Balsera, Jürgen Soll and Bob B. Buchanan, "Protein Import in Chloroplasts: An Emerging Regulatory Role for Redox". It appears now pretty clear that the import mechanisms are under redox control, Toc 35, Tic 55 and Tic 110 being the leading candidates.

The next three chapters are devoted to glutaredoxin and glutathionylation in plants. Shutian Li and Sabine Zachgo (Glutaredoxins in Development and Stress Responses of Plants) present remarkable results that implicate glutaredoxins of the CC type in flowering. These findings were absolutely unexpected and the ROXY (CC-type glutaredoxins) story once again shows the combined power of genetics and biochemistry. The contribution from the group of Stéphane D. Lemaire (Gao *et al.*, Glutathionylation in Photosynthetic Organisms) describes the latest developments concerning protein glutathionylation in plants, including the detection of glutathionylated proteins by proteomics and the role of glutaredoxins in deglutathionylation. Finally, the contribution of the group of Nicolas Rouhier (a chapter by Selles *et al.* "Glutaredoxin: The Missing Link Between Thiol-Disulfide Oxidoreductases and Iron Sulfur Enzymes") shows that both iron–sulfur containing (ISC) proteins and glutaredoxins contain CxxC motives in their primary sequence. These motives or their CGFS derivatives are instrumental in glutaredoxins having the capacity to bind ISCs and also to transfer them to apoproteins. These developments are highly exciting as they connect those two separate protein worlds together.

The two last chapters describe thioredoxins from cereals. Studies in the laboratory of Bob Buchanan have shown earlier that thioredoxin is important in germination. In this volume, Pablo Pulido, Fernando Domínguez and Francisco Javier Cejudo have written a chapter entitled "Oxidative Stress and Thiol-Based Antioxidants in Cereal Seeds". They describe the oxidative stress generated during germination and the thiol-based antioxidant system of cereals formed by 1-Cys peroxiredoxin (1-Cys Prx), thioredoxin *h* (Trx *h*) and NADPH-dependent thioredoxin reductase (NTR). Finally, the last chapter is a contribution by the group of Birte Svensson (Hägglund *et al.*, Molecular Recognition in NADPH-Dependent Plant Thioredoxin Systems – Catalytic Mechanisms, Structural Snapshots and Target Identifications). This chapter describes the molecular properties of thioredoxins from barley, with emphasis on structure and proteomic methods that allow the detection of protein targets.

Together with the above-mentioned colleagues, we have tried in this volume to cover many important aspects of ROS impact on plants (generation, signalling, deleterious effects and destruction mechanisms). Quite often, this area of research has led to interesting and unexpected new developments (flowering and glutaredoxins, glutaredoxin–thioredoxin interconnections, redox control of protein import, glutaredoxins and iron–sulfur assembly, etc.). Obviously many other developments are still to come and the field is getting ever more exciting. We hope that this volume will prove useful to researchers and students in this area. The last comment concerning this volume is that coincidentally, as we were closing the receipt of the contributions, Dr. Naoko Ohkama-Ohtsu has given birth to a baby boy, Makoto, on 1 July 2009. All of the authors join me to congratulate the mom and wish a happy and successful life to the new born and to his older sister Yuka.

Jean-Pierre Jacquot

CONTENTS OF VOLUMES 35–51

Series Editor (Volumes 35–44)

J.A. CALLOW
School of Biosciences, University of Birmingham, Birmingham, United Kingdom

Contents of Volume 35

Contents of Volume 36

Contents of Volume 37

ANTHOCYANINS IN LEAVES

Edited by **K. S. Gould and D. W. Lee**

Contents of Volume 38

Contents of Volume 39

Contents of Volume 40

Contents of Volume 41

Contents of Volume 42

Contents of Volume 43

Contents of Volume 44

Series Editors (Volume 45–)

JEAN-CLAUDE KADER
Laboratoire Physiologie Cellulaire et Moléculaire des Plantes, CNRS, Université de Paris, Paris, France
MICHEL DELSENY
Laboratoire Génome et Développement des Plantes, CNRS IRD UP, Université de Perpignan, Perpignan, France

Contents of Volume 45

RAPESEED BREEDING

Contents of Volume 46

Contents of Volume 47

Contents of Volume 48

Contents of Volume 49

Contents of Volume 50

Contents of Volume 51

Oxidation of Proteins in Plants—Mechanisms and Consequences

LEE J. SWEETLOVE*[,1] AND IAN M. MØLLER[†]

*Department of Plant Sciences, University of Oxford, South Parks Road, Oxford OX1 3RB, United Kingdom
[†]Department of Genetics and Biotechnology, Faculty of Agricultural Sciences, Aarhus University, Forsøgsvej 1, DK-4200 Slagelse, Denmark

ABSTRACT

The production of reactive oxygen and reactive nitrogen species in plant cells can lead to a variety of modifications of proteins through oxidation of amino acid side groups. The widespread occurrence of such modifications is becoming appreciated as new proteomic approaches allow their systematic identification. Oxidized amino acid

[1]Corresponding author: E-mail: Lee.sweetlove@plants.ox.ac.uk

Advances in Botanical Research, Vol. 52

0065-2296/09 $35.00
DOI: 10.1016/S0065-2296(09)52001-0

residues can be identified directly by mass spectrometry if the modification is stable, but it is more common to covalently tag the oxidized group by reaction with a marker molecule. The marker molecule generally allows visualization through immunodetection and isolation of modified proteins by affinity purification. Although there are several technical caveats with such approaches, they have been useful in documenting the extent of oxidative modification of proteins and have highlighted a number of proteins where oxidative modification is critical for protein function. A view that such modifications could have signalling ramifications is emerging. However, in many cases there is a lack of information as to the effect of oxidation on protein activity or function. Severe protein oxidation is costly to the cell since oxidatively damaged proteins need to be degraded by specific proteases or damaged cellular components recycled via the autophagy pathway. Avoiding this cost is clearly advantageous, and it has been proposed that proteins may have an over-representation of easily oxidizable amino acids on their surface to act as decoy or sacrificial residues, thus preventing or postponing oxidation of residues more important for the function of the protein.

I. INTRODUCTION

Oxidative stress occurs when the rate of production of reactive oxygen (ROS) and/or reactive nitrogen species (RNS) is greater than the capacity of the cell's antioxidant defences to detoxify them. As a consequence, the extent to which ROS and RNS oxidize key cellular macromolecules is dramatically increased. Some of these oxidation events may prevent the normal functioning of the target macromolecules, and the resulting change in cellular homeostasis is referred to as oxidative stress. Along with lipids, proteins are the key class of macromolecules in the cell that can be oxidized in a way that contributes to cellular oxidative stress. The aim of this chapter is to review recent advances in our understanding of the process of protein oxidation. We will briefly introduce the production of ROS and RNS, summarize the main mechanisms by which proteins can be oxidized, before reviewing recent work providing an overview of the proteins that are targets of oxidation, and the importance of specific oxidation events in the regulation of cellular processes. We will also consider the cost of protein oxidation and describe our current understanding of the mechanisms by which oxidized proteins are processed and removed.

II. THE FORMATION OF ROS AND RNS

The biochemistry of formation of ROS and RNS has been extensively reviewed and we do not wish to reproduce this exhaustive information here. Our aim in this section is to briefly outline the main points and provide the context for the rest of the chapter. For a more detailed account of the chemistry and biochemistry of ROS and RNS formation, the reader should consult Halliwell and Gutteridge's excellent textbook (Halliwell and Gutteridge, 2007). In addition, there are numerous relevant plant-specific reviews that we wholeheartedly recommend (Apel and Hirt, 2004; Delledonne, 2005; del Rio *et al.*, 2006; Halliwell, 2006; Møller, 2001; Møller *et al.*, 2007; Noctor and Foyer, 1998; Noctor *et al.*, 2007; Rinalducci *et al.*, 2008).

ROS are produced either by partial, single-electron reduction of oxygen to generate superoxide, hydrogen peroxide and hydroxyl radicals or by alteration of oxygen electron spin states by photoactivation to generate singlet oxygen (Fig. 1). The latter happens exclusively in the chloroplast via

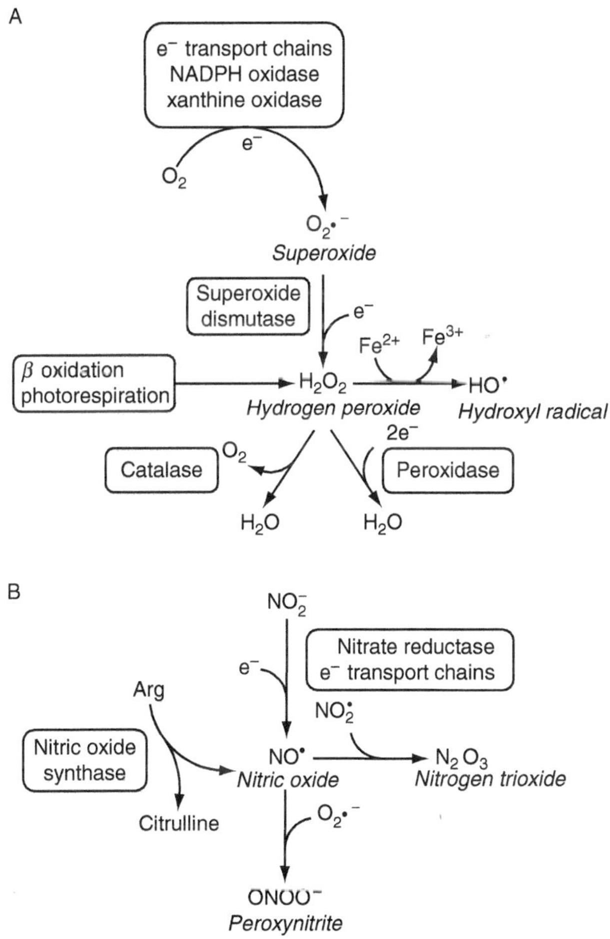

Fig. 1. Formation of the most important reactive oxygen species (A) and reactive nitrogen species (B).

a photosensitization reaction involving photosystem II. In the light, the chloroplast is also the main source of superoxide (Foyer and Noctor, 2003) as a result of 'electron leakage' from the photosynthetic electron transport chain. However, significant quantities of superoxide are also generated as a by-product of the mitochondrial electron transport chain (Foyer and Noctor, 2003; Møller, 2001) and in the apoplast as a consequence of NADPH oxidase activity in response to biotic (Sagi and Fluhr, 2001) and abiotic (Achard *et al.*, 2008) stresses. Superoxide will rapidly chemically dismutate to form hydrogen peroxide, a reaction that is accelerated manyfold by the presence of superoxide dismutases in most subcellular compartments. Hydrogen peroxide is also generated directly in very large quantities in peroxisomes as a by-product of photorespiratory metabolism and the β-oxidation of fatty acids. In the presence of reduced transition metals, Fenton chemistry reduces hydrogen peroxide to the extremely reactive hydroxyl radical.

Different ROS vary significantly in terms of their properties and reactivity; the order of reactivity being hydroxyl radical > superoxide > hydrogen peroxide. Their reactivity sets limits on how far different ROS can propagate from their site of production. The hydroxyl radical is so reactive that it will react more or less indiscriminately with the first molecule it encounters. In contrast, the relatively low reactivity of hydrogen peroxide means that it can accumulate to significant concentrations and can diffuse as far as 1 μm from its site of production (Møller *et al.*, 2007). Superoxide will travel a shorter distance (up to 30 nm) and moreover, as a charged species at cellular pH (p*K*a 4.8—Halliwell and Gutteridge, 2007), it is confined to the subcellular compartment in which it is produced. The superoxide formed by the NADPH oxidase outside the plasma membrane, where the pH is normally significantly lower than inside the cell, could be substantially protonated and may find it easier to enter the cell by crossing the plasma membrane as a neutral molecule.

Plants also produce RNS, particularly nitric oxide ($NO^{\bullet}$) and peroxynitrite ($ONOO^{-}$). It remains unclear what is the most significant source of NO in plant cells. Efforts to uncover a canonical nitric oxide synthase, analogous to that found in mammals, have been beset with controversy (Zemojtel *et al.*, 2006). A putative nitric oxide synthase in *Arabidopsis* (Guo and Crawford, 2005) was ultimately revealed to be incapable of NO synthesis from arginine and instead was established as a plastid-localized GTPase (Gas *et al.*, 2009). This protein is still linked with $NO^{\bullet}$ production (knockouts reduce $NO^{\bullet}$ levels) but the mechanism is unclear. There are two other potential sources of $NO^{\bullet}$ in plants: nitrate reductase and the mitochondrial electron transport chain under anoxia. Nitrate reductase can reduce nitrite to $NO^{\bullet}$. This could occur *in vivo* when the nitrite accumulates (the preferred substrate of nitrate reductase is, of course, nitrate) such as during anoxia when nitrite reductase is inhibited (Meyer *et al.*, 2005). Anoxia is also a prerequisite for $NO^{\bullet}$ production by mitochondria, allowing nitrite to serve as an alternative terminal electron acceptor in the electron transport chain. Because $NO^{\bullet}$ inhibits complex IV of the respiratory chain, but not the alternative oxidase, $NO^{\bullet}$ production has been proposed as a mechanism for controlling respiratory balance under low oxygen (Benamar *et al.*, 2008; Borisjuk *et al.*, 2007). Whatever the mechanisms of $NO^{\bullet}$ synthesis, there are reliable measurements demonstrating its presence in a range of plant tissues. Although $NO^{\bullet}$ is a radical, its reactivity with proteins is limited. However, $NO^{\bullet}$ does react extremely readily with superoxide to form peroxynitrite and this anion is more significant in terms of protein oxidation.

III. MECHANISMS OF OXIDATION OF PROTEINS

The following protein amino acids contain side groups that can be oxidized by different ROS and RNS leading to stable covalent modifications: Cys, Met, His, Arg, Lys, Pro, Tyr and Trp. The reactions are summarized in Fig. 2. Most of these reactions are essentially irreversible, although in the specific case of oxidation of thiols, enzyme-catalyzed re-reduction is possible (Bechtold *et al.*, 2004; Rouhier *et al.*, 2006). Considering only the oxidation by ROS, most of the oxidation reactions shown in Fig. 2 are found to be only triggered by the highly reactive hydroxyl radical or singlet oxygen. This means that such oxidation events are only likely to occur in proteins that are localized extremely close to the site of production of these radicals.

Hydroxyl radical formation from hydrogen peroxide requires the presence of a reduced transition metal such as Fe^{2+}. To avoid inadvertent formation of hydroxyl radicals, the free metal ion concentration is maintained at extremely low levels (Jeong and Guerinot, 2009). Metal ion coordination sites in proteins may, therefore, represent the main source of metal ions for Fenton chemistry and could form the key sites of oxidation (Stadtman, 1990). As well as true metalloproteins that require metal ion co-factors for their structure and function, it is apparent that a more generic form of association of metal ions with the surface of proteins is possible (Szpunar, 2005). Certain metal ion co-factors, such as Fe-S, are particularly susceptible to oxidation. For example, the TCA cycle enzyme aconitase contains a cubane $[4Fe–4S]^{2+}$ cluster that is essential for its activity. Only three of the four irons are ligated to Cys residues. The fourth iron is exposed to the solvent and is open to attack from superoxide, which causes a one-electron

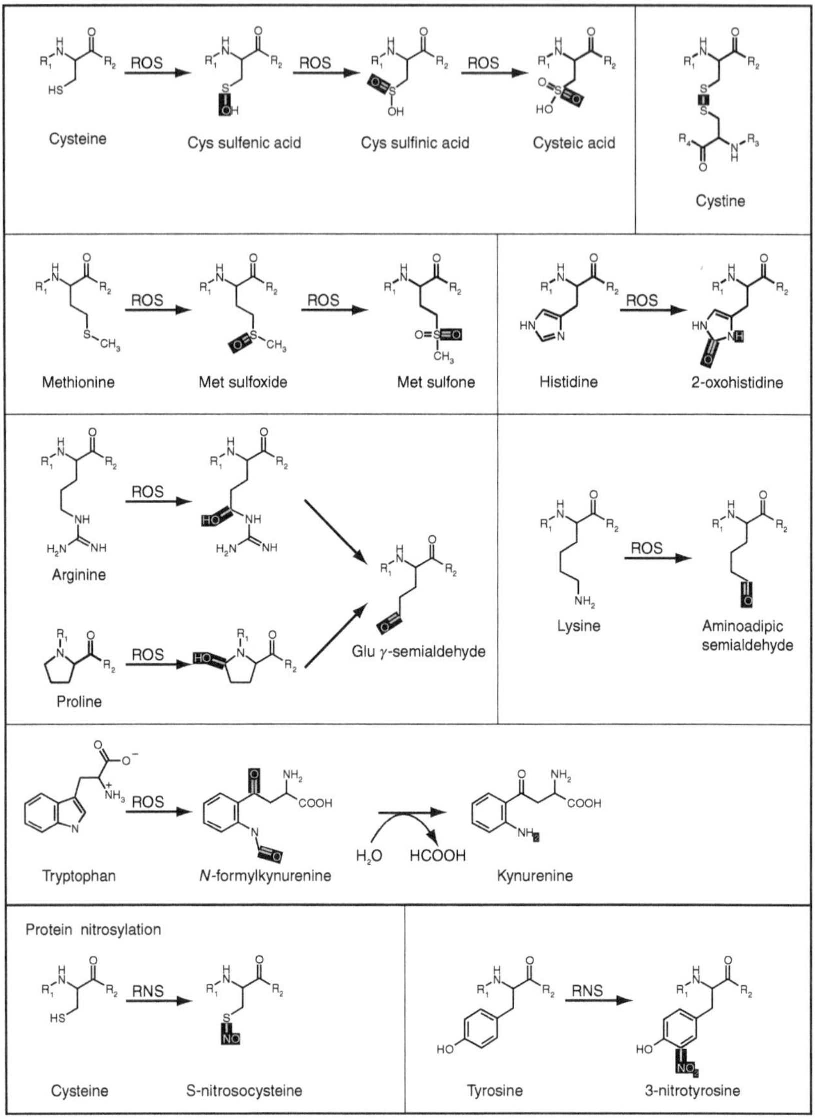

Fig. 2. Reactions leading to common oxidation of amino acids.

oxidation of the iron-sulphur cluster, releasing the exposed iron in the ferrous state and inactivating the enzyme (Kennedy *et al.*, 1983).

Cys thiols are the most readily oxidized of all the amino acid side groups, a fact that is exploited in the well-known regulatory thiol-disulfide transition. While the latter is a deliberate, enzyme-catalyzed event, ROS-triggered oxidation of the Cys thiol progressively to sulfenic, sulfinic and cysteic acid can also occur (Fig. 2). This is sometimes referred to as 'over-oxidation' of Cys, to distinguish it from regulatory thiol-disulfide exchanges. Cys sulphur atoms can also be oxidized by RNS to form a Cys nitrosyl group (Fig. 2).

IV. METHODS FOR MEASURING PROTEIN OXIDATION

The detection and identification of oxidized proteins is a key experimental goal in the field of oxidative stress. Global oxidation of proteins is frequently used as a marker of oxidative stress. Monitoring of specific oxidation events on a protein-by-protein basis is important to deepen our understanding of the significance of protein oxidation by answering key questions such as: Are certain proteins more susceptible to oxidation than others? Are certain amino acid side groups more commonly oxidized? Does protein oxidation necessarily have a negative impact on protein function?

Broadly speaking, there are two main approaches used in the identification of oxidized amino acid side groups: specific reaction of the oxidized group with a chemical tag that can be readily visualized (e.g., using antibodies to the chemical tag) or direct identification of the oxidized group by mass spectrometry. The latter approach is only suitable for relatively stable modifications such as oxidation of Trp to *N*-formylkynurenine. In both approaches, the advent of proteomic approaches has been a key advance, allowing the identification of proteins containing specific oxidative modifications. There have been a number of such studies over the last few years and these will form the focus of subsequent sections of this chapter. However, before describing these studies, we wish to highlight some important limitations to the methodologies for detecting oxidized protein residues.

Carbonylated amino acid residues in proteins are tagged by reacting with dinitrophenylhydrazine (DNPH) and the conjugate is recognized by a DNP-specific peroxidase-linked antibody (Berlett and Stadtman, 1997; Dean *et al.*, 1997) (Fig. 3). The amount of DNP-labelling can then be quantified by the standard immunological techniques and values around 0.2 carbonyl group per 50 kDa protein have been reported for plant tissues. Much higher values (10–100-fold) have also been reported, but these are most likely overestimates (Møller *et al.*, 2007), indicating that the technique might not be straightforward to use. The DNP-labelled proteins can also be identified, for example, by affinity purification followed by mass spectrometry (Fig. 3). Using this technique, 20 mitochondrial proteins were shown to be carbonylated probably *in vivo* while a further 32 could be carbonylated by

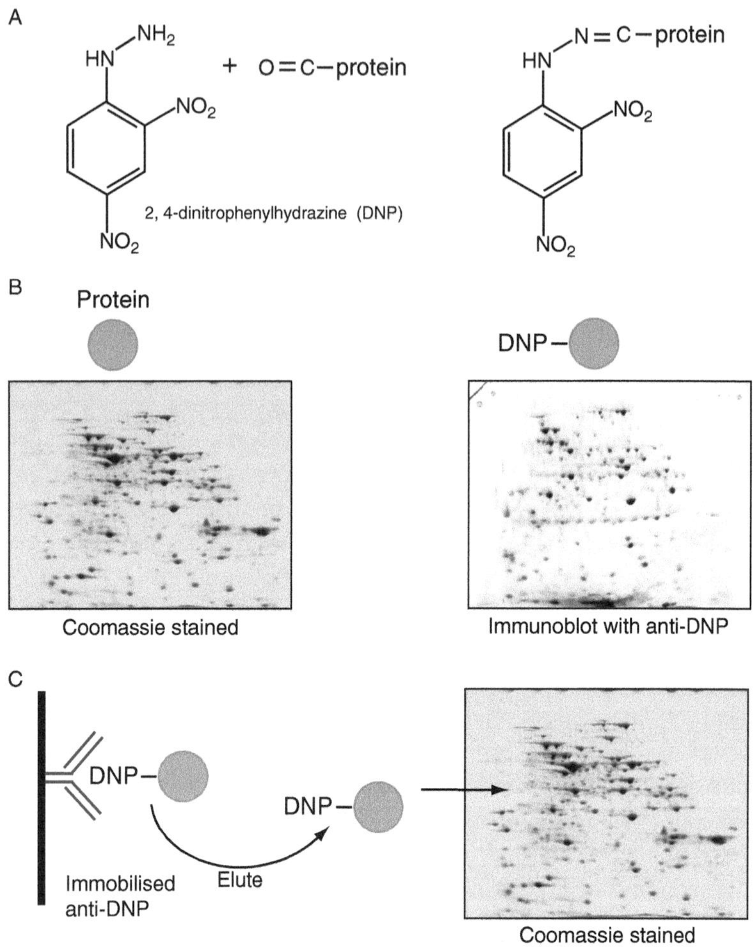

Fig. 3. Methods of identifying proteins with carbonyl groups. (A) Reaction of 2,4-dinitrophenylhydrazine (DNPH) with protein carbonyls. (B) Identification of protein carbonyls by two-dimensional gel electrophoresis and immunoblotting with antibodies specific to DNP. (C) Identification of protein carbonyls by immunoaffinity purification using immobilized antibodies specific to DNP.

an *in vitro* oxidative treatment (Kristensen *et al.*, 2004). One limitation of this technique is that it is unclear whether DNPH reacts with all carbonylated amino acids; thus *N*-formylkynurenine was not found in any of the DNP-labelled proteins identified by Kristensen *et al.* (2004).

Another approach that is increasingly commonly used is the 'biotin-switch' assay to detect protein *S*-nitrosylation (Jaffrey *et al.*, 2001; Sell *et al.*, 2008). The assay is based on the selective reduction of *S*-nitroso groups

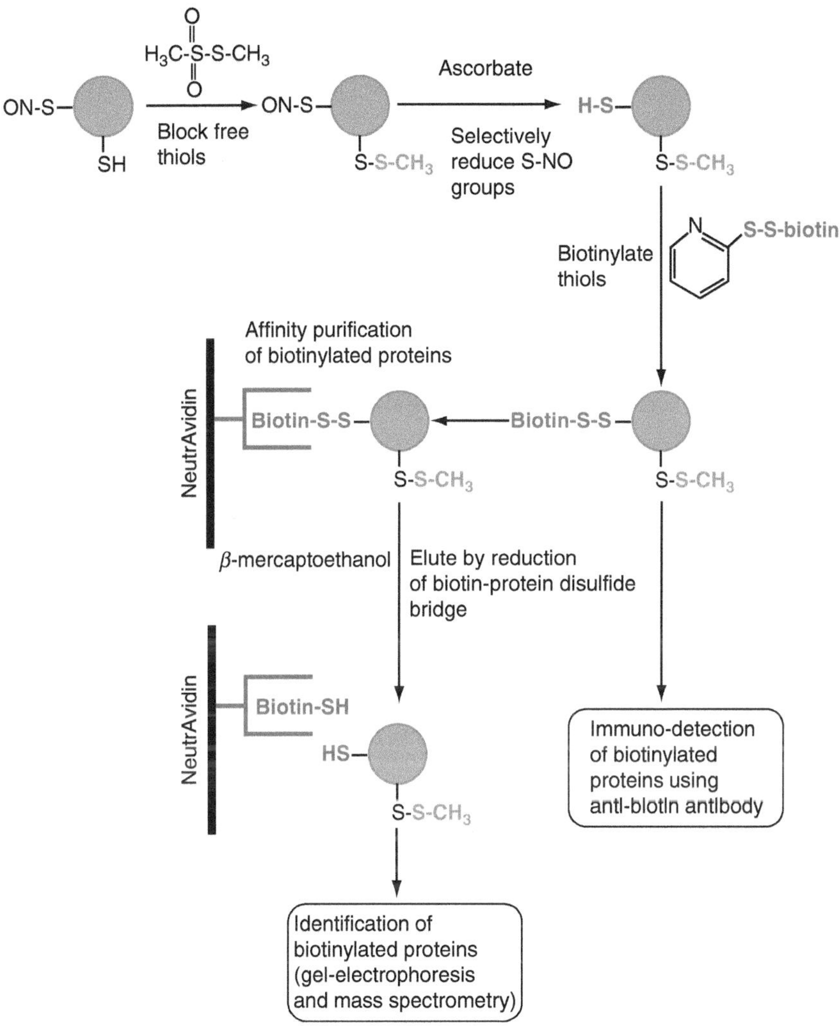

Fig. 4. The biotin-switch method. Free protein thiols are blocked by alkylation using *S*-methyl methanethiosulfonate (shown) or *N*-ethyl maleimide. Nitroso-thiols (S-NO) are then selectively reduced using ascorbate and the resulting thiols tagged with biotin using Biotin-HPDP (*N*-(6-(Biotinamido)hexyl)-3′-(2′-pyridyldithio)-propionamide. Biotinylated proteins can be detected directly using anti-biotin antibodies or can be affinity purified using NeutrAvidin prior to identification by mass spectrometry.

to thiols using ascorbate (having first blocked free thiols by alkylation) and detection of the revealed thiols using a biotinylation reagent (Fig. 4). The biotin group can be detected and quantified by western blotting or, alternatively, used to affinity-purify modified proteins for subsequent proteomic identification (Lindermayr *et al.*, 2005). The relatively simple, yet powerful, nature of this technique has led to its widespread adoption. However, recently, questions have been raised as to the specificity of the assay. One of the key concerns regards the possibility of promiscuous reduction of oxidized thiols other than *S*-nitroso groups by ascorbate. For example, it has been found that pre-reduced bovine serum albumin can be biotinylated during the biotin-switch procedure and even disulfide bridges were reduced by ascorbate (Huang and Chen, 2006). This phenomenon was traced to an ascorbate- and light-dependent modification of the biotinylation reagent, biotin-HPDP (*N*-(6-(Biotinamido)hexyl)-3′-(2′-pyridyldithio)-propionamide), to biotin-SH which is responsible for non-specific protein biotinylation via thiol/disulfide exchange at blocked Cys (Forrester *et al.*, 2007). Provided the procedure is carried out in the absence of indirect sunlight, this problem can be avoided (Sell *et al.*, 2008).

Nevertheless, there remain some further concerns. Generally, a concentration of 1 mM ascorbate is recommended in the biotin-switch assay. However, some *S*-nitroso-proteins react slowly with ascorbate and higher concentrations (10–50 mM) may be required to efficiently reduce all *S*-nitroso groups. This may create other problems: a recent report has revealed that at higher concentrations, ascorbate may lead to light-independent reduction of disulfides (Giustarini *et al.*, 2008). Moreover, given that plant tissues contain millimolar concentrations of ascorbate, one may question how it is possible to extract proteins with *S*-nitroso groups at all? Surely, most *S*-nitroso groups would have been reduced to thiols *in vivo* by intracellular ascorbate? Maybe, *S*-nitroso modifications are introduced to proteins artefactually during extraction of proteins from the tissue? Clearly, the biotin-switch assay must be used with some care and the results interpreted with caution. Ideally, the occurrence of protein *S*-nitrosylation should be confirmed by an independent means.

A number of new redox probes have been developed recently, which, although they do not directly measure target protein oxidation, may be helpful in characterizing the local redox environment. The best characterized of these is the roGFP, a modified GFP containing two additional surface Cys that form an intramolecular disulfide in oxidizing conditions, altering the spectroscopic properties of the fluorophore (Hanson *et al.*, 2004). This roGFP has been successfully used in plants (Jiang *et al.*, 2006) where it reports the redox status of the glutathione pool (Meyer *et al.*, 2007). The roGFP can be targeted to different subcellular compartments and, when used correctly, can provide precise quantitative information on the glutathione redox status (Schwarzlander *et al.*, 2008). Recent work using roGFP in *Arabidopsis* leaves has demonstrated that abiotic stresses can have differential effects in different subcellular compartments and suggests that the mitochondrion may be more susceptible to oxidation than other subcellular locations (Schwarzlander *et al.*, 2009). Because oxidation of roGFP is based on Cys oxidation, the probe gives a good indication of the likelihood of general oxidation of protein Cys. However, it provides a rather indirect measure of the oxidative load which would be best estimated by measurement of the steady-state levels of superoxide and hydrogen peroxide. Although there are several widely used 'ROS-sensor' dyes (e.g., 2′,7′-dichlorodihydrofluorescein diacetate; H_2DCFDA), these are, at best, semi-quantitative and lack specificity for individual ROS (Kristiansen *et al.*, 2009). More promising, are genetically encoded probes for superoxide (Wang *et al.*, 2008) and hydrogen peroxide but these are yet to be tested in plants.

V. WHICH PROTEINS, WHICH OXIDATIONS?

Despite the technical limitations described in section IV, the proteomic methods to detect specific protein oxidation events have been used with some success to characterize the extent and importance of protein oxidation. In this section, we will review some of these studies and summarize the overall impact of different types of side-group oxidations. This topic has been reviewed in detail recently, so we will mainly focus on very recent work while still attempting to provide a general overview.

A. PROTEIN CARBONYLATION

Formation of carbonyls is one of the most widespread protein oxidative modifications affecting Arg, His, Lys, Pro, Thr and Trp amino acid residues (Fig. 2). Because of the large number of amino acids that can be oxidized in this way, carbonylation can occur on more or less any protein and is often used as a general indicator of oxidative stress. That said, proteomic studies have revealed that the extent of carbonylation is not uniform across the proteome and there is some evidence of proteins that are more sensitive to carbonylation than others (Johansson *et al.*, 2004; Kristensen *et al.*, 2004). Moreover, the extent of protein carbonylation is very uneven during the developmental cycle of *Arabidopsis*, with a pronounced fall in protein carbonyls in vegetative tissues just prior to bolting (Johansson *et al.*, 2004). This is in distinct contrast to the situation in animals where more protein carbonyls are observed with age. Nevertheless, a carbonyl-proteomic study has demonstrated that the loss of viability of *Arabidopsis* seeds associated with age is accompanied by a pronounced increase in specific protein carbonyls (Rajjou *et al.*, 2008). Oxidation of proteins such as heat shock proteins and LEA (late embryogenesis abundant) proteins is proposed to lead to a reduced capacity to maintain protein function in the dehydrated seed. However, the role of protein oxidation and oxidative stress in seed viability is not as straightforward as this suggests; the breaking of seed dormancy after the initial dehydration is dependent on a burst of ROS (Oracz *et al.*, 2009) and is associated with *increased* carbonylation of a specific set of proteins in the embryo axes (Oracz *et al.*, 2007). Increased protein carbonylation is also surprisingly associated with elevated atmospheric CO_2 (Qiu *et al.*, 2008), an observation that has substantial implications for the predictions of plant productivity in future global atmospheric compositions.

At the subcellular level, the bioenergetic organelles contain the highest levels of carbonylated proteins (Bartoli *et al.*, 2004; Johansson *et al.*, 2004), consistent with the high rates of ROS production in mitochondria and chloroplasts.

B. OXIDATION OF SULPHUR-CONTAINING AMINO ACIDS

The sulphur-containing groups of both Met and Cys are relatively easily oxidized causing a range of modifications that are, on the whole, reversible (Møller *et al.*, 2007). The oxidation of Cys thiols to disulphides is one of the most well-recognized regulatory post-translational modifications, allowing protein activity to be linked to redox-active processes such as photosynthesis. However, Cys thiols can also be oxidized in a number of other ways (Fig. 2) and it is becoming increasingly apparent that these Cys modifications are also of regulatory importance. *S*-nitrosylation has emerged as perhaps the most important non-disulphide modification of Cys (Wang *et al.*, 2006).

An initial proteomic survey of *Arabidopsis* nitroso-proteins using the biotin-switch technique revealed that nitrosylation is widespread (Lindermayr *et al.*, 2005). Some 25 nitroso-proteins were identified in control leaves. After treatment with $NO^{\bullet}$ or an $NO^{\bullet}$ donor, this number increased to 41 and 57, respectively. Proteins in a range of functional classes were affected including redox-related, stress-related and metabolic proteins. However, given the specificity issues of the biotin-switch method and the non-physiological nature of the treatments, this list of proteins requires further investigation to establish the physiological and functional significance of the putative nitrosylation events. A more physiological study used the same biotin-switch/proteomic approach to study the protein nitrosylation during pathogen infection (Romero-Puertas *et al.*, 2008). *Arabidopsis* plants were infiltrated with *Pseudomonas syringae pv. Tomato*, inducing a hypersensitive response. The nitrosylation pattern of 16 proteins was altered after pathogen infection. Most of these proteins were enzymes of central carbon metabolism (photosynthesis, glycolysis, TCA cycle) as well as redox-active enzymes. Some of these proteins, such as the glycolytic enzyme GAPDH, are known to be inhibited by nitrosylation of an active site Cys (Padgett and Whorton, 1995).

In this study, no attempt was made to independently confirm nitrosylation of the identified proteins or establish to what extent nitrosylation alters their functional state. Moreover, for most of the identified proteins there is no direct link to either a general pathogen response or to the cell-death component of the hypersensitive response. However, two possible mechanistic links between protein nitrosylation and pathogen response have been uncovered in two independent studies. First, it was found that *Arabidopsis* metacaspase 9, one of the key cell-death-related proteases, is nitrosylated *in vivo* (Belenghi *et al.*, 2007). Moreover, this nitrosylation event affects both the autoprocessing and the proteolytic activity, suggesting that nitrosylation could be an important regulatory event in the control of cell death. Secondly, it was found that another important protein in plant-pathogen responses, salicylic acid-binding protein 3 (SABP3), is nitrosylated in *Arabidopsis* infected with *P. syringae pv. Tomato* (Wang *et al.*, 2009). As well as binding the immune activator, salicylic acid, SABP3 has a carbonic anhydrase activity that is believed to be required for the expression of resistance in the host. Nitrosylation suppresses both these activities of SABP3, possibly contributing to a negative feedback loop that modulates the plant defence response.

Perhaps the most intriguing of the recent *S*-nitrosylation discoveries involves PrxIIE, a chloroplastic member of the peroxiredoxin family (Gama *et al.*, 2008). Peroxiredoxins are classified as alkyl hydroperoxide reductases, reflecting their broad specificity towards a range of hydroperoxide substrates including hydrogen peroxide (Dayer *et al.*, 2008). Recently, the list of substrates for PrxIIE was extended to include peroxynitrite (Romero-Puertas *et al.*, 2007). What is particularly interesting about this observation is that in the same work, Romero-Puertas *et al.* demonstrated that PrxIIE activity is inhibited by *S*-nitrosylation of a catalytic Cys. Based on this observation, the authors suggested that through nitrosylation and inhibition of the peroxynitrite detoxification reaction of PrxIIE, nitric oxide could influence the formation of its own secondary toxic ions such as peroxynitrite. However, it should be pointed out that this is a feed-forward loop that would lead to uncontrolled peroxynitrite formation, a situation that could be considerably damaging to the plant cell. The extent to which this would be a problem would depend on the importance of PrxIIE for peroxynitrite detoxification. It is likely that, in addition to PrxIIE, there are other enzymes capable of reducing peroxynitrite. This is an area that will benefit from further research.

Another potentially important modification of sulphur-containing amino acids is the formation of mixed disulphides, for example, with glutathione (glutathionylation) (Rouhier *et al.*, 2008). Several enzymes have been identified that are glutathionylated *in vitro* (Dixon *et al.*, 2002) and *in vivo* (Ito *et al.*, 2006), with concomitant effects on the target enzyme activity. A more extensive study using biotinylated glutathione identified 10 proteins glutathionylated *in vivo*, including primary metabolic enzymes, a HSP70 and structural proteins such as actin and tubulin (Dixon *et al.*, 2005). Glutathionylation occurs via reaction of a protein thiol with GSSG and thereby connects the glutathione redox state to post-translational modification of target proteins (Rouhier *et al.*, 2008). As well as directly modifying protein function, glutathionylation may also serve to protect against irreversible over-oxidation of thiols.

C. OTHER PROTEIN MODIFICATIONS

Although not as prevalent as carbonylation or thiol oxidation, there are a number of other types of protein oxidation that may have biological significance. These include oxidation of tryptophan (Møller and Kristensen, 2006) and the covalent modification of proteins by reactive aldehydes such as hydroxynonenal (Winger *et al.*, 2007). So far, these modifications have only been experimentally tracked in mitochondrial proteins, but it is likely that they occur in proteins from other subcellular localizations. As well as modification of specific amino acid side groups, protein function can also be affected through oxidative loss of co-factors. The best known example of this is the loss of the Fe-S cluster of aconitase, although several other metalloenzymes are known to be vulnerable to oxidative inactivation. Other co-factors such as lipoic acid (required in key mitochondrial dehydrogenases) can also be lost leading to enzyme inactivation. Modification of lipoic acid occurs relatively readily, accounting for the sensitivity to environmental stress of processes such as photosynthesis and respiration that are dependent on lipoic acid-containing enzymes (Taylor *et al.*, 2005).

VI. DOES PROTEIN OXIDATION MEAN PROTEIN DYSFUNCTION?

The proteomic studies that we have described thus far provide a broad-ranging survey of protein oxidations that can occur in response to a variety of oxidizing treatments, stresses and during development. However, there is not necessarily a straightforward relationship between the occurrence of oxidized amino acid residues and functional inactivation of the protein. In order to understand the biological significance of protein oxidation, it is vital that the proteomic studies are followed up with an equally comprehensive analysis of the activities of the target proteins. To date, this information is not available. Although there have been several studies of the metabolic consequences of altered redox status (Kolbe *et al.*, 2006; Morgan *et al.*, 2008), these were not specifically linked to protein oxidation. Generally speaking, the discussions of the significance of protein oxidation fall back on a small number of well-known examples (such as inactivation of aconitase by superoxide and GAPDH by nitrosylation) that were originally established in the mammalian literature. An additional problem is that many of the treatments used to investigate protein oxidation may be non-physiological, relying on the application of exogenous oxidizing agents at concentrations that generate unknown internal concentrations of these molecules.

While specific modification of catalytic Cys can generally be shown to have functional consequence (Romero-Puertas *et al.*, 2007), the functional effects of oxidation of non-catalytic Cys and oxidation of other residues by carbonylation are not known. An investigation of metal-catalyzed carbonylation of peroxisomal proteins has, however, established that there is no simple relationship between carbonylation and impairment of enzyme activity (Nguyen and Donaldson, 2005). Carbonylation caused the inactivation of four enzymes: malate synthase, isocitrate lyase, catalase and malate dehydrogenase. However, the protein that contained the highest proportion of carbonyls (catalase) exhibited the least inhibition of the four enzymes. Conversely, isocitrate lyase was completely inactivated and yet had the least carbonylation.

The explanation for this variation between the degree of oxidation and the degree of functional inactivation clearly lies in the functional importance of the oxidized residues. It is likely that most residues that become oxidized are surface-accessible residues, many of which may have little role in the overall protein structure or function. In fact, it is possible that proteins are organized with 'sacrificial' amino acids on their surface to permit a certain degree of oxidation without necessarily impairing the protein function. This concept of sacrificial or decoy residues has some support from a study of animal mitochondrial proteins (Bender *et al.*, 2008).

In animal mitochondria, there is an alternative genetic code in which the AUA codon is reassigned from Ile to Met. This results in a significantly higher proportion of Met in mitochondrial-encoded respiratory proteins than in nuclear-encoded proteins. It is argued that, as an easily oxidized residue, Met provides sacrificial protection for a group of proteins that are exposed to a high oxidative pressure. Two pieces of evidence are provided in support of this hypothesis. First, the AUA-encoded Met residues are predominantly located on the surface of mitochondrial-encoded respiratory proteins. And secondly, the application of two model methionyl compounds to isolated rat liver mitochondria provided significant protection against iron/hydrogen peroxide toxicity (Bender *et al.*, 2008). The 6% Met residues found in mitochondrial proteins is equivalent to 600 nmol Met/mg proteins, and this should be compared to the steady-state level of 4 nmol carbonyl groups/mg mitochondrial proteins observed. It would, therefore, appear that the Met residues in mitochondrial proteins have the capacity to absorb a significant proportion of the oxidative damage in mitochondria? However, it is an open question how many oxidized Met residues a protein molecule can tolerate—even if located on its surface—before losing its function.

VII. REMOVAL AND PROCESSING OF OXIDIZED PROTEINS

Oxidized proteins are rapidly removed from the cell by proteolytic degradation. The exact mechanisms by which oxidized proteins are targeted for degradation are not clear, but one feasible possibility is that oxidation causes exposure of hydrophobic residues that are recognized by proteases. Indeed, some proteases have been shown to have a preference for oxidized protein substrates (Bota and Davies, 2002). Carbonylation, in particular, has been proposed as a 'marker' of protein quality control (Nystrom, 2005). Proteases of a variety of classes are present in all subcellular compartments. In some cases, mutation of a specific protease (e.g., mitochondrial FtsH4) has been shown to lead to an accumulation of protein carbonyls, suggesting that the protease is responsible for processing of oxidized proteins (Gibala *et al.*, 2009). Moreover, the lack of this mitochondrial protease leads to developmental abnormalities, emphasizing the importance of dealing with oxidized proteins. Mitochondrial proteases seem to be particularly important, with mutation in a different class of mitochondrial protease (a lon protease) leading to arrested early development and heat sensitivity of *Arabidopsis* seedlings (Rigas *et al.*, 2009).

Although plants possess an extensive protease network, oxidized proteins can nevertheless accumulate under severe stress conditions, despite the induction of protease activity (Penaa *et al.*, 2006; Sweetlove *et al.*, 2002). Extreme oxidative damage can lead to the formation of protein aggregates that cannot be processed by proteases. The most likely route for dealing with such aggregates is via the autophagy pathway. Autophagy involves the formation of specific intracellular vesicular structures, called autophagosomes, which trap cytoplasmic material and deliver it to the vacuole for degradation (Bassham, 2007; Patel *et al.*, 2006; Thompson and Vierstra, 2005). This can include whole cell components (e.g., organelles) and is thought to be a mechanism for the removal of dysfunctional organelles and to facilitate the nutrient recycling during starvation by sacrifice of cell components. Recently, it has been shown that the induction of oxidative stress by application of hydrogen peroxide to *Arabidopsis* plants can induce autophagy (Xiong *et al.*, 2007). Moreover, autophagy-deficient mutants are more sensitive to oxidative stress and can accumulate higher levels of carbonylated proteins (Xiong *et al.*, 2007). This suggests that autophagy is an important mechanism by which the cell processes oxidized proteins.

The peptide products that result from processing of oxidized proteins may have a role in signalling during oxidative stress. It is often considered that ROS themselves are the most likely signals of oxidative stress. However, as discussed earlier, superoxide cannot cross membranes and the hydroxyl radical is too reactive to act as a specific signal. This leaves hydrogen peroxide as the most likely ROS messenger (Neill *et al.*, 2002), partly because of its relative stability and partly because it can cross membranes through aquaporins (Bienert *et al.*, 2007). However, the disadvantage with hydrogen peroxide as a gene regulator is that it lacks specificity. This is where the degradation of oxidatively damaged proteins may contribute to retrograde signalling (Koppen and Langer, 2007). Peptides from oxidized mitochondrial proteins especially peptides containing oxidized amino acids as such peptides could not derive from normal turnover of the protein—would have the specificity to activate specific genes not activated by ROS-generated messengers from other sources such as the plasma membrane NADPH oxidase or the chloroplasts. If peptides are indeed retrograde signals, we need to identify the peptide-specific receptors in the nucleus.

VIII. THE COST OF PROTEIN OXIDATION

The removal and replacement of oxidatively damaged proteins is quite costly. It has been estimated that 2–20% of the ATP produced by a plant mitochondrion may be used for replacing its proteins (Møller *et al.*, 2007). The estimate was based on turnover halflives of 20 to >100 h for mammalian mitochondrial proteins since no data was available for plant mitochondria. Using inducible RNAi knockdown of three mitochondrial proteins (UCP, MnSOD and PrxIIF) in rosette leaves of *Arabidopsis*, the complete turnover of these proteins has recently been measured to be from 1 day (UCP) up to 12 days (MnSOD and PrxIIF), equivalent to halflives of 6–72 h, respectively

(M. Schwarzlander, M. Morgan and L. J. Sweetlove, unpublished data). This may indicate that protein turnover in plant mitochondria is faster than in mammalian mitochondria. However, these rates were estimated in the absence of any experimentally induced oxidative stress and we have, as yet, no indication of the degree to which protein oxidation contributed to the turnover.

IX. SUMMARY

Proteins can be modified by ROS and RNS in a large number of ways. Oxidation of cysteines and carbonylation of several other amino acids are the most common modifications, but nitrosylation is also potentially important. Some of the modifications are regulatory, viz the oxidation of cysteinyl disulfides, while others are damaging to the function of the protein, viz the oxidation of the iron-sulphur cluster in aconitase. Oxidized proteins can be removed by proteases found in all cellular compartments and it is possible that peptides derived from such degradation can act as messengers. The combined synthesis and breakdown of oxidized proteins results in a steady-state level of oxidized proteins which increases under stress conditions. The turnover of proteins, whether necessitated by protein oxidation or by regulatory mechanisms, is quite expensive and may consume a significant proportion of the cellular ATP production.

REFERENCES

Achard, P., Renou, J. P., Berthome, R., Harberd, N. P. and Genschik, P. (2008). Plant DELLAs restrain growth and promote survival of adversity by reducing the levels of reactive oxygen species. *Current Biology* **18**, 656–660.

Apel, K. and Hirt, H. (2004). Reactive oxygen species: Metabolism, oxidative stress, and signal transduction. *Annual Review Plant Biology* **55**, 373–399.

Bartoli, C. G., Gomez, F., Martinez, D. E. and Guiamet, J. J. (2004). Mitochondria are the main target for oxidative damage in leaves of wheat (*Triticum aestivum L.*). *Journal of Experimental Botany* **55**, 1663–1669.

Bassham, D. C. (2007). Plant autophagy – More than a starvation response. *Current Opinion in Plant Biology* **10**, 587–593.

Bechtold, U., Murphy, D. J. and Mullineaux, P. M. (2004). *Arabidopsis* peptide methionine sulfoxide reductase2 prevents cellular oxidative damage in long nights. *Plant Cell* **16**, 908–919.

Belenghi, B., Romero-Puertas, M. C., Vercammen, D., Brackenier, A., Inze, D., Delledonne, M. and Van Breusegem, F. (2007). Metacaspase activity of *Arabidopsis thaliana* is regulated by *S*-nitrosylation of a critical cysteine residue. *Journal of Biological Chemistry* **282**, 1352–1358.

Benamar, A., Rolletschek, H., Borisjuk, L., Avelange-Macherel, M. H., Curien, G., Mostefai, H. A., Andriantsitohaina, R. and Macherel, D. (2008). Nitrite-nitric oxide control of mitochondrial respiration at the frontier of anoxia. *Biochimica et Biophysica Acta* **1777**, 1268–1275.

Bender, A., Hajieva, P. and Moosmann, B. (2008). Adaptive antioxidant methionine accumulation in respiratory chain complexes explains the use of a deviant genetic code in mitochondria. *Proceedings of the National Academy of Sciences of the United States of America* **105**, 16496–16501.

Berlett, B. S. and Stadtman, E. R. (1997). Protein oxidation in aging, disease, and oxidative stress. *Journal of Biological Chemistry* **272**, 20313–20316.

Bienert, G. P., Møller, A. L., Kristiansen, K. A., Schulz, A., Møller, I. M., Schjoerring, J. K. and Jahn, T. P. (2007). Specific aquaporins facilitate the diffusion of hydrogen peroxide across membranes. *Journal of Biological Chemistry* **282**, 1183–1192.

Borisjuk, L., Macherel, D., Benamar, A., Wobus, U. and Rolletschek, H. (2007). Low oxygen sensing and balancing in plant seeds: A role for nitric oxide. *New Phytologist* **176**, 813–823.

Bota, D. A. and Davies, K. J. (2002). Lon protease preferentially degrades oxidized mitochondrial aconitase by an ATP-stimulated mechanism. *Nature Cell Biology* **4**, 674–680.

Dayer, R., Fischer, B. B., Eggen, R. I. and Lemaire, S. D. (2008). The peroxiredoxin and glutathione peroxidase families in *Chlamydomonas reinhardtii*. *Genetics* **179**, 41–57.

Dean, R. T., Fu, S., Stocker, R. and Davies, M. J. (1997). Biochemistry and pathology of radical-mediated protein oxidation. *Biochemical Journal* **324**(Pt. 1), 1–18.

Delledonne, M. (2005). NO news is good news for plants. *Current Opinion in Plant Biology* **8**, 390–396.

del Rio, L. A., Sandalio, L. M., Corpas, F. J., Palma, J. M. and Barroso, J. B. (2006). Reactive oxygen species and reactive nitrogen species in peroxisomes. Production, scavenging, and role in cell signaling. *Plant Physiology* **141**, 330–335.

Dixon, D. P., Davis, B. G. and Edwards, R. (2002). Functional divergence in the glutathione transferase superfamily in plants. Identification of two classes with putative functions in redox homeostasis in *Arabidopsis thaliana*. *Journal of Biological Chemistry* **277**, 30859–30869.

Dixon, D. P., Skipsey, M., Grundy, N. M. and Edwards, R. (2005). Stress-induced protein *S*-glutathionylation in *Arabidopsis*. *Plant Physiology* **138**, 2233–2244.

Forrester, M. T., Foster, M. W. and Stamler, J. S. (2007). Assessment and application of the biotin switch technique for examining protein *S*-nitrosylation under conditions of pharmacologically induced oxidative stress. *Journal of Biological Chemistry* **282**, 13977–13983.

Foyer, C. H. and Noctor, G. (2003). Redox sensing and signalling associated with reactive oxygen in chloroplasts, peroxisomes and mitochondria. *Physiologia Plantarum* **119**, 355–364.

Gama, F., Brehelin, C., Gelhaye, E., Meyer, Y., Jacquot, J. P., Rey, P. and Rouhier, N. (2008). Functional analysis and expression characteristics of chloroplastic Prx IIE. *Physiologia Plantarum* **133**, 599–610.

Gas, E., Flores-Perez, U., Sauret-Gueto, S. and Rodriguez-Concepcion, M. (2009). Hunting for plant nitric oxide synthase provides new evidence of a central role for plastids in nitric oxide metabolism. *Plant Cell* **21**, 18–23.

Gibala, M., Kicia, M., Sakamoto, W., Gola, E. M., Kubrakiewicz, J., Smakowska, E. and Janska, H. (2009). The lack of mitochondrial AtFtsH4 protease alters *Arabidopsis* leaf morphology at the late stage of rosette development under short day photoperiod. *Plant Journal.* DOI: 10.1111/j.1365-313X.2009.03907.x.

Giustarini, D., Dalle-Donne, I., Colombo, R., Milzani, A. and Rossi, R. (2008). Is ascorbate able to reduce disulfide bridges? A cautionary note. *Nitric Oxide* **19**, 252–258.

Guo, F. Q. and Crawford, N. M. (2005). *Arabidopsis* nitric oxide synthase1 is targeted to mitochondria and protects against oxidative damage and dark-induced senescence. *Plant Cell* **17**, 3436–3450.

Halliwell, B. (2006). Reactive species and antioxidants. Redox biology is a fundamental theme of aerobic life. *Plant Physiology* **141**, 312–322.

Halliwell, B. and Gutteridge, J. M. C. (2007). Free radicals in biology and medicine 4th edn., Oxford University Press, Oxford.

Hanson, G. T., Aggeler, R., Oglesbee, D., Cannon, M., Capaldi, R. A., Tsien, R. Y. and Remington, S. J. (2004). Investigating mitochondrial redox potential with redox-sensitive green fluorescent protein indicators. *Journal of Biological Chemistry* **279**, 13044–13053.

Huang, B. and Chen, C. (2006). An ascorbate-dependent artifact that interferes with the interpretation of the biotin switch assay. *Free Radicals in Biology and Medicine* **41**, 562–567.

Ito, J., Heazlewood, J. L. and Millar, A. H. (2006). Analysis of the soluble ATP-binding proteome of plant mitochondria identifies new proteins and nucleotide triphosphate interactions within the matrix. *Journal of Proteome Research* **5**, 3459–3469.

Jaffrey, S. R., Erdjument-Bromage, H., Ferris, C. D., Tempst, P. and Snyder, S. H. (2001). Protein *S*-nitrosylation: A physiological signal for neuronal nitric oxide. *Nature Cell Biology* **3**, 193–197.

Jeong, J. and Guerinot, M. L. (2009). Homing in on iron homeostasis in plants. *Trends in Plant Science* **14**, 280–285.

Jiang, K., Schwarzer, C., Lally, E., Zhang, S., Ruzin, S., Machen, T., Remington, S. J. and Feldman, L. (2006). Expression and characterization of a redox-sensing green fluorescent protein (reduction-oxidation-sensitive green fluorescent protein) in *Arabidopsis*. *Plant Physiology* **141**, 397–403.

Johansson, E., Olsson, O. and Nyström, T. (2004). Progression and specificity of protein oxidation in the life cycle of *Arabidopsis thaliana*. *Journal of Biological Chemistry* **279**, 22204–22208.

Kennedy, M. C., Emptage, M. H., Dreyer, J. L. and Beinert, H. (1983). The role of iron in the activation-inactivation of aconitase. *Journal of Biological Chemistry* **258**, 11098–11105.

Kolbe, A., Oliver, S. N., Fernie, A. R., Stitt, M., van Dongen, J. T. and Geigenberger, P. (2006). Combined transcript and metabolite profiling of *Arabidopsis* leaves reveals fundamental effects of the thiol-disulfide status on plant metabolism. *Plant Physiology* **141**, 412–422.

Koppen, M. and Langer, T. (2007). Protein degradation within mitochondria: Versatile activities of AAA proteases and other peptidases. *Critical Reviews in Biochemistry and Molecular Biology* **42**, 221–242.

Kristensen, B. K., Askerlund, P., Bykova, N. V., Egsgaard, H. and Møller, I. M. (2004). Identification of oxidised proteins in the matrix of rice leaf mitochondria by immunoprecipitation and two-dimensional liquid chromatography-tandem mass spectrometry. *Phytochemistry* **65**, 1839–1851.

Kristiansen, K. A., Jensen, P. E., Møller, M. I. and Schulz, A. (2009). Monitoring reactive oxygen species formation and localization in living cells by use of the fluorescent probe CM-H_2DCFDA and confocal laser microscopy. *Physiologia Plantarum* **136**, 369–383.

Lindermayr, C., Saalbach, G. and Durner, J. (2005). Proteomic identification of *s*-nitrosylated proteins in *Arabidopsis*. *Plant Physiology* **137**, 921–930.

Meyer, C., Lea, U. S., Provan, F., Kaiser, W. M. and Lillo, C. (2005). Is nitrate reductase a major player in the plant NO (nitric oxide) game? *Photosynthesis Research* **83**, 181–189.

Meyer, A. J., Brach, T., Marty, L., Kreye, S., Rouhier, N., Jacquot, J. P. and Hell, R. (2007). Redox-sensitive GFP in *Arabidopsis thaliana* is a quantitative biosensor for the redox potential of the cellular glutathione redox buffer. *Plant Journal* **52**, 973–986.

Møller, I. M. (2001). Plant mitochondria and oxidative stress: Electron transport, NADPH turnover, and metabolism of reactive oxygen species. *Annual Review of Plant Physiology and Plant Molecular Biology* **52**, 561–591.

Møller, I. M. and Kristensen, B. K. (2006). Protein oxidation in plant mitochondria detected as oxidized tryptophan. *Free Radicals in Biology and Medicine* **40**, 430–435.

Møller, I. M., Jensen, P. E. and Hansson, A. (2007). Oxidative modifications to cellular components in plants. *Annual Review of Plant Biology* **58**, 459–481.

Morgan, M. J., Lehmann, M., Schwarzlander, M., Baxter, C. J., Sienkiewicz-Porzucek, A., Williams, T. C., Schauer, N., Fernie, A. R., Fricker, M. D., Ratcliffe, R. G., Sweetlove, L. J. and Finkemeier, I. (2008). Decrease in manganese superoxide dismutase leads to reduced root growth and affects TCA cycle flux and mitochondrial redox homeostasis. *Plant Physiology* **147**, 101–114.

Neill, S., Desikan, R. and Hancock, J. (2002). Hydrogen peroxide signalling. *Current Opinion in Plant Biology* **5**, 388.

Nguyen, A. T. and Donaldson, R. P. (2005). Metal-catalyzed oxidation induces carbonylation of peroxisomal proteins and loss of enzymatic activities. *Archives of Biochemistry and Biophysics* **439**, 25–31.

Noctor, G. and Foyer, C. H. (1998). Acorbate and glutathione: Keeping active oxygen under control. *Annual Review of Plant Physiology and Plant Molecular Biology* **49**, 249–279.

Noctor, G., De Paepe, R. and Foyer, C. H. (2007). Mitochondrial redox biology and homeostasis in plants. *Trends in Plant Science* **12**, 125–134.

Nyström, T. (2005). Role of oxidative carbonylation in protein quality control and senescence. *EMBO Journal* **24**, 1311–1317.

Oracz, K., El-Maarouf Bouteau, H., Farrant, J. M., Cooper, K., Belghazi, M., Job, C., Job, D., Corbineau, F. and Bailly, C. (2007). ROS production and protein oxidation as a novel mechanism for seed dormancy alleviation. *Plant Journal* **50**, 452–465.

Oracz, K., El-Maarouf-Bouteau, H., Kranner, I., Bogatek, R., Corbineau, F. and Bailly, C. (2009). The Mechanisms involved in seed dormancy alleviation by hydrogen cyanide unravel the role of reactive oxygen species as key factors of cellular signaling during germination. *Plant Physiology* **150**, 494–505.

Padgett, C. M. and Whorton, A. R. (1995). *S*-nitrosoglutathione reversibly inhibits GAPDH by *S*-nitrosylation. *American Journal of Physiology* **269**, C739–C749.

Patel, S., Caplan, J. and Dinesh-Kumar, S. P. (2006). Autophagy in the control of programmed cell death. *Current Opinion in Plant Biology* **9**, 391–396.

Penaa, L. B., Pasquinia, L. A., Tomaroa, M. L. and Gallego, S. M. (2006). Proteolytic system in sunflower (*Helianthus annuus L.*) leaves under cadmium stress. *Plant Science* **171**, 531–537.

Qiu, Q. S., Huber, J. L., Booker, F. L., Jain, V., Leakey, A. D., Fiscus, E. L., Yau, P. M., Ort, D. R. and Huber, S. C. (2008). Increased protein carbonylation in leaves of *Arabidopsis* and soybean in response to elevated [CO_2]. *Photosynthesis Research* **97**, 155–166.

Rajjou, L., Lovigny, Y., Groot, S. P., Belghazi, M., Job, C. and Job, D. (2008). Proteome-wide characterization of seed aging in *Arabidopsis*: A comparison between artificial and natural aging protocols. *Plant Physiology* **148**, 620–641.

Rigas, S., Daras, G., Laxa, M., Marathias, N., Fasseas, C., Sweetlove, L. J. and Hatzopoulos, P. (2009). Role of Lon1 protease in post-germinative growth and maintenance of mitochondrial function in *Arabidopsis thaliana*. *New Phytologist* **181**, 588–600.

Rinalducci, S., Murgiano, L. and Zolla, L. (2008). Redox proteomics: Basic principles and future perspectives for the detection of protein oxidation in plants. *Journal of Experimental Botany* **59**, 3781–3801.

Romero-Puertas, M. C., Laxa, M., Matte, A., Zaninotto, F., Finkemeier, I., Jones, A. M., Perazzolli, M., Vandelle, E., Dietz, K. J. and Delledonne, M. (2007). *S*-nitrosylation of peroxiredoxin II E promotes peroxynitrite-mediated tyrosine nitration. *Plant Cell* **19**, 4120–4130.

Romero-Puertas, M. C., Campostrini, N., Matte, A., Righetti, P. G., Perazzolli, M., Zolla, L., Roepstorff, P. and Delledonne, M. (2008). Proteomic analysis of *S*-nitrosylated proteins in *Arabidopsis thaliana* undergoing hypersensitive response. *Proteomics* **8**, 1459–1469.

Rouhier, N., Vieira Dos Santos, C., Tarrago, L. and Rey, P. (2006). Plant methionine sulfoxide reductase A and B multigenic families. *Photosynthesis Research* **89**, 247–262.

Rouhier, N., Lemaire, S. D. and Jacquot, J. P. (2008). The role of glutathione in photosynthetic organisms: Emerging functions for glutaredoxins and glutathionylation. *Annual Review of Plant Biology* **59**, 143–166.

Sagi, M. and Fluhr, R. (2001). Superoxide production by plant homologues of the gp91(phox) NADPH oxidase. Modulation of activity by calcium and by tobacco mosaic virus infection. *Plant Physiology* **126**, 1281–1290.

Schwarzlander, M., Fricker, M. D., Muller, C., Marty, L., Brach, T., Novak, J., Sweetlove, L. J., Hell, R. and Meyer, A. J. (2008). Confocal imaging of glutathione redox potential in living plant cells. *Journal of Microscopy* **231**, 299–316.

Schwarzlander, M., Fricker, M. D. and Sweetlove, L. J. (2009). Monitoring the *in vivo* redox state of plant mitochondria: Effect of respiratory inhibitors, abiotic stress and assessment of recovery from oxidative challenge. *Biochimica et Biophysica Acta* **1787**, 468–475.

Sell, S., Lindermayr, C. and Durner, J. (2008). Identification of *S*-nitrosylated proteins in plants. *Methods in Enzymology* **440**, 283–293.

Stadtman, E. R. (1990). Covalent modification reactions are marking steps in protein turnover. *Biochemistry* **29**, 6323–6331.

Sweetlove, L. J., Heazlewood, J. L., Herald, V., Holtzapffel, R., Day, D. A., Leaver, C. J. and Millar, A. H. (2002). The impact of oxidative stress on *Arabidopsis* mitochondria. *Plant Journal* **32**, 891–904.

Szpunar, J. (2005). Advances in analytical methodology for bioinorganic speciation analysis: Metallomics, metalloproteomics and heteroatom-tagged proteomics and metabolomics. *Analyst* **130**, 442–465.

Taylor, N. L., Heazlewood, J. L., Day, D. A. and Millar, A. H. (2005). Differential impact of environmental stresses on the pea mitochondrial proteome. *Molecular and Cellular Proteomics* **4**, 1122–1133.

Thompson, A. R. and Vierstra, R. D. (2005). Autophagic recycling: Lessons from yeast help define the process in plants. *Current Opinion in Plant Biology* **8**, 165–173.

Wang, Y., Yun, B. W., Kwon, E., Hong, J. K., Yoon, J. and Loake, G. J. (2006). *S*-nitrosylation: An emerging redox-based post-translational modification in plants. *Journal of Experimental Botany* **57**, 1777–1784.

Wang, W., Fang, H., Groom, L., Cheng, A., Zhang, W., Liu, J., Wang, X., Li, K., Han, P., Zheng, M., Yin, J. Wang, W. *et al.* (2008). Superoxide flashes in single mitochondria. *Cell* **134**, 279–290.

Wang, Y. Q., Feechan, A., Yun, B. W., Shafiei, R., Hofmann, A., Taylor, P., Xue, P., Yang, F. Q., Xie, Z. S., Pallas, J. A., Chu, C. C. and Loake, G. J. (2009). *S*-nitrosylation of AtSABP3 antagonizes the expression of plant immunity. *Journal of Biological Chemistry* **284**, 2131–2137.

Winger, A. M., Taylor, N. L., Heazlewood, J. L., Day, D. A. and Millar, A. H. (2007). The cytotoxic lipid peroxidation product 4-hydroxy-2-nonenal covalently modifies a selective range of proteins linked to respiratory function in plant mitochondria. *Journal of Biological Chemistry* **282**, 37436–37447.

Xiong, Y., Contento, A. L., Nguyen, P. Q. and Bassham, D. C. (2007). Degradation of oxidized proteins by autophagy during oxidative stress in *Arabidopsis*. *Plant Physiology* **143**, 291–299.

Zemojtel, T., Frohlich, A., Palmieri, M. C., Kolanczyk, M., Mikula, I., Wyrwicz, L. S., Wanker, E. E., Mundlos, S., Vingron, M., Martasek, P. and Durner, J. (2006). Plant nitric oxide synthase: A never-ending story? *Trends in Plant Science* **11**, 524–525.

Reactive Oxygen Species: Regulation of Plant Growth and Development

HYUN-SOON KIM,* YOON-SIK KIM,*,† KYU-WOONG HAHN,† HYOUK JOUNG* AND JAE-HEUNG JEON*,[1]

*Plant Systems Engineering Center, Korea Research Institute of Bioscience and Biotechnology (KRIBB), Daejon 305-333, South Korea
†Department of Biological Sciences, Hannam University, Daejeon 306-791, Korea

ABSTRACT

In plants, reactive oxygen species (ROS) are continuously produced from aerobic metabolic processes such as the photosynthetic and respiratory reactions. The cellular accumulation of ROS, which are highly reactive, is highly cytotoxic. Therefore, all the aerobic organisms have been evolved to develop efficient ROS-scavenging mechanisms. In recent years, the role of ROS in the regulation of plant growth and development has been identified. Increased ROS production is functionally coupled to the effects of plant growth regulators. The specific ROS that are related to specific plant hormones may control plant growth and development. The recent discovery of a

[1]Corresponding author: E-mail: jeonjh@kribb.re.kr

Advances in Botanical Research, Vol. 52

0065-2296/09 $35.00
DOI: 10.1016/S0065-2296(09)52002-2

tip-high, Ca^{2+}-interdependent, ROS gradient produced by NADPH oxidase and its close association with polarized growth will provide information on the dual role of ROS in plants, as both the toxic byproducts of aerobic metabolism and the key regulators of growth and developmental pathways.

ABBREVIATIONS

ABA	abscisic acid
ABRE	ABA-responsive elements
ACC	1-aminocyclopropane-1-carboxylate
ACO	ACC oxidase
APX	ascrobate peroxidase
BAP	6-benzylaminopurine
CAT	catalase
Cu/ZnSOD	Cu/Zn superoxide dismutase
DPI	diphenylene iodonium
GAs	gibberellins
MAPK	mitogen-activated protein kinase
MV	methyl viologen
NOX	NADPH oxidase
ROS	reactive oxygen species

I. INTRODUCTION

Plants use molecular oxygen for efficient energy production; as a result, reactive oxygen species (ROS) are also produced. The cellular accumulation of ROS, which are highly reactive, is highly cytotoxic. Consequently, the evolution of all aerobic organisms is dependent on the development of efficient ROS-scavenging mechanisms. The distinct subcellular localization and biochemical properties of antioxidant enzymes, their differential inducibility at the enzyme and gene expression level and the plethora of nonenzymatic scavengers render the antioxidant systems with very versatile and flexible units that can control ROS accumulation temporally and spatially (Shao *et al.*, 2008).

The growth and development of plants are profoundly affected by environmental conditions, because plants are unable to migrate from extreme environmental conditions. To survive stressful conditions, plants must slow their rate of growth. A wide range of adaptive strategies to cope with adverse environmental conditions has evolved. ROS production under abiotic stress is kept under tight control by versatile and supportive antioxidant systems that modulate intracellular ROS concentrations and maintain system homeostasis. ROS signaling is controlled by the production and scavenging of ROS (Fig. 1). Different developmental or environmental signals feed into the ROS signaling network and perturb ROS homeostasis in a compartment-specific or even cell-specific manner (Mittler *et al.*, 2004).

Recently, a new role for ROS in the regulation of plant growth and development has been identified. Increased ROS production is functionally coupled to the effects of plant growth regulators. There are many reports regarding the relationship between ROS and plant hormones for regulation of plant growth and development. The specific ROS that are related to specific plant hormones may control plant growth and development. We summarized the recent discovery of a tip-high, Ca^{2+}-interdependent, ROS gradient produced by NADPH oxidase and its close association with polarized growth, such as pollen tubes and root hairs (Monshausen *et al.*, 2008). It provides information on the dual role of ROS in plants, as both toxic by-products of aerobic metabolism and key regulators of growth and developmental pathways.

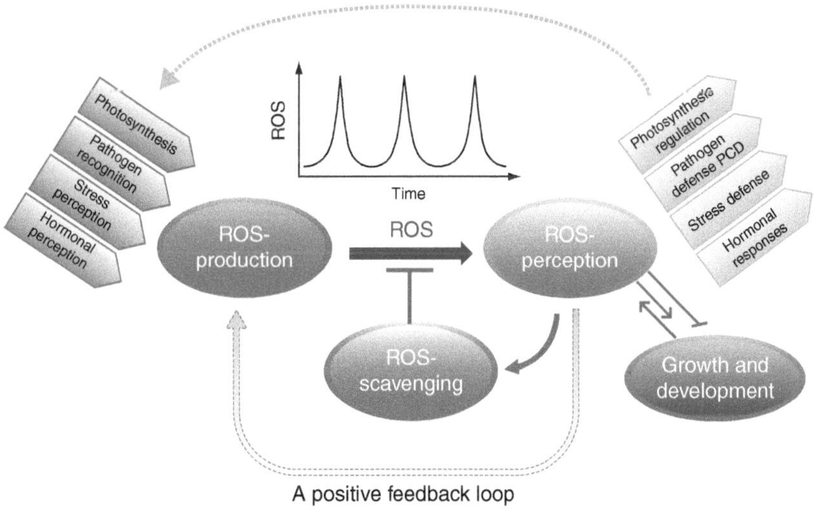

Fig. 1. Modulation of ROS signaling by the reactive oxygen gene network of plants. Different cellular signals result in the enhanced production of ROS in cells by the ROS-producing pathways of the network. ROS are perceived by different ROS sensors and activate cellular responses. The intensity, duration, and localization of the ROS signals are determined by interplay between the ROS-producing and the ROS-scavenging pathways. This figure was originally published in Mittler *et al.* (2004) and is reproduced by permission from Elsevier Ltd.

II. REDOX REGULATION

Higher plants require oxygen for the efficient production of energy, as do other aerobic organisms. As a consequence of aerobic metabolism, ROS, namely superoxide radicals ($O_2^{\bullet-}$), hydrogen peroxide (H_2O_2), and hydroxyl radicals ($OH^\bullet$) are formed by the partial reduction of molecular oxygen. An increase in the production of ROS is an inevitable by-product of the reduction of O_2 during many of these processes (Apel and Hirt, 2004). Since the introduction of molecular oxygen into our atmosphere by O_2-producing photosynthetic organisms, ROS have been the unwelcome companions of aerobic life (Halliwell, 2006). A common feature among different ROS types is their ability to cause oxidative damage to proteins, DNA, and lipids. The cellular accumulation of ROS, which are highly reactive, is extremely cytotoxic (Apel and Hirt, 2004). Consequently, the evolution of all aerobic organisms is dependent on the development of efficient ROS-scavenging mechanisms (Foyer and Noctor, 2005a). Higher plants possess very efficient enzymatic and nonenzymatic antioxidant defense systems that allow ROS-scavenging and protect plant cells from oxidative damage (Shao *et al.*, 2008). The distinct subcellular localization and biochemical properties of antioxidant enzymes, their differential inducibility at the enzyme and gene expression level, and the plethora of nonenzymatic scavengers render the antioxidant systems with a very versatile and flexible unit that can control ROS accumulation temporally and spatially (Foyer and Noctor, 2005a; Shao *et al.*, 2008).

ROS are formed by partial reduction of molecular oxygen. The simultaneous presence of both oxidized and reduced forms of electron carriers, known as redox poising, involves a continuous flux of electrons to molecular oxygen from multiple sites in the photosynthetic and respiratory electron transport chains (Foyer and Noctor, 2005b). Both enzymatic and nonenzymatic antioxidants, such as glutathione, ascorbate, tocopherol, proline, betaine, and others, can protect higher plant cells from oxidative damage by scavenging ROS. In addition to their crucial roles in defense systems and as enzyme cofactors, antioxidants influence higher plant growth and development by modifying a variety of processes, ranging from mitosis and cell elongation to senescence and death. Most importantly, antioxidants provide essential information on the cellular redox state and regulate expression of genes associated with biotic and abiotic stress responses to optimize defense and survival. Special attention has been paid to the interaction between ROS and ROS-antioxidants as a metabolic interface for different types of signals derived from metabolism and from the changing environment (Foyer and Noctor, 2005b).

As open systems, living cells require a constant flux of energy for continuous biomass production and consumption, and they depend on cellular homeostasis to maintain all functions. This, in turn, can only be achieved when the relatively small pools of ATP/ADP, NAD(P)H/NAD(P), and other redox carriers, as well as cellular pH, remain in balance. This allows input and withdrawal of both energy and reducing equivalents at the required rates and keeps the system in homeostasis (Scheibe *et al.*, 2005). Cellular homeostasis can be maintained as long as the capacity for acclimation is sufficient. If an impact is too rapid, and acclimation on the level of gene expression cannot occur, cell damage and death are initiated (Scheibe *et al.*, 2005).

The extent to which ROS accumulate is determined by the antioxidant defense system, which enables organisms to maintain proteins and other cellular components in an active state for metabolism. Like all other aerobic organisms, plants maintain most cytoplasmic thiols in the reduced (–SH) state because of the low thiol-disulfide redox potential imposed by millimolar amounts of the thiol buffer glutathione. Unlike many animal cells, plant cells synthesize high concentrations of ascorbate (vitamin C), which is an additional hydrophilic redox buffer that provides robust protection against oxidative challenge. Redox homeostasis is governed by the presence of large pools of these antioxidants that absorb and buffer reductants and oxidants (Foyer and Noctor, 2005b). Plants also make tocopherols (vitamin E) that act as important liposoluble redox buffers. Although tocopherol is considered a major singlet oxygen scavenger, it is also an effective scavenger of other ROS. The reduced scavenging form of tocopherol is regenerated by ascorbate. Moreover, because the tocopherol redox couple has a more positive midpoint potential than that of the ascorbate pool, the range of effective superoxide scavenging is further enhanced. The ability of the ascorbate, glutathione, and tocopherol pools to act as redox buffers in plant cells is one of their most important attributes (Foyer and Noctor, 2005b). Pathways of ROS signaling are made possible by homeostatic regulation, achieved by antioxidant redox buffering. Because antioxidants continuously process ROS, they determine the lifetime and specificity of the ROS signal.

Higher plants can sense, transduce, and translate ROS signals into the appropriate cellular responses. This process requires the presence of redox-sensitive proteins that can undergo reversible oxidation/reduction and thereby switch on and off, depending on the cellular redox state (Foyer and Noctor, 2003). ROS can oxidize redox-sensitive proteins directly or indirectly via ubiquitous redox-sensitive molecules, such as glutathione (GSH) or thioredoxins, which control the cellular redox state in higher plants (Foyer and Noctor, 2005a). There are many recent reports regarding the molecular mechanisms of redox-sensitive regulation of protein functions. Redox-sensitive metabolic enzymes may directly modulate corresponding cellular metabolism. Another molecular mechanism involves redox-sensitive signaling proteins, which execute their function via downstream signaling components, such as kinases, phosphatases, and transcription factors (Foyer and Noctor, 2005b). Redox regulation of protein function plays an important role in many biological processes, such as enzymatic activity, protein–protein interactions, and even DNA-binding activity (Maya-Ampudia and Bernal-Lugo, 2006). Recently, many proteins related to cell signaling pathways have been identified as redox sensitive, such as phospholipase C (Kamata and Hirata, 1999), heterotrimeric G-proteins (Nishida *et al.*, 2000; Pfannschmidt *et al.*, 2003), protein tyrosine phosphatases (Chiarugi and Cirri, 2003), and Ca^{2+}-mobilizing proteins, for example, Ca^{2+} channels and Ca^{2+}-ATPases (Suzuki and Ford, 1999).

ROS are involved in redox regulation because they reversibly oxidize cysteine residues. This regulation modulates and integrates different cellular responses to extracellular stimuli. The best characterized redox signal transduction system in higher plants is the stromal ferredoxin–thioredoxin system, which functions in the regulation of photosynthetic carbon metabolism. Signal transmission involves disulfide-thiol conversion in target enzymes and is probably achieved by a light-induced decrease in the thioredoxin redox potential (Barnes and Mayfield, 2003; Schürmann and Buchanan, 2008; Schürmann and Jacquot, 2000; Yabuta *et al.*, 2004). Alternative reduction systems are glutaredoxins, some of which are also located in plastids/ chloroplasts (Rouhier *et al.*, 2008). In higher plants, ROS utilize and/or interfere with other signaling pathways or molecules, forming a signaling network (Foyer and Noctor, 2003). The network of redox signals from energy-generating organelles orchestrates metabolism to adjust energy production to utilization, interfacing with hormone signaling to respond to environmental change at every stage of plant development (Foyer and Noctor, 2003).

III. PLANT HORMONES AND ROS

To survive under stressful conditions, plants slow their rate of growth. This strategy enables plants to limit the risk of heritable damage, while conserving energy for defense purposes (May *et al.*, 1998). The oxidation–reduction (redox) cascades of the photosynthetic and respiratory electron transport chains not only provide the driving forces for metabolism but also generate redox signals that both participate in and regulate every aspect of plant biology, ranging from gene expression and translation to enzyme chemistry. ROS play an important signaling role in these adaptive processes as ubiquitous messengers of stress responses. Plants respond to changes in their environment by varying the amount of ROS, thereby delivering environmental signals to the nucleus for transcription of specific genes. The role of ROS as signal transducers has been studied in the areas of plant physiology, defense, and gene expression (Apel and Hirt, 2004). There are many reports regarding the relationship between ROS and plant hormones for regulation of plant growth and development (Cui *et al.*, 1999; Joo *et al.*, 2001). ROS are asymmetrically generated in roots by gravistimulation in regions of reduced growth (Cui *et al.*, 1999; Joo *et al.*, 2001). A function for ROS in root curvature was discovered by inhibition of cell growth, thus contributing to tropisms. In a previous study, auxin also induced ROS production in roots, while the auxin-transport inhibitor, *N*-1-naphthylphthlamic acid, did not inhibit hydrogen peroxide (H_2O_2)-induced root curvature, leading to the hypothesis that ROS play a role downstream of transport in auxin signaling and gravitropism (Cui *et al.*, 1999; Joo *et al.*, 2001). Generally, plant hormones are located downstream of the ROS signal. H_2O_2 induces accumulation of stress hormones, such as salicylic acid (SA) and ethylene (Chamnongpol *et al.*, 1998). In addition, H_2O_2-mediated stomatal closure was completely disrupted in the loss-of-function mutant of the ethylene receptor *etr1-7*, suggesting a role for ETR1 in guard cell ROS signaling and stomatal closure (Desikan *et al.*, 2005). Interestingly, in another recent study, ethylene was proposed to counteract stomatal closure (Tanaka *et al.*, 2005). Abscisic acid (ABA)-induced stomatal closure was inhibited by ethylene or the hormone precursor 1-aminocyclopropane-1-carboxylic acid and by the ethylene-overproducing mutation *eto1-1* (Tanaka *et al.*, 2005). In winter squash, ROS stimulate ethylene production by activating the expression of *ACC* synthase and diphenylene iodonium (DPI), which is an inhibitor of ROS generation, thereby blocking ethylene production (Watanabe and Sakai, 1998). Elevated ROS (H_2O_2) production via overexpression of the chloroplastic Cu/ZnSOD gene triggered ethylene synthesis by activating *ACO* gene expression in transgenic potatoes (Kim *et al.*, 2008).

Abiotic stress adversely affects plant growth and development. Several plant growth regulators, such as SA, ABA, jasmonic acid, and ethylene, play a crucial role in altering plant morphology in response to various stresses (Jia *et al.*, 2002; Larkindale and Knight, 2002; Swiątek *et al.*, 2002). Among them, ABA is an important regulator that coordinates plant development with environmental stress conditions, such as drought, high salinity, and cold temperature. ABA regulates many important plant developmental processes and induces tolerance to different stresses. Reportedly, ABA causes generation of ROS (Hu *et al.*, 2005; Jiang and Zhang, 2001; Kwak *et al.*, 2003; Pei *et al.*, 2000), which are important intermediate components of the ABA-induced antioxidant defense system (Hu *et al.*, 2005; Jiang and Zhang, 2004).

ABA production is increased in tissues during stressful conditions, and this causes a variety of physiological effects, including water potential regulation. Reportedly, the increase in the level of ABA plays a major role in iron-induced biosynthesis of ferritin (Lobreaux *et al.*, 1993). Inducibility of ferritin mRNA accumulation by iron is dramatically decreased in the maize ABA-deficient mutant *vp2*, and it can be rescued by addition of exogenous ABA, confirming the involvement of ABA in the iron response in plants. Guan and Scandalios (1998) demonstrated that SOD-related genes may respond to ABA and osmotic stress via alternate pathways. They hypothesized that the increase in Sod4 and Sod4A mRNA in response to ABA is due in part to ABA-mediated metabolic changes that alter oxygen free radical levels and induce the antioxidant defense system (Guan and Scandalios, 1998). Recently, Kempa *et al.* reported that salt stress induces complex readjustment of carbohydrate metabolism and that ABA triggers the initial steps of carbon mobilization (Kempa *et al.*, 2008). Moreover, ABA prevents a series of events that culminate in programmed cell death, whereas GAs initiate apoptosis. ROS are key elements in aleurone programmed cell death. GA-treated barley aleurone cells lose their ability to scavenge ROS, and this loss ultimately results in the oxidative damage and cell death. ABA-treated cells, on the other hand, maintain their ability to scavenge ROS and remain viable (Fath *et al.*, 2001). ABA biosynthesis is regulated by osmotic stress at multiple steps. Both ABA-dependent and ABA-independent osmotic stress signaling first modify constitutively expressed transcription factors, leading to the expression of early response transcriptional activators, which then activate downstream stress tolerance effector genes (Zhu, 2002). ABA modifies gene expression, and the analysis of ABA-responsive promoters has revealed a diverse collection of potential *cis*-acting regulatory elements. ABA regulates the expression of numerous stress-responsive genes. Based on various promoter analyses, ABA-responsive elements (ABREs) have been determined and a number of ABRE binding factors isolated (Kang *et al.*, 2002). The protein kinase family is involved in both the primary ABA signaling pathway and in early ABA perception in plants. Receptor-like kinase1 (RPK1), a Leu-rich repeat (LRR) receptor kinase in the plasma membrane, is upregulated by ABA in *Arabidopsis thaliana* (Osakabe *et al.*, 2005). The WRKY proteins are a superfamily of regulators that control diverse developmental and physiological processes. Also, the annexins (AnnAt1 and AnnAt4) play an important role in both osmotic stress and ABA signaling in a Ca^{2+}-dependent manner (Lee *et al.*, 2004).

ABA-induced leaf stomatal closure during drought stress has been well characterized. The synthesis of H_2O_2 by NADPH oxidase in guard cells is activated by ABA, and H_2O_2 mediates stomatal closure by activating plasma membrane Ca^{2+} channels (Kwak *et al.*, 2003; Pei *et al.*, 2000). There is some evidence that H_2O_2 is an endogenous component of ABA signaling in *Arabidopsis* guard cells. The synthesis of H_2O_2 by NADPH oxidase in guard cells is activated by ABA, and H_2O_2 mediates stomatal closure by activating plasma membrane Ca^{2+} channels (Kwak *et al.*, 2003; Pei *et al.*, 2000). Sarath *et al.* (2007) reported that the ABA activity requires ROS, and that seed germination is impacted by ROS and ABA (Sarath *et al.*, 2007).

ROS play a vital role in the guard cell ABA signaling network (Desikan *et al.*, 2004). ROS signaling during drought stress acts both downstream of stomatal closure and upstream in the ABA signaling network. Recent data have shown that *Arabidopsis* glutathione peroxidases, which in plants are in reality thioredoxin-dependent peroxidases (Navrot *et al.*, 2006) may also have distinct dual roles: a general scavenger and an oxidative signal transducer that relays the H_2O_2 signal in ABA and drought stress signaling. The T-DNA insertion mutant of *Arabidopsis* glutathione peroxidase3 (ATGPX3) exhibited a greater rate of water loss under drought stress, higher sensitivity to H_2O_2 treatment during seed germination and seedling development, and enhanced production of H_2O_2 in guard cells. The *atgpx3* mutation disrupted the ABA activation of calcium channels and the expression of ABA- and stress-responsive genes (Miao *et al.*, 2006). Furthermore, transgenic plants overexpressing ATGPX6 were more resistant to drought stress and recovered better than the wild type. ATGPX6 was suggested to have a dual role, as both an important scavenger of H_2O_2 and an essential element of the ABA signaling pathway, mediating stomatal regulation in response to drought stress (Miao *et al.*, 2006).

Mitogen-activated protein kinase (MAPK) cascades play an important role in mediating stress responses in eukaryotic organisms (Jonak *et al.*, 2002). MAPK and its immediate upstream activators, MAPK kinase (MAPKK) and MAPKK kinase, constitute a functionally interlinked MAPK cascade. MAPKs play a role in regulation of plant growth, development, and response to the environment (Jonak *et al.*, 2002; Mittler *et al.*, 2004; Moon *et al.*, 2003; Samuel and Ellis, 2002). The ABA signal is transmitted to the transcriptional apparatus through MAPK signaling (Lu *et al.*, 2002). It has been suggested that MAPK is involved in ABA-induced antioxidant defense, and that ROS are required for MAPK activation and induction of antioxidant defense during ABA signaling. A recent study showed that inhibition of MAPK signaling reduced the ABA-induced H_2O_2 production (Zhang *et al.*, 2006).

OsMAPK5 can positively regulate drought, salt, and cold tolerance and negatively modulate PR gene expression and broad-spectrum disease resistance (Xiong and Yang, 2003). Kim *et al.* identified *CIPK3* as a molecular link between stress- and ABA-induced calcium signaling and gene expression in plant cells. Because the cold signaling pathway is largely independent of endogenous ABA production, *CIPK3* represents a cross-talk node between the ABA-dependent and ABA-independent pathways during stress responses (Kim *et al.*, 2003). Using a conditional gain-of-function transgenic system, Liu and Zhang demonstrated that activation of SIPK, a tobacco stress-responsive MAPK, and MPK6, the *A. thaliana* ortholog of tobacco SIPK, regulates plant stress responses. DRE/CRT, a *cis*-acting element, is apparently involved in ABA-independent gene expression in response to dehydration and low-temperature stress (Liu and Zhang, 2004).

GAs are tetracyclic diterpenoid hormones that play an essential role in plant growth and development. They control a variety of growth responses in higher plants, including stem elongation, fruit ripening, flowering, and seed germination (Davies, 1995). In potatoes, GAs have an inhibitory effect on tuberization. Gibberellin (GA) activity decreased under conditions that promoted tuberization, such as short days (Kumar and Wareing, 1974), and increased in plants subjected to conditions that inhibited tuberization (Menzel, 1983). Transgenic potato plants, with elevated ROS (H_2O_2) production due to overexpression of the chloroplastic Cu/ZnSOD gene, show delayed plant growth and enhanced microtuberization. In addition to the difference in radical-scavenging capacity, the sense (SS4) and antisense (SA1) transgenic potato plants showed different morphological characteristics and growth patterns during tissue culture (Park *et al.*, 2006). A higher concentration of $O_2^{\bullet -}$ was detected in the antisense transgenic potato plants (SA1). These plants showed an enhanced plant growth, delayed microtuberization, and a higher level of all GAs tested (Fig. 2). These results suggest that a certain concentration of $O_2^{\bullet -}$ is necessary for stem elongation and that $O_2^{\bullet -}$ might activate shoot growth in potatoes. A specific ROS ($O_2^{\bullet -}$) acts as a signal transducer via the GA biosynthetic pathway for regulation of plant growth and tuber development in potatoes (Kim *et al.*, 2007).

A cysteine-rich, GA-induced gene (GAST1), the expression of which coincides with stem elongation, was identified in tomatoes. To date, numerous GAST1 homologs have been identified in various plant species. All of them encode small proteins with a putative signal peptide at the N terminus and a conserved C-terminal region comprising approximately 60 amino acids with 12 conserved cysteine residues. The corresponding genes are expressed during diverse developmental processes, including shoot and petal cell elongation, cessation of cell elongation, cell division, and root growth. GA induced a GAST1-like gene in petunia, GIP1–5. GIP2 is expressed in elongating zones, and its suppression in transgenic petunia plants inhibits stem elongation. The catalytic disulfide bonds (putative redox-active cysteines) in these proteins were involved in redox regulation. Analysis of GIP2, GIP4,

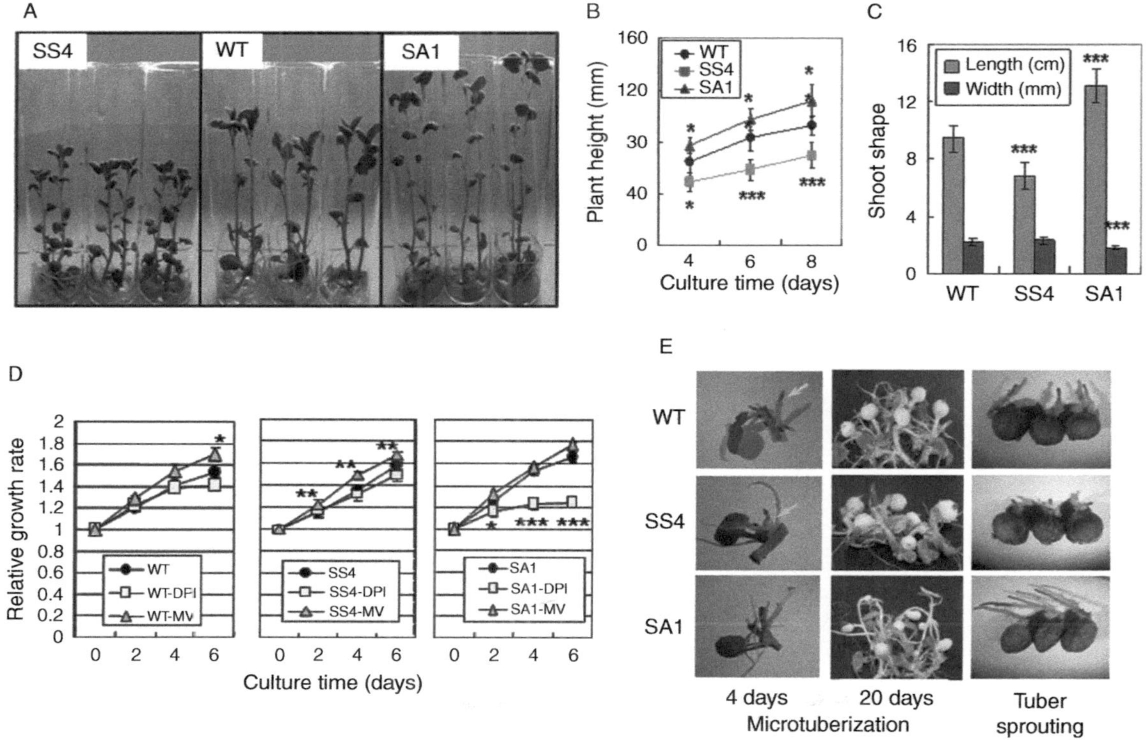

Fig. 2. Specific ROS regulate plant growth and tuber development of potato. Growth patterns of *in vitro* cultured wild-type (WT) and transgenic plants, SS4 (higher concentrations of H_2O_2) and SA1 (higher concentrations of $O_2^{\bullet -}$) were checked. (A) Photographs of growing potato shoot and growth curve (B) of WT and transgenic potatoes in liquid vial culture system. (C) Length and width of shoots were checked

and GIP5 expression revealed that these genes are induced by H_2O_2. Transgenic petunia plants expressing GIP2 showed reduced H_2O_2 levels in leaves following wounding. GIP2 expression also reduced the levels of H_2O_2 in guard cells following osmotic stress and ABA treatment, suppressing stomatal closure. In addition, the transgene promoted stem and corolla elongation. Thus, GIP2 apparently affects growth by regulating the levels of ROS (Wigoda *et al.*, 2006).

Seed germination is an important developmental change in the plant life cycle. Upon imbibition, cells switch from quiescence to very active metabolism, allowing biosynthesis of macromolecules and repair activities required for efficient germination (Gidrol *et al.*, 1994). Activation of a protective enzyme (SOD) upon imbibition was observed during germination of low-viability soybean seeds. A relationship between the zeatin riboside, one of the most abundant plant hormones, and ROS has been reported. It is probably directly related to the increased production of superoxide anion, which may also lead to the oxidation of endogenous cytokinin, one of the most abundant plant hormones, as suggested by the *in vitro* data. The mRNA encoding Fe-SOD was decreased following the treatment of cytokinin-starved, soybean cell suspension cultures with cytokinin. In general, cytokinin efficiently protects cells against oxidative stress, as demonstrated in *Escherichia coli* cells (Gidrol *et al.*, 1994). An *Agrobacterium tumefaciens* strain with mutations in the cytokine biosynthetic gene induced delayed tumorigenesis, and smaller, abnormal tumors, which underwent significantly reduced superoxide stress compared with tumors produced by the wild-type strain (Jia *et al.*, 1996). In *Kalanchoe laciniata*, T-DNA inserted into the plant genome may create oxidative stress in the infected tissue. Toxic superoxide anions produced by this stress deregulate cell division, resulting in oncogenesis. Transgenic plants expressing high levels of SOD activity exhibited a 55% reduction in $O_2^{\bullet-}$ production compared with nontransgenic plants. These results suggest that the phytohormones involved in superoxide stress may deregulate cell division, oncogenicity, and tumor formation (Jia *et al.*, 1996). Recent

for 10 days after transferring to vial. (D) Effect of treatments that modify ROS concentration on the growth of transgenic potatoes. To investigate the growth rate, plant heights were measured at 0, 2, 4, and 6 days after transferring to liquid culture media containing 10 μM DPI, an inhibitor of NADPH oxidase, and 5 μM MV, a contact herbicide that causes the massive, light-mediated accumulation of $O_2^{\bullet-}$ in photosynthetic tissues. (E) Large differences in *in vitro* microtuber formation and sprouting were observed. A specific ROS ($O_2^{\bullet-}$) acts as a signal transducer via the GA biosynthetic pathways for regulation of plant growth and tuber development in potatoes. This figure was originally published in Kim *et al.* (2007) and is reproduced by permission from Springer Publishing.

results also show that the cytokinin BAP reduces levels of ROS, and enhances the activity of antioxidant enzymes (CAT, APX). This result suggests that BAP protects the cell membrane and photosynthetic machinery from oxidative damage during delayed senescence in the dark. Increased oxidative stress displayed during dark-senescence of wheat leaves (*Triticum aestivum* L.) is caused by both the increased levels of radicals and the loss of antioxidant capacity. BAP reduced the degradation of the light-harvesting chlorophyll *a*/*b* binding protein (LHCP-2), and the large (LSU) and small subunits (SSU) of Rubisco. BAP maintains Chl levels by preventing degradation, rather than inducing Chl biosynthesis (Zavaleta-Mancera *et al.*, 2007).

IV. POLARIZED CELL GROWTH AND DEVELOPMENT

In recent years, a new role for ROS has been identified: the control and regulation of biological processes, such as growth, cell cycle, programmed cell death, hormone signaling, biotic and abiotic stress responses, and development (Foreman *et al.*, 2003). These studies suggest a dual role for ROS in plant biology, as both toxic byproducts of aerobic metabolism and key regulators of growth and developmental pathways. The use of ROS as signaling molecules by plant cells suggests that, during the course of evolution, plants were able to achieve a high degree of control over ROS toxicity and have evolved using ROS as signaling molecules. Controlling ROS toxicity, while enabling ROS to act as signaling molecules, appears to require a large reactive oxygen gene network composed of at least 152 genes in *Arabidopsis* (Mittler *et al.*, 2004).

ROS are emerging as important regulators of plant development. There is now abundant evidence that ROS play roles in cell growth and that spatial regulation of ROS production is an important determinant of plant form (Gapper and Dolan, 2006). The huge volumetric increase in plant cells as they transit from the founder population in and around meristems to the differentiated cells of the mature organ indicates that cell growth is an important component of plant development (Sugimoto-Shirasu and Roberts, 2003). Plant plasma membrane (PM)-localized NADPH oxidases (NOXs), which are partly homologous to the gp91phox (NOX2) catalytic subunit of the mammalian phagocyte NOX (Keller *et al.*, 1998), have been implicated in ROS formation during both the oxidative burst caused by pathogen infection (Torres *et al.*, 2002) and ABA-induced stomatal closure (Kwak *et al.*, 2003). Many recent discoveries suggest that ROS control development via regulation of cell growth. ROS that play a role in development are produced by NOXs that generate the superoxide radical ($O_2^{\bullet -}$), using NADPH as an electron donor. Moreover, additional studies have identified Rho-related GTPases as positional regulators of Nox function (Carol and Dolan, 2006; Jones *et al.*, 2007).

Studies of the localized accumulation of ROS in the apical region of *Aspergillus nidulans* hyphae strongly suggest that ROS play a key role in the spatial regulation of polar growth in fungal systems. In addition, NOX or related flavoproteins are responsible for the generation of ROS at hyphal tips and Rac1 activates NOX, whereas NOXR and Cdc42 may function together in a parallel pathway that regulates Nox localization (Semighini and Harris, 2008). In zygotes of the alga *Fucus serratus*, a tip-high, Ca^{2+}-interdependent, ROS gradient was associated with the polarized growth in *F. serratus* zygotes. Suppression of the ROS gradient inhibits polarized zygotic growth; conversely, exogenous ROS generation can redirect zygotic polarization following inhibition of endogenous ROS. These findings support a model in which localized production of ROS at the rhizoid tip stimulates formation of a localized tip-high, $[Ca^{2+}]$cyt gradient. Modulation of intracellular $[Ca^{2+}]$ cyt signals by ROS is a common motif in many plant and algal systems. A recent study extends this mechanism to embryogenesis (Coelho *et al.*, 2008). Consideration of the ways in which $[Ca^{2+}]$cyt signals are generated in other tip-growing systems may shed light on the mechanisms responsible for the generation of the embryogenic $[Ca^{2+}]$cyt gradient. In the two best-studied examples—pollen tubes (Hepler *et al.*, 2001; Potocky *et al.*, 2007) and root hairs (Foreman *et al.*, 2003)—the tip-high, $[Ca^{2+}]$cyt gradient was closely associated with the generation of ROS by NADPH oxidase. This ROS-$[Ca^{2+}]$cyt link is a well established and evolutionarily conserved signaling motif that has been observed in a number of plant tissues (Bothwell and Ng, 2005; Mori and Schroeder, 2004).

In *Arabidopsis*, this class of genes is referred to as *Arabidopsis* respiratory burst oxidase homologs (Atrboh) (Keller *et al.*, 1998; Torres *et al.*, 2002). Recently, a requirement for ROS in root-hair growth was demonstrated in *A. thaliana* using a loss-of-function knockout of *AtrbohC/RHD2* (Foreman *et al.*, 2003). *AtrbohC/RHD2* encodes a superoxide ($O_2^{\bullet -}$)-producing NOX, and the mutant reduced the ROS formation at the tip of very short root hairs (Foreman *et al.*, 2003). Because root-hair growth also requires tip-localized ROS production *via* the *AtrbohC/RHD2* NOX homolog (Foreman *et al.*, 2003), these results show that NOX is involved in ROS production during tip growth of plant cells. Furthermore, transgenic tomato plants expressing plant NOX RNAi transgenes show a wide range of pleiotropic developmental abnormalities (Sagi *et al.*, 2004). ROS derived from NOX are involved in polarized growth of pollen tubes. Localized ROS formation is a general feature of tip growth in plant cells, even though the cell wall composition of pollen tubes and root hairs differ considerably (Potocky *et al.*, 2007). ROS have been indirectly involved, by use of NOX-inhibitors such as DPI, in the control of the deposition of secondary cell walls in both developing cotton fibers (Potikha *et al.*, 1999) and in maize leaf growth (Rodriguez *et al.*, 2002).

The tip-high, Ca^{2+} gradient is thought to regulate the activity of components of the growth machinery, including the cytoskeleton, Ca^{2+}-dependent regulatory proteins, and the secretory apparatus during tip growth. In pollen tubes, both the Ca^{2+} gradient and cell elongation show oscillatory behavior, reinforcing the link between elongation of the root hairs and the associated tip-focused Ca^{2+} gradient (Monshausen *et al.*, 2008). Elevated apical Ca^{2+} is thought to provide a spatial determinant for growth by facilitating membrane fusion at the tip and regulating a host of Ca^{2+}-dependent proteins required for tip growth (Monshausen *et al.*, 2008). Indeed, the structures of the apical actin cytoskeleton and its regulatory proteins, such as villins, gelsolins, and actin-depolymerizing factors, are thought to be regulated by the tip-high Ca^{2+} (Yokota *et al.*, 2005). Similarly, calmodulin and protein kinases (Yoon *et al.*, 2006), which also play a role in sustaining tip growth, are regulated via the cytosolic Ca^{2+} gradient.

In some systems, tip growth oscillates, with periods of rapid expansion alternating with slower growth rates. While this has been independently confirmed for different species of pollen tubes by several research groups (Hwang *et al.*, 2005; Messerli and Robinson, 1997; Watahiki *et al.*, 2004), oscillating growth in root hairs has only recently been reported (Monshausen *et al.*, 2007). One of the most extensively studied tip-growing systems is that of lily (*Lilium longiflorum*) pollen tubes. In this system, growth oscillation was accompanied by oscillations in the tip-focused Ca^{2+} gradient (Messerli and Robinson, 1997; Watahiki *et al.*, 2004). The Ca^{2+} channel that sustains the tip-focused Ca^{2+} gradient in root hairs is thought to be gated by membrane voltage and ROS (Foreman *et al.*, 2003). During oscillating growth, dynamic increases in extracellular ROS and pH that oscillate with a similar frequency as growth but lagged growth oscillations by 7–8 s were observed (Monshausen *et al.*, 2007). These extracellular changes are thought to play a role in restricting growth at the tip (pH) and along the shank immediately behind the tip (ROS). The oscillations in extracellular ROS and pH lag the oscillatory increases in the intracellular Ca^{2+} gradient. Therefore, the oscillatory nature of the cytosolic Ca^{2+} gradient might be linked to the extracellular changes in ROS and pH as part of a system that limits growth after the initial burst of elongation has occurred, with elevations in Ca^{2+} being driven by each growth pulse and then triggering subsequent ROS and pH responses to limit additional expansion. This model is consistent with the likely Ca^{2+}-dependence of NADPH oxidases, which contain an EF hand-like

Ca^{2+}-binding domain that appear critical for tip growth. Recent data suggest that ROS- and Ca^{2+}-regulation of growth form a feedback loop to sustain tip growth (Takeda *et al.*, 2008). The spatial and temporal aspects of these three oscillatory parameters in relation to growth are depicted in the model shown in Fig. 3 (Monshausen *et al.*, 2008). Cells develop polarized shapes by generating and maintaining localized sites of growth. Elevation of cytoplasmic Ca^{2+} concentrations and the localized production of ROS by ROOT HAIR DEFECTIVE2 (RHD2) NADPH oxidase are important for root hair growth in *A. thaliana* (Carol and Dolan, 2006; Foreman *et al.*, 2003; Hepler *et al.*, 2001). There is evidence that a two-part mechanism determines the shape of root hair cells. A prerequisite is the apical localization of the RHD2 protein in the hair cell tip. Once localized to the tip, a positive feedback loop

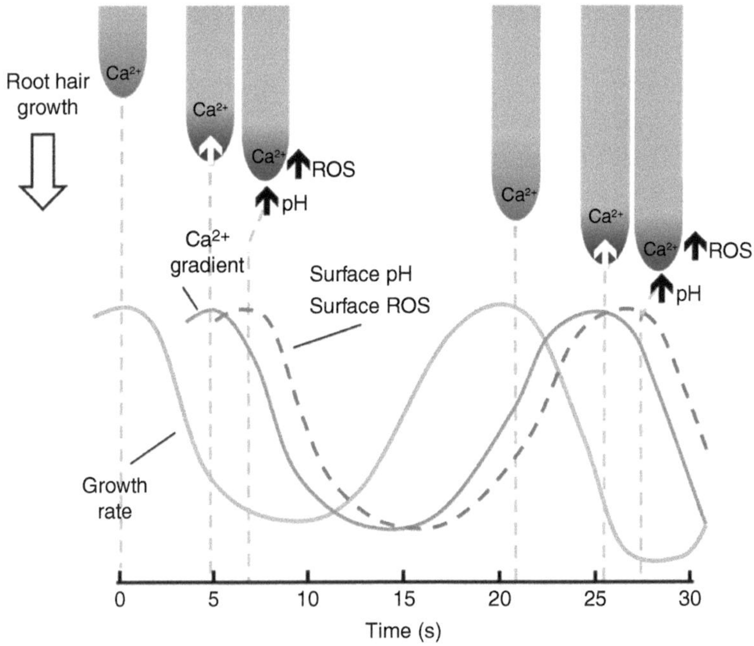

Fig. 3. Oscillation model of tip growth of *Arabidopsis* root hairs. Oscillations of the Ca^{2+} gradient lagged oscillations in growth by 5 s. Dynamic increases in extracellular ROS and pH that oscillate with a similar frequency as growth but lag growth oscillations by 7–8 s. The maximal oscillatory increase in cytosolic Ca^{2+} is triggered by cell expansion associated with tip growth and plays a role in the subsequent restriction of growth. These three oscillation parameters, maximal oscillatory increases in cytosolic Ca^{2+}, extracellular ROS, and pH, may be linked as part of a system that limits growth after the initial burst of elongation has occurred. This figure was originally published in Monshausen *et al.* (2008) and is reproduced by permission from *Plant Physiology*®/ *The Plant Cell.* (See color plate section at the back of the book.)

involving RHD2 is initiated. ROS derived from RHD2 NADPH oxidase may activate hyperpolarization-activated Ca^{2+} channels that transport Ca^{2+} into the cells, thereby activating the RHD2 NADPH oxidase activity through its EF hand and Ca^{2+}-dependent protein kinase activity. This positive feedback system, in concert with an independent mechanism for localizing RHD2 protein to the tip of the cell, provides a robust mechanism that explains how cells, such as root hairs, maintain polarity during morphogenesis (Takeda *et al.*, 2008).

ACKNOWLEDGMENTS

This research was supported by grants from the KRIBB Research Initiative Program and the Plant Diversity Research Center of the 21st Century Frontier Research Program, which is funded by the Ministry of Science and Technology of the Korean government.

REFERENCES

Apel, K. and Hirt, H. (2004). Reactive oxygen species: Metabolism, oxidative stress, and signal transduction. *Annual Review of Plant Biology* **55**, 373–399.

Barnes, D. and Mayfield, S. P. (2003). Redox control of posttranslational process in the chloroplast. *Antioxidants & Redox Signaling* **5**, 89–94.

Bothwell, J. and Ng, C. (2005). The evolution of Ca^{2+} signalling in photosynthetic eukaryotes. *New Phytologist* **166**, 21–38.

Carol, R. J. and Dolan, L. (2006). The role of reactive oxygen species in cell growth: Lessons from root hairs. *Journal of Experimental Botany* **57**, 1829–1834.

Chamnongpol, S., Willekens, H., Moeder, W., Langebartels, C., Sandermann, H., Jr., Van Montagu, M., Inze, D. and Van Camp, W. (1998). Defense activation and enhanced pathogen tolerance induced by H_2O_2 in transgenic plants. *Proceedings of the National Academy of Sciences of the United States of America* **95**, 5818–5823.

Chiarugi, P. and Cirri, P. (2003). Redox regulation of protein tyrosine phosphatases during receptor tyrosine kinase signal transduction. *Trends in Biochemical Science* **28**, 509–514.

Coelho, S., Brownlee, C. and Bothwell, J. (2008). A tip-high, Ca^{2+}-interdependent, reactive oxygen species gradient is associated with polarized growth in *Fucus serratus* zygotes. *Planta* **227**, 1037–1046.

Cui, K., Xing, G., Liu, X., Xing, G. and Wang, Y. (1999). Effect of hydrogen peroxide on somatic embryogenesis of *Lycium barbarum* L. *Plant Science* **146**, 9–16.

Davies, P. J. (1995). The plant hormones: Their nature, occurrence and functions. *In* "Plant Hormones: Physiology, Biochemistry and Molecular Biology", (P. J. Davies, ed.), pp. 1–12. Kluwer Academic Publisher, Dordrecht, The Netherlands.

Desikan, R., Cheung, M., Bright, J., Henson, D., Hancock, J. and Neil, S. (2004). ABA, hydrogen peroxide and nitric oxide signalling in stomatal guard cells. *Journal of Experimental Botany* **395**, 205–212.

Desikan, R., Hancock, J. T., Bright, J., Harrison, J., Weir, I., Hooley, R. and Neill, S. J. (2005). A role for ETR1 in hydrogen peroxide signaling in stomatal guard cells. *Plant Physiology* **137**, 831–834.

Fath, A., Bethke, P. and Jones, R. (2001). Enzymes that scavenge reactive oxygen species are down-regulated prior to gibberellic acid-induced programmed cell death in barley aleurone. *Plant Physiology* **126**, 156–166.

Foreman, J., Demidchik, V., Bothwell, J. H. F., Mylona, P., Miedema, H., Angel Torres, M., Linstead, P., Costa, S., Brownlee, C., Jones, J. D. G., Davies, J. M. and Dolan, L. (2003). Reactive oxygen species produced by NADPH oxidase regulate plant cell growth. *Nature* **422**, 442–446.

Foyer, C. H. and Noctor, G. (2003). Redox sensing and signaling associated with reactive oxygen in chloroplasts, peroxisomes and mitochondria. *Physiologia Plantarum* **119**, 355–364.

Foyer, C. H. and Noctor, G. (2005a). Oxidant and antioxidant signalling in plants: A re-evaluation of the concept of oxidative stress in a physiological context. *Plant, Cell and Environment* **28**, 1056–1071.

Foyer, C. H. and Noctor, G. (2005b). Redox homeostis and antioxidant signaling: A metabolic interface between stress perception and physiological responses. *The Plant Cell* **17**, 1866–1875.

Gapper, C. and Dolan, L. (2006). Control of plant development by reactive oxygen species. *Plant Physiology* **141**, 341–345.

Gidrol, X., Lin, W., Degousee, N., Yip, S. and Kush, A. (1994). Accumulation of reactive oxygen species and oxidation of cytokinin in germinating soybean seeds. *European Journal of Biochemistry* **224**, 21–28.

Guan, L. and Scandalios, J. (1998). Two structurally similar maize cytosolic superoxide dismutase genes, *Sod4* and *Sod4A*, respond differentially to abscisic acid and high osmoticum. *Plant Physiology* **117**, 217–224.

Halliwell, B. (2006). Reactive species and antioxidants. Redox biology is a fundamental theme of aerobic life. *Plant Physiology* **141**, 312–322.

Hepler, P., Vidali, L. and Cheung, A. (2001). Polarized cell growth in higher plants. *Annual Review of Cell Developmental Biology* **17**, 159–187.

Hu, X., Jiang, M., Zhang, A. and Lu, J. (2005). Abscisic acid-induced apoplastic H_2O_2 accumulation up-regulates the activities of chloroplastic and cytosolic antioxidant enzymes in maize leaves. *Planta* **223**, 57–68.

Hwang, J., Gu, Y., Lee, Y. and Yang, Z. (2005). Oscillatory ROP GTPase activation leads the oscillatory polarized growth of pollen tubes. *Molecular Biology of the Cell* **16**, 5385–5399.

Jia, S., Kumar, P. and Kush, A. (1996). Oxidative stress in Agrobacterium-induced tumors on Kalanchoe plants. *The Plant Journal* **10**, 545–551.

Jia, W., Wang, Y., Zhang, S. and Zhang, J. (2002). Salt-stress-induced ABA accumulation is more sensitively triggered in roots than in shoots. *Journal of Experimental Botany* **53**, 2201–2206.

Jiang, M. and Zhang, J. (2001). Effect of abscisic acid on active oxygen species, antioxidative defence system and oxidative damage in leaves of maize seedlings. *Plant and Cell Physiology* **42**, 1265–1273.

Jiang, M. Y. and Zhang, J. H. (2004). Abscisic acid and antioxidant defense in plant cells. *Acta Botanica Sinica* **46**, 1–9.

Jonak, C., Ökrész, L., Bögr, E. L. and Hirt, H. (2002). Complexity, cross talk and integration of plant MAP kinase signaling. *Current Opinion in Plant Biology* **5**, 415–424.

Jones, M., Raymond, M., Yang, Z. and Smirnoff, N. (2007). NADPH oxidase-dependent reactive oxygen species formation required for root hair growth depends on ROP GTPase. *Journal of Experimental Botany* **58**, 1261–1270.

Joo, J. H., Bae, Y. S. and Lee, J. S. (2001). Role of auxin-induced reactive oxygen species in root gravitropism. *Plant Physiology* **126**, 1055–1060.

Kamata, H. and Hirata, H. (1999). Redox regulation of cellular signalling. *Cellular Signaling* **11**, 1–14.

Kang, J.-Y., Choi, H.-I., Im, M.-Y. and Kim, S. (2002). *Arabidopsis* basic leucine zipper proteins that mediate stress-responsive abscisic acid signaling. *The Plant Cell* **14**, 343–357.

Keller, T., Damude, H. G., Werner, D., Doerner, P., Dixon, R. A. and Lamb, C. (1998). A plant homolog of the neutrophil NADPH oxidase gp91phox subunit gene encodes a plasma membrane protein with Ca^{2+} binding motifs. *The Plant Cell* **10**, 255–266.

Kempa, S., Krasensky, J., Santo, S., Kopka, J. and Jonak, C. (2008). A central role of abscisic acid in stress-regulated carbohydrate metabolism. *PLoS ONE* **3**, e3935.

Kim, K.-N., Cheong, Y., Grant, J., Pandey, G. and Luan, S. (2003). *CIPK3*, a calcium sensor-associated protein kinase that regulates abscisic acid and cold signal transduction in *Arabidopsis*. *The Plant Cell* **15**, 411–423.

Kim, M. S., Kim, H. S., Kim, Y. S., Baek, K. H., Oh, H. W., Hahn, K. W., Bae, R. N., Lee, I. J., Joung, H. and Jeon, J. H. (2007). Superoxide anion regulates plant growth and tuber development of potato. *Plant Cell Reports* **26**, 1717–1725.

Kim, Y., Kim, H., Lee, Y., Kim, M., Oh, H., Joung, H. and Jeon, J. (2008). Elevated H2O2 production via overexpression of a chCu/ZnSOD gene of lily triggers ethylene synthesis in transgenic potato. *Plant Cell Reports* **27**, 973–983.

Kumar, D. and Wareing, P. F. (1974). Studies on tuberization of *Solanum andigena*. *New Phytologist* **73**, 833–840.

Kwak, J., Mori, I., Pei, Z., Leonhardt, N., Torres, M., Dangl, J., Bloom, R., Bodde, S., Jones, J. and Schroeder, J. (2003). NADPH oxidase AtrbohD and AtrbohF genes function in ROS-dependent ABA signaling in *Arabidopsis*. *The EMBO Journal* **22**, 2623–2633.

Larkindale, J. and Knight, M. (2002). Protection against heat stress-induced oxidative damage in *Arabidopsis* involves calcium, abscisic acid, ethyelen, and salicylic acid. *Plant Physiology* **128**, 682–695.

Lee, S., Lee, E., Yang, E., Lee, J., Park, A., Song, W. and Park, O. (2004). Proteomic identification of annexins, calcium-dependent membrane binding proteins that mediate osmotic stress and abscisic acid signal transduction in *Arabidopsis*. *The Plant Cell* **16**, 1378–1391.

Liu, Y. and Zhang, S. (2004). Phosphorylation of 1-aminocyclopropane-1-carboxylic acid synthase by MPK6, a stress-responsive mitogen-activated protein kinase, induces ethylene biosynthesis in *Arabidopsis*. *The Plant Cell* **16**, 3386–3399.

Lobreaux, S., Hardy, T. and Briat, J.-F. (1993). Abscisic acid is involved in the iron-induced synthesis of maize ferritin. *The EMBO Journal* **12**, 651–657.

Lu, C., Han, M.-H., Guevara-Garcia, A. and Fedoroff, N. (2002). Mitogen-activated protein kinase signaling in postgermination arrest of development by abscisic acid. *Proceedings of the National Academy of Sciences of the United States of America* **99**, 15812–15817.

May, M. J., Vernoux, T., Leaver, C., Van Montagu, M. and Inze, D. (1998). Glutathione homeostasis in plants: implications for environmental sensing and plant development. *Journal of Experimental Botany* **49**, 647–667.

Maya-Ampudia, V. and Bernal-Lugo, I. (2006). Redox-sensitive target detection in gibberellic acid-induced barley aleurone layer. *Free Radical Biology & Medicine* **40**, 1362–1368.

Menzel, B. M. (1983). Tuberization in potato (*Solanum tuberosum* cultivar Sebago) at high temperatures: Gibberellin content and transport from buds. *Annals of Botany* **52**, 697–702.

Messerli, M. and Robinson, K. (1997). Tip localized Ca^{2+} pulses are coincident with peak pulsatile growth rates in pollen tubes of *Lilium longiflorum*. *Journal of Cell Science* **110**, 1269–1278.

Miao, Y., Lv, D., Wang, P., Wang, X., Chen, J., Miao, C. and Song, C. (2006). An *Arabidopsis* glutathione peroxidase functions as both a redox transducer and a scavenger in abscisic acid and drought stress responses. *The Plant Cell* **18**, 2749–2766.

Mittler, R., Vanderauwera, S., Gollery, M. and Van Breusegem, F. (2004). Reactive oxygen gene network of plants. *Trends in Plant Science* **9**, 490–498.

Monshausen, G. B., Bibikova, T. N., Messerli, M. A., Shi, C. and Gilroy, S. (2007). Oscillations in extracellular pH and reactive oxygen species modulate tip growth of *Arabidopsis* root hairs. *Proceedings of the National Academy of Sciences of the United States of America* **104**, 20996–21001.

Monshausen, G. B., Messerli, M. A. and Simon Gilroy, S. (2008). Imaging of the yellow cameleon 3.6 indicator reveals that elevations in cytosolic Ca^{2+} follow oscillating increases in growth in root hairs of *Arabidopsis*. *Plant Physiology* **147**, 1690–1698.

Moon, H., Lee, B., Choi, G., Shin, D., Theertha Prasad, D., Lee, O., Kwak, S. S., Hoon Kim, D., Nam, J., Bahk, J., Chan Hong, J. Yeol Lee, S. *et al.* (2003). NDP kinase 2 interacts with two oxidative stress-activated MAPKs to regulate cellular redox state and enhances multiple stress tolerance in transgenic plants. *Proceedings of the National Academy of Sciences of the United States of America* **100**, 358–363.

Mori, I. and Schroeder, J. (2004). Reactive oxygen species activation of plant Ca^{2+} channels. A signaling mechanism in polar growth, hormone transduction, stress signaling, and hypothetically mechanotransduction. *Plant Physiology* **135**, 702–708.

Navrot, N., Collin, V., Gualberto, J., Gelhaye, E., Hirasawa, M., Rey, P., Knaff, D. B., Issakidis, E., Jacquot, J.-P. and Rouhier, N. (2006). Plant glutathione peroxidases are functional peroxiredoxins distributed in several subcellular compartments and regulated during biotic and abiotic stresses. *Plant Physiology* **142**, 1364–1379.

Nishida, M., Maruyama, Y., Tanaka, R., Kontani, K., Nagao, T. and Kurose, H. (2000). $G\alpha i$ and $G\alpha o$ are target proteins of reactive oxygen species. *Nature* **408**, 492–495.

Osakabe, Y., Maruyama, K., Seki, M., Satou, M., Shinozaki, K. and Yamaguchi-Shinozaki, K. (2005). Leucine-rich repeat receptor-like kinasel is a key membrane-bound regulator of abscisic acid early signaling in *Arabidopsis*. *The Plant Cell* **17**, 1105–1119.

Park, J. Y., Kim, H. S., Youn, J. W., Kim, M. S., Kim, K. S., Joung, H. and Jeon, J. H. (2006). Cloning of superoxide dismutase (SOD) gene of lily 'Marcopolo' and expression in transgenic potatoes. *Agricultural Chemistry and Biotechnology* **49**, 1–7.

Pei, Z., Murata, Y., Benning, G., Thomine, S., Klusener, B., Allen, G., Grill, E. and Schroeder, J. (2000). Calcium channels activated by hydrogen peroxide mediate abscisic acid signalling in guard cells. *Nature* **406**, 731–734.

Pfannschmidt, T., Schutze, K., Fey, V., Sherameti, I. and Oelmuller, R. (2003). Chloroplast redox control of nuclear gene expression-A new class of plastid signals in interorganellar communication. *Antioxidants & Redox Signaling* **5**, 95–101.

Potikha, T., Collins, C., Johnson, D., Delmer, D. and Levine, A. (1999). The involvement of hydrogen peroxide in the differentiation of secondary walls in cotton fibers. *Plant Physiology* **119**, 849–858.

Potocky, M., Jones, M., Bezvoda, R., Smirnoff, N. and Zarsky, V. (2007). Reactive oxygen species produced by NADPH oxidase are involved in pollen tube growth. *New Phytologist* **174**, 742–751.

Rodriguez, A. A., Grunberg, K. A. and Taleisnik, E. L. (2002). Reactive oxygen species in the elongation zone of maize leaves are necessary for leaf extension. *Plant Physiology* **129**, 1627–1632.

Rouhier, N., Lemaire, S. and Jacquot, J. P. (2008). The role of glutathione in photosynthetic organisms: Emerging functions for glutaredoxins and glutathionylation. *Annual Reviews of Plant Biology* **59**, 143–166.

Sagi, M., Davydov, O., Orazova, S., Yesbergenova, Z., Ophir, R., Stratmann, J. W. and Fluhr, R. (2004). Plant respiratory burst oxidase homologs impinge on wound responsiveness and development in *Lycopersicon esculentum. The Plant Cell* **16**, 616–628.

Samuel, M. and Ellis, B. (2002). Double jeopardy: Both overexpression and suppression of a redox-activated plant mitogen-activated protein kinase render tobacco plants ozone sensitive. *The Plant Cell* **14**, 2059–2069.

Sarath, G., Hou, G., Baird, L. and Mitchell, R. (2007). ABA, ROS and NO are key players during switchgrass seed germination. *Plant Signaling & Behavior* **2**, 492–493.

Scheibe, R., Backhausen, J. E., Emmerlich, V. and Holtgrefe, S. (2005). Strategies to maintain redox homeostasis during photosynthesis under changing condition. *Journal of Experimental Botany* **56**, 1481–1489.

Schürmann, P. and Buchanan, B. B. (2008). The ferredoxin/thioredoxin system of oxygenic photosynthesis. *Antioxidants and Redox Signaling* **10**, 1235–1274.

Schürmann, P. and Jacquot, J. P. (2000). Plant thioredoxin systems revisited. *Annual Review of Plant Physiology and Plant Molecular Biology* **51**, 371–400.

Semighini, C. P. and Harris, S. D. (2008). Regulation of apical dominance in *Aspergillus nidulans* hyphae by reactive oxygen species. *Genetics* **179**, 1919–1932.

Shao, H. B., Chu, L. Y., Lu, Z. H. and Kang, C. M. (2008). Primary antioxidant free radical scavenging and redox signaling pathways in higher plant cells. *International Journal of Biological Sciences* **4**, 8–14.

Sugimoto-Shirasu, K. and Roberts, K. (2003). "Big it up": Endoreduplication and cell-size control in plants. *Current Opinion in Plant Biology* **6**, 544–553.

Suzuki, Y. J. and Ford, G. D. (1999). Redox regulation of signal transduction in cardiac and smooth muscle. *Journal of Molecular and Cellular Cardiology* **37**, 345–353.

Swiątek, A., Lenjou, M., van Bockstaele, D., Inzé, D. and van Onckelen, H. (0327). Differential effect of jasmonic acid and abscisic acid on cell cycle progression in tobacco BY-2 cells. *Plant Physiology* **128**, 201–211.

Takeda, S., Gapper, C., Kaya, H., Bell, E., Kuchitsu, K. and Dolan, L. (2008). Local positive feedback regulation determines cell shape in root hair cells. *Science* **319**, 1241–1244.

Tanaka, Y., Sano, T., Tamaoki, M., Nakajima, N., Kondo, N. and Hasezawa, S. (2005). Ethylene inhibits abscisic acid-induced stomatal closure in *Arabidopsis. Plant Physiology* **138**, 2337–2343.

Torres, M., Dangl, J. and Jones, J. (2002). *Arabidopsis* gp91phox homologues AtrbohD and AtrbohF are required for accumulation of reactive oxygen intermediates in the plant defense response. *Proceedings of the National Academy of Sciences United States of America* **99**, 517–522.

Watahiki, M., Trewavas, A. and Parton, R. (2004). Fluctuations in the pollen tube tip-focused calcium gradient are not reflected in nuclear calcium level: A comparative analysis using recombinant yellow cameleon calcium reporter. *Sexual Plant Reproduction* **17**, 125–130.

Watanabe, T. and Sakai, S. (1998). Effects of active oxygen species and methyl jasmonate on expression of the gene for a wound-inducible 1-aminocyclo-propane-1-carboxylate synthase in winter squash (*Cucurbita maxima*). *Planta* **206**, 570–576.

Wigoda, N., Ben-Nissan, G., Granot, D., Schwartz, A. and Weiss, D. (2006). The gibberellin-induced, cysteine-rich protein GIP2 from *Petunia hybrida* exhibits in planta antioxidant activity. *The Plant Journal* **48**, 796–805.

Xiong, L. and Yang, Y. (2003). Disease resistance and abiotic stress tolerance in rice are inversely modulated by an abscisic acid-inducible mitogen-activated protein kinase. *The Plant Cell* **15**, 745–759.

Yabuta, Y., Maruta, T. and Yoshimura, K. (2004). Two distinct redox signaling pathways for cytosolic APX induction under photooxdidative stress. *Plant and Cell Physiology* **45**, 1586–1594.

Yokota, E., Tominaga, M., Mabuchi, I., Tsuji, Y., Staiger, C., Oiwa, K. and Shimmen, T. (2005). Plant villin, lily P-135-ABP, possesses G-actin binding activity and accelerates the polymerization and depolymerization of actin in a Ca^{2+}-sensitive manner. *Plant and Cell Physiology* **46**, 1690–1703.

Yoon, G., Dowd, P., Gilroy, S. and McCubbin, A. (2006). Calcium-dependent protein kinase isoforms in Petunia have distinct functions in pollen tube growth, including regulating polarity. *The Plant Cell* **18**, 867–878.

Zavaleta-Mancera, H., Lopez-Delgado, H., Loza-Tavera, H., Mora-Herrera, M., Trevilla-Garcia, C., Vargas-Suares, M. and Ougham, H. (2007). Cytokinin promotes catalase and ascorbate peroxidase activites and preserves the chloroplast integrity during dark-senescence. *Journal of Plant Physiology* **164**, 1572–1582.

Zhang, A., Jiang, M., Zhang, J., Tan, M. and Hu, X. (2006). Mitogen-activated protein kinase is involved in abscisic acid-induced antioxidant defense and acts downstream of reactive oxygen species production in leaves of maize plants. *Plant Physiology* **141**, 475–487.

Zhu, J.-K. (2002). Salt and drought stress signal transduction in plants. *Annual Review of Plant Biology* **53**, 247–273.

Ultraviolet-B Induced Changes in Gene Expression and Antioxidants in Plants

S. B. AGRAWAL,[1] SURUCHI SINGH
AND MADHOOLIKA AGRAWAL

Laboratory of Air Pollution and Global Climate Change, Ecology Research Circle, Department of Botany, Banaras Hindu University, Varanasi 221005, India

ABSTRACT

The depletion of the stratospheric ozone layer leads to an increase in the level of ultraviolet-B radiations reaching the Earth's surface. UV-B radiations are known to have damaging effects on all forms of life. In plants, the UV-B exposure leads to the generation of reactive oxygen species (ROS), eventually resulting in oxidative stress. ROS induce lipid peroxidation of biological membranes, destroy the natural lipid-soluble antioxidants, and alter the expression of several genes through nonspecific signaling pathways. The integration of the thylakoid membrane appears to be much more sensitive than the activities of the photosynthetic components bound within.

[1]Corresponding author: Email: sbagrawal56@gmail.com

Advances in Botanical Research, Vol. 52

0065-2296/09 $35.00
DOI: 10.1016/S0065-2296(09)52003-4

However, the decrease of mRNA transcripts in the photosynthetic complexes and other chloroplast proteins are among the early events of UV-B damage. Other genes, encoding defense-related proteins are rapidly upregulated under UV-B irradiation. UV-B radiation induced production of ROS, increased the antioxidant capacity and thus, minimized the magnitude of negative impact of UV-B on plants. Specific signaling pathway includes the UVR8 component that regulates the expression of a set of genes essential for the protection of plant against UV-B. This chapter comprises information regarding the UV-B perception, signal transduction, regulation of gene expression, ROS formation, and its metabolism from various studies performed under growth chamber, green house, and field conditions.

I. INTRODUCTION

Depletion of the stratospheric ozone layer by human activity produced ozone-depleting substances has been recognized as a global environmental hazard for more than three decades. The electromagnetic radiation emitted from the sun in the ultraviolet (UV) range (200–400 nm) constitutes about 7% of the total radiation. As it passes through the atmosphere, the total flux transmitted is greatly reduced and the composition of UV radiation is modified (Frohnmeyer and Staiger, 2003). Increases in the UV-B radiation have been estimated to continue until 2050s in the boreal and subarctic regions (Weatherhead *et al.*, 2005). Owing to its high energy, the impact of UV-B on metabolic processes of plants can be very harmful (Hollosy, 2002; Kakani *et al.*, 2003). UV-B radiations induce oxidative stress (Singh *et al.*, 2009; Panagopoulos *et al.*, 1990); however, the mechanism of formation of reactive oxygen species (ROS) is not well known (Rao *et al.*, 1996). UV-B can induce damage to DNA, protein, membrane, and photosynthetic apparatus (Julkunen-Titto *et al.*, 2005). To keep this damage to a minimum, plants induce enzymatic and nonenzymatic antioxidative defense systems.

Targets of UV-mediated photomodification and photosensitization reactions (Greenberg *et al.*, 1997) include nucleotides, amino acids, lipids, and pigments (Jordan, 1996). DNA is a potentially sensitive target molecule for UV-B, because it absorbs UV-B efficiently and undergoes transformation that leads to the formation of the cyclobutane pyrimidine dimers (CPD) (Dany *et al.*, 2001) and the pryimidine (6-4) pyrimidinone photoproduct, both of which are formed by covalent bonding of adjacent pyrimidines (Nakajima *et al.*, 1998). These DNA lesions, if not repaired, may interfere with DNA transcription and replication, and can lead to misreading of the genetic code and ultimately cause mutations, growth inhibition, and potential death (Giordano *et al.*, 2004; Jiang *et al.*, 1997). Levels of UV-B tolerance differ considerably between genera, species, and even closely related cultivars. Plants somewhat tolerant to UV-B are found in areas having high UV-B influx like lower latitudes or high altitudes (Sullivan *et al.*, 1992). UV-B radiation is potentially damaging to plants, impairing gene transcription and translation, as well as photosynthesis (Jansen *et al.*, 1998). The biological impact of UV-B radiation depends on a number of factors, including the ratio of UV-B and photosynthetically active radiations (PAR), the spectral distribution within the UV-B wavelength band, genetic factors, and the exposure history of the plant (Frohnmeyer and Staiger, 2003; Jansen *et al.*, 1998). The chloroplasts are highly vulnerable to photooxidation owing to a high content in polyunsaturated fatty acids (PUFA) in their membrane system (Chow *et al.*, 1992a,b), thus significantly increasing the production of ROS (Salin, 1991).

II. UV-B PERCEPTION

It is essential that before inducing a particular cellular response, UV-B must be perceived by some kind of photoreceptors. This perception is coupled to the terminal response by signal transduction mechanisms (Fig. 1). Even though a lot of information is available on various light-sensing systems, the existence of UV-B receptors is questionable (Frohnmeyer and Staiger, 2003). Little is known about the nature of the UV-B receptors that are thought to mediate these responses (Ulm and Nagy, 2005). However, the

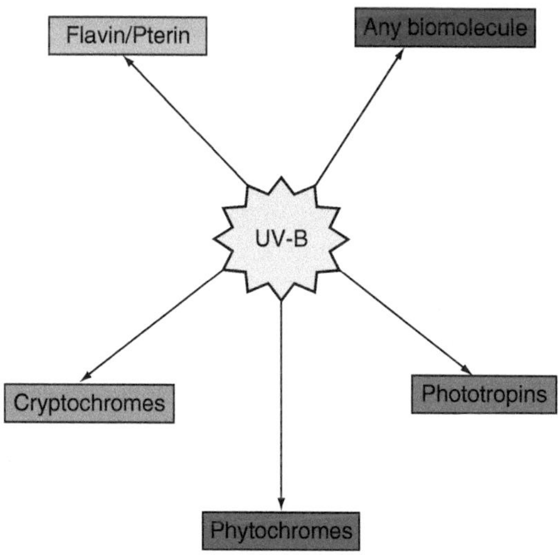

Fig. 1. Possible UV-B photoreceptors involved in signaling.

perception of UV-B radiation can be attributed to the action of phytochromes and cryptochromes as they absorb UV-B to some extent (Beggs *et al.*, 1986). To cope with the changes in the composition of light, different photoreceptor classes have evolved, such as specific UV-B photoreceptors (Nagy and Schäfer, 2000), cryptochromes, and phototropins, monitoring the blue/Ultraviolet-A (B/UV-A) region of the spectrum, whereas phytochromes monitor primarily red (R) and far-red (FR) wavelengths. The most well characterized receptors are phytochromes. The cryptochromes and phytochromes control growth and developmental responses to variations in the wavelength, intensity, and diurnal variation of the irradiation (Smith *et al.*, 2000). The phototropins primarily control the direction of growth in response to light and/or intracellular chloroplast movement in response to light intensity (Sakai *et al.*, 2001).

The cryptochromes are flavoproteins localized in the nucleus, each carrying two chromophores, a pterin, or a diazaflavin at one site and a FAD at the other. Photoreception by each of the three classes of receptors triggers specific intracellular signaling pathways that induce changes in gene expression, which drives various growth and developmental responses (Tepperman *et al.*, 2001). The hypothesis that phytochromes and cryptochromes serve as putative UV-B receptors has been disproven. Studies dealing with mutants lacking these receptors demonstrate that UV-B radiation independently affects the hypocotyl elongation response (Suesslin and Frohnmeyer, 2003). Five different isoforms of phytochromes (phytochrome A, B, C, D, and E) and four different blue light receptors (cryptochromes CRY1 and CRY2 and two membrane-activated phototropins) have been identified in *Arabidopsis thaliana* (Kagawa *et al.*, 2001). According to Brosche and Strid (2003), phytochromes do not act as UV-B receptors that control changes in gene expression. In fact, mutants of *A. thaliana* (Phy A, Phy B, and Phy AB), lacking functional phytochrome A and/or B were found to maintain UV-B induction of CHS gene expression in wild plants (Wade *et al.*, 2001). They also concluded that phytochromes do not act as primary photoreceptors, but they may mediate various responses of UV-B. Frohnmeyer *et al.* (1998) proposed that phytochrome A signaling involves the activation of one or more heteromeric G proteins and the subsequent participation of three different pathways depend upon calcium and/or cGMP. DNA can also be a UV-B receptor and a number of responses are related to the UV-B absorption by DNA as they are stimulated maximally by wavelengths between 250 and 280 nm (Frohnmeyer and Staiger, 2003). Action spectra revealed a maximal stimulation between wavelengths 290 and 310, whereas radiation below 290 nm inhibited these responses (Herrlich *et al.*, 1997).

Moreover, genome-wide analyses of gene expression suggest the involvement of more than one UV-B receptor (Ulm *et al.*, 2004). Although UV-B receptor(s) are still unknown, it is clear that they are different from phytochromes, cryptochromes, and phototropins (Brosche and Strid, 2003; Ulm *et al.*, 2004). Several attempts were made to identify the UV-B receptors, but failed due to lack of bioassays for mutant screening. There is a large agreement that UV-B receptor consists of a protein with a bound pterin or flavin (Galland and Senger, 1988). In fact, any biomolecule which interacts with UV-B photons and induces the specific stress responses should be quite different from the commonly observed photoreceptors (Brosche and Strid, 2003).

III. UV-B INDUCED SIGNAL TRANSDUCTION

After UV-B perception, a signal pathway must be established to bring about changes in gene expression. Molecular absorption of UV-B radiation through multiple transduction pathways leads to changes in the transcriptional machinery (Brosche and Strid, 2003; Mackerness, 2000). To identify UV-B signaling components, it is important to characterize genome-wide changes in gene expression that are generated by UV-B exposure (Ulm and Nagy, 2005).

UV-B signaling includes induction of alkalinization response, activation of NADPH oxidase, ion fluxes, and activation of mitogen-activated protein kinases (MAPKs) (Stratman, 2003). A MAPK module consists of a MAPKKK–MAPKK–MAPK that is linked in different ways to upstream receptors and downstream targets. Receptor-mediated activation of a MAPKKK can occur through physical interaction and/or phosphorylation by the receptor itself, intermediate bridging factors, or interlinking MAPKKKKs. MAPKs are serine/threonine kinases that phosphorylate a variety of substrates including CHS (Jenkins *et al.*, 1997) and pathogen-related genes (Mackerness, 2000). Activation of the MAP kinase cascade (Kalbin and Strid, 2006) and calcium release (Christie and Jenkins, 1996) are other responses. Upon UV-B exposure, the transcription of genes of the phenylpropanoid pathway is induced (Jenkins *et al.*, 2001).

A number of *Arabidopsis* mutants, which are hypersensitive to UV-B radiations show defects in several types of cellular functions, such as flavonoid biosynthesis (Rao and Ormrod, 1995) and DNA repair (Britt *et al.*, 1993). Also, ribosomal proteins, important for protein synthesis, are induced upon UV-B irradiation and an upregulation of ribosomal protein transcripts seems to be important for the maintenance of general cellular functions (Izaguirre *et al.*, 2003). UV-B is known to damage the ribosomes by cross-linking cytosolic ribosomal proteins to RNA (Casati and Walbot, 2004). Casati and Walbot (2003) have also reported an induction in several chaperones under UV-B radiation in maize genotypes, indicating that damaged proteins are recognized, repaired, and recycled.

UV-B radiation effects on photosynthesis have been clearly demonstrated with multiple sites of inhibition (Strid *et al.*, 1994). Downregulation of photosynthetic genes, both nuclear- and chloroplast-encoded, may cause substantial loss of protein content and activity leading to decreased photosynthetic function. The decline in mRNA transcripts seems to be more rapid for nuclear-encoded genes than for chloroplastic genes. The mRNAs for the nuclear CAB genes, encoding the chlorophyll *a*/*b*-binding proteins for light harvesting antenna of PSII, are more rapidly degraded than the mRNA for the plastid localized psbA encoding D1 protein of PSII (Jordan *et al.*, 1991). The mRNA for nuclear-encoded atpC gene corresponding to the gamma subunit of the ATP synthase declines more rapidly than the mRNA transcripts for the atpB and atpE encoding for β and subunits of ATPase (Zhang *et al.*, 1994). The mRNA for the small subunit of Rubisco, encoded by the nuclear rbcS gene, declines ahead of the mRNA for the large subunit encoded by rbcL gene in the chloroplast (Jordan *et al.*, 1992). Taylor (1989) observed that nuclear genes are regulated mainly at the transcriptional level, whereas the plastid-encoded genes are subjected to considerable posttranscriptional regulation.

Huang *et al.* (1997) have elucidated the involvement of calcium and calmodulin as secondary messengers in many physiological processes in plant cells under UV-B exposure (Fig. 2). The importance of Ca^{2+} in modulating plant responses to external stimuli of biotic and abiotic origin is now well established (Kiegle *et al.*, 2000). Ca^{2+}-dependent modulation of cellular processes occurs via intracellular Ca^{2+}-binding proteins, also known as Ca^{2+} sensors. The calcium-dependent pathway regulates the expression of genes, such as CAB (encoding chlorophyll *a*,*b*-binding proteins) and is able to direct partial chloroplast development. The cGMP-dependent pathway regulates the expression of CHS (chalcone synthase) and production of anthocyanin pigments. High levels of cGMP pathway can negatively regulate the two calcium-dependent pathways and high levels of Ca^{2+}. Ca^{2+}-activated calmodulin can negatively regulate the cGMP pathway that controls the CHS gene expression. Frohnmeyer *et al.* (1997) gave evidence that Ca^{2+} and calmodulin are negative regulators for the phytochrome control of CHS expression while they act as positive regulators for UV-mediated CHS gene expression. Long and Jenkins (1998) suggested that perception of UV-B initiates redox processes in the plasma membrane.

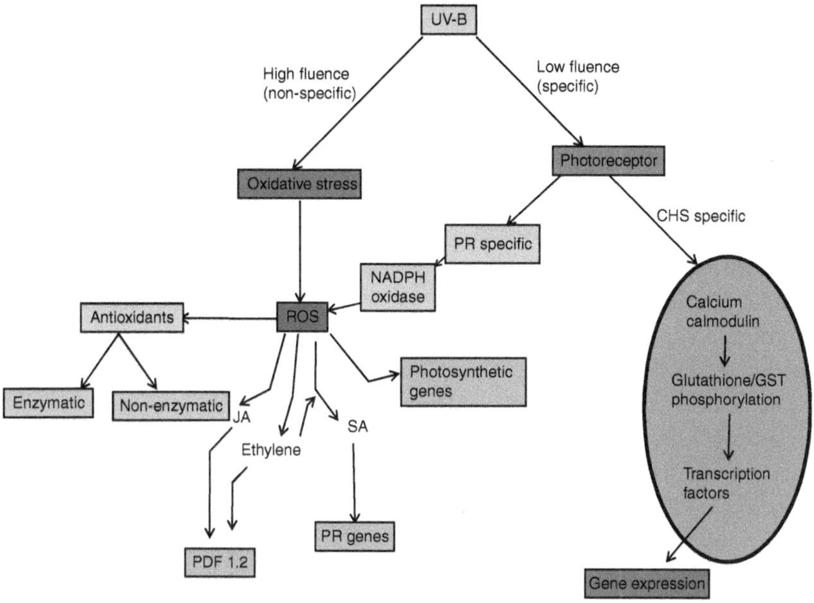

Fig. 2. Signal transduction and multiple signaling pathways mediating responses to UV-B (redrawn with changes after Mackerness, 2000; Mackerness *et al.*, 1999). PR, pathogenesis related; JA, jasmonic acid; SA, salicylic acid; CHS, chalcone synthase; ROS, reactive oxygen species; PDF 1.2, plant defensin gene. (See color plate section at the back of the book.)

IV. REGULATION OF GENE EXPRESSION BY UV-B

UV-B induced modification in gene expression is very complex and specific. Gene expression is both up- and downregulated by the UV-B exposure (Jordan *et al.*, 1994). UV-B affects the gene expression at different levels—transcriptional, translational, and posttranslational (Mackerness *et al.*, 1997). Ulm *et al.* (2004) postulated the interaction of at least two UV-B perception and signaling pathways, one pathway controlled by shorter wavelengths of UV-B and the other controlled by longer wavelengths of UV-B, the former negatively interfering with the latter. They also described for the first time a whole genome expression analysis of transcripts of *Arabidopsis* after UV-B exposure and identified a robust set of early low-level UV-B responsive genes of which more than 20% shared transcription factors. UV-B regulates sets of defense-related and other genes that are activated via different signaling pathways involving ROS, salicylic acid (SA), jasmonic acid (JA), and ethylene (C_2H_4) (Broschė *et al.*, 2002; Green and Fluhr, 1995; Mackerness *et al.*, 2001) (Fig. 2). Examples include PR-1, PR-2, PR-5, the defense gene PDF 1.2, and proteinase inhibitor genes.

Surplus *et al.* (1998) have clearly demonstrated differences in gene activity response of *Arabidopsis* mutants that are insensitive to SA, JA, and ethylene (NahG, jar1, and etr 1-1, respectively). An increase in expression of two pathogen-related genes PR-1 and PDF 1.2 were found to be dependent on SA and ethylene or JA and ethylene, respectively. In contrast, downregulation of RNA transcripts for photosynthetic proteins was dependent on all three compounds (Jordan, 1996). Exposure of plants to low fluence of UV-B promoted the expression of a range of genes involved in UV-B protection, including genes responsible for the production of flavonoids and several phenolic compounds working as sunscreen against UV-B (Casati and Walbot, 2003; Ulm *et al.*, 2004). Brown *et al.* (2005) reported UV RESISTANCE LOCUS 8 (UVR 8) protein in *Arabidopsis* working as a UV-B specific signaling component controlling the expression of a range of genes essential for UV-B protection. UVR8 regulates the expression of the transcription factor HY5.

Salicyclic acid is a component of the signal transduction pathway that leads to the regulation of PR genes in response to pathogen attack and various abiotic stress factors (Dempsey and Klessig, 1994). Surplus *et al.* (1998) observed that changes in ROS and SA, in response to UV-B exposure, are primarily a consequence of cellular damage and lesion formation resulting from extreme and prolonged UV-B treatment. SA was shown to play a role in the mobilization of defense pathways leading to an upregulation of three acidic-type pathogenesis-related (PR) genes in response to UV-B radiation (Surplus *et al.*, 1998). Green and Fluhr (1995) have shown that UV-B radiation resulted in an increase of PR-1 mRNA and protein levels in tobacco. It is possible that PR-1 protein induced by UV-B has a role in protecting cells from the damaging effects of UV-B radiation. They have also elucidated some of the components of the signal transduction cascade between UV-B and PR-1. In addition, ROS have been shown to be involved in the induction of PR-1 by SA. SA has recently been shown to increase the intracellular hydrogen peroxide (H_2O_2) concentration, probably by inhibiting the catalase (CAT) activity (Chen *et al.*, 1993).

Concentrations of JA frequently increase in response to wounding (Blechert *et al.*, 1995) and pathogen infection (Vijayan *et al.*, 1998). Exogenous application of JA has been shown to enhance the expression of an array of stress-related genes, such as thionin (Epple *et al.*, 1995) and defensins in *Arabidopsis* (Clarke *et al.*, 1998) and proteinase inhibitors in tomato (Farmer *et al.*, 1992). In contrast, elevated levels of JA can downregulate genes encoding proteins required for photosynthesis (Reinbothe *et al.*, 1994). Studies on tomato indicated that a number of wound-inducible genes are also upregulated in response to UV-B radiation and JA was shown to be involved in this response (Conconi *et al.*, 1996).

The gaseous plant hormone, ethylene has also been identified as a signaling component in wounding and defense responses (Morgan and Drew, 1997). Ethylene biosynthesis is promoted by many stress factors, including wounding (O' Donnel *et al.*, 1996), pathogen infection (Hammond-Kosack and Jones, 1996), and UV-B radiation (Predieri *et al.*, 1995). Exogenously applied ethylene induces transcription of a number of defense-associated genes, such as many basic PR genes (Potter *et al.*, 1993).

High concentrations of ROS can lead to phytotoxicity, whereas relatively low levels can influence signaling and gene expression (Dat *et al.*, 2000). ROS control various biological programs (Apel and Hert, 2004). Being small and able to diffuse over short distances, ROS are ideally suited to act as signaling molecules. Among different ROS, only H_2O_2 can cross plant membrane, and therefore, can directly function in cell-to-cell signaling. Nitric oxide (NO) and H_2O_2 act as signaling molecules in plants, this being essential in response to environmental stresses (He *et al.*, 2004). Pharmacological data have suggested that NO is important in regulating gene expression in response to UV-B (Mackerness *et al.*, 2001). He *et al.* (2005) showed using an epidermal strip bioassay and laser-scanning confocal microscopy that generation of H_2O_2 and NO are required for the UV-B induced stomatal closure. According to Mackerness *et al.* (2001), H_2O_2, NO, and SA act as second messengers mediating responses of specific genes to UV-B radiation. In addition, they also reported that under UV-B radiation, the increase in PR-1 transcript and decrease in Lhcb transcript were mediated through pathways involving H_2O_2 derived from superoxide, but the upregulation of CHS was not controlled by ROS scavengers and reduced by NO synthase. In contrast, upregulation of PDF 1.2 transcript was mediated through pathways directly involving $O_2^{-\bullet}$.

UV-B induced gene expression has been shown to occur via H_2O_2, as exposure of *Arabidopsis* plants to UV-B in the presence of antioxidants led to the downregulation of the UV-induced gene PDF 1.2 (Mackerness *et al.*, 1999). In soybean, H_2O_2 induced the expression of defense-related genes, glutathione *S*-transferase (GST) and glutathione peroxidase (GPX) (Levine *et al.*, 1994). In *Arabidopsis* suspension cultures, H_2O_2 induced the expression of GST and phenylalanine ammonia lyase (PAL) (Desikan *et al.*, 1998a). H_2O_2 can induce the expression of genes potentially involved in its synthesis (Desikan *et al.*, 1998b), and also of those encoding proteins involved in its degradation, implying a complex mechanism for cellular regulation of oxidative status. Recent studies have also shown that H_2O_2 regulates stomatal movement through the activation of Ca^{2+} channels (Köhler *et al.*, 2003).

Reduction of root, hypocotyl, and coleoptile growth under UV-B is likely to result from cell wall stiffening processes related to the formation of cross-links among cell wall polymers (Fry, 1986) which are formed due to H_2O_2 (Schopfer, 1996). It is widely known that H_2O_2 triggers the expression of a set of genes including various antioxidant enzymes, such as APX (ascorbate peroxidase), SOD (superoxide dismutase), GR (glutathione reductase), and CAT related to plant defense (Neil *et al.*, 2002). H_2O_2 also helps in the induction of a subset of defense genes, including proteinase inhibitors and polyphenol oxidase (Orozco-Cárdenas *et al.*, 2001). It is a relatively stable ROS, uncharged at physiological pH, small-sized, and hence freely diffusible across membranes. Depending upon its compartmentalization, H_2O_2 concentration varies from micromolar to low millimolar range (Cheeseman, 2006). It causes oxidative protein modification at distal area from its place of production (Scandalios *et al.*, 1997). Mitochondrial H_2O_2 is generated through the dismutation of $O_2^{-\bullet}$ (Forman and Boveris, 1982). The mitochondrial membranes are an important source of intracellular H_2O_2 steady state levels, via the mitochondrial generation of $O_2^{-\bullet}$ (Rich and Bonner, 1978).

To prevent over reduction of the electron-transport chain (ETC) under conditions that limit CO_2 fixation, higher plants have evolved the photorespiratory pathway to regenerate $NADP^+$ (Shao and Chu, 2005), H_2O_2 is one of the byproduct produced in peroxisomes, whereas it can also be formed as a byproduct of β-oxidation of fatty acids. The targets of H_2O_2 are Calvin cycle enzymes, iron containing enzymes, D1/D2 proteins, and Mn clusters in PSII. H_2O_2 is a potent oxidant of an enzyme of thiol groups; its inhibitory effect on CO_2 fixation is due to the inactivation of thiol-regulated enzymes. The involvement of brassinosteroids (BR) in signaling events during UV-B stress was investigated by Sävenstrand *et al.* (2004). BR are growth regulators involved in growth, development, and stress tolerance. Reduced levels of gene expression of CHS, PYROA, MEB 5.2, and PR-5 were observed in BR-deficient *A. thaliana* mutants, indicating the need for a complete BR pathway for proper UV-B-dependent gene expression (Sävenstrand *et al.*, 2004).

V. SOURCES OF ROS

It is widely accepted that various ROS are involved in the responses of plants to UV-B, both as signaling and damaging agents (Šnyrychová *et al.*, 2007). As a result, increased antioxidant activity (Jansen *et al.*, 2008) and higher amounts of oxidative membrane damage products (Malanga *et al.*, 1997) are observed. Ascorbate radicals (Hideg *et al.*, 1997), long-lived chlorophyll radicals (Kumagai *et al.*, 1999) as well as oxygen-, carbon-, and nitrogen-centered free radicals (He *et al.*, 2005) were detected with electron paramagnetic resonance (EPR) spectroscopy in a variety of plants upon UV-B exposure. The main sources of ROS in plants are ETC in chloroplast and mitochondria, some peroxidases and oxidases (NADPH oxidase, xanthine oxidase, lipoxygenase, glycolate oxidase, amine oxidase, etc.), photosensitizers, such as chlorophyll molecules (Dat *et al.*, 2000) and peroxisomes (Foyer and Noctor, 2003). He *et al.* (2006) reported that UV-B mediated H_2O_2 inhibited the pollen germination and tube growth of *Paeonia suffruticosa* and *Paulownia tomentosa*. Environmental stress conditions reduce $NADP^+$ regeneration by the Calvin cycle, consequently the photosynthetic ETC is over reduced, forming superoxide radical ($O_2^{-\bullet}$) and singlet oxygen ($^1O_2$) in the chloroplast (Krause, 1994). Essential UV-B targets in photosynthetic organisms include photosystem II (PSII), whose electron transport is inhibited and its D1 and D2 subunits damaged (Vass *et al.*, 1996). Within PSII, QA and QB quinone electron acceptors (Greenberg *et al.*, 1989), Tyr-Z and Tyr-D redox active tyrosines (Yerkes *et al.*, 1990) as well as the catalytic Mn cluster of the water oxidizing complex are damaged (Renger *et al.*, 1989).

The ETC in chloroplast operates in an O_2 rich environment, such that leakage of electrons from overloaded ETC will lead to ROS production. The e-flow from excited PS centers is directed to NADPH, this then enters the Calvin cycle and reduces the final electron acceptor, CO_2. In conditions of overloading of ETC, a part of the electron flow is diverted from ferredoxin to O_2, reducing it to superoxide free radical. Dai *et al.* (1997) observed that excessive accumulation of ROS in leaves following UV-B treatment strongly inhibited the photosynthetic electron activity. The outlet of e^-(s) from ferredoxin to O_2 is called the Mehler reaction. The rate of H_2O_2 production in the Mehler reaction is sufficiently high to cause an accumulation of 10 M H_2O_2 within 0.5 s which leads up to 50% inhibition of CO_2 fixation when the scavenging enzymatic system of chloroplast does not function. The acceptor side of ETC in PSII also provides sites (QA, QB) of electron leakage to O_2 producing $O_2^{-\bullet}$ (Dat *et al.*, 2000). Once superoxide anions are produced, they follow different pathways. On the internal lumen membrane surface, $O_2^{-\bullet}$ may be protonated to H_2O_2 which initiates lipid peroxidation. On the external membrane surface, $O_2^{-\bullet}$ is enzymatically or spontaneously dismutated to H_2O_2 and O_2. In dismutation, however, the scavenging of one ROS type, $O_2^{-\bullet}$ yields another ROS type, H_2O_2 (Foyer *et al.*, 1994). At the level of Fe-S centers where Fe^{2+} is available, H_2O_2 is transformed through the Fenton reaction into $^\bullet OH$. Conditions unfavorable for CO_2 fixation will inevitably lead to an increased ROS accumulation, as more O_2 molecules will be used as electron acceptors. Also, input of light energy to O_2 produces highly reactive singlet oxygen which has a hazardous effect on chloroplast pigment–protein complexes (Niyogi, 1999).

Relatively higher contents of PUFA are found in thylakoid membranes which facilitates photosynthesis (He and Hader, 2002). It has been suggested that exposure to UV-B results in the generation of ROS within the chloroplasts, as thylakoid membranes are very rapidly perturbed upon exposure to UV-B radiation (Chow *et al.*, 1992a). PUFAs, that is, linolenic acid and levulonic acid are the major fatty acids in the plant membrane, galactolipids (thylakoid membrane), and phospholipids (all other membranes). PUFAs are particularly susceptible to attack by $_1O^2$ and $^{\bullet}OH$ forming complex mixtures of lipid hydroperoxides. Among the ROS, the superoxide anion ($O_2^{-\bullet}$) plays a central role in the peroxidation of lipids via the formation of more active species, such as hydroxyl radical and singlet oxygen that react directly with unsaturated fatty acids to generate lipid peroxides. PUFA peroxidation decreases the fluidity of the membrane, increases the leakiness, and causes secondary damage to the membranes. The lipid peroxidation, which is indicative of damage to cellular membranes, interferes with the function of membranes (Kramer *et al.*, 1991; Panagopoulos *et al.*, 1990), and ultimately results in bleaching of chlorophyll (Elstner, 1982). Lipid hydroperoxides, formed as a result of lipid peroxidation, can affect membrane properties by increasing the membrane hydrophilicity of the internal side of the bilayer (Frenkel, 1991). The phenomenon is very important for the termination of lipid peroxidation, since increased hydrophilicity of the membrane favors the generation of tocopherol by ascorbate.

In order to stop peroxidative chain reactions, α-tocopherol must be located in the direct pathway of propagation of lipid radical reaction (Scholz *et al.*, 1990). Lipid peroxides can also be detoxified through conjugation with glutathione. Aldehydes like 4-hydroxy-2-nonenal (HNE) and malondialdehyde (MDA) as well as hydroxyl and keto fatty acids are formed due to lipid peroxidation. The MDA content has been reported to increase under UV-B radiation in cucumber leaves (Kramer *et al.*, 1991). MDA formed under UV-B radiation increases with increasing UV-B doses (Table I). Takeuchi *et al.* (1995) reported growth inhibition due to lipid peroxidation caused by UV-B exposure. Carletti *et al.* (2003) reported oxidative effect of UV-B radiation on biological membranes of maize seedlings under 8.35 kJ $m^{-2}d^{-1}$ fluence rate. Plant mitochondria too are considered an important source of superoxide and peroxide, when electron transport through the cytochrome part of the respiratory chain is restricted due to stress-induced physical changes in the membrane components (Wagner and Moore, 1997). Under these circumstances, both the high reduction of respiratory chain components before the cytochromes and the increased oxygen

TABLE I

Effects of UV-B on Various Enzymatic and Nonenzymatic Antioxidants, MDA Contents and Lipid Peroxides of Various Plant Species

Plant species	UV-B dose	Growth conditions	Parameters	Change	Reference(s)
Cucumis sativus L.	0.2 W m^{-2} s^{-1}	Growth chamber	APX	+ 5.25 fold	Kondo and Kawashima (2000)
			SOD	+ 4.5 fold	
			GR	+ 1.55 fold	
	147 mW m^{-2} s^{-1}	Growth chamber	MDA content	+ 2.64 fold	Yao *et al.* (2006)
	152.7		MDA content	+ 1.21 fold	
	3.4 kJ m^{-2} d^{-1}	Growth chamber	MDA contents	+ 1.04 fold	Teklemariam and Blake (2003)
	5.5 kJ m^{-2} d^{-1}		MDA contents	+ 1.6 fold	
	10.6 kJ m^{-2} d^{-1}		MDA contents	+ 1.8 fold	
Glycine max	30 kJ m^{-2} s^{-1}	Green house	α-tocopherol	No change	Galatro *et al.* (2001)
			Ascorbic acid	− 0.97 fold	
	60 kJ m^{-2} s^{-1}		α-tocopherol	− 0.88 fold	
			Ascorbic acid	− 0.81 fold	
	147 mW m^{-2} s^{-1}	Growth chamber	MDA content	Increased	Yao *et al.* (2006)
	152.7 mW m^{-2} s^{-1}		MDA content	Increased	
	Solar UV-B	Green house	CAT	+ 0.96 fold	Xu *et al.* (2008)
			APX	+ 1.2 fold	
			POX	No change	
			SOD	− 0.62 fold	
			GR	+ 1.4 fold	
	5 kJ $m^{-2}$$d^{-1}$	Field	SOD	− 0.98 fold	Yanqun *et al.* (2003)
			MDA contents	+ 1.35 fold	
	A+7.1 kJ m^{-2}	Field	CAT	− 0.76 fold	Ambasht and Agrawal (2003a)
			AA	− 1.17 fold	
			POX	+ 1.2 fold	

(*continues*)

TABLE I (*continued*)

Plant species	UV-B dose	Growth conditions	Parameters	Change	Reference(s)
Triticum aestivum L.	4.2 kJ $m^{-2}d^{-1}$	Growth chambers	SOD	+ 1.8 fold	Yang *et al.* (2007)
			CAT	− 1.25 fold	
			GR	− 1.66 fold	
			APX	− 1.12 fold	
	10.3 kJ $m^{-2}d^{-1}$		SOD	+ 2.27 fold	
			CAT	− 1.45 fold	
			GR	+ 2 fold	
			APX	− 1.4 fold	
	75 W m^{-2}	Growth chamber	SOD	+ 2.4 fold	Dawar *et al.* (1998)
			CAT	+ 1.7 fold	
			POX	No difference	
			MDA	+ 1.2 fold	
	49 kJ m^{-2} d^{-1}	Growth chamber	CAT	+ 1.9 fold	Alexieva *et al.* (2001)
			POX	− 0.66 fold	
			SOD	+ 1.4 fold	
	A+7.1 kJ m^{-2}	Field	CAT	+ 2.8 fold	Ambasht and Agrawal (2003b)
			POX	− 0.76 fold	
			AA	− 0.33 fold	
	A+7.1 kJ m^{-2}	Field	AA	− 0.66 fold	Agrawal and Rathore (2007)
			CAT	+ 1.32 fold	
			SOD	+ 1.04 fold	
			POX	+ 0.71 fold	
Pisum sativum L.	49 kJ m^{-2} d^{-1}	Growth chamber	CAT	+ 1.4 fold	Alexieva *et al.* (2001)
			POX	+ 1.41 fold	
			SOD	+ 1.31 fold	
	A+7.1 kJ m^{-2}	Field	SOD	+ 1.2 fold	Agrawal and Mishra (2009)
			POX	+ 1.6 fold	
			CAT	− 0.72 fold	
			AA	− 0.89 fold	
			LPO	+ 1.5 fold	

Crotalaria juncea L	1 kJ m^{-2} d^{-1}	Field	SOD	+ 1.1 fold	Selvakumar (2008)
	1.4			+ 1.5 fold	
	4.7			+ 2 fold	
	6			+ 3 fold	
	1 kJ m^{-2} d^{-1}		MDA content	+ 1.2 fold	
	1.4			+ 1.4 fold	
	4.7			+ 1.4 fold	
	6			+ 1.8 fold	
	1 kJ m^{-2} d^{-1}		CAT	− 0.62 fold	
	1.4			− 0.69 fold	
	4.7			− 0.76 fold	
	6			− 0.84 fold	
	12.2 kJ m^{-2} d^{-1}	Growth chamber	CAT	− 0.69 fold	Balakrishnan *et al.* (2005)
			POX	+ 1.64 fold	
			SOD	+ 1.54 fold	
Capsicum annuum L.	5.8 W m^{-2}	Green house	POX	+ 10.8 fold	Madhavian *et al.* (2008)
			APX	+ 3.3 fold	
			CAT	+ 2.5 fold	
			GR	+ 2.6 fold	
Helianthus annuus L.	15 kJ m^{-2} s^{-1}	Growth chamber	CAT	+ 1.2 fold	Yannarelli *et al.* (2006)
			APX	− 0.82 fold	
			GPX	+ 1.35 fold	
	30 kJ m^{-2} s^{-1}		CAT	+ 1.2 fold	
			APX	− 0.88 fold	
			GPX	+ 1.38 fold	
Helianthus annuus L.	15 kJ m^{-2} s^{-1}	Growth chamber	SOD	− 0.45 fold	Costa *et al.* (2002)
			AA	+ 1.53 fold	
			APX	− 0.77 fold	
			CAT	+ 1.2 fold	
			GR	+ 1.15 fold	

(*continues*)

TABLE I (*continued*)

Plant species	UV-B dose	Growth conditions	Parameters	Change	Reference(s)
	30 kJ m^{-2} s^{-1}		SOD	− 0.36 fold	
			AA	+ 1.39 fold	
			APX	− 0.88 fold	
			CAT	+ 1.2 fold	
			GR	+ 1.02 fold	
Lycopersicum esculentum L.	6.3 $kJm^{-2}d^{-1}$	Field	MDA	+ 1.32 fold	Balakumar *et al.* (1997)
			SOD	+ 2.24 fold	
			CAT	+ 1.5 fold	
Zea mays L.	8.35 kJ m^{-2} d^{-1}	Growth chamber	γ-tocopherol	− 0.97 fold	Carletti *et al.* (2003)
			α-tocopherol	− 0.84 fold	
			Ascorbate	+ 1.01 fold	
			Proline	+ 1.12 fold	
Gunnera magellanica L.	2 kJ m^{-2} d^{-1}	Green house	Lipid peroxides	No change	Giordano *et al.* (2004)
	4 kJ m^{-2} d^{-1}		Lipid peroxides	No change	
	6.5 kJ m^{-2} d^{-1}		Lipid peroxides	No change	
Populus kangdingensis L.	4.4 kJ m^{-2} s^{-1}	Green house	SOD	+ 1.05 fold	Ren *et al.* (2008)
			APX	+ 1.93 fold	
			CAT	+ 1.65 fold	
P. cathayana L.	4.4 kJ m^{-2} s^{-1}	Green house	SOD	+ 1.12 fold	
			APX	+ 7.56 fold	
			CAT	− 0.60 fold	
Crepis capillaries L.	3 kJ m^{-2}	Growth chamber	SOD	+ 1.4 fold	Rancélienë *et al.* (2005)
	9 kJ m^{-2}		SOD	+ 2 fold	
Hordeum vulgare L.	21 kJ m^{-2} d^{-1}	Growth chamber	APX	+ 1.22 fold	Zancan *et al.* (2008)
			CAT	− 0.64 fold	
Spinacia oleracea L.	A+7.1 kJ m^{-2}	Field	CAT	− 0.67 fold	Mishra and Agrawal (2006)
			POX	+ 1.67 fold	
			AA	− 0.62 fold	
			MDA	+ 1.17 fold	

Abelmoschus esculentum L.	A+1.8 kJ m^{-2} d^{-1}	Field	SOD	+ 1.38 fold	Kumari *et al.* (2009)
			APX	+ 1.07 fold	
			MDA	+ 2.94 fold	
			POX	− 0.72 fold	
Vigna unguiculata L.	1 kJ m^{-2} d^{-1}	Field	SOD	+ 1.3 fold	Selvakumar (2008)
	1.4			+ 1.5 fold	
	4.7			+ 2.5 fold	
	6			+ 2 fold	
	1 kJ m^{-2} d^{-1}		MDA content	+ 1.1 fold	
	1.4			+ 1.2 fold	
	4.7			+ 1.55 fold	
	6			+ 2 fold	
	1 kJ m^{-2} d^{-1}		CAT	− 0.88 fold	
	1.4			− 0.64 fold	
	4.7			− 0.51 fold	
	6			− 0.63 fold	
Picea asperata L.	A+3.31 kJ m^{-2} d^{-1}	Field	MDA content	+ 1.2 fold	Yao and Liu (2007)
			POX	+ 1.6 fold	
			APX	+ 1.7 fold	
			SOD	+ 3 fold	
			CAT	+ 1.4 fold	
			GR	+ 2.9 fold	
Acer mono Maxim	A+14.33 kJ m^{-2} d^{-1}	Field	POD	+ 2.28 fold	Yao and Liu (2006)
			SOD	+ 1.7 fold	
			CAT	+ 2.4 fold	
			APX	− 0.95 fold	
			GR	− 0.96 fold	
Hippophae rhamnoides L.	A+5.30 kJ m^{-2} d^{-1}	Field	MDA contents	+ 1.01 fold	Yang and Yao (2008)
			AA	− 0.74 fold	
Vacciniummyrtillus L.	Not available	Field	AA	+ 1.11 fold	Taulavouri *et al.* (1998)
			GSH	− 0.96 fold	

(*continues*)

TABLE I (*continued*)

Plant species	UV-B dose	Growth conditions	Parameters	Change	Reference(s)
Cassia auriculata L.	7.5 kJ m^{-2}	Growth chambers	LPO	+ 1.22 fold	Agarwal (2007)
			AA	+ 2.6 fold	
			DHA	+ 2.8 fold	
			GSH	+ 12.2 fold	
			SOD	+ 1.5 fold	
			CAT	+ 1.9 fold	
			POX	+ 1.4 fold	
	15 kJ m^{-2}		LPO	+ 1.34 fold	
			AA	+ 2.1 fold	
			DHA	+ 2.2 fold	
			GSH	+ 5.2 fold	
			SOD	+ 1.8 fold	
			CAT	+ 1.7 fold	
			POX	+ 1.6 fold	
Sorghum vulgare L.	A+7.1 kJ m^{-2}	Field	AA	− 0.84 fold	Ambasht and Agrawal (1998)
			POX	+ 1.2 fold	
			CAT	− 0.8 fold	
Vigna radiata L.	A+7.1 kJ m^{-2}		AA	− 0.8 fold	Agrawal and Rathore (2007)
			CAT	− 0.95 fold	
			SOD	+ 1.31 fold	
			POX	+ 0.27 fold	
Helianthus annuus L.	8.6 W m^{-2}	Green house	MDA content	+ 2.9 fold	Cechin *et al.* (2008)
Triticosecale	2.6 kJ m^{-2} d^{-1}	Growth chamber	POX	+ 1.08 fold	Skorska and Szwarc (2007)
			CAT	− 0.87 fold	
Oryza sativa L.	6 kJ m^{-2}	Green house	MDA content	+ 1.13 fold	Dai *et al.* (1997)
	13 kJ m^{-2}		MDA content	+ 1.31 fold	

Changes in enzymatic and nonenzymatic antioxidants, MDA content and lipid peroxides when compared to their respective controls. (A: ambient level; +/− represents increase/decrease).

level in the cell due to lower respiratory rate enhance the production of $O_2^{-\bullet}$ and H_2O_2. ROS are also generated in plants at the plasma membrane level or extracellularly in the apoplast. NADPH-dependent oxidase (NADPH oxidase) of plasma membrane has recently been regarded as a source of ROS for the oxidative burst (Lamb and Dixon, 1997). Chemical inhibitors of NADPH oxidase, such as diphenylene iodonium (DPI) have been shown to block or severely reduce ROS production upon biotic or abiotic stresses (Orozco-Cárdenas and Ryan, 1999). UV-B induces NADPH-oxidase activity which leads to peroxide formation (Rao *et al.*, 1996). Mackerness *et al.* (2001) provided evidence to show that UV-B exposure induced NADPH oxidase and cell wall peroxidases mediated ROS synthesis in the leaves of *Arabidopsis* suggesting that there are multiple sources of H_2O_2 production in response to UV-B radiation. It may be possible that the plants recognize the UV-B radiation through mechanisms identical to those involved in the pathogen infection.

Highly energetic photons of the UV-B range are absorbed by the chromophore groups of many biologically important molecules, such as chlorophyll, phycobiliproteins, and quinones. These molecules can act as photosensitizers for the production of ROS (Caldwell *et al.*, 1998; Jordan, 1996). Under normal growth condition, the energy of the excited chlorophylls or phycobilins is utilized efficiently for photosynthesis. However, the inhibition of photosynthesis or ETC under excess of UV-B may lead to photosensitization process as well as the formation of ROS. Normally, the excited singlet state of the chlorophyll serves to transfer energy or electrons. To emit energy, chlorophyll uses fluorescence or conversion to the triplet state which leads to the formation of singlet oxygen (Arora *et al.*, 2002). The formation of singlet oxygen via photsensitization is known to play a crucial role in damaging the D1 protein (Hideg *et al.*, 1994). UV-B chromophores, such as aromatic amino acids, NADH, and phenolic compounds can also be activated by the absorption of UV-B radiations and react with molecular oxygen to form singlet oxygen and superoxide anion. Barta *et al.* (2004) observed that the dominant ROS produced under UV-B stress was $O_2^{-\bullet}$ whereas ${}_1O^2$ was a minor contributor (Hideg *et al.*, 2002).

Upregulation of antioxidant system and increased expression of genes related to oxidative stress are found in plants grown under lower near-field intensities of UV-B (Brosche and Strid, 2003). Under UV-B stress, the inhibition of the ETC due to the degradation of the D1 protein of PSII may promote the energy transfer from triplet chlorophyll to oxygen to form singlet oxygen (Jordan, 1996). The imbalance between the light phase and the Calvin cycle, probably due to the decreased activity of ribulose-1,5-bisphosphate carboxylase/oxygenase (Rubisco) by UV radiation promotes the formation of superoxide radicals at the level of ferredoxin in photosystem I (PSI) (Bischof *et al.*, 2000). Another important source of ROS formation, especially of H_2O_2, is the photorespiration in the peroxisomes. During CO_2 fixation, Rubisco uses CO_2 to carboxylate RuBP. This enzyme can also use molecular O_2 to oxygenate RuBP. During this reaction, glycolate is formed and transported from chloroplast to peroxisomes. The glycolate is then oxidized and H_2O_2 is formed as a byproduct. Oxygen reduction absorbs significant levels of the photosynthetic electron flux not only through its role in photorespiration, but also by its direct reduction by PSI (Asada, 1999).

Higher plants can sense, transduce, and translate ROS signals into specific cellular responses, the mechanism is dependent on the presence of redox-sensitive proteins that can undergo reversible oxidation/reduction and may switch "on" and "off" depending on the cellular redox state. ROS can oxidize the redox-signaling proteins directly or indirectly via the ubiquitous redox-sensitive molecules, such as glutathione (GSH) or thioredoxin, which control the cellular redox state in higher plants (Shao *et al.*, 2005). Activation of pH-dependent cell wall peroxidases takes place under alkaline pH, and in the presence of a reductant H_2O_2. Alkalinization of the apoplast upon elicitor recognition proceeds the oxidative burst and the production of H_2O_2 by pH-dependent cell wall peroxidases (Wojtaszek, 1997).

The UV-B radiation-induced inhibition of PSII photochemistry results in excessive excitation energy, which if not dissipated safely may damage PSII due to over reduction of reaction centers (Demmig-Adams, 2003). The alternative way to dissipate this excessive energy is either directly through the Mehler reaction or indirectly through photorespiration which favors the production of $O_2^{-\bullet}$ and H_2O_2 (Asada, 1999). The ROS thus produced disturb metabolic balance (Galatro *et al.*, 2001). Furthermore, ROS are known to activate genes, the products of which can in turn affect the expression of other genes (Mackerness *et al.*, 1999). The direct damage to the key enzymes involved in photosynthesis and respiratory pathways may also promote ROS formation (Jordan, 1996).

VI. METABOLISM OF ROS

Solar UV-B radiation produces ROS, eventually producing an oxidative stress (Brosche and Strid, 2003; Surplus *et al.*, 1998). Plants have developed complex antioxidant defense systems involving several enzymes and metabolites, to scavenge excess ROS produced under UV-B stress (Jansen *et al.*, 2008). The enzymatic antioxidants include SOD (EC 1.15.1.1), CAT

(EC 1.11.1.6), peroxidase (POD; EC 1.11.1.7), APX (EC 1.11.1.11), GR (EC 1.6.4.2), dehydroascorbate reductase (DHAR; EC 1.8.5.1), monodehydroascorbate reductase (MDHAR; EC 1.6.5.4), GPX (EC 1.11.1.9) and nonenzymatic antioxidant systems include reduced glutathione (GSH), ascorbic acid (AsA), α-tocopherol, and carotenoids, etc. The major ROS-scavenging pathways of plants include SODs found in all cellular compartments, the water–water cycle in chloroplasts, cytosol, mitochondria, apoplast and peroxisomes, GPX and CAT in peroxisomes. Transcripts of key enzymes of antioxidative enzyme system, such as APX, SOD, POD, CAT are induced by UV-B radiation (Agrawal and Rathore, 2007; Jansen *et al.*, 1998; Kumari *et al.*, 2009; Willekens *et al.*, 1994). Genes encoding scavenging enzymes are differentially expressed in response to UV-B, when the transcript levels of GR and GPX rise, while those of SOD remain unaltered or even dropped (Strid *et al.*, 1994; Willekens *et al.*, 1994).

SOD is a metallo-enzyme containing either Cu and Zn or Mn which catalyzes the dismutation reaction of superoxide anion ($O_2^{-\bullet}$) into H_2O_2 and O_2. In plants, three different SODs are present: in cytosol (Cu/Zn-SOD), in mitochondria (Mn-SOD), and in chloroplasts (Cu/Zn-SOD) (Fig. 3). These SODs can be easily differentiated on the basis of mRNA transcripts, as well as activity levels with *in situ* staining technique on gel

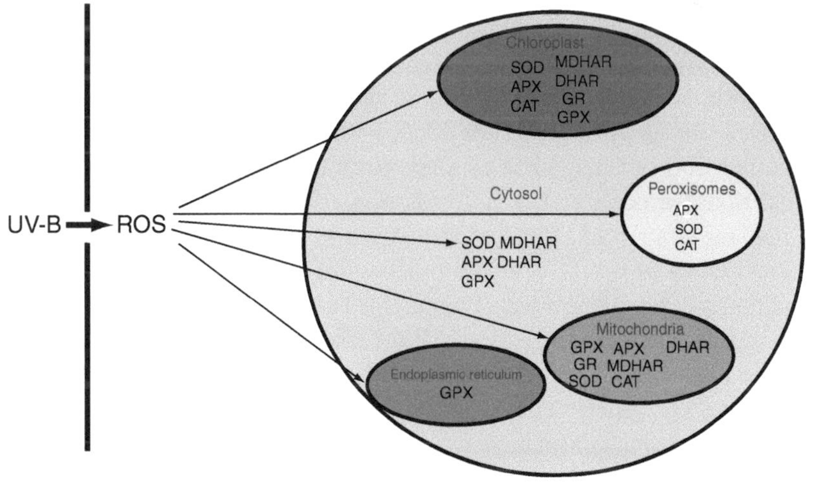

Fig. 3. Showing the production of SOD, APX, GR, GPX, CAT, MDHAR, and DHAR in different cellular compartments by UV-B induced ROS. SOD, superoxide dismutase; APX, ascorbate peroxidase; GR, glutathione reductase; GPX, glutathione peroxidase; CAT, catalase; MDHAR, monodehydroascorbate reductase; DHAR, dehydroascorbate reductase.

(Rao *et al.*, 1996). H_2O_2 is generated as a result of direct UV-B photochemical reaction in plants (Larson, 1988) and also due to the increase in SOD activity under UV-B irradiation (Balakumar *et al.*, 1997). Xu *et al.* (2008) and Yanqun *et al.* (2003), however, reported decrease in SOD activity in UV-B sensitive *Glycine max* (Table I). Rao *et al.* (1996) showed induction of cytosolic as well chloroplastic Cu, Zn-SOD, due to preferential expression of Cu, Zn-SOD-3, -4, and -5 isoforms, while UV-B exposure did not significantly affect Mn-SOD. Strid *et al.* (1994) reported a decrease in mRNA transcript of chloroplastic SOD of *Pisum sativum* grown under supplemental UV-B. Increase in SOD activity is a general response indicating oxidative stress in plants under elevated UV-B radiation in growth chamber, glass house, and field experiments (Table I). Selvakumar (2008) reported linear increase in SOD activity in *Crotalaria juncea* with corresponding increase in UV-B dose. Increase in SOD activity was found to be associated with concurrent decline in CAT activity, suggesting accumulation of H_2O_2 in the plant cells exposed to UV-B (Selvakumar, 2008).

APX isoenzymes are distributed in at least four distinct cellular compartments: stromal APX (sAPX) and thylakoid-membrane bound (tAPX) in chloroplasts, microbody (including glyoxysomes and peroxisome), membrane-bound APX (mAPX), and cytosolic APX (cAPX) (Chen and Asada, 1989) (Fig. 3). A fifth APX isoenzyme (mit APX) occurs in mitochondria as a membrane-bound form (Leonardis *et al.*, 2000) (Fig. 3). The isoforms are encoded by distinct genes and differ in size, specificity for their electron donors, and sensitivity to inactivation (Chen and Asada, 1989). In *Arabidopsis*, a total of five genes have been identified as coding for various isoforms in the chloroplast (APX4 and APX5), cytosol (APX1 and APX2), and microbodies (APX3) (Santos *et al.*, 1996). This represents a full gene set but other isoforms may arise by some form of posttranscriptional or post-translational processes. UV-B irradiation increased the APX activity in *A. thaliana* (Rao *et al.*, 1996). Acceleration in APX activity, in response to elevated levels of UV-B, is a common response to foliar tissues under field conditions (Table I). However, Yannarelli *et al.* (2006) and Costa *et al.* (2002) reported no significant effect on APX activity of cotyledons of sunflower exposed in growth chamber and Dai *et al.* (1997) in rice leaves grown under greenhouse conditions. Yao and Liu (2006) also reported decline in APX activity in *Acer mono* Maxim leaves at a high UV-B dose under field conditions (Table I). The balance between SOD and APX or APX activities in cells is crucial for determining the steady state level of $O_2^{-\bullet}$ and H_2O_2 (Bowler *et al.*, 1991). ROS production can also be decreased by the alternative channelling of electrons in the ETC of the chloroplast and mitochondria by a group of enzymes called alternative oxidases (AOXs).

The ascorbate–glutathione cycle is the major defense system against ROS in chloroplasts, cytosol, mitochondria, apoplast, and peroxisomes. The ascorbate–glutathione cycle in the chloroplast is also referred as "Asada-Foyer-Halliwell pathway" (Fig. 4). The cycle involves several enzymes (APX; MDHAR and DHAR, GR), ascorbate (AsA), and glutathione (GSH) as oxidoreductants, H_2O_2 as an electron acceptor, and NADPH as an H-donor (Fig. 4). APX uses two molecules of AsA to reduce H_2O_2 with generation of two molecules of monodehydroascorbate (MDHA); MDHA can be reduced to AsA, in a reaction catalyzed by MDHAR. AsA can also be nonenzymatically regenerated from MDHA. DHA is always produced during the rapid disproportionation of MDHA radical and DHA is then reduced to AsA by the action of DHAR using GSH as the reducing substrate. This results in the generation of glutathione disulfide (GSSG), which is regenerated to GSH by GR. Thus, the ascorbate–glutathione cycle is involved in the full scavenging of H_2O_2, the utilization of reducing NADPH units, and the continous supply of $NADP^+$ as well as in the dissipation of excess excitation energy. In this way, the cycle minimizes the overloading of the ETC and contributes to normalization of the redox status in chloroplast (Asada, 1992). The hydroxyl radical ($^{\bullet}OH$) cannot be subjected to enzymatic breakdown but can be

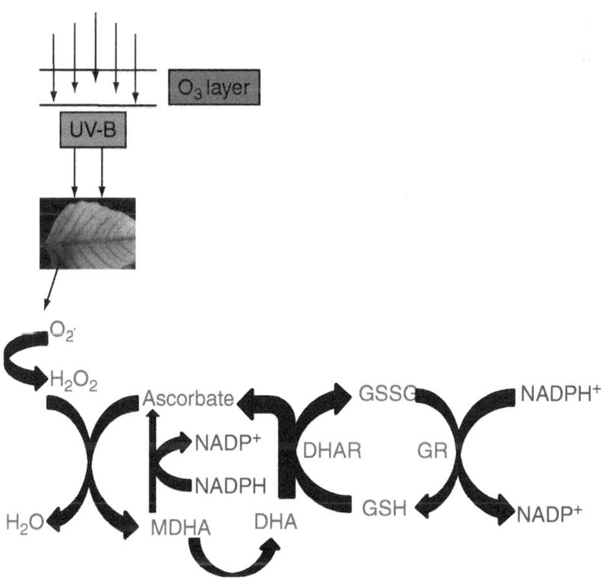

Fig. 4. UV-B induced ROS generation followed by Asada–Halliwell Pathway of oxyradicals scavenging and involvement of various antioxidant enzymes.

scavenged by ascorbate, tocopherol, and glutathione (Larson, 1988; Niyogi, 1999; Noctor and Foyer, 1998). Mechanisms involving interference with ˙OH generation seem to be more effective (Mittler, 2002). Ferritin, a Fe^{2+} binding protein, blocks the Fenton reaction and hence the ˙OH formation.

POD are monomeric hemoprotein that catalyze the oxidation of a range of substrates by H_2O_2. PODs are involved in physiological processes like phenol-oxidation (Kobayashi *et al.*, 1994), cross-linking of phenolic compounds to proteins and polysaccharides and/or deposition of polyphenols and lignin (Lagrimini, 1991), suberization (Bernards and Lewis, 1998), pathogen resistance (Bestwick *et al.*, 1998), and the oxidative degradation of the major endogenous auxin (Gazaryan *et al.*, 1998). UV-B is known to alter the distribution of POD isoforms, (Murali *et al.*, 1988) and have a number of potential roles in plant growth, development, and differentiation (Gaspar *et al.*, 1991). PODs that use glutathione as a cosubstrate have been rarely identified in plants, but PODs specific for ascorbate have been often observed (Chen and Asada, 1989). In addition, PODs are believed to metabolize H_2O_2 by using phenols as a cosubstrate through an ascorbate-dependent pathway (Otter and Polle, 1994). Balakrishnan *et al.* (2005) reported increased POD activity under elevated UV-B radiation and the increasing trend reached a maximum (64.5%) on the fourth day in treated seedlings (Table I). Kumari *et al.* (2009) reported decline in POD activity in *Abelmoschus esculentus* L. in field conditions under 1.8 kJ m^{-2} d^{-1} UV-B dose above ambient (Table I).

The role of POD in IAA catabolism in plants has been demonstrated by the decrease of IAA levels in transgenic *Nicotiana sylvestris* over expressing the anionic POD (Jansen, 2002). Anionic PODs are believed to utilize phenolic compounds, such as coniferyl alcohol and H_2O_2 to initiate the chain reaction that leads to lignification (Polle *et al.*, 1994). Using RT-PCR analysis, Kim *et al.* (2007) reported the responses of 10 POD genes from cell cultures of sweet potato to treatment with UV-B and found that four anionic POD genes swpa1, swpa2, swpa3, and swpa4 were highly induced by UV-B, while other genes were not expressed. A link between POD activity and UV tolerance was also found in *Spirodela punctata* (Jansen *et al.*, 2001). The plant PODs in the apoplast catalyze the formation of aromatic oxyl radicals from several aromatic compounds (Takahama, 2004) and POD-dependent production of such organic radicals often results in the generation of ROS (Kagan *et al.*, 1990).

CATs are tetrameric heme-containing enzymes that use H_2O_2 as a substrate and convert it to H_2O and O_2, thus protecting the cells from the damaging effects of H_2O_2 accumulation (Sanchez-Casas and Klesseg, 1994). CATs are present in peroxisomes, glyoxysomes, and related organelles where H_2O_2-generating enzymes, such as glycolate oxidase are found (Fig. 3).

There are three main isoforms: CAT1, CAT2, and CAT3. Induction of Cat1 and Cat2 transcripts, as well as of the enzyme activities by UV-B radiation, is well known (Alexieva *et al.*, 2001; Willekens *et al.*, 1994). A specific isozyme CAT3 is present in maize mitochondria. Willekens *et al.* (1994) reported that under UV-B treatment, Cat1 is highly expressed in photosynthesizing cells of *Nicotiana plumbaginifolia* where it can scavenge the H_2O_2 that is produced during photorespiration. Cat3 is most abundant in seeds and, therefore likely to be linked with glyoxysomal functions, whereas Cat2 is found to be uniformly distributed in plants with a particular preference for vascular tissues. In maize, the expression and accumulation of Cat2 and Cat3 CAT genes is induced by UV-B suggesting that both the genes may act together to scavenge ROS generated by UV-B to protect the plants from oxidative damage (Boldt and Scandalios, 1997). It is known that the limited protective action of CAT is attributed to its poor affinity for its substrate, its sensitivity to light-induced inactivation (Foyer *et al.*, 1994), and inhibition of its activity by high $O_2^{-\bullet}$ or H_2O_2 concentrations (Lardinois, 1995). When the production of ROS exceeds the capacity of antioxidant metabolism to remove them, oxidative damage to cellular macromolecules and structure occurs, which if unchecked leads to cell death. The studies conducted on CAT, in response to UV-B, showed both increase and decrease in its activity (Table I). Experiments conducted in growth chambers mostly show induction of CAT activity upon UV-B exposure (Table I). CAT activity mostly increases in leaves of relatively resistant plants, such as *Lycopersicum esculentum*, *Picea asperata*, *A. mono* Maxim, *Triticum aestivum*, under natural field conditions (Agrawal and Rathore, 2007; Ambasht and Agrawal, 2003a; Balakumar *et al.*, 1997; Yao and Liu, 2006). In sensitive leguminous plants, however, CAT activity declines when UV-B irradiation increases under field conditions (Agrawal and Mishra, 2009). The different affinities of APX (M range) and CAT (mM range) for H_2O_2 suggest that they belong to two different classes of H_2O_2-scavenging enzymes, where APX might be responsible for the removal of excess ROS during stress.

Peroxisomes are not only the site of ROS detoxification by CAT but also the site of ROS production by glycolate oxidase and fatty acid β-oxidation. In addition, peroxisomes might be one of the cellular sites for NO biosynthesis (Corpas *et al.*, 2001). NO has been shown be involved in ROS-induced cell death in plants (Delledonne *et al.*, 2001). Generally, plants with suppressed APX production induce SOD, CAT, and GR to compensate for the loss of APX, whereas plants with suppressed CAT production induce APX, GPX, and mitochondrial AOX (Willekens *et al.*, 1997). Pyro A expression increases following exposure to acute UV-B dose, and this has been associated with the singlet oxygen scavenging properties of pyridoxine (Broschė *et al.*, 2002).

Pyro A expression is downregulated in plants acclimated to low chronic UV-B (Hectors *et al.*, 2007).

Ascorbate and glutathione act in the aqueous phase, whereas the lipophilic antioxidants tocopherols and carotenoids are active in the membrane environment. Tocopherols are lipid-soluble plant antioxidants and are precursors of Vit. E. In plants, tocopherols are involved in the reduction of PUFA radicals that are formed in UV-B stressed plants (DeLong and Steffen, 1998). Acute exposure of UV-B leads to decrease in α-tocopherol levels in plants (Carletti *et al.*, 2003; DeLong and Steffen, 1998; Galatro *et al.*, 2001; Jain *et al.*, 2003) (Table I), reflecting reactions with lipid radicals. In the thylakoid membrane, α-tocopherol protects the structure and function of photosynthetic membrane by efficiently scavenging ROS and lipid alkyl and peroxyl radicals (Hess, 1993). Incorporation of α-tocopherol into phosphatidylcholine liposomes has been shown to prevent oxidative degradation during UV-B exposure (Pelle *et al.*, 1990). It has been shown that α-tocopheroxyl radicals are readily generated in UV-B irradiated liposomes, resulting in the immediate loss of α-tocopherol antioxidative function which is not regained unless reducing agents, such as ascorbate/thiols are present (Kagan *et al.*, 1992). Reducing agents that donate electrons to α-tocopheroxyl radicals recycle α-tocopherol to its reduced form and thus sustain its antioxidative function. UV-B induced decreases in α-tocopherol levels have been reported to be paralleled by an increase in ascorbate levels (Jain *et al.*, 2003).

Levels of glutathione and ascorbate are upregulated in response to UV-B (Takeuchi *et al.*, 1996). AsA and glutathione are involved in the neutralization of secondary products of ROS reactions (Conklin *et al.*, 2000) and are found at high concentrations in chloroplast and other cellular compartments (5–20 mM AsA and 1–5 mM glutathione). An analysis of *Arabidopsis* single and double mutant plants have shown that decreases in the levels of one of three main plant antioxidants (tocopherols, ascorbate, or glutathione) result in increases in the remaining antioxidants. Reduced to oxidized ratios of AsA and glutathione are essential for the proper scavenging of ROS in cells. Transporters for AsA and glutathione are likely to be the central in determining the specific concentrations of these compounds and the redox potential in different cellular compartments (Noctor and Foyer, 1998). Responses of AsA levels in plants to elevated UV-B are both negative and positive. Most of the studies conducted in tropical field conditions show a decline in the AsA content. Taulavouri *et al.* (1998) found increases in AsA content with a decline in GSH content in *Vaccinium myrtillus* L. grown in field conditions at elevated UV-B (Table I). This suggests a greater GR activity in these plants to regenerate AsA at the expense of GSH. Generally, GR activity increases under UV-B exposure (Table I) but Yang *et al.* (2007) have shown its reduction at lower dose whereas higher doses increase GR activity in growth chamber studies. Yao and Liu (2006) have also shown reduction in GR activity at high UV-B dose.

Glutathione (GSH), a disulfide reductant protects thiols of enzymes and reacts with singlet oxygen, H_2O_2, and $^{\bullet}OH$ (Millar *et al.*, 2003). GSH has been detected virtually in all cell compartments, such as cytosol, chloroplast, endoplasmic reticulum, vacuoles, and mitochondria (Millar *et al.*, 2003). The change in the ratio of reduced GSH to oxidized GSSG during the degradation of H_2O_2 is essential in certain redox-signaling pathways. A reduction in the ratio of reduced glutathione/total glutathione in vtc1 mutant as compared to the wild type was observed during the first day of UV-B treatment in *A. thaliana* (Gao and Zhang, 2008). Enhanced glutathione biosynthesis in chloroplasts can result in oxidative damage to cells rather than their protection, possibly by altering the overall redox state of chloroplasts (Creissen *et al.*, 1999). Reduced glutathione content is shown to increase in *Cassia auriculata* L. grown under growth chambers at 7.5 and 15 kJ m^{-2} UV-B radiation (Table I).

VII. CONCLUSION

UV-B acts as a signal to induce changes in gene expression distal from its origin of perception. However, the nature of the UV-B photoreceptor is not completely known. UV-B photoreceptor would probably be a protein with a flavin and/or pterin chromophores. Studies on signal transduction intermediates conducted through combinations of cell physiology, biochemical, and genetic approaches indicated ROS as the best characterized signaling intermediates generated by UV-B. UV-B signaling includes induction of MAPK, ion fluxes, and NADPH oxidase. UV-B induced modifications in gene expression are precise with a downregulation of photosynthetic genes and upregulation of defense genes. Expression of genes is affected by UV-B at different levels from transcription, translation, and posttranslational modification. ROS obviously lead to cellular damage, however, they act as signal transducer for the expression of certain defense-related genes involved with various antioxidants and enzymes and other genes involving SA, ethylene, and JA. UVR8 protein is a UV-B specific signaling component controlling the expression of a range of genes essential for UV-B protection. H_2O_2, NO, and SA act as second messengers mediating responses of specific genes to UV-B radiation. Induction of a number of defense mechanisms, such as production of UV-B screening pigments, increase in antioxidant enzymes, and induction of PR proteins are also mediated at the level of gene expression. UV-B induced formation of ROS proceeds through multiple pathways and increased with increasing UV-B doses. UV-B induced antioxidants are important in providing tolerance to plants against UV-B. The molecular mechanisms behind UV-B responses are poorly understood. It is particularly important to trace the mechanism of UV-B perception and signal transduction pathways involved in various metabolic responses and the biochemical changes related to ROS. Identification of the receptors involved in perception of UV-B would be of great significance in suggesting the way for manipulating UV-B resistance without affecting responses of other stresses.

REFERENCES

Agarwal, S. (2007). Increased antioxidant activity in *Cassia* seedlings under UV-B radiation. *Biologia Plantarum* **51**(1), 157–160.

Agrawal, S. B. and Mishra, S. (2009). Effects of supplemental ultraviolet-B and cadmium on growth, antioxidants and yield of *Pisum sativum* L. *Ecotoxicology and Environmental Safety* **72**, 610–618.

Agrawal, S. B. and Rathore, D. (2007). Changes in oxidative stress defense system in wheat (*Triticum aestivum* L.) and mung bean (*Vigna radiata* L.) cultivars grown with and without mineral nutrients and irradiated by supplemental ultraviolet-B. *Environmental and Experimental Botany* **59**(1), 21–33.

Alexieva, V., Sergiev, I., Mapelli, S. and Karanov, E. (2001). The effect of drought and ultraviolet radiation on growth and stress markers in pea and wheat. *Plant Cell and Environment* **24**, 1337–1344.

Ambasht, N. K. and Agrawal, M. (1998). Physiological and biochemical responses of *Sorghum vulgare* plants to supplemental ultraviolet-B radiation. *Canadian Journal of Botany* **76**, 1290–1294.

Ambasht, N. K. and Agrawal, M. (2003a). Interactive effects of ozone and ultraviolet-B radiation on physiological and biochemical characteristics of soybean plants. *Journal of Plant Biology* **30**, 37–45.

Ambasht, N. K. and Agrawal, M. (2003b). Effects of enhanced UV-B radiation and tropospheric ozone on physiological and biochemical characteristics of field grown wheat. *Biologia Plantarum* **47**, 625–628.

Apel, K. and Hert, H. (2004). Reactive oxygen species: Metabolism, oxidative stress and signal transduction. *Annual Review of Plant Biology* **55**, 373–399.

Arora, A., Sairam, R. K. and Srivastava, G. C. (2002). Oxidative stress and antioxidative system in plants. *Current Science* **82**, 1227–1238.

Asada, K. (1992). Ascorbate peroxidase—A hydrogen peroxide-scavenging enzyme in plants. *Physiologia Plantarum* **85**, 235–241.

Asada, K. (1999). The water–water cycle in chloroplasts: Scavenging of active oxygen and dissipation of excess photons. *Annual Review of Plant Physiology and Plant Molecular Biology* **50**, 601–639.

Balakrishnan, V., Venkatesan, K., Ravindran, K. C. and Kulandaivelu, G. (2005). Protective mechanism in UV-B treated *Crotalaria juncea* L. seedlings. *Plant Protection Science* **41**, 115–120.

Balakumar, T., Gayathri, B. and Anbudurai, P. R. (1997). Oxidative stress injury in tomato plants induced by supplemental UV-B radiation. *Biologia Plantarum* **39**, 215–221.

Barta, C., Kalai, T., Hideg, K., Vass, I. and Hideg, E. (2004). Differences in the ROS-generating efficacy of various ultraviolet wavelengths in detached spinach leaves. *Functional Plant Biology* **31**, 23–28.

Beggs, C. J., Schneider-Ziebert, U. and Wellman, E. (1986). UV-B radiation and adaptive mechanisms in plants. *In* "Stratospheric Ozone Reduction Solar Ultraviolet Radiation and Plant Life" (R. C. Worrest and C. C. Caldwell, eds.), pp. 235–250. Springer, Berlin.

Bernards, M. A. and Lewis, N. G. (1998). The macromolecular aromatic domain in suberized tissue: A changing paradigm. *Phytochemistry* **47**, 915–933.

Bestwick, C. S., Brown, I. R. and Mansfield, J. W. (1998). Localized changes in peroxidase activity accompany hydrogen peroxide generation during the development of a nonhost hypersensitive reaction in lettuce. *Plant Physiology* **118**, 1067–1078.

Bischof, K., Hanelt, D. and Wiencke, C. (2000). Effects of ultraviolet radiation on photosynthesis and related enzyme reactions of marine macroalgae. *Planta* **211**, 555–562.

Blechert, S., Brodschelm, W., Holder, S., Kammerer, L., Kutchan, T. M., Mueller, M. J., Xia, Z.-Q. and Zenk, M. H. (1995). The octadecanoid pathway: Signal molecules for the regulation of secondary pathways. *Proceedings of the National Academy of Sciences of the United States of America* **92**, 4099–4105.

Boldt, R. and Scandalios, J. G. (1997). Influence of UV-light on the expression of Cat2 and Cat3 catalase genes in maize. *Free Radicals Biology and Medicine* **23**, 505–514.

Bowler, C., Slooten, L., Vandenbranden, S., De Rycke, R., Botterman, J., Sybesma, C., Van Montagu, M. and Inze, D. (1991). Manganese superoxide dismutase can reduce in transgenic plants. *The EMBO Journal* **10**, 1723–1732.

Britt, A. B., Chen, J. J., Wykoff, D. and Mitchell, D. (1993). A UV-sensitive mutant of *Arabidopsis* defective in the repair of pyrimidine-pyrimidinone (6–4) dimmers. *Science* **261**, 1571–1574.

Brosche, M. and Strid, A. (2003). Molecular events following perception of ultraviolet-B radiation by plants. *Physiologia Plantarum* **117**, 1–10.

Broschė, M., Schuler, M., Kalbina, I., Connor, L. and Strid, A. (2002). Gene regulation by low level UV-B radiation: Identification by DNA array analysis. *Photochemical and Photobiological Sciences* **1**, 656–664.

Brown, B. A., Cloix, C., Jiang, G. H., Kaiserli, E., Herzyk, P., Kliebenstein, D. J. and Jenkins, G. I. (2005). A UV-B-specific signaling component orchestrates plant UV protection. *Proceedings of National Academy of Sciences of the United States of America* **102**, 18225–18230.

Caldwell, M. M., Bjorn, L. O., Bornman, J. F., Flint, S. D., Kulandaivelu, G., Teramura, A. H. and Tevini, M. (1998). Effects of increased solar ultraviolet radiation on terrestrial plants. *Ambio* **24**, 166–173.

Carletti, P., Masi, A., Wonisch, A., Grill, D., Tausz, M. and Ferretti, M. (2003). Changes in antioxidant and pigment pool dimensions in UV-B irradiated maize seedlings. *Environmental and Experimental Botany* **50**, 149–157.

Casati, P. and Walbot, V. (2003). Gene expression profiling in response to ultraviolet radiation in maize genotypes with varying flavonoids content. *Plant Physiology* **132**, 1739–1754.

Casati, P. and Walbot, V. (2004). Crosslinking of ribosomal proteins to RNA in maize ribosomes by UV-B and its effect on translation. *Plant Physiology* **136**, 3319–3332.

Cechin, I., Corniani, N., Fumis, T. D. and Cataneo, A. C. (2008). Ultraviolet-B and water stress effects on growth, gas exchange and oxidative stress in sunflower plants. *Radiation Environmental and Biophysics* **47**, 405–413.

Cheeseman, J. M. (2006). Hydrogen peroxide concentrations in leaves under natural conditions. *Journal of Environmental Botany* **57**, 2435–2444.

Chen, G. X. and Asada, K. (1989). Ascorbate peroxidase in tea leaves: Occurrence of two isozymes and the differences in their enzymatic and molecular properties. *Plant and Cell Physiology* **30**, 987–998.

Chen, Z., Silva, H. and Klesseg, D. (1993). Active oxygen species in the induction of plant systemic required resistance by salicyclic acid. *Science* **262**, 1883–1886.

Chow, W. S., Strid, A. and Anderson, J. M. (1992a). Recovery of photosynthetic functions and components after supplementary UV-B irradiation. *Photosynthetic Research* **34**, 224.

Chow, W. S., Strid, A. and Anderson, J. M. (1992b). Short term treatment of pea plants with supplementary ultraviolet-B radiation: Recovery time-courses of some photosynthetic functions and components. *In* "Research in Photosynthesis, Vol. IV" (N. Murata, ed.), pp. 361–364. Kluwer Academic Publishers, Dordrecht, The Netherlands.

Christie, J. M. and Jenkins, G. I. (1996). Distinct UV-B and UV-A/blue light signal transduction pathways induce chalcone synthase gene expression in *Arabidopsis* cells. *Plant Cell* **8**, 1555–1567.

Clarke, J. D., Liu, Y., Klessig, D. F. and Dong, X. (1998). Uncoupling PR gene expression from NPR1 and bacterial resistance: Characterization of the dominant *Arabidopsis* cpr6–1 mutant. *Plant Cell* **10**, 557–569.

Conconi, A., Smerdon, M. J., Howe, G. A. and Ryan, C. A. (1996). The octadecanoid signalling pathway in plants mediates a response to UV radiation. *Nature* **383**, 826–829.

Conklin, P. L., Saracco, S. A., Norris, S. R. and Last, R. L. (2000). Identification of ascorbic acid-deficient *Arabidopsis thaliana* mutants. *Genetics* **154**, 847–856.

Corpas, F. J., Barroso, J. B. and del Rio, L. A. (2001). Peroxisomes as a source of reactive oxygen species and nitric oxide signal molecules in plant cells. *Trends of Plant Sciences* **6**, 145–150.

Costa, H., Gallego, S. M. and Tomaro, S. L. (2002). Effect of UV-B radiation on antioxidant defense system in sunflower cotyledons. *Plant Science* **162**, 939–945.

Creissen, G., Firmin, J., Fryer, M., Kular, B., Leyland, M., Reynolds, H., Pastori, G., Wellburn, F., Baker, N. R., Wellburn, A. and Mullineaux, P. (1999). Elevated glutathione biosynthetic capacity in the chloroplasts of transgenic tobacco paradoxically causes increased oxidative stress. *Plant Cell* **11**, 1277–1291.

Dai, Q., Yan, B., Huang, S., Liu, X., Peng, S., Miranda, M. L. L., Chavez, A. Q., Vergara, B. S. and Olszyk, D. H. (1997). Response of oxidative stress defense systems in rice (*Oryza sativa* L.) leaves with supplemental UV-B radiation. *Physiologia Plantarum* **101**, 301–308.

Dany, A. L., Douki, T., Triantaphylides, C. and Cadet, J. (2001). Repair of the main UV-induced thymine dimeric lesions within *Arabidopsis thaliana* DNA: Evidence for the major involvement of photoreactivation pathways. *Journal of Photochemistry and Photobiology, B: Biology* **65**, 127–135.

Dat, J., Vandenabeele, S., Vranova, E., Van Montage, M., Inze, D. and Breusegem Van, F. (2000). Dual action of the active oxygen species during plant stress. *Cellular and Molecular Life Sciences* **57**, 779–795.

Dawar, S., Vani, T. and Singhal, G. S. (1998). Stimulation of antioxidant enzymes and lipid peroxidation by UV-B irradiation in thylakoid membrane of wheat. *Biologia Plantarum* **41**, 65–73.

Delledonne, M., Zeier, J., Marocco, A. and Lamb, C. (2001). Signal interactions between nitric oxide and reactive oxygen intermediates in the plant hypersensitive disease resistance. *Nature* **394**, 585–588.

DeLong, J. M. and Steffen, K. L. (1998). Lipid peroxidation and α-tocopherol content in α-tocopherol—Supplemented membranes during UV-B exposure. *Environmental and Experimental Botany* **39**, 177–185.

Demmig-Adams, B. (2003). Linking the xanthophylls cycle with thermal energy dissipation. *Photosynthetic Research* **76**, 73–80.

Dempsey, D. A. and Klessig, D. F. (1994). Salicylic acid, active oxygen species and systemic acquired resistance in plants. *Trends in Cell Biology* **4**, 334–338.

Desikan, R., Burnett, E. C., Hancock, J. T. and Neill, S. J. (1998a). Harpin and hydrogen peroxide induce the expression of a homologue of *gp 91-phox* in *Arabidopsis thaliana* suspension cultures. *Journal of Experimental Botany* **49**, 1767–1771.

Desikan, R., Reynolds, A., Hancock, J. T. and Neill, S. J. (1998b). Harpin and hydrogen peroxide both initiate programmed cell death but have differential effects on defense gene expression in *Arabidopsis* suspension cultures. *Biochemical Journal* **330**, 115–120.

Elstner, E. F. (1982). Oxygen activation and oxygen toxicity. *Annual Review of Plant Physiology* **33**, 73–96.

Epple, P., Apel, K. and Bohlmann, H. (1995). An *Arabidopsis thaliana* thionin gene is inducible via a signal transduction pathway different from that for PR proteins. *Plant Physiology* **109**, 813–820.

Farmer, E. E., Johnson, R. R. and Ryan, C. A. (1992). Regulation of expression of proteinase inhibitor genes by methyl jasmonate and jasmonic acid. *Plant Physiology* **98**, 995–1002.

Forman, H. J. and Boveris, A. (1982). Superoxide radical and hydrogen peroxide in mitochondria. *In* "Free Radicals in Biology" (W. A. Prior, ed.), pp. 65–90. Academic Press, Boca Raton, FL.

Foyer, C. H. and Noctor, G. (2003). Redox sensing and signalling associated with reactive oxygen in chloroplasts, peroxisomes and mitochondria. *Physiologia Plantarum* **119**, 355–364.

Foyer, C. H., Descourvieres, P. and Kunert, K. J. (1994). Protection against oxygen radicals: an important defence mechanism studied in transgenic plants. *Plant Cell and Environment* **17**, 507–523.

Frenkel, C. (1991). Disruption of macromolecular hydration—A possible origin of chilling destabilization of biopolymers. *Trends in Food Science and Technology* **2**, 39–41.

Frohnmeyer, H. and Staiger, D. (2003). Ultraviolet-B radiation mediated responses in plants. Balancing damage and protection. *Plant Physiology* **133**, 1420–1428.

Frohnmeyer, H., Bowler, C. and Schafer, E. (1997). Evidence for some signal transduction elements involved in UV-light dependent responses in parsley protoplasts. *Journal of Experimental Botany* **48**, 739–750.

Frohnmeyer, H., Bowler, C., Zhu, J. K., Yamagata, H., Schäfer, E. and Chuan, N. H. (1998). Different roles for calcium and calmodulin in phytochrome- and UV-regulated expression of chalcone synthase. *Plant Journal* **13**, 763–772.

Fry, S. C. (1986). Cross-linking of matrix polymers in the growing cells of angiosperms. *Annual Review of Plant Physiology* **37**, 165–186.

Galatro, A., Simontacchi, M. and Puntarulo, S. (2001). Free radical generation and antioxidant content in chloroplasts from soybean leaves exposed to ultraviolet-B. *Physiologia Plantarum* **113**, 564–570.

Galland, P. and Senger, H. (1988). The role of pterins in the photoreception and metabolism in plants. *Photochemistry and Photobiology* **48**, 811–820.

Gao, Q. and Zhang, L. (2008). Ultraviolet-B-induced oxidative stress and antioxidant defense system responses in ascorbate-deficient vtc1 mutants of *Arabidopsis thaliana*. *Journal of Plant Physiology* **165**, 138–148.

Gaspar, T. H., Penel, C., Hagega, D. and Greppin, H. (1991). Peroxidases in plant growth differentiation and development processes. *In* "Biochemical, Molecular and Physiological Aspects of Plant Peroxidase" (J. Lobarzewski, H. Gneppin, C. Penel and T. H. Gaspar, eds.), pp. 244–250. University de Geneve, Switzerland.

Gazaryan, I. G., Lagrimini, L. M., Mellon, F. A., Naldrett, M. J., Ashby, G. A. and Thorneley, R. N. F. (1998). Identification of the skatolyl hydroperoxide and its role in the peroxidase-catalyzed oxidation of indol-3-yl acetic acid. *Biochemical Journal* **333**, 223–232.

Giordano, C. V., Galatro, A., Puntarulo, S. and Ballar, C. L. (2004). The inhibitory effects of UV-B radiation (280–315 nm) on *Gunnera magellanica* growth correlate with increased DNA damage but not with oxidative damage to lipids. *Plant Cell and Environment* **27**, 1415–1423.

Green, R. and Fluhr, R. (1995). UV-B-induced PR1 accumulation is mediated by active oxygen species. *Plant Cell* **7**, 203–212.

Greenberg, B. M., Gaba, V., Canaani, O., Malkin, S., Mattoo, A. K. and Edelman, M. (1989). Separate photosensitizers mediate degradation of 32kDa photosystem II reaction center polypeptide in the visible and UV spectral regions. *Proceedings of the National Academy of Sciences of the United States of America* **86**, 6617–6620.

Greenberg, B. M., Wilson, M. I., Huang, X., Duxbury, C. L., Gerhardt, K. E. and Gensemer, R. W. (1997). The effect of ultraviolet-B radiation on higher plants. *In* "Plants for Environmental Studies" (W. Wang, J. W. Gorsuch and J. S. Hughes, eds.), pp. 1–35. Lewis Publishers, Boca Raton, FL.

Hammond-Kosack, K. E. and Jones, J. D. G. (1996). Resistance gene-dependent plant defence responses. *Plant Cell* **8**, 1773–1791.

He, V. and Hader, D. P. (2002). Reactive oxygen species and UV-B: effect on cyanobacteria. *Photochemistry and Photobiology* **1**, 729–736.

He, Y., Tang, R. H., Hao, Y., Stevens, R. D., Cook, C. W., Ahn, S. M., Jing, L., Yang, Z., Chen, L., Guo, F., Fiorani, F. Jackson, R. B. *et al.* (2004). Nitric oxide represses the *Arabidopsis* floral transition. *Science* **305**, 1968–1971.

He, J. M., Xu, H., She, X. P., Song, X. G. and Zhao, W. M. (2005). The role and the interrelationship of hydrogen peroxide and nitric oxide in the UV-B-induced stomatal closure in broad bean. *Functional Plant Biology* **32**, 237–247.

He, J. M., Liu, Z. H., Xu, H., She, X. P. and Huang, C. (2006). The involvement of hydrogen peroxide in UV-B inhibited pollen germination and tube growth of *Paeonia suffruticosa* and *Paulownia tomentosa in vitro*. *Plant Growth Regulation* **49**, 199–208.

Hectors, K., Prinsen, E., De Coen, W., Jansen, M. A. K. and Guisez, Y. (2007). *Arabidopsis thaliana* plants acclimated to low dose rates of ultraviolet-B radiation show specific changes in morphology and gene expression in the absence of stress symptoms. *New Phytologist* **175**, 255–270.

Herrlich, P., Blattner, C., Knebel, A., Bender, K. and Rahmsdorf, H. J. (1997). Nuclear and non-nuclear targets of genotoxic agents in the induction of gene expression: Shared principles in yeast, rodents, man and plants. *Journal of Biological Chemistry* **378**, 1217–1229.

Hess, J. L. (1993). Vitamin E, α-tocopherol. *In* "Antioxidants of Higher Plants" (R. Alscher and J. L. Hess, eds.), pp. 111–134. CRC Press, Boca Raton, FL.

Hideg, E., Spetea, C. and Vass, I. (1994). Singlet oxygen production in thylakoid membranes during photoinhibition as detected by ESR spectroscopy. *Photosynthetic Research* **39**, 191–199.

Hideg, E., Mano, J., Ohno, C. and Asada, K. (1997). Increased levels of monodehydroascorbate radical in UV-irradiated broad bean leaves. *Plant Cell Physiology* **25**, 684–690.

Hideg, E., Barta, C., Kálai, T., Vass, I., Hideg, K. and Asada, K. (2002). Detection of singlet oxygen and superoxide with fluorescent sensors in leaves under stress by photoinhibition or UV radiation. *Plant Cell Physiology* **43**, 1154–1164.

Hollosy, F. (2002). Effects of ultraviolet radiation on plant cells. *Micron* **33**, 179–197.

Huang, F., Philosoph-Hadas, S., Meir, S., Callaham, D., Sabato, R., Zelcer, A. and Helper, P. (1997). Increases in cytosolic Ca^{2+} in parsley mesophyll cells correlate with leave senescence. *Plant Physiology* **115**, 51–60.

Izaguirre, M. M., Scopel, A. L., Baldwin, I. T. and Ballare, C. L. (2003). Convergent responses to stress solar ultraviolet-B radiation and *Manduca sexta* herbivory elicit overlapping transcriptional responses in field grown plants of *Nicotiana longiflora*. *Plant Physiology* **132**, 1755–1767.

Jain, K., Kataria, S. and Guruprasad, K. N. (2003). Changes in antioxidant defenses of cucumber cotyledons in response to UV-B and to the free radicals generating compound AAPH. *Plant Science* **165**, 551–557.

Jansen, M. A. K. (2002). Ultraviolet-B radiation effects on plants: Induction of morphogenic responses. *Physiologia Plantarum* **116**, 423–429.

Jansen, M. A. K., Gaba, V. and Greenberg, B. M. (1998). Higher plants and UV-B radiation: Balancing, damage, repair and acclimation. *Trends in Plant Science* **3**, 131–135.

Jansen, M. A. K., vanden Noort, R. E., Tan, M. Y. A., Prinsen, E., Lagrimini, L. M. and Thornley, R. N. F. (2001). Phenol-oxidizing peroxidases contribute to the protection of plants from ultraviolet radiation stress. *Plant Physiology* **126**, 1012–1023.

Jansen, M. A. K., Hectors, K., O'Brien, N. M., Guisez, Y. and Potters, G. (2008). Plant stress and human health: Do human consumers benefit from UV-B acclimated crops? *Plant Science* j.plantsci2008.04.010/j.plantsci2008.04.010.

Jenkins, G. I., Fuglevand, G. and Christie, J. M. (1997). UV-B perception and signal transduction. *In* Plants and UV-B: Responses to Environmental Change, (P. Lumsden, ed.), pp. 135–156. Cambridge University Press, UK.

Jenkins, G. I., Long, J. C., Wade, H. K., Shenton, M. R. and Bibikova, T. N. (2001). UV and blue light signaling: Pathways regulating chalcone synthase gene expression in *Arabidopsis*. *New Phytologist* **151**, 121–131.

Jiang, C. Z., Yee, L., Mitchell, D. L. and Britt, A. B. (1997). Photorepair mutants of *Arabidopsis*. *Proceedings in National Academy of Science of the United States of America* **94**, 7441–7445.

Jordan, B. R. (1996). The effects of ultraviolet-B radiation on plants: A molecular perspective. *Advances in Botanical Research* **22**, 97–162.

Jordan, B. R., Chow, W. S., Strid, A. and Anderson, J. M. (1991). Reduction in cab and psbA RNA transcripts in response to supplementary UV-B radiation. *FEBS Letters* **284**, 5–8.

Jordan, B. R., He, J., Chow, W. S. and Anderson, J. M. (1992). Changes in mRNA levels and polypeptide subunits of ribulose bisphosphate carboxylase in response to supplemental UV-B radiation. *Plant Cell and Environment* **15**, 91–98.

Jordan, B. R., James, P., Strid, A. and Anthony, R. (1994). The effect of supplementary UV-B radiation on gene expression and pigment composition in etiolated and green pea leaf tissue: UV-B-induced changes are gene specific and dependent upon the developmental stage. *Plant Cell Environment* **17**, 45–54.

Julkunen-Titto, R., Häggman, H., Aphalo, P. J., Lavola, A., Tegelberg, R. and Veteli, T. (2005). Growth and defense in deciduous trees and shrubs under UV-B. *Environmental Pollution* **137**, 415–427.

Kagan, Y. E., Serbinova, E. A. and Packer, L. (1990). Generation and recycling of radicals from phenolic antioxidants. *Archives of Biochemistry and Biophysics* **280**, 33–39.

Kagan, V. E., Witt, E., Goldman, R., Scita, G. and Packer, L. (1992). Ultraviolet light-induced generation of vitamin E radicals and their recycling. A possible photosensitizing effect of vitamin E in skin. *Free Radicals Research Communications* **16**, 51–64.

Kagawa, T., Sakai, T., Suetsugu, N., Oikawa, K., Ishiguro, S., Kato, T., Tabata, S., Okada, K. and Wada, M. (2001). *Arabidopsis* NPL1: A phototropin homolog controlling the chloroplast high-light avoidance response. *Science* **291**, 2138–2141.

Kalbin, I. and Strid, Å. (2006). Supplementary ultraviolet-B irradiation reveals differences in stress responses between *Arabidopsis thaliana* ecotypes. *Plant, Cell and Environment* **29**, 754–763.

Kakani, V. G., Reddy, K. R., Zhao, D. and Sailaja, K. (2003). Field crop response to ultraviolet-B radiation: a review. *Agricultural and Forest Meteorology* **120**, 191–218.

Kiegle, E., Moore, C. A., Haseloff, J., Tester, M. A. and Knight, M. R. (2000). Cell type-specific calcium responses to drought, salt and cold in the *Arabidopsis* root. *Plant Journal* **23**, 267–278.

Kim, Y. H., Lim, S., Han, S. H., Lee, J. C., Song, W. K., Bang, J. W., Kwon, S. Y., Lee, H. S. and Kwak, S. S. (2007). Differential expression of 10 sweetpotato peroxidases in response to sulfur dioxide, ozone and ultraviolet radiation. *Plant Physiology and Biochemistry* **45**, 908–914.

Kobayashi, A., Kaguchi, Y., Kanzaki, H., Kajeyama, S. and Kawazu, K. (1994). A new type of antimicrobial phenolics produced by plant peroxidase and its possible role in the chemical defense systems against plant pathogens. *Zeitschrift für Naturforschung* **49**, 411–414.

Köhler, B., Hills, A. and Blatt, M. R. (2003). Control of guard cell ion channels by hydrogen peroxide and abscisic acid indicates their action through alternate signaling pathways. *Plant Physiology* **131**, 385–388.

Kondo, N. and Kawashima, M. (2000). Enhancement of the tolerance to oxidative stress in cucumber (*Cucumis sativus* L.) seedlings by UV-B irradiation: Possible involvements of phenolic compounds and antioxidative enzymes. *Journal of Plant Research* **113**, 311–317.

Kramer, G. F., Norman, H. A., Krizek, D. T. and Mirecki, R. M. (1991). Influence of UV-B radiation on polyamines, lipid peroxidation and membrane lipids in cucumber. *Phytochemistry* **30**, 2101–2108.

Krause, G. H. (1994). The role of oxygen in photoinhibition of photosynthesis. *In* "Causes of Photooxidative Stress and Amelioration of Defence Systems in Plants" (C. H. Foyer and P. M. Mullineaux, eds.), pp. 43–76. CRC Press, Boca Raton, FL.

Kumagai, J., Katoh, H., Miyazaki, T., Hidema, J. and Kumagai, T. (1999). Differences in the sensitivity to UVB radiation of two cultivars of rice (*Oryza sativa* L.) based on observation of long-lived radicals. *Journal of Radiation Research* **40**, 303–310.

Kumari, R., Singh, S. and Agrawal, S. B. (2009). Combined effects of Psoralens and ultraviolet-B on growth, pigmentation and biochemical parameters of *Abelmoschus esculentus* L. *Ecotoxicology and Environmental safety* **72**, 1129–1136.

Lagrimini, L. M. (1991). Wound-induced deposition of polyphenols in transgenic plants overexpressing peroxidase. *Plant Physiology* **96**, 577–583.

Lamb, C. and Dixon, R. A. (1997). The oxidative burst in plant disease resistance. *Annual Review of Plant Physiology and Plant Molecular Biology* **48**, 251–275.

Lardinois, O. M. (1995). Reactions of bovine liver catalase with superoxide radicals and hydrogen-peroxide. *Free Radical Research* **22**, 251–274.

Larson, R. A. (1988). The antioxidants of higher plants. *Phytochemistry* **27**, 969–978.

Leonardis, S. D., Dipierro, N. and Dipierro, S. (2000). Purification and characterization of an ascorbate peroxidase from potato tuber mitochondria. *Plant Physiology and Biochemistry* **38**, 773–779.

Levine, A., Tenhaken, R., Dixon, R. and Lamb, C. (1994). H_2O_2 from the oxidative burst orchestrates the plant hypersensitive disease resistance response. *Cell* **79**, 583–593.

Long, J. C. and Jenkins, G. I. (1998). Involvement of plasma membrane redox activity and calcium homeostasis in the UV-B and UV-A/blue light induction of gene expression in *Arabidopsis*. *Plant Cell* **12**, 1939–1950.

Mackerness, S. A.-H. (2000). Plant responses to ultraviolet-B (UV-B: 280-320 nm) stress: What are the key regulators? *Plant Growth Regulator* **37**, 27–39.

Mackerness, S. A.-H., Jordan, B. R. and Thoman, B. (1997). UV-B effects on the expression of genes encoding proteins involved in photosynthesis. *In* "Plants and UV-B" (P. Lumsden, ed.), pp. 113–134. University Press, Cambridge.

Mackerness, S. A.-H., Surplus, S. L., Blake, P., John, C. F., Buchanan-Wollaston, V., Jordan, B. R. and Thomas, B. (1999). Ultraviolet-B—Induced stress and changes in gene expression in *Arabidopsis thaliana*: Role of signalling pathways controlled by jasmonic acid, ethylene and reactive oxygen species. *Plant Cell and Environment* **22**, 1413–1423.

Mackerness, S. A.-H., John, C. F., Jordan, B. and Thomas, B. (2001). Early signaling components in ultraviolet-B responses: Distinct roles for different reactive oxygen species and nitric oxide. *FEBS Letter* **489**, 237–242.

Madhavian, K., Ghorbanli, M. and Kalantari, K. M. (2008). The effects of ultraviolet radiation on some antioxidant components and enzymes in *Capsicum annuum* L. *Turkish Journal of Botany* **32**, 129–134.

Malanga, G., Calmanovici, G. and Puntarulo, S. (1997). Oxidative damage to chloroplasts from *Chlorella vulgaris* exposed to ultraviolet-B radiation. *Physiologia Plantarum* **101**, 455–467.

Millar, A. H., Mittova, V., Kiddle, G., Heazlewood, J. L., Bartoli, C. G., Theodoulou, F. L. and Foyer, C. H. (2003). Control of ascorbate synthesis by respiration and its implications for stress responses. *Plant Physiology* **133**, 443–447.

Mishra, S. and Agrawal, S. B. (2006). Interactive effects between supplemental UV-B radiation and heavy metals on growth and biochemical characteristics of *Spinacia oleracea* L. *Brazilian Journal of Plant Physiology* **18**(1), 1–8.

Mittler, R. (2002). Oxidative stress, antioxidant and stress tolerance. *Trends Plant Science* **7**, 405–410.

Morgan, P. W. and Drew, M. C. (1997). Ethylene and plant responses to stress. *Physiologia Plantarum* **100**, 620–630.

Murali, N. S., Teramura, A. H. and Randall, S. K. (1988). Response differences between two soybean cultivars with contrasting UV-B radiation sensitivities. *Photochemistry and Photobiology* **48**, 653–657.

Nagy, F. and Schäfer, E. (2000). Nuclear and cytosolic events of light-induced, phytochrome-regulated signaling in higher plants. *EMBO Journal* **19**, 157–163.

Nakajima, S., Sugiyama, M., Iwai, S., Hitomi, K., Otoshi, E., Kim, S. T., Jiang, C. Z., Todo, T., Britt, A. B. and Yamamoto, K. (1998). Cloning and characterization of a gene (UVR3) required for photorepair of 6–4 photoproducts in *Arabidopsis thaliana. Nucleic Acids Research* **26**, 638–644.

Neil, S., Desikan, R., Clarke, A., Hurst, R. D. and Hancock, J. T. (2002). Hydrogen peroxide and nitric oxide as signaling molecules in plants. *Journal of Experimental Botany* **53**, 1237–1247.

Niyogi, K. K. (1999). Photoprotection revisited: Genetic and molecular approaches. *Annual Review of Plant Physiology and Plant Molecular Biology* **50**, 333–359.

Noctor, G. and Foyer, C. H. (1998). Ascorbate and glutathione: Keeping active oxygen under control. *Annual Review of Plant Physiology and Plant Molecular Biology* **49**, 249–279.

O' Donnel, P. J., Calvert, C., Atzorn, R., Wasrwenack, C., Leyser, H. M. O. and Boulles, D. J. (1996). Ethylene as a signal mediating the wound response of tomato plants. *Nature* **274**, 1914–1917.

Orozco-Cárdenas, M. L. and Ryan, C. A. (1999). Hydrogen peroxide is generated systematically in plant leaves by wounding and systemin via the octadecanoid pathway. *Proceedings of National Academy of Science of the United States of America* **96**, 6553–6557.

Orozco-Cárdenas, M. L., Narváez-Vásquez, J. and Ryan, C. A. (2001). Hydrogen peroxide acts as a second messenger for the induction of defense genes in tomato plants in response to wounding, systemin and methyl jasmonate. *Plant Cell* **13**, 179–191.

Otter, T. and Polle, A. (1994). The influence of apoplastic ascorbate on the activities of cell-wall associated peroxidase and NADH-oxidases in needles of Norway spruce (*Picea abies* L.). *Plant and Cell Physiology* **35**, 1231–1238.

Panagopoulos, I., Bornman, J. F. and Bjorn, L. O. (1990). Effects of ultraviolet radiation and visible light on growth, fluorescence induction, ultraweak luminescence and peroxidase activity in sugar beet plants. *Journal of Photochemistry and Photobiology B: Biology* **8**, 73–87.

Pelle, E., Maes, D., Padulo, G. A., Kim, E. K. and Smith, W. P. (1990). An *in vitro* model to test relative antioxidant potential: Ultraviolet-induced lipid peroxidation in liposomes. *Archives of Biochemistry and Biophysics* **283**, 234–240.

Polle, A., Otter, T. and Serfert, F. (1994). Apoplastic peroxidases and lignification in needles of Norway spruce (*Picea abies* L.). *Plant Physiology* **116**, 53–60.

Potter, S., Ukness, S., Lawton, K., Winter, A. M., Chandler, D., Dimaio, J., Novitzky, R., Ward, E. and Ryals, J. (1993). Regulation of hevein-like protein in *Arabidopsis. Molecular Plant-Microbe Interactions* **6**, 680–681.

Predieri, S., Norman, H. A., Krizek, D. T., Pillai, P., Mirecki, R. M. and Zimmerman, R. H. (1995). Influence of UV-B radiation on membrane lipid composition and ethylene evolution in doyenne d'hiver pear shoots grown *in vitro* under different photosynthetic photon fluxes. *Environmental and Experimental Botany* **35**, 151–160.

Rancélienë, V., Vyoniauskienë, R., Janëys, Z. and Dlekytë, K. (2005). Action of UV-B on *Crepis capillaris* (L.) Wallr. Plants in controlled environmental conditions. *Biologija* **3**, 74–80.

Rao, M. V. and Ormrod, D. P. (1995). Impact of UV-B and ozone on oxygen free radical scavenging system in *Arabidopsis thaliana* genotypes differing in flavonoid biosynthesis. *Photochemistry and Photobiology* **62**, 719–726.

Rao, M. V., Paliyath, G. and Ormrod, D. P. (1996). Ultraviolet-B and ozone-induced biochemical changes in antioxidant enzymes of *Arabidopsis thaliana*. *Plant Physiology* **110**, 125–136.

Reinbothe, S., Mollenhauer, B. and Reinbothe, C. (1994). JIPs and RIPs: The regulation of plant gene expression by jasmonates in response to environmental cues and pathogens. *Plant Cell* **6**, 1197–1209.

Ren, J., Yao, Y., Yang, Y., Korpelainen, H., Junttila, O. and Li, C. (2008). Growth and physiological responses to supplemental UV-B radiation of two contrasting poplar species. *Tree Physiology* **26**, 665–672.

Renger, G., Völker, M., Eckert, H. J., Fromme, R., HohmVeit, S. and Graber, P. (1989). On the mechanism of photosystem II deterioration by UV-B irradiation. *Photochemistry and Photobiology* **49**, 97–105.

Rich, P. R. and Bonner, W. D. (1978). The sites of superoxide anion generation in higher plant mitochondria. *Archives of Biochemistry and Biophysics* **188**, 206–213.

Sakai, T., Kagawa, T., Kasahara, M., Swartz, T. E., Christie, J. M., Briggs, W. R., Wada, M. and Okada, K. (2001). *Arabidopsis* nph1 and npl1: Blue-light receptors that mediate both phototropism and chloroplast relocation. *Proceedings of National Academy of Science of the United States of America* **48**, 6969–6974.

Salin, M. L. (1991). Chloroplasts and mitochondrial mechanisms for protection against oxygen toxicity. *Free Radical Research Communications* **12**, 851–858.

Sanchez-Casas, P. and Klesseg, D. F. (1994). A salicylic acid-binding activity and a salicylic acid-inhibitable catalase activity are present in a variety of plant species. *Plant Physiology* **106**, 1675–1679.

Santos, M., Gousseau, H., Lister, C., Foyer, C., Creissen, G. and Mullineaux, P. (1996). Cytosolic ascorbate peroxidase from *Arabidopsis thaliana* L. is encoded by a small multigene family. *Planta* **198**, 64–69.

Sävenstrand, H., Brosch, M. and Strid, Å. (2004). Ultraviolet-B signalling: *Arabidopsis* brassinosteroids mutants are defective in UV-B regulated defence gene expression. *Plant Physiology and Biochemistry* **42**(9), 687–694.

Scandalios, J. G., Guan, L. and Polidoros, A. N. (1997). Catalase in plants:gene structure, properties, regulation and expression. *In* "Oxidative Stress and Molecular Biology of Antioxidant Defenses" (J. G. Scandalios, ed.), pp. 343–406. Cold Spring Harbor Laboratory Press, Cold Spring Harbor, New York.

Scholz, R. W., Graham, K. S., Wynn, M. K. and Reddy, C. C. (1990). Interaction of glutathione and α-tocopherol in the inhibition of lipid peroxidation in rat liver microsomes. *In* "Biological Oxidation Systems" (C. C. Reddy, G. A. Hamilton and K. N. Madyastha, eds.), pp. 841–867. Academic Press, New York.

Schopfer, P. (1996). Hydrogen peroxide-mediated cell-wall stiffening *in vitro* in maize coleoptiles. *Planta* **199**, 43–49.

Selvakumar, V. (2008). Ultraviolet-B radiation (280–315 nm) invoked antioxidant defense systems in *Vigna unguiculata* (L.) Walp and *Crotalaria juncea* L. *Photosynthetica* **46**(1), 98–106.

Shao, H. B. and Chu, L. V. (2005). Plant molecular biology in China: Opportunities and challenges. *Plant Molecular Biology Reporter* **23**, 345–358.

Shao, H. B., Liang, Z. S. and Shao, M. S. (2005). Changes of some anti-oxidative enzymes under soil water deficits among 10 wheat genotypes at maturation stage. *Coloids and Surface B: Biointerfaces* **45**, 7–13.

Singh, S., Mishra, S., Kumari, R. and Agrawal, S. B. (2009). Response of supplemental ultraviolet-B and nickel on pigments, metabolites, and antioxidants of *Pisum sativum*. L. Journal of environmental biology (in press), 30.

Skorska, E. and Szwarc, W. (2007). Influence of UV-B radiation on young triticale plants with different waz cover. *Biologia Plantarum* **51**(1), 189–192.

Smith, J. L., Buritt, D. J. and Bannister, P. (2000). Shoot dry weight, chlorophyll and UV-B absorbing compounds as indicators of a plant's sensitivity to UV-B radiation. *Annals of Botany* **86**, 1057–1063.

Šnyrychová, I., Kós, P. B. and Hideg, E. (2007). Hydroxyl radicals are not the protagonists of UV-B-induced damage in isolated thylakoid membranes. *Functional Plant Biology* **34**, 1112–1121.

Stratman, J. (2003). Ultraviolet-B radiation co-opts defense signaling pathways. *Trends Plant Science* **8**, 526–533.

Strid, A., Chow, W. S. and Anderson, J. M. (1994). UV-B damage and protection at the molecular level in plants. *Photosynthesis Research* **39**, 475–489.

Suesslin, C. and Frohnmeyer, H. (2003). An *Arabidopsis* mutant defective in UV-B light-mediated responses. *Plant Journal* **33**, 591–601.

Sullivan, J. H., Teramura, A. H. and Ziska, L. H. (1992). Variation in UV-B sensitivity in plants from a 3000 m or elevational gradient in Hawaii. *American Journal of Botany* **79**, 737–743.

Surplus, S. L., Jordan, B. R., Murphy, A. M., Carr, J. P., Thomas, B. and Mackerness, S. A.-H. (1998). UV-B induced response in *Arabidopsis thaliana*: Role of salicylic acid and ROS in the regulation of transcripts and acidic PR proteins. *Plant Cell and Environment* **21**, 685–694.

Takahama, U. (2004). Oxidation of vacuolar and apoplastic phenolic substrates by peroxidase: Physiological significance of the oxidation reactions. *Phytochemistry Review* **3**, 207–219.

Takeuchi, Y., Fukeimoto, R., Hirokazu, K., Sakaki, T. and Kitao, M. (1995). Peroxidation of lipids and growth inhibition induced by UV-B radiation. *Plant Cell Reports* **14**(9), 566–570.

Takeuchi, Y., Kubo, H., Kasahara, H. and Sakaki, T. (1996). Adaptive alterations in the activities of scavengers of active oxygen in cucumber cotyledons irradiated with UV-B. *Journal of Plant Physiology* **147**, 589–592.

Taulavouri, E., Bäckman, M., Taulavuori, K., Gwynn-Jones, D., Johanson, U., Laine, K., Callaghan, T., Sonesson, M. and Björn, L. O. (1998). Long-term exposure to enhanced ultraviolet-B radiation in the sub-arctic does not cause oxidative stress in *Vaccinium myrtillus*. *New Phytologist* **140**, 691–697.

Taylor, W. C. (1989). Regulatory interactions between nuclear and plastid genomes. *Annual Review in Plant Physiology and Plant Molecular Biology* **40**, 211–233.

Teklemariam, T. and Blake, T. J. (2003). Effects of UV-B preconditioning on heat tolerance of cucumber (*Cucumis sativus* L.). *Environmental and Experimental Botany* **50**, 169–182.

Tepperman, J. M., Zhu, T., Chang, H. S., Wang, X. and Quail, P. H. (2001). Multiple transcription-factor genes are early targets of phytochrome A signalling. *Proceedings of the National Academy of Sciences of the United States of America* **98**, 9437–9442.

Ulm, R. and Nagy, F. (2005). Signaling and gene regulation in response to ultraviolet light. *Current Opinion in Plant Biology* **8**, 477–482.

Ulm, R., Baumann, A., Oravecz, A., Mate, Z., Adam, E., Oakeley, E. J., Scahfer, E. and Nagy, F. (2004). Genome wide analysis of gene expression reveals function of the bZIP transcription factor HY5 in the UV-B response of

Arabidopsis. *Proceedings of the National Academy of Sciences of the United States of America* **101**, 1397–1402.

Vass, I., Sass, L., Spetea, C., Bakou, A., Ghanotakis, D. F. and Petrouleas, V. (1996). UV-B induced inhibition of photosystem II electron transport studied by ESR and chlorophyll fluorescence. Impairment of donor and acceptor side components. *Biochemistry* **35**, 8964–8973.

Vijayan, P., Shockey, J., Levesque, C. A., Cook, R. J. and Browse, J. (1998). A role for jasmonate in pathogen defence of *Arabidopsis*. *Proceedings of the National Academy of Sciences of the United States of America* **95**, 7209–7214.

Wade, H. K., Bibikova, T. N., Valentine, W. J. and Jenkins, G. I. (2001). Interactions within a network of phytochrome, crptochrome and UV-B phototransduction pathways regulate chalcone synthase gene expression in *Arabidopsis* leaf tissue. *Plant Journal* **25**, 675–685.

Wagner, A. M. and Moore, A. L. (1997). Structure and function of the plant alternative oxidase: Its putative role in the oxygen defence mechanism. *Bioscience Reports* **17**, 319–333.

Weatherhead, B., Tanskanen, A. and Stevermer, A. (2005). Ozone and ultraviolet radiation. *In* "Arctic Climate Impact Assessment Scientific Report" (C. Symon, L. Arris and B. Heal, eds.), pp. 151–182. Cambridge University Press, New York.

Willekens, H., Camp, W. V., Van Montagu, M., Inze, D., Langebartels, C. H. and Sandermann, H. (1994). Ozone, sulphur dioxide and ultraviolet-B have similar effects on mRNA accumulation of antioxidant genes in *Nicotiana plumbaginifolia* L. *Plant Physiology* **106**, 1007–1014.

Willekens, H., Chamnogopol, S., Davey, M., Schraudner, M. and Langebartels, C. (1997). Catalase is a sink for H_2O_2 and is indispensable for stress defense in C3 plants. *EMBO Journal* **16**, 4806–4816.

Wojtaszek, P. (1997). Oxidative burst: An early plant response to pathogen infection. *Biochemical Journal* **322**, 681–692.

Xu, C., Natarajan, S. and Sullivan, J. H. (2008). Impact of solar ultraviolet-B radiation on the antioxidant defense system in soybean lines differing in flavonoid contents. *Environmental and Experimental Botany* **63**, 39–48.

Yang, Y. Q. and Yao, Y. (2008). Photosynthetic responses to solar UV-A and UV-B radiation in low and high-altitude populations of *Hippophae rhamnoides*. *Photosynthetica* **46**(2), 307–311.

Yang, S. H., Wang, L. J., Li, S. H., Duan, W., Loescher, W. and Liang, Z. C. (2007). The effects of UV-B radiation on photosynthesis in relation to photosystem II photochemistry, thermal dissipation and antioxidant defenses in winter wheat (*Triticum aestivum* L.) seedlings at different growth temperatures. *Functional Plant Biology* **34**, 907–917.

Yannarelli, G. G., Gallego, S. M. and Tomaro, M. L. (2006). Effect of UV-B radiation on the activity and isoforms of enzymes with peroxidase activity in sunflower cotyledons. *Environmental and Experimental Botany* **56**, 174–181.

Yanqun, Z., Yuan, L., Haryan, C. J. and Ianjun, C. (2003). Intraspecific differences in physiological response to 20 soybean cultivars to enhanced ultraviolet-B radiation under field conditions. *Environmental and Experimental Botany* **50**, 87–97.

Yao, X. and Liu, Q. (2006). Changes in morphologal, photosynthetic and physiological responses of Mono Maple seedlings to enhanced UV-B and to nitrogen addition. *Plant Growth Regulation* **50**, 165–177.

Yao, X. and Liu, Q. (2007). Changes in photosynthesis and antioxidant defenses of *Picea asperata* seedlings to enhanced ultraviolet-B radiation and to nitrogen supply. *Physiologia Plantarum* **129**, 364–374.

Yao, Y., Yang, Y., Ren, J. and Li, C. (2006). UV-Spectra dependence of seedling injury and photosynthetic pigment change in *Cucumis sativus* and *Glycine max*. *Environmental and Experimental Botany* **57**, 160–167.

Yerkes, C. T., Kramer, D. M., Fenton, J. M. and Crofts, A. R. (1990). UV-photoinhibition: Studies *in vitro* and in intact plants. *In* "Current Research in Photosynthesis" (M. Baltscheffsky, ed.), pp. II.6.381–II.6.384. Kluwer Academic Publishers, Dordrecht, The Netherlands.

Zancan, S., Suglia, I., Rocca, N. L. and Ghisi, R. (2008). Effects of UV-B radiation on antioxidant parameters of iron-deficient barley plants. *Environmental and Experimental Botany* **63**, 71–79.

Zhang, J. E., Hu, X., Henkow, L., Jordan, B. R. and Strid, A. (1994). The effects of ultraviolet-B radiation on CF_o F_1-ATPASE. *Biochimica Biophysica Acta* **1185**, 295–302.

Roles of γ-Glutamyl Transpeptidase and γ-Glutamyl Cyclotransferase in Glutathione and Glutathione-Conjugate Metabolism in Plants

NAOKO OHKAMA-OHTSU,*[,1] KEIICHI FUKUYAMA[†]
AND DAVID J. OLIVER[‡]

*RIKEN Plant Science Center, 1-7-22 Suehiro-cho, Tsurumi-ku, Yokohama-City, Kanagawa 230-0045, Japan
[†]Department of Biological Sciences, Graduate School of Science, Osaka University, Toyonaka, Osaka 560-0043, Japan
[‡]Department of Genetics, Development and Cell Biology, Iowa State University, Ames, IA 50011, USA

[1]Corresponding author: Email: ohkama_ohtsu@psc.riken.jp

Advances in Botanical Research, Vol. 52 0065-2296/09 $35.00

DOI: 10.1016/S0065-2296(09)52004-6

ABSTRACT

In the mammalian γ-glutamyl cycle, γ-glutamyl transpeptidase (GGT) is responsible for extracellular glutathione (GSH) degradation leading to the subsequent reabsorption of the constituent amino acids into the cell. The physiological functions of the plant GGTs have recently been discovered in *Arabidopsis thaliana*. In *Arabidopsis* GGTs are located in the apoplast and the vacuole. The GGT in the apoplast plays an important role in the prevention of oxidative stress by degrading the oxidized form of glutathione (GSSG). The vacuolar GGT is required for degradation of GSH conjugates in the vacuole allowing their further metabolism. Amino acid sequences with high similarity to *Arabidopsis* GGTs were found in various plant species, and these sequences suggest that the apoplastic and vacuolar functions of GGTs are broadly conserved in plants. Three-dimensional structures of bacterial GGTs and sequence alignment indicated that the overall folding of the *Arabidopsis* GGTs was very similar to that of bacterial enzymes and that the spatial arrangement of the residues involved in the recognition of the γ-glutamyl moiety was conserved in all GGTs. While plant GGT function is limited to the apoplast or vacuole, most of GSH exists in the cytosol and is degraded through a GGT-independent pathway initiated by γ-glutamyl cyclotransferase.

I. INTRODUCTION

Glutathione (GSH), a tripeptide of γ-Glu-Cys-Gly, performs redox reactions through its thiol residue, enabling it to be an important molecule for redox homeostasis in plant cells (Foyer and Noctor, 2005, 2009). As a component of the ascorbate–GSH cycle GSH detoxifies photosynthetically generated H_2O_2 (Foyer and Noctor, 2009; Noctor and Foyer, 1998). The reversible glutathionylation of cysteine residues in proteins modulated by glutaredoxins can control enzyme activity and acts as part of the signal transduction mechanism as plants sense and respond to oxidative stress (see Foyer and Noctor, 2009; Rouhier *et al.*, 2008 for recent reviews). GSH, either directly or through glutaredoxins, is essential for the function of type II peroxiredoxins which are important for the reduction of hydrogen peroxide and alkyl hydroperoxides (Rouhier *et al.*, 2008). GSH is also involved in the detoxification of both heavy metals and xenobiotics. Toxic xenobiotics, including herbicides, are conjugated with GSH by GSTs and these GSH conjugates are sequestered into the vacuole (Marrs, 1996). GSH is also polymerized to form phytochelatins [(γ-Glu-Cys)$_{2-11}$-Gly]. Phytochelatins chelate several heavy metals including Cd^{2+}, Cu^{2+}, arsenate, and Ag^{2+} that are then transported into the vacuole where they are metabolically inactive (Cobbett and Goldsbrough, 2002). Furthermore, GSH controls cell size and root development by regulating the cell cycle (Veroux *et al.*, 2000; Xiang *et al.*, 2001).

The cellular concentration of GSH is high, probably several millimolar in plants. When GSH synthesis is blocked more than 80% of the GSH is degraded within one day (Ohkama-Ohtsu *et al.*, 2008), suggesting a relatively high rate of GSH turnover in plants even under low stress conditions. GSH is the major storage and transport form of organic sulfur, a major source for cysteine for protein synthesis, and a precursor for metabolite synthesis (Leustek *et al.*, 2000). The rate of all these processes may be determined by the rate of GSH turnover. Because of the roles of GSH in protecting plants from environmental stress and as a reservoir for central biochemicals, particularly cysteine, it is important to understand GSH breakdown.

GSH is synthesized from Glu, Cys, and Gly in two reactions both of which use ATP. First γ-glutamyl cysteine (γ-EC) is synthesized from Glu and Cys by γ-EC synthetase (Hell and Bergmann, 1990). Then Gly is attached to the γ-EC by GSH synthetase (Wang and Oliver, 1996). In *Arabidopsis thaliana*, γ-EC synthetase and GSH synthetase are each encoded by single genes, *GSH1* (May and Leaver, 1994) and *GSH2* (Wang and Oliver, 1996). The expression of both genes is regulated transcriptionally by heavy metals and jasmonic acid, and *GSH1* translation is controlled by oxidative stress (Xiang and Oliver, 1998). The activity of γ-EC synthetase is regulated by the redox homeostasis affected by GSH (Gromes *et al.*, 2008; Hell and Bergmann, 1990; Jez *et al.*, 2004) and possibly γ-EC (Pasternak *et al.*, 2008) levels. γ-EC synthetase is exclusively localized in the plastids. GSH synthetase is predominantly cytosolic although it is also found in chloroplasts (Wachter *et al.*, 2005). Recently Pasternak *et al.* (2008) reported that restricting GSH synthesis to the cytosol is sufficient for normal plant development.

While several groups have studied the mechanism and control of GSH synthesis in plants relatively little is known about its degradation. The γ-glutamyl bond between Glu and Cys is resistant to standard peptidases (Taniguchi and Ikeda, 1998). The only enzyme that was known to break this linkage was γ-glutamyl transpeptidase (GGT, EC 2.3.2.2, also known as γ-glutamyl transferase). Studying GGT, therefore, provided a unique opportunity for learning about GSH degradation. GGT catalyzes the transfer of the γ-glutamyl group of GSH to a variety of acceptors (Tate and Meister, 1985). When water is the acceptor the γ-glutamyl bond is hydrolyzed to form Glu and the dipeptide cysteinylglycine (Gys-Gly). When an amino acid or a peptide is used as the acceptor, the transpeptidase reaction forms the γ-glutamyl amino acid or a γ-glutamyl peptide and Cys-Gly.

This chapter describes the mammalian, bacterial, and plant GGTs and particularly focuses on their recently discovered functions in plants. It also describes a newly discovered GSH degradation pathway in plants where the first enzyme for GSH degradation is γ-glutamyl cyclotransferase (GGC).

II. CHARACTERISTICS OF GGTs

The mammalian GGT is a glycosylated heterodimeric transmembrane protein that is derived from a proprotein by processing (Kinlough *et al.*, 2005). The N-terminal targeting sequence of the mammalian protein is not cleaved, acting as an anchor domain in the plasma membrane. The *Escherichia coli* GGT is also a heterodimer, but the protein is not glycosylated and processing of the proprotein to the large and small subunits is autocatalytic (Suzuki and Kumagai, 2002). The *E. coli* protein has a signal peptide as its N-terminus that is cleaved, releasing the protein in a soluble form in the periplasmic space (Suzuki *et al.*, 1986).

The plant GGTs have been partially characterized (Kawasaki *et al.*, 1982; Martin and Slovin, 2000; Nakano *et al.*, 2006a,b,c; Shaw *et al.*, 2005; Steinkamp and Rennenberg, 1984; Storozhenko *et al.*, 2002). Analysis of the gene sequence and the resulting protein from *Arabidopsis* (Storozhenko *et al.*, 2002), *Allium cepa* (onion) (Shaw *et al.*, 2005), and *Raphanus sativus* (radish) (Nakano *et al.*, 2006a,b) suggests an initial proprotein of about 60 kDa that, upon processing, yields a subunit of about 40 kDa, which would be equivalent to the large subunit of the mammalian enzyme, and a smaller subunit of about 20 kDa. Since the enzyme binds to a ConA-Sepharose column it is likely that it is glycosylated and therefore extracellular (Shaw *et al.*, 2005). This is supported by the fractionation of the protein with membranes and cell-wall components under low-salt conditions (Martin and Slovin, 2000; Nakano *et al.*, 2004, 2006b; Storozhenko *et al.*, 2002). Release of the enzyme from these components with high salt suggests that the enzyme is bound by ionic interactions to the membrane or the cell wall. Various γ-glutamyl dipeptides, reduced GSH, oxidized glutathione (GSSG), and GSH conjugates are used as substrates by plant GGTs (Martin and Slovin, 2000; Nakano *et al.*, 2004, 2006b; Storozhenko *et al.*, 2002).

III. PHYSIOLOGICAL FUNCTIONS OF GGT IN ANIMALS

In mammals, GGT is a component of the γ-glutamyl cycle (Meister and Larsson, 1995) (Fig. 1A). In this cycle, extracellular GSH is broken down by GGT that is located on the plasma membrane with its active site in the extracellular space. GGT transfers the γ-linked Glu from GSH to water or another amino acid producing either Glu or a γ-glutamyl amino acid, respectively. The resulting Glu and γ-glutamyl amino acids are transported back into the cell. There is some disagreement on the role of water as compared

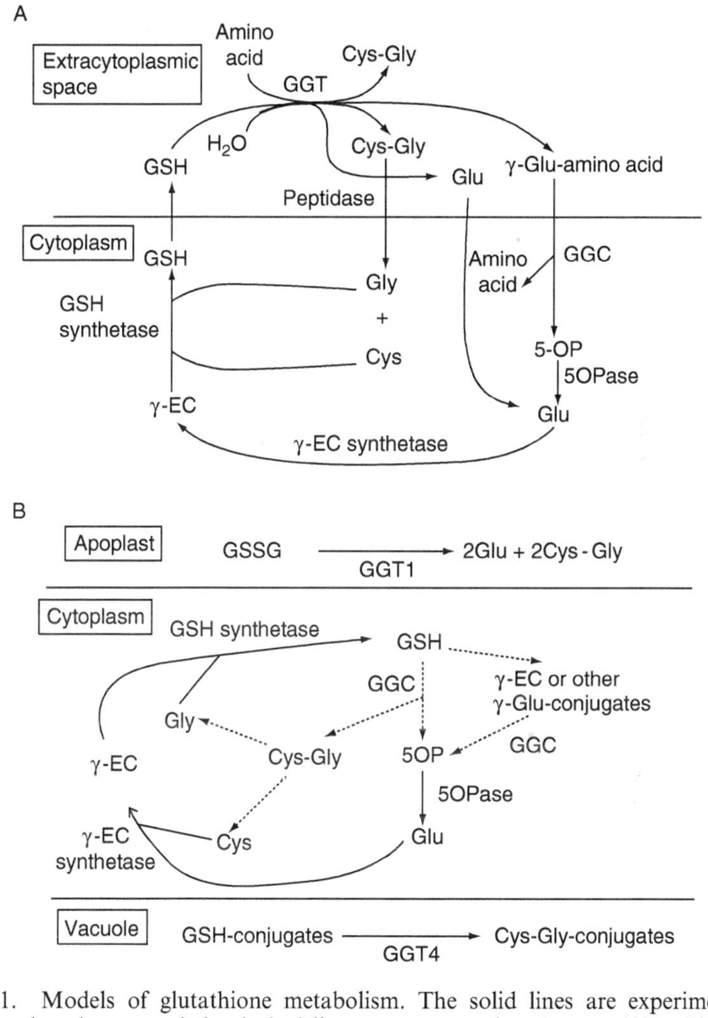

Fig.1. Models of glutathione metabolism. The solid lines are experimentally confirmed pathways and the dashed lines are proposed pathways. (A) γ-Glutamyl cycle in mammals. Glutathione is degraded in extracellular space by GGT and the constituent amino acids are transported back to the cytoplasm to resynthesize GSH. The extracellular and intracellular reactions may occur in different tissues and/or organs. (B) Glutathione metabolism proposed in *Arabidopsis*. Roles of GGT are limited to the extracytosolic spaces such as the degradation of oxidized GSH (GSSG) by GGT1 in the apoplast and the degradation of GSH conjugates by GGT4 in the vacuole. Cytosolic GSH is degraded by a sequential reaction of GGC and 5OPase without involvement of GGT.

to amino acids as cosubstrates with GSH. It was originally assumed that γ-glutamyl amino acids were imported into the cell and that this might be important in amino acid uptake (Meister, 1973). More recently, experiments have indicated that the normal product of the GGT reaction is free Glu

(Hanigan and Pitot, 1985). Irrespective of the glutamate acceptor, the reaction is key to initiating the breakdown of extracellular GSH with the ultimate uptake of the essential amino acid Cys back into the cell.

Inside the cell, γ-glutamyl amino acids are converted into 5-oxoproline (5OP, synonyms: pyroglutamate, pyrrolidone carboxylate) and free amino acids by GGC. 5-Oxoprolinase (5OPase) hydrolyzes 5OP to Glu in an ATP-dependent reaction. Outside the cell Cys-Gly produced from GSH by the GGT reaction is hydrolyzed to Cys and Gly by a dipeptidase. These amino acids then reenter the cell. γ-EC synthetase and GSH synthetase reform GSH from the resulting Glu, Cys, and Gly. Activities of GGT, GGC, and 5OPase are highest in secretory tissues including those in the kidney. In this organ, GSH is filtered from the blood and deposited into the excretory stream. GGT triggers the reabsorption of this GSH by breaking the γ-glutamyl bond thus initiating uptake of the component amino acids. Mice with a mutation in GGT lost GSH through the excretory system resulting in glutathionuria, cysteine deficiency, and growth and reproductive defects that can be prevented by feeding the Cys precursor acetylcysteine (Harding *et al.*, 1997; Kumar *et al.*, 2000; Lieberman *et al.*, 1996).

IV. PHYSIOLOGICAL FUNCTIONS OF GGTs IN PLANTS

Some plants accumulate γ-glutamyl amino acids that are apparently produced by GGT (Kasai *et al.*, 1982; Kawasaki *et al.*, 1982). GGT in onion bulb is involved in producing flavor compounds by hydrolyzing precursors occurring as γ-glutamyl conjugates (Shaw *et al.*, 2005). Other physiological functions of GGT in plants have recently been studied in the model plant *Arabidopsis*. There are four *Arabidopsis* genes whose translated sequences are homologous to animal GGTs, *GGT1* (AT4g39640), *GGT2* (At4g39650), *GGT3* (AT1g69820), and *GGT4* (AT4g29210). *GGT4* (At4g29210), originally called GGT3 in our earlier papers (Ohkama-Ohtsu *et al.*, 2007a,b), was renamed later (Ohkama-Ohtsu *et al.*, 2008) to be in accordance with the nomenclature of Martin *et al.* (2007) who simultaneously discovered its function (Grzam *et al.*, 2007). Amino acid sequences of GGT1 and GGT2 showed high identity (82%), while GGT4 showed 46% and 48% identity to GGT1 and GGT2, respectively. *GGT3* has a short open reading frame that is similar to the C-terminal encoding portion of *GGT1* or *GGT2*. There is no expression sequence tag (EST) corresponding to *GGT3* and mRNA originating from *GGT3* was not detected by reverse transcriptase (RT)-PCR (Martin *et al.*, 2007; Ohkama-Ohtsu *et al.*, 2007a). Thus *GGT3* was assumed to be a pseudo gene. This was supported by the observation that in *ggt1*, *ggt2*, and *ggt4* single and double mutants all of the GGT activity was accounted for by the three major genes and that in the leaves of *ggt1/ggt4* double mutant there were no detectable GGT activity (Ohkama-Ohtsu *et al.*, 2007b). Martin *et al.* (2007), however, suggested that GGT3 may be functional because there might be low levels of expression detected in microarray experiments (GENEVESTIGATOR; https://www.genevestigator.ethz.ch/gv/index.jsp) and with *GGT3*: *β*-glucuronidase (GUS) transgenic plants. They also showed that the *ggt3* knockout plant had reduced inflorescence height and seed yield, supporting functionality of GGT3 (Martin *et al.*, 2007). Three mRNAs originating from *GGT3* were also identified in deep sequencing experiments with *Arabidopsis* directly confirming its expression at ultralow abundance (Todd Mockler, Oregon State University—personal communications). It is not clear if the gene encodes an active protein and if so what its specific function might be.

GGT activity assays in the *Arabidopsis ggt* knockout mutants showed that GGT1 comprised over 80% of the total enzyme activity in most tissues. Two exceptions are siliques where GGT1 and GGT2 each contribute about half of the total activity and roots where the activity is evenly divided between GGT1 and GGT4 (Martin *et al.*, 2007; Ohkama-Ohtsu *et al.*, 2007a,b).

A. EXTRACELLULAR GGT1 AND GGT2 IN *ARABIDOPSIS*

GGT1: GUS transgenic plants that express the GUS reporter gene under control of the *GGT1* promoter showed that *GGT1* was expressed in vascular tissue, specifically in the phloem of the midrib and minor veins of leaves, roots, and flowers (Fig. 2A–E) (Martin *et al.*, 2007; Ohkama-Ohtsu *et al.*, 2007a). This localization was confirmed in leaves by laser microdissection where GGT1 mRNA was 17-fold higher in the vascular than the mesophyll tissues (Fig. 2F) (Ohkama-Ohtsu *et al.*, 2007a). *GGT2*:GUS fusion constructs showed that expression of *GGT2* was limited to embryos, endosperm, outer integument, and a small portion of the funiculus in developing siliques (Fig. 3) (Ohkama-Ohtsu *et al.*, 2007a). Martin *et al.* (2007) also detected *GGT2*:GUS expression in immature trichomes and pollen. Masi *et al.* (2007) has developed an enzyme histochemical method that allowed them to demonstrate GGT in the leaf vascular tissue, stomata, and root meristems in maize. However, it is not clear if the chromogen will penetrate to the vacuole in these tissue slices and measure GGT activity there.

Experimental evidence strongly suggests that GGT1 and GGT2 are extracellular. GGT1 overexpressed in tobacco showed activity in the apoplastic space (Storozhenko *et al.*, 2002). GGT activity derived from *GGT1* and

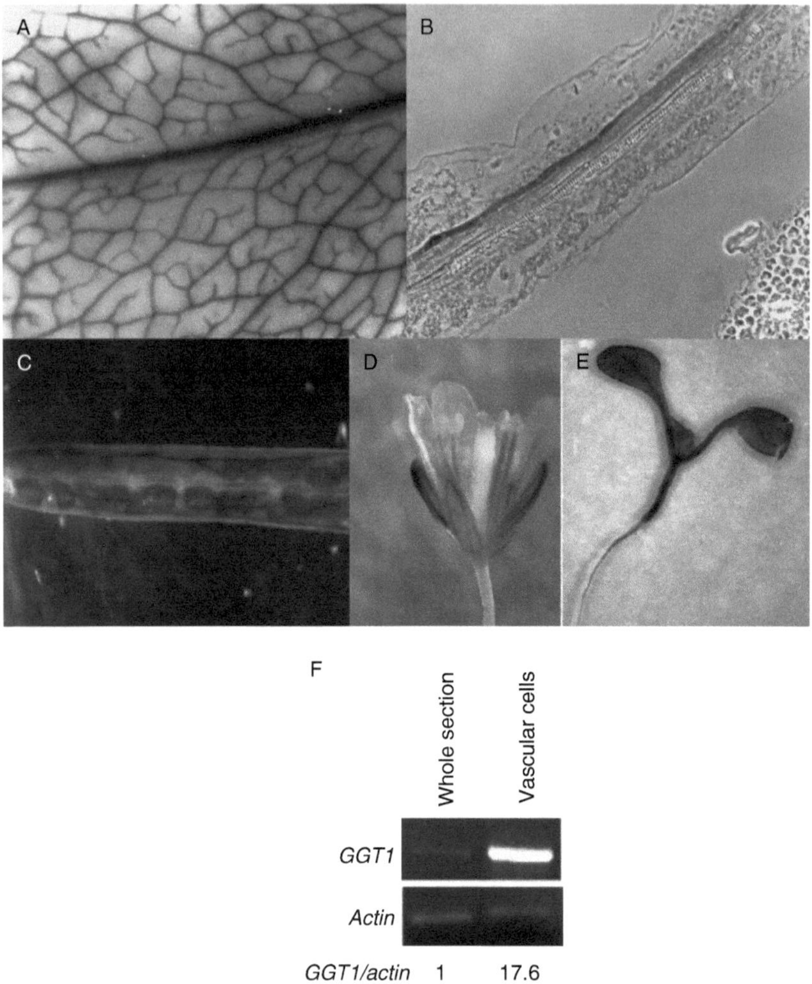

Fig. 2. GUS staining of *GGT1:: GUS* fusion. (A) GUS staining of a leaf. (B) Longitudinal section of a petiole. (C) Maturing silique showing staining in vascular tissue but no staining in developing seeds. (D) Flower with GUS staining in the vascular tissue of sepals, petals, and filaments of anthers. (E) One-week-old plant showing staining in the roots. (F) GGT1 transcript accumulation in vascular cells. Vascular cells in leaf cross sections were collected by laser microdissection. Total RNA extracted from vascular cells or whole sections was subjected to RT-PCR analysis. Values below the gel images are the levels of GGT1 mRNA relative to that of *ACTIN8* quantified using real-time PCR. These figures were originally published in Ohkama-Ohtsu *et al.* (2007a) and are reproduced with permission from Blackwell Publishing. (See color plate section at the back of the book.)

GGT2 was also extracted with high salt (Martin *et al.*, 2007; Ohkama-Ohtsu *et al.*, 2007a). Half of GGT1 activity was detected on the surface of protoplast (Ohkama-Ohtsu *et al.*, 2007a), suggesting that GGT1 is apoplastic and

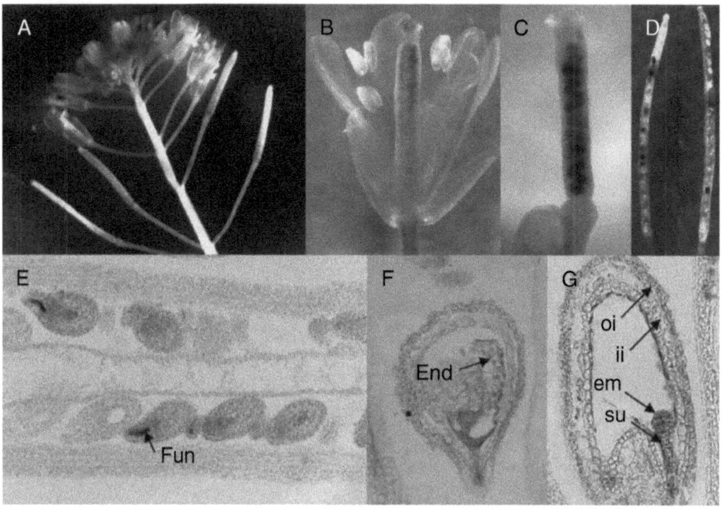

Fig. 3. GUS staining of *GGT2::GUS* fusion. (A) *Arabidopsis* shoot. (B) Flower stained before anthers open. (C) Pistil two days after pollination. (D) Mature siliques from a line that was still segregating for the *GGT2*::GUS insert. (E) Longitudinal section of a 5-day-old silique to illustrate staining in the funiculus (fun). (F) Longitudinal section of a seed three days after pollination (end, endosperm). (G) Longitudinal section of a seed 5 days after fertilization (oi, outer integument; ii, inner integument; em, embryo; su, suspensor). These figures were originally published in Ohkama-Ohtsu *et al.* (2007a) and are reproduced with permission from Blackwell Publishing. (See color plate section at the back of the book.)

maybe associated with the plasma membrane and cell wall. The extracellular localization was confirmed in maize root tips by immunohistochemical techniques where the maize GGT1 equivalent was localized to the cell-wall space (Ferretti *et al.*, 2009; Masi *et al.*, 2007). A specific Cys-Gly dipeptidase has been isolated from radish that may function to complete extracellular GSH metabolism (Kumada *et al.*, 2007).

The function of GGT1 in the apoplast was elucidated by analyzing *ggt1* knockout plants. The *ggt1* knockout plants show yellowing in older leaves (Martin *et al.*, 2007; Ohkama-Ohtsu *et al.*, 2007a). This phenotype was likely to be caused by oxidative stress because lipid peroxidation in the *ggt1* mutant leaves was higher than that in the wild-type leaves (Ohkama-Ohtsu *et al.*, 2007a) and because shading the leaves decreased chlorosis. Concentration of oxidized glutathione (GSSG) in the apoplastic space of leaves from *ggt1* mutant was sevenfold higher than that in leaves from wild-type plants, while concentration of reduced GSH in *ggt1* mutant was not different from that in the wild-type plants. The apoplast is an important space for the detoxification of reactive oxygen species (ROS) (Hernández *et al.*, 2001;

Lamb and Dixon, 1997; Luwe, 1996; Pignocchi and Foyer, 2003; Vanacker *et al.*, 1998a,b; Vanacker *et al.*, 1998c). Our results demonstrated that GGT1 mitigates oxidative stress by degrading GSSG in the apoplast (Ohkama-Ohtsu *et al.*, 2007a). It is not clear if the disruption of GSSG degradation in *ggt1* is directly blocking ROS destruction in the apoplast and thus causing the oxidative stress or if the altered GSH/GSSG ratio is initiating a signal that results in stress. The turnover of GSH in the apoplast is small enough that *ggt1* knockouts do not show altered total GSH levels (Ohkama-Ohtsu *et al.*, 2007a). This would imply either a role for GSH/GSSG signaling or that scavenging of relatively small amounts of extracellular ROS has large physiological consequences. The *Arabidopsis* gene microarray expression database GENEVESTIGATOR revealed that *GGT1* is induced by ozone or salt treatment, both of which cause oxidative stress. This supports the role of GGT1 in mitigating oxidative stress and suggests that plants may upregulate expression of the gene in response to reactive oxygen or stress-inducing conditions.

Expression of GGT1 in vascular tissues and GGT2 in siliques led us to postulate that GGT1 and GGT2 may also function in transporting GSH from its source (leaf) to sink (seed) tissues. Although a statistically significant decrease in ^{35}S-GSH transport into siliques in the *ggt2* knockout was not detected (Ohkama-Ohtsu *et al.*, 2007a), this idea is consistent with the high levels of GSH in the funiculus noted by Cairns *et al.* (2006). This colocalization of high levels of GGT2 expression and high GSH levels may indicate that GGT2 is acting as part of a GSH pump in this tissue.

B. VACUOLAR GGT4 IN *ARABIDOPSIS*

GGT4 has a longer N-terminal sequence than GGT1 and GGT2; thus, this protein was expected to show a different subcellular localization. GGT4 fused to green fluorescent protein localized inside the vacuole (Fig. 4) (Grzam *et al.*, 2007; Ohkama-Ohtsu *et al.*, 2007b). The monobromobimane-GSH conjugate was used as a model compound to examine whether GGT4 is involved in the degradation of GSH conjugates in the vacuole. Monobromobimane reacts with GSH in the cytosol in a reaction largely catalyzed by GST and the GSH-bimane derivative is transported to the vacuole within 1 h (Fricker *et al.*, 2000). In the wild-type plants the GSH-bimane was nearly all degraded to Cys-bimane within 24 h (Ohkama-Ohtsu *et al.*, 2007b). In roots where GGT4 contributes 50% of the total GGT activity, degradation of the GSH-bimane was completely inhibited in the *ggt4* knockout mutants. Thus in roots vacuolar GGT4 seems to be primarily responsible for degrading all of this GSH conjugate. In leaves, where GGT4

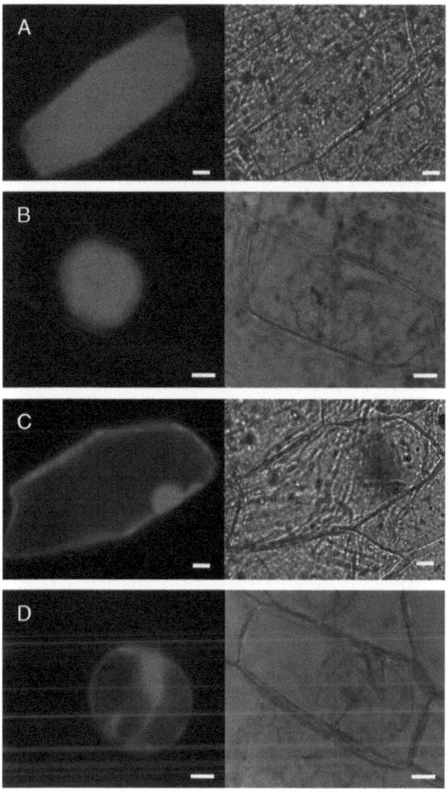

Fig. 4. Fluorescent micrographs of onion cells transiently expressing GGT4 ORF. N-terminal 100 amino acids of GGT4 fused to GFP (A and B), and unmodified GFP (C and D). In (B) and (D) cells were put in 10 mM mannitol solution. Images of the same cells taken under white light are on the right. Bars = 20 μm. These figures were originally published in Ohkama-Ohtsu *et al.* (2007b) and are reproduced with permission from Blackwell Publishing. (See color plate section at the back of the book.)

contributes only for 10% of the total GGT activity (the other 90% is GGT1), degradation of GSH-bimane was only 50% inhibited in the *ggt4* knockout mutants compared to the wild-type plants (Ohkama-Ohtsu *et al.*, 2007b). In leaves while half of the GSH-bimane is metabolized in the vacuole by GGT4, some GSH-conjugate degradation also appears to happen in the apoplast using GGT1. Involvement of GGT4 in GSH-bimane degradation was confirmed in *Arabidopsis* by Grzam *et al.* (2007) and in radish by Nakano *et al.* (2006c).

Various xenobiotics, including herbicides, may be detoxified by being conjugated to GSH and sequestered in the vacuole (Coleman *et al.*, 1997; Dixon *et al.*, 1998). The conjugations are catalyzed by GSTs

(Edwards *et al.*, 2000; Marrs, 1996), and some GSH conjugates are transported into the vacuole by ABC transporters on the tonoplast membrane (Rea, 1999). In the vacuole, GSH-conjugated xenobiotics are degraded into Cys conjugates and further metabolized to such compounds as malonylcysteine or *S*-methyl-derivatives (Rennenberg and Lamoureux, 1990). It is thought that derivatization of Cys conjugates is a mechanism to prevent further metabolism and to render derivatives transport inactive (Rea *et al.*, 1998; Rennenberg and Lamoureux, 1990). Thus GGT4 plays an essential role in the sequestration and further metabolism of GSH conjugates in the vacuole.

The potential involvement of GGT4 in herbicide resistance was tested with *ggt4* knockout plants (Ohkama-Ohtsu *et al.*, 2007b); however, sensitivity to herbicides was not different between *ggt4* mutant and wild-type plants. This suggested that in these cases degradation of GSH-herbicide conjugates by GGT4 was not essential for herbicide resistance. The steps for conjugation to GSH by GSTs or transportation out of the cytosol are likely to be more essential for herbicide detoxification as shown in other crop species (Marrs, 1996).

The hypothesis that GGT4 scavenges the constituent amino acid of GSH for their transport from the vacuole back into the cytosol was also tested (Ohkama-Ohtsu *et al.*, 2007b). *ggt4* mutant plants grown under nitrogen deficiency and herbicide treatment did not show any difference either in appearance or in Gly and Glu concentrations compared to wild-type plants. This suggested that nitrogen flux through GSH-conjugate degradation in the vacuole is minor under these conditions. An analysis of *GGT4* expression using GENEVESTIGATOR, however, revealed that *GGT4* mRNA is induced under low nitrogen conditions. GGT4 may therefore have roles in nitrogen scavenging although we were unable to document an increase in nitrogen deficiency stress in the *ggt4* mutant.

V. THREE-DIMENSIONAL STRUCTURES OF GGTs FROM BACTERIA AND *ARABIDOPSIS*

GGT is known to catalyze hydrolysis and/or transpeptidation of a γ-glutamyl compound like GSH in two steps. In the first step GGT cleaves the peptide bond at the γ-position of the glutamyl moiety of the substrate to form a γ-glutamyl-enzyme intermediate. In the second step the glutamyl moiety in the intermediate is transferred to a water molecule (hydrolysis) or to an amino acid or peptide (transpeptidation). The molecular mechanism of the GGT reaction has been studied in detail for the bacterial GGTs using their three-dimensional structures as determined by X-ray crystallography (Boanca *et al.*, 2007; Okada *et al.*, 2006). The mature forms of GGTs from *E. coli* and *Helicobacter pylori* are both heterodimeric with large and small subunits that are generated by posttranslational processing. In the mature form the small subunit is surrounded by the large subunit (Fig. 5A). GGT has the $\alpha\beta\beta\alpha$ fold, as seen in members of the N-terminal nucleophile hydrolase superfamily (Brannigan *et al.*, 1995; Oinonen and Rouvinen, 2000). The central β-sheets, which comprise β-strands from both subunits, are flanked by the two α-helical domains. The structures of the two bacterial GGTs are very similar to each other.

In *E. coli* GGT the active site residue Thr-391, the N-terminus of the small subunit, is located in a groove on the enzyme surface. The detailed manner of GSH recognition by GGT as well as the mechanistic implication for the GGT reaction were provided by X-ray analysis of the short-lived γ-glutamyl-enzyme intermediate that was prepared by flash cooling the crystal soaked in a GSH solution (Okada *et al.*, 2006). The carbonyl carbon atom of the γ-glutamyl moiety derived from GSH is linked to the Oγ atom of Thr-391 through a covalent bond. The glutamyl moiety is held by conserved residues through extensive hydrogen bonds (Fig. 5B). The residues characterized as being essential for the catalytic activity of the mammalian GGTs (Ikeda *et al.*, 1993, 1995a,b) are mostly in direct interaction with the α-amino and α-carboxyl groups and the carbonyl oxygen of the moiety. This carbonyl oxygen atom is in close proximity to the NH groups of Gly-483 and Gly-484, suggesting that these two residues stabilize the tetrahedral transition state occurring in the catalytic reaction. The crystallographic analysis of the intermediate located a water molecule (W2) that may attack the carbonyl carbon atom of the γ-glutamyl moiety for hydrolysis. The crystal structure of GGT in complex with glutamate, the product, demonstrated that the mode of interaction with glutamate is similar to that was observed in the γ-glutamyl-enzyme intermediate. In both the intermediate and the product complex, the γ-glutamyl moiety is shielded from solvent by a lid loop (residues 438–449). However, the area in the bacterial GGTs that is involved in the binding of either the Cys-Gly moiety of GSH or any peptide to which the γ-glutamyl moiety is transferred has so far been poorly characterized.

How the structural change occurs in the maturation process of this enzyme was shown by a structural comparison between the Thr391Ala mutant protein, a mimic of the precursor that is unable to undergo autoprocessing, and the mature form of GGT (Okada *et al.*, 2007). As expected from the long distance (>30 Å) between the C-terminus of the large subunit and the N-terminus of the small subunit in the mature GGT, marked structural changes occur upon cleavage of the Gln390-Thr391 bond. That is, the

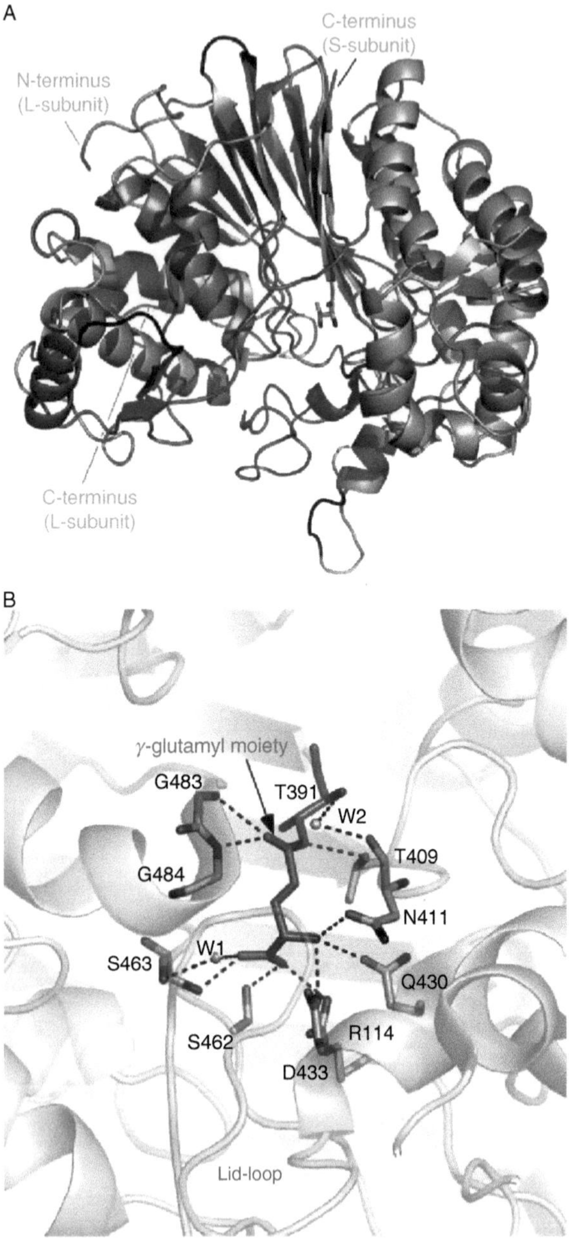

Fig. 5. (A) Ribbon drawing of *E. coli* GGT. Brown and green show the large subunit and the small subunit, respectively. The stick model at the center is the active residue Thr391, the N-terminus of the small subunit. The segments shown in black correspond to the regions in which large insertion/deletion of amino acid residues occur in the *Arabidopsis* GGTs. (B) Close-up view of the active site pocket of the γ-glutamyl-enzyme intermediate. The γ-glutamyl moiety is bound to Thr391 and the

segment Ile378-Gln390 in the mutant protein moves away from the active site pocket. This allows the movement of one side of the walls comprising the pocket in such a way that the pocket size is appropriate for entry of the γ-glutamyl moiety. The removal of the Ile378-Gln390 segment from the pocket also allows the lid-loop to cover the pocket. The flexible nature of the lid loop, which was demonstrated by the X-ray analyses of the mutant protein and the substrate-free form of GGT, implies that the lid loop will open and close the pocket during the course of the catalytic reaction.

The sequence alignment of the plant GGTs with typical bacterial and mammalian GGTs is given in Fig. 6. When the amino acid sequences of *Arabidopsis* GGT1, GGT2, and GGT4 are compared with those of the bacterial GGTs, the identities are in the range of 31–35%. In the *Arabidopsis* GGTs several residues are deleted before $\alpha 3$ and inserted between $\beta 19$ and $\alpha 19$ and between $\beta 21$ and $\beta 22$ (segments shown in black in Fig. 5A). None of these are included in the core of the GGT structure. In addition, the insertion and the deletion of residues in the plant GGTs occur in the regions of the bacterial GGTs that correspond to the loops between the secondary structures, indicating that the folds of these three plant GGTs are very similar to those of the bacterial GGTs.

The residues involved in the recognition of the γ-glutamyl moiety as well as the catalytic threonine residue (Thr391 in *E. coli* GGT) are highly conserved between the bacterial and plant GGTs. Among these residues only one substitution, Gln at the 430 position in *E. coli* GGT by Glu, may scarcely affect the substrate recognition; in the plant GGTs the O atom of this Glu may occupy the corresponding site of the O atom of Gln in *E. coli* GGT and is expected to be involved in the interaction with the α-amino group of the γ-glutamyl moiety. Taken together, all the three *Arabidopsis* GGTs may adopt the necessary tertiary and quaternary structures for the catalytic reaction of GGT. In contrast to GGT1 and GGT2, the sequence alignment shows that GGT4 has a long peptide segment at the N-terminus relative to *E. coli* GGT that is probably responsible for targeting to the vacuole. Structural information for the long N-terminal segment in GGT4 is unavailable from the structures of the bacterial GGTs.

GGT3 appears to be missing a portion of the N-terminus found in the closely related proteins GGT1 and GGT2 (Martin *et al.*, 2007). While the C-terminal active sites are retained in this protein, the segments

---

residues that interact with the moiety are colored. Hydrogen bonds are shown by broken lines. The lid loop (yellow) covers on the pocket. Except for the residues involved in the interaction with the γ-glutamyl moiety, individual atoms are omitted for clarity. PDB ID: 2DBW. (See color plate section at the back of the book.)

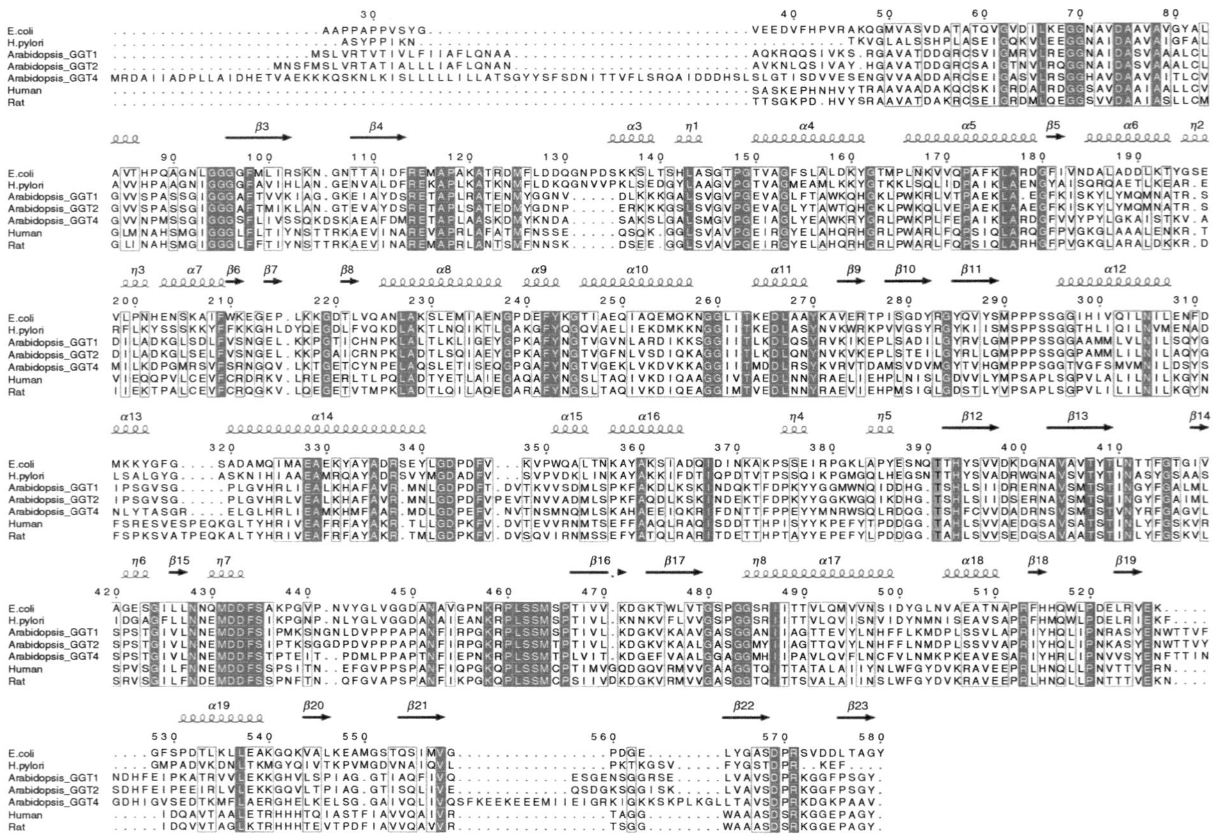
E.coli
H.pylori
Arabidopsis_GGT1
Arabidopsis_GGT2
Arabidopsis_GGT4
Human
Rat

corresponding to the α-helices and β-strands of the large subunit in *E. coli* GGT are missing. It is difficult to imagine how this protein could fold into the correct conformation.

GGT from *E. coli* contains an endogenous peptidase activity that cleaves that proprotein into the catalytically active large and small subunits. It is not known if the plant enzyme is autocatalytic. The catalytic Thr residue that is responsible for the catalysis is conserved in the plant sequences and as with the *E. coli* protein ends up as the N-terminal of the small subunit in radish (Nakano *et al.*, 2006b).

VI. GGT-LIKE PROTEINS IN OTHER PLANTS THAN *ARABIDOPSIS*

Protein sequences homologous to *Arabidopsis* GGTs are found in various plant species, including as *R. sativus* (radish, Nakano *et al.*, 2006a), *A. cepa* (onion, Shaw *et al.*, 2005), *Populus tremula x Populus alba* (gray poplar, GenBank Acc. No. ABB59573), *Zea mays* (maize, ACG27955), *Physcomitrella patens* (moss, XP_001772768), *Vitis vinifera*, (wine grape, CAO65418), and *Oryza sativa* (rice, TIGR's LOC_Os ID, Os01g05810, Os01g05820, and Os04g38450, http://rice.plantbiology.msu.edu/index.shtml).

A phylogenetic tree of the rice and *Arabidopsis* protein sequences as well as the known sequences for radish and maize is shown in Fig. 7. The *Arabidopsis* GGT1, GGT2, and GGT3 proteins are all similar and cluster with the radish GGT1 and GGT2 proteins as well as rice Os04g38450 and maize ACG27955. Given the apoplastic localization of the *Arabidopsis* GGT1 and GGT2 this same localization is expected for the rice and maize proteins. This correspondence between Os04g38450 and the apoplastic GGT1 and GGT2 is supported by iPSORT (http://hc.ims.u-tokyo.ac.jp/iPSORT/) which predicts that this one of the three rice proteins has a signal peptide protein at its N-terminus although neither WoLF PSORT (http://wolfpsort.org/) nor TargetP (http://www.cbs.dtu.dk/services/TargetP/) support this localization. Immunohistochemical methods have demonstrated that there is an apoplastic GGT in maize (Ferretti *et al.*, 2009; Masi *et al.*, 2007), so one presumably also occurs in rice and these are probably the specific proteins identified. Two

Fig. 6. Sequence alignment of bacterial, plant, and mammalian GGTs. The secondary structures observed in *E. coli* GGT are shown above the sequences. The residue numbers are for *E. coli* GGT. Thr391 shown in orange is the N-terminus of the small subunit of the mature GGT and the catalytic residue not only in the GGT reaction but also in the autoprocessing of the precursor protein. (See color plate section at the back of the book.)

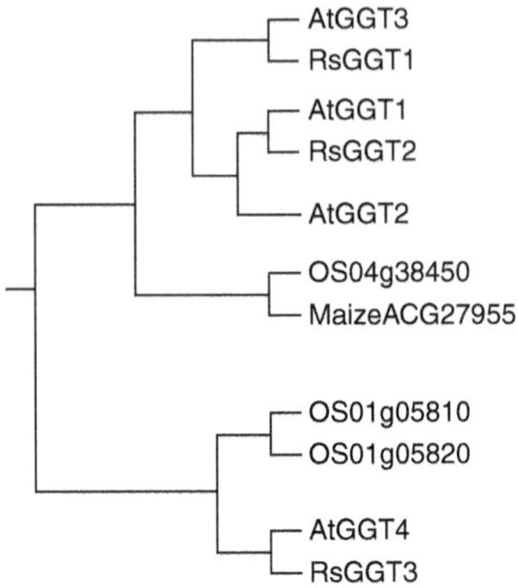

Fig. 7. Phylogenetic relationship between the four GGT proteins encoded in the *Arabidopsis* genome (AtGGT1, AtGGT2, AtGGT3, and AtGGT4), the three proteins encoded by the rice genome (Os01g05810, Os01g05820, and Os04g38450), and the proteins that have been identified from radish (RsGGT1, RsGGT2, and RsGGT3) and maize (Maize ACG27955). The sequences were compared by using ClustalW. The apoplastic proteins exemplified by *Arabidopsis* GGT1 and GGT2 clearly separate from the group that contains the vacuolar *Arabidopsis* GGT4.

of the rice proteins (Os01g05810 and Os01g05820) cluster with the vacuolar *Arabidopsis* GGT4 and radish RsGGT3 suggesting a similar subcellular localization. Interestingly neither of the rice proteins is predicted to target to that organelle by the subcellular localization programs. Clearly, more work needs to be done to confirm the role of vacuolar GGTs in monocots.

VII. THE PATHWAY FOR GSH DEGRADATION IN THE CYTOSOL IN PLANTS

In plants most of GSH exists in the cytoplasm. The GSH concentration in the mitochondria is reported to be the highest in the cytoplasm, 2.5-fold higher than that in the cytosol (Zechmann *et al.*, 2008). However considering that cytosol occupies about 15-fold larger volume of the cytoplasm compared to the mitochondria (Douce, 1985), the largest portion of the total GSH is in the cytosol in plants. Cytosolic GSH is degraded by enzymes other than GGTs because all of the functional GGTs in *Arabidopsis* are extracytosolic, with GGT1 and GGT2 in the apoplast and GGT4 in the vacuole. This idea was supported by the observation that GSH concentrations in *ggt* knockout mutants were not different from those in wild-type plants (Martin *et al.*, 2007; Ohkama-Ohtsu *et al.*, 2007a,b).

A GGT-independent degradation pathway of cytosolic GSH in plants is described by Ohkama-Ohtsu *et al.* (2008). As shown in Fig. 1B, cytoplasmic GSH is metabolized by the sequential reaction of GGC and 5OPase, both cytosolic enzymes (Rennenberg *et al.*, 1981; Steinkamp and Rennenberg, 1987). This pathway was suggested by following observations.

In *Arabidopsis*, a single gene (At5g37830) shows high homology to 5OPase from animals and was named *OXP (OXOPROLINASE) 1*. 5OPase catalyzes 5OP conversion to Glu. While wild-type plants show 5OPase activity it was not detected in the *oxp1* knockout plants. The *oxp1* knockout plants accumulate much more 5-oxoproline (5OP) and less Glu than wild-type plants suggesting substantial metabolite flux through 5OP and that 5OP is a major contributor to steady-state Glu levels. 5OP formation is strongly correlated with the level of GSH in *Arabidopsis* plants. Decreasing the GSH concentrations by treating with buthionine sulfoximine (BSO), an inhibitor for γ-EC synthetase, or using the *cad2* mutant with low γ-EC synthetase levels resulted in a coordinated decrease in 5OP concentrations. Conversely, increasing GSH levels in feeding experiments resulted in elevated 5OP concentrations. These results suggest that GSH is the predominant precursor of 5OP. This agrees with the previous report that GSH fed to tobacco was converted to 5OP (Rennenberg *et al.*, 1980).

Ohkama-Ohtsu *et al.* (2008) further confirmed that 5OP concentrations were not changed in the *ggt* mutants, including a *ggt1/ggt4/oxp1* triple mutant, and that GSH degradation rates were identical between *ggt* knockouts and wild-type plants. These results showed that GGT is not a major contributor to both 5OP production and GSH degradation.

Conversion of GSH to 5OP is catalyzed by the cytosolic plant enzyme GGC. Protein extracts from *Arabidopsis* seedlings showed GGC activity with GSH or γ-EC as a substrate (Ohkama-Ohtsu *et al.*, 2008). K_m and V_{max} values were 1.28 mM and 0.18 nmol mg^{-1} protein min^{-1} for GSH, and 4.24 mM and 0.52 nmol mg^{-1} protein min^{-1} for γ-EC. GGC reaction rates were calculated using the best available estimates of cytoplasmic GSH or γ-EC concentrations (Fricker *et al.*, 2000). The cytoplasmic GSH concentration they measured was 2–3 mM. Assuming the same ratio of γ-EC to GSH in the cytoplasm as is found in intact tissue, the cytoplasmic γ-EC concentration would be about 0.1 mM. Using these values with the measured kinetic constants in the Michaelis–Menten equation suggests that the GGC reaction rates are 0.11 nmol mg^{-1} protein min^{-1} for GSH and 0.02 nmol mg^{-1} protein min^{-1} for γ-EC. This calculated *in vivo* rate agrees very well with the measured rates of GSH degradation and 5OP accumulation in the *ggt1/ggt4/oxp1* triple mutant (Ohkama-Ohtsu *et al.*, 2008). These values strongly support the mutant studies that nearly all of GSH metabolism involves its conversion to 5OP by the enzyme GGC. While we have generally presented the conversion of GSH to 5OP by GGC as a one-step process it is also possible to postulate an intermediate. The higher V_{max} with γ-EC might support the notion that γ-EC or some other γ-glutamyl-amino acids could be preferred GGC substrates and that a two-step conversion occurs.

The proteins or genes corresponding to GGC have not been identified from any plants. Recently a protein showing GGC activity was identified in humans (Oakley *et al.*, 2008). The human GGC does not use GSH as a substrate *in vitro*. Homology searches using the human GGC sequence did not identify any similar proteins in plants (Oakley *et al.*, 2008) although GGC activity has been detected in tobacco (Steinkamp and Rennenberg, 1985, 1987) and *Arabidopsis* (Ohkama-Ohtsu *et al.*, 2008). This suggests that the plant and animal enzymes are too highly divergent to be recognized by this method. Human GGC contains a *BtrG*-like fold that might be a signature feature of GGC. A small family of *Arabidopsis* proteins with unknown function also contains this fold (de la Cruz *et al.*, 2008) and may include the *Arabidopsis* GGC. Specific identification of the *Arabidopsis* gene encoding GGC and analysis of the mutant are necessary to verify its function and to test our hypothesis that it is the major source of GSH metabolism *in vivo*.

VIII. DIFFERENCES IN THE GSH DEGRADATION PATHWAYS BETWEEN ANIMALS AND PLANTS

Fig. 1B shows a model of the degradation pathway for GSH and GSH conjugates in plants. GSH degradation occurs in the cytoplasm by GGC with only minor GGT involvement in total GSH turnover. The role of the GGTs is limited to the extracytosolic space, the degradation of oxidized glutathione (GSSG) by GGT1 in the apoplast, and the metabolism of GSH conjugates by GGT4 in the vacuole.

The plant pathway is different from the traditional γ-glutamyl cycle reported in mammals (Fig. 1A). In mammals the major site of GSH breakdown is extracellular and the reaction is catalyzed by GGT. This cycle is most active in the kidneys and important for triggering reabsorption of the constituent amino acids of GSH into the cell. The differences between the plant and animal pathways may reflect the lack of a secretory system in plants. Alternatively, GGT may catalyze only extracellular GSH degradation in mammals and, as in plants, another as yet unidentified enzyme is responsible for GSH turnover in the cytosol. Patients with GGT deficiency had increased GSH concentrations in plasma and urine but the cellular levels are normal, possibly supporting this idea (Goodman *et al.*, 1971; Hammond *et al.*, 1995;

O'Daly, 1968; Ristoff and Larsson, 1998; Wright *et al.*, 1980). Having demonstrated this variation of the γ-glutamyl cycle in plants where 5OP is formed from GSH by GGC and not GGT, it is interesting to ask if a GGC activity fills a similar role in animals *in vivo*. We were only able to prove this pathway in plants because of the capability of creating mutant plants that lacked both GGT and 5OPase activity. This has not been undertaken in animals.

IX. PERSPECTIVE

Plant GGT detoxifies ROS in the apoplast by metabolizing GSSG. This was discovered by analyses of *Arabidopsis ggt1* mutant plants, which showed oxidative stress and accumulated GSSG in the apoplast. Plant GGT also functions in GSH-conjugates degradation in the vacuole in *Arabidopsis*. While herbicides are sequestered in the vacuole in the form of GSH conjugates, endogenous GSH conjugates have not yet been identified. Some flavonoids and phytoalexins have been shown to conjugate with GSH *in vitro* (Dean *et al.*, 1995; Li *et al.*, 1997; Marrs *et al.*, 1995; Walczak and Dean, 2000); however, it is not known if they also form conjugates *in vivo*. It should be possible to identify endogenous GSH conjugates by metabolomic analysis using the *ggt4* mutant which is deficient in vacuolar GSH-conjugate degradation. It will be interesting to know how GSH contributes to vacuolar transport and metabolism of plant metabolites.

Although plant GGTs are important for GSH metabolism in the apoplast and vacuole, they are not involved in the major GSH degradation pathway in the cytoplasm. Cytoplasmic GSH is proposed to be degraded by the sequential reaction of GGC and 5OPase. Considering that GSH exists in mM concentration, substantial metabolite flux may occur through GSH degradation. Particularly GSH is the major storage and transport form for Cys and it must be broken down in order to release the amino acid. Specific identification of plant GGC and analysis of the mutant will provide insights into physiological role of GSH degradation in plant metabolism. In addition it will be important to study how GSH synthesis and degradation, which appear to be in the same cellular compartment, are regulated.

ACKNOWLEDGMENTS

This study is supported by grants from the U.S. National Science Foundation (MCB-0841528) and the RIKEN Special Postdoctoral Researcher Program. We thank Dr. Kei Wada, Osaka University, for his aid with preparation of the figures.

REFERENCES

Boanca, G., Sand, A., Okada, T., Suzuki, H., Kumagai, H., Fukuyama, K. and Barycki, J. J. (2007). Autoprocessing of *Helicobacter pylori* γ-glutamyl transpeptidase leads to the formation of a threonine-threonine catalytic triad. *Journal of Biological Chemistry* **282**, 534–541.

Brannigan, J. A., Dodson, G., Duggleby, H. J., Moody, P. C. E., Smith, J. L., Tomchick, D. R. and Murzin, A. G. (1995). A protein catalytic frameworth with an N-terminal nucleophile is capable of self-activation. *Nature* **378**, 416–419.

Cairns, N. G., Pasternak, M., Wachter, A., Cobbett, C. S. and Meyer, A. J. (2006). Maturation of *Arabidopsis* seeds is dependent on glutathione biosynthesis within the embryo. *Plant Physiology* **141**, 446–455.

Cobbett, C. and Goldsbrough, P. (2002). Phytochelatins and metallothioneins: roles in heavy metal detoxification and homeostasis. *Annual Review of Plant Biology* **53**, 159–182.

Coleman, J., Blake-Kalff, M. and Davies, E. (1997). Detoxification of xenobiotics by plants: Chemical modification and vacuolar compartmentation. *Trends in Plant Science* **2**, 144–151.

Dean, J. V., Devarenne, T. P., Lee, I. S. and Orlofsky, L. E. (1995). Properties of a maize glutathione *S*-transferase that conjugates coumaric acid and other phenylpropanoids. *Plant Physiology* **108**, 985–994.

de la Cruz, N. B., Peterson, F. C. and Volkman, B. F. (2008). Solution structure of At3g28950 from *Arabidopsis thaliana*. *Proteins* **71**, 546–551.

Dixon, D. P., Cummins, L., Cole, D. J. and Edwards, R. (1998). Glutathione-mediated detoxification systems in plants. *Current Opinion in Plant Biology* **1**, 258–266.

Douce, R. (1985). Mitochondria in Higher Plants: Structure, Function, and Biogenesis Academic Press, New York, p. 5.

Edwards, R., Dixon, D. P. and Walbot, V. (2000). Plant glutathione *S*-transferases: Enzymes with multiple functions in sickness and in health. *Trends in Plant Science* **5**, 193–198.

Ferretti, M., Destro, T., Tosatto, S. C. E., La Rocca, N., Rascio, N. and Masi, A. (2009). Glutamyl transferase in the cell wall participates in extracellular glutathione salvage from the root apoplast. *New Phytologist* **181**, 115–126.

Foyer, C. H. and Noctor, G. (2005). Redox homeostasis and antioxidant signaling: A metabolic interface between stress perception and physiological responses. *The Plant Cell* **17**, 1866–1875.

Foyer, C. H. and Noctor, G. (2009). Redox regulation in photosynthetic organisms: Signaling, acclimation, and practical implications. *Antioxidants and Redox Signaling* **11**, 861–905.

Fricker, M. D., May, M., Meyer, A. J., Sheard, N. and White, N. S. (2000). Measurement of glutathione levels in intact roots of *Arabidopsis*. *Journal of Microscopy* **198**, 162–173.

Goodman, S. I., Mace, J. W. and Pollack, S. (1971). Serum-glutamyl transpeptidase deficiency. *The Lancet* **1**, 234–235.

Gromes, R., Hothorn, M., Lenherr, E. D., Rybin, V., Scheffzek, K. and Rausch, T. (2008). The redox switch of γ-glutamylcysteine ligase via a reversible monomer-dimer transition is a mechanism unique to plants. *The Plant Journal* **54**, 1063–1075.

Grzam, A., Martin, M. N., Hell, R. and Meyer, A. J. (2007). Glutamyl transpeptidase GGT4 initiates vacuolar degradation of glutathione *S*-conjugates in *Arabidopsis*. *FEBS Letters* **581**, 3131–3138.

Hammond, J. W., Potter, M., Wilken, B. and Truscott, R. (1995). Siblings with γ-glutamyl transferase deficiency. *Journal of Inherited Metabolic Disease* **18**, 82–83.

Hanigan, M. H. and Pitot, H. C. (1985). γ-Glutamyl transpeptidase—Its role in hepatocarcinogenesis. *Carcinogenesis* **6**, 165–172.

Harding, C. O., Williams, P., Wagner, E., Chang, D. S., Wild, K., Colwell, R. E. and Wolff, J. A. (1997). Mice with genetic γ-glutamyl transpeptidase deficiency exhibit glutathionuria, severe growth failure, reduced life spans, and infertility. *Journal of Biological Chemistry* **272**, 12560–12567.

Hell, R. and Bergmann, L. (1990). γ-Glutamylcysteine synthetase in higher plants: Catalytic properties and subcellular localization. *Planta* **180**, 603–612.

Hernández, J. A., Ferrer, M. A., Jimenez, A., Barcelo, A. R. and Sevilla, F. (2001). Antioxidant systems and O_2^-/H_2O_2 production in the apoplast of pea leaves. Its relation with salt-induced necrotic lesions in minor veins. *Plant Physiology* **127**, 817–831.

Ikeda, Y., Fujii, J. and Taniguchi, N. (1993). Significance of Arg-107 and Glu-108 in the catalytic mechanism of human γ-glutamyl transpeptidase: Identification by site-directed mutagenesis. *Journal of Biological Chemistry* **268**, 3980–3985.

Ikeda, Y., Fujii, J., Anderson, M. E., Taniguchi, N. and Meister, A. (1995a). Involvement of Ser-451 and Ser-452 in the catalysis of human γ-glutamyl transpeptidase. *Journal of Biological Chemistry* **270**, 22223–22228.

Ikeda, Y., Fujii, J., Taniguchi, N. and Meister, A. (1995b). Human γ-glutamyl transpeptidase mutants involving conserved aspartate residues and the unique cysteine residues of the light subunit. *Journal of Biological Chemistry* **270**, 12471–12475.

Jez, J. M., Cahoon, R. E. and Chen, S. (2004). *Arabidopsis thaliana* glutamate-cysteine ligase: Functional properties, kinetic mechanism, and regulation of activity. *Journal of Biological Chemistry* **279**, 33463–33470.

Kasai, T., Ohmiya, A. and Sakamura, S. (1982). Glutamyltranspeptidases in the metabolism of γ-glutamyl peptides in plants. *Phytochemistry* **21**, 1233–1239.

Kawasaki, Y., Ogawa, T. and Sasaoka, K. (1982). Occurrence and some properties of the novel γ-glutamyl transferase responsible for the synthesis of γ-L-glutamyl-D-alanine in pea seedlings. *Biochimica et Biophysica Acta* **716**, 194–200.

Kinlough, C. L., Poland, P. A., Bruns, J. B. and Hughey, R. P. (2005). γ-Glutamyl-transpeptidase: Disulfide bridges, propeptide cleavage, and activation in the endoplasmic reticulum. *Methods in Enzymology* **401**, 426–449.

Kumada, H.-O., Koizumi, Y. and Sekiya, J. (2007). Purification and characterization of dipeptidase hydrolyzing L-cysteinylglycine from radish cotyledon. *Bioscience, Biotechnology and Biochemistry* **71**, 3102–3104.

Kumar, T. R., Wiseman, A. L., Kala, G., Kala, S. V., Matzuk, M. M. and Lieberman, M. W. (2000). Reproductive defects in γ-glutamyl transpeptidase-deficient mice. *Endocrinology* **141**, 4270–4277.

Lamb, C. and Dixon, R. A. (1997). The oxidative burst in plant disease resistance. *Annual Review of Plant Physiology and Plant Molecular Biology* **48**, 251–275.

Leustek, T., Martin, M. N., Bick, J. A. and Davies, J. P. (2000). Pathways and regulation of sulfur metabolism revealed through molecular and genetic studies. *Annual Review and Plant Physiology and Plant Molecular Biology* **51**, 141–165.

Li, Z. S., Alfenito, M., Rea, P. A., Walbot, V. and Dixon, R. A. (1997). Vacuolar uptake of the phytoalexin medicarpin by the glutathione conjugate pump. *Phytochemistry* **45**, 689–693.

Lieberman, M. W., Wiseman, A. L., Shi, Z. Z., Carter, B. Z., Barrios, R., Ou, C. N., Chevez-Barrios, P., Wang, Y., Habib, G. M., Goodman, J. C., Huang, S. L. Lebovitz, R. M. *et al.* (1996). Growth retardation and cysteine deficiency in γ-glutamyl transpeptidase-deficient mice. *Proceedings of the National Academy of Sciences of the United States of America* **93**, 7923–7926.

Luwe, M. (1996). Antioxidants in the apoplast and symplast of beech (*Fagus sylvatica* L.) leaves: Seasonal variations and responses to changing ozone concentrations in air. *Plant Cell and Environment* **19**, 321–328.

Marrs, K. A. (1996). The functions and regulation of glutathione *S*-transferases in plants. *Annual Review of Plant Physiology and Plant Molecular Biology* **47**, 127–158.

Marrs, K. A., Alfenito, M. R., Lloyd, A. M. and Walbot, V. (1995). A glutathione *S*-transferase involved in vacuolar transfer encoded by the maize gene *Broneze-2*. *Nature* **375**, 397–400.

Martin, M. N. and Slovin, J. P. (2000). Purified γ-glutamyl transpeptidases from tomato exhibit high affinity for glutathione and glutathione *S*-conjugates. *Plant Physiology* **122**, 1417–1426.

Martin, M. N., Saladores, P. H., Lambert, E., Hudson, A. O. and Leustek, T. (2007). Localization of members of the γ-glutamyl transpeptidase family identifies sites of glutathione and glutathione *S*-conjugate hydrolysis. *Plant Physiology* **144**, 1715–1732.

Masi, A., Destro, T., Turetta, L., Varotto, S., Caporale, G. and Ferretti, M. (2007). Localization of γ-glutamyl transferase activity in *Zea mays* organs and tissues. *Journal of Plant Physiology* **164**, 1527–1535.

May, M. J. and Leaver, C. J. (1994). *Arabidopsis thaliana* γ-glutamylcysteine synthetase is structurally unrelated to mammalian, yeast, and *Escherichia coli* homologs. *Proceedings of the National Academy of Sciences of the United States of America* **91**, 10059–10063.

Meister, A. (1973). On the enzymology of amino acid transport. *Science* **180**, 33–39.

Meister, A. and Larsson, A. (1995). Glutathione synthetase deficiency and other disorders of the γ-glutamyl cycle. *In* "The Metabolic and Molecular Bases of Inherited Disease, Vol. 1" (C. R. Scriver, A. L. Beaudet, W. S. Sly and D. Valle, eds.), pp. 1461–1495. McGraw-Hill, New York.

Nakano, Y., Okawa, S., Yamaguchi, T., Koizumi, Y. and Sekiya, J. (2004). Occurrence of two forms of γ-glutamyl transferase in radish plant. *Plant Biotechnology* **3**, 243–246.

Nakano, Y., Prieto, R., Hara, M. and Sekiya, J. (2006a). Cloning and molecular analysis of radish (*Raphanus sativus* L.) cDNAs encoding heterodimeric γ-glutamyltransferase. *Plant Biotechnology* **23**, 419–424.

Nakano, Y., Okawa, S., Yamauchi, T., Koizumi, Y., Sekiya, J. and Sekiya, J. (2006b). Purification and properties of soluble and bound γ-glutamyltransferases from radish cotyledon. *Bioscience, Biotechnology and Biochemistry* **70**, 369–376.

Nakano, Y., Okawa, S., Prieto, R. and Sekiya, J. (2006c). Subcellular localization and possible functions of γ-glutamyltransferase in the radish (*Raphanus sativus* L.) plant. *Bioscience Biotechnology and Biochemistry* **70**, 1790–1793.

Noctor, G. and Foyer, C. H. (1998). ASCORBATE AND GLUTATHIONE: Keeping active oxygen under control. *Annual Review of Plant Physiology and Plant Molecular Biology* **49**, 249–279.

Oakley, A. J., Yamada, T., Liu, D., Coggan, M., Clark, A. G. and Board, P. G. (2008). The identification and structural characterization of C7orf24 as γ-glutamyl cyclotransferase. An essential enzyme in the γ-glutamyl cycle. *Journal of Biological Chemistry* **283**, 22031–22042.

O'Daly, S. (1968). An abnormal sulphydryl compound in urine. *Irish Journal of Medical Science* **7**, 578–579.

Ohkama-Ohtsu, N., Radwan, S., Peterson, A., Zhao, P., Badr, A. F., Xiang, C. and Oliver, D. J. (2007a). Characterization of the extracellular γ-glutamyl transpeptidases, GGT1 and GGT2, in *Arabidopsis*. *The Plant Journal* **49**, 865–877.

Ohkama-Ohtsu, N., Zhao, P., Xiang, C. and Oliver, D. J. (2007b). Glutathione conjugates in the vacuole are degraded by γ-glutamyl transpeptidase GGT3 in *Arabidopsis*. *The Plant Journal* **49**, 878–888.

Ohkama-Ohtsu, N., Oikawa, A., Zhao, P., Xiang, C., Saito, K. and Oliver, D. J. (2008). A γ-glutamyl transpeptidase-independent pathway of glutathione catabolism to glutamate via 5-oxoproline in *Arabidopsis*. *Plant Physiology* **148**, 1603–1613.

Oinonen, C. and Rouvinen, J. (2000). Structural comparison of Ntn-hydrolases. *Protein Science* **9**, 2329–2337.

Okada, T., Suzuki, H., Wada, K., Kumagai, H. and Fukuyama, K. (2006). Crystal structures of γ-glutamyltranspeptidase from *Escherichia coli*, a key enzyme in glutathione metabolism, and its reaction intermediate. *Proceedings of the National Academy of Sciences of the United States of America* **103**, 6471–6476.

Okada, T., Suzuki, H., Wada, K., Kumagai, H. and Fukuyama, K. (2007). Crystal structure of the γ-glutamyltranspeptidase precursor protein from *Escherichia coli*: Structural changes upon autocatalytic processing and implications for the maturation mechanism. *Journal of Biological Chemistry* **282**, 2433–2439.

Pasternak, M., Lim, B., Wirtz, M., Hell, R., Cobbett, C. S. and Meyer, A. J. (2008). Restricting glutathione biosynthesis to the cytosol is sufficient for normal plant development. *The Plant Journal* **53**, 999–1012.

Pignocchi, C. and Foyer, C. H. (2003). Apoplastic ascorbate metabolism and its role in the regulation of cell signaling. *Current Opinion in Plant Biology* **6**, 379–389.

Rea, P. A. (1999). MRP subfamily ABC transporters from plants and yeast. *Journal of Experimental Botany* **50**, 895–913.

Rea, P. A., Li, Z. S., Lu, Y. P., Drozdowicz, Y. M. and Martinoia, E. (1998). From vacuolar GS-X pumps to multispecific ABC transporters. *Annual Review of Plant Physiology and Plant Molecular Biology* **49**, 727–760.

Rennenberg, H. and Lamoureux, G. L. (1990). Physiological processes that modulate the concentration of glutathione in plant cells. *In* "Sulfur Nutrition and Assimilation in Higher Plants" (H. Rennenberg, C. Brunold, L. J. De Kok and I. Stulen, eds.), pp. 53–65. SPB Academic Publishing, The Hague, The Netherlands.

Rennenberg, H., Steinkamp, R. and Polle, A. (1980). Evidence for the participation of a 5-oxo-prolinase in degradation of glutathione in *Nicotiana tabacum*. *Zeitschrift für Naturforschung. Section C: Biosciences* **35**, 708–711.

Rennenberg, H., Steinkamp, R. and Kesselmeier, J. (1981). 5-oxoprolinase in *Nicotiana tabacum*: Catalytic properties and subcellular localization. *Physiologia Plantarum* **62**, 211–216.

Ristoff, E. and Larsson, A. (1998). Patients with genetic defects in the γ-glutamyl cycle. *Chemico-Biological Interactions* **111–112**, 113–121.

Rouhier, N., Lemaire, S. D. and Jacquot, J.-P. (2008). The role of glutathione in photosynthetic organisms: Emerging functions for glutaredoxins and glutathionylation. *Annual Review of Plant Biology* **59**, 143–166.

Shaw, M. L., Pither-Joyce, M. D. and McCallum, J. A. (2005). Purification and cloning of a γ-glutamyl transpeptidase from onion (*Allium cepa*). *Phytochemistry* **66**, 515–522.
Steinkamp, R. and Rennenberg, H. (1984). γ-Glutamyl-transpeptidase in tobacco suspension cultures: Catalytic properties and subcellular localization. *Physiologia Plantarum* **61**, 251–256.
Steinkamp, R. and Rennenberg, H. (1985). Degradation of glutathione in plant cells: Evidence against the participation of a γ-glutamyl transpeptidase. *Zeitschrift für Naturforschung. Section C: Biosciences* **40c**, 29–33.
Steinkamp, R. and Rennenberg, H. (1987). γ-Glutamylcyclotransferase in tobacco suspension cultures: Catalytic properties and subcellular localization. *Physiologia Plantarum* **69**, 499–503.
Storozhenko, S., Belles-Boix, E., Babiychuk, E., Herouart, D., Davey, M. W., Slooten, L., Van Montagu, M., Inze, D. and Kushnir, S. (2002). γ-Glutamyl transpeptidase in transgenic tobacco plants. Cellular localization, processing, and biochemical properties. *Plant Physiology* **128**, 1109–1119.
Suzuki, H. and Kumagai, H. (2002). Autocatalytic processing of γ-glutamyltranspeptidase. *Journal of Biological Chemistry* **277**, 43536–43543.
Suzuki, H., Kumagai, H. and Tockikura, T. (1986). γ-Glutamyltranspeptidase from *Escherichia coli* K-12: Formation and localization. *Journal of Bacteriology* **168**, 1332–1335.
Taniguchi, N. and Ikeda, Y. (1998). γ-Glutamyl transpeptidase. *In* "Handbook of Proteolytic Enzymes" (A. Barrett, N. Rawlings and J. Woessner, eds.), pp. 512–516. Academic Press, London.
Tate, S. S. and Meister, A. (1985). γ-Glutamyl transpeptidase from kidney. *Methods in Enzymology* **113**, 400–419.
Vanacker, H., Carver, T. L. and Foyer, C. H. (1998a). Pathogen-induced changes in the antioxidant status of the apoplast in barley leaves. *Plant Physiology* **117**, 1103–1114.
Vanacker, H., Foyer, C. H. and Carver, T. L. M. (1998b). Changes in apoplastic antioxidants induced by powdery mildew attack in oat genotypes with race non-specific resistance. *Planta* **208**, 444–452.
Vanacker, H., Harbinson, J., Ruisch, J., Carver, T. L. W. and Foyer, C. H. (1998c). Antioxidant defenses of the apoplast. *Protoplasma* **205**, 129–140.
Veroux, T., Wilson, R. C., Seeley, K. A., Reichheld, J. P., Muroy, S., Brown, S., Maughan, S. C., Cobbett, C. S., Van Montagu, M., Inze, D., May, M. J. and Sung, Z. R. (2000). The *ROOT MERISTEMLESS1/CADMIUM SENSITIVE2* gene defines a glutathione-dependent pathway involved in initiation and maintenance of cell division during postembryonic root development. *The Plant Cell* **12**, 97–110.
Wachter, A., Wolf, S., Steininger, H., Bogs, J. and Rausch, T. (2005). Differential targeting of GSH1 is achieved by multiple transcription initiation: Implications for the compartmentation of glutathione biosynthesis in the Brassicaceae. *The Plant Journal* **41**, 15–30.
Walczak, H. A. and Dean, J. V. (2000). Vacuolar transport of the glutathione conjugate of trans-cinnamic acid. *Phytochemistry* **53**, 441–446.
Wang, C. L. and Oliver, D. J. (1996). Cloning of the cDNA and genomic clones for glutathione synthetase from *Arabidopsis thaliana* and complementation of a *gsh2* mutant in fission yeast. *Plant Molecular Biology* **31**, 1093–1104.
Wright, E. C., Stern, J., Ersser, R. and Patrick, A. D. (1980). Glutathionuria: γ-Glutamyl transpeptidase deficiency. *Journal of Inherited Metabolic Disease* **2**, 3–7.

Xiang, C. and Oliver, D. J. (1998). Glutathione metabolic genes coordinately respond to heavy metals and jasmonic acid in *Arabidopsis*. *The Plant Cell* **10**, 1539–1550.

Xiang, C., Werner, B. L., Christensen, E. M. and Oliver, D. J. (2001). The biological functions of glutathione revisited in *Arabidopsis* transgenic plants with altered glutathione levels. *Plant Physiology* **126**, 564–574.

Zechmann, B., Mauch, F., Sticher, L. and Muller, M. (2008). Subcellular immunocytochemical analysis detects the highest concentrations of glutathione in mitochondria and not in plastids. *Journal of Experimental Botany* **59**, 4017–4027.

The Redox State, a Referee of the Legume–*Rhizobia* Symbiotic Game

DANIEL MARINO,* CHIARA PUCCIARIELLO,† ALAIN PUPPO* AND PIERRE FRENDO*,[1]

**Interactions Biotiques et Santé Végétale UMR INRA 1301-CNRS 6243-Université de Nice-Sophia Antipolis, 400 Route des Chappes, BP167, 06903 Sophia Antipolis Cedex, France*
†Plant & Crop Physiology Laboratory, Scuola Superiore Santa Anna, Via Mariscoglio 34, 56124 Pisa, Italy

[1]Corresponding author: Email: frendo@unice.fr

Advances in Botanical Research, Vol. 52 0065-2296/09 $35.00

DOI: 10.1016/S0065-2296(09)52005-8

ABSTRACT

The symbiosis between legumes and *Rhizobia* leads to the formation of a new organ, the nodule, where nitrogen fixation takes place. The plant provides bacteria with energy and a micro-aerobic environment compatible with nitrogenase activity. In exchange, bacteria provide a nitrogen supply to the plant. Therefore, nodules represent a unique model for the study of developmental processes, plant–microbe, and carbon/nitrogen/oxygen metabolism interactions. Reactive oxygen species (ROS) have been described to be produced at every step of the symbiotic association: during the symbiosis establishment, during the nitrogen fixation and during the senescence of the nodule. ROS can be harmful provoking oxidative damage or can have a positive role in signalling processes. To deal with ROS production, nodules are fitted with a large panel of enzymatic and non-enzymatic antioxidant mechanisms. This review summarizes the present knowledge about ROS production and scavenging systems in legume nodules, and about their functional roles during the different steps of the symbiotic interaction.

I. INTRODUCTION

The acquisition of mineral nutrients is one of the major challenges for plant survival. After carbon, hydrogen and oxygen, nitrogen (N) is the most important element for plant growth, because of its presence in major biomolecules. Although N is very abundant on Earth with most of it found in the atmosphere in the form of dinitrogen (N_2), soils are often deficient in N. Only a reduced group of prokaryotes, known as diazotrophes or N_2-fixing organisms, are able to reduce N_2 to NH_4^+ (Graham and Vance, 2003). These microorganisms are able to fix N_2 in a free-living state or in symbiotic association with plants (Zehr *et al.*, 2003). The N_2-fixing symbiosis (NFS) can be established in non-leguminous plants between the filamentous bacterial genus *Frankia* and actinorhizal plants (Wall, 2000), but this association will not be covered in this review. With legumes as host plants, the NFS occurs with bacteria generally known as *Rhizobia*. Nowadays it is known that nodulation with legumes is not restricted to the Rhizobiaceae family but may involve other genera such as *Burkholderia*, *Ralstonia* or *Methylobacterium* (Sprent, 2007). On the other hand, *Rhizobia* can also nodulate the non-leguminous plant genus *Parasponia* (Trinick, 1973). In this review, we will focus on the importance of redox metabolism in NFS.

II. PRODUCTION OF REACTIVE OXYGEN SPECIES DURING LEGUME–*RHIZOBIA* SYMBIOSIS

A. ESTABLISHMENT OF NFS

The establishment of symbiosis involves a complex molecular dialogue between host and symbiont leading to infection of root hairs by the bacteria and ultimately to the formation of root nodules, where the nitrogen fixation process takes place (Oldroyd and Downie, 2008). This symbiotic interaction is beneficial for both partners: the plant supplies energy and carbon to the bacteria in the form of photo-assimilates and the bacteria provide the plant with usable forms of N.

The initial step of the symbiotic interaction is a chemical cross-talk between both partners (Fig. 1). Plants secrete (iso)flavonoids that are recognized by the bacteria, resulting in the induction of nodulation genes (nod genes), which encode proteins that synthesize and export specific lipochito-oligosaccharides, the so-called Nod factors (NFs) (Oldroyd and Downie, 2008). The species-specific modifications of NFs have a key role in defining the specificity of the interaction between *Rhizobia* and their plant hosts (Gage, 2002). NFs activate a number of processes leading to root-hair infection after recognition by the plant. One of the first plant responses to NFs is an increase in the intracellular calcium level in root hairs, followed by marked calcium spiking and alteration of the root-hair cytoskeleton (Sieberer *et al.*, 2005). This results in the reorientation of root-hair tip growth leading to its curling around bacteria. Simultaneously, NFs also trigger induction of cell division in inner

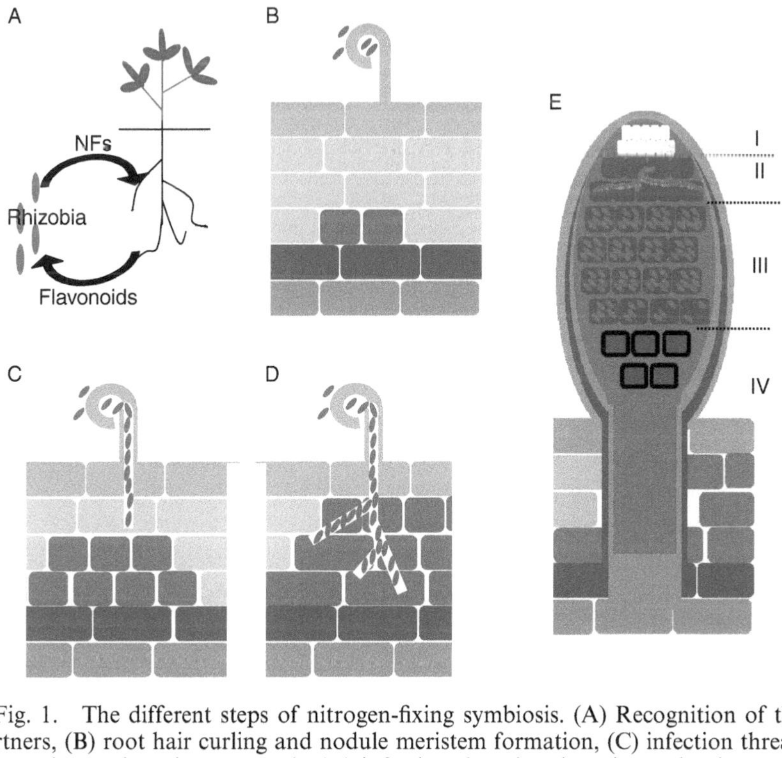

Fig. 1. The different steps of nitrogen-fixing symbiosis. (A) Recognition of the partners, (B) root hair curling and nodule meristem formation, (C) infection thread penetration and meristem growth, (D) infection thread and meristem development, (E) indeterminate nodule with meristematic (I), infection (II), fixing (III), and senescent (IV) zones. (See color plate section at the back of the book.)

cortical cells leading to a nodule primordium formation. Then, infection of root hairs takes place through the progressive formation of an inwardly growing tunnel known as the infection thread (IT). *Rhizobia* enter and divide within the IT, which subsequently traverses the entire root hair and outer cortical cells in order to reach the inner cortex (Gage, 2002). The path followed by the IT in the cortex is anticipated by the formation of transcellular cytoplasmic bridges known as pre-IT (Timmers *et al.*, 1999; van Brussel *et al.*, 1992). Recently, it has been shown that IT elongation anticipates bacterial-IT colonization indicating that this process is primarily host driven (Fournier *et al.*, 2008). In the inner cortex, bacteria are released by endocytosis into the host cells. Each bacterium is endocytosed individually into a host cell of the inner cortex in an unwalled compartment. This organelle-like structure is called the symbiosome, in which bacteria differentiate into bacteroids. The co-differentiation of the two symbiotic partners ends in the formation of the functional N_2-fixing nodule (Oldroyd and Downie, 2008).

Nodules can be of two types: determinate or indeterminate (Hirsch, 1992). Determinate nodules are formed in legumes of tropical origin (bean, soybean or the model *Lotus japonicus*), are spherical, and lack a persistent meristem. In the determinate nodules, cell expansion, rather than cell division, is responsible for nodule growth (Franssen *et al.*, 1992). Indeterminate nodules are formed in legumes of subtropical and temperate origins (pea, clover, or the model *Medicago truncatula*). These nodules have a persistent meristem that often yields a cylindrical or branched structure. Five zones can be distinguished in these nodules: zone I made of meristematic cells; zone II where the ITs are present and bacteria are endocytosed; zone III where bacteria become competent to perform symbiotic N_2-fixation (SNF); zone IV where the symbiotic partnership is lost and senescence is evident; and zone V where bacteria are essentially free living (Timmers *et al.*, 2000).

B. ROS AND ESTABLISHMENT OF SYMBIOSIS

Reactive oxygen species (ROS) are now described as secondary messengers in many plant signalling processes (Apel and Hirt, 2004). At the early stage of the symbiotic interaction, the oxidation of nitro blue tetrazolium (NBT) can be detected in ITs, showing that superoxide anions ($O_2^{\bullet -}$) are produced during the infection process (Ramu *et al.*, 2002; Santos *et al.*, 2001). Interestingly, this production was not observed when *M. truncatula* plants were inoculated with a *Sinorhizobium meliloti* nodD1ABC mutant that is impaired for NFs' production, therefore suggesting a role for NFs in ROS production (Ramu *et al.*, 2002). Furthermore, it has been shown that hydrogen peroxide (H_2O_2) accumulation occurs during the infection process (Jamet *et al.*, 2007;

Rubio *et al.*, 2004; Santos *et al.*, 2001). In this framework, a peroxidase (*rip1*), that has sequence motifs with homology to ROS responsive cis elements, is induced in the presence of NFs in *M. truncatula–S. meliloti* interaction (Cook *et al.*, 1995; Ramu *et al.*, 2002). Besides, *rip1* transcription and ROS production co-localize in the same root cells (Ramu *et al.*, 2002). Similarly, a high induction of a peroxidase (*Srprx1*) expression has been shown after inoculation or NFs' treatment in *Sesbania rostrata* (Den Herder *et al.*, 2007). Moreover, a transient increase in ROS level has been shown within a few seconds after treatment with NFs in growing *Phaseolus vulgaris* root hairs (Cardenas *et al.*, 2008). This increase was suppressed when the plants were treated with diphenyleneiodonium (DPI), an inhibitor of NADPH oxidases. This result suggests that the activation of NADPH oxidases and the consequent ROS production is closely linked to the perception and signalling of NFs (Cardenas *et al.*, 2008). The authors proposed that this specific transient ROS signature was the earliest downstream signal after the perception of a symbiotic or pathogenic molecular stimulus by the plant. NADPH oxidases (NOX in animals, RBOH in plants, from Respiratory Burst Oxidative Homolog) are membrane-bound proteins which reduce O_2 to $O_2^{\bullet-}$. They have been shown to play many critical roles in plants during the hypersensitive response, stomatal closure, root-hair development, cell death, or response to abiotic stresses (Foreman *et al.*, 2003; Jones *et al.*, 2007; Kwak *et al.*, 2003; Torres *et al.*, 2005; Yoshioka *et al.*, 2003). The inhibition of ROS production by pharmacological approaches was also shown to inhibit root-hair curling and IT formation (Peleg-Grossman *et al.*, 2007). The importance of ROS production has been confirmed by using a catalase (CAT) overexpressing *S. meliloti* strain, acting as a H_2O_2 sink, which showed a delayed nodulation combined to an enlargement of ITs (Jamet *et al.*, 2007).

A number of tropical flooding-tolerant legumes, such as the semiaquatic *S. rostrata*, provide bacteria with special ways of invading roots to establish NFS. On well-aerated roots, root-hair curling invasion is used, whereas in flooded conditions where root hairs are few, the symbiosis takes places through the fissure region at the base of lateral or adventitious roots by cortical, intercellular crack entry. The formation of infection pockets has been shown to be associated with localized cell death and the production of H_2O_2 (D'Haeze *et al.*, 2003; Goormachtig *et al.*, 2004). Pharmacological experiments have also showed that ROS and ethylene are involved in the NFs-induced signal cascade leading to nodule development (D'Haeze *et al.*, 2003). All these data indicate that ROS production may be involved in the development of a proper symbiotic interaction.

In contrast with these results, it has been shown that after 20–30 min of exposure to NFs, H_2O_2 efflux reduces to 60% of the steady-state level in

M. truncatula (Shaw and Long, 2003). This down-regulation of early ROS efflux was not observed in plants deficient for genes encoding proteins involved in the symbiotic interaction, like a Ca^{2+}-dependent protein kinase (CDPK1) or the putative NFs receptor, Nod factor perception (NFP). These results suggest that ROS variations are linked to the NFs signalling transduction pathway (Ivashuta *et al.*, 2005). The reduction of ROS efflux after treatment with NFs has been correlated to a transient decrease of transcript abundance of two NADPH oxidase homologs in *M. truncatula MtRboh2* and *MtRBoh3* 1 h after NFs treatment (Lohar *et al.*, 2007). This transient down-regulation of both *Rboh* was not observed in the nfp mutant, showing that the early stage of NFs perception is required for the regulation of *MtRboh2* and *3* (Lohar *et al.*, 2007). The apparent contradictory results concerning the production of ROS and its involvement in the early stages of the interaction may be reconciled assuming that ROS production is differentially modulated at different time points of the symbiotic interaction.

As for plant–pathogen interactions, ROS may be important actors in the initial interaction between the two partners, which leads to the recognition of the bacteria as a friend or as a foe. Thus, plant pathogenesis and symbiosis have been proposed as variations on a common theme (Baron and Zambryski, 1995). There is also some evidence suggesting that *Rhizobia* might be recognized by the plant as useful partners or be treated as intruders which trigger the plant defence response. Indeed, several plant defence responses are induced during root-nodule ontogeny (Gamas *et al.*, 1998; Parniske *et al.*, 1990; Vasse *et al.*, 1993). In fact, the abortion of IT in *Medicago* has been associated to a hypersensitive response triggered in the root cortex (Penmetsa and Cook, 1997; Vasse *et al.*, 1993). Moreover, a *S. meliloti* mutant strain deficient in NFs production and thus, in plant infection, induced H_2O_2 accumulation during the first hours post-inoculation compared with wild-type *S. meliloti* (Bueno *et al.*, 2001). The rhizobial poly- and oligo-saccharides have also been proposed to play an important role in the process of establishing a successful symbiosis through the suppression or avoidance of ROS production (Albus *et al.*, 2001; Scheidle *et al.*, 2005; Tellstrom *et al.*, 2007). A *S. meliloti* mutant which is defective in plant invasion has also an impaired exopolysaccharide production. This mutant triggers the induction of a large number of genes belonging to the plant defence category in *M. truncatula*, which may be linked to ROS production (Jones *et al.*, 2008). Different studies of *Rhizobia*–legume associations showed that mutant strains defective in this carbohydrate production are impaired in the successful formation of nodules, with the concomitant induction of the plant defence reaction (Campbell *et al.*, 2002; Jones *et al.*, 2008; Niehaus *et al.*, 1993; Perotto *et al.*, 1994; van Workum *et al.*, 1998).

C. ROS AND SYMBIOSIS FUNCTIONING

ROS are also likely to be produced in functioning mature nodules (Fig. 2). The latter need high amounts of energy to fuel the nitrogenase with enough ATP to reduce the atmospheric N_2 (16 mol of ATP for each mol of N_2 reduced), which results in a high flux of O_2 into the nodules. Paradoxically, the nitrogenase enzymatic complex is extremely sensitive to O_2. The reason of the sensitivity of nitrogenase to O_2 is complex and one hypothesis is that nitrogenase has the capacity to reduce O_2, leading to the production of ROS which rapidly destroy it (Dalton, 1995). To deal with this dilemma, nodules have developed different strategies to protect the nitrogenase from O_2: (i) microsymbionts exhibit high respiration rates (Dalton, 1995), (ii) uninfected cell layers in the inner cortex act as an O_2 diffusion barrier (Tjepkema and Yocum, 1974) and (iii) leghemoglobin (Lb), a myoglobin-like 16-kDa O_2-binding protein present at millimolar concentrations in the cytoplasm of infected cells, delivers O_2 efficiently to mitochondria and bacteroids for respiration while decreasing free O_2 to the nanomolar range. The essential

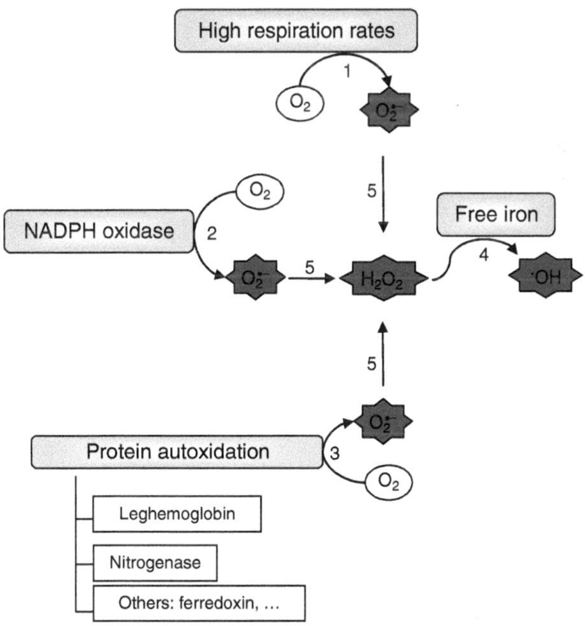

Fig. 2. Putative reactive oxygen species sources in nodules. (1) $O_2^{\bullet-}$ production associated to O_2 reduction during the electron transport chain; (2) enzymatic $O_2^{\bullet-}$ production by NADPH oxidases; (3) $O_2^{\bullet-}$ release during protein autooxidation; (4) H_2O_2 reduction to $\cdot OH$ through Fenton reaction; (5) $O_2^{\bullet-}$ dismutation to H_2O_2 by SODs.

role of Lb was demonstrated in nodules of RNAi *L. japonicus* lacking Lb leading to an increase in free O_2 and a loss of nitrogenase and N_2 fixation (Ott *et al.*, 2005).

The high respiration rates needed for a N_2-fixation energy supply are certainly the main source of ROS in nodules. The way to produce ROS as associated by-products to the electron respiration chain will not be covered in this review (Apel and Hirt, 2004). Amongst the ROS-producing proteins, ferredoxin, the proximal electron donor to nitrogenase in the bacteroids, is well known to reduce O_2 to $O_2^{\bullet-}$ (Becana *et al.*, 2000). Bacteroid hydrogenase is also an O_2-labile enzyme, whose inactivation also involves $O_2^{\bullet-}$ production, which may self-inactivate the protein through destruction of the Fe-S cluster, as also suggested for nitrogenase (Becana *et al.*, 2000). Uricase, which catalyzes the oxidation of uric acid to allantoine in ureide-exporting legumes is also O_2 dependent and susceptible to auto-oxidation with the consequent formation of $O_2^{\bullet-}$ (Dalton *et al.*, 1991).

Lb is certainly another source of ROS in N_2-fixing nodules (Fig. 3). Ferro-Lb (Lb-Fe^{2+}), the oxygenated active form of Lb, is subject to auto-oxidation in which $O_2^{\bullet-}$ is released when changing to the Lb-Fe^{3+} inactive form (Fridovich, 1986; Puppo *et al.*, 1981). $O_2^{\bullet-}$ can disproportionate to H_2O_2 and both can produce the highly reactive hydroxyl radical ($^{\bullet}OH$) by the metal-catalyzed Fenton reaction. This reaction is greatly accelerated in nodules, since H_2O_2 attacks the heme group of Lb, releasing free iron

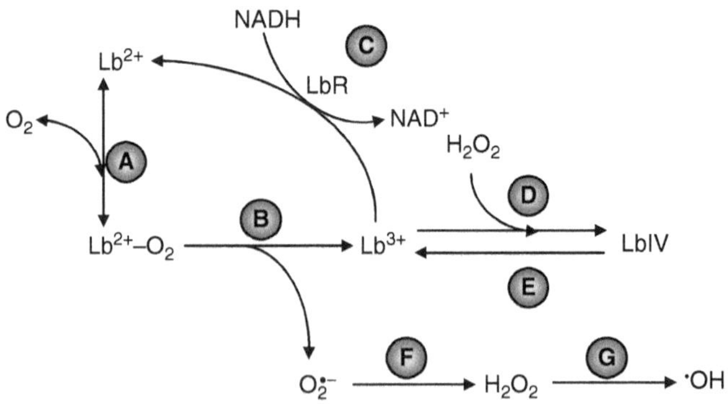

Fig. 3. Metabolic pathways involving Lb and formation of ROS. A) Reversible oxygenation of Lb^{2+}; B) Autoxidation of Lb^{2+}-O_2 to Lb^{3+} with release of $O_2^{\bullet-}$; C) Lb^{3+} reduction by ferric Lb reductase (LbR); D) H_2O_2 reaction with Lb^{3+} to generate the inactive LbIV (ferryl) form; E) LbIV reduction to Lb^{3+} by ASA or thiols; other ROS can also be generated from $O_2^{\bullet-}$, by its dismutation to H_2O_2 (F) and by H_2O_2 reduction to $^{\bullet}OH$ through Fenton reaction (G).

which then reacts to create ·OH (Puppo and Halliwell, 1988). H_2O_2 also directly reacts with both the ferrous and the ferric forms of Lb and oxidizes them to the inactive ferryl (LbIV) form (Aviram *et al.*, 1978; Puppo *et al.*, 1993). Thus, at least three major forms of ROS are generated by Lb in a self-perpetuating cycle. The oxidized Lb, that results from the release of $O_2^{\bullet -}$, is unable to bind O_2 and thus requires the presence of an NADH-dependent Lb reductase to restore Lb to its physiologically active form (Becana and Klucas, 1990). The hypothesis of Lb being one of the major ROS sources in nodules has been recently confirmed in RNAi *L. japonicus* lines completely lacking Lb: a significant decrease of H_2O_2 levels and an increase in reduced ascorbate was observed in these plants (Gunther *et al.*, 2007).

Nevertheless, the lack of ROS detection in the N_2-fixing zone of the nodules may be attributed to a strong antioxidant defence present in the nodules (see later), or to the fast reaction of ROS with biological compounds present in the nodule (Rubio *et al.*, 2004; Santos *et al.*, 2001). This lack of detection might be also due to the limitations of ROS *in situ* detection techniques. In contrast, ROS have been detected around the senescing symbiosomes during nodule senescence, suggesting their involvement in this process (Alesandrini *et al.*, 2003; Rubio *et al.*, 2004). A combination of multiple factors including ROS may be involved in the control of the events leading to the rupture of the symbiotic interaction (Puppo *et al.*, 2005).

III. INVOLVEMENT OF ANTIOXIDANT SYSTEMS IN THE LEGUME–*RHIZOBIUM* SYMBIOSIS

A large amount of ROS is generated during the whole life of the nodule, as described earlier. Thus, a large panel of enzymatic and non-enzymatic antioxidant mechanisms is present in the symbiotic partners both to scavenge harmful ROS, protecting the cell from oxidative damage, and to regulate the steady-state level of ROS which also act as signal molecules (Fig. 4).

A. ANTIOXIDANT SYSTEMS IN NODULE FORMATION AND FUNCTIONING

The main soluble non-enzymatic antioxidant molecules identified in nodule tissue are ascorbate (ASA), glutathione (GSH) and its legume-specific homolog, homoglutathione (hGSH) (Price, 1957). In plants, three pathways for ASA biosynthesis have been proposed: through L-galactose (Wheeler *et al.*, 1998), through galacturonic acid (Agius *et al.*, 2003), and from *myo*-inositol (Lorence *et al.*, 2004). The best known function of ASA is related to its antioxidant property, as a direct scavenger of ROS and a

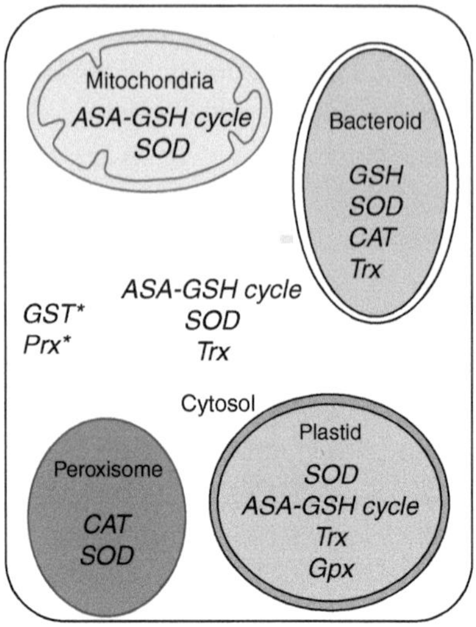

Fig. 4. Localization of different antioxidant systems in the symbiotic cell. Asterisk (*) indicates a localization not indicated in the literature. CAT, catalase; SOD, superoxide dismutase; Gpx, glutathione peroxidase; Trx, thioredoxin; GST, glutathione-*S*-transferase; Prx, peroxiredoxin; ASA–GSH cycle, ascorbate–glutathione cycle.

substrate for ascorbate peroxidase (APX) in the ASA–GSH H_2O_2-scavenging cycle (Noctor and Foyer, 1998). ASA is also involved in other critical processes of the plant, such as the regulation of mitosis, quiescence and cell growth (Potters *et al.*, 2000, 2004). The multiple functions of ASA identified in several papers suggest its involvement in the signalling of complex regulation processes as biotic and abiotic stresses, in addition to its antioxidant and redox buffer properties (Conklin and Barth, 2004; Pastori *et al.*, 2003; Smirnoff and Wheeler, 2000). In the root nodule, ASA was found at a concentration range of 0.5–2 mM (Dalton *et al.*, 1986; Gogorcena *et al.*, 1997; Matamoros *et al.*, 1999a). Recently, Matamoros *et al.* (2006) showed that the genes involved in ASA biosynthesis are expressed in nodules of indeterminate and determinate legumes and that an active galactono-1,4-lactone dehydrogenase (GalLDH), the enzyme catalyzing the last step in the main pathway of ASA biosynthesis, is present in mitochondria of the infected nodule tissue. Moreover, GalLDH activity was found to be abundant in the meristematic region of indeterminate nodules (Matamoros *et al.*, 2006). The importance of ASA in the nodulation process was suggested as ASA exogenous application induced an increased nitrogenase activity and plant nodule number as well as a delayed senescence in the soybean nodule

(Bashor and Dalton, 1999; Swaraj and Garg, 1970). However, ASA or its precursor galactono-1,4-lactone exogenous supply did not prevent water stress effects on SNF in pea nodules (Zabalza *et al.*, 2008). Nevertheless, whereas the presence of ASA in the N_2-fixing zone of nodules can suggest its involvement in antioxidant defence, the high levels found in the apex of indeterminate nodules underline the possibility of some additional roles (Matamoros *et al.*, 2006). The requirement of ASA for cell division and elongation suggests its involvement, together with thiols, in meristematic and invasion zones having high mitotic activity in nodules (Matamoros *et al.*, 2006).

GSH is the most abundant low molecular weight thiol in plants. It is involved in multiple aspects of plant defence, acting directly as an antioxidant and as part of the ASA–GSH cycle (Noctor and Foyer, 1998). It is also involved in the detoxification of xenobiotics (Dixon *et al.*, 2002), heavy metals (Cobbett and Goldsbrough, 2002) and in the tolerance to (a)biotic stresses (Rausch and Wachter, 2005). GSH has been shown to be involved in the regulation of cell division (Sanchez-Fernandez *et al.*, 1997; Vernoux *et al.*, 2000) and post-translational modification of proteins (Dixon *et al.*, 2005; Michelet *et al.*, 2005). In certain legumes, homoglutathione (hGSH) is present instead of, or in addition to, GSH (Frendo *et al.*, 1999; Klapheck, 1988; Matamoros *et al.*, 1999b), but its specific role, if any, in nodule functioning is not yet known. These compounds share an equivalent chemical structure, suggesting similar functions in plants (Harrison *et al.*, 2003). In the root nodule, (h)GSH was found in the millimolar concentration range. Previous studies have shown that (h)GSH is more abundant in the nodules than in other organs of the plants (Frendo *et al.*, 1999; Matamoros *et al.*, 1999b). The highest levels are observed in the infection zones of both determinate and indeterminate nodules (Matamoros *et al.*, 1999b). The biosynthesis pathway of (h)GSH is a two-step ATP-dependent process, catalyzed by γ-glutamylcysteine synthetase (γECS) and (h)GSH synthetase (Fig. 5). The hGSH synthetase (hGSHS) is derived from GSH synthetase (GSHS) by a gene duplication event which occurred after the divergence of plant orders. This gene duplication may have happened via a tandem duplication (Frendo *et al.*, 2001). The importance of GSH in nodule meristem formation (Frendo *et al.*, 2005), together with its role in the regulation of postembryonic root meristem (Vernoux *et al.*, 2000), suggests a role of GSH in initiation and maintenance of cell division in the nodule meristematic area. The observations that the GSH level increases in soybean nodules in the first 30 days of the plant life (Dalton *et al.*, 1986), that its concentration in effective nodules is higher than in ineffective ones (Dalton *et al.*, 1993b) and that a strong correlation exists between the nodule GSH content and its nitrogen fixation capacity

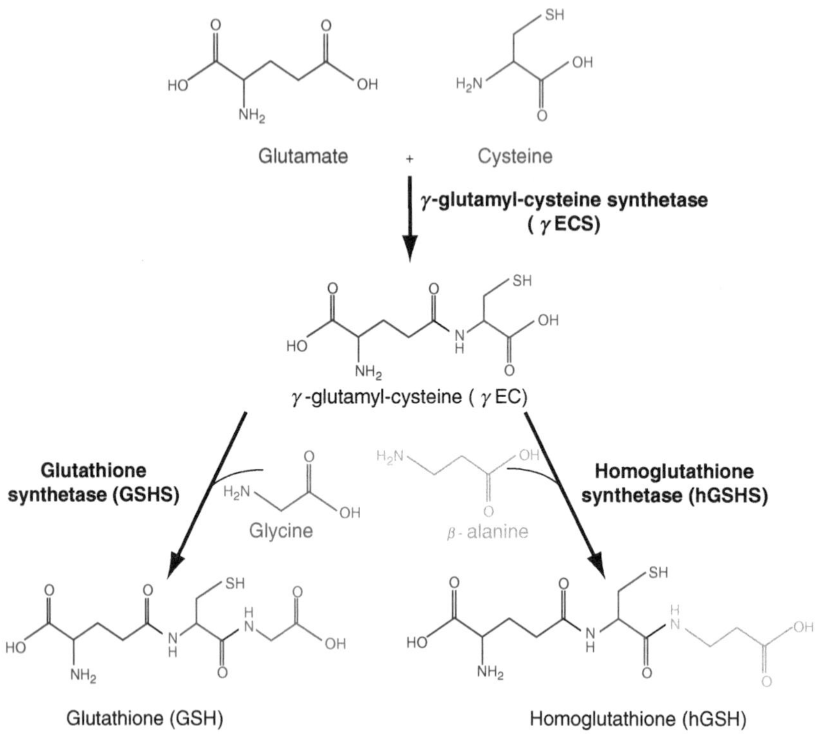

Fig. 5. Biosynthesis pathway of GSH and hGSH. All enzymes are shown in bold. (See color plate section at the back of the book.)

(Dalton *et al.*, 1993b; Groten *et al.*, 2005) suggest a role for GSH in maintaining the nodule physiological functions. This role is probably related to the GSH antioxidant and scavenging capacity in relation to the strong reducing conditions required by the N_2 fixation. This hypothesis is strengthened by the evidence that a high O_2 partial pressure in the nodule leads to an increase in GSH content, showing the importance of antioxidant defence in nodule functioning (Dalton *et al.*, 1991). GSH is also important as a substrate of glutathione *S*-transferases (GST) which catalyze the conjugation of toxic xenobiotics and oxidatively produced compounds to GSH. At least 14 isoforms of GST are expressed in soybean nodules with *GST9* being the most prevalent isoform. Reduction of the GST9 level by RNAi in transgenic composite plants leads to a decrease in nitrogenase activity and an increase in oxidatively damaged proteins indicating that GSTs are involved in antioxidant defence and are critical for supporting SNF (Dalton *et al.*, 2009).

The inability of *S. meliloti* to synthesize GSH leads to an impaired nodule phenotype, characterized by early senescence during symbiosis with *Medicago sativa* showing its crucial role in the symbiotic process (Harrison *et al.*, 2005). The GSHS-deficient *S. meliloti* strain (SmgshB) showed an increased oxidized γEC pool correlated to a higher CAT activity, suggesting that the mutant strain is under oxidative stress. The drastic phenotype observed during nodulation compared to the absence of a marked phenotype during free-living growth may reflect the need for a higher antioxidant defence during symbiosis. Moreover, a GSH-deficient mutant *Rhizobium tropici* strain is outcompeted by the wild-type strain in co-inoculation experiments and nodules induced by the mutant bacteria presented an early senescent pattern, which was associated with increased levels of $O_2^{\bullet -}$ accumulation (Muglia *et al.*, 2008; Riccillo *et al.*, 2000).

The ASA–GSH cycle is one of the main antioxidant systems in plants; it detoxifies H_2O_2 via its reduction to water. The reaction is catalyzed by APX, using the reducing power of ASA that is regenerated by either an NADH-dependent monodehydroascorbate reductase (MDHAR) or coupled reactions that include dehydroascorbate reductase (DHAR). DHAR uses the reduction power of GSH that is oxidized to GSSG, and its regeneration is obtained by an NADPH-dependent GSH-reductase (GR) (Asada, 1999; Noctor and Foyer, 1998). All the components of the ASA–GSH cycle have been found in the cytosol of the nodule cells (Dalton, 1995; Dalton *et al.*, 1986, 1992) and some enzymes were also present in plastids and mitochondria (Dalton, 1995; Dalton *et al.*, 1993a; Iturbe-Ormaetxe *et al.*, 2001). Several results strongly suggest that the ASA–GSH cycle is critical for nodule functioning. A higher content of the ASA–GSH pool has been found in the nodule than in uninfected roots, and the presence of the cycle is correlated with nodule effectiveness (Dalton *et al.*, 1993b). Moreover, the activities of some of the enzymes and component concentrations of the ASA–GSH cycle have been found to decline with nodule age (Dalton *et al.*, 1986; Groten *et al.*, 2005). APX, the key enzyme of the cycle, is very abundant in nodule-infected cells (Dalton *et al.*, 1993a, 1998), where it has probably a function in H_2O_2 scavenging. The protective role of ASA, GSH, and the enzymes of the cycle may be crucial because nitrogenase, Lb, and other nodule proteins are particularly subjected to oxidation by ROS (Dalton *et al.*, 1986). In an *in vitro* model system consisting of bacteria or soybean nodule bacteroids, APX has been shown to enhance N_2 fixation up to 4.5-fold; in parallel, Lb was maintained in a favourable oxygenation state (Ross *et al.*, 1999). Moreover, the reduction of N_2 fixation in parallel with the reduction of activity of some of the ASA–GSH cycle enzymes occurs after nodule treatment with urea, linking antioxidant defence efficiency and N_2 fixation (Dalton *et al.*, 1991). APX is also present in nodule parenchyma cells (Dalton *et al.*, 1998), where it may have a role in the O_2 diffusion barrier which controls O_2 entry in the infected zone. Dalton *et al.* (1998) proposed that O_2 entrance is regulated by a respiratory activity adjustment with APX functioning as scavengers of the high H_2O_2 level generated by respiration. On the basis of this idea, aerobic respiration in the parenchyma may restrain the diffusion of O_2 into the nodule (Becana *et al.*, 2000). Finally, the ASA–GSH needs high amounts of reducing power for its functioning. Recent research has been focused on the role of NAD(P)H-regenerating mechanisms in the ROS scavenging network. Proteins that function to reduce NAD(P) are essential for plant tolerance to oxidative stress (Noctor, 2006). In this context, NADP-dependent isocitrate dehydrogenase has been proposed to play a key role in pea nodule antioxidant defence (Marino *et al.*, 2007b).

Antioxidant enzymes such as superoxide dismutase (SOD) and CAT are also abundant in nodules (Dalton *et al.*, 1998; Rubio *et al.*, 2004). The SOD family is composed of metalloenzymes that catalyze the dismutation of $O_2^{\bullet -}$ to H_2O_2 and O_2. The family is divided into three different classes characterized by different metal ions (CuZn, Mn and Fe) present in the active site. All of them were found in the nodule (Becana *et al.*, 1989; Puppo *et al.*, 1987). In higher plants, CuZnSODs have been localized in the cytosol, chloroplasts and apoplast; MnSODs in mitochondria and peroxisomes; and FeSODs in chloroplasts (Bowler *et al.*, 1994; Ogawa *et al.*, 1997; Sandalio *et al.*, 1987; Van Camp *et al.*, 1990). Interestingly, the vuFeSOD has been found in the nodule of cowpea (*Vigna unguiculata*), with an unusual extra-plastidial localization in the cytosol (Moran *et al.*, 2003). The vuFeSOD protein was found to be more abundant in mature and senescing nodules, suggesting that this enzyme is induced by the free Fe released by Lb degradation and is then involved in antioxidant protection during senescence (Moran *et al.*, 2003). A detailed microscopic analysis from Rubio and co-workers (2004) focused on the localization of SODs in pea and alfalfa indeterminate nodules showed that transcripts and proteins corresponding to cytosolic CuZnSOD were present in the nodule, most abundantly in the meristematic and invasion zones, interzone II-III, and distal parts of the N_2-fixing zone. Instead, the transcripts and proteins of MnSOD were mainly present in zone III, especially in the infected cells, suggesting different and specific functions for the two isoforms in the nodules. Moreover, a large amount of H_2O_2 was found in association with CuZnSOD, suggesting it as a nodule source of H_2O_2 localized in the cytosol adjacent to cell walls, in the IT, and in the cortex adjacent to zones I and II (Rubio *et al.*, 2004). The authors proposed that an NADPH oxidase could be the source of $O_2^{\bullet -}$ and be involved in cell wall growth and the progression of IT. One should also note that the micro-symbiont CuZn-SOD is expressed during infection, suggesting a bacterial response to the ROS produced by the plant (Ampe *et al.*, 2003).

CAT decomposes H_2O_2 to O_2 and water without using reducing power, an energy-efficient mechanism that functions in the peroxisome where it is present in high concentrations (Scandalios *et al.*, 1997). In nodules, CAT activity was detected in the peroxisome and its presence in mitochondria is unclear (Iturbe-Ormaetxe *et al.*, 2001; Puppo *et al.*, 1982). Early studies showed that effective nodules have greater CAT activities than ineffective ones in white clover and soybean (Francis and Alexander, 1972). Results from Dalton and co-workers (1998) on common bean and cowpea nodules reported that CAT activity was markedly higher in the central zone than in the periphery of both legume nodules. The authors indicated that this localization was consistent with the primary role of CAT in the elimination of H_2O_2, generated as a consequence of ureide synthesis in the central zone. The importance of CAT during symbiosis is supported by the finding that free-living *S. meliloti* have three CAT isoforms differentially expressed during nodule formation: two monofunctional (KatA and KatB) and one bifunctional (KatC). Whereas KatB is expressed during all the symbiotic steps, KatC is expressed in the IT and in the infection zone and KatA is restricted to the fixation zone of the mature nodule (Jamet *et al.*, 2003). Though single mutants did not show any marked phenotype, the double mutants katB/katC and katA/katC had impaired nodule formation and N_2-fixation activity (Jamet *et al.*, 2003). Recent results from Jamet *et al.* (2007) reported also that the katB overexpressing strain of *S. meliloti* RmkatB^{++} was characterized by a delayed symbiosis and an IT enlargement, suggesting that H_2O_2 is required for its optimal progression through the root-hair and plant cell layers.

Thioredoxins (Trxs) and glutaredoxins (Grxs) are ubiquitous disulfide reductases that regulate the redox status of target proteins. They are involved in the reduction of peroxiredoxins (Prxs) and glutathione peroxidases (Gpxs), two families of thiol-based peroxidases (Rouhier *et al.*, 2008a; Tripathi *et al.*, 2009). Contrary to their common denomination, in plants Gpxs catalyze the reduction of lipid peroxides and other organic peroxides using not glutathione as their name wrongly indicates, but rather thioredoxins as the preferred electron donors and they belong in fact to the family of thioredoxin peroxidases (Koh *et al.*, 2007; Navrot *et al.*, 2006; Rouhier and Jacquot, 2005). Expression of six Gpx genes has been found in *L. japonicus* nodules, with LjGpx3 being the most abundantly expressed (Ramos *et al.*, 2009). Immunogold labelling revealed the presence of Gpx in amyloplasts of nodules from *L. japonicus*, alfalfa and *S. rostrata* (Ramos *et al.*, 2009). Prxs are also an important component of the plant antioxidant network as they detoxify alkyl hydroperoxides, H_2O_2 and peroxynitrite. Cytosolic (Prx II A) and mitochondrial (Prx II F) Prxs have been detected in pea nodules and the abundance of the cytosolic isoform has been shown to be modulated by ASA (Groten *et al.*, 2006).

Trxs and Grxs are also implicated in the post-translational regulation of protein activity and in redox signalling transduction pathways via the regulation of disulfide bonds (Dietz, 2008; Meyer *et al.*, 2008; Rouhier *et al.*, 2008b). There are multiple indications that Trxs also play an important role in the establishment of the N_2-fixing symbiotic process. Recently, a detailed analysis of Trxs in *M. truncatula* has established the existence of two isoforms that do not belong to any of the previously described types (Alkhalfioui *et al.*, 2008). Interestingly, these novel isoforms seem to be specifically dedicated to the symbiotic interaction according to their expression pattern. In *Glycine max*, a Trx was found to be involved in nodulation, as plants deficient in this isoform have a reduced number of nodules (Lee *et al.*, 2005). The bacterial partner, a *Bradyrhizobium japonicum* mutant deficient in *tlpA* (a gene encoding a protein having a Trx domain) was notably defective in the development of a N_2-fixing endosymbiosis and exhibited a strong decrease in oxidase activity compared to wild-type bacteria (Loferer *et al.*, 1993). Similarly, the Trx of *S. meliloti* CE52G is required for melanin production and SNF (Castro-Sowinski *et al.*, 2007).

Transcriptomic analyses further showed the importance of antioxidant systems in symbiosis. Colebatch and colleagues (2004) isolated transcripts particularly abundant in *L. japonicus* nodule in comparison to the root. Eight hundred and sixty genes were more expressed in *L. japonicus* nodule than in roots, and several metabolic pathways appeared to be co-ordinately up-regulated in the nodules, including glycolysis, CO_2 fixation, amino acid biosynthesis, and purine, haem, and redox metabolisms. APX, GPX, glucose-6-phosphate dehydrogenase, Prx, chloroplastic Trx of the m type and guanosine 5′-diphosphate-mannose pyrophosphorylase, which is involved in ASA biosynthesis (Smirnoff *et al.*, 2001), were found to be more expressed in nodules than in roots.

Taken together, these data emphasize the role of the antioxidant defence and redox state regulation in the functioning of the N_2-fixation metabolism during the symbiosis between legumes and *Rhizobia*.

B. ANTIOXIDANT SYSTEMS IN THE ESTABLISHMENT OF SYMBIOSIS AND PLANT DEFENCE REGULATION

For the establishment of a successful symbiotic association, plant defence responses need to be properly regulated (Mithofer, 2002). During the first phase of nodulation, the regulation of *Rhizobia* infection by the host shows similarities with incompatible pathogenic interactions (see Section II-B).

Large-scale expression profile studies are very helpful to evaluate the induction of defence related genes in the early events of nodulation. An expression profiling analysis on *M. truncatula* roots inoculated by *S. meliloti* identified a large collection of downregulated genes encoding pathogenesis and defence-related proteins (El Yahyaoui *et al.*, 2004). In spite of this, several genes involved in the general stress response were strongly up-regulated, suggesting the necessity to balance the need of *Rhizobium* infection and the control of its extension (El Yahyaoui *et al.*, 2004). Lohar and co-workers (2006) analyzed the regulation of transcripts at early nodulation events in *M. truncatula*, showing that the root response to *S. meliloti* was characterized by an induction of defence and disease response genes, 1 h post-inoculation, followed by their suppression coincident with IT penetration into root hairs (Lohar *et al.*, 2006). Similar results showing defence response genes with a transient increase in expression during the early symbiotic stage and then a subsequent decline were reported also in response to *Glomus versiforme* arbuscular mycorrhiza root colonization of *M. truncatula* (Liu *et al.*, 2003).

Plants with a reduced level of (h)GSH are partially inhibited in the formation of nodules showing the importance of these molecules during the first step of symbiosis between *M. truncatula* and *S. meliloti* (Frendo *et al.*, 2005). A transcript profile analysis of these (h)GSH-depleted plants inoculated with *S. meliloti* revealed a significant up-regulation of defence-related genes, indicating that the early plant defence reaction occurring after *Rhizobium* inoculation is affected. Moreover, amongst these genes, up-regulation of salicylic acid (SA)-regulated genes was observed, suggesting a possible SA metabolism modification under these conditions (Pucciariello *et al.*, in press). Therefore, these results suggest the existence of a link between redox state regulation, SA, and the efficiency of the nodulation process.

SA has been extensively described as a hormone involved in plant defence against pathogens (Durrant and Dong, 2004; Loake and Grant, 2007). Its involvement in controlling the symbiotic process has been suggested in reports showing that exogenous SA treatments were capable of inhibiting indeterminate-type nodule formation (Martinez-Abarca *et al.*, 1998; van Spronsen *et al.*, 2003). Furthermore, the inoculation of alfalfa plants with either incompatible *Rhizobium* or a *Rhizobium* mutant blocked in NFs' synthesis led to an accumulation of SA in roots, in contrast to plants inoculated with a wild-type compatible *S. meliloti* strain (Martinez-Abarca *et al.*, 1998; Stacey *et al.*, 2006; van Spronsen *et al.*, 2003). Finally, expression of *NahG* (salicylate hydroxylase), a bacterial gene degrading SA, in both *M. truncatula* and *L. japonicus* plants, led to enhanced nodulation and infection showing the important role for SA-mediated plant pathways in controlling nodule formation in both determinate and indeterminate nodule-forming hosts (Stacey *et al.*, 2006). In this context, NONEXPRESSOR OF PATHOGENESIS-RELATED GENES1 (NPR1), a master regulator of SA-mediated defence genes, has been shown to be regulated by the redox state (Dong, 2004). NPR1 is sequestered in the cytoplasm as an oligomer through intermolecular disulfide bonds. *S*-nitrosylation of NPR1 by *S*-nitrosoglutathione (GSNO) facilitates its oligomerization, which maintains protein homeostasis upon SA induction. Conversely, the SA-induced NPR1 oligomer-to-monomer transition is governed by Trxs (Tada *et al.*, 2008). Thus, the regulation of NPR1, through the opposite actions of GSNO and Trx, may be a link between redox state, defence reaction, and the regulation of the legume–*Rhizobium* interactions.

SA, ethylene and jasmonic acid are all hormones involved in plant response to biotic stresses and important players in the symbiotic process (Penmetsa and Cook, 1997; Penmetsa *et al.*, 2008). The interplay between these different hormones is well established during plant–pathogen interactions (Lopez *et al.*, 2008). The modulation of the redox state is a key regulator of this interplay as it is involved in many regulatory pathways involved in biotic and abiotic stress responses (Fujita *et al.*, 2006; Wiermer *et al.*, 2005). However, the interactions between the redox state and these different hormones during the symbiotic process still need to be clarified.

C. ANTIOXIDANT SYSTEMS DURING NODULE SENESCENCE

The N_2-fixing capacity of the nodule starts to decline 3–5 weeks after *Rhizobium* inoculation. A hallmark of nodule senescence is the activation of proteolytic enzymes with the final degradation of bacteroids and nodule cells (Pladys and Vance, 1993). Many results indicate a shift in the redox homeostasis of the nodule from establishment to senescence. The amounts of antioxidant enzymes such as FeSOD increase in senescent nodules in parallel with Lb degradation (Moran *et al.*, 2003). ASA and GSH levels decrease very early during nodule ageing, in parallel to N_2 fixation (Evans *et al.*, 1999; Groten *et al.*, 2005). Senescence causes a 50% decrease in concentration of hGSH in the soybean and pea nodule content and 82% decrease in GSH in pea nodules (Evans *et al.*, 1999; Groten *et al.*, 2005; Matamoros *et al.*, 2003). Puppo *et al.* (2005) suggested that ASA and GSH have a central role in nodule senescence. These authors have put forward two hypotheses to explain how they could be involved. In the simplest one, the progressive decline in GSH and ASA content should lead to an increased level of oxidative stress, triggering nodule senescence. In the second hypothesis, ASA and GSH declining content during ageing is accompanied by a similar decrease of ROS and maintenance of the redox balance. The decline of ASA may contribute to nodule ageing both inhibiting mitosis and meristem activity and increasing ABA synthesis, with the mobilization of proteolytic activities (Puppo *et al.*, 2005). In this context, a transcript profiling analysis of senescent *M. truncatula* nodules demonstrated the activation of defence and stress-response transcripts and showed also a high degree of overlap with leaf senescence in terms of the gene families involved (Van de Velde *et al.*, 2006). In spite of this, De Michele and co-workers (2009), crossing their cDNA-AFLP results on senescent leaves of *M. truncatula* with the nodule data from Van de Velde *et al.* (2006), showed a higher degree of difference than previously supposed. However, these authors have suggested that since the two analyses were conducted with two different *M. truncatula* lines, a part of the transcriptional discrepancy could be due to the different genotypes. Groten and co-workers (2005), analyzing *Pisum sativum* inoculated with *Rhizobium leguminosarum*, did not find any evidence for oxidative stress or programmed cell death induced under senescence, even though ASA and GSH levels decreased. Recently, *S. meliloti* cells were engineered to overexpress *Anabaena variabilis* flavodoxin, a protein that is involved in the response to oxidative stress (Redondo *et al.*, 2008). In *M. sativa* plants nodulated by this *S. meliloti* strain, the decline of nitrogenase activity and the nodule structural and ultrastructural alterations that are associated with nodule senescence were significantly delayed compared to plants nodulated with wild-type *Rhizobia*. In parallel, substantial changes in the nodule antioxidant metabolism and significant lower lipid peroxidation were observed in the flavodoxin-expressing nodules. The authors suggested that the observed amelioration of the oxidative balance and the delay in nodule senescence was most likely due to a role of the protein in ROS detoxification (Redondo *et al.*, 2008).

IV. REDOX CONTROL OF NFS UNDER ENVIRONMENTAL STRESSES

Higher plants survive in a changing environment by inducing modifications in their metabolism and development. The association between legumes and *Rhizobia* is particularly sensitive to inadequate environmental conditions. In most types of stress, the breakdown of the association is characterized by the reduction of the N_2-fixing capacity associated to the senescence of the nodule. SNF is particularly sensitive to stresses which alter photosynthesis (defoliation or continuous darkness), water content (drought or salt stress), and nitrogen metabolism (nitrate fertilization). Taking into account (i) the numerous processes able to generate ROS in the nodule (see Section II) and (ii) the high susceptibility of the nitrogenase protein complex to O_2 and ROS, it appears that any stress-induced modification of O_2 and/or ROS concentration may have a dramatic effect on nodule functioning.

One of the most general effects of stresses associated with the diminution of SNF is the modification of the nodule O_2 diffusion barrier. Indeed, modification of the nodule O_2 permeability associated to the change of nodule respiration is observed early during stress conditions. During nitrate fertilization, the O_2 diffusion resistance is increased and respiration is reduced in the pea nodule (Escuredo *et al.*, 1996). A similar increase in the O_2 diffusion barrier resistance has been observed in bean, *Vigna radiata* and soybean plants (Minchin *et al.*, 1989; Vessey *et al.*, 1988). Similarly, other stresses such as drought, salt stress, dark treatment, or cold stress also modify the nodule internal O_2 concentration by reducing the permeability of the O_2 diffusion barrier or decreasing the respiration rate in bean, soybean, alfalfa and pea plants (Matamoros *et al.*, 1999a; Naya *et al.*, 2007; Serraj and Sinclair, 1996; Serraj *et al.*, 1994; van Heerden *et al.*, 2008). In contrast to these results, *M. truncatula* seems to respond differently to salt stress with a higher nodule conductance in sensitive lines (Aydi *et al.*, 2004). Taken together, these results show that under stress conditions O_2 availability is modified in the nodule. Thus, the production of ROS also could be altered.

A second element which plays a crucial role in the modification of the redox balance through a higher ROS production is the presence of a high Lb concentration. This Fe-containing protein is present in a millimolar concentration range in nodules (Appleby, 1984). During stress conditions, the quantity of Lb decreases and the amount of catalytic iron increases, suggesting a degradation of the protein and the release of the iron fraction (Gogorcena *et al.*, 1995, 1997; Moran *et al.*, 2003). The augmentation of the free iron concentration may be involved in the production of ROS through the Fenton reaction. The oxidation of Lb from ferrous Lb^{2+} to ferric Lb^{3+} has been detected in soybean nodules during dark and flooding stresses (Lee *et al.*, 1995). Moreover, exposure of young and mature nodules to oxidant stress results in appearance of absorptions similar to those from senescent nodules in the electron paramagnetic resonance spectra (Mathieu *et al.*, 1998). These data should be linked to the degradation effect detected for Lb radicals on the peribacteroid membrane (Moreau *et al.*, 1996) and the increased iron-dependent free radical production during nodule senescence (Becana and Klucas, 1992). The higher expression of ferritin, the main intracellular iron storage protein which keeps free Fe in a safe and bioavailable form, observed in the nodule during natural senescence or stress conditions also points to the modification of the iron metabolism under these conditions (Clement *et al.*, 2008; Matamoros *et al.*, 1999b; Strozycki *et al.*, 2007).

Taken together, these results show that Lb degradation is potentially involved in the higher production of radicals during stress conditions.

Alteration of the O_2 metabolism in the nodule and the change of the Lb/Fe pools observed during stress strongly suggest a higher production of ROS, protein and lipid radicals altering the cell redox balance. In this context, the antioxidant compounds, GSH/hGSH, ASA and NADPH, involved in the maintenance of the redox state should be present in a more oxidized state under environmental stresses. Indeed, multiple reports showed that the nodule stress-induced response is associated with a decrease of the antioxidant molecules and a higher content of their oxidized forms (Escuredo *et al.*, 1996; Gogorcena *et al.*, 1997; Marino *et al.*, 2007a; Naya *et al.*, 2007). Moreover, antioxidant enzyme activity has been shown to decrease during stress conditions (Gogorcena *et al.*, 1995; Jebara *et al.*, 2005; Rubio *et al.*, 2002; Tejera Garcia *et al.*, 2007). Taken together, these results indicate that the redox balance may shift towards a more oxidized state. In contrast to these results showing a general down-regulation of the antioxidant defence of the nodule under stress conditions, other articles have suggested that the nodule antioxidant defence is stable or up-regulated under stress conditions (Loscos *et al.*, 2008; Naya *et al.*, 2007). This discrepancy may be linked to the analyzed parameter (antioxidant compound, enzymatic activity, or gene expression analysis), the differential intensity of the stress, the resistance of the plant to treatments, or the growth conditions which may modify the impact of the stress for the nodule.

A modification of the redox metabolism in the stressed nodule has thus been clearly shown. However, it is not fully clear whether the modification of the redox metabolism plays a role in the adaptation of the nodule to stress as it is observed in the leaves (Gechev *et al.*, 2006; Noctor, 2006) or a consequence of the cellular imbalance leading to the degradation of the nodule and to the plant/bacteroid cell death as observed during natural senescence (Rubio *et al.*, 2004; Santos *et al.*, 2001; Van de Velde *et al.*, 2006). To test whether oxidative stress is a result of or a prerequisite for the modulation of SNF, Marino *et al.* (2006) have applied on pea plants a direct oxidative stress through the application of paraquat (PQ) which generates $O_2^{\bullet -}$. This endogenous production of ROS, which induces the alteration of the redox balance of GSH and ASA, provokes the decrease in SNF. This decrease of SNF is concomitant with a Lb content diminution as observed under environmental stress conditions. Moreover, this diminution of SNF is preceded by a decrease in sucrose synthase (SS) activity under moderate oxidative stress. This decrease of the nodule carbon metabolism is a feature of the nodule under environmental stress conditions in pea, soybean, and common bean plants (Galvez *et al.*, 2005; Gogorcena *et al.*, 1997; Gordon *et al.*, 1997). The authors concluded that the oxidative stress may be part of the signal transduction pathway involved in SNF regulation under environmental constraints. Interestingly, the diminution of SS activity is correlated to the down-regulation of SS gene expression, and *in vitro* experiments showed that oxidizing agents were able to inhibit SS activity and that this inhibition was completely reversed by the addition of the strong reducing agent dithiothreitol (Marino *et al.*, 2008). These results are consistent with a regulation model of nodule SS exerted by the cellular redox state at both the transcriptional and the post-translational levels, suggesting that SS may be strongly regulated by the redox state under stress conditions. In contrast, a recent article showed that exposure of alfalfa plants to a moderate drought caused the inhibition of SNF but had no effect on SS activity, suggesting that a down-regulation of SS activity under stress conditions may not be a general nodule feature (Naya *et al.*, 2007).

Another potential involvement of redox regulation in the response of the nodule to environmental stresses implicates the action of abscissic acid (ABA), the concentration of which is strongly increased in soybean nodules under drought stress (Clement *et al.*, 2008). The five-fold accumulation of ABA observed in stressed nodules compared to stressed roots 5 days after stress application suggests that ABA synthesis is much higher in nodules than in roots under drought conditions. The potential role of ABA in the nodule response to environmental stresses has been shown by exogenous treatment (Gonzalez *et al.*, 2001). ABA treatment affected the SNF, with a reduction of 70% within 5 days and Lb content declined in parallel with the SNF. However, SS activity was not affected by ABA treatment showing that ABA may not directly regulate SS activity. The authors suggested that ABA is involved in a Lb/O_2-related control of the SNF. The importance of ABA in the nodulation process has been recently genetically demonstrated (Ding *et al.*, 2008). These authors characterized a novel locus of *M. truncatula*, SENSITIVITY TO ABA, which dictates the sensitivity of the plant to ABA and impacts the regulation of nodulation. Moreover, the genetic inhibition of ABA signalling through the use of a dominant-negative allele of ABSCISIC ACID INSENSITIVE1 (*ABI1*) led to a hypernodulation phenotype. *abi1* and *abi2* are two negative regulators involved in the signal transduction pathways of ABA (Grill and Himmelbach, 1998; Naya *et al.*, 2007). ABI2 and in a lesser extent ABI1 have been shown to physically interact with the glutathione peroxidase 3 (AtGPX3) in *Arabidopsis* (Miao *et al.*, 2006). The reduced form of ABI2 was converted to the oxidized form by the addition of oxidized AtGPX3 *in vitro*, which might mediate ABA and oxidative signalling. Thus, ATGPX3 might specifically relay the H_2O_2 signal as an oxidative signal transducer in ABA and drought stress signalling. In this context, the production of ROS may not only induce ABA production in the nodule but may also control its level of response through different check points of the transduction pathway.

V. CONCLUSIONS AND PERSPECTIVES

This review highlights that ROS and the antioxidant defence are involved in numerous aspects of the NFS, ranging from the establishment of the interaction to the functioning of the nodule under unfavourable environmental conditions. Multiple reports have shown that the bacterial antioxidant defence plays an important role in the establishment and the functioning of the SNF. Conversely, whereas a lot of literature describes the redox modifications occurring during the symbiosis process, very few reports have provided evidence concerning the importance of the plant redox metabolism in the regulation of the symbiotic process (Table I). Plants with modified ROS production and antioxidant content will be very useful tools to address this question. Indeed, tilling for single ethyl methanesulfonate (EMS) mutations, retrotransposon, T-DNA tagging, and RNA interference (RNAi) are reverse genetic approaches which can be performed in model plants like *M. truncatula* or *L. japonicus* (Sato and Tabata, 2006; Tadege *et al.*, 2005). Moreover, promoters controlling timing and tissue specificity may be used to allow a specific modulation of gene expression through RNAi.

It appears that the regulation of bacterial infection during the nodulation process may involve a modification of the redox state which balances the plant defence reaction. However, the developmental processes which lead to the cell dedifferentiation, the nodule meristem formation, and the cell differentiation leading to the specific SNF metabolism are still an open field in which redox regulation may have an important function. Moreover, the specific metabolism involved in the SNF as hypoxic conditions, high respiration, and plant–*Rhizobia* interaction may implicate specific regulation processes. In this framework, the characterization of the cross-talk between the plant and the bacteria will be very interesting to understand how this could be implicated in the global regulation of the nodule functioning. Besides, the bacteroid response to environmental conditions has been studied in much less detail than the response of the plant partner. Nevertheless, the fine characterization of the multiple pathways involving the redox state in the regulation of the symbiotic process will be important to decipher the interaction between this regulatory process and other master players such as

TABLE I

Description of the Different Antioxidant Systems in Nodules

Antioxidant	Activity	Involvement in legume–*Rhizobia* symbiosis demonstrated by genetical approaches
ASA	Substrate of APX in the ASA–GSH cycle	
GSH	Substrate of Gpx	Initiation and maintenance of cell division in nodule meristem (Frendo *et al.*, 2005)
	Substrate of GST	
	Reducing power for DHAR, and substrate of GR in the ASA–GSH cycle	
	Detoxification of xenobiotics and heavy metals	Bacteria protection from oxidative stress and delayed senescence (Harrison *et al.*, 2005; Muglia *et al.*, 2008)*
	Post-translational protein regulation	
ASA–GSH cycle enzymes	Detoxification of H_2O_2	
SODs	Dismutation of $O_2^{\bullet -}$ to H_2O_2	
CATs	Detoxification of H_2O_2	Essential for correct nodule formation and SNF activity (Jamet *et al.*, 2003)*
		Infection thread progression (Jamet *et al.*, 2007)*
Trxs	Post-translational protein regulation	Nodule number control (Lee *et al.*, 2005)
		Essential for a correct symbiosis establishment and SNF performance (Castro-Sowinski *et al.*, 2007; Loferer *et al.*, 1993)*
Grxs	Post-translational protein regulation	
GSTs	Detoxification of xenobiotics and reduction of oxidized molecules	Essential antioxidant defense for SNF functioning (Dalton *et al.*, 2009)
Prxs	Detoxification of H_2O_2, peroxynitrate, and alkyl hydroperoxides	
Gpxs	Reduction of lipid peroxides and other organic peroxides	

ASA, ascorbate; GSH, glutathione; ASA–GSH cycle, ascorbate–glutathione cycle; DHAR, dehydroascorbate reductase; GR, glutathione reductase; SODs, superoxide dismutases; CAT, catalases; Trxs, thioredoxins; Grxs, glutaredoxins; GSTs, glutathione-*S*-transferases; Prxs, peroxiredoxins; Gpxs, glutathione peroxidases.

*Indicates results obtained with genetically modified *Rhizobia*.

hormones. The finding of legume orthologs for regulatory genes defined in *Arabidopsis* will be useful to carry out this task.

The redox regulation of metabolism is not only performed through a transcriptional regulation but also via post-translational modifications of proteins. The glutathionylation, the oxidative/nitrosative modifications of proteins, and the redox modification of the proteins by the redoxins are important regulatory processes which have not been studied during symbiotic interactions. A global analysis of redox-modified proteins will be important to pinpoint the implication of the redox regulation in the N_2-fixing symbiotic process. In this context, adaptation of proteomic approaches to identify redox-modulated proteins will be very useful (Dietz, 2008; Gao *et al.*, 2009; Wait *et al.*, 2005).

Legume–*Rhizobia* symbiosis involves plant–microorganism interactions, plant development processes, and a specific anaerobic metabolism which makes it an interesting working model which can provide information of general interest to plant biologists. From a more environmental point of view, since legumes are unique as they do not need N_2-fertilizers and display a high protein content, they have a crucial role to play in the development of more sustainable agriculture. As such, the comprehension of the regulatory processes leading to an efficient SNF will be crucial in the future.

ACKNOWLEDGMENTS

We gratefully acknowledge Julie Hopkins and Terence Neil Ledger for a critical reading of the manuscript.

REFERENCES

Agius, F., Gonzalez-Lamothe, R., Caballero, J. L., Munoz-Blanco, J., Botella, M. A. and Valpuesta, V. (2003). Engineering increased vitamin C levels in plants by overexpression of a D-galacturonic acid reductase. *Nature Biotechnology* **21**, 177–181.

Albus, U., Baier, R., Holst, O., Pühler, A. and Niehaus, K. (2001). Suppression of an elicitor-induced oxidative burst reaction in *Medicago sativa* cell cultures by *Sinorhizobium meliloti* lipopolysaccharides. *New Phytologist* **151**, 597–606.

Alesandrini, F., Mathis, R., Van de Sype, G., Hérouart, D. and Puppo, A. (2003). Possible roles for a cysteine protease and hydrogen peroxide in soybean nodule development and senescence. *New Phytologist* **158**, 131–138.

Alkhalfioui, F., Renard, M., Frendo, P., Keichinger, C., Meyer, Y., Gelhaye, E., Hirasawa, M., Knaff, D. B., Ritzenthaler, C. and Montrichard, F. (2008). A novel type of thioredoxin dedicated to symbiosis in legumes. *Plant Physiology* **148**, 424–435.

Ampe, F., Kiss, E., Sabourdy, F. and Batut, J. (2003). Transcriptome analysis of *Sinorhizobium meliloti* during symbiosis. *Genome Biology* **4**, R15.

Apel, K. and Hirt, H. (2004). Reactive oxygen species: Metabolism, oxidative stress, and signal transduction. *Annual Review of Plant Biology* **55**, 373–399.

Appleby, C. A. (1984). Leghemoglobin and rhizobium respiration. *Annual Review of Plant Physiology and Plant Molecular Biology* **35**, 443–478.

Asada, K. (1999). The water–water cycle in chloroplasts: Scavenging of active oxygens and dissipation of excess photons. *Annual Review of Plant Physiology and Plant Molecular Biology* **50**, 601–639.

Aviram, I., Wittenberg, A. and Wittenberg, J. B. (1978). The reaction of ferrous leghemoglobin with hydrogen peroxide to form leghemoglobin(IV). *The Journal of Biological Chemistry* **253**, 5685–5689.

Aydi, S., Drevon, J. J. and Abdelly, C. (2004). Effect of salinity on root-nodule conductance to the oxygen diffusion in the *Medicago truncatula–Sinorhizobium meliloti* symbiosis. *Plant Physiology and Biochemistry* **42**, 833–840.

Baron, C. and Zambryski, P. C. (1995). The plant response in pathogenesis, symbiosis, and wounding: Variations on a common theme? *Annual Review of Genetics* **29**, 107–129.

Bashor, C. J. and Dalton, D. A. (1999). Effects of exogenous application and stem infusion of ascorbate on soybean (*Glycine max*) root nodules. *New Phytologist* **142**, 19–26.

Becana, M. and Klucas, R. V. (1990). Enzymatic and nonenzymatic mechanisms for ferric leghemoglobin reduction in legume root nodules. *Proceedings of the National Academy of Sciences of the United States of America* **87**, 7295–7299.

Becana, M. and Klucas, R. V. (1992). Transition metals in legume root nodules. Iron dependant free radical production increases during nodule senescence. *Proceedings of the National Academy of Sciences of the United States of America* **87**, 7295–7299.

Becana, M., Paris, F. J., Sandalio, L. M. and del Río, L. A. (1989). Isoenzymes of superoxide dismutase in nodules of *Phaseolus vulgaris* L., *Pisum sativum* L., and *Vigna unguiculata* (L.) Walp. *Plant Physiology* **90**, 1286–1292.

Becana, M., Dalton, D. A., Moran, J. F., Iturbe-Ormaetxe, I., Matamoros, M. A. and Rubio, M. C. (2000). Reactive oxygen species and antioxidants in legume nodules. *Physiologia Plantarum* **109**, 372–381.

Bowler, C., Van Camp, W., Van Montagu, M. and Inzé, D. (1994). Superoxide dismutase in plants. *Critical Review in Plant Science* **13**, 199–218.

Bueno, P., Soto, M. J., Rodriguez-Rosales, M. P., Sanjuan, J., Olivares, J. and Donaire, J. P. (2001). Time-course of lipoxygenase, antioxidant enzyme activities and H_2O_2 accumulation during early stages of *Rhizobium* legume symbiosis. *New Phytologist* **152**, 91–96.

Campbell, G. R., Reuhs, B. L. and Walker, G. C. (2002). Chronic intracellular infection of alfalfa nodules by *Sinorhizobium meliloti* requires correct lipopolysaccharide core. *Proceedings of the National Academy of Sciences of the United States of America* **99**, 3938–3943.

Cardenas, L., Martinez, A., Sanchez, F. and Quinto, C. (2008). Fast, transient and specific intracellular ROS changes in living root hair cells responding to Nod factors (NFs). *The Plant Journal* **27**, 802–813.

Castro-Sowinski, S., Matan, O., Bonafede, P. and Okon, Y. (2007). A thioredoxin of *Sinorhizobium meliloti* CE52G is required for melanin production and symbiotic nitrogen fixation. *Molecular Plant Microbe Interactions* **20**, 986–993.

Clement, M., Lambert, A., Herouart, D. and Boncompagni, E. (2008). Identification of new up-regulated genes under drought stress in soybean nodules. *Gene* **426**, 15–22.

Cobbett, C. and Goldsbrough, P. (2002). Phytochelatins and metallothioneins: Roles in heavy metal detoxification and homeostasis. *Annual Review of Plant Biology* **53**, 159–182.

Colebatch, G., Desbrosses, G., Ott, T., Krusell, L., Montanari, O., Kloska, S., Kopka, J. and Udvardi, M. K. (2004). Global changes in transcription orchestrate metabolic differentiation during symbiotic nitrogen fixation in *Lotus japonicus*. *The Plant Journal* **39**, 487–512.

Conklin, P. L. and Barth, C. (2004). Ascorbic acid, a familiar small molecule intertwined in the response of plants to ozone, pathogens and the onset of senescence. *Plant, Cell and Environment* **27**, 959–970.

Cook, D., Dreyer, D., Bonnet, D., Howell, M., Nony, E. and Vandenbosh, K. (1995). Transient induction of a peroxidase gene in *Medicago truncatula* precedes infection by *Rhizobium meliloti*. *The Plant Cell* **7**, 43–55.

Dalton, D. A. (1995). Antioxidant defenses of plants and fungi. *In* "Oxidative Stress and Antioxidant Defenses in Biology" (Chapman and Hall, ed.), pp. 298–355. Chapman and Hall, New York, USA.

Dalton, D. A., Russell, S. A., Hanus, F. J., Pascoe, G. A. and Evan, H. J. (1986). Enzymatic reactions of ascorbate and glutathione that prevent peroxide damage in soybean root nodules. *Proceedings of the National Academy of Sciences of the United States of America* **83**, 3811–3815.

Dalton, D. A., Post, C. J. and Langeberg, L. (1991). Effects of ambient oxygen and of fixed nitrogen on concentrations of glutathione, ascorbate, and associated enzymes in soybean root nodules. *Plant Physiology* **96**, 812–818.

Dalton, D. A., Langeberg, L. and Robbins, M. (1992). Purification and characterization of monodehydroascorbate reductase from soybean root nodules. *Archives of Biochemistry and Biophysics* **292**, 281–286.

Dalton, D. A., Bard, L. M., Langeberg, L., Taugher, C. Y., Anyan, W. R., Vance, C. P. and Sarath, G. (1993a). Subcellular localization of oxygen defense enzymes in soybean (*Glycine max* (L.) Meer) root nodules. *Plant Physiology* **102**, 481–489.

Dalton, D. A., Langeberg, L. and Treneman, N. C. (1993b). Correlations between the ascorbate–glutathione pathway and effectiveness in legume root nodules. *Physiologia Plantarum* **84**, 365–370.

Dalton, D. A., Joyner, S. L., Becana, M., Iturbe-Ormaetxe, I. and Chatfield, J. M. (1998). Antioxidant defenses in the peripheral cell layers of legume root nodules. *Plant Physiology* **116**, 37–43.

Dalton, D. A., Boniface, C., Turner, Z., Lindahl, A., Kim, H. J., Jelinek, L., Govindarajulu, M., Finger, R. E. and Taylor, C. G. (2009). Physiological roles of glutathione *S*-transferases in soybean root nodules. *Plant Physiology* **11**, 521–530.

De Michele, R., Formentin, E., Todesco, M., Toppo, S., Carimi, F., Zottini, M., Barizza, E., Ferrarini, A., Delledonne, M., Fontana, P. and Lo Schiavo, F. (2009). Transcriptome analysis of *Medicago truncatula* leaf senescence: Similarities and differences in metabolic and transcriptional regulations as compared with *Arabidopsis*, nodule senescence and nitric oxide signalling. *New Phytologist* **181**, 563–575.

Den Herder, J., Lievens, S., Rombauts, S., Holsters, M. and Goormachtig, S. (2007). A symbiotic plant peroxidase involved in bacterial invasion of the tropical legume *Sesbania rostrata*. *Plant Physiology* **144**, 717–727.

D'Haeze, W., De Rycke, R., Mathis, R., Goormachtig, S., Pagnotta, S., Verplancke, C., Capoen, W. and Holsters, M. (2003). Reactive oxygen species and ethylene play a positive role in lateral root base nodulation of a semiaquatic legume. *Proceedings of the National Academy of Sciences of the United States of America* **100**, 11789–11794.

Dietz, K.-J. (2008). Redox signal integration: From stimulus to networks and genes. *Physiologia Plantarum* **133**, 459–468.

Ding, Y., Kalo, P., Yendrek, C., Sun, J., Liang, Y., Marsh, J. F., Harris, J. M. and Oldroyd, G. E. (2008). Abscissic acid coordinates nod factor and cytokinin signaling during the regulation of nodulation in *Medicago truncatula. Plant Cell* **20**, 2681–2695.

Dixon, D. P., Lapthorn, A. and Edwards, R. (2002). Plant glutathione transferases. *Genome Biology* **3**, Review S3004.

Dixon, D. P., Skipsey, M., Grundy, N. M. and Edwards, R. (2005). Stress-induced protein *S*-glutathionylation in *Arabidopsis. Plant Physiology* **138**, 2233–2244.

Dong, X. (2004). NPR1, all things considered. *Current Opinion in Plant Biology* **7**, 547–552.

Durrant, W. E. and Dong, X. (2004). Systemic acquired resistance. *Annual Review of Phytopathology* **42**, 185–209.

El Yahyaoui, F., Kuster, H., Ben Amor, B., Hohnjec, N., Puhler, A., Becker, A., Gouzy, J., Vernie, T., Gough, C., Niebel, A., Godiard, L. and Gamas, P. (2004). Expression profiling in *Medicago truncatula* identifies more than 750 genes differentially expressed during nodulation, including many potential regulators of the symbiotic program. *Plant Physiology* **136**, 3159–3176.

Escuredo, P. R., Minchin, F. R., Gogorcena, Y., Iturbe-Ormaetxe, I., Klucas, R. V. and Becana, M. (1996). Involvement of activated oxygen in nitrate-induced senescence of pea root nodules. *Plant Physiology* **110**, 1187–1195.

Evans, P. J., Gallesi, D., Mathieu, C., Hernandez, M. J., de Felipe, M., Halliwell, B. and Puppo, A. (1999). Oxidative stress occurs during soybean nodule senescence. *Planta* **208**, 73–79.

Foreman, J., Demidchik, V., Bothwell, J. H., Mylona, P., Miedema, H., Torres, M. A., Linstead, P., Costa, S., Brownlee, C., Jones, J. D., Davies, J. M. and Dolan, L. (2003). Reactive oxygen species produced by NADPH oxidase regulate plant cell growth. *Nature* **422**, 442–446.

Fournier, J., Timmers, A. C., Sieberer, B. J., Jauneau, A., Chabaud, M. and Barker, D. G. (2008). Mechanism of infection thread elongation in root hairs of *Medicago truncatula* and dynamic interplay with associated rhizobial colonization. *Plant Physiology* **148**, 1985–1995.

Francis, A. J. and Alexander, M. (1972). Catalase activity and nitrogen fixation in legume root nodules. *Canadian Journal of Microbiology* **18**, 861–864.

Franssen, H. J., Vijn, I., Yang, W. C. and Bisseling, T. (1992). Developmental aspects of the *Rhizobium*–legume symbiosis. *Plant Molecular Biology* **19**, 89–107.

Frendo, P., Gallesi, D., Turnbull, R., Van de Sype, G., Hérouart, D. and Puppo, A. (1999). Localisation of glutathione and homoglutathione in *Medicago truncatula* is correlated to a differential expression of genes involved in their synthesis. *The Plant Journal* **17**, 215–219.

Frendo, P., Jimenez, M. J., Mathieu, C., Duret, L., Gallesi, D., Van de Sype, G., Herouart, D. and Puppo, A. (2001). A *Medicago truncatula* homoglutathione synthetase is derived from glutathione synthetase by gene duplication. *Plant Physiology* **126**, 1706–1715.

Frendo, P., Harrison, J., Norman, C., Hernandez Jimenez, M. J., Van de Sype, G., Gilabert, A. and Puppo, A. (2005). Glutathione and homoglutathione play a critical role in the nodulation process of *Medicago truncatula. Molecular Plant Microbe Interactions* **18**, 254–259.

Fridovich, I. (1986). Biological effects of the superoxide radical. *Archives in Biochemistry and Biophysics* **247**, 1–11.

Fujita, M., Fujita, Y., Noutoshi, Y., Takahashi, F., Narusaka, Y., Yamaguchi-Shinozaki, K. and Shinozaki, K. (2006). Crosstalk between abiotic and biotic stress responses: A current view from the points of convergence in the stress signaling networks. *Current Opinion in Plant Biology* **9**, 436–442.

Gage, D. J. (2002). Analysis of infection thread development using Gfp- and DsRed-expressing *Sinorhizobium meliloti*. *Journal of Bacteriology* **184**, 7042–7046.

Galvez, L., Gonzalez, E. M. and Arrese-Igor, C. (2005). Evidence for carbon flux shortage and strong carbon/nitrogen interactions in pea nodules at early stages of water stress. *Journal of Experimental Botany* **56**, 2551–2561.

Gamas, P., de Billy, F. and Truchet, G. (1998). Symbiosis-specific expression of two *Medicago truncatula* nodulin genes, MtN1 and MtN13, encoding products homologous to plant defense proteins. *Molecular Plant Microbe Interactions* **11**, 393–403.

Gao, X.-H., Bedhomme, M., Veyel, D., Zaffagnini, M. and Lemaire, S. D. (2009). Methods for analysis of protein glutathionylation and their application to photosynthetic organisms. *Molecular Plant* **2**, 218–235.

Gechev, T. S., Van Breusegem, F., Stone, J. M., Denev, I. and Laloi, C. (2006). Reactive oxygen species as signals that modulate plant stress responses and programmed cell death. *Bioessays* **28**, 1091–1101.

Gogorcena, Y., Iturbe-Ormaetxe, I., Escuredo, P. R. and Becana, M. (1995). Antioxidant defenses against activated oxygen in pea nodules subjected to water stress. *Plant Physiology* **108**, 753–759.

Gogorcena, Y., Gordon, A. J., Escuredo, P. R., Minchin, F. R., Witty, J. F., Moran, J. F. and Becana, M. (1997). N_2 fixation, carbon metabolism and oxidative damage in nodules of dark stressed common bean plants. *Plant Physiology* **113**, 1193–1201.

Gonzalez, E. M., Galvez, L. and Arrese-Igor, C. (2001). Abscissic acid induces a decline in nitrogen fixation that involves leghaemoglobin, but is independent of sucrose synthase activity. *Journal of Experimental Botany* **52**, 285–293.

Goormachtig, S., Capoen, W. and Holsters, M. (2004). Rhizobium infection: Lessons from the versatile nodulation behaviour of water-tolerant legumes. *Trends in Plant Science* **9**, 518–522.

Gordon, A. J., Minchin, F. R., Skot, L. and James, C. L. (1997). Stress-induced declines in soybean N_2 fixation are related to nodule sucrose synthase activity. *Plant Physiology* **114**, 937–946.

Graham, P. H. and Vance, C. P. (2003). Legumes: Importance and constraints to greater use. *Plant Physiology* **131**, 872–877.

Grill, E. and Himmelbach, A. (1998). ABA signal transduction. *Current Opinion in Plant Biology* **1**, 412–418.

Groten, K., Vanacker, H., Dutilleul, C., Bastian, F., Bernard, S., Carzaniga, R. and Foyer, C. H. (2005). The roles of redox processes in pea nodule development and senescence. *Plant Cell and Environment* **28**, 1293–1304.

Groten, K., Dutilleul, C., van Heerden, P. D., Vanacker, H., Bernard, S., Finkemeier, I., Dietz, K. J. and Foyer, C. H. (2006). Redox regulation of peroxiredoxin and proteinases by ascorbate and thiols during pea root nodule senescence. *FEBS Letters* **580**, 1269–1276.

Gunther, C., Schlereth, A., Udvardi, M. and Ott, T. (2007). Metabolism of reactive oxygen species is attenuated in leghemoglobin-deficient nodules of *Lotus japonicus*. *Molecular Plant Microbe Interactions* **20**, 1596–1603.

Harrison, J., Puppo, A. and Frendo, P. (2003). The synthesis and the roles of glutathione and homoglutathione in legumes. *Advances in Plant Physiology* **6**, 385–412.

Harrison, J., Jamet, A., Muglia, C. I., Van de Sype, G., Aguilar, O. M., Puppo, A. and Frendo, P. (2005). Glutathione plays a fundamental role in growth and symbiotic capacity of *Sinorhizobium meliloti*. *Journal of Bacteriology* **187**, 168–174.

Hirsch, A. M. (1992). Developmental biology of legume nodulation. *New Phytologist* **122**, 211–237.

Iturbe-Ormaetxe, I., Matamoros, M. A., Rubio, M. C., Dalton, D. A. and Becana, M. (2001). The antioxidants of legume nodule mitochondria. *Molecular Plant-Microbe Interactions* **14**, 1189–1196.

Ivashuta, S., Liu, J., Liu, J., Lohar, D. P., Haridas, S., Bucciarelli, B., VandenBosch, K. A., Vance, C. P., Harrison, M. J. and Gantt, J. S. (2005). RNA interference identifies a calcium-dependent protein kinase involved in *Medicago truncatula* root development. *Plant Cell* **17**, 2911–2921.

Jamet, A., Sigaud, S., Van de Sype, G., Puppo, A. and Herouart, D. (2003). Expression of the bacterial catalase genes during *Sinorhizobium meliloti–Medicago sativa* symbiosis and their crucial role during the infection process. *Molecular Plant Microbe Interactions* **16**, 217–225.

Jamet, A., Mandon, K., Puppo, A. and Herouart, D. (2007). H_2O_2 is required for optimal establishment of the *Medicago sativa/Sinorhizobium meliloti* symbiosis. *Journal of Bacteriology* **189**, 8741–8745.

Jebara, S., Jebara, M., Limam, F. and Aouani, M. E. (2005). Changes in ascorbate peroxidase, catalase, guaiacol peroxidase and superoxide dismutase activities in common bean (*Phaseolus vulgaris*) nodules under salt stress. *Journal of Plant Physiology* **162**, 929–936.

Jones, M. A., Raymond, M. J., Yang, Z. and Smirnoff, N. (2007). NADPH oxidase-dependent reactive oxygen species formation required for root hair growth depends on ROP GTPase. *Journal of Experimental Botany* **58**, 1261–1270.

Jones, K. M., Sharopova, N., Lohar, D. P., Zhang, J. Q., VandenBosch, K. A. and Walker, G. C. (2008). Differential response of the plant *Medicago truncatula* to its symbiont *Sinorhizobium meliloti* or an exopolysaccharide-deficient mutant. *Proceedings of the National Academy of Sciences of the United States of America* **105**, 704–709.

Klapheck, S. (1988). Homoglutathione: Isolation, quantification and occurrence in legumes. *Physiologia Plantarum* **74**, 727–732.

Koh, C. S., Didierjean, C., Navrot, N., Panjikar, S., Mulliert, G., Rouhier, N., Jacquot, J. P., Aubry, A., Shawkataly, O. and Corbier, C. (2007). Crystal structures of a poplar thioredoxin peroxidase that exhibits the structure of glutathione peroxidases: Insights into redox-driven conformational changes. *Journal of Molecular Biology* **370**, 512–529.

Kwak, J. M., Mori, I. C., Pei, Z. M., Leonhardt, N., Torres, M. A., Dangl, J. L., Bloom, R. E., Bodde, S., Jones, J. D. and Schroeder, J. I. (2003). NADPH oxidase AtrbohD and AtrbohF genes function in ROS-dependent ABA signaling in *Arabidopsis*. *The EMBO Journal* **22**, 2623–2633.

Lee, K., Shearman, L. L., Erickson, B. K. and Klucas, R. V. (1995). Ferric leghemoglobin in plant-attached leguminous nodules. *Plant Physiology* **109**, 261–267.

Lee, M. Y., Shin, K. H., Kim, Y. K., Suh, J. Y., Gu, Y. Y., Kim, M. R., Hur, Y. S., Son, O., Kim, J. S., Song, E., Lee, M. S. Nam, K. H. *et al.* (2005). Induction of thioredoxin is required for nodule development to reduce reactive oxygen species levels in soybean roots. *Plant Physiology* **139**, 1881–1889.

Liu, J., Blaylock, L., Endre, G., Cho, J., Town, C. D. A. V. K. and Harrison, M. J. (2003). Transcript profiling coupled with spatial expression analyses reveals genes involved in distinct developmental stages of an arbuscular mycorrhizal symbiosis. *Plant Cell* **15**, 2106–2123.

Loake, G. and Grant, M. (2007). Salicylic acid in plant defence: The players and protagonists. *Current Opinion in Plant Biology* **10**, 466–472.

Loferer, H., Bott, M. and Hennecke, H. (1993). *Bradyrhizobium japonicum* TlpA, a novel membrane-anchored thioredoxin-like protein involved in the biogenesis of cytochrome aa3 and development of symbiosis. *The EMBO Journal* **12**, 3373–3383.

Lohar, D. P., Sharopova, N., Endre, G., Penuela, S., Samac, D., Town, C., Silverstein, K. A. and VandenBosch, K. A. (2006). Transcript analysis of early nodulation events in *Medicago truncatula*. *Plant Physiology* **140**, 221–234.

Lohar, D. P., Haridas, S., Gantt, J. S. and Vandenbosch, K. A. (2007). A transient decrease in reactive oxygen species in roots leads to root hair deformation in the legume–rhizobia symbiosis. *New Phytologist* **173**, 39–49.

Lopez, M. A., Bannenberg, G. and Castresana, C. (2008). Controlling hormone signaling is a plant and pathogen challenge for growth and survival. *Current Opinion in Plant Biology* **11**, 420–427.

Lorence, A., Chevone, B. I., Mendes, P. and Nessler, C. L. (2004). *myo*-Inositol oxygenase offers a possible entry point into plant ascorbate biosynthesis. *Plant Physiology* **134**, 1200–1205.

Loscos, J., Matamoros, M. A. and Becana, M. (2008). Ascorbate and homoglutathione metabolism in common bean nodules under stress conditions and during natural senescence. *Plant Physiology* **146**, 1282–1292.

Marino, D., Gonzalez, E. M. and Arrese-Igor, C. (2006). Drought effects on carbon and nitrogen metabolism of pea nodules can be mimicked by paraquat: Evidence for the occurrence of two regulation pathways under oxidative stresses. *Journal of Experimental Botany* **57**, 665–673.

Marino, D., Frendo, P., Ladrera, R., Zabalza, A., Puppo, A., Arrese-Igor, C. and Gonzalez, E. M. (2007a). Nitrogen fixation control under drought stress. Localized or systemic? *Plant Physiology* **143**, 1968–1974.

Marino, D., Gonzalez, E. M., Frendo, P., Puppo, A. and Arrese-Igor, C. (2007b). NADPH recycling systems in oxidative stressed pea nodules: A key role for the NADP+-dependent isocitrate dehydrogenase. *Planta* **225**, 413–421.

Marino, D., Hohnjec, N., Kuster, H., Moran, J. F., Gonzalez, E. M. and Arrese-Igor, C. (2008). Evidence for transcriptional and post-translational regulation of sucrose synthase in pea nodules by the cellular redox state. *Molecular Plant Microbe Interactions* **21**, 622–630.

Martinez-Abarca, F., Herrera-Cervera, J. A., Bueno, P., Sanjuan, J., Bisseling, T. and Olivares, J. (1998). Involvement of salicylic acid in the establishment of the *Rhizobium meliloti*–alfalfa symbiosis. *Molecular Plant Microbe Interactions* **11**, 153–155.

Matamoros, M. A., Baird, L. M., Escuredo, P. R., Dalton, D. A., Minchin, F. R., Iturbe-Ormaetxe, I., Rubio, M. C., Moran, J. F., Gordon, A. J. and Becana, M. (1999a). Stress induced legume root nodule senescence. Physiological, biochemical and structural alterations. *Plant Physiology* **121**, 97–112.

Matamoros, M. A., Moran, J. F., Iturbe-Ormaetxe, I., Rubio, M. C. and Becana, M. (1999b). Glutathione and homoglutathione synthesis in legume root nodules. *Plant Physiology* **121**, 879–888.

Matamoros, M. A., Clemente, M. R., Sato, S., Asamizu, E., Tabata, S., Ramos, J., Moran, J. F., Stiller, J., Gresshoff, P. M. and Becana, M. (2003). Molecular analysis of the pathway for the synthesis of thiol tripeptides in the model legume *Lotus japonicus*. *Molecular Plant Microbe Interactions* **16**, 1039–1046.

Matamoros, M. A., Loscos, J., Coronado, M. J., Ramos, J., Sato, S., Testillano, P. S., Tabata, S. and Becana, M. (2006). Biosynthesis of ascorbic acid in legume root nodules. *Plant Physiology* **141**, 1068–1077.

Mathieu, C., Moreau, S., Frendo, P., Puppo, A. and Davies, M. J. (1998). Direct detection of radicals in intact soybean nodules: Presence of nitric oxide leghemoglobin complexes. *Free Radical Biology & Medicine* **24**, 1242–1249.

Meyer, Y., Siala, W., Bashandy, T., Riondet, C., Vignols, F. and Reichheld, J. P. (2008). Glutaredoxins and thioredoxins in plants. *Biochimica Biophysica Acta* **1783**, 589–600.

Miao, Y., Lv, D., Wang, P., Wang, X. C., Chen, J., Miao, C. and Song, C. P. (2006). An *Arabidopsis* glutathione peroxidase functions as both a redox transducer and a scavenger in abscissic acid and drought stress responses. *The Plant Cell* **18**, 2749–2766.

Michelet, L., Zaffagnini, M., Marchand, C., Collin, V., Decottignies, P., Tsan, P., Lancelin, J. M., Trost, P., Miginiac-Maslow, M., Noctor, G. and Lemaire, S. D. (2005). Glutathionylation of chloroplast thioredoxin f is a redox signaling mechanism in plants. *Proceedings of the National Academy of Sciences of the United States of America* **102**, 16478–16483.

Minchin, F. R., Becana, M. and Sprent, J. I. (1989). Short-term inhibition of legume N_2 fixation by nitrate. II. Nitrate effects on nodule oxygen diffusion. *Planta* **180**, 46–52.

Mithofer, A. (2002). Suppression of plant defence in rhizobia–legume symbiosis. *Trends in Plant Science* **7**, 440–444.

Moran, J. F., James, E. K., Rubio, M. C., Sarath, G., Klucas, R. V. and Becana, M. (2003). Functional characterization and expression of a cytosolic iron-superoxide dismutase from cowpea root nodules. *Plant Physiology* **133**, 773–782.

Moreau, S., Davies, M. J., Mathieu, C., Herouart, D. and Puppo, A. (1996). Leghemoglobin-derived radicals. Evidence for multiple protein-derived radicals and the initiation of peribacteroid membrane damage. *The Journal of Biological Chemistry* **271**, 32557–32562.

Muglia, C., Comai, G., Spegazzini, E., Riccillo, P. M. and Aguilar, O. M. (2008). Glutathione produced by *Rhizobium tropici* is important to prevent early senescence in common bean nodules. *FEMS Microbiology Letters* **23**, 191–198.

Navrot, N., Collin, V., Gualberto, J., Gelhaye, E., Hirasawa, M., Rey, P., Knaff, D. B., Issakidis, E., Jacquot, J. P. and Rouhier, N. (2006). Plant glutathione peroxidases are functional peroxiredoxins distributed in several subcellular compartments and regulated during biotic and abiotic stresses. *Plant Physiology* **142**, 1364–1379.

Naya, L., Ladrera, R., Ramos, J., Gonzalez, E. M., Arrese-Igor, C., Minchin, F. R. and Becana, M. (2007). The response of carbon metabolism and antioxidant defenses of alfalfa nodules to drought stress and to the subsequent recovery of plants. *Plant Physiology* **144**, 1104–1114.

Niehaus, K., Kapp, D. and Pühler, A. (1993). Plant defence and delayed infection of alfalfa pseudonodules induced by a exopolysaccharide (EPS I)-deficient *Rhizobium meliloti* mutant. *Planta* **190**, 415–425.

Noctor, G. (2006). Metabolic signalling in defence and stress: The central roles of soluble redox couples. *Plant Cell &Environment* **29**, 409–425.

Noctor, G. and Foyer, C. (1998). Ascorbate and glutathione: Keeping active oxygen under control. *Annual Review of Plant Physiology and Plant Molecular Biology* **49**, 249–279.

Ogawa, K., Kanematsu, S. and Asada, K. (1997). Generation of superoxide anion and localization of CuZn-superoxide dismutase in the vascular tissue of spinach hypocotyls: Their association with lignification. *Plant and Cell Physiology* **38**, 1118–1126.

Oldroyd, G. E. and Downie, J. A. (2008). Coordinating nodule morphogenesis with rhizobial infection in legumes. *Annual Review of Plant Biology* **59**, 519–546.

Ott, T., van Dongen, J. T., Gunther, C., Krusell, L., Desbrosses, G., Vigeolas, H., Bock, V., Czechowski, T., Geigenberger, P. and Udvardi, M. K. (2005). Symbiotic leghemoglobins are crucial for nitrogen fixation in legume root nodules but not for general plant growth and development. *Current Biology* **15**, 531–535.

Parniske, M., Zimmermann, C., Cregan, P. B. and Werner, D. (1990). Hypersensitive reaction of nodule cells in the *Glycine max sp./Bradyrhizobium japonicum* symbiosis occurs at the genotype specific level. *Botanica Acta* **103**, 318–320.

Pastori, G. M., Kiddle, G., Antoniw, J., Bernard, S., Veljovic-Jovanovic, S., Verrier, P. J., Noctor, G. and Foyer, C. H. (2003). Leaf vitamin C contents modulate plant defense transcripts and regulate genes that control development through hormone signaling. *The Plant Cell* **15**, 939–951.

Peleg-Grossman, S., Volpin, H. and Levine, A. (2007). Root hair curling and *Rhizobium* infection in *Medicago truncatula* are mediated by phosphatidylinositide-regulated endocytosis and reactive oxygen species. *Journal of Experimental Botany* **58**, 1637–1649.

Penmetsa, R. V. and Cook, D. R. (1997). A legume ethylene-insensitive mutant hyperinfected by its rhizobial symbiont. *Science* **275**, 527–530.

Penmetsa, R. V., Uribe, P., Anderson, J., Lichtenzveig, J., Gish, J. C., Nam, Y. W., Engstrom, E., Xu, K., Sckisel, G., Pereira, M., Baek, J. M. Lopez-Meyer, M. *et al.* (2008). The *Medicago truncatula* ortholog of *Arabidopsis* EIN2, sickle, is a negative regulator of symbiotic and pathogenic microbial associations. *The Plant Journal* **55**, 580–595.

Perotto, S., Brewin, N. J. and Kannenberg, E. L. (1994). Cytological evidence for a host defence response that reduces cell and tissue invasion in pea nodules by lipopolysaccharide-defective mutants of *Rhizobium leguminosarum* strain. *Molecular Plant Microbe Interactions* **7**, 99–112.

Pladys, D. and Vance, C. P. (1993). Proteolysis during development and senescence of effective and plant gene-controlled ineffective alfalfa nodules. *Plant Physiology* **103**, 379–384.

Potters, G., Horemans, N., Caubergs, R. J. and Asard, H. (2000). Ascorbate and dehydroascorbate influence cell cycle progression in a tobacco cell suspension. *Plant Physiology* **124**, 17–20.

Potters, G., Horemans, N., Bellone, S., Caubergs, R. J., Trost, P., Guisez, Y. and Asard, H. (2004). Dehydroascorbate influences the plant cell cycle through a glutathione-independent reduction mechanism. *Plant Physiology* **134**, 1479–1487.

Price, C. (1957). A new thiol in legumes. *Nature* **180**, 148–149.

Puppo, A. and Halliwell, B. (1988). Generation of hydroxyl radicals by soybean nodule leghaemoglobin. *Planta* **173**, 405–410.

Puppo, A., Rigaud, J. and Job, D. (1981). Role of superoxide anion in leghemoglobin autoxidation. *Plant Science Letters* **22**, 353–360.

Puppo, A., Dimitrijevic, L. and Rigaud, J. (1982). Possible involvement of nodule superoxide dismutase and catalase in leghemoglobin protection. *Planta* **156**, 374–379.

Puppo, A., Dimitrijevic, L. and Rigaud, J. (1987). O_2 consumption and superoxide dismutase content in purified mitochondria from soybean root nodules. *Plant Science* **50**, 3–11.

Puppo, A., Monny, C. and Davies, M. J. (1993). Glutathione-dependent conversion of ferryl leghemoglobin into the ferric form—A potential protective process in soybean (*Glycine max*) root-nodules. *Biochemical Journal* **289**, 435–438.

Puppo, A., Groten, K., Bastian, F., Carzaniga, R., Soussi, M., Lucas, M. M., de Felipe, M. R., Harrison, J., Vanacker, H. and Foyer, C. H. (2005). Legume nodule senescence: roles for redox and hormone signalling in the orchestration of the natural aging process. *New Phytologist* **165**, 683–701.

Ramos, J., Matamoros, M. A., Naya, L., James, E. K., Rouhier, N., Sato, S., Tabata, S. and Becana, M. (2009). The glutathione peroxidase gene family of *Lotus japonicus*: Characterization of genomic clones, expression analyses and immunolocalization in legumes. *New Phytologist* **181**, 103–114.

Ramu, S. K., Peng, H. M. and Cook, D. R. (2002). Nod factor induction of reactive oxygen species production is correlated with expression of the early nodulin gene *rip1* in *Medicago truncatula*. *Molecular Plant Microbe Interactions* **15**, 522–528.

Rausch, T. and Wachter, A. (2005). Sulfur metabolism: A versatile platform for launching defence operations. *Trends in Plant Science* **10**, 503–509.

Redondo, F. J., de la Pena, T. C., Morcillo, C. N., Lucas, M. M. and Pueyo, J. J. (2008). Overexpression of flavodoxin in bacteroids induces changes in antioxidant metabolism leading to delayed senescence and starch accumulation in alfalfa root nodules. *Plant Physiology* **19**, 1166–1178.

Riccillo, P. M., Muglia, C. I., de Bruijn, F. J., Roe, A. J., Booth, I. R. and Aguilar, O. M. (2000). Glutathione is involved in environmental stress responses in *Rhizobium tropici*, including acid tolerance. *Journal of Bacteriology* **182**, 1748–1753.

Ross, E. J., Kramer, S. B. and Dalton, D. A. (1999). Effectiveness of ascorbate and ascorbate peroxidase in promoting nitrogen fixation in model systems. *Phytochemistry* **52**, 1203–1210.

Rouhier, N. and Jacquot, J. P. (2005). The plant multigenic family of thiol peroxidases. *Free Radical Biology and Medicine* **38**, 1413–1421.

Rouhier, N., Koh, C. S., Gelhaye, E., Corbier, C., Favier, F., Didierjean, C. and Jacquot, J. P. (2008a). Redox based anti-oxidant systems in plants: Biochemical and structural analyses. *Biochimica Biophysica Acta* **1780**, 1249–1260.

Rouhier, N., Lemaire, S. D. and Jacquot, J. P. (2008b). The role of glutathione in photosynthetic organisms: Emerging functions for glutaredoxins and glutathionylation. *Annual Review of Plant Biology* **59**, 143–166.

Rubio, M. C., Gonzalez, E. M., Minchin, F. R., Webb, K. J., Arrese-Igor, C., Ramos, J. and Becana, M. (2002). Effects of water stress on antioxidant enzymes of leaves and nodules of transgenic alfalfa overexpressing superoxide dismutases. *Physiologia Plantarum* **115**, 531–540.

Rubio, M. C., James, E. K., Clemente, M. R., Bucciarelli, B., Fedorova, M., Vance, C. P. and Becana, M. (2004).*Localization of superoxide dismutases and hydrogen peroxide in legume root nodules* **17**, 1294–1305.

Sanchez-Fernandez, R., Fricker, M., Corben, L. B., White, N. S., Sheard, N., Leaver, C. J., Van Montagu, M., Inzé, D. and May, M. J. (1997). Cell proliferation and hair tip growth in the *Arabidopsis* root are under mechanistically different forms of redox control. *Proceedings of the National Academy of Sciences of the United States of America* **94**, 2745–2750.

Sandalio, L. M., Palma, J. M. and del Rio, L. A. (1987). Localization of manganese superoxide dismutase in peroxisomes isolated from *Pisum sativum* L. *Plant Science* **51**, 1–8.

Santos, R., Herouart, D., Sigaud, S., Touati, D. and Puppo, A. (2001). Oxidative burst in alfalfa-*Sinorhizobium meliloti* symbiotic interaction. *Molecular Plant Microbe Interactions* **14**, 86–89.

Sato, S. and Tabata, S. (2006). *Lotus japonicus* as a platform for legume research. *Current Opinion in Plant Biology* **9**, 128–132.

Scandalios, J. G., Guan, L. and Polidoros, A. N. (1997). Catalases in plants: Gene structure, properties, regulation, and expression. *In* "Oxidative Stress and the Molecular Biology of Antioxidant Defenses" (J. G. Scandalios, ed.), pp. 343–406. Cold Spring Harbor, New York, NY.

Scheidle, H., Groß, A. and Niehaus, K. (2005). The lipidA substructure of the *Sinorhizobium meliloti* lipopolysaccharide is sufficient to suppress the oxidative burst in host plant. *New Phytologist* **165**, 559–566.

Serraj, R. and Sinclair, T. R. (1996). Inhibition of nitrogenase activity and oxygen permeability by water deficit. *Journal of Experimental Botany* **47**, 1067–1073.

Serraj, R., Roy, G. and Drevon, J. J. (1994). Salt stress induces a decrease in the oxygen uptake of soybean nodules and in their permeability to oxygen diffusion. *Physiologia Plantarum* **91**, 161–168.

Shaw, S. L. and Long, S. R. (2003). Nod factor inhibition of reactive oxygen efflux in a host legume. *Plant Physiology* **132**, 2196–2204.

Sieberer, B. J., Timmers, A. C. and Emons, A. M. (2005). Nod factors alter the microtubule cytoskeleton in *Medicago truncatula* root hairs to allow root hair reorientation. *Molecular Plant Microbe Interactions* **18**, 1195–1204.

Smirnoff, N. and Wheeler, G. L. (2000). Ascorbic acid in plants: Biosynthesis and function. *Critical Reviews in Biochemistry and Molecular Biology* **35**, 291–314.

Smirnoff, N., Conklin, P. L. and Loewus, F. A. (2001). Biosynthesis of ascorbic acid in plants: A renaissance. *Annual Review of Plant Physiology and Plant Molecular Biology* **52**, 437–467.

Sprent, J. I. (2007). Evolving ideas of legume evolution and diversity: A taxonomic perspective on the occurrence of nodulation. *New Phytologist* **174**, 11–25.

Stacey, G., McAlvin, C. B., Kim, S. Y., Olivares, J. and Soto, M. J. (2006). Effects of endogenous salicylic acid on nodulation in the model legumes *Lotus japonicus* and *Medicago truncatula*. *Plant Physiology* **141**, 1473–1481.

Strozycki, P. M., Szczurek, A., Lotocka, B., Figlerowicz, M. and Legocki, A. B. (2007). Ferritins and nodulation in *Lupinus luteus*: Iron management in indeterminate type nodules. *Journal of Experimental Botany* **58**, 3145–3153.

Swaraj, K. and Garg, O. P. (1970). The effect of ascorbic acid, when applied to the rooting medium, on nodulation and nitrogen fixation in gram (*Cicer arietinum*). *Plant Physiology* **23**, 889–897.

Tada, Y., Spoel, S. H., Pajerowska-Mukhtar, K., Mou, Z., Song, J., Wang, C., Zuo, J. and Dong, X. (2008). Plant immunity requires conformational charges of NPR1 via *S*-nitrosylation and thioredoxins. *Science* **321**, 952–956.

Tadege, M., Ratet, P. and Mysore, K. S. (2005). Insertional mutagenesis: A Swiss Army knife for functional genomics of *Medicago truncatula*. *Trends in Plant Science* **10**, 229–235.

Tejera Garcia, N. A., Iribarne, C., Palma, F. and Lluch, C. (2007). Inhibition of the catalase activity from *Phaseolus vulgaris* and *Medicago sativa* by sodium chloride. *Plant Physiology and Biochemistry* **45**, 535–541.

Tellstrom, V., Usadel, B., Thimm, O., Stitt, M., Kuster, H. and Niehaus, K. (2007). The lipopolysaccharide of *Sinorhizobium meliloti* suppresses defense-associated gene expression in cell cultures of the host plant *Medicago truncatula*. *Plant Physiology* **143**, 825–837.

Timmers, A. C., Auriac, M. C. and Truchet, G. (1999). Refined analysis of early symbiotic steps of the *Rhizobium–Medicago* interaction in relationship with microtubular cytoskeleton rearrangements. *Development* **126**, 3617–3628.

Timmers, A. C., Soupene, E., Auriac, M. C., de Billy, F., Vasse, J., Boistard, P. and Truchet, G. (2000). Saprophytic intracellular rhizobia in alfalfa nodules. *Molecular Plant Microbe Interactions* **13**, 1204–1213.

Tjepkema, J. D. and Yocum, C. S. (1974). Measurement of oxygen partial pressure within soybean nodules by oxygen microelectrodes. *Planta* **119**, 351–360.

Torres, M. A., Jones, J. D. and Dangl, J. L. (2005). Pathogen-induced, NADPH oxidase-derived reactive oxygen intermediates suppress spread of cell death in *Arabidopsis thaliana*. *Nature Genetics* **37**, 1130–1134.

Trinick, M. J. (1973). Symbiosis between *Rhizobium* and the non-legume *Trema aspera*. *Nature* **244**, 459–460.

Tripathi, B. N., Bhatt, I. and Dietz, K. J. (2009). Peroxiredoxins: A less studied component of hydrogen peroxide detoxification in photosynthetic organisms. *Protoplasma* **235**, 3–15.

van Brussel, A. A., Bakhuizen, R., van Spronsen, P. C., Spaink, H. P., Tak, T., Lugtenberg, B. J. and Kijne, J. W. (1992). Induction of pre-infection thread structures in the leguminous host plant by mitogenic lipo-oligosaccharides of *Rhizobium*. *Science* **257**, 70–72.

Van Camp, W., Bowler, C., Villarroel, R., Tsang, E. W. T., Van Montagu, M. and Inzé, D. (1990). Characterization of iron superoxide dismutase cDNAs from plants obtained by genetic complementation in *Escherichia coli*. *Proceedings of the National Academy of Sciences of the United States of America* **87**, 9903–9907.

Van de Velde, W., Guerra, J. C., De Keyser, A., De Rycke, R., Rombauts, S., Maunoury, N., Mergaert, P., Kondorosi, E., Holsters, M. and Goormachtig, S. (2006). Aging in legume symbiosis. A molecular view on nodule senescence in *Medicago truncatula*. *Plant Physiology* **141**, 711–720.

van Heerden, P. D., Kiddle, G., Pellny, T. K., Mokwala, P. W., Jordaan, A., Strauss, A. J., de Beer, M., Schluter, U., Kunert, K. J. and Foyer, C. H. (2008). Roles for the regulation of respiration and the oxygen diffusion barrier in soybean in the protection of symbiotic nitrogen fixation from chilling-induced inhibition and shoots from premature senescence. *Plant Physiology* **148**, 316–327.

van Spronsen, P. C., Tak, T., Rood, A. M., van Brussel, A. A., Kijne, J. W. and Boot, K. J. (2003). Salicylic acid inhibits indeterminate-type nodulation but not determinate-type nodulation. *Molecular Plant Microbe Interactions* **16**, 83–91.

van Workum, W. A. T., van Slageren, S., van Brussel, A. A. N. and Kijne, J. W. (1998). Role of exopolysaccharides of *Rhizobium leguminosarum* bv. *Vicae* as host plant-specific molecules required for infection thread formation during nodulation of *Vicia faba*. *Molecular Plant Microbe Interactions* **11**, 1233–1241.

Vasse, J., de Billy, F. and Truchet, G. (1993). Abortion of infection during the *Rhizobium meliloti*–alfalfa symbiotic interaction is accompanied by a hypersensitive reaction. *The Plant Journal* **4**, 555–566.

Vernoux, T., Wilson, R. C., Seeley, K. A., Reichheld, J. P., Muroy, S., Brown, S., Maughan, S. C., Cobbett, C. S., Van Montagu, M., Inze, D., May, M. J. and Sung, Z. R. (2000). The ROOT MERISTEMLESS1/CADMIUM SENSITIVE2 gene defines a glutathione-dependent pathway involved in initiation and maintenance of cell division during postembryonic root development. *The Plant Cell* **12**, 97–110.

Vessey, J. K., Walsh, K. B. and Layzell, D. B. (1988). Oxygen limitation of nitrogen fixation in stem-girdled and nitrate-treated soybean. *Physiologia Plantarum* **73**, 113–121.

Wait, R., Begum, S., Brambilla, D., Carabelli, A. M., Conserva, F., Rocco Guerini, A., Eberini, I., Ballerio, R., Gemeiner, M., Miller, I. and Gianazza, E. (2005). Redox options in two-dimensional electrophoresis. *Amino Acids* **28**, 239–272.

Wall, L. G. (2000). The actinorhizal symbiosis. *Journal of Plant Growth Regulation* **19**, 167–182.

Wheeler, G. L., Jones, M. A. and Smirnoff, N. (1998). The biosynthetic pathway of vitamin C in higher plants. *Nature* **393**, 365–369.

Wiermer, M., Feys, B. J. and Parker, J. E. (2005). Plant immunity: The EDS1 regulatory node. *Current Opinion in Plant Biology* **8**, 383–389.

Yoshioka, H., Numata, N., Nakajima, K., Katou, S., Kawakita, K., Rowland, O., Jones, J. D. and Doke, N. (2003). *Nicotiana benthamiana* gp91phox homologs NbrbohA and NbrbohB participate in H_2O_2 accumulation and resistance to *Phytophthora infestans*. *The Plant Cell* **15**, 706–718.

Zabalza, A., Galvez, L., Marino, D., Royuela, M., Arrese-Igor, C. and Gonzalez, E. M. (2008). The application of ascorbate or its immediate precursor, galactono-1, 4-lactone, does not affect the response of nitrogen-fixing pea nodules to water stress. *Journal of Plant Physiology* **165**, 805–812.

Zehr, J. P., Jenkins, B. D., Short, S. M. and Steward, G. F. (2003). Nitrogenase gene diversity and microbial community structure: A cross-system comparison. *Environmental Microbiology* **5**, 539–554.

Reactive Oxygen Species in *Phanerochaete chrysosporium*: Relationship Between Extracellular Oxidative and Intracellular Antioxidant Systems

MÉLANIE MOREL,[1] ANDREW A. NGADIN, JEAN-PIERRE JACQUOT AND ERIC GELHAYE

IFR 110 Ecosystèmes Forestiers, Agroressources, Bioprocédés et Alimentation, Unité Mixte de Recherches INRA UHP 1136 Interaction Arbres Microorganismes, Université Nancy I BP 239, 54506 Vandoeuvre-lès-Nancy Cedex, France

[1]Corresponding author: Email: mmorel@scbiol.uhp-nancy.fr

Advances in Botanical Research, Vol. 52

0065-2296/09 $35.00
DOI: 10.1016/S0065-2296(09)52006-X

ABSTRACT

The basidiomycete *Phanerochaete chrysosporium* is a model of ligninolytic fungus which has been studied for a long time. The lignin degradation mediated by this fungus occurs through oxidative processes involving a large set of extracellular enzymes including lignin oxidases and lignin-degrading auxiliary enzymes. In this context, the production of reactive oxygen species (ROS) by this fungus occurs in physiological conditions, that is, during the wood degradation. Ligninolytic basidiomycetes have thus had to develop strategies to protect themselves against oxidative damages induced during lignin oxidation. The excretion of extracellular ligninolytic enzymes is indeed linked at least partially to the fungal intracellular redox state, suggesting a relationship between the intracellular antioxidant system and the production of extracellular ROS by this fungus. This review describes the extracellular systems involved in ROS production, the intracellular systems protecting against ROS, as well as the relationship between them.

I. EXTRACELLULAR REACTIVE OXYGEN SPECIES (ROS) FORMATION

Lignin plays a key role in the carbon cycle as the most abundant aromatic compound in nature, providing the protective matrix surrounding the cellulose microfibrils of plant cell walls. This amorphous and insoluble polymer lacks stereoregularity and, in contrast to cellulose and hemicellulose, it is not susceptible to hydrolytic attack. Lignin is made up of phenylpropanoid units which are linked by a variety of carbon–carbon and carbon–oxygen bonds, making it very difficult to degrade. However, its degradation by certain fungi was recognized and described nearly 125 years ago. Collectively referred to as white-rot fungi (since they degrade brown lignin and leave behind white cellulose), these are the only microbes capable of efficient depolymerization and mineralization of lignin. All are basidiomycetes, a fungal group that includes both edible mushrooms and plant pathogens such as smuts and rust. They secrete an array of peroxidases and oxidases that act non-specifically via the generation of lignin-free radicals, which then undergo spontaneous cleavage reactions. The non-specific nature and exceptional oxidation potential of the enzymes has attracted considerable interest for application in bioprocesses such as organopollutant degradation and fibre bleaching.

The enzymes which are part of the ligninolytic system have been first discovered in *Phanerochaete chrysosporium* (Tien and Kirk, 1983). This system is composed of lignin peroxidases (LiPs), manganese peroxidases (MnPs) and glyoxal oxidase (Glenn *et al.*, 1983; Kersten and Kirk, 1987; Tien and Kirk, 1983). LiPs as well as MnPs represent each 1% of the *Phanerochaete* secretome, and total oxidases represent 11% (Bouws *et al.*, 2008). Additionally, many white fungi, with the notable exception of

P. chrysosporium, produce extracellular laccases, which catalyze the one-electron oxidation of phenols to phenoxy radicals. Several strains produce laccase and MnP but apparently not LiP, suggesting that they degrade lignin by an oxidative mechanism somewhat different from that of *P. chrysosporium* (Gold and Alic, 1993). Such diversity is also observed in the genetic repertoire of ligninolytic enzymes: *P. chrysosporium* possesses over a dozen of different peroxidase genes but no true laccase sequences, while *Coprinopsis cinerea* encodes an abundant set of laccases but only one peroxidase (Kilaru *et al.*, 2006).

A recent study has classified the enzymes potentially involved in lignin catabolism into sequence-based families and integrated them into a newly developed database, designated fungal oxidative lignin enzymes (FOLy) (Levasseur *et al.*, 2008). These ligninolytic families were divided into the categories of lignin oxidases (LO) and lignin-degrading auxiliary (LDA) enzymes according to their direct or indirect action on lignin degradation. Laccases, LiPs, MnPs and versatile oxidases, which represent a third type of oxidases combining catalytic properties of both LiPs and MnPs enzymes, were found in the LO group. Most enzymes classified as LDAs are H_2O_2 producers. Among H_2O_2-generating enzymes, aryl-alcohol oxidase and glyoxal oxidase are thought to be the main enzymes responsible for the production of H_2O_2 (Kersten and Cullen, 2007). This category regroups also vanillyl-alcohol oxidase, pyranose oxidase, galactose oxidase and glucose oxidase (Fig. 1). *P. chrysosporium* exhibits 16 peroxidases from LO group, 3 aryl-alcohol oxidases, 1 glyoxal oxidase, 1 pyranose oxidase and 1 glucose oxidase.

A. LIGNIN OXIDASES

The LiPs are encoded by a family of 10 structurally related genes designed A to J (Gaskell *et al.*, 1994). Even though the repeated pattern of genomic organization indicates that the LiP family probably arose via a series of duplication events, their differential regulation in response to medium composition suggests specific biological role for individual isozymes (Stewart and Cullen, 1999). In *P. chrysosporium*, one of these proteins has been purified to homogeneity by ion-exchange chromatography. The protein contains one protohaem IX per molecule. It catalyzes, non-stereospecifically, several oxidations in the alkyl side chains of lignin-related compounds: $C\alpha$–$C\beta$ cleavage in lignin-related compounds of the type aryl-$C\alpha$HOH–$C\beta$H–R–$C\gamma H_2OH$ (R = -aryl or -*O*-aryl), oxidation of benzyl alcohols to aldehydes or ketones, intradiol cleavage in phenylglycol structures and hydroxylation of benzylic methylene groups. It also catalyzes oxidative coupling of phenols, perhaps

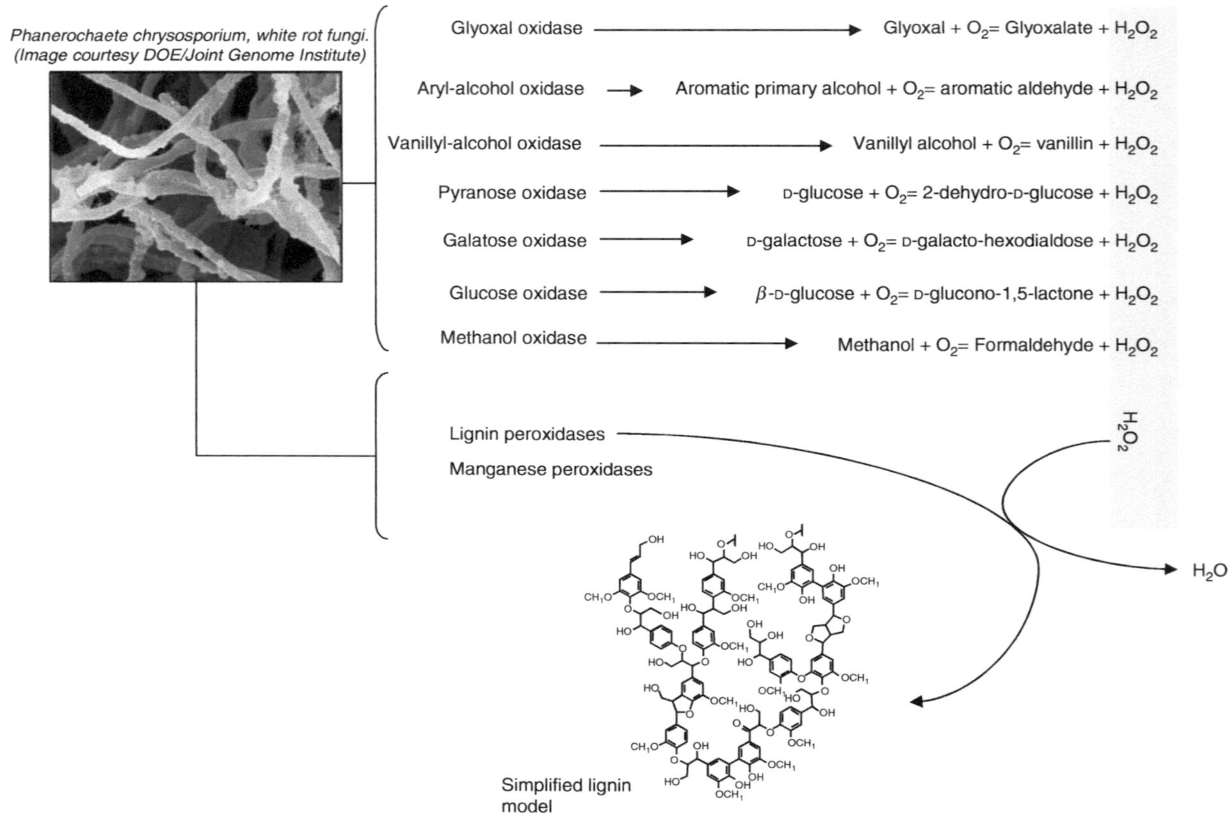

Fig. 1. Extracellular enzymes involved in lignin degradation in *Phanerochaete chrysosporium*. The fungus is able to excrete a large array of oxidases in the extracellular medium. These enzymes generate hydrogen peroxide, which is required for peroxidase activity to cleave lignin linkages.

explaining the long-recognized association between phenol oxidation and lignin degradation. All reactions require H_2O_2. The $C\alpha$–$C\beta$ cleavage and the methylene hydroxylation reactions involve substrate oxygenation, the oxygen atom arising from O_2 and not H_2O_2 (Tien and Kirk, 1984). A second lignin peroxidase has been characterized by heterologous production in insect cells (Johnson *et al.*, 1992). The recombinant enzyme was purified to near homogeneity and is capable of oxidizing veratryl alcohol, iodide, and, to a lesser extent, guaiacol.

MnP oxidizes Mn^{2+} to Mn^{3+}, using H_2O_2 as oxidant. In addition to previously characterized MnP genes (*mnp1*, *mnp2*, *mnp3*) (Alic *et al.*, 1997; Orth *et al.*, 1994; Pease *et al.*, 1989; Pribnow *et al.*, 1989), genome analysis revealed two others MnP genes (*mnp4* and *mnp5*). If MnPs do have a major role in lignin degradation, one possibility is that subsequent reactions of Mn^{3+} may generate other oxidants that can cleave non-phenolic structures. For example, Mn^{3+} oxidizes oxalate that chelates it to generate CO_2 and a formate anion radical, which reacts with O_2 to give another molecule of CO_2 and also superoxide ($O_2^{\bullet -}$). At the low pH values in wood undergoing white rot, most of this $O_2^{\bullet -}$ will occur in its protonated form as the perhydroxyl radical ($HOO^{\bullet}$), a strong oxidant that can abstract hydrogen atoms from donors such as unsaturated fatty acids. The result of this chemistry would be lipid peroxidation that generates potentially ligninolytic peroxyl radicals through a radical chain reaction. In agreement with this hypothesis, it has been shown that MnPs catalyze lipid peroxidation in the presence of chelated Mn^{2+} and H_2O_2, that non-phenolic synthetic lignin is cleaved *in vitro* as a result, and that white-rot fungi produce extracellular lipids that could provide the necessary peroxidizable substrates in wood. A role for peroxyl radicals in these reactions is also suggested by data showing that other peroxyl-radical-generating systems are involved in the cleavage of non-phenolic lignin structures (Hammel and Cullen, 2008; Kersten and Cullen, 2007).

The sites of MnP production were localized at different stages of cultivation by an immunolabelling procedure. MnP was mainly concentrated in the chlamydospore-like cells and principally distributed in Golgi-like vesicles located at the periphery of the cytoplasm. The apices of hyphae in the outer layer of the pellets were apparently minor sites of MnP production. Maximal MnP release into the culture supernatant coincided with apparent autolysis of the chlamydospore-like cells (Jimenez-Tobon *et al.*, 2003).

Versatile peroxidases represent a third type of peroxidase, combining catalytic properties of both Lip and MnP enzymes (Heinfling *et al.*, 1998). Other peroxidases were reported in the FOLy database: chloroperoxidases are potential chlorinators of lignin and may thus account for some of the high-molecular-weight organochlorine residues. Cellobiose dehydrogenase plays a role in carbohydrate metabolism, and some studies also suggest a role in lignin degradation. This haem-containing protein can generate hydroxyl radicals by Fenton-type reactions, thereby oxidizing lignin (Levasseur *et al.*, 2008).

B. LIGNIN-DEGRADING AUXILIARY ENZYMES

Peroxide is required as an oxidant in peroxidative reactions, and several oxidases have been proposed to play a role in this regard. Glyoxal oxidase (GLX) is an extracellular glycoprotein of 68 kDa with two isozymic forms (pI 4.7 and 4.9). It is a copper radical oxidase that catalyzes the oxidation of several simple aldehyde-, α-hydroxycarbonyl- and α-dicarbonyl compounds, coupled to the reduction of O_2 to H_2O_2. During the reaction, GLX is activated by LiP, suggesting a possible extracellular circuit for the control of H_2O_2 production by GLX, and control of Lip activity by H_2O_2 (Kersten and Cullen, 2007; Singh and Chen, 2008). An analysis of the *P. chrysosporium* draft genome had identified six sequences with significant similarity to GLX and designated them *cro1* through *cro6* (Vanden Wymelenberg *et al.*, 2006b). The predicted mature protein sequences diverge substantially from one another, but the residues coordinating copper and constituting the radical redox site are conserved. Transcript profiles, microscopic examination and lignin analysis of inoculated thin wood sections are consistent with differential regulation as decay advances. However, the role of the diverse *cro* genes in lignocellulose degradation is not clearly established.

Fungal glucose oxidase catalyzes the oxidation of D-glucose to β-D-gluconolactone and H_2O_2 in the presence of molecular oxygen. One glucose oxidase was isolated from *P. chrysosporium*. It is a flavoprotein containing two identical polypeptides with a molecular weight of 80 kDa each. The flavin analysis data revealed 1.5 mol of flavin per mol of purified glucose oxidase from *P. chrysosporium* (Kelley and Reddy, 1986). Very few studies have concerned this enzyme and its role remains to be determined. While no evidence supports a role of glucose oxidase in wood degradation, an important role has been proposed for pyranose oxidase (POx) (Daniel *et al.*, 1994; Giffhorn, 2000). POx is preferentially localized in the periplasmic space and the associated membranous materials. The protein has been purified to apparent homogeneity from mycelium extracts of *P. chrysosporium*. It is a homotetrameric flavoprotein with subunits of about 65 kDa, which is not glycosylated compared to POx from other fungi (Artolozaga *et al.*, 1997). A K_m value for O_2 of 0.13 mM has been determined for *P. chrysosporium* POx (Artolozaga *et al.*, 1997).

Structurally related to glucose and pyranose oxidases, three aryl-alcohol oxidases have been identified in *P. chrysosporium* (Varela *et al.*, 2001). The precise role of these enzymes remains uncertain, but they may support a redox cycle by supplying extracellular peroxide, perhaps coupled to intracellular aryl-alcohol dehydrogenase (Kersten and Cullen, 2007). Among H_2O_2-generating enzymes, aryl-alcohol oxidases and glyoxal oxidase are thought to be the main enzymes responsible for the production of H_2O_2 (Kersten and Cullen, 2007). However, many other oxidases, such as methanol oxidase or vanillyl-alcohol oxidase, could be responsible for providing H_2O_2 for the ligninolytic systems.

Some transcriptomic and proteomic analyses have revealed that MnPs, LiPs and a copper radical oxidase are secreted in the medium when *P. chrysosporium* is cultivated on softwood (Ravalason *et al.*, 2008). In ligninolytic cultures corresponding to C-limited and N-limited conditions, peptides corresponding to LiPs, MnPs and glyoxal oxidase were identified in the culture medium (Vanden Wymelenberg *et al.*, 2006a). Moreover, two MnP coding genes, one LiP coding gene and one glucose-methanol-choline oxidoreductase coding gene, were up-regulated four-fold or greater during the initiation of ligninolytic enzymes production (Minami *et al.*, 2007).

II. INTRACELLULAR ROS FORMATION

The major part of oxygen consumed by aerobic cells is converted into water in mitochondria via a four-electron reduction reaction, catalyzed by cytochrome *c* oxidase (respiratory complex IV). A much smaller part is converted into H_2O_2 as a result of two-electron reduction, catalyzed by a number of enzymes. It should be emphasized that in addition to the most active ensemble of oxidoreductases, known as the respiratory chain, intact mitochondria contain a number of oxidoreductases both in the matrix and in the outer mitochondrial membrane potentially capable of superoxide production. Thus, attributing the generation of hydrogen peroxide solely to the respiratory chain components, something common in the current literature, should be done with caution (Grivennikova and Vinogradov, 2006). Nothing is really known concerning *P. chrysosporium* mitochondrial functioning. Most of the published studies concern mammalian cells; however, it was shown that the structure of the different respiratory complexes is very similar in all forms of life (Joseph-Horne *et al.*, 2001).

A. ROS PRODUCTION IN THE MITOCHONDRIAL INNER MEMBRANE AND MATRIX

In mitochondria, there are several reactions which can potentially produce ROS. The proteins involved contain either a flavin or an iron–sulphur centre with low redox potentials, responsible for the superoxide ion production. It is assumed that the respiratory chain is the major source of ROS in mitochondria. Indeed, the number of electrons passing through the respiratory chain is much larger than in any other cellular redox system, so large quantities of ROS could be obtained even if a small portion of this electron flow results in formation of $O_2^{-\bullet}$ rather than H_2O (Skulachev, 2006).

Most of the studies have been carried out on mammalian systems and they show that complexes I and III are the components capable of the univalent reduction of oxygen into superoxide (Muller *et al.*, 2004).

Complex I is the entry point for electrons from NADH into the respiratory chain. It is a ca. 1-MDa complex comprising 45 polypeptides in mammalian (Sazanov, 2007). An FMN cofactor accepts electrons from NADH and passes them through a chain of seven iron–sulphur centres to the Coenzyme Q reduction site. The mechanism of $O_2^{-\bullet}$ production by isolated complex I is now reasonably well understood (Murphy, 2009). The isolated complex produces $O_2^{-\bullet}$ from the reaction of O_2 with the fully reduced FMN, and the proportion of the FMN that is fully reduced is set by the $NADH/NAD^+$ ratio. Consequently, inhibition of the respiratory chain by damage, mutation, ischemia, loss of cytochrome *c* or by the buildup of $NADH/NAD^+$ ratio will lead to $O_2^{-\bullet}$ formation (Fig. 2A). In contrast, for most situations where mitochondria are respiring normally on NADH-linked substrates and the $NADH/NAD^+$ ratio is relatively low, only small amounts of $O_2^{-\bullet}$ are produced from complex I. The other mechanism by which complex I produces large amounts of $O_2^{-\bullet}$ is during reverse electron transport. Reverse electron transport occurs for mitochondria when electron supply reduces the CoQ pool, which in the presence of a significant Δp, forces electrons back from $CoQH_2$ into complex I, and can reduce NAD^+ to NADH at the FMN site (Fig. 2B). Although the site of $O_2^{-\bullet}$ production during reverse electron transport is not known, the rate of $O_2^{-\bullet}$ production linked to this electron flow reversal seems to be the highest that occurs in mitochondria (Hurd *et al.*, 2007; Lambert and Brand, 2004a,b).

Complex III is an enzyme complex oxidizing coenzyme Q using cytochrome *c* as an electron acceptor. In mammalian, the monomer is about 240 kDa and comprises 11 polypeptides, three haems, and an FeS centre (Iwata *et al.*, 1998). When supplied with $CoQH_2$ and when the Qi site is inhibited by antimycin, complex III produces large amounts of $O_2^{-\bullet}$ from the reaction of O_2 with an ubisemiquinone bound at the Qo site. This $O_2^{-\bullet}$ is released from complex III to both sides of the inner membrane (Fig. 2C). However, in the absence of antimycin, the Qo site ubisemiquinone is not stabilized and $O_2^{-\bullet}$ production by complex III is low. Although complex III can be induced to produce $O_2^{-\bullet}$ with the inhibitor antimycin, its production in mitochondria under physiological conditions is far lower and is negligible

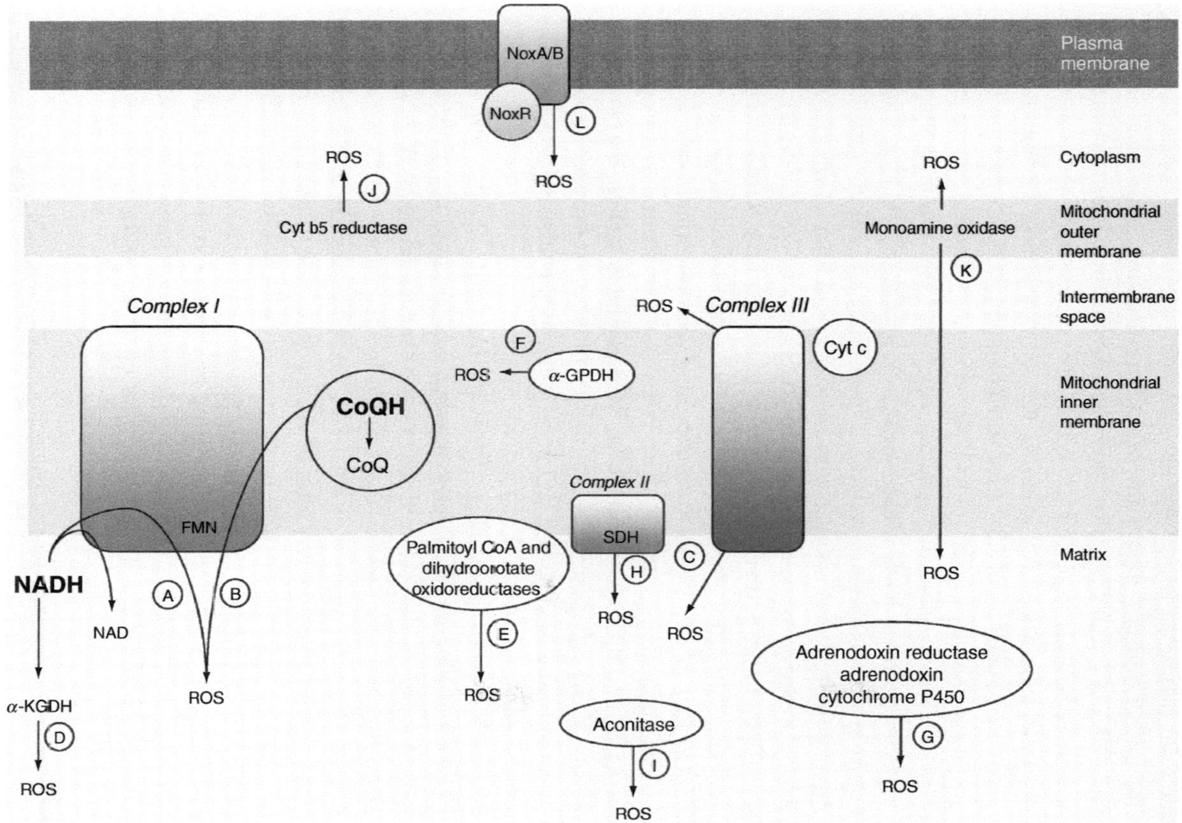

Fig. 2. (Continued)

compared with the maximum rate of $O_2^{-\bullet}$ production from complex I. Moreover, complex I produces superoxide exclusively into the mitochondrial matrix, whereas complex III produces superoxide at both the matrix and cytosolic sides (Muller *et al.*, 2004).

Within mitochondria, other sites of $O_2^{-\bullet}$ production have been listed (Murphy, 2009). Most of these sites have been divided into sites that interact with the matrix NADH pool and those that are connected to the CoQ pool within the inner membrane. For example, the combination of 2-oxoglutarate with a reduced NADH pool may lead to significant ROS production from α-ketoglutarate dehydrogenase (α-KGDH) (Fig. 2D). Indeed, one component of α-KGDH is dihydrolipoamide dehydrogenase, which contains a flavin that can produce ROS when its electron acceptor NAD^+ is limiting (Bunik and Sievers, 2002; Starkov *et al.*, 2004; Tretter and Adam-Vizi, 2004). Some other potential sites interact with the CoQ pool. Oxidation of palmitoyl-CoA or dihydroorotate may produce ROS (Eaton, 2002; Forman and Kennedy, 1976). In particular, in the absence of its natural electron acceptor, CoQ, reduced dihydroorotate dehydrogenase can produce H_2O_2 *in vitro* (Fig. 2E) (Loffler *et al.*, 1996). Moreover, α-glycerophosphate dehydrogenase is a FAD-containing enzyme which takes electrons from α-glycerophosphate to CoQ, and this is associated with ROS production (Fig. 2F) (Tretter *et al.*, 2007). Many other enzymes that can be induced to produce ROS are not connected to the NADH or CoQ pools, such as the adrenodoxin reductase/adrenodoxin/cytochrome P450 system in the matrix that receives electrons from NADPH pool (Fig. 2G) (Hanukoglu, 2006). Succinate dehydrogenase belonging to complex II, when incorporated into liposomes, can produce ROS via its FAD reduced in the absence of electron acceptor (Fig. 2H) (Zhang *et al.*, 1998). Aconitase is located in the mitochondrial matrix. It catalyzes conversion of citrate to isocitrate as part of the tricarboxylic acid cycle. Upon inactivation of the enzyme, due to oxidation of its iron–sulphur cluster by superoxide, production of hydroxyl radical mediated by released Fe^{2+} is induced (Fig. 2I) (Vasquez-Vivar *et al.*, 2000).

Most of the described studies concern mammalian cells; however, it was shown that the structure of the different complexes is very similar in all forms of life (Joseph-Horne *et al.*, 2001). An *in silico* analysis has described, using a

Fig. 2. Reactive oxygen species (ROS) production sites. (A–I) ROS production sites in the mitochondrial inner membrane and matrix. (J, K) ROS production sites in the mitochondrial outer membrane. (L) ROS production by NADPH oxidases (NOX) at the plasma membrane. α-KGDH, α-ketoglutarate dehydrogenase; α-GPDH, α-glycerophosphate dehydrogenase; CoQ, Coenzyme Q; SDH, succinate dehydrogenase; cyt *c*, cytochrome *c*; FMN, flavin mononucleotide. (See color plate section at the back of the book.)

comparative genomic approach, the oxidative phosphorylation system of many fungi, including *P. chrysosporium* (Lavin *et al.*, 2008). This study demonstrated that *P. chrysosporium* possesses the five complexes of the respiratory chain including complex I, one alternative oxidase (AOX) and three alternative NAD(P)H dehydrogenases. This is not the case for some other fungi. For example, *Saccharomyces cerevisiae, Candida glabrata, Kluyveromyces lactis, and Eremothecium gossypii* do not possess complex I or the AOX.

B. ROS PRODUCTION IN THE MITOCHONDRIAL OUTER MEMBRANE

Cytochrome *b*5 reductase oxidizes cytoplasmic NAD(P)H and reduces cytochrome *b*5 in the outer membrane (Fig. 2J). It was reported that mitochondrial cytochrome *b*5 reductase may produce superoxide at a high rate of ca. 300 nmol/min/mg protein (Whatley *et al.*, 1998).

Monoamine oxidase is a flavoprotein ubiquitously expressed in higher eukaryotic organisms, which catalyzes oxidative deamination of biogenic amines. This enzyme contributes to an increase in the steady-state concentrations of reactive species within both the mitochondrial matrix and the cytosol (Fig. 2K) (Cadenas and Davies, 2000).

C. ROS PRODUCTION BY NADPH OXIDASES (NOX) AT THE PLASMA MEMBRANE

Three different subfamilies of NOX are found in the fungal kingdom (Aguirre *et al.*, 2005). Specific isoforms have been shown by genetic analysis to be required for various cellular differentiations, including development of sexual fruiting bodies and ascospore germination (Aguirre *et al.*, 2005; Lara-Ortiz *et al.*, 2003; Malagnac *et al.*, 2004). ROS production catalyzed by NOX has also been proposed to serve for defence against other fungi (Fig. 2L) (Haedens *et al.*, 2005; Silar, 2005). Taken together these findings indicate that ROS production by NOX is a universal signalling system among multicellular organisms. The diffusible nature of superoxide and H_2O_2 makes them ideal second messengers for signalling within the cell, and in the case of H_2O_2, which can cross the cell membrane, inter-cellular signalling. A survey of available fungi genomes revealed considerable variation in NOX gene composition. This gene is completely absent in some species (*S. cerevisiae*, *Schizosaccharomyces pombe*, *Ustilago maydis*, and *Rhizopus oryzae*), but there are up to four copies in *Fusarium solani*, reflecting the diverse morphologies and life cycles of fungal species (Takemoto *et al.*, 2007). Because NOX genes were found in a wide range of fungi from

Chytridomycota to Ascomycota, it is likely that NOX was an ancestral enzymatic function that has been lost during evolution. *P. chrysosporium* possesses the three NOX genes (NOXA, NOXB and NOXR) but not NOXC, which was found in only seven ascomycota fungi (Takemoto *et al.*, 2007).

It should have been implicit that the true source of oxidative stress is not the generation *per se* but spatiotemporal imbalance of ROS production and detoxification. Indeed it is generally accepted that levels of ROS are fine-tuned by the rates of their formation and decomposition by enzymatic and non-enzymatic systems.

III. HOW TO DEAL WITH INTRACELLULAR ROS?

A. SMALL ANTIOXIDANT MOLECULES

1. The glutathione

Glutathione (γ-L-glutamyl-L-cysteinyl-glycine) is synthesized through the activity of two enzymes, the γ-L-glutamyl-L-cysteine ligase (GCL) and the glutathione synthetase (GS). In *P. chrysosporium*, one GCL (PcGCL) and also one GS (PcGS) are predicted from the sequenced genome (Table I). In addition, glutathione has been detected in this fungus (Belinky *et al.*, 2003). Glutathione plays a major role in ROS detoxification, reacting directly with different compounds such as hydrogen peroxide or acting as electron donor to various peroxidases, a point which will be discussed later. In addition, glutathione plays a major role in different detoxification pathways particularly against electrophilic compounds, and also in heavy metal detoxification as a phytochelatin precursor (Noctor and Foyer, 1998; Rouhier *et al.*, 2008). Furthermore, glutathione is involved in post-translational modifications called glutathionylation, this mechanism being involved in different processes such as protein protection against oxidation (Rouhier *et al.*, 2008). These different points will be discussed in the following sections dealing with thiol-dependent systems.

2. Other molecules

L-Ascorbic acid is an important ROS scavenger produced by higher plants (Smirnoff and Wheeler, 2000) and in mammals with few exceptions (Linster and Van Schaftingen, 2007). In fungi, ascorbate is mainly replaced by D-erythroascorbate, a C_5 ascorbate analogue similar in structure and physico-chemical properties to ascorbate (Baroja-Mazo *et al.*, 2005). D-Erythroascorbate appears to be the naturally occurring ascorbate analogue in ascomycetes and basidiomycetes, so it is presumed that it may have similar functions to

TABLE I
Proteins involved in the P. chrysosporium antioxidant systems

	Identification number	Localization	Expression studies
Superoxide dismutase			
PcMnSOD1	9267	**M**	Yes
PcMnSOD2	94129	S	Yes
PcMnSOD3	131933	C	No
PcMnSOD4	1323	C	No
PcCu/ZnSOD	128732	S	No
Catalase			
PcCat1	128306	P	Yes
PcCat2	134956	C	Yes
PcCat3	127288	C	Yes
PcCat4	124398	**Mb**	Yes
PcCat5	127266	C	Yes
γ-L-Glutamyl-L cysteine synthase			
PcGCL	133953	C	No
Glutathione synthetase			
PcGS	130608	C	No
Glutathione peroxidase			
PcGpx	130274	C	Yes
Peroxiredoxin			
Pc2-Cys Prx	8807	C	Yes
Pc1-Cys Prx1	10009	**Mb**	Yes
Pc1-Cys Prx2	126313	**Mb**	Yes
PcPrxQ1	6867	C	No
PcPrxQ2	124208	N	No
PcPrxII1	125657	**Mb**	Yes
PcPrxII2	133514	C	No
Thioredoxin			
PcTrx1	7266	C	Yes
PcTrx2	7498	S/**Mb**	Yes
PcTrx3	44586	C	Yes
PcTrx4	122495	M	Yes
NADPH thioredoxin reductase			
PcNTR	8527	C/**M**	Yes
Glutaredoxin			
PcGrx1	10446	N/C	No
PcGrx2	124106	S	No
PcGrx3	127248	M	No
PcGrx4	135692	N	No
Glutathione reductase			
PcGR1	876	M/**C**	Yes
PcGR2	10525	M/C	Yes
PcGR3	135167	C	Yes

(*continues*)

TABLE I (*continued*)

	Identification number	Localization	Expression studies
Methionine sulphoxide reductase			
PcMsrA	122315	M	No
PcMsrB	137333	M	No
Glutathione-*S*-transferase			
PcGTO1	EU791894	M	No
PcGTO2	126388	M	No
PcGTO3	EU791893	C	No
PcGTO4	7168	C	No
PcGTO5	7169	M	No
PcGTO6	3911	C	No
PcGTO7	6880	C	No
PcGTO8	6881	M/C	No
PcURE2p1	503	M	No
PcURE2p2	140156	C	No
PcURE2p3	140271	M	No
PcURE2p4	137250	C	No
PcURE2p5	128511	M	No
PcURE2p6	2269	C	No
PcURE2p7	2266	C	No
PcURE2p8	2268	N	No
PcURE2p9	140259	C	No
PcGTE1	5118	C	No
PcGTE2	5119	C	No
PcGTE3	5122	C	No
PcGTE4	5300	S	No
PcGTE5	7058	M	No
PcGTT2.1	6766	C	No
PcGTT2.2	6683	M/C	No
PcGTT2.2	7971	M	Yes
PcMAK16	137531	N	No
PcEFBγ	39727	C	No
Benzoquinone reductase			
PcBQR1	10307	S/C/M	Yes
PcBQR2	121028	C	Yes
PcBQR3	129887	S/C	Yes
PcBQR4	139901	C	Yes
Alternative dehydrogenase			
PcNDI	3743	M	No
PcNDE1.1	134180	C	No
PcNDE1.2	123031	M	No
PcNDE2	6157	M	No
Alternative oxidase			
PcAOX	40093	M	No

Localizations in bold have been determined according to experimental studies; the others are based on software predictions (psort, target and mitoprot). M, mitochondria; C, cytosol; N, nucleus; S, secreted; Mb, membrane.

L-ascorbate in these organisms. Usually, erythroascorbate in fungi appears to be glycosylated at the C_5 position by a range of sugars including glucose, galactose and xylose. In *P. chrysosporium*, metabolomic experiments have failed to identify both ascorbate and erythroascorbate in this fungus; however, threonate and erythronate were identified as oxygen-stress responsive metabolites (Miura *et al.*, 2004). The significance and the role of this metabolite accumulation during oxidative stress remain unexplained. In the same experiments, intracellular accumulation of veratryl alcohol (VA) has been reported when *P. chrysosporium* was submitted to oxidative stress (Miura *et al.*, 2004). VA is a well-characterized secondary metabolite involved in LiP protection and lignin degradation (Kersten and Cullen, 2007). Again, the significance of this accumulation remains to be explored.

B. ENZYMATIC ROS DETOXIFICATION

Concerning *P. chrysosporium* and more generally basidiomycetes, studies devoted to enzymes involved in ROS detoxification are particularly scarce. By comparison with other fungi and by mining genomic, transcriptomic and proteomic databases, different genes coding for proteins involved in the cellular redox balance could be identified in *P. chrysosporium* (Table I). The predicted proteins are putatively distributed in several subcellular compartments.

1. Superoxide dismutases

Superoxide dismutases (SODs) have an antioxidant function by catalyzing the disproportionation of superoxide anion to hydrogen peroxide. The SOD activity requires the presence of active metal ions in the active site allowing to classify them as Cu/ZnSODs and MnSODs. In *P. chrysosporium*, the presence of at least three MnSOD genes has been reported: MnSOD1 and two additional putative MnSOD genes: MnSOD2, which is located in scaffold 8 (nucleotides 478804 to 479683) and MnSOD3, which is located in scaffold 9 (nucleotides 1861632 to 1862495) (Matityahu *et al.*, 2008). On the other hand, the same authors indicated in a previous study that no Cu/ZnSOD activity or homologous sequence has been detected in this organism (Belinky *et al.*, 2002; Matityahu *et al.*, 2008). Nevertheless, additional sequences encoding putative SODs could be found in *P. chrysosporium* genome: one related to a Cu/ZnSOD located in scaffold 8 (nucleotides 464475 to 465167) and one related to a Mn SOD in scaffold 2 (nucleotides 883075 to 884217) that we named MnSOD4. MnSOD1 has been shown to be localized in mitochondria (Belinky *et al.*, 2002), whereas the Cu/ZnSOD and MnSOD2 are predicted to be secreted. As confirmed by RNAi experiments, the main SOD activity observed in *P. chrysosporium* cultivated in liquid conditions is due to MnSOD1 (Belinky *et al.*, 2002; Matityahu *et al.*, 2008). The expression of MnSOD1 has been shown to be downregulated at the transcriptional level by the presence of exogenous benzoic acid (Matsuzaki *et al.*, 2008). In addition, MnSOD2 has also been shown to be down-regulated during initiation of ligninolytic enzyme production (Minami *et al.*, 2007). MnSOD1 expression has been particularly studied in Mn-depletion conditions, since these culture conditions induce an oxidative stress in *P. chrysosporium* and also induce lignin peroxidase production (Belinky *et al.*, 2003; Rothschild *et al.*, 1999). This relationship between oxidative stress, MnSOD1, and lignin peroxidase production is discussed in Section IV.

2. *Catalases*

Catalases reduce hydrogen peroxide using the redox properties of a haem group. It has been shown for long time that *P. chrysosporium* possesses catalases, four different isoforms having been detected (Kwon and Anderson, 2001). At least five encoding genes: PcCat1, PcCat2, PcCat3, PcCat4 and PcCat5 could be found in the sequenced genome (Table I). PcCat2, PcCat3 and PcCat 5 are clearly related together and belong to clade 2 as defined for *Histoplasma capsulatum* catalases (Johnson *et al.*, 2002). These three proteins are predicted to be cytosolic. On the other hand PcCat1 is related to small-subunit catalase protein with greatest similarity to known peroxisomal catalases (Johnson *et al.*, 2002).

Previous cytological studies have shown that a catalase activity is present in the periplasmic space and is up-regulated during ligninolytic metabolism (Forney *et al.*, 1982). More recently, PcCat4 has been shown to be the main isoform associated with the fungal outer membrane. The production of this isoform is strongly induced under ligninolytic conditions (up to 35-fold), suggesting a role in surveillance against extracellular ROS (Shary *et al.*, 2008). Furthermore, this isoform has also been shown to be upregulated after addition of vanillin in the culture medium (Shimizu *et al.*, 2005). Nevertheless, the detection of catalase activity in the whole cells does not seem to be directly correlated with the production of LiPs (Kwon and Anderson, 2001). It is particularly true during induction of LiPs by intracellular oxidative stress due to Mn depletion (Belinky *et al.*, 2003), suggesting that the regulation of intracellular isoforms is not linked to the lignin metabolism.

3. *Thiol peroxidases*

SODs and catalases are peroxidases which use metal ions to catalyze the reduction of ROS. Cells have also developed other enzymatic systems able to reduce oxidative molecules in the absence of a metal. These proteins possess often cysteinyl residue(s) in their catalytic site, these latter being involved in the reduction of peroxides to the corresponding alcohols. The catalytic efficiently of such enzymes is lower in comparison to metal-containing enzymes. Depending on the system used to regenerate these cysteine-containing enzymes, two classes of thiol peroxidases can be distinguished: the glutathione peroxidases and peroxiredoxins.

a. Glutathione peroxidases. Even though they are named Glutathione peroxidases (Gpxs), only mammalian Gpxs used glutathione for their regeneration. Their bacterial, fungal, and plant counterparts use thioredoxins as reductants (Rouhier and Jacquot, 2005). In *P. chrysosporium*, only one gene related to Gpx could be found, the corresponding protein (PcGpx) being predicted as cytosolic (Table I) (Morel *et al.*, 2008). This protein is related to the phospholipid hydroperoxide glutathione peroxidase 3 (Gpx3) from yeast and could have a function in the reduction of lipid hydroperoxides. Furthermore, the yeast Gpx3 acts as a sensor of oxidative stress and is involved in the oxidation of the AP1 transcriptional factor (Delaunay *et al.*, 2002). PcGpx has been shown to be slightly up-regulated after addition of exogenous vanillin (Shimizu *et al.*, 2005). Such aromatic compounds induce oxidative stress suggesting a function of this protein in the cellular redox control.

Besides the so-called Gpx, other fungal proteins use glutathione to reduce organic and non-organic peroxides. These proteins are related to glutathione-*S*-transferases (GSTs). In yeast, at least five isoforms are involved in cellular redox control (Herrero *et al.*, 2008): ScGTT1, ScGTT2 and three GST belonging to the omega class. The function(s) of these proteins remain unclear, but ScGTT1 is able to reduce *in vitro* organic hydroperoxides, whereas omega GSTs act as 1-cys thiol transferase (Garcera *et al.*, 2006). In *P. chrysosporium*, at least 27 genes encoding GST-related proteins have been detected (Morel *et al.*, 2009). First attempts have been performed in our laboratory to characterize this superfamily at the protein level. Particularly, PcGTT2, an isoform related to ScGTT2, has been shown to reduce very efficiently cumene hydroperoxides ($k_{cat} = 141\ s^{-1}$) (Morel *et al.*, unpublished data). This efficiency suggests that this isoform could be involved in the reduction of oxidized lipids and also in the membrane repair after lipid peroxidation.

b. Peroxiredoxins. Among the thiol-peroxidase superfamily, four groups of peroxiredoxins (Prxs) are present in fungi based on sequence homology and also on biochemical data : 2-Cys Prxs, 1-Cys Prx, Prx Q and type II Prx (Rouhier and Jacquot, 2005). In *P. chrysosporium*, the distribution of Prxs is the following one: one 2-Cys Prx (Pc2-CysPrx), two 1-Cys Prx (Pc1-CysPrx1 and Pc1-CysPrx2), two Prx Q (PcPrxQ1 and PcPrxQ2) and two type II Prx

(PcPrxII1 and PcPrxII2) (Table I) (Morel *et al.*, 2008). To date, biochemical and physiological data concerning these proteins remain particularly scarce. The 2-Cys peroxiredoxin has been heterologously produced and the resulting purified protein exhibits a typical activity against hydrogen peroxide. Nevertheless, potential physiological reductants (i.e., thioredoxins) have not been tested in this study (Jiang *et al.*, 2005). The 2-Cys Prx has been identified in the cytosolic fraction by proteomics (Shimizu *et al.*, 2005) and shown to be down-regulated in the presence of benzoic acid (Matsuzaki *et al.*, 2008). PcPrxII1, Pc1-CysPrxs and Pc2-CysPrx have been found in *P. chrysosporium* microsomal preparations (Shary *et al.*, 2008). The localization of these proteins in microsomal fraction is in accordance with the hypothesis that peroxiredoxins and more generally thiol peroxidases are involved in ROS sensing and signalling mechanisms linked in particular to lipid peroxidation (Rouhier and Jacquot, 2005). Among the Prxs not identified in this proteomic study, one is predicted nuclear (PcPrx Q2) by protein localization software. It is interesting to note that no Prx is predicted as mitochondrial in *P. chrysosporium* (Morel *et al.*, 2008).

C. REPAIR OF OXIDATIVE PROTEIN DAMAGE

Besides the enzymatic network involved in ROS detoxification, cells possess also enzymatic mechanisms involved in repairing ROS-damaged proteins. The sulphur containing amino acids, cysteine and methionine, are the major targets of oxidation, their level of oxidation being also used by the cells as signalization mechanisms. Two thiol-related systems are mainly involved in the redox regulation/protein repair mechanisms, namely the thioredoxin system (Gelhaye *et al.*, 2005) and the glutathione/glutaredoxin system (Rouhier *et al.*, 2004). In this volume, the papers by Hagglund *et al.*, Nishiyama and Hisabori, Li and Zachgo, Selles *et al.* and Gao *et al.* describe various aspects of the thioredoxin and glutaredoxin systems in plants and cyanobacteria and they are a complement to this study in fungi.

1. The thioredoxin system

Thioredoxins are small proteins, which possess two cysteines in their catalytic site. They catalyze dithiol/disulphide exchange interacting with more or less specific targets as Gpxs or Prxs (Gelhaye *et al.*, 2005), and besides their function in redox signalization/ detoxification, they are involved in various cellular processes (Gelhaye *et al.*, 2005; Herrero *et al.*, 2008). In *P. chrysosporium*, four genes encoding thioredoxins have been identified: PcTrx1, PcTrx2, PcTrx3 and PcTrx4 (Morel *et al.*, 2008). PcTrx4 is predicted to be mitochondrial, whereas PcTrx1 and PcTrx3 are probably cytosolic (Table I). A putative secretion signal has been detected in the N-terminal part of PcTrx2 (Morel *et al.*, 2008). Interestingly, PcTrx2 has been shown to be present in *P. chrysosporium* microsomal preparations (Shary *et al.*, 2008). Since phylogenetic studies suggest that the presence of PcTrx2 orthologs is restricted to basidiomycetes (Morel *et al.*, 2008), further work will be required to functionally characterize this isoform. PcTrx1 and PcTrx4 have been produced heterologously in our laboratory; the resulting proteins are able to reduce efficiently dithionitrobenzene, a compound classically used to characterize thioredoxins (Gelhaye *et al.*, unpublished data). The regeneration of Trxs is usually catalyzed by NADPH thioredoxin reductase (NTR). In *P. chrysosporium*, only one gene encoding an NTR could be found (PcNTR). Nevertheless, an N-terminal extension recognized as a transit peptide has been predicted suggesting an alternative splicing leading to either a cytosolic or a mitochondrial targeting of the resulting proteins (Morel *et al.*, 2008,). This dual targeting has been previously shown for instance in the case of *Cryptococcus neoformans* NTR or for the *Arabidopsis* protein (Missall and Lodge, 2005; Reichheld *et al.*, 2005). After production of PcNTR in *Escherichia coli*, the resulting flavoprotein was able to reduce PcTrx1 and PcTrx4, confirming that *P. chrysosporium* possesses a functional thioredoxin system (Gelhaye *et al.*, unpublished data). In the presence of aromatic hydrocarbons, which have been shown to increase the mitochondrial production of ROS, PcNTR production was up-regulated at the protein level (Shimizu *et al.*, 2005). In contrast, the induction of the Trx system does not seem to be not correlated with the lignin metabolism, since the expression of PcTrx1 is downregulated at the transcriptional level during the initiation of ligninolytic enzyme production (Minami *et al.*, 2007). Similar results have been obtained in our laboratory when the whole Trx system expression was studied under Mn depletion, conditions which also trigger LiP production (Morel *et al.*, unpublished data).

2. The glutathione/glutaredoxin system

Glutaredoxins are also small proteins involved in various cellular reactions mainly due to the presence of one or two cysteinyl residues in their active site (Rouhier *et al.*, 2008). In *P. chrysosporium*, four genes encoding glutaredoxins have been reported: PcGrx1, PcGrx2, PcGrx3 and PcGrx4 (Table I) (Morel *et al.*, 2008). Based on phylogenetic analysis and sequence comparison with previously characterized proteins, it has been suggested that PcGrx3 should be involved in mitochondrial formation of Fe–S cluster (Morel *et al.*, 2008; Rouhier *et al.*, 2004). Concerning PcGrx4 and PcGrx1, a nuclear localization is predicted for both proteins. Nevertheless, PcGrx1 could be also targeted to the cytosol (Morel *et al.*, 2008). On the other hand, PcGrx2 is predicted to be secreted. To date, among the different transcriptomic and proteomic studies (Matsuzaki *et al.*, 2008; Minami *et al.*, 2007; Shary *et al.*, 2008; Shimizu *et al.*, 2005), no glutaredoxin has been identified. Grxs could act as thiol oxidoreductases, reducing protein disulphides or glutathione–protein mixed disulphides, with reduced glutathione acting as hydrogen donor. The system includes also NADPH and glutathione reductase (GR). In *P. chrysosporium*, three genes encoding GRs have been found: PcGR1, PcGR2 and PcGR3. PcGR1 and PcGR2 are predicted to be mitochondrial, whereas PcGR3 should be cytosolic (Table I) (Morel *et al.*, 2008). A putative alternative translation initiation site containing the consensus sequence AXXAUG is found in the genes encoding GR 876 and GR 10525 (Morel *et al.*, 2008). The dual targeting has been confirmed for GR 876 which has been found in the cytosolic fraction (Shimizu *et al.*, 2005). Overall, five isoforms (two mitochondrial and three cytosolic) of GR should be present in *P. chrysosporium*, PcGR1, PcGR2 and PcGR3 genes being expressed (Morel *et al.*, 2008). The physiological significance of such high number of isoforms remains unclear and requires further investigations.

3. Methionine sulphoxide reductases

Methione sulphoxide reductases (Msr) are enzymes able to reduce methione sulphoxide (MetSO) resulting from methionine oxidation occurring under oxidative conditions (Agbas and Moskovitz, 2009; Tarrago *et al.*, 2009; Vlamis-Gardikas, 2008). Two classes of Msr could be identified, MsrA and MsrB, specific for the *S*- and *R*-diastereomers of MetSO, respectively. In *P. chrysosporium*, one MsrA (PcMsrA) and one MsrB (PcMsrB) have been identified, both proteins being predicted as mitochondrial (Table I) (Morel *et al.*, 2008). No additional data are available about expression and activity of these proteins.

D. PREVENTING ROS FORMATION AND SUBSEQUENT OXIDATIVE DAMAGES

Various compounds are able to initiate redox reactions leading to the production of ROS. Among them, phenolic compounds are known as being able to produce toxic ROS molecules. In fact, these aromatic compounds could be oxidized to *ortho*- and *para*-related benzoquinones. Quinones, beside being potential strong electrophilic molecules (Huyen *et al.*, 2009), also act as catalysts in the generation of ROS. In fact, quinones can undergo one-electron reduction resulting in the generation of semiquinone radicals, which reduce molecular oxygen, producing ROS such as superoxide anion and then leading to redox-cycling reactions. Since white-rot fungi and particularly *P. chrysosporium* are able to completely mineralize lignin, they are exposed to various quinones resulting in particular from the lignin oxidation. Since benzoquinones are generated by the peroxidase-catalyzed oxidation of lignin and appear to be key intermediates in the degradation of aromatic compounds (Valli and Gold, 1991; Valli *et al.*, 1992), *P. chrysosporium* is also thought to possess extensive systems for preventing ROS formation from quinones. Besides the quinone catabolic pathways (Valli and Gold, 1991; Valli *et al.*, 1992), which will be not discussed here, we will give later an overview of benzoquinone reductases (BQR) and conjugation systems present in *P. chrysosporium.* In addition, alternative pathways preventing ROS production in mitochondria, mainly due to incomplete reduction of ubiquinone or menaquinone by NADH oxidoreductases will be discussed in this section.

1. Benzoquinone reductases

As previously discussed, a wide variety of oxidized metabolic intermediates are generated during the oxidative degradation of lignin, including in particular substituted quinones and hydroquinones. Some quinones are potent redox active compounds generating a redox cycling with their corresponding semiquinone radical, thus producing superoxide anion radicals. To prevent this first one-electron reduction step, the presence of intracellular enzymes involved in the reduction of benzoquinones has been shown in *P. chrysosporium* (Akileswaran *et al.*, 1999; Brock and Gold, 1996; Brock *et al.*, 1995). These enzymes are not able to directly reduce ROS, but they prevent ROS generation. Mining the genome suggests that at least four genes encode 1,4 BQR-related proteins: PcBQR1, PcBQR2, PcBQR3 and PcBQR4 (Table I). PcBQR1 has been previously thoroughly studied and has been shown to be cytosolic (Akileswaran *et al.*, 1999; Brock and Gold, 1996). Nevertheless, an export peptide signal is present in the N-terminal part of the protein suggesting the possibility of alternative splicing (Akileswaran *et al.*, 1999). PcBQR3 is predicted to be also secreted, whereas PcBQR2 and PcBQR4 seem to be cytosolic. The expression of PcBQR1 has been shown to be highly regulated both at the transcriptional and the translational levels induced by the presence of benzoquinone-related compounds (Akileswaran *et al.*, 1999; Shimizu *et al.*, 2005). As other BQRs (Andrade *et al.*, 2007; Lee *et al.*, 2007; Turley and Taliercio, 2008), PcBQR1 is a NADPH-dependent quinone reductase using FMN as cofactor, able to reduce various quinone-containing compounds (Brock *et al.*, 1995). Moreover, quinones can be reduced via other flavoproteins as NTR (Bironaite *et al.*, 1998).

*2. Glutathione-*S*-transferases*

Aromatic phenolic compounds, which result from the activation of xenobiotics by various oxidative enzymes as P450 monooxygenases and peroxidases, can be conjugated to biomolecules, that is, sugars via glycosyltransferases or to the tripeptide glutathione via glutathione *S*-transferases (GSTs). Since nothing is known about glycosyltransferases in *P. chrysosporium*, only an overview of GSTs is given as follows.

The GST main chemistry is to catalyze the conjugation of the tripeptide glutathione with compounds containing an electrophilic centre, in particular on quinone-derived compounds, to form more soluble, non-toxic peptide derivatives, ready to be degraded, excreted, or compartmentalized by phase III enzymes (Hayes *et al.*, 2005). In addition, during oxidative stress, lipid peroxidation leads to the formation of electrophilic compounds as 4-hydroxy-2-nonenal (4-HNE) (Forman *et al.*, 2008). 4-HNE can bind covalently to proteins forming stable adducts and leading to their inactivation. GSTs catalyze Michael addition of GSH to these various substrates and are important in preventing ROS damages.

In *P. chrysosporium*, at least 27 genes encoding related-GSTs have been detected in the genome (Table I) and six classes could be identified. The URE2p, GTT2, EFBγ, MAK16 and Omega classes have been identified in agreement with a previous study on ascomycetes (McGoldrick *et al.*, 2005). A phylogenetic analysis has revealed a new class, enclosing proteins which are related to bacterial LIG proteins and which could also be involved in the degradation of internalized lignin-derived compounds (Masai *et al.*, 2007). *P. chrysosporium* possesses five isoforms of this class that we named GTE (glutathione transferase etherase-related) (Morel *et al.*, 2009). The functions of these different GSTs remain to be elucidated; nevertheless, some isoforms have been shown to be involved in pentachlorophenol catabolism pathway (Valli *et al.*, 1992). The first attempts made in our laboratory to characterize *P. chrysosporium* GSTs have shown that some isoforms (omega class), harbouring a cysteine in their catalytic site, are able to catalyze deglutathionylation reactions cleaving in particular C–S links. Further studies will be required to determine the structural features of such enzymes and the physiological significance of such activities.

3. *Alternative pathways in mitochondria*

a. Alternative dehydrogenases. In most animals, complex I provides the sole mechanism for entry of electrons from NADH into the respiratory chain. By contrast, in fungi as in plants it seems that alternatives to complex I are widely distributed. Three classes of NADH:ubiquinone oxidoreductases are now recognized. Complex I is assigned to class 1, while alternative dehydrogenases of plant and fungal mitochondria belong to class 2. In class 3, electron transfer is coupled to Na^+ pumping, but this is restricted to bacteria (Joseph-Horne *et al.*, 2001). *S. cerevisiae* lacks complex I. Instead, reducing equivalents are delivered to the respiratory chain via three NADH dehydrogenases (NADH-dh): NDE1, NDE2 (both external) and NDI1 (internal) (De Vries *et al.*, 1992). The other fungus in which alternative NADH-dh have been extensively studied is *Neurospora crassa* (Joseph-Horne *et al.*, 2001). These enzymes use NAD(P)H from either the cytosol (external enzymes) or the mitochondrial matrix (internal enzymes). They do not pump protons and may be useful as a system that keeps reducing equivalents at physiological levels, but their precise role is not completely clear. The variation in number and specificity among species suggests that they fulfil specific needs of different organisms.

The proton-pumping complex I and the alternative NADH-dh have overlapping roles and both activities are probably required for the optimal functioning of the cells. In *N. crassa*, it seems that complex I and the NDE1 protein are constitutively expressed throughout the fungal life cycle (Melo *et al.*, 2001). The complexity of the enzyme, its involvement in several biological processes, such as fungal and plant development, and different metabolic pathways in microorganisms and, its contribution to energy conservation argue against the possibility of the complete substitution of complex I with alternative enzymes (Archer *et al.*, 1993; Claas *et al.*, 2000; Dupuis *et al.*; 1998, Laval-Favre *et al.*, 1997; Rasmusson *et al.*, 1998; Videira, 1998; Zambrano and Kolter, 1993). For instance, the introduction of the *E. coli NDH-2* gene in *Pseudomonas denitrificans* allowed the disruption of complex I genes in the latter organism (Finel, 1996). Moreover, the *NDI1* gene of *S. cerevisiae* was used to complement complex I defects in mammalian cells (Seo *et al.*, 1998, 1999, 2000).

Four sequences have been identified as putative NADH-dh in *P. chrysosporium*. One sequence (PcNDI) exhibits homology with internal NADH-dh, and three sequences could correspond to external NADH-dh (PcNDE1.1, PcNDE1.2, PcNDE2) (Table I). However nothing is known concerning their regulation.

b. Alternative oxidase. AOX is present in the mitochondria of plants, fungi and many types of yeast. This protein exists in the inner membrane as a homodimer. Two distinct states of the dimer can be identified: an oxidized state in which the dimer is covalently cross-linked by a disulphide bridge and a reduced state which is maintained through non-covalent interactions (Sluse and Jarmuszkiewicz, 1998). The reduced form can be four-to five-fold more active than the oxidized form and the ratio of oxidized to reduced protein varies considerably between species and tissues (Umbach and Siedow, 1993). This enzyme transfers electrons from the ubiquinol pool directly to oxygen without contributing to the proton transfer across the mitochondrial membrane. Electron flow through the alternative pathway bypasses two of the three sites along the cytochrome chain where electron transport is coupled to ATP synthesis.

AOX is involved in stress responses, programmed cell death, and maintenance of the cellular redox balance. Its activity is controlled by several parameters among which we can distinguish regulatory events and substrate availability. Regulation of AOX activity can occur at different levels: (i) gene expression that affects the amount of the protein in the membrane and differential gene expression that modifies the ratio between isoforms; (ii) post-translational modifications of the protein (i.e. its redox status that affects the nature of the dimer); (iii) the action of allosteric effectors like pyruvate (Sluse and Jarmuszkiewicz, 1998).

Interestingly, stimulation of AOX by AMP and respiratory rates obtained after inhibition of the cytochrome pathway showed that fungal/protist AOX is activated under low-energy conditions, in contrast to plant AOX, which is activated under high-energy conditions (Juarez *et al.*, 2006).

In *Aspergillus fumigatus*, both AOX activity and mRNA expression were induced with menadione or paraquat, suggesting an important role of AOX under oxidative stress (Magnani *et al.*, 2007). Similarly, knock-out AOX mutants are more susceptible both to an imposed *in vitro* oxidative stress condition and to macrophage killing, suggesting that AOX is required for the *A. fumigatus* pathogenicity, mainly for the survival of the fungus conidia during host infection and resistance to ROS generated by macrophages (Magnani *et al.*, 2008). It has been proposed that AOX may serve a more general function by limiting mitochondrial ROS formation (Maxwell *et al.*, 1999).

P. chrysosporium exhibits one sequence coding for AOX (PcAOX). Nothing is known concerning its regulation in *P. chrysosporium*.

IV. RELATIONSHIP BETWEEN INTRACELLULAR ROS AND LIGNIN DEGRADATION

The formation of LiP is particularly dependent on exposure of cultures to high oxygen tensions; indeed, cultures of *P. chrysosporium* which are oxygen starved at the pellet centre produce much lower levels of LiPs and MnPs (Dosoretz *et al.*, 1990). It has been thus proposed that a high partial pressure of oxygen in the culture headspace is needed to make sufficient oxygen available to the submerged hyphae (Michel *et al.*, 1992). Oxygen has, however, the potential to give rise to toxic oxygen-free radicals that are capable of oxidizing, fragmenting, and cross-linking proteins, carbohydrates, lipids, and nucleic acids. Zacchi *et al.* (2000) suggested that cultures of

P. chrysosporium exposed to O_2 to trigger LiP synthesis are subjected to oxygen toxicity, which leads to disorganization of the cellular ultrastructure and chlamydospore development, probably in response to accumulating ROS. This ROS accumulation has been also confirmed by ROS concentration measurements and the enhancement of the antioxidant defence system in *P. chrysosporium* (Belinky *et al.*, 2003). It has been thus proposed that the induction of LiP by oxygen may result from reactions of ROS with the cell surface or from the entry of these species into the cells.

The LiP formation normally induced by high oxygen concentrations can be replaced by manganese deficiency, and this substitution is effective under both nitrogen limitation and excess (Rothschild *et al.*, 1999). The question arising from this result concerns the mechanism by which Mn deficiency can

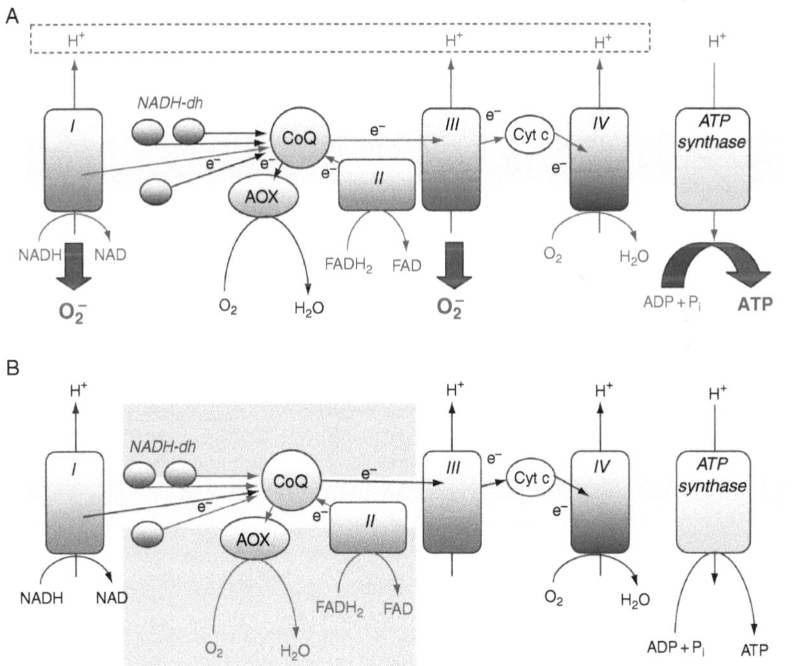

Fig. 3. Regulation of the mitochondrial electron transport chain of *P. chrysosporium* grown in a ligninolytic condition compared to a control condition. (A) When the intracellular oxidative stress is low, all the classical mitochondrial complexes are involved in creating the proton gradient required to produce ATP. Complexes I and III are responsible for most of the superoxide production. To maintain a nontoxic amount of superoxide inside the cell, some detoxifying enzymes function in their degradation. (B) When there is a high oxidative stress, another level of regulation involves alternative mitochondrial pathways to bypass complexes I and III and limit ROS production. Red arrows indicate the preferential pathway in conditions (A) and (B). (See color plate section at the back of the book.)

replace the high level of oxygen needed for LiP formation. A hypothetical explanation may be that both Mn deficiency and a high level of oxygen can increase the level of oxygen radicals generated within the fungal cell. The fact that no Mn-dependent superoxide dismutase (Mn-SOD) activity was observed when there was a Mn deficiency is consistent with results reported for other non-fungal organisms (Borrello *et al.*, 1992; Hassan and Schrum, 1994). Because of the importance of Mn-SOD and high Mn^{2+} levels as antioxidants, it may be assumed that Mn-deficient cultures of *P. chrysosporium* are more susceptible to oxidative stress than non-deficient cultures. An antioxidant role of LiP has consequently been proposed (Morpeth, 1987), since LiP formation was initiated only after a decrease in the levels of the antioxidant enzymes, catalase and SOD.

Another aspect of LiP induction has been studied. We focused on the bioenergetic state of LiP production in *P. chrysosporium*. Although no significant difference in total respiration has been observed between ligninolytic and non-ligninolytic conditions, some differences in mitochondrial functioning have been highlighted (Fig. 3) (Morel *et al.*, unpublished data). The participation of complex I is reduced in the ligninolytic conditions and all or almost all the activity is assured by the NADH-dehydrogenases (NADH-dh) (Fig. 3B). Moreover, oxidative activity is rather due to AOX than to complex IV. By reducing the activity of complexes I, III, and IV, the cells reduce the proton gradient required for ATP synthesis, leading to a lower energetic state of the fungus grown in a ligninolytic conditions. This supports the hypothesis that a reduced bioenergetic state of fungal cells could be involved in LiP induction (Morel *et al.*, unpublished data).

REFERENCES

Agbas, A. and Moskovitz, J. (2009). The role of methionine oxidation/reduction in the regulation of immune response. *Current Signal Transduction Therapy* **4**, 46–50.

Aguirre, J., Rios-Momberg, M., Hewitt, D. and Hansberg, W. (2005). Reactive oxygen species and development in microbial eukaryotes. *Trends in Microbiology* **13**, 111–118.

Akileswaran, L., Brock, B. J., Cereghino, J. L. and Gold, M. H. (1999). 1, 4-benzoquinone reductase from *Phanerochaete chrysosporium*: cDNA cloning and regulation of expression. *Applied and Environmental Microbiology* **65**, 415–421.

Alic, M., Akileswaran, L. and Gold, M. H. (1997). Characterization of the gene encoding manganese peroxidase isozyme 3 from *Phanerochaete chrysosporium*. *Biochimica et Biophysica Acta* **1338**, 1–7.

Andrade, S. L. A., Patridge, E. V., Ferry, J. G. and Einsle, O. (2007). Crystal structure of the NADH: Quinone oxidoreductase WrbA from *Escherichia coli*. *Journal of Bacteriology* **189**, 9101–9107.

Archer, C. D., Wang, X. and Elliott, T. (1993). Mutants defective in the energy-conserving NADH dehydrogenase of *Salmonella typhimurium* identified by a decrease in energy-dependent proteolysis after carbon starvation. *Proceedings of the National Academy of Sciences of the United States of America* **90**, 9877–9881.

Artolozaga, M. J., Kubatova, E., Volc, J. and Kalisz, H. M. (1997). Pyranose 2-oxidase from *Phanerochaete chrysosporium*—Further biochemical characterisation. *Applied Microbiology and Biotechnology* **47**, 508–514.

Baroja-Mazo, A., del Valle, P., Rua, J., de Cima, S., Busto, F., de Arriaga, D. and Smirnoff, N. (2005). Characterisation and biosynthesis of D-erythroascorbic acid in *Phycomyces blakesleeanus*. *Fungal Genetics and Biology* **42**, 390–402.

Belinky, P. A., Goldberg, D., Krinfeld, B., Burger, M., Rothschild, N., Cogan, U. and Dosoretz, C. G. (2002). Manganese-containing superoxide dismutase from the white-rot fungus *Phanerochaete chrysosporium*: Its function, expression and gene structure. *Enzyme and Microbial Technology* **31**, 754–764.

Belinky, P. A., Flikshtein, N., Lechenko, S., Gepstein, S. and Dosoretz, C. G. (2003). Reactive oxygen species and induction of lignin peroxidase in *Phanerochaete chrysosporium*. *Applied and Environmental Microbiology* **69**, 6500–6506.

Bironaite, D., Anuseviciu, Z., Jacquot, J. P. and Cenas, N. (1998). Interaction of quinones with *Arabidopsis thaliana* thioredoxin reductase. *Biochimica et Biophysica Acta* **1383**, 82–92.

Borrello, S., De Leo, M. E., Wohlrab, H. and Galeotti, T. (1992). Manganese deficiency and transcriptional regulation of mitochondrial superoxide dismutase in hepatomas. *FEBS Letters* **310**, 249–254.

Bouws, H., Wattenberg, A. and Zorn, H. (2008). Fungal secretomes—Nature's toolbox for white biotechnology. *Applied and Microbiology and Biotechnology* **80**, 381–388.

Brock, B. J. and Gold, M. H. (1996). 1,4-benzoquinone reductase from the basidiomycete *Phanerochaete chrysosporium*: Spectral and kinetic analysis. *Archives of Biochemistry and Biophysics* **331**, 31–40.

Brock, B. J., Rieble, S. and Gold, M. H. (1995). Purification and characterization of a 1,4 benzoquinone reductase from the basidiomycete *Phanerochaete chrysosporium*. *Applied and Environmental Microbiology* **61**, 3076–3081.

Bunik, V. I. and Sievers, C. (2002). Inactivation of the 2-oxo acid dehydrogenase complexes upon generation of intrinsic radical species. *European Journal of Biochemistry* **269**, 5004–5015.

Cadenas, E. and Davies, K. J. (2000). Mitochondrial free radical generation, oxidative stress, and aging. *Free Radical Biology and Medicine* **29**, 222–230.

Claas, K., Weber, S. and Downs, D. M. (2000). Lesions in the nuo operon, encoding NADH dehydrogenase complex I, prevent PurF-independent thiamine synthesis and reduce flux through the oxidative pentose phosphate pathway in *Salmonella enterica* serovar typhimurium. *Journal of Bacteriology* **182**, 228–232.

Daniel, G., Volc, J. and Kubatova, E. (1994). Pyranose oxidase, a major source of H(2)O(2) during wood degradation by *Phanerochaete chrysosporium*, *Trametes versicolor*, and *Oudemansiella mucida*. *Applied and Environmental Microbiology* **60**, 2524–2532.

Delaunay, A., Pflieger, D., Barrault, M. B., Vinh, J. and Toledano, M. B. (2002). A thiol peroxidase is an H_2O_2 receptor and redox-transducer in gene activation. *Cell* **111**, 471–481.

De Vries, S., Van Witzenburg, R., Grivell, L. A. and Marres, C. A. (1992). Primary structure and import pathway of the rotenone-insensitive NADH-ubiquinone

Pease, E. A., Andrawis, A. and Tien, M. (1989). Manganese-dependent peroxidase from *Phanerochaete chrysosporium*. Primary structure deduced from cDNA sequence. *Journal of Biological Chemistry* **264**, 13531–13535.

Pribnow, D., Mayfield, M. B., Nipper, V. J., Brown, J. A. and Gold, M. H. (1989). Characterization of a cDNA encoding a manganese peroxidase, from the lignin-degrading basidiomycete *Phanerochaete chrysosporium*. *Journal of Biological Chemistry* **264**, 5036–5040.

Rasmusson, A. G., Heiser, V. V., Zabaleta, E., Brennicke, A. and Grohmann, L. (1998). Physiological, biochemical and molecular aspects of mitochondrial complex I in plants. *Biochimica et Biophysica Acta* **1364**, 101–111.

Ravalason, H., Jan, G., Molle, D., Pasco, M., Coutinho, P. M., Lapierre, C., Pollet, B., Bertaud, F., Petit-Conil, M., Grisel, S., Sigoillot, J. C. Asther, M. *et al.* (2008). Secretome analysis of *Phanerochaete chrysosporium* strain CIRM-BRFM41 grown on softwood. *Applied Microbiology and Biotechnology* **80**, 719–733.

Reichheld, J. P., Meyer, E., Khafif, M., Bonnard, G. and Meyer, Y. (2005). AtNTRB is the major mitochondrial thioredoxin reductase in *Arabidopsis thaliana*. *FEBS Letters* **579**, 337–342.

Rothschild, N., Levkowitz, A., Hadar, Y. and Dosoretz, C. G. (1999). Manganese deficiency can replace high oxygen levels needed for lignin peroxidase formation by *Phanerochaete chrysosporium*. *Applied and Environmental Microbiology* **65**, 483–488.

Rouhier, N. and Jacquot, J. P. (2005). The plant multigenic family of thiol peroxidases. *Free Radical Biology and Medicine* **38**, 1413–1421.

Rouhier, N., Gelhaye, E. and Jacquot, J. P. (2004). Plant glutaredoxins: Still mysterious reducing systems. *Cellular and Molecular Life Sciences* **61**, 1266–1277.

Rouhier, N., Lemaire, S. D. and Jacquot, J. P. (2008). The role of glutathione in photosynthetic organisms: Emerging functions for glutaredoxins and glutathionylation. *Annual Review of Plant Biology* **59**, 143–166.

Sazanov, L. A. (2007). Respiratory complex I: Mechanistic and structural insights provided by the crystal structure of the hydrophilic domain. *Biochemistry* **46**, 2275–2288.

Seo, B. B., Kitajima-Ihara, T., Chan, E. K., Scheffler, I. E., Matsuno-Yagi, A. and Yagi, T. (1998). Molecular remedy of complex I defects: Rotenone-insensitive internal NADH-quinone oxidoreductase of *Saccharomyces cerevisiae* mitochondria restores the NADH oxidase activity of complex I-deficient mammalian cells. *Proceedings of the National Academy of Sciences of the United States of America* **95**, 9167–9171.

Seo, B. B., Matsuno-Yagi, A. and Yagi, T. (1999). Modulation of oxidative phosphorylation of human kidney 293 cells by transfection with the internal rotenone-insensitive NADH-quinone oxidoreductase (NDI1) gene of *Saccharomyces cerevisiae*. *Biochimica et Biophysica Acta* **1412**, 56–65.

Seo, B. B., Wang, J., Flotte, T. R., Yagi, T. and Matsuno-Yagi, A. (2000). Use of the NADH-quinone oxidoreductase (NDI1) gene of *Saccharomyces cerevisiae* as a possible cure for complex I defects in human cells. *Journal of Biological Chemistry* **275**, 37774–37778.

Shary, S., Kapich, A. N., Panisko, E. A., Magnuson, J. K., Cullen, D. and Hammel, K. E. (2008). Differential expression in *Phanerochaete chrysosporium* of membrane-associated proteins relevant to lignin degradation. *Applied and Environmental Microbiology* **74**, 7252–7257.

Shimizu, M., Yuda, N., Nakamura, T., Tanaka, H. and Wariishi, H. (2005). Metabolic regulation at the tricarboxylic acid and glyoxylate cycles of the lignin-degrading basidiomycete *Phanerochaete chrysosporium* against exogenous addition of vanillin. *Proteomics* **5**, 3919–3931.

Silar, P. (2005). Peroxide accumulation and cell death in filamentous fungi induced by contact with a contestant. *Mycological Research* **109**, 137–149.

Singh, D. and Chen, S. (2008). The white-rot fungus *Phanerochaete chrysosporium*: Conditions for the production of lignin-degrading enzymes. *Applied Microbiology and Biotechnology* **81**, 399–417.

Skulachev, V. P. (2006). Bioenergetic aspects of apoptosis, necrosis and mitoptosis. *Apoptosis* **11**, 473–485.

Sluse, F. E. and Jarmuszkiewicz, W. (1998). Alternative oxidase in the branched mitochondrial respiratory network: An overview on structure, function, regulation, and role. *Brazilian Journal of Medical and Biological Research* **31**, 733–747.

Smirnoff, N. and Wheeler, G. L. (2000). Ascorbic acid in plants: Biosynthesis and function. *Critical Reviews in Biochemistry and Molecular Biology* **35**, 291–314.

Starkov, A. A., Fiskum, G., Chinopoulos, C., Lorenzo, B. J., Browne, S. E., Patel, M. S. and Beal, M. F. (2004). Mitochondrial alpha-ketoglutarate dehydrogenase complex generates reactive oxygen species. *Journal of Neuroscience* **24**, 7779–7788.

Stewart, P. and Cullen, D. (1999). Organization and differential regulation of a cluster of lignin peroxidase genes of *Phanerochaete chrysosporium. Journal of Bacteriology* **181**, 3427–3432.

Takemoto, D., Tanaka, A. and Scott, B. (2007). NADPH oxidases in fungi: Diverse roles of reactive oxygen species in fungal cellular differentiation. *Fungal Genetics and Biology* **44**, 1065–1076.

Tarrago, L., Laugier, E. and Rey, P. (2009). Protein-repairing methionine sulfoxide reductases in photosynthetic organisms: Gene organization, reduction mechanisms, and physiological roles. *Molecular Plant* **2**, 202–217.

Tien, M. and Kirk, T. K. (1983). Lignin-degrading enzyme from the hymenomycete *Phanerochaete chrysosporium* burds. *Science* **221**, 661–663.

Tien, M. and Kirk, T. K. (1984). Lignin-degrading enzyme from *Phanerochaete chrysosporium*: Purification, characterization, and catalytic properties of a unique H(2)O(2)-requiring oxygenase. *Proceedings of the National Academy of Sciences of the United States of America* **81**, 2280–2284.

Tretter, L. and Adam-Vizi, V. (2004). Generation of reactive oxygen species in the reaction catalyzed by alpha-ketoglutarate dehydrogenase. *Journal of Neuroscience* **24**, 7771–7778.

Tretter, L., Takacs, K., Hegedus, V. and Adam-Vizi, V. (2007). Characteristics of alpha-glycerophosphate-evoked H2O2 generation in brain mitochondria. *Journal of Neurochemistry* **100**, 650–663.

Turley, R. B. and Taliercio, E. (2008). Cotton benzoquinone reductase: Up-regulation during early fiber development and heterologous expression and characterization in *Pichia pastoris*. *Plant Physiology and Biochemistry* **46**, 780–785.

Umbach, A. L. and Siedow, J. N. (1993). Covalent and noncovalent dimers of the cyanide-resistant alternative oxidase protein in higher plant mitochondria and their relationship to enzyme activity. *Plant Physiology* **103**, 845–854.

Valli, K. and Gold, M. H. (1991). Degradation of 2,4 dichlorophenol by the lignin-degrading fungus *Phanerochaete chrysosporium. Journal of Bacteriology* **173**, 345–352.

Valli, K., Brock, B. J., Joshi, D. K. and Gold, M. H. (1992). Degradation of 2,4 dinitrotoluene by the lignin-degrading fungus *Phanerochaete chrysosporium. Applied and Environmental Microbiology* **58**, 221–228.

Vanden Wymelenberg, A., Minges, P., Sabat, G., Martinez, D., Aerts, A., Salamov, A., Grigoriev, I., Shapiro, H., Putnam, N., Belinky, P., Dosoretz, C. Gaskell, J. *et al.* (2006a). Computational analysis of the *Phanerochaete chrysosporium* v2.0 genome database and mass spectrometry identification of peptides in ligninolytic cultures reveal complex mixtures of secreted proteins. *Fungal Genetics and Biology* **43**, 343–356.

Vanden Wymelenberg, A., Sabat, G., Mozuch, M., Kersten, P. J., Cullen, D. and Blanchette, R. A. (2006b). Structure, organization, and transcriptional regulation of a family of copper radical oxidase genes in the lignin-degrading basidiomycete *Phanerochaete chrysosporium. Applied and Environmental Microbiology* **72**, 4871–4877.

Varela, E., Guillen, F., Martinez, A. T. and Martinez, M. J. (2001). Expression of *Pleurotus eryngii* aryl-alcohol oxidase in *Aspergillus nidulans*: Purification and characterization of the recombinant enzyme. *Biochimica et Biophysica Acta* **1546**, 107–113.

Vasquez-Vivar, J., Kalyanaraman, B. and Kennedy, M. C. (2000). Mitochondrial aconitase is a source of hydroxyl radical. An electron spin resonance investigation. *Journal of Biological Chemistry* **275**, 14064–14069.

Videira, A. (1998). Complex I from the fungus *Neurospora crassa. Biochimica et Biophysica Acta* **1364**, 89–100.

Vlamis-Gardikas, A. (2008). The multiple functions of the thiol-based electron flow pathways of *Escherichia coli*: Eternal concepts revisited. *Biochimica et Biophysica Acta-General Subjects* **1780**, 1170–1200.

Whatley, S. A., Curti, D., Das Gupta, F., Ferrier, I. N., Jones, S., Taylor, C. and Marchbanks, R. M. (1998). Superoxide, neuroleptics and the ubiquinone and cytochrome *b*5 reductases in brain and lymphocytes from normals and schizophrenic patients. *Molecular Psychiatry* **3**, 227–237.

Zacchi, L., Morris, I. and Harvey, P. J. (2000). Disordered ultrastructure in lignin-peroxidase-secreting hyphae of the white-rot fungus *Phanerochaete chrysosporium. Microbiology* **146**, 759–765.

Zambrano, M. M. and Kolter, R. (1993). *Escherichia coli* mutants lacking NADH dehydrogenase I have a competitive disadvantage in stationary phase. *Journal of Bacteriology* **175**, 5642–5647.

Zhang, L., Yu, L. and Yu, C. A. (1998). Generation of superoxide anion by succinate-cytochrome *c* reductase from bovine heart mitochondria. *Journal of Biological Chemistry* **273**, 33972–33976.

Physiological Impact of Thioredoxin- and Glutaredoxin-Mediated Redox Regulation in Cyanobacteria

YOSHITAKA NISHIYAMA* AND TORU HISABORI†,[1]

*Department of Biochemistry and Molecular Biology, Graduate School of Science and Engineering, Saitama University, 255 Shimo-Okubo, Sakura-ku, Saitama 338-8570, Japan
†Chemical Resource Laboratory, Tokyo Institute of Technology, 4259 Nagatsuta, Midori-Ku, Yokohama 226-8503, Japan

ABSTRACT

Cyanobacteria are photosynthetic bacteria, which are thought to be derived from ancestral oxygen-evolving photosynthetic organisms. Recent progress in proteomics using redox-protein affinity chromatography, two-dimensional electrophoresis and mass spectrometry has improved our understanding of the complicated redox-regulation networks that exist in photosynthetic organisms, and studies with cyanobacteria have

[1]Corresponding author: Email: thisabor@res.titech.ac.jp

Advances in Botanical Research, Vol. 52

0065-2296/09 $35.00
DOI: 10.1016/S0065-2296(09)52007-1

made a great contribution to this area. Moreover, a number of remarkable differences relating to redox-regulated proteins between higher plants and cyanobacteria have also been uncovered as a result of these studies. In addition, novel redox-regulation systems that govern gene expression and protein synthesis have also been recently described for cyanobacteria. The redox-regulation system is an important multiphasic control system that ensures cell viability for this photosynthetic organism.

I. INTRODUCTION: THE REDOX-BALANCING SYSTEM IN CYANOBACTERIA

The architecture of the photosynthetic apparatus suggests that the ancestral oxygen-evolving photosynthetic organisms were closely related to extant cyanobacteria. Photosynthetic organisms have evolved the capability to use water as an electron donor for photosynthesis, leading to the synthesis of molecular oxygen. Over one billion years, oxygen concentrations in the earth's atmosphere increased, enabling the evolution of a new metabolic pathway, respiration. The organisms that were subjected to these novel atmospheric conditions rapidly evolved a set of new defence systems which allowed protection against oxidative damage caused by reactive oxygen species (ROS), a by-product of the presence of molecular oxygen within the cell. Two major defence systems that these ancestral organisms evolved are the redox-balancing system, which involves proteins such as thioredoxin (Trx) and glutaredoxin (Grx), and the anti-oxidative stress system, which involves proteins such as catalases, superoxide dismutases, peroxiredoxins and peroxidases. Moreover, the cellular redox-balancing system has become an efficient system which controls the metabolic activity of photosynthetic organisms allowing them to adapt to the alternating light–dark conditions.

Trx, the key player of the redox-balancing system, is a ubiquitous protein which is found across an extremely wide range of living organisms. Genomic analysis of various organisms indicates the presence of a strikingly wide variety of isoforms of Trx and Grx proteins. Even in the cyanobacterium *Synechocystis* sp. PCC 6803, whose genome size is only 3.57 Mb, four Trx isoform genes are present (Kaneko *et al.*, 1996a,b), and these genes are all expressed in physiological growth conditions (Florencio *et al.*, 2006; Hishiya *et al.*, 2008). These findings suggest that Trx is a critical protein, especially for photosynthetic organisms. In this review, we describe the significance of the redox-balancing system in cyanobacterial cells and discuss the newly identified functions of the redox-balancing system in sustaining cyanobacterial cell viability.

II. SYNCHRONIZATION BETWEEN REDOX EQUILIBRIUM AND PHOTOSYNTHESIS

The significance of the redox-balancing system in photosynthetic organisms is very often emphasized by the relevance between the reducing equivalents that are generated by the electron transfer reactions and the reduction levels of redox-sensitive chloroplast enzymes that are important for the carbon assimilation pathway. This relation is already defined for the thiol enzymes in chloroplasts, which work in ATP synthesis (Mills and Mitchell, 1982) and the Calvin–Benson cycle (Buchanan, 1991). The thiol enzymes involved in these systems are directly reduced and activated by Trx in chloroplasts in order to promote efficient photosynthesis. The redox-regulation system therefore seems to be a reasonable switching system for photosynthetic organisms to maintain the efficiency of metabolic pathways under light conditions and to avoid futile reverse reactions under dark conditions (Schurmann, 2003).

When the *Synechocystis* sp. PCC 6803 genome was fully sequenced (Kaneko *et al.*, 1996a,b), the existence of four genes for Trx (*sll1057*, *slr0233*, *slr0623* and *slr1139*) and three genes for Trx-like proteins (*sll0685*, *sll1980* and *slr1796*) was revealed. In addition, the genes for two Trx reductases, ferredoxin-Trx reductase (FTR, the heterodimer consisting of the gene products of *sll0554* and *ssr0330*) and NADPH-Trx reductase (NTR, the gene product of *slr0600*), were also identified. These two reduction pathways for Trx clearly perform different functions: disruption of the *ntr* gene results in cells which are very sensitive to oxidative stress, whereas disruption of the *ftr-v* gene caused slow cell growth but did not affect cell viability under oxidative stress conditions (Hishiya *et al.*, 2008). The reduction levels of Trx isoforms were also shown to vary within the gene disruptants described earlier. In addition, these gene disruptions strongly affected the expression levels of other redox-balancing system proteins, such as Trx and other Trx reductases, suggesting that the redox level in the cyanobacterial cells is an important determinant in the regulation of the expression of certain specific proteins, as mentioned elsewhere in this review.

The identification of novel proteins capable of interacting with Trx isoforms in photosynthetic organisms has been significantly enhanced by way of proteomic approaches, such as two-dimensional gel electrophoresis analysis (Yano *et al.*, 2001) and Trx-affinity chromatography (Balmer *et al.*, 2003; Motohashi *et al.*, 2001; Yamazaki *et al.*, 2004). More than 100 proteins in higher plant chloroplasts are currently reported to be potential Trx-interacting partners. Although many of the proteins in the list have not actually been confirmed biochemically as Trx target proteins (Hisabori *et al.*, 2007), the data accumulated certainly enhance our understanding of the redox-balancing system in chloroplasts and in cells of photosynthetic organisms (Montrichard *et al.*, 2009).

Another well-studied key protein in the redox-balancing system is Grx. Grx is related to Trx in terms of its three-dimensional structure, also possessing the so-called Trx fold, and the cysteine residues that form a similar active site. In *Synechocystis* sp. PCC 6803, two Grx genes, *slr1562* and *ssr2061*, have been identified (Kaneko *et al.*, 1996a,b): both proteins contain the conserved CPFC motif, a nearly classical active site sequence for Grx proteins. However, based on the growth phenotype of their disruptants, both proteins appear to be dispensable for cell growth under normal growth conditions (Marteyn *et al.*, 2009). In particular, the *grx1* gene disruptant (*slr1562*) showed no difference in growth compared to the wild type. In contrast, the *grx2* (*ssr2061*) disruptant did show some sensitivity to oxidative stress. Protein–protein interaction experiments have shown that the redox exchange reaction between Grx1 and Grx2 is observed *in vivo* as well as *in vitro* (Marteyn *et al.*, 2009). Furthermore, Grx1 seems to be reduced by the NTR system, suggesting the existence of crosstalk between the Trx system and the Grx system in cyanobacterial cells. Currently, the function of Grx is believed to be very diverse in various organisms, ranging from activation of ribonucleotide reductase (Holmgren, 1976), reduction of dehydroascorbate (Wells *et al.*, 1990), regulation of transcription factors, to protection of cells against apoptosis and cellular defence system for ROS (Holmgren, 2000). Grx is also involved in reversible protein inactivation by glutathionylation of the cysteine residues of the target proteins (Rouhier *et al.*, 2008).

Grx target proteins in *Synechocystis* sp. PCC 6803 were recently extensively surveyed by a proteomic approach, using monothiol Grx2 mutant (Li *et al.*, 2007), and these results can be compared to a similar study carried out with a Grx from poplar, a land plant (Rouhier *et al.*, 2005). In *Synechocystis*, 42 proteins were captured as potential target proteins, including anti-oxidative stress proteins, such as catalase and peroxiredoxin; several Calvin cycle enzymes, which are well known as thiol enzymes like phosphoribulokinase, glyceraldehyde 3-phosphate dehydrogenase and fructose 1,6-bisphosphatase; molecular chaperone; and elongation factor Tu. Of these suggested Grx target proteins, 13 proteins were included in the list of proteins captured by affinity chromatography using monothiol Trx in *Synechocystis* cells. Critical information relating to the existence of conserved cysteines in the molecule must be evaluated in order to discuss the overlap of the target proteins for Grx and Trx. As discussed in this review, there are certainly remarkable differences in the conserved cysteines among cyanobacteria, green algae and higher plants. For the comprehensive evaluation of these newly proposed redox pathways, individual biochemical experiments concerning the proteins listed as Trx or Grx targets are essential in order to further define the redox network system in the cells of photosynthetic organisms.

III. PHYSIOLOGICAL PHENOMENA CONTROLLED BY REDOX: GENE EXPRESSION

As mentioned earlier, light is an important signal that controls the synthesis of proteins in photosynthetic organisms. Indeed, synthesis of various proteins involved in photosynthesis is induced in light conditions. The light-dependent synthesis of proteins is regulated at several steps that include transcription, post-transcriptional modification of mRNA (e.g., RNA editing, splicing, processing), translation and post-translational modification (e.g., phosphorylation, formation of disulfide bonds). Primarily, light absorbed by chlorophylls is used to drive electron transport in the photosynthetic machinery located in and around the thylakoid membrane. Thus, the majority of light signals are converted to redox signals in photosynthetic organisms. However, limited information is available regarding the mechanisms of redox regulation of gene expression and protein synthesis.

Initially the redox state of plastoquinone, an electron carrier that connects photosystem II (PSII) and the cytochrome b_6/f complex, was proposed to be important for the transcriptional regulation of photosynthesis-related genes in plants and algae (reviewed in Allen and Pfannschmidt, 2000; Li *et al.*, 2008; Oelze *et al.*, 2008; Pfannschmidt, 2003). In fact, the redox state of the plastoquinone pool regulates the transcription of the genes of the light-harvesting complex proteins in unicellular green algae *Dunaliella tertiolecta* and *Chlamydomonas reinhardtii* (Chen *et al.*, 2004; Durnford and Falkowski, 1997; Durnford *et al.*, 2003; Escoubas *et al.*, 1995) and the *psbA* gene for the D1 protein, a reaction centre protein of PSII, and *psaAB* genes for reaction centre proteins of photosystem I (PSI) in mustard (Pfannschmidt *et al.*, 1999). In contrast, studies with barley and transgenic tobacco plants have demonstrated that the redox state of the plastoquinone pool is not involved in the expression of genes for the light-harvesting proteins or for photoacclimative responses (Anderson *et al.*, 1997; Montane *et al.*, 1998).

Cyanobacteria constitute a suitable photosynthetic organism in which to thoroughly clarify the role of the redox signal. Recent DNA microarray analysis using inhibitors of the photosynthetic electron transport has revealed that, in *Synechocystis* sp. PCC 6803, the redox state of components located downstream of plastoquinone is more critical for transcriptional regulation than that of the plastoquinone pool (Hihara *et al.*, 2003).

DNA microarray analysis in *Arabidopsis* also suggests that the redox state of the components on the acceptor side of PSI is important for the light-dependent regulation of the expression of nuclear-encoded genes (Piippo *et al.*, 2006).

In *Synechocystis* sp. PCC 6803, a small LuxR-type regulator, PedR, has been identified as a component that is regulated by the redox state of photosynthetic electron transport (Nakamura and Hihara, 2006). This transcription factor activates the expression of *chlL*, *chlN*, *chlB* and *slr1957* and represses the expression of *ndhD2*, *rpe* and the *pedR-sll0296* operon, when the activity of the photosynthetic electron transport is low. Under high light conditions, the supply of reducing equivalents increases as a result of the stimulation of the photosynthetic electron transport, and PedR is transiently inactivated with a concomitant conformational change. Thus, redox regulation of PedR enables transient activation or repression of the target genes in response to rapid changes in light conditions. Recent *in vitro* and *in vivo* studies of PedR have revealed that its activity is regulated by Trx (Y. Hihara, personal communication). Thus, Trx-mediated redox signals must be critical for the regulation of the light-dependent gene expression in cyanobacteria.

IV. PHYSIOLOGICAL PHENOMENA CONTROLLED BY REDOX: PROTEIN SYNTHESIS

The most striking example of proteins whose synthesis in photosynthetic organisms is controlled by redox conditions is the D1 protein of PSII. The synthesis of the D1 protein is markedly induced during the shift from dark to light conditions in chloroplasts and cyanobacteria. The light-dependent synthesis of the D1 protein occurs at the translational level in chloroplasts of plants (Klein and Mullet, 1987; Zhang *et al.*, 2002) and algae (Trebitsh and Danon, 2001), while it occurs at both the transcriptional and the translational levels in cyanobacteria (Tyystjarvi *et al.*, 2001, 2004).

In *C. reinhardtii*, the translational initiation of *psbA* mRNA for the D1 protein is regulated in a redox-dependent manner (reviewed in Marin-Navarro *et al.*, 2007). Under light conditions, a complex of four proteins, namely, RB47, RB38, RB60 and RB55, is bound to the 5′ untranslated region (5′ UTR) of the *psbA* mRNA where they promote translational initiation (Kim and Mayfield, 1997). Within the protein complex, RB47 and RB60 are redox-active proteins: RB47 interacts with the 5′ UTR of *psbA* mRNA under reducing conditions, while RB60, a protein disulfide isomerase, regulates the redox state of RB47 (Alergand *et al.*, 2006; Kim and Mayfield, 1997). Although it has been suggested that RB60 receives reducing equivalents from

PSI in a redox pathway via ferredoxin and Trx (Trebitsh *et al.*, 2000), whether RB60 is directly reduced by Trx *in vitro* and *in vivo* remains to be clarified.

In contrast, light-dependent synthesis of the D1 protein is mainly regulated at the elongation step of translation in higher plants (Edhofer *et al.*, 1998; Kim *et al.*, 1991). The translational elongation of the product of *psbA* mRNA with its concomitant insertion into the PSII complex is activated by reducing equivalents from PSI (Kuroda *et al.*, 1996; Zhang *et al.*, 2000) or by a proton gradient across the thylakoid membrane, which is generated by photosynthetic electron transport (Muhlbauer and Eichacker, 1998). Binding of some trans-acting factors to the 5′ UTR of *psbA* mRNA in a redox-dependent manner *in vitro* in *Arabidopsis* was reported (Shen *et al.*, 2001), suggesting that translation might also be regulated at the initiation step as well as at the elongation step. However, detailed mechanisms of the redox-dependent regulation of translation in chloroplasts remain to be elucidated.

Recently, remarkable progress has been made towards understanding the mechanisms of redox regulation of protein synthesis in cyanobacteria. Oxidative stress enhances PSII photoinhibition by inhibiting the repair of PSII (Nishiyama *et al.*, 2001, 2004; reviewed in Nishiyama *et al.*, 2006). This inhibition is initially induced due to ROS-induced suppression of the synthesis of proteins required for repair, such as the D1 protein, at elongation step of translation (Nishiyama *et al.*, 2001, 2004). A study with an *in vitro* translation system, derived from *Synechocystis* sp. PCC 6803, has revealed that elongation factor G (EF-G), a key protein for translational elongation, is a primary target of ROS-induced inactivation within the translational machinery (Kojima *et al.*, 2007).

Inactivation of EF-G has been shown to be attributable to the oxidation of two specific cysteine residues and formation of a disulfide bond (Kojima *et al.*, 2009). The disulfide bond in the oxidized EF-G is reduced by Trx and the resulting reduced form of EF-G regains its activity to mediate translation *in vitro* (Kojima *et al.*, 2009). Furthermore, the reduction and the subsequent activation of EF-G by Trx have been observed *in vivo* (Kojima *et al.*, 2009). This phenomenon might also explain aspects of the light-dependent control of translation. Activation of the synthesis of the D1 protein requires reducing equivalents derived from PSI in *Synechocystis* sp. PCC 6803 (Allakhverdiev *et al.*, 2005) as well as in plants (Kuroda *et al.*, 1996; Zhang *et al.*, 2000). Thus, it is likely that the translational machinery is regulated by the redox state of EF-G, which is oxidized by ROS and reduced by reducing equivalents that are generated by the photosynthetic electron transport and mediated by Trx. Figure 1 depicts the control of D1 translation by EF-G itself reduced by the Trx system.

Overexpression of EF-G in cyanobacteria resulted in the enhancement of not only the synthesis of the D1 protein, but also that of almost all proteins on

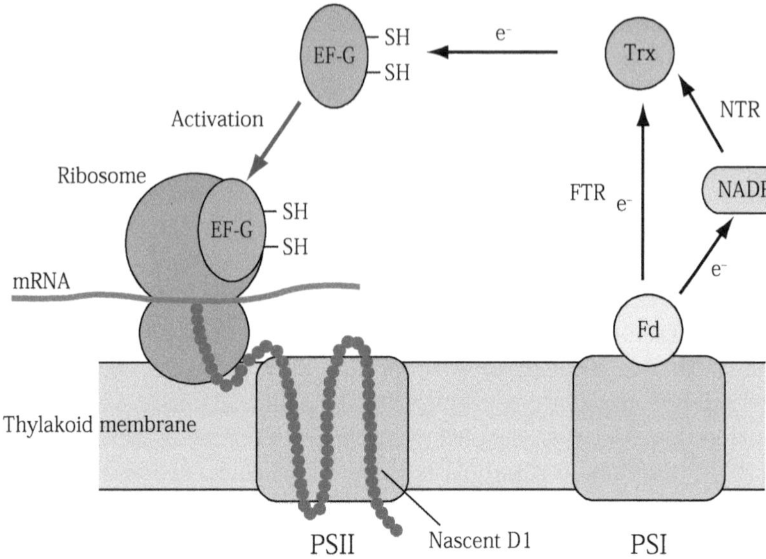

Fig. 1. A model for the mechanism of the redox-regulation of the translational machinery in cyanobacteria. Reducing equivalents that are generated at PSI as a result of the photosynthetic electron transport are transmitted to EF-G in a Trx-mediated redox pathway. The resultant reduction of EF-G activates the light-dependent synthesis of proteins, such as the D1 protein. Thus, the photosynthetic machinery and the translational machinery are interconnected via redox regulation. Black arrows indicate the pathways of reducing equivalents. Fd, ferredoxin. (See color plate section at the back of the book.)

thylakoid membranes under photo-oxidative conditions (Kojima *et al.*, 2007). This observation suggests that many of the light-induced proteins might be regulated by redox signals that are mediated by Trx and EF-G in cyanobacteria.

EF-G proteins of *Synechocystis* sp. PCC 6803 (Lindahl and Florencio, 2003) and spinach chloroplasts (Balmer *et al.*, 2003) have been captured by Trx-affinity column as potential targets of Trx. Since specific cysteine residues that are targets of Trx are conserved in EF-G proteins of cyanobacteria and plant chloroplasts (Kojima *et al.*, 2009), it is possible that the translational machinery in chloroplasts might also be regulated by Trx-mediated redox signals.

V. THE PROTEOMIC APPROACH REVEALS A VARIETY OF Trx TARGET PROTEINS

As mentioned earlier, the target proteins of Trx and its related proteins have been largely revealed by the proteomic approach carried out within this decade. Presently, approximately 500 proteins are described as potential targets for the cytosolic, mitochondrial and chloroplast Trx proteins in higher plants (Balmer *et al.*, 2003, 2004a,b, 2006; Marchand *et al.*, 2004; Motohashi *et al.*, 2001; Yamazaki *et al.*, 2004). In cyanobacteria, the Trx target proteins have been mainly studied in *Synechocystis* sp. PCC 6803, and 82 proteins in total have already been assigned as target proteins, including membrane-integrated and soluble proteins (Florencio *et al.*, 2006; Hosoya-Matsuda *et al.*, 2005; Lindahl and Florencio, 2003; Perez-Perez *et al.*, 2006). Trx targets have also been studied in the green algae *C. reinhardtii* (Lemaire *et al.*, 2004), and the anaerobic photosynthetic bacteria *Chlorobaculum tepidum* (Hosoya-Matsuda *et al.*, 2009). Recently, Montrichard *et al.* described all the Trx target proteins that have been reported to date in photosynthetic organisms (Montrichard *et al.*, 2009). The comparison of the target proteins from higher plants, green algae and cyanobacteria in the article indicates that 60% of the cyanobacterial target proteins of the list are unique to this organism. In order to obtain the global picture of conserved cysteines present in these unique target protein candidates in cyanobacteria, the cysteine residues in each of the listed proteins have been checked. To this end, we have categorized the conserved cysteines as 'cyano-Cys', which are only observed in the target proteins and their homologs in four cyanobacteria, *Synechocystis* sp. PCC 6803, *Anabaena* sp. PCC 7120, *Thermosynechococcus elongatus* BP-1 and *Gloeobacter violaceus* PCC 7421, and also as 'global-Cys', which is observed in the target proteins and their homologs in these four cyanobacteria, plus the higher plant *Arabidopsis thaliana*. When given Cys residues are only observed in these cyanobacteria and in the target proteins in *C. reinhardtii*, these were categorized as 'unicell-Cys'. Although the number of the analyzed proteins is limited in our analysis, we found that 18 unique target proteins observed only in cyanobacteria possess one or multiple 'cyano-Cys' residues, and 14 proteins did not have 'global-Cys' (Table I). In total, the ratios of the 'cyano-Cys' in the conserved cysteines are 71% for the target proteins unique in cyanobacteria, and 30% for the target proteins observed among cyanobacteria, green algae and higher plants. These results suggest that the potential target disulfide for Trx on the target proteins unique in cyanobacteria may be mainly formed by the 'cyano-Cys' residues. In contrast, phosphoglucomutase (only one Cys in the molecule) and sugar-nucleotide epimerase (seven Cys in the molecule) do not have 'cyano-Cys', although they were observed as Trx targets in *Synechocystis*, suggesting that the reported interaction between Trx and these proteins might be unique in *Synechocystis*, or simply due to non-specific interaction.

In addition to these specific Cys residues in cyanobacterial enzymes linked to redox regulation, there has been high interest concerning the insertion of peptide sequences containing redox-sensitive Cys in eukaryotic

TABLE I

Thioredoxin target proteins revealed by proteomics studies

Category[a]	Trx target proteins	*Synechocystis* sp. PCC6803	Cyano-Cys	Global-Cys	Unicell-Cys	References
(1)	Acetolactate synthase, small subunit	sll0065	81			Mata-Cabana *et al.* (2007)
	Aspartyl-tRNA synthetase	slr1720			6, 222, 568	Mata-Cabana *et al.* (2007)
	Carboxysomal protein	sll1031	217, 279			Mata-Cabana *et al.* (2007), Lindahl and Florencio (2003)
	ClpB1	slr1641	312			Mata-Cabana *et al.* (2007)
	ClpC	sll0535	10, 13, 32, 35, 39, 417			Mata-Cabana *et al.* (2007)
	Ferredoxin sulfite reductase	slr0963	569	139, 451, 497, 501		Mata-Cabana *et al.* (2007), Lindahl and Florencio (2003)
	GDP-mannose dehydratase	sll1212	184			Lindahl and Florencio (2003)
	Glucan branching enzyme	sll0158	75, 353, 657, 682, 740			Lindahl and Florencio (2003)
	Glycogen phosphorylase	sll1356	85, 309, 832 (slr1367)	144, 782		Perez-Perez *et al.* (2006)
	Glycogen synthase 2	sll0945	127, 169, 400	360		Lindahl and Florencio (2003)
	Heme oxygenase 1	sll1184	28			Mata-Cabana *et al.* (2007)
	Lysyl-tRNA synthetase	slr1550	365			Perez-Perez *et al.* (2006)
	Oxyanion-translocating ATPase, ArsA	sll0086	25, 293			Mata-Cabana *et al.* (2007)
	PAPS sulfotransferase	slr1791		230		Schmidt and Christen (1978)

	Phosphoglucomutase	sll0726	None			Lindahl and Florencio (2003)
	Photosystem I protein PsaD	slr0737		65		Mata-Cabana *et al.* (2007)
	Polyribonucleotide nucleotidyltransferase	sll1043		455		Perez-Perez *et al.* (2006)
	Porphobilinogen synthase	sll1994	119, 121, 129			Lindahl and Florencio (2003)
	RNA polymerase, subunit α	sll1818		8, 261		Mata-Cabana *et al.* (2007)
	RNA polymerase subunit β	sll1787		413, 620		Lindahl and Florencio (2003)
	RNA polymerase subunit β'	sll1789	214, 286, 293, 296, 652, 999			Lindahl and Florencio (2003)
	Serine-*O*-acetyl transferase	slr1348	28, 234			Mata-Cabana *et al.* (2007)
	Sugar-nucleotide epimerase	sll0576	None			Lindahl and Florencio (2003)
	Sulfate adenylyltransferase	slr1165	213, 243, 285, 332			Lindahl and Florencio (2003)
	Universal stress protein-family, Usp1	slr0244	215, 227	87, 274		Mata-Cabana *et al.* (2007)
	Valyl-tRNA synthetase	slr0557	40, 179, 272, 404	671		Lindahl and Florencio (2003)
(2)	Argininosuccinate synthetase	slr0585	20	120		Lindahl and Florencio (2003)
(3)	ADP-glucose pyrophosphorylase	slr1176		55, 325, 330		Lindahl and Florencio (2003)
	Argininosuccinate lyase	slr1133	274		347	Perez-Perez *et al.* (2006)
	Carbonic anhydrase, type β	slr1347		76, 138		Mata-Cabana *et al.* (2007)
	FtsH	sll1463		268		Mata-Cabana *et al.* (2007)

(*continues*)

TABLE I (*continued*)

Category[a]	Trx target proteins	*Synechocystis* sp. PCC6803	Cyano-Cys	Global-Cys	Unicell-Cys	References
	Glyceraldehyde 3-phosphate dehydrogenase 2	sll1342	75	19, 154, 158		Mata-Cabana *et al.* (2007), Perez-Perez *et al.* (2006)
	Isocitrate dehydrogenase (NADP)	slr1289		131		Papen *et al.* (1983)
	1-Cys peroxiredoxin	slr1198	167	45		Mata-Cabana *et al.* (2007), Lindahl and Florencio (2003), , Hosoya-Matsuda *et al.* (2005)
	Peroxiredoxin II (YLR109-homolog)	sll1621		56, 80		Lindahl and Florencio (2003), Hosoya-Matsuda *et al.* (2005)
	Phosphoglycerate kinase	slr0394		97, 216	313	Perez-Perez *et al.* (2006)
	Pyruvate dehydrogenase component E1, subunit α	slr1934	39, 163, 178			Mata-Cabana *et al.* (2007)
	Pyruvate dehydrogenase component E1, subunit β	sll1721				Mata-Cabana *et al.* (2007)
	Transketolase	sll1070	155, 368, 423	570		Perez-Perez *et al.* (2006)
(4)	ATP synthase, subunit α	sll1326		194		Mata-Cabana *et al.* (2007)
	ATP synthase, subunit β	slr1329		53		Mata-Cabana *et al.* (2007)
	Chaperonin 1 60 kDa GroEL	slr2076	344, 519			Mata-Cabana *et al.* (2007), Lindahl and Florencio (2003)
	Elongation factor G	slr1463	105, 388, 547V[b]	257G[b]		Lindahl and Florencio (2003)

Elongation factor Tu	sll1099		82	Mata-Cabana *et al.* (2007), Lindahl and Florencio (2003)
Fructose 1,6-bisphosphate aldolase, class II	sll0018	91, 145, 316		Perez-Perez *et al.* (2006)
Glutamate synthase GOGAT (Fdx)	sll1499	1247	27, 53, 60, 192, 500, 673, 681, 691, 897, 1163, 1169, 1174, 1390, 1427, 1428	Lindahl and Florencio (2003)
GST	sll1545	None		Mata-Cabana *et al.* (2007)
Type 2 NADH dehydrogenase, NdbC	sll1484	146		Mata-Cabana *et al.* (2007)
Phosphoribulokinase	sll1525		19, 41, 229, 235	Mata-Cabana *et al.* (2007), Perez-Perez *et al.* (2006)
RubisCO large subunit	slr0009	312	79, 167, 187, 242, 279, 422, 454	Mata-Cabana *et al.* (2007), Lindahl and Florencio (2003)

[a]Category: (1) Trx target proteins observed only in cyanobacterium *Synechocystis* sp. PCC 6803; (2) Trx target proteins observed in both cyanobacterium *Synechocystis* sp. PCC 6803 and *Chlamydomonas reinhardtii*; (3) Trx target proteins observed in both cyanobacterium *Synechocystis* sp. PCC 6803 and *Arabidopsis thaliana*; (4) Trx target proteins observed in these three organisms.

[b]Conserved cysteine in three cyanobacteria but not in Synechocystis PCC6803 was shown as the original amino acid.

photosynthetic enzymes. There is a large variety of Cys insertion motives in redox-regulated enzymes, and they were summarized by Jacquot *et al.* (1997). So far, there is no general rule as to how the Cys insertion links the change in properties of redox-sensitive enzymes compared to redox-insensitive enzymes. Thus, biochemical analysis of the proteins listed from the proteomic studies will be absolutely required in order to draw definitive conclusions on the interaction with Trx, the critical disulfide bond as a target of Trx, the physiological significance of the suggested interaction and the molecular evolution of the redox-regulated enzymes.

VI. PERSPECTIVES

Our knowledge concerning the Trx and Grx target proteins, and the glutathionylated proteins in the cells has dramatically increased during the past decade, and we now have to figure out the very complex picture of the redox-related protein networks, particularly in photosynthetic organisms. However, biochemical and physiological studies on the listed proteins are still required in order to determine whether the suggested interaction between the target proteins and Trx/Grx is of physiological significance. Particularly enigmatic are the mechanisms by which redox-balancing system proteins can detect the change in redox balance within cells. From the study of the cyanobacterial-disruptant strains of NTR and FTR proteins, the average reduction levels of Trx isoforms of these disruptants were found to undergo significant changes as compared to those of the wild-type cells (Hishiya *et al.*, 2008). For instance, the reduction level of Trx-*m* in the *ntr* disruptant was decreased to about 25% of that of the wild-type cells, whereas that in the *ftr* mutant was about 65%. These results indicate that levels of the reduced form Trx-*m* in the *ntr* mutant cells decrease dramatically, although a portion of this protein is still reduced. Since certain amounts of the reducing equivalents seem to be required to maintain the function of the anti-oxidative stress system proteins, it is possible that the observed decrease of the reduced form Trx-*m* directly affects the cell viability under oxidative stress conditions. In contrast, the decrease of the reduced form of Trx-*m* was not so significant in the case of the *ftr* mutant, although the mutant showed an obvious delay in cell growth. This phenomenon could not be explained just by the redox regulation of protein synthesis, which is deeply linked to cell growth rate. A threshold of the redox level of the susceptible proteins, the ratio of reduced form within the whole Trx proteins and/or the amounts of reduced forms, may be present, constituting an important component in allowing them to exert their function. To evaluate the significance of this hypothetical threshold, the change of the redox level of the desired protein under the various conditions, for example, under high light, or in the dark, and also in the presence of various chemicals, must be examined. In addition, studies on the redox changes of the thiol enzymes in the *ntr*- and *ftr*-disruptant cells under various conditions that directly affect the redox balance in the cells will be useful to help understand the whole redox control network which operates within cyanobacterial cells.

ACKNOWLEDGMENTS

This study is supported by the grants-in-aid for science research to T. H. (No. 17GS0316) from the Japan Society for the Promotion of Science.

REFERENCES

Alergand, T., Peled-Zehavi, H., Katz, Y. and Danon, A. (2006). The chloroplast protein disulfide isomerase RB60 reacts with a regulatory disulfide of the RNA-binding protein RB47. *Plant and Cell Physiology* **47**, 540–548.

Allakhverdiev, S. I., Nishiyama, Y., Takahashi, S., Miyairi, S., Suzuki, I. and Murata, N. (2005). Systematic analysis of the relation of electron transport and ATP synthesis to the photodamage and repair of photosystem II in *Synechocystis*. *Plant Physiology* **137**, 263–273.

Allen, J. F. and Pfannschmidt, T. (2000). Balancing the two photosystems: Photosynthetic electron transfer governs transcription of reaction centre genes in chloroplasts. *Philosophical Transactions of the Royal Society of London, Series B: Biological Sciences* **355**, 1351–1359.

Anderson, J., Price, D., Chow, W., Hope, A. and Badger, M. (1997). Reduced levels of cytochrome *bf* complex in transgenic tobacco leads to marked photochemical reduction of the plastoquinone, without significant change in acclimation to irradiance. *Photosynthesis Research* **53**, 215–227.

Balmer, Y., Koller, A., Val, G. D., Manieri, W., Schurmann, P. and Buchanan, B. B. (2003). Proteomics gives insight into the regulatory function of chloroplast thioredoxins. *Proceedings of the National Academy of Sciences of the United States of America* **100**, 370–375.

Balmer, Y., Koller, A., Val, G. D., Schurmann, P. and Buchanan, B. B. (2004a). Proteomics uncovers proteins interacting electrostatically with thioredoxin in chloroplasts. *Photosynthesis Research* **79**, 275–280.

Balmer, Y., Vensel, W. H., Tanaka, C. K., Hurkman, W. J., Gelhaye, E., Rouhier, N., Jacquot, J. P., Manieri, W., Schurmann, P., Droux, M. and Buchanan, B. B. (2004b). Thioredoxin links redox to the regulation of fundamental processes of plant mitochondria. *Proceedings of the National Academy of Sciences of the United States of America* **101**, 2642–2647.

Balmer, Y., Vensel, W. H., Hurkman, W. J. and Buchanan, B. B. (2006). Thioredoxin target proteins in chloroplast thylakoid membranes. *Antioxidants & Redox Signaling* **8**, 1829–1834.

Buchanan, B. B. (1991). Regulation of CO2 assimilation in oxygenic photosynthesis: The ferredoxin/thioredoxin system. Perspective on its discovery, present status, and future development. *Archives of Biochemistry and Biophysics* **288**, 1–9.

Chen, Y. B., Durnford, D. G., Koblizek, M. and Falkowski, P. G. (2004). Plastid regulation of Lhcb1 transcription in the chlorophyte alga *Dunaliella tertiolecta*. *Plant Physiology* **136**, 3737–3750.

Durnford, D. and Falkowski, P. (1997). Chloroplast redox regulation of nuclear gene transcription during photoacclimation. *Photosynthesis Research* **53**, 229–241.

Durnford, D., Price, J., Mckim, S. and Sarchfield, M. (2003). Light-harvesting complex gene expression is controlled by both transcriptional and post-transcriptional mechanisms during photoacclimation in *Chlamydomonas reinhardtii*. *Physiologia Plantarum* **118**, 193–205.

Edhofer, I., Muhlbauer, S. K. and Eichacker, L. A. (1998). Light regulates the rate of translation elongation of chloroplast reaction center protein D1. *European Journal of Biochemistry* **257**, 78–84.

Escoubas, J. M., Lomas, M., LaRoche, J. and Falkowski, P. G. (1995). Light intensity regulation of cab gene transcription is signaled by the redox state of the plastoquinone pool. *Proceedings of the National Academy of Sciences of the United States of America* **92**, 10237–10241.

Florencio, F. J., Perez-Perez, M. E., Lopez-Maury, L., Mata-Cabana, A. and Lindahl, M. (2006). The diversity and complexity of the cyanobacterial thioredoxin systems. *Photosynthesis Research* **89**, 157–171.

Hihara, Y., Sonoike, K., Kanehisa, M. and Ikeuchi, M. (2003). DNA microarray analysis of redox-responsive genes in the genome of the cyanobacterium *Synechocystis* sp. strain PCC 6803. *Journal of Bacteriology* **185**, 1719–1725.

Hisabori, T., Motohashi, K., Hosoya-Matsuda, N., Ueoka-Nakanishi, H. and Romano, P. G. N. (2007). Towards a functional dissection of thioredoxin networks in plant cells. *Photochemistry and Photobiology* **83**, 145–151.

Hishiya, S., Hatakeyama, W., Mizota, Y., Hosoya-Matsuda, N., Motohashi, K., Ikeuchi, M. and Hisabori, T. (2008). Binary reducing equivalent pathways using NADPH-thioredoxin reductase and ferredoxin-thioredoxin reductase in the cyanobacterium *Synechocystis* sp. strain PCC 6803. *Plant and Cell Physiology* **49**, 11–18.

Holmgren, A. (1976). Hydrogen donor system for *Escherichia coli* ribonucleoside-diphosphate reductase dependent upon glutathione. *Proceedings of the National Academy of Sciences of the United States of America* **73**, 2275–2279.

Holmgren, A. (2000). Antioxidant function of thioredoxin and glutaredoxin systems. *Antioxidant & Redox Signaling* **2**, 811–820.

Hosoya-Matsuda, N., Motohashi, K., Yoshimura, H., Nozaki, A., Inoue, K., Ohmori, M. and Hisabori, T. (2005). Anti-oxidative stress system in cyanobacteria: Significance of type II peroxiredoxin and the role of 1-Cys peroxiredoxin in *Synechocystis* sp. strain PCC 6803. *Journal of Biological Chemistry* **280**, 840–846.

Hosoya-Matsuda, N., Inoue, K. and Hisabori, T. (2009). Roles of thioredoxin in the obligate anaerobic green sulfur photosynthetic bacterium *Chlorobaculum tepidum*. *Molecular Plant* **2**, 336–343.

Jacquot, J. P., Lancelin, J. M. and Meyer, Y. (1997). Thioredoxins: Structure and function in plant cells. *The New Phytologist* **136**, 543–570.

Kaneko, T., Sato, S., Kotani, H., Tanaka, A., Asamizu, E., Nakamura, Y., Miyajima, N., Hirosawa, M., Sugiura, M., Sasamoto, S., Kimura, T. Hosouchi, T. *et al.* (1996a). Sequence analysis of the genome of the unicellular cyanobacterium *Synechocystis* sp. strain PCC6803. II. Sequence determination of the entire genome and assignment of potential protein-coding regions. *DNA Research* **3**, 109–136.

Kaneko, T., Sato, S., Kotani, H., Tanaka, A., Asamizu, E., Nakamura, Y., Miyajima, N., Hirosawa, M., Sugiura, M., Sasamoto, S., Kimura, T. Hosouchi, T. *et al.* (1996b). Sequence analysis of the genome of the unicellular cyanobacterium *Synechocystis* sp. strain PCC6803. II. Sequence determination of the entire genome and assignment of potential protein-coding regions (supplement). *DNA Research* **3**, 185–209.

Kim, J. and Mayfield, S. P. (1997). Protein disulfide isomerase as a regulator of chloroplast translational activation. *Science* **278**, 1954–1957.

Kim, J., Klein, P. G. and Mullet, J. E. (1991). Ribosomes pause at specific sites during synthesis of membrane-bound chloroplast reaction center protein D1. *Journal of Biological Chemistry* **266**, 14931–14938.

Klein, R. R. and Mullet, J. E. (1987). Control of gene expression during higher plant chloroplast biogenesis. Protein synthesis and transcript levels of psbA, psaA-psaB, and rbcL in dark-grown and illuminated barley seedlings. *Journal of Biological Chemistry* **262**, 4341–4348.

Kojima, K., Oshita, M., Nanjo, Y., Kasai, K., Tozawa, Y., Hayashi, H. and Nishiyama, Y. (2007). Oxidation of elongation factor G inhibits the synthesis of the D1 protein of photosystem II. *Molecular Microbiology* **65**, 936–947.

Kojima, K., Motohashi, K., Morota, T., Oshita, M., Hisabori, T., Hayashi, H. and Nishiyama, Y. (2009). Regulation of translation by the redox state of elongation factor G in the cyanobacterium *Synechocystis* sp. PCC6803. *Journal of Biological Chemistry* **284**, 18685–18691.

Kuroda, H., Kobashi, K., Kaseyama, H. and Satoh, K. (1996). Possible involvement of a low redox potential component(s) downstream of photosystem I in the translational regulation of the D1 subunit of the photosystem II reaction center in isolated pea chloroplasts. *Plant and Cell Physiology* **37**, 754–761.

Lemaire, S. D., Guillon, B., Le Marechal, P., Keryer, E., Miginiac-Maslow, M. and Decottignies, P. (2004). New thioredoxin targets in the unicellular photosynthetic eukaryote *Chlamydomonas reinhardtii*. *Proceedings of the National Academy of Sciences of the United States of America* **101**, 7475–7480.

Li, M., Yang, Q., Zhang, L., Li, H., Cui, Y. and Wu, Q. (2007). Identification of novel targets of cyanobacterial glutaredoxin. *Archives of Biochemistry and Biophysics* **458**, 220–228.

Li, Z., Wakao, S., Fischer, B. B. and Niyogi, K. K. (2008). Sensing and responding to excess light. *Annual Review of Plant Biology* **60**, 239–260.

Lindahl, M. and Florencio, F. J. (2003). Thioredoxin-linked processes in cyanobacteria are as numerous as in chloroplasts, but targets are different. *Proceedings of the National Academy of Sciences of the United States of America* **100**, 16107–16112.

Marchand, C., Le Marechal, P., Meyer, Y., Miginiac-Maslow, M., Issakidis-Bourguet, E. and Decottignies, P. (2004). New targets of *Arabidopsis* thioredoxins revealed by proteomic analysis. *Proteomics* **4**, 2696–2706.

Marin-Navarro, J., Manuell, A. L., Wu, J. and Mayfield, S. P. (2007). Chloroplast translation regulation. *Photosynthesis Research* **94**, 359–374.

Marteyn, B., Domain, F., Legrain, P., Chauvat, F. and Cassier-Chauvat, C. (2009). The thioredoxin reductase-glutaredoxins-ferredoxin crossroad pathway for selenate tolerance in *Synechocystis* PCC6803. *Molecular Microbiology* **71**, 520–532.

Mata-Cabana, A., Florencio, F. and Lindahl, M. (2007). Membrane proteins from the cyanobacterium *Synechocystis* sp. PCC 6803 interacting with thioredoxin. *Proteomics* **7**, 3953–3963.

Mills, J. D. and Mitchell, P. (1982). Modulation of coupling factor ATPase activity in intact chloroplasts. Reversal of thiol modulation in the dark. *Biochimica et Biophysica Acta* **679**, 75–83.

Montane, M. H., Tardy, F., Kloppstech, K. and Havaux, M. (1998). Differential control of xanthophylls and light-induced stress proteins, as opposed to light-harvesting chlorophyll *a/b* proteins, during photosynthetic acclimation of barley leaves to light irradiance. *Plant Physiology* **118**, 227–235.

Montrichard, F., Alkhalfioui, F., Yano, H., Vensel, W. H., Hurkman, W. J. and Buchanan, B. B. (2009). Thioredoxin targets in plants: The first 30 years. *Journal of Proteomics* **72**, 452–474.

Motohashi, K., Kondoh, A., Stumpp, M. T. and Hisabori, T. (2001). Comprehensive survey of proteins targeted by chloroplast thioredoxin. *Proceedings of the National Academy of Sciences of the United States of America* **98**, 11224–11229.

Muhlbauer, S. K. and Eichacker, L. A. (1998). Light-dependent formation of the photosynthetic proton gradient regulates translation elongation in chloroplasts. *Journal of Biological Chemistry* **273**, 20935–20940.

Nakamura, K. and Hihara, Y. (2006). Photon flux density-dependent gene expression in *Synechocystis* sp. PCC 6803 is regulated by a small, redox-responsive, LuxR-type regulator. *Journal of Biological Chemistry* **281**, 36758–36766.

Nishiyama, Y., Yamamoto, H., Allakhverdiev, S. I., Inaba, M., Yokota, A. and Murata, N. (2001). Oxidative stress inhibits the repair of photodamage to the photosynthetic machinery. *EMBO Journal* **20**, 5587–5594.

Nishiyama, Y., Allakhverdiev, S. I., Yamamoto, H., Hayashi, H. and Murata, N. (2004). Singlet oxygen inhibits the repair of photosystem II by suppressing the translation elongation of the D1 protein in *Synechocystis* sp. PCC 6803. *Biochemistry* **43**, 11321–11330.

Nishiyama, Y., Allakhverdiev, S. I. and Murata, N. (2006). A new paradigm for the action of reactive oxygen species in the photoinhibition of photosystem II. *Biochimica et Biophysica Acta* **1757**, 742–749.

Oelze, M. L., Kandlbinder, A. and Dietz, K. J. (2008). Redox regulation and overreduction control in the photosynthesizing cell: Complexity in redox regulatory networks. *Biochimica et Biophysica Acta* **1780**, 1261–1272.

Papen, H., Neuer, G., Refaian, M. and Bothe, H. (1983). The isocitrate dehydrogenase from cyanobacteria. *Archives of Microbiology* **134**, 73–79.

Perez-Perez, M. E., Florencio, F. J. and Lindahl, M. (2006). Selecting thioredoxins for disulphide proteomics: Target proteomes of three thioredoxins from the cyanobacterium *Synechocystis* sp. PCC 6803. *Proteomics* **6**(Suppl. 1), S186–S195.

Pfannschmidt, T. (2003). Chloroplast redox signals: How photosynthesis controls its own genes. *Trends in Plant Science* **8**, 33–41.

Pfannschmidt, T., Nilsson, A. and Allen, T. (1999). Photosynthetic control of chloroplast gene expression. *Nature* **397**, 625–628.

Piippo, M., Allahverdieva, Y., Paakkarinen, V., Suoranta, U. M., Battchikova, N. and Aro, E. M. (2006). Chloroplast-mediated regulation of nuclear genes in *Arabidopsis thaliana* in the absence of light stress. *Physiological Genomics* **25**, 142–152.

Rouhier, N., Villarejo, A., Srivastava, M., Gelhaye, E., Keech, O., Droux, M., Finkemeier, I., Samuelsson, G., Dietz, K. J., Jacquot, J. P. and

Wingsle, G. (2005). Identification of plant glutaredoxin targets. *Antioxidants & Redox Signaling* **7**, 919–929.

Rouhier, N., Lemaire, S. and Jacquot, J. P. (2008). The role of glutathione in photosynthetic organisms: Emerging functions for glutaredoxins and glutathionylation. *Annual Reviews of Plant Biology* **59**, 143–166.

Schmidt, A. and Christen, U. (1978). A factor-dependent sulfotransferase specific for 3′ -phosphoadenosine-5′ -phosphosulfate (PAPS) in the cyanobacterium *Synechococcus* 6301. *Planta* **140**, 239–244.

Schurmann, P. (2003). Redox signaling in the chloroplast: The ferredoxin/thioredoxin system. *Antioxidant & Redox Signaling* **5**, 69–78.

Shen, Y., Danon, A. and Christopher, D. A. (2001). RNA binding-proteins interact specifically with the Arabidopsis chloroplast psbA mRNA 5′ untranslated region in a redox-dependent manner. *Plant and Cell Physiology* **42**, 1071–1078.

Trebitsh, T. and Danon, A. (2001). Translation of chloroplast psbA mRNA is regulated by signals initiated by both photosystems II and I. *Proceedings of the National Academy of Sciences of the United States of America* **98**, 12289–12294.

Trebitsh, T., Levitan, A., Sofer, A. and Danon, A. (2000). Translation of chloroplast psbA mRNA is modulated in the light by counteracting oxidizing and reducing activities. *Molecular and Cellular Biology* **20**, 1116–1123.

Tyystjarvi, T., Herranen, M. and Aro, E. M. (2001). Regulation of translation elongation in cyanobacteria: Membrane targeting of the ribosome nascent-chain complexes controls the synthesis of D1 protein. *Molecular Microbiology* **40**, 476–484.

Tyystjarvi, T., Sirpio, S. and Aro, E. M. (2004). Post-transcriptional regulation of the psbA gene family in the cyanobacterium *Synechococcus* sp. PCC 7942. *FEBS Letters* **576**, 211–215.

Wells, W. W., Xu, D. P., Yang, Y. F. and Rocque, P. A. (1990). Mammalian thioltransferase (glutaredoxin) and protein disulfide isomerase have dehydroascorbate reductase activity. *Journal of Biological Chemistry* **265**, 15361–15364.

Yamazaki, D., Motohashi, K., Kasama, T., Hara, Y. and Hisabori, T. (2004). Target proteins of the cytosolic thioredoxins in *Arabidopsis thaliana*. *Plant and Cell Physiology* **45**, 18–27.

Yano, H., Wong, J. H., Lee, Y. M., Cho, M. J. and Buchanan, B. B. (2001). A strategy for the identification of proteins targeted by thioredoxin. *Proceedings of the National Academy of Sciences of the United States of America* **98**, 4794–4799.

Zhang, L., Paakkarinen, V., van Wijk, K. J. and Aro, E. M. (2000). Biogenesis of the chloroplast-encoded D1 protein: Regulation of translation elongation, insertion, and assembly into photosystem II. *Plant Cell* **12**, 1769–1782.

Zhang, N., Kallis, R. P., Ewy, R. G. and Portis, A. R., Jr. (2002). Light modulation of Rubisco in *Arabidopsis* requires a capacity for redox regulation of the larger Rubisco activase isoform. *Proceedings of the National Academy of Sciences of the United States of America* **99**, 3330–3334.

Use of Transgenic Plants to Uncover Strategies for Maintenance of Redox Homeostasis During Photosynthesis

GUY T. HANKE, SIMONE HOLTGREFE, NICOLAS KÖNIG, INGA STRODTKÖTTER, INGO VOSS AND RENATE SCHEIBE[1]

Department of Plant Physiology, University of Osnabrueck, D-49069 Osnabrueck, Germany

[1]Corresponding author: Email: scheibe@biologie.uni-osnabrueck.de

Advances in Botanical Research, Vol. 52

0065-2296/09 $35.00
DOI: 10.1016/S0065-2296(09)52008-3

ABSTRACT

Plant cells encounter a spectacular variation in the supply and consumption of redox components, due to changes in photosynthesis caused by the environment. To prevent these huge fluxes from causing catastrophic oxidative damage, there is an extensive network of compensatory, buffering mechanisms. These must be integrated with signaling cascades in a greater redox network, to ensure that short-term responses are adequate and that, if buffering capacity is exceeded, there is a response at the transcript level. Transgenic approaches have been fundamental in identifying the interconnections between redox fluxes, buffering, and signaling networks. In this review we discuss how this has shaped current understanding, and how transgenics might be used in the future to unravel the complex network required for energy metabolism, redox homeostasis, autotrophic growth and development under changing conditions. There are obvious problems associated with describing a highly interconnected network in the linear format of a written review, but we attempt this by first describing how redox poise is maintained in electron transport chains, move on to buffering pathways throughout the cell, and finally describe the mechanisms that detect signals, leading to interpretation of these changes at the level of altered transcription.

ABBREVIATIONS

AGPase	ADPglucose pyrophosphorylase
AOX	alternative oxidase
CET	cyclic electron flow
Cyt b_6f	cytochrome b_6f complex
Fd	ferredoxin
FNR	ferredoxin-NADP reductase
GAPDH	glyceraldehyde 3-P dehydrogenase
G6PDH	glucose 6-P dehydrogenase
Grx	glutaredoxin
LET	linear electron flow
MDH	malate dehydrogenase

NPQ	nonphotochemical quenching
NTR	NADP-thioredoxin reductase
OPPP	oxidative pentose-phosphate pathway
PET	photosynthetic electron transport chain
PQ	plastoquinone
Prx	peroxiredoxin
PSI	photosystem I
PSII	photosystem II
ROS	reactive oxygen species
Trx	thioredoxin

I. INTRODUCTION: STUDYING CONTROL OF REDOX NETWORKS

A. EARLY AND PRESENT-DAY TRANSGENIC APPROACHES

Redox metabolism and the use of redox mechanisms in signaling are ubiquitous in living cells. However, along with the photosynthetic bacteria, plants are exceptional in redox status due to their ability to harvest energy in the form of light, through photosynthetic electron transport (PET). This allows the production of large amounts of ATP and reducing compounds such as NADPH, enabling the complex metabolism whereby inorganic carbon, nitrogen, and sulfur can be assimilated. Perhaps the most remarkable part of this photosynthetic process is not PET itself, but the careful buffering mechanisms, which control the destructive reactive species that are inevitably generated, and the exquisite sensitivity of the redox-signaling networks that allow the plant to respond to redox changes and prevent damage. The extremely high redox challenges faced by land plants have resulted in evolution of many systems to guarantee survival under extreme conditions (Foyer and Noctor, 2009; Oelze *et al.*, 2008), many of which involve the free transfer of reductant between cellular compartments, for dissipation elsewhere, for example, by the malate valve (Fig. 1). In this review, we will discuss studies on terrestrial angiosperms, as they are more subject to dramatic changes in environmental conditions that affect redox metabolism than those in aqueous systems. Readers are referred to other, excellent reviews for similar work on cyanobacteria and algae (see Hosoya-Matsuda *et al.*, 2005; Lemaire *et al.*, 2004; Sommer *et al.*, 2003).

The greatly increased understanding of redox networks that we have derived from transgenic work has been mostly generated as a by-product of functional genomic approaches. The first antisense experiments were

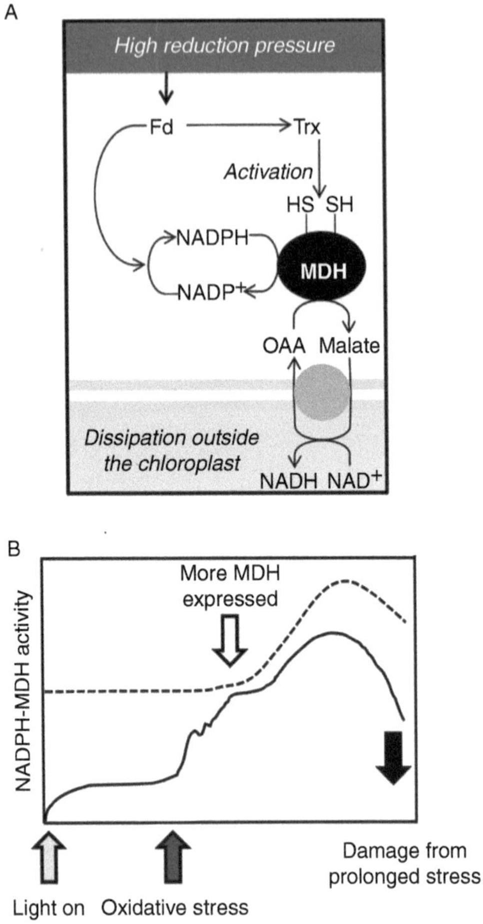

Fig. 1. The malate valve. (A) Reducing power is transferred across membranes by the malate valve. The chloroplast enzyme NADP-dependent malate dehydrogenase (NADP-MDH) allows the chloroplast to balance the ATP and NADPH outputs of PET, by oxidizing excess NADPH, generating malate that is exported in return for the substrate, oxaloacetate (OAA). Activity of the enzyme is regulated by the ferredoxin (Fd):thioredoxin (Trx) system and by the NADPH/NADP ratio, ensuring that under conditions of excess reducing pressure from PET, the enzyme is activated to relieve this pressure. (B) Regulation of NADP-MDH at the protein and transcript level by stress, based on biochemical data (Becker *et al.*, 2006). Enzyme activity of chloroplast NADP-MDH is shown as total enzyme capacity (black dashes) and the enzyme fully activated *in situ* by the Fd:Trx system (solid line). Dark grey upward arrow indicates initiation of stress (in this case high light causing increased reductive pressure from PET), white downward arrow indicates increased stress inducing higher MDH capacity, black downward arrow indicates irreversible oxidative damage when the buffering limits of the system are reached.

performed by Van der Kroll *et al.* (1988) with chalcone synthetase (CHS) resulting in changed flower pigmentation. This suggested that it might be possible to unravel the role of a gene/protein by identifying the disruption of a given property in its absence. In the early 1990s, but most effectively following the availability of the *Arabidopsis* genome in 2000, much effort went into the identification of a function for each gene, that is, "functional genomics," as an approved and well-supported way of studying plant metabolism and development (see Radhamony *et al.*, 2005). In later studies, multilevel analysis, using transcriptomics, proteomics, and metabolomics, has often revealed multiple effects upon knockout of a single gene, rather than disruption of a specific function. In addition to transgenic reduction of chloroplast proteins, redox perturbations are also caused by knockout of genes encoding proteins targeted to other compartments, such as the mitochondria (Millar *et al.*, 2001; Noctor *et al.*, 2004, 2007). This illustrates the integrated response necessary to balance the overall redox state and to protect the cell.

These antisense and knockout approaches have solved many questions relating to known pathways, and identified new pathways and regulatory properties (for review, see Thorneycroft *et al.*, 2001). Interestingly, protein content frequently does not correlate with functional disruption. For example when RubisCO content is reduced, the flux control coefficient is highly dependent on growth conditions (light, CO_2, N-nutrition), and can range from 0.1 to 0.8 (Stitt and Sonnewald, 1995). In addition to gene disruption techniques, a constructive approach to determine a minimal set of essential genes (e.g., genes required for normal embryo development) is also being taken, and this information is accumulating in the SeedGene database (Meinke *et al.*, 2008; Tzafrir *et al.*, 2004). Finally, manipulation of primary metabolism in many transgenics has resulted in major developmental changes, revealing the role of metabolite levels in controlling gene expression (Raines and Paul, 2006).

B. THE DIFFERENT CHALLENGES OF ANALYZING REDOX METABOLISM, BUFFERING/POISING, AND SIGNALING

Severe phenotypes arise from disruption of major metabolic pathways, such as PET, where the plants must make large morphological adjustments, simply to survive (Hansson *et al.*, 2007; Schwenkert *et al.*, 2006; Umate *et al.*, 2007; Voss *et al.*, 2008). In these situations, the challenge lies in dissecting the impact of the absence of a single protein, from the adaptive response to its loss. Such plants form an important resource to study the impact (on the cell) of severe alterations in redox metabolism.

By contrast, identifying the physiological role of proteins involved in buffering changes in cellular redox status can be problematic due to the numerous compensatory mechanisms. One particular example of this is the chloroplast enzyme NADP-malate dehydrogenase (NADP-MDH), an enzyme which oxidizes excess NADPH to maintain the redox status of the chloroplast and is activated under conditions of reductive pressure (Fig. 1A). Figure 1B illustrates our understanding of the role NADP-MDH plays in conditions of increasing reductant output from PET (such as high light, low temperature, or lack of the acceptor CO_2). Initially, the activation state of the enzyme is moderate, but rises to almost 100% of the capacity under conditions of high reductive pressure in the stroma. When 100% activation state is finally reached, a further increase in total enzyme capacity is detected, indicative of an adaptive response at the transcription level, yielding more NADP-MDH to combat the stress. Damage will only occur if environmental changes are too rapid for a timely adaptive response in gene expression and translation, or so severe that they exceed the capacity of the upregulated system. Intriguingly, transgenics lacking NADP-MDH partially or completely have no visible phenotype under nonstressed growth conditions (Backhausen *et al.*, 1998; Faske *et al.*, 1997; I. Strodtkötter and R. Scheibe, unpublished results) (see later discussion).

This lack of obvious phenotype is also true for several other transgenic plants lacking specific proteins involved in balancing redox homeostasis. For example, *Arabidopsis* plants lacking the main isoform of alternative oxidase (AOX1A) are indistinguishable from wild-type plants when grown under standard conditions (Giraud *et al.*, 2008; Strodtkötter *et al.*, 2009). Only when additional pressure, such as cold, drought, or antimycin A, challenges the plants do the effects become visible as retarded growth. Even challenging the plant with specific stress may not reveal a phenotype, as transgenics compensate for the absence of a specific mechanism by upregulating alternative pathways. Such redundancy between pathways demonstrates the importance of antioxidant systems to plant metabolism for maintaining redox balance. Where compensatory metabolic pathways are operating, specific growth conditions or treatments may resolve a transgenic phenotype.

Realization of the impact that changes in redox metabolism can have on signaling pathways has often resulted serendipitously from gene-knockout studies. Interference with redox-related metabolism in chloroplasts, the cytosol, peroxisomes, and mitochondria has frequently led to interesting, but also contrasting results indicating changes in the signaling pathways of the cell. In some cases, knockout of genes that were initially assumed to code for redundant proteins, due to the presence of multiple isoforms (e.g., thioredoxins, glutaredoxins, peroxiredoxins, and the various thioredoxin reductases) resulted in phenotypes with altered developmental and morphological properties (Chi *et al.*, 2008; Lepistö *et al.*, 2009; Li *et al.*, 2009; Oelze *et al.*, 2008; Reichheld *et al.*, 2007; Xing *et al.*, 2005).

The absence of one of the cytosolic isoforms of glyceraldehyde 3-P dehydrogenase (GapC1) or of the nonphosphorylating GAPDH led to changes at multiple levels. These include changes in the expression of transcription factors, which cause widespread changes in protein expression and development (Rius *et al.*, 2006, 2008). Visible phenotypes will often also depend on the growth conditions, such as photoperiod, temperature, and nutrient availability. Ambiguity relating to chosen growth conditions for transgenics was demonstrated in early experiments with plants possessing decreased RubisCO (Lauerer *et al.*, 1993; Quick *et al.*, 1991a,b). A growth phenotype could only be observed under nonlimiting conditions for N supply or at high light, when the full capacity of RubisCO was required. These broad pleiotropic effects, that vary depending on the growth conditions, make identifying the specific function of the knocked out gene highly problematic. However, such plants can give us valuable insights into the signaling networks involved in maintaining redox homeostasis, and the feedback mechanisms that control PET, its buffering mechanisms, and the redox-poising apparatus of the whole cell.

II. BALANCING REDOX NETWORKS WITHIN PET

Plants contain both the PET chain in chloroplasts and the mitochondrial respiratory electron transfer chain. It is now clear that in green leaves the mitochondrial pathway also makes a large contribution to the light-dependent redox metabolism of the plant cell (Raghavendra and Padmasree, 2003). This has been extensively reviewed (Noctor *et al.*, 2004, 2007), and we will therefore devote this study primarily to discussing redox changes derived from photosynthetic electron transfer, and how these are integrated into the cellular redox network. Other cellular compartments will be considered only in cases of obvious metabolic contributions for optimizing and protecting photosynthesis.

A. PET IS BOTH ESSENTIAL AND DANGEROUS

PET is the major source of reducing power in the plant cell, and causes redox fluxes that dominate the chloroplast and control gene expression in both the chloroplast and the nucleus. Much of the redox metabolism of the chloroplast, and to a lesser extent the cell, is dedicated to buffering any variation in PET output against capacity for ATP/NADPH consumption. In addition, PET is the source of signals, for example, reactive oxygen species (ROS), that change rapidly in response to environmental conditions, and initiate widespread changes in protein activity and gene expression.

Land plants must endure extreme variations in parameters such as illumination, temperature, and water availability, which cause great changes in PET, and also the sink capacity for its redox output. In order to survive over this spectrum of conditions, they must maintain a balance between harvesting light energy for optimal growth and adapting to suboptimal conditions, as well as protecting themselves from irreversible damage under excess light. Since changes in conditions can either be subtle or abrupt, and may be expected (day/night) or unexpected and random (sun flecks), the system for control should be both sensitive and robust. This is necessary to maintain stable energy and metabolic fluxes, while avoiding detrimental effects and irreversible damage under extreme conditions. Transgenic approaches to analyzing the function of PET genes are well reviewed by Pesaresi *et al.* (2001), but there have been many subsequent developments, which can be constructively used to discuss the relationship between PET and cellular redox networks.

B. PREVIOUS AIMS OF TRANSGENIC INVESTIGATIONS OF PET

To date, the majority of transgenic studies on PET have been dedicated to the identification of new components of the pathway (DalCorso *et al.*, 2008; Joet *et al.*, 2002; Shikanai *et al.*, 1998; Takabayashi *et al.*, 2002; Wu *et al.*, 1999), PET protein complex assembly (Hashimoto *et al.*, 2003; Lennartz *et al.*, 2001, Stockel and Oelmüller, 2004), and dissection of the roles of highly homologous proteins (Hanke and Hase, 2008; Ishihara *et al.*, 2007; Lintala *et al.*, 2009; Lundin *et al.*, 2007), and complementary parts of the pathway (DalCorso *et al.*, 2008; Munekage *et al.*, 2002).

Transgenic approaches have been particularly fruitful when probing genes of unknown function in photosynthetic regulons, such as those identified by Biehl *et al.* (2005) and when investigating the function of gene products where there are no homologs (Maiwald *et al.*, 2003). Another powerful use of transgenics has been to search for unknown proteins involved in a defined function. By predicting the phenotype that will occur when a specific process has been disrupted, mutant populations can be screened for such a phenotype to identify the proteins involved (Hashimoto *et al.*, 2003; Miles, 1994; Munekage *et al.*, 2002). Finally, protein function has been probed by transforming knockout plants with altered replacement genes (Rosgaard *et al.*, 2005). This approach is also related to the fact that many PET genes are encoded on the plastome, making replacement by homologous recombination with site-directed mutants a powerful tool for probing their function (Schwenkert *et al.*, 2006; Umate *et al.*, 2007).

A range of transgenics with disrupted PET genes and coordinately altered redox metabolism have been generated in these studies. These PET transgenics will provide powerful tools to investigate how the output of PET is interpreted by the plant and the redox networks that contribute to balancing such heavy disruption. Indeed, some studies have already examined changes at the level of transcription (Ihnatowicz *et al.*, 2004; Lintala *et al.*, 2009; Maiwald *et al.*, 2003) and protein activity (Lundin *et al.*, 2007; Schwenkert *et al.*, 2006) in transgenic plants with altered PET.

C. REDUNDANCY WITHIN PET THROUGH HOMOLOGOUS PROTEINS

Disruption of PET will result in obvious redox imbalances and problems for plant metabolism. Perhaps for this reason, several PET genes are present as multiple isoforms, which are capable of compensating for each other's function and preventing disruption of cellular redox poise. In order to generate a loss-of-function phenotype transgenically, double and sometimes triple mutants are necessary. Examples include PgrL1 (DalCorso *et al.*, 2008), LHCI light-harvesting complex (LHC) proteins (Ihalainen *et al.*, 2005), and PsaD1/PsaD2 (Ihnatowicz *et al.*, 2004).

Antisense experiments with conserved nucleotide sequences can sometimes be used to generate loss-of-function mutants against a multiple gene homolog background. For example an antisense approach in potato has yielded plants reduced in both ferredoxin isoproteins (Holtgrefe *et al.*, 2003), and the same approach in tobacco has reduced total ferredoxin reductase (FNR) (Hajirezaei *et al.*, 2002). These phenotypes are more profound than when single isoproteins are absent in *Arabidopsis* (Hanke and Hase, 2008; Hanke *et al.*, 2008; Lintala *et al.*, 2009; Voss *et al.*, 2008). PSI-D also provides a good example of how an alternative RNA interference (RNAi) approach can be used to reduce transcript of two homologous genes (Haldrup *et al.*, 2003), where one construct was used to silence two genes.

Comparing reciprocal knockout mutants has proved a powerful tool in finding out whether the gene products have compensatory function, or diverse roles. For example, *Arabidopsis* plants contain three homologs of the photosystem II protein PSBP, (PSBP, PPL1, and PPL2). A comparison of mutants showed that while PSBP and PPL1 are involved in optimizing water splitting at PSII, as revealed by stress treatment, PPL2 is a thylakoid protein involved in assembly of the NDH complex (which is greatly reduced in the mutants) (Ishihara *et al.*, 2007). *Arabidopsis* also contains two separate genes for leaf-type Fd:NADPH reductase (FNR), and transgenic approaches seem to suggest that they can compensate each other's basic function (Hanke *et al.*, 2008; Lintala *et al.*, 2007, 2009). Analysis of transcripts in the respective knockouts was required to uncover a difference between knockout of FNR1 and FNR2, revealing a proportionally greater role for FNR2 in coordinating stress responses (Lintala *et al.*, 2009).

D. REDOX BALANCING WITHIN PET BY ALTERNATIVE PATHWAYS

The relative demand of plant metabolism for ATP and reducing equivalents can vary greatly depending on environmental conditions. One of the most dynamic and effective ways plants can respond to these variations is by altering the pathway of electron flow through thylakoid-membrane complexes. The fate of the final reduced products of PET, reduced ferredoxin (Fd) and NADPH, can be regulated to promote either linear electron flow (LET), in which electrons are delivered to sinks in the stroma, or cyclic electron flow (CET), in which electrons are returned to PET with no net generation of reductant, while retaining the generation of a proton gradient to drive ATP synthesis (Fig. 2). There are two independent pathways of CET mediated either by electron transfer from NADPH to the thylakoid-bound NADPH-dehydrogenase (NDH) complex, or electron transfer by reduced Fd through an unidentified PQ reductase, which may comprise or involve the cytochrome b_6f (Cyt b_6f) complex and FNR (Joliot and Joliot, 2006). Transgenic work has been fundamental in revealing how these two CET pathways are essential to maintaining redox balance around PET, and how they compensate for each other.

CET allows much greater passage of electrons through the Cyt b_6f complex, generating a pH gradient more rapidly than in LET. This pH gradient is necessary to initiate the xanthophyll cycle, which provides the majority of nonphotochemical quenching (NPQ) in higher plants. NPQ is the mechanism whereby excess excitation energy is diverted out of PET to prevent the generation of damaging reductive pressure and subsequently ROS formation. Initiation of NPQ at high light intensities is therefore dependent on CET, and this fact has been cleverly exploited to identify mutants of CET by screening for mutants with retarded NPQ in a reverse genetics approach. This enabled identification of the proton gradient reduced (*pgr*) mutants (Munekage *et al.*, 2002), which form the basis of our developing understanding of Fd-dependent CET. Reverse genetics has also enabled identification of genes essential to the alternative, NDH-dependent pathway. Earlier, forward genetics approaches had established that this particular branch of CET was uniquely responsible for the return of electrons to the PQ pool following

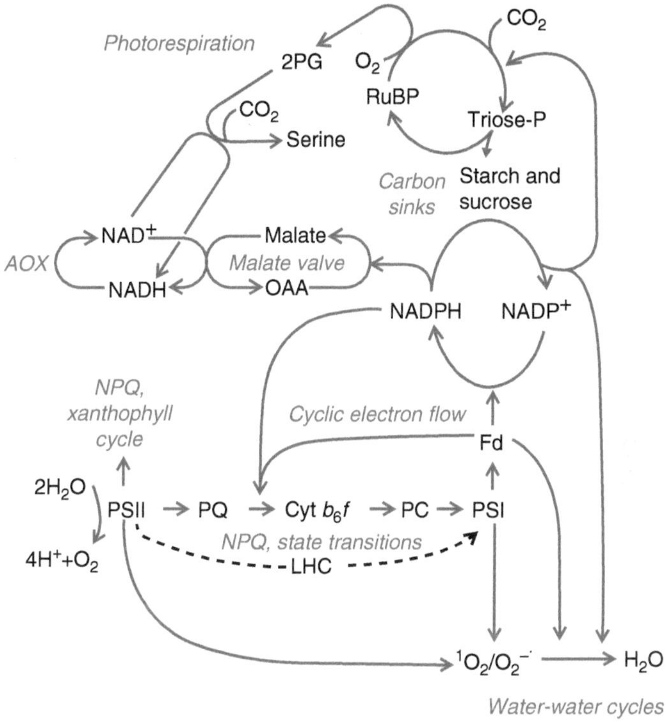

Fig. 2. Simplified picture showing redox balancing and buffering of photosynthetic electron flow (PET). Light excitation of photosystem II (PSII) enables the splitting of water to yield high-energy electrons and protons. These electrons are passed via plastoquinone (PQ) and the cytochrome b_6f complex (Cyt b_6f) to photosystem I (PSI), further generating the proton gradient used to drive ATP synthesis (not shown for reasons of clarity). Excited PSI donates electrons to ferredoxin (Fd), which can then be transferred to the NADPH that fuels triose phosphate (triose-P) generation in the Calvin cycle, eventually ending up in carbon sinks, such as starch and sucrose. Under conditions where high light, or other stress, causes PSII to generate oxidative species, the excess reductive pressure can be relieved by multiple pathways. Within PET, excess electrons can be released through NPQ. This phenomenon can include electron donation to the xanthophyll cycle and the movement of light-harvesting complexes (LHC) from PSII to PSI. Excitation of PSI then results in cyclic electron flow, either via ferredoxin (Fd) or NADPH. These soluble electron carriers are also used in reactions to remove excess oxidative species released from PSI and PSII by the Mehler reaction in water–water cycles. The ratio of $NADP^+$ to NADPH is exquisitely balanced in the chloroplast stroma by a number of compensatory pathways. NADPH can be oxidized in the Calvin cycle, where, depending on RuBisCO activity, it can fix either CO_2 or O_2 in photorespiration, or used to reduce oxaloacetate (OAA) to malate in the malate valve. Both photorespiration and the malate valve then result in export of the reduced products to other cellular compartments (not shown for clarity) to regenerate intermediates and remove excess reducing equivalents. A major contributor is the alternative oxidase of mitochondria (AOX), which can oxidize NADH without generating a proton gradient. (See color plate section at the back of the book.)

transfer to darkness (visible as fluorescence of PSII as electrons flow back from PQ). Screening for perturbation in this PSII fluorescence allowed identification of the proteins involved in assembly of the NDH complex (Hashimoto *et al.*, 2003; Takabayashi *et al.*, 2009), such as CRR2.

The two mechanisms of CET have subtly different roles in photosynthesis, which are still not fully understood, and can compensate for one another, meaning that severe phenotypes are not visible until both pathways are disrupted. These processes are well reviewed (Shikanai, 2007), but it is worth discussing them briefly against the background of the cellular redox network. Knockouts of NDH subunits in tobacco (Burrows *et al.*, 1998; Sazanov *et al.*, 1998; Shikanai *et al.*, 1998) and *Arabidopsis* (Takabayashi *et al.*, 2002, 2009) grow well if not subjected to stress, demonstrating that the Fd pathway can adequately compensate for any loss in activity, although some disruption in cyclic flow can be measured by fluorescence. CET is most active in the dark-to-light transition, and under stressful conditions; therefore, probing for phenotypes must involve screening under these conditions. Indeed, a chlorotic phenotype of NDH mutants can be revealed by high light stress (Endo *et al.*, 1999). The only transgenics with disrupted Fd-dependent electron flow are the *pgr5* (Munekage *et al.*, 2002) and *pgrL1* (DalCorso *et al.*, 2008) mutants. Mutants of *pgr5* and double mutants of *pgrL1* show a more severe phenotype than the *ndh* mutants, but plants still grow under nonstressful conditions. Combining mutants of *pgr5* with *ndh* mutants results in a very severe phenotype (Munekage *et al.*, 2004). Pgr5 is now thought to regulate the transition between LET and CET, rather than have a direct involvement (Nandha *et al.*, 2007). Transgenic approaches have therefore established that CET plays an irreplaceable role in balancing redox homeostasis and identified two compensatory pathways whose differential function we do not yet fully understand.

PSII is readily damaged by excess reduction pressure, and specific mechanisms ensure deregulation of the photosystem under these conditions. This deregulation can be measured as a decrease in PSII chlorophyll fluorescence, and because this does not involve transfer of electrons into PET, it is known as NPQ. Under nondamaging conditions, NPQ is composed of two major contributions: dissipation of excess PSII electrons, which are channeled into the xanthophyll cycle, or downregulation of PSII by relocation of the LHC antenna from PSII to PSI in a state transition. Increased flux into the xanthophyll cycle has been measured under conditions of limiting electron acceptors at PSI (Takahashi *et al.*, 2006). In the state transition LHCs are phosphorylated by a redox-sensitive mechanism, which causes their dissociation from PSII and association with PSI, thus downregulating PSII activity while enhancing PSI activity (Aro and Ohad, 2003; Zer and Ohad, 2003).

This is important in balancing the excitation of the photosystems in response to environmental changes and metabolic demands (Kanervo *et al.*, 2005). Transgenic approaches have been integral in identifying molecular determinants of state transition changes. For example, the importance in state transitions of LHC binding to PSI was established through transgenic programs aimed at identifying functions for peripheral PSI subunits. These studies showed that subunits PSI-H (Lunde *et al.*, 2000; Varotto *et al.*, 2002) and PSI-G (Jensen *et al.*, 2002) are essential for LHC binding and the ability to perform state transitions. Both these NPQ mechanisms are clearly important for protecting PSII and ensuring balanced excitation of the photosystems, but *npq* mutants show relatively little photoinhibition (and phenotypes can only be measured by chlorophyll fluorescence) (Sarvikas *et al.*, 2006). This indicates that in the absence of NPQ, other mechanisms can readily compensate to relieve excess electron pressure. The main candidates for alternative mechanisms are electron transport to O_2 and removal of ROS by the various water–water cycles, which will be addressed in Section IV, since the resulting compounds are likely to function also as signals for induction of other pathways.

III. BUFFERING OF REDOX POISE BY COORDINATED AND COMPENSATORY PATHWAYS

A. PET PRODUCTS NOT REQUIRED BY SINKS MUST BE BUFFERED AND ACCEPTORS REGENERATED

In addition to the thylakoid-membrane-based compensatory mechanisms that enable plants to maintain redox homeostasis under a wide range of conditions, there is a broad network of redox-buffering metabolism, extending throughout the cell. The interaction between this redox-poising network and PET is complex as, in addition to being the major generator of damaging ROS, PET is also the source of much of the reductive power that supports the buffering mechanisms.

In PET, light energy is used to generate electrons, the transfer of which through thylakoid-membrane complexes leads to $NADP^+$ photoreduction and establishes the proton gradient required for ATP synthesis. PET-dependent production of ATP and NADPH therefore represents the main source of metabolic energy in photosynthetic organisms. The continuous provision of ATP, NADPH, and NADH is essential for metabolism and all other endergonic processes (Fig. 3), and so catabolic energy reconversion from storage compounds is also necessary in nonphotosynthetic conditions.

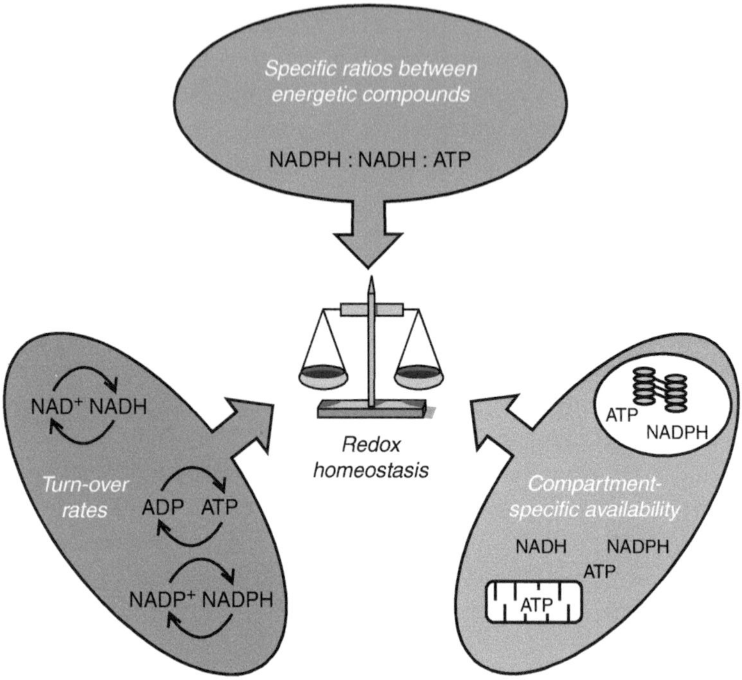

Fig. 3. Fine-tuning energy production and consumption, and the ratios between energy-carrying molecules. Energy generation and consumption in a plant cell needs to be tightly coupled with respect to species, turnover rate, ratio, and subcellular compartmentation of the energy carriers. The $NADP^+$/NADPH, NAD^+/NADH, and ADP/ATP ratios throughout the cell must be stably maintained to ensure acceptor molecules are available as sinks, and reductant and energy supply for metabolism is adequate. Energy-generating pathways produce these molecules in specific ratios, while demands for them in different cellular compartments can vary dramatically according to the metabolic demands of cell type, developmental stage, and environmental conditions. $NADP^+$/NADPH, NAD^+/NADH, and ADP/ATP ratios must be buffered to account for their rate of formation, consumption in specific compartments, and the ratio between the separate electron carriers and the adenylate pool. (See color plate section at the back of the book.)

Energetic pathways, which provide and consume reducing power, are located in different compartments of the plant cell. Transfer of the reducing equivalents (NADPH, NADH) and of energy in the form of ATP (or GTP) between compartments is therefore necessary to maintain metabolism in a multicompartmented cell. This transfer is not direct in most cases, since these compounds are not transported easily across membranes, but rather operates by indirect shuttle systems (Ebbighausen *et al.*, 1987; Scheibe, 2004).

Generation and consumption of ATP, NADPH, and NADH is tightly controlled, because their pool sizes are relatively small, and in addition to providing energy, in their dephosphorylated and oxidized forms, respectively, they must also function as phosphate and electron acceptors. Since the stoichiometric ratio between ATP and reducing equivalents produced may be considerably different from the metabolic requirements of a particular compartment, shuttle systems must also act to remove reductive excess and regenerate electron acceptors. Figure 3 illustrates how energy conversion and energy consumption must be finely tuned in each of the cellular compartments in order to ensure optimal ADP/ATP, $NADP^+$/NADPH, and NAD^+/NADH ratios, and avoid excess reduction pressure, overenergization, and subsequent production of radicals and oxidative stress (Scheibe *et al.*, 2005). Avoidance of such oxidative stress is achieved through a network of buffering pathways that provide multiple alternative sinks to store excess reductive power or energy and absorb the reactive species that may be generated (Fig. 2). Therefore, from the plant's perspective, these great fluctuations in redox metabolism are anticipated and buffered, so that previous ideas of "stress" are now outdated. In fact, oxidative stress has recently been redefined as "an imbalance between oxidants and antioxidants, leading to disruption of redox signaling and control and/or molecular damage" (Jones, 2006).

B. BUFFERING REDOX FLUCTUATION THROUGH ANTIOXIDANT ENZYMES

In photosynthetic metabolism, robustness is achieved through complex metabolic networks to avoid perturbations in this dynamic process (Luo *et al.*, 2009). Antioxidant systems are present in all cellular compartments as a rapid buffering system to absorb excess ROS before oxidative damage occurs: ROS generation from PET increases greatly when electron supply exceeds consumption, and the dissipatory systems (see Fig. 2), are overloaded, resulting in electron transfer to O_2 (Backhausen *et al.*, 2000). Energy transfer to O_2 from PSII results in generation of singlet oxygen ($^1O_2$), and energy transfer from Fd to O_2 at the acceptor side of PSI generates superoxide radical anions ($O_2^{\cdot -}$) (Apel and Hirt, 2004; Asada, 2006; Foyer and Noctor, 2000, 2005a; Halliwell, 2006). In the subsequent water–water cycle, $O_2^{\cdot -}$ is dismutated to H_2O_2 and O_2 by superoxide dismutase (SOD), and H_2O_2 in turn can be converted to H_2O by catalase (CAT) or ascorbate peroxidase (APX) using ascorbate as an electron donor (Asada, 2006). Regeneration of ascorbate by reduced glutathione (GSH) requires glutathione reductase (GR) and NADPH, but is also possible through electron donation from Fd (Miyake and Asada, 1994; Sano *et al.*, 2005). There are large pools of the substrates of the water–water cycle (reduced glutathione and ascorbate) present in chloroplasts, enabling very effective short-term buffering of fluctuations in the redox output of PET.

Recently, an additional, related antioxidant system has been identified using a transgenic approach. NTRC is an NADPH-dependent thioredoxin (Trx) reductase, which also contains a Trx domain (Serrato *et al.*, 2004). The protein transfers electrons from NADPH to regenerate reduced chloroplast 2-Cys peroxiredoxin (2-CP) when oxidized by H_2O_2. Knockout mutants of NTRC are smaller, pale green, and sensitive to oxidative stress (Pérez-Ruiz *et al.*, 2006).

Transgenic approaches have amply demonstrated the importance of maintaining balanced GSH and ascorbate pools to buffering the redox state of plant cells. For example, plants with suppressed expression of the thylakoid-bound Cu/Zn-SOD showed heavy growth inhibition, indicating that the water–water cycle is poorly replaced by other systems (Rizhsky *et al.*, 2003). Also, mutation of a chloroplast copper transporter resulted in a severe phenotype when electron acceptors were limiting, a defect which was explained as decreased water–water cycle capacity due to reduced SOD activity (Higuchi *et al.*, 2009). There are multilayer antioxidant systems distributed over different compartments, which all contain a variety of scavenging mechanisms. For example, APX exists as a small gene family, with three isoforms present in chloroplasts and two in the cytosol, CAT is represented by three peroxisomal isoproteins, and cytosolic and chloroplastic GR isoforms have been described (Edwards *et al.*, 1990; Frugoli *et al.*, 1996; Jespersen *et al.*, 1997; Narendra *et al.*, 2006; Santos *et al.*, 1996).

Such a multiplicity of systems and isoenzymes in variable compartments means that plant cells possess an extensive network of compensatory mechanisms to ensure that ascorbate and GSH levels are adequately buffered. For example, there is evidence from transgenic studies that the classical ascorbate- and GSH-dependent water–water cycle may compensate in part for lack of the NTRC-dependent pathway (Pérez-Ruiz *et al.*, 2006). Single and even double knockouts of thylakoid and stromal *APX* genes show no obvious phenotype under nonstressing growth conditions, due to compensating increases in the alternative antioxidant enzymes 2-CP, cytosolic APX, and CAT (Davletova *et al.*, 2005; Dietz *et al.*, 2006; Giacomelli *et al.*, 2007; Kangasjärvi *et al.*, 2008). Correspondingly, antisense lines in which both *2-CP* transcripts were reduced resulted in a severe phenotype early in development, but this was alleviated when plants reached maturity, through increased peroxidase and SOD expression and activity (Baier and Dietz, 1999) and increased upregulation of genes for chloroplast APX (Baier *et al.*, 2000). Intriguingly, compensation is even possible within the GSH regeneration pathway. In *gr1* knockout mutants, where the cytosolic GR is absent, a Trx *h3* and NADPH:thioredoxin reductase (NTRA) combination is thought to directly reduce GSSG to GSH (Marty *et al.*, 2009).

Due to these compensating mechanisms, a phenotypic difference from wild type can sometimes only be shown through specific growth conditions or targeted saturation of the GSH and ascorbate pool buffering systems. For example, a phenotype for chloroplast *apx* mutants was only visible when double mutants were subject to abrupt high light stress, after growth under low light (Kangasjärvi *et al.*, 2008). The same principle was used when analyzing *apx* double mutants that had been crossed with the *vtc*-null, ascorbate-deficient line, in order to observe oxidative stress and photobleaching (Giacomelli *et al.*, 2007).

Antisense reduction in the total transcript levels of GR by 30–70% also yielded increased sensitivity to oxidative stress (Ding *et al.*, 2009; Foyer *et al.*, 1995). These plants showed adaptive responses typical of mutations that disrupt the antioxidant system. Cellular glutathione levels were increased, while the redox state shifted to the oxidized form. In addition, ascorbate contents were decreased, and there was accumulation of H_2O_2. Interestingly, double antisense mutants lacking APX and CAT are more resistant to oxidative stress than the single mutants (Davletova *et al.*, 2005; Rizhsky *et al.*, 2002). This protection seems to be mediated by increased oxidative pentose-phosphate pathway (OPPP) activity and downregulation of PET (Rizhsky *et al.*, 2002).

Because the water water cycle depends on reductants supplied by PET (as reduced Fd and NADPH), the interaction of PET, ROS production, and removal is highly complex. Indeed, a strong correlation has been established between FNR content and stress responses (Palatnik *et al.*, 1997, 2003; Rodriguez *et al.*, 2007), possibly due to interaction with stress-relieving metabolism of ascorbate (Lintala *et al.*, 2009).

C. THE ROLE OF PHOTORESPIRATION AND RESPIRATION IN REDOX POISING OF C3 PLANTS

The enzyme RubisCO has dual specificity for CO_2 and O_2, and the photorespiratory pathway detoxifies the glycollate generated by this inefficient, competing oxygenase activity (Fig. 2). One consequence of the multicompartment location of the enzymes of photorespiration is the generation of glycine and serine in the mitochondria, enabling electrons to be fed into the mitochondrial electron transport chain. This has led to our understanding that photorespiration can also act as a valve mechanism to remove excess reductant generated in high light and maintain redox homeostasis (Foyer *et al.*, 2009; Niyogi, 2000).

Work with transgenic plants has been central to the investigation of photorespiration (see Foyer *et al.*, 2009 for a recent review). In these studies, the initial goal to identify the function of a single gene could not always be achieved, but the importance of photorespiration was elegantly demonstrated, with severe phenotypes under photorespiratory (low CO_2) conditions.

Glycine decarboxylase (GDC) is a critical multisubunit enzyme in photorespiration, and homozygous barley mutants for the GDC H-protein (containing only 1% of the wild-type GDC activity) showed no obvious differences compared to the wild type under nonphotorespiratory conditions. However, in high O_2 and low CO_2 these plants exhibited a severe phenotype with early leaf senescence (Blackwell *et al.*, 1990; Igamberdiev *et al.*, 2001; Wingler *et al.*, 1997). By contrast, double knockout mutants of both peroxisomal MDH (pMDH) isoforms (Cousins *et al.*, 2008), and mutants of hydroxypyruvate reductase isoforms HPR1 and HPR2 (Timm *et al.*, 2008) did not result in severe phenotypes, even under photorespiratory conditions. This implies that photorespiration might be readily compensated for by other metabolism, but other results indicate this is not the case, and so there is likely an, as yet unidentified, upregulated alternative to HPR in these plants. The essential role of photorespiration has been well demonstrated in other knockouts of photorespiration genes, which showed compromised PSII D1 repair leading to photoinhibition (Boldt *et al.*, 2005; Somerville and Ogren, 1982; Somerville and Somerville, 1983; Takahashi *et al.*, 2007), and a double mutant lacking both transcripts coding for the P-subunit of GDC in *Arabidopsis thaliana* did not survive for longer than a few weeks even when conditions were nonphotorespiratory (Engel *et al.*, 2007).

In plants, mitochondria are not only a site of energy production, but they also contribute to redox homeostasis and optimizing photosynthesis (Raghavendra and Padmasree, 2003), as well as serving in stress protection and defense. Plant mitochondria possess several poising features, including internal and external NAD(P)H dehydrogenases, uncoupling proteins (UCP), and AOX, allowing the flexible removal of excess electrons by transfer to O_2 without the generation of ATP. Some of these mechanisms are unique to plants, emphasizing their specific role in adjusting redox status in cells subject to the great variations in redox state caused by PET. This means that transgenics with perturbed mitochondrial electron transport often show no visible phenotype under standard growth conditions. In the case of AOX, illustrated in Fig. 4, a severe phenotype for knockout mutants of a specific AOX isoprotein could only be observed when the compensating phosphorylating pathway was disrupted by an inhibitor (Fig. 4D; Strodtkötter *et al.*, 2009).

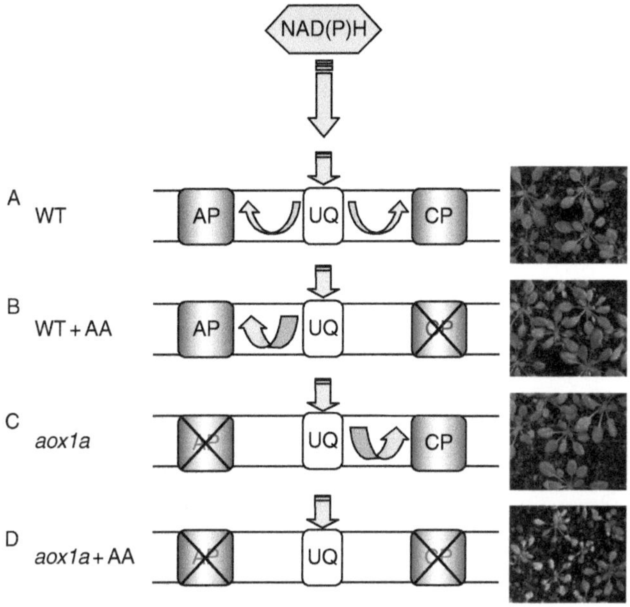

Fig. 4. Revealing a phenotype in mutants of the alternative oxidase (AOX1A) against a background of compensatory pathways. (A) Wild-type mitochondria possess two distinct pathways for the transfer of electrons from reduced ubiquinone to molecular oxygen: the cytochrome pathway (CP) via cytochrome oxidase (COX) that is linked to ATP production and the alternative pathway (AP) via alternative oxidase (AOX) without concomitant ATP production. (B) Under normal conditions, the loss of AOX1A by transgenic knockout (*aox1a*) does not yield an obvious phenotype, as it is compensated by the CP. (C) CP can be inhibited by antimycin a (AA), but on application to wild type this does not result in a stressed phenotype because increased AOX activity compensates for the inhibition. (D) After application of AA a phenotype for the *aox1a* mutant is finally visible due to elimination of the compensating CP resulting in profound stress, thus resolving the function of AOX in relieving oxidative stress. (See color plate section at the back of the book.)

Prior to the transgenic era, inhibitors were the tool of choice for examining the effects of loss of function on a plant, and they remain a powerful option, particularly in combination with transgenics, as shown here. The advantage presented by transgenics is that a gene coding for a specific isoprotein, in this case *AOX1A*, can be knocked out to dissect its function from that of other gene products. Other examples of transgenic perturbation of mitochondrial redox metabolism include the CMS II mutants of mitochondrial electron transport, which lack a functional complex I for NADH oxidation, but still reach the same final biomass as the wild type (Noctor *et al.*, 2004). The function of these nonphosphorylating alternatives to the cytochrome pathway may be revealed by rapid changes in redox status or inhibition of compensating pathways. Knockout of one of the various UCP isoforms (UCP1) resulted in localized oxidative stress in *A. thaliana* (Sweetlove *et al.*, 2006), while knockout of the alternative oxidase (AOX1A) led to increased susceptibility to low temperature (Watanabe *et al.*, 2008), to antimycin A-treatment, and to combined light and drought stress (Giraud *et al.*, 2008; Strodtkötter *et al.*, 2009). Multiple effects became apparent due to oxidative stress in tobacco cells lacking AOX, causing upregulation of the antioxidant defense system and resulting in programmed cell death (Robson and Vanlerberghe, 2002).

D. COMPENSATING PATHWAYS ENSURE OPTIMAL REDOX HOMEOSTASIS

Several pathways enable correct poising of the plant cellular redox state and balance fluctuations in PET output caused by changes in environmental conditions (Fig. 2). These include the combination of AOX in mitochondria and the chloroplast NADP-dependent malate dehydrogenase (NADP-MDH)-based malate valve which allows the export of reducing equivalents from the chloroplast (Scheibe, 2004; Fig. 1), and their dissipation through nonproton-gradient-generating oxidation in the mitochondria (Niyogi, 2000; Fig. 3). Clearly, maintaining redox poise is of paramount importance to the plant, and a high degree of redundancy is desirable, but this diversity of pathways has made transgenic approaches especially problematic. The failure to find significant phenotypes in mutants lacking proteins involved in these pathways is common and demonstrates that alternative systems for energy dissipation are readily used to prevent excess reduction pressure and ROS generation. In many cases, a phenotype can only be detected by using treatments that cause oxidative stress.

For example, NADP-MDH was decreased by an antisense approach that left little residual activity, but no significant phenotype was detected (tobacco: Faske *et al.*, 1997; potato: Backhausen *et al.*, 1998; Laisk *et al.*, 2007). Knockout mutants of *A. thaliana* with no NADP-MDH activity also do not exhibit any visible growth retardation, and this seems to be partly due to upregulation of the water–water cycle and photorespiration, as *NTRC* and *GDC* transcripts are elevated in these plants (I. Strodtkötter and R. Scheibe, unpublished results). Conversely, when photorespiratory enzyme transcripts were decreased, NADP-MDH capacity was found to increase in compensation (Igamberdiev *et al.*, 2001). In plants lacking a chloroplast peroxiredoxin (PrxFII), no phenotype was visible under normal growth conditions, but upregulation of several antioxidant systems suggested compensation.

However, this compensation could not overcome oxidative stress induced by cadmium or the AOX inhibitor SHAM (Dietz *et al.*, 2006).

Because of the redundancy within redox-poising pathways, their importance for homeostasis can often only be determined through very specific growth conditions or treatments. This may also be partly related to their physiological role in buffering redox stresses under exceptional conditions, and the extremely mild and constant conditions under which most transgenic plants are cultivated for the experiments. Finally, transgenic experiments on these pathways have enabled the identification of alternative, compensatory pathways, as mutants with mild or no obvious phenotype are screened to identify genes that are upregulated, and to find out what kind of metabolism could substitute for the antisense reduction or knockout.

E. BUFFERING REDOX FLUCTUATION BY SINK CAPACITY

Energy conversion in the light reactions is highly linked to energy consumption during assimilation of C, N, and S. It has been shown that knockout of a specific FNR isoprotein affects the ratio of electron channeling into C and N metabolism (Hanke *et al.*, 2008), possibly due to altered $NADP^+$ photoreduction capacity and/or CET. It follows that changes in the content of enzymes involved in CO_2 fixation, Calvin cycle, starch biosynthesis, and export of carbohydrates will be reflected to some extent in the redox poise of the chloroplast (Figs. 2 and 3). Transgenic studies to knockout enzymes of carbon assimilation have had mixed results in terms of altering carbon fixation. Early antisense studies aimed to determine limiting steps in order to improve photosynthetic yield (Raines, 2003). It was found that reduced RubisCO content in *rbcS* antisense-transformed tobacco resulted in decreased photosynthesis only when CO_2 or light were not limiting (Lauerer *et al.*, 1993; Quick *et al.*, 1991a), and also led to decreased nitrate assimilation (Quick *et al.*, 1991b). Similarly, plants with reduced phosphoribulokinase (Banks *et al.*, 1999) and NADP-glyceraldehyde 3-P dehydrogenase content (Price *et al.*, 1995) were able to balance carbon, nitrogen, and energy metabolism. Surprisingly, phosphoglucomutase-antisense plants with reduced plastidic and cytosolic isoforms achieved wild-type levels of starch synthesis (Fernie *et al.*, 2002), suggesting a new pathway circumventing both reactions. Thus, the absence of proteins seemingly essential to carbon metabolism can be largely compensated under nonsaturating conditions. Other examples include knockout of the chloroplast triose-phosphate/phosphate translocator (TPT), which does not lead to visible phenotypes (Schneider *et al.*, 2002). Even a knockout of ADP-glucose pyrophosphorylase (AGPase), with a starchless phenotype, seems able to compensate readily for this metabolic disturbance (Lin *et al.*, 1988). These results indicate that plants readily alternate and substitute channeling of carbon into either sucrose or starch, possibly to maintain a constant flow of photosynthetic power into photosynthate under favorable conditions.

Later transgenic approaches did identify enzymes which dramatically impact on carbon-fixation rates, and some of these studies provide strong evidence that disrupting photosynthetic sink capacity has profound effects on the redox poise of the cell and in turn on PET. For example decreases in plastid transketolase (Henkes *et al.*, 2001) and sedoheptulose 1,7-bisphosphatase (Lawson *et al.*, 2006) resulted in major effects on photosynthesis, phenylpropanoid metabolism, and development. A moderate decrease in plastid aldolase also resulted in severe growth inhibition (Haake *et al.*, 1998). Häusler *et al.* (2009) demonstrated that a double mutant lacking any TPT and AGPase activities (and therefore inefficient at both exporting triose phosphate for sucrose production and at synthesizing starch in the chloroplast) develops a severe phenotype. Notably, there is an 80% decrease in PET, and a massive increase in photorespiration, as the plants seek to relieve reduction pressure in the thylakoids by using O_2 as a sink. Evidence is also presented for the upregulation of other compensatory pathways in these plants, including the water–water cycle, chlororespiration, and NDH-dependent CET.

Interestingly, enzymes normally associated with the reconversion of carbon into reducing power in the dark also appear to have a strong effect on cellular redox poise. Stromal glucose 6-phosphate dehydrogenase (G6PDH) generates NADPH using Glc 6-P for the OPPP. In plants with greatly reduced leaf Fd content, the OPPP functions in the light to provide the NADPH required to detoxify ROS by NTRC and 2-CP (Voss *et al.*, 2008). However, in an antisense line of tobacco with decreased activity of the plastid P2-isoform of G6PDH, any redox imbalance appeared to be compensated, and the plants were even more resistant toward oxidative stress (Debnam *et al.*, 2004).

IV. CHANGES IN REDOX STATE ARE TRANSLATED INTO SIGNALING CASCADES TO ADJUST METABOLISM

A. TRANSGENIC STUDIES OF REDOX SIGNALING

The metabolism that helps to buffer the redox status of plant cells is inextricably linked to signaling cascades at many levels, forming a network that ensures induction of short-term and long-term responses to changing environmental and developmental conditions in order to maintain redox homeostasis. For example, antisense and knockout mutants for photorespiratory enzymes accumulate metabolites (malate, glycine) that in turn affect the activities of mitochondrial respiration (Bykova *et al.*, 2005), photosynthesis (Heineke *et al.*, 2001), and GS2/GOGAT (Schjoerring *et al.*, 2006) at the posttranscriptional and posttranslational level. Maintaining buffering pathways and balancing PET source and sink capacity under these conditions requires responses at the levels of gene expression, protein translation and modification, and regulation of protein location and activity. Such networks must regulate short-term changes in metabolism to adapt to transient change and also initiate large-scale changes in gene expression when buffering capacity is exceeded upon sustained stress.

Because severe fluctuations in the redox status of the cell are damaging, the redox poise is heavily buffered. Any signaling network must therefore be extremely sensitive, in order to detect changes below the threshold of potential damage, and anticipate induction of the compensating metabolism (Apel and Hirt, 2004; Foyer and Noctor, 2005b, 2009). In addition, when buffering capacity is exceeded, for example due to sudden, excess reducing pressure from PET, this must also be detected. Under these conditions signaling cascades induce downregulation of metabolism that is connected to stress generation (e.g., expression of the proteins involved in PET), and in extreme cases even promote controlled cell death, rather than allow an unregulated buildup of excess oxidative species (Gadjev *et al.*, 2008).

Study of the redox-signaling cascades linking the redox state of the plant cell to gene expression is a developing field, and many of the genes involved remain to be identified. Our understanding of these processes has been greatly enhanced by studies on transgenic plants where the redox network has been disturbed, and concurrent changes in nuclear gene expression have been detected (e.g., Queval *et al.*, 2007; Voss *et al.*, 2008; Wan *et al.*, 2009). In an alternative approach to identify new redox-signaling cascades, transgenic lines were screened against a genetic background of the luciferase reporter gene under the 2-CP promoter (which mediates redox-state-dependent expression) (Heiber *et al.*, 2007). This enabled identification of various redox-imbalanced (*rimb*) mutants by their weak reporter gene expression. In another recent approach, transcript analysis of knockout plants lacking a redox-responsive transcription factor (RRTF1) and grown under various types of redox stress revealed subsets of genes involved in the redox imbalance response network (Khandelwal *et al.*, 2008).

B. DIFFERENT SOURCES OF REDOX SIGNALS

Short-term robustness in photosynthetic metabolism is achieved by complex metabolic networks that avoid perturbations in a dynamic process (Luo *et al.*, 2009; Fig. 2). In contrast, upon exposure to sustained stress, redox-homeostasis and optimal cell function is maintained by adjusting gene expression. This is complicated by the membrane barriers that separate the source of most changes in redox status (PET and mitochondrial respiration) from the site of most gene expression, in the nucleus. The sources of these signals from different pathways has been the subject of intense speculation, with current models focusing on ROS-based signals originating in the PQ pool (Durnford and Falkowski, 1997; Escoubas *et al.*, 1995). The actual signal from the PQ pool remains unclear, but it has been speculated that it may be H_2O_2. Indeed, when H_2O_2 is slightly enhanced in double mutants of *apx*, the chloroplast redox state is modulated and in turn activates retrograde signaling from the chloroplast to the nucleus (Kangasjärvi *et al.*, 2008). By comparing transcript differences from the wild type when either the Cyt b_6f complex or the chloroplast ATPase complex was transgenically eliminated, genes that are differentially regulated by reduction or oxidation of the PQ pool have been identified (Maiwald *et al.*, 2003).

C. IMPACT OF ALTERED THIOL CASCADES ON PROTEIN ACTIVITY

As the location of PET, chloroplasts are able to rapidly detect changes in PET acceptor redox state, which can be interpreted through the Fd–Trx system (Buchanan, 1984; Dietz *et al.*, 2002). This coordinates fine regulation of light/dark-modulated chloroplast enzymes by reducing regulatory cysteines (Scheibe, 1991), meaning that changes in redox state lead to rapid changes in enzyme activities, contributing to short-term acclimation of photosynthesizing cells (Scheibe *et al.*, 2005; Fig. 1). Redox regulation at cysteine residues has also been shown to control metabolism and signaling in other cellular compartments (Buchanan and Balmer, 2005). The reversible formation of mixed disulfides between proteins and GSH is frequently involved in such processes, particularly in oxidative conditions (Dalle-Donne *et al.*, 2009; Giustarini *et al.*, 2004). It is noteworthy that there is an extremely large number (over 100) of redox-active proteins of the redoxin superfamily in plants (Meyer *et al.*, 2008).

Transgenic approaches to alter the constituents of PET have also given us new information regarding stromal signaling pathways to control protein activity. For example, when electron transport from PET to the stroma is disrupted by decreased availability of the stromal electron receptor Fd

(Voss *et al.*, 2008), or by decreased capacity to release electrons from PSI (Haldrup *et al.*, 2003), then the Fd:Trx regulatory system fails to activate the enzymes of the Calvin cycle. This leads to a massive decrease in carbon fixation, which results in severe phenotypes. The plants seem to compensate in part by using catabolic metabolism to temporarily increase the reduction potential of the Trx pool enough to activate the Calvin-cycle enzymes (Voss *et al.*, 2008).

Reverse genetic approaches using gene-silencing or knockout technology have been used to try and identify specific functions of the many members of the Trx family. Because the role of Trx *m* in light/dark modulation of chloroplast enzymes was considered well understood, it was rather unexpected to find a severe effect on chloroplast development and growth in RNAi rice lacking all Trx *m* forms (Chi *et al.*, 2008), indicating an additional role in plant development. It seems likely that this dramatic phenotype is due to the recently identified role of Trx *m* in reducing HCF164, a thylakoid-membrane thioredoxin-like protein that regulates assembly of the Cyt b_6f complex (Motohashi and Hisabori, 2009). Knockout mutants of *HCF164* are severely deficient in Cyt b_6f accumulation (Lennartz *et al.*, 2001).

D. IMPACT OF ALTERED THIOL CASCADES ON SIGNAL TRANSDUCTION

Redox-active proteins such as Trx, glutaredoxins (Grx), and peroxiredoxins (Prx) are also involved in defense responses (Dietz *et al.*, 2006; Gelhaye *et al.*, 2005; Meyer *et al.*, 2008; Rouhier and Jacquot, 2002; Rouhier *et al.*, 2004) and could interpret the redox state of the cell to drive developmental processes. There are multiple isoforms of the superfamily of redox-signaling proteins, or redoxins, some of which have unknown, or poorly understood functions. These have presented attractive potential targets for transgenic approaches to investigate the possible action of thiol cascades in regulating developmental processes.

Knockout of *h*-type Trxs did not have any effect upon development or sensitivity toward oxidative stress treatment, even when fivefold mutants for Trx *h1, h2, h3, h4*, and *h5* were generated (Meyer *et al.*, 2005). This indicates that some Trx proteins are largely redundant, and can be compensated for by other pathways, or may only have physiological roles under very specific conditions. In our laboratory, we have isolated 18 T-DNA insertion lines in separate Trx genes (Fig. 5), none of which has any obvious phenotype under nonstressful growth conditions (S. Holtgrefe, unpublished results). In another example, *A. thaliana* knockout lines for sulfiredoxin were characterized by an increased level of oxidized Prx and a visible phenotype under photooxidative stress, but under standard culture conditions no difference from

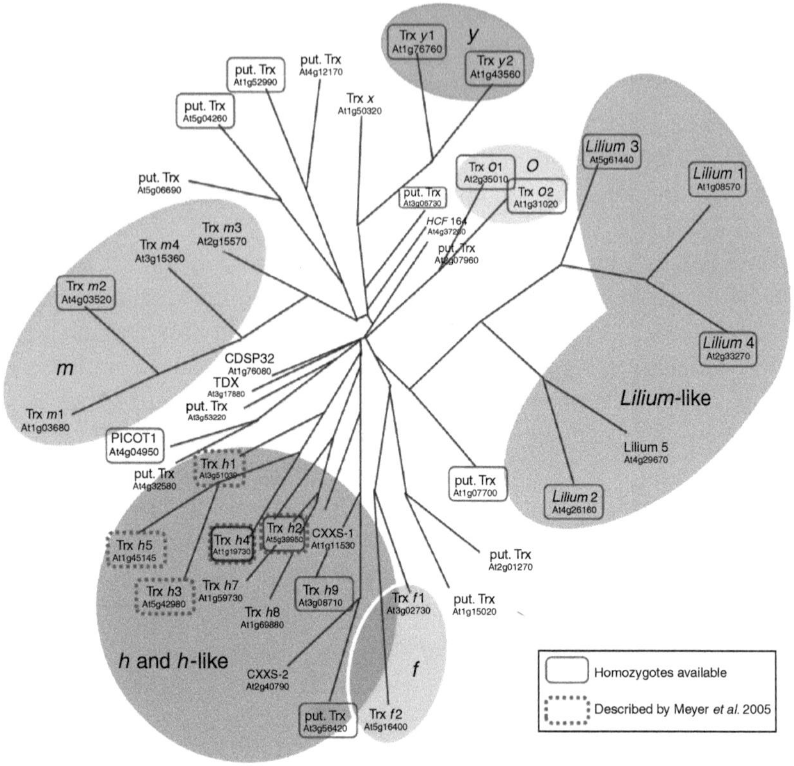

Fig. 5. Evolution and diversity of thioredoxin family genes in *Arabidopsis*. The tree was constructed with the PHYLIP program package (Felsenstein, 1993), using the Neighbor-Joining method with 1000 bootstraps. Confirmed homozygous knockout plants which have been grown in our laboratory are highlighted in the boxes. Those in boxes with dotted lines were described by Meyer *et al.* (2005).

wild-type plants was visible (Rey *et al.*, 2007). These results indicate that the gene products have very specific roles, related to redox-stress conditions. Another possible explanation for the lack of an obvious phenotype may be that, early in seedling development, compensatory pathways are upregulated, so that the original function of the gene is masked in the null mutant. Knockout of Trx reductases has revealed an important role in development. For example, the lack of NTRC in chloroplasts results in changed photoperiodic development (Lepistö *et al.*, 2009).

The developmental or stress-induced changes visible in these mutants result from disturbing multistep signaling cascades and thus represent pleiotropic effects that cannot be attributed to direct action of a specific gene product. They rather reflect the presence of a complex network of cellular control at many levels of regulation.

E. Impact of Altered ROS Status on Signal Transduction

The redox state of the glutathione and ascorbate pools and the concentration of ROS have been heavily implicated in coordinating redox signaling cascades. These findings have been extensively reviewed elsewhere (Foyer and Noctor, 2009; Maughan and Foyer, 2006; Sharma and Dietz, 2009), and therefore only a brief description will be attempted here. Transgenic disruption of genes encoding enzymes involved in eliminating ROS, such as CAT, APX, GR, and SOD, disturbs cellular redox poise, initiating signals that result in gene expression changes to maintain redox homeostasis. The redox state, as well as the total amount of glutathione has been found to affect gene expression, impacting on many developmental processes (May *et al.*, 1998) such as root development (Espunya *et al.*, 2006; Morgan *et al.*, 2008), and xylem differentiation (Henmi *et al.*, 2005). Moreover, petal development was shown to be altered upon knockout of one of the Grx genes (Li *et al.*, 2009; Wang *et al.*, 2009).

Many redox signaling cascades function to interpret environmental conditions, and so the growth of transgenic plants that will be used for experiments to examine the effects of eliminating signaling protein components must take place under carefully controlled conditions. This was made especially clear in work on the *cat2* knockout mutants (Queval *et al.*, 2007), where dramatically different phenotypes were observed in short-day (smaller rosettes, increased total glutathione, and proportionally more oxidized glutathione) and long-day (lesions indicating activation of an executor pathway) growth conditions. Also, an increase in the malate-valve enzyme NADP-MDH was only apparent in short-day conditions in wild-type *Arabidopsis*, while long-day photoperiod resulted in the increase of antioxidant enzymes and in flowering induction (Becker *et al.*, 2006).

Clearly, disturbances in signaling cascades will have far-reaching long-term consequences for global gene expression, and simple comparisons of mutant and wild-type transcripts will also detect many pleiotropic effects. This can be overcome by comparing complementary transgenics, for example two mutants, both incapable of photoautotrophy, but one with very reduced PQ and the other with very oxidized PQ to identify genes differentially regulated by reduction or oxidation of the PQ pool (Maiwald *et al.*, 2003). Alternatively, further conditions can be applied to transgenics, such as high light or DCMU treatment when investigating a knockout of the redox-regulated transcription factor 1 (*rrtf1*) (Khandelwal *et al.*, 2008) to separate short-term gene induction events controlled by the transcription factor from those that were pleiotropically induced.

It is likely that when microarray experiments on gene-knockout and/or stressed plants detect up- or downregulation of genes of many different pathways as well as of transcription factors, these often represent long-term responses to the genotype background, influencing alternative pathways (Mittler *et al.*, 2004). Signal transduction pathways analyzed in other contexts, such as phytohormone action, often interact with components of the redox network, and development is frequently found to be governed by redox state. For example, ROS and abscissic acid signaling appear to be interconnected (Christmann *et al.*, 2006; Miao *et al.*, 2006). This means that both morphology and life cycle are responding to changing conditions, demonstrating the high flexibility necessary for land plants to compete and propagate under a wide range of conditions. The regulatory network integrates many signals coming from the environment with those generated by the endogenous genetic program.

F. FEEDBACK OF REDOX SIGNALS TO ADJUST PET

Severe disruption in PET by transgenic knockout of a single gene has been found to lead to a signaling cascade which results in downregulation of other PET genes. Examples include the PETC subunit of Cyt b_6f (Maiwald *et al.*, 2003) and the PsaD1 subunit of PSI (Ihnatowicz *et al.*, 2004), presumably as the plant attempts to reduce the capacity for excess ROS production when certain steps in PET become restricted. Moreover, in many cases transgenic disruption of pathways buffering redox poise also results in downregulation of the proteins involved in PET. It is highly likely that in the absence of appropriate mechanisms to fully buffer the redox fluctuations caused by variable PET activity, signaling cascades occur which result in the downregulation of PET (Häusler *et al.*, 2009; Pérez-Ruiz *et al.*, 2006; Reiser *et al.*, 2004). Thus, the plant will sacrifice almost any pathway, even photosynthesis, rather than allow the redox poise of the cell to be disrupted.

Indeed, extreme perturbation of redox poise appears to lead to programmed cell death (Gadjev *et al.*, 2008), possibly to avoid large-scale production and spread of oxidative species. A more oxidized cellular redox poise is known to induce senescence and cell death, both during normal development and upon stress or pathogen attack (see Foyer and Noctor, 2005a). This imbalance can arise through disruption of redox-poising mechanisms or indirectly as a result of disturbed signaling cascades. For example, knockout of *cpr5* (constitutive pathogen response), a transmembrane protein of unknown function, resulted in fivefold increase in antioxidant systems, indicating that the mutant plants suffered oxidative stress (Jing *et al.*, 2008). This altered cellular redox poise leads to early leaf senescence, but the other components involved are not yet identified.

V. CONCLUSIONS

A. CONTRIBUTIONS OF TRANSGENIC APPROACHES TO UNDERSTANDING THE REDOX NETWORK

The study of plant metabolism has depended on the comparison of different genotypes for enlightenment since the time of Gregor Mendel. At first, variants that had arisen naturally and as a consequence of breeding programs were studied, to further understanding of genetics and metabolism, and give rise eventually to the "one gene–one function" hypothesis that drove the initial functional genomics projects. When it became clear that great information about the function of a gene product could be derived from studying mutations in a gene for loss of function (reverse genetics), and identifying the genes whose disruption resulted in specific genetic and metabolic traits (forwards genetics), it was desirable to generate populations of more plants with mutations in separate genes, as research tools. This was primarily done using chemical mutagens (e.g. ethane methosulfate) or radiation, to induce base changes, but was revolutionized by exploiting the invasive properties of *Agrobacterium tumefaciens* to disrupt nuclear genes with modified versions of the T-DNA cassette and use of transposons. To complement this, homologous recombination could be used to delete specific plastome genes. Alternative approaches to decreasing or extinguishing a gene product have been applied at the RNA level, through antisense and RNAi techniques. These provide the advantage of allowing a comparison to be made between the degree of reduction in a gene product and the severity of the resultant phenotype, and also allow the simultaneous knockout of similar genes. Inhibitors can also be used to disrupt activity at the protein level in a nonspecific way, but the potential for inhibitor side effects will always make transgenics the primary tool of choice for studying gene product function. In the future, establishing inducible knockout technology represents an ideal, since it enables the separation of the immediate effects of the loss of one gene product from long-term, adaptive response to its absence.

Great progress in our understanding of redox networks and their high level of complexity has therefore come through transgenic studies, and this can be visualized as shown in Fig. 6. Initial studies assumed a clear relationship between input (changed gene activity) and output (effect on a specific property), as shown in Fig. 6A. It rapidly became apparent that metabolic pathways are

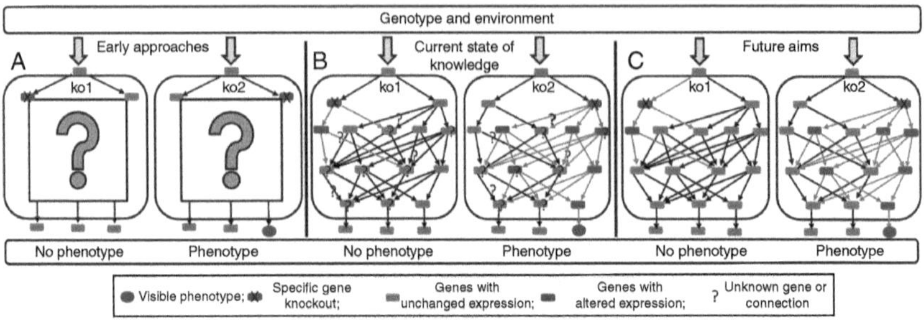

Fig. 6. Increased understanding of redox systems through transgenic work. (A) Early approaches. It was previously hoped that a function for each gene could be resolved by identifying loss of a single function in knockout plants (ko1 and ko2). For proteins in the redox network, in some cases (ko1) there was no detectable effect upon the manipulated pathway, while in others (ko2) the expected phenotypic change occurred. (B) Current state of knowledge. Approaches to examine changes in transgenics at the transcriptome and metabolome level, upon specific environmental changes, reveal clusters, hubs, and nodes in the redox network related to the knocked out gene. In some cases, identification of the compensating pathways that act to prevent an obvious phenotype is possible (ko1), while in others unexpected phenotypes occur (ko2). There are still many genes of unknown function and uncharacterized interactions, which must now be clarified with classical biochemical and molecular biology approaches. (C) Future aims. We hope that comprehensive knowledge of all the components of the redox network, their functions, and integrated responses to various developmental and environmental changes will enable the construction of dynamic computer models. In this way, predictions of the response to various natural and transgenic variations will be possible, revolutionizing plant breeding and transgenic technology.

interconnected and changes at one point can result in multiple effects on outputs (Fig. 6B). Since many effects were initially unexpected, we have come to understand more about the redox system, in which strongly interlinked metabolic and gene expression processes enable adaptation to knockout situations and growth adjustment. These networks contain circular pathways and hubs that interconnect different clusters of reactions, some of which initially seemed unrelated to redox systems. The redox network is also subject to great influence from differences in cell type, development, and environmental cues such as time, nutrition, and stress. As the number of transgenic studies increases and complex changes can be monitored using improved methodology, a more complete picture of the multidimensional network in plant cells will emerge (Fig. 6C). One ultimate aim is to be able to predict the effect of any change posed upon the redox system. Computational methods, such as the program MapMan (Urbanczyk-Wochniak *et al.*, 2006), will be essential in attempting to integrate the bewildering complexity of these multiple, interlinked processes in a systems biology approach.

B. INTEGRATION OF TRANSGENIC RESULTS INTO A MODELING-BASED SYSTEMS BIOLOGY APPROACH

One possibility for the study of complex redox networks is through a modeling approach, and this has yielded some significant results. Application of the theory of metabolic control analysis (MCA) has revealed that control is often spread over many steps of a pathway and is not achieved by one, single regulatory step (Fell, 1997). Rather, metabolic pathways are relatively stable, and there is little change in flux even when large portions of one enzyme are removed (Fridlyand and Scheibe, 2000; Westerhoff *et al.*, 1995), and increase or decrease of a "limiting" enzyme may not lead to large alteration in fluxes (Fridlyand *et al.*, 1999). However, in models the complexity of the system is generally not sufficiently understood, and we are a long way from the situation represented in Fig. 6C, since not all pathways have been identified or characterized, and many proteins have multiple homologs. For example, the optimal distribution of photosynthetically generated electrons to various acceptors in order to poise the ATP/NADPH ratio and avoid ROS formation is highly flexible (Fridlyand and Scheibe, 1999). Eventual inputs to this kind of modeling of the redox system will have to include data on pathway fluxes, enzyme constants and activation states, concentrations of enzymes, metabolites, ions and H^+, protein modifications, crowding, and channeling effects.

C. FUTURE APPLICATION OF TRANSGENICS TO STUDY METABOLISM, SIGNALING, AND OVERALL CONTROL

The transgenic elimination or reduction of single, or multiple gene products related to the redox network can be regarded as a kind of stress that the plant must overcome with compensatory pathways in order to reproduce. This requires intimate integration of redox metabolism and buffering with signaling cascades, often involving ROS, and resulting in responses at gene expression level. The result may be a morphological phenotype, such as those described in "growing out of trouble" (Potters *et al.*, 2007), or may only be visible at the metabolic level of the pathways that are compensating for the absence of the gene product(s). Future studies of redox systems will require the use of transgenics to comprehensively identify which steps are critical, and establish the compensatory pathways for those which are not, through transcriptomics, metabolomics, interactomics, and mathematical models.

Examples of how transgenics can be used to generate the kind of data necessary for the construction of such modeling approaches include a cross between mutants of stromal PSI subunits (which have increased CET) with double mutants of pgrL1 (which are unable to switch to CET). The resulting severe phenotype demonstrates that cyclic flow was compensating for the deficiency of NADP photoreduction when PSI activity was decreased (DalCorso *et al.*, 2008). A powerful use of PET mutants in the future will also be to identify the source and mechanism of chloroplast-derived redox signals controlling nuclear gene expression, which have previously been investigated using inhibitors (Fey *et al.*, 2005; Schütze *et al.*, 2008). Papers to date with gene-expression profiling of PET-disrupted plants include transgenic knockouts of the ATPase and Cyt b_6f (Maiwald *et al.*, 2003), PSI (Ihnatowicz *et al.*, 2004, 2008), and FNR (Lintala *et al.*, 2009).

It will also be of interest to use PET mutants to examine how different redox fluxes contribute to regulation of protein components within the chloroplast. In this respect, studies have already been conducted on the impact of highly reduced PQ (caused by knockout of PsaE and PsaD), on phosphorylation of Lhc-4 of PSI (Ihnatowicz *et al.*, 2008), reduction in the redox-dependent phosphorylation of PSII reaction center proteins D1 and D2 in knockouts of PsbO2 due to altered D-protein turnover (Lundin *et al.*, 2007), and differential redox regulation of the PSII D1 and D2 proteins (downregulated) and the LHCII (upregulated) in mutants of PsbI (Schwenkert *et al.*, 2006).

In addition to being essential for photosynthesis, PET also creates a unique set of problems for plants. Due to the huge redox fluxes involved, this situation is particularly challenging in terms of redox homeostasis. Photosynthetic organisms therefore employ diverse and extensive mechanisms for buffering deviations from the steady state, and for preventing oxidative damage. Against this background, land plants must also interpret different combinations of metabolic, developmental, and environmental triggers. These cause multiple, parallel, and overlapping clusters of multilevel responses, and it is now clear that the redox poise of the cell is crucial to the initiation of such responses. Primary signals result in subtle shifts in redox balance, acting as mediators promoting a wide range of responses, either directly, or through further signaling systems involving concentrations of primary and secondary metabolites and calcium. Although the initial hope of the "Functional Genomics Initiatives" to identify every gene function using reverse genetics may not be easily reached, tremendous progress has been made using this approach, especially in the field of redox metabolism. The emerging picture is not simple, but every new study provides pieces of the puzzle that contribute to our understanding of the complex network that enables a living plant to grow and compete.

ACKNOWLEDGMENTS

Work performed in the authors' lab over many years in the framework of a research unit has led to the ideas put forward in this review; Renate Scheibe gratefully acknowledges financial support given by the DFG (FOR 387, EM 166/1). The authors thank Heike Schwiderski for her help with the preparation of the manuscript.

REFERENCES

Apel, K. and Hirt, H. (2004). Reactive oxygen species: Metabolism, oxidative stress, and signal transduction. *Annual Reviews of Plant Biology* **55**, 373–399.

Aro, E.-M. and Ohad, I. (2003). Redox regulation of thylakoid protein phosphorylation. *Antioxidants and Redox Signaling* **5**, 55–67.

Asada, K. (2006). Production and scavenging of reactive oxygen species in chloroplasts and their functions. *Plant Physiology* **141**, 391–396.

Backhausen, J. E., Emmerlich, A., Holtgrefe, S., Horton, P., Nast, G., Rogers, J. J. M., Müller-Röber, B. and Scheibe, R. (1998). Transgenic potato plants with altered expression levels of chloroplast NADP-malate dehydrogenase: Interactions between photosynthetic electron transport and malate metabolism in leaves and in isolated intact chloroplasts. *Planta* **207**, 105–114.

Backhausen, J. E., Kitzmann, C., Horton, P. and Scheibe, R. (2000). Electron acceptors in isolated intact spinach chloroplasts act hierarchically to prevent over-reduction and competition for electrons. *Photosynthesis Research* **64**, 1–13.

Baier, M. and Dietz, K.-J. (1999). Protective function of chloroplast 2-cysteine peroxiredoxin in photosynthesis. Evidence from transgenic *Arabidopsis*. *Plant Physiology* **119**, 1407–1414.

Baier, M., Noctor, G., Foyer, C. H. and Dietz, K. J. (2000). Antisense suppression of 2-cysteine peroxiredoxin in *Arabidopsis* specifically enhances the activities and expression of enzymes associated with ascorbate metabolism but not glutathione metabolism. *Plant Physiology* **124**, 823–832.

Banks, F. M., Driscoll, S. P., Parry, M. A. J., Lawlor, D. W., Knight, J. S., Gray, J. C. and Paul, M. J. (1999). Decrease in phosphoribulokinase activity by antisense RNA in transgenic tobacco. Relationship between photosynthesis growth, and allocation at different nitrogen levels. *Plant Physiology* **119**, 1125–1136.

Becker, B., Holtgrefe, S., Jung, S., Wunrau, C., Kandlbinder, A., Baier, M., Dietz, K.-J., Backhausen, J. E. and Scheibe, R. (2006). Influence of the photoperiod on redox regulation and stress responses in *Arabidopsis thaliana* L. (Heynh.) plants under long- and short-day conditions. *Planta* **224**, 380–393.

Biehl, A., Richly, E., Noutsos, C., Salamini, F. and Leister, D. (2005). Analysis of 101 nuclear transcriptomes reveals 23 distinct regulons and their relationship to metabolism, chromosomal gene distribution and co-ordination of nuclear and plastid gene expression. *Gene* **344**, 33–41.

Blackwell, R. D., Murray, A. J. S. and Lea, P. J. (1990). Photorespiratory mutants of the mitochondrial conversion of glycine to serine. *Plant Physiology* **94**, 1316–1322.

Boldt, R., Edner, C., Kolukisaoglu, U., Hagemann, M., Weckwerth, W., Wienkoop, S., Morgenthal, K. and Bauwe, H. (2005). D-Gylcerate 3-kinase, the last unknown enzyme in the photorespiratory cycle in *Arabidopsis*, belongs to a novel kinase family. *Plant Cell* **17**, 2413–2420.

Buchanan, B. B. (1984). The ferredoxin/thioredoxin system: A key element in the regulatory function of light in photosynthesis. *Bioscience* **34**, 378–383.

Buchanan, B. B. and Balmer, Y. (2005). Redox regulation: A broadening horizon. *Annual Reviews of Plant Biology* **56**, 187–220.

Burrows, P. A., Sazanov, L. A., Svab, Z., Maliga, P. and Nixon, P. J. (1998). Identification of a functional respiratory complex in chloroplasts through analysis of tobacco mutants containing disrupted plastid ndh genes. *The EMBO Journal* **17**, 868–876.

Bykova, N. V., Keerberg, O., Pärnik, T., Bauwe, H. and Gardeström, P. (2005). Interaction between photorespiration and respiration in transgenic potato plants with antisense reduction in glycine decarboxylase. *Planta* **222**, 130–140.

Chi, Y. H., Moon, J. C., Park, J. H., Kim, H.-S., Zulfugarov, I. S., Fanata, W. I., Jang, H. H., Lee, J. R., Lee, Y. M., Kim, S. T., Chung, Y.-Y. Lim, C. O. *et al.* (2008). Abnormal chloroplast development and growth inhibition in rice thioredoxin *m* knock-down plants. *Plant Physiology* **148**, 808–817.

Christmann, A., Moes, D., Himmelbach, A., Yang, Y., Tang, Y. and Grill, E. (2006). Integration of abscissic acid signalling into plant responses. *Plant Biology* **8**, 314–325.

Cousins, A. B., Pracharoenwattana, I., Zhou, W., Smith, S. M. and Badger, M. R. (2008). Peroxisomal malate dehydrogenase is not essential for photorespiration in *Arabidopsis* but its absence causes an increase in the stoichiometry of photorespiratory CO_2 release. *Plant Physiology* **148**, 786–795.

DalCorso, G., Pesaresi, P., Masiero, S., Aseeva, E., Schunemann, D., Finazzi, G., Joliot, P., Barbato, R. and Leister, D. (2008). A complex containing PGRL1 and PGR5 is involved in the switch between linear and cyclic electron flow in *Arabidopsis*. *Cell* **132**, 273–285.

Dalle-Donne, I., Rossi, R., Colombo, G., Giustarini, D. and Milzani, A. (2009). Protein *S*-glutathionylation: A regulatory device from bacteria to humans. *Trends in Biochemical Science* **34**, 85–96.

Davletova, S., Rizhsky, L., Liang, H., Shengqiang, Z., Oliver, D. J., Coutu, J., Shulaev, V., Schlauch, K. and Mittler, R. (2005). Cytosolic ascorbate peroxidase 1 is a central component of the reactive oxygen gene network of *Arabidopsis*. *The Plant Cell* **17**, 268–281.

Debnam, P. M., Fernie, A. R., Leisse, A., Golding, A., Bowsher, C. G., Grimshaw, C., Knight, J. S. and Emes, M. J. (2004). Altered activity of the P2 isoform of plastidic glucose 6-phosphate dehydrogenase in tobacco (*Nicotiana tabacum* cv. Samsun) causes changes in carbohydrate metabolism and response to oxidative stress in leaves. *The Plant Journal* **38**, 49–59.

Dietz, K.-J., Link, G., Pistorius, K. and Scheibe, R. (2002). Redox regulation in oxygenic photosynthesis. *Progress in Botany* **63**, 207–245.

Dietz, K.-J., Jacob, S., Oelze, M.-L., Laxa, M., Tognetti, V., de Miranda, S. M. N., Baier, M. and Finkemeier, I. (2006). The function of peroxiredoxins in plant organelle redox metabolism. *Journal of Experimental Botany* **57**, 1697–1709.

Ding, S., Lu, Q., Zhang, Y., Yang, Z., Wen, X., Zhang, L. and Lu, C. (2009). Enhanced sensitivity to oxidative stress in transgenic tobacco plants with decreased glutathione reductase activity leads to a decrease in ascorbate pool and ascorbate redox state. *Plant Molecular Biology* **69**, 577–592.

Durnford, D. G. and Falkowski, P. G. (1997). Chloroplast redox regulation of nuclear gene transcription during photoacclimation. *Photosynthesis Research* **53**, 229–241.

Ebbighausen, H., Hatch, M.D, Lilley, R. Mc C., Krömer, S., Stitt, M. and Heldt, H. W. (1987). On the function of malate-oxalacetate shuttles in plant cell. *In* "Plant Mitochondria: Structural, Functional and Physiological Aspects" (A. L. Moore and R. B. Beechey eds.), pp. 171–180. Plenum Press, New York.

Edwards, E. A., Rawsthorne, S. and Mullineaux, P. M. (1990). Subcellular distribution of multiple forms of glutathione reductase in leaves of pea (*Pisum sativum* L.). *Planta* **180**, 278–284.

Endo, T., Shikanai, T., Takabayashi, A., Asada, K. and Sato, F. (1999). The role of chloroplastic NAD(P)H dehydrogenase in photoprotection. *FEBS Letters* **457**, 5–8.

Engel, N., van den Daele, K., Kolukisaoglu, Ü., Morgenthal, K., Weckwerth, W., Pärnik, T., Keerberg, O. and Bauwe, H. (2007). Deletion of glycine decarboxylase in *Arabidopsis* is lethal under non-photorespiratory conditions. *Plant Physiology* **144**, 1328–1335.

Escoubas, J. M., Lomas, M., Laroche, J. and Falkowski, P. G. (1995). Light-intensity regulation of cab gene-transcription is signaled by the redox state of the plastoquinone pool. *Proceedings of the National Academy of Sciences of the United States of America* **92**, 10237–10241.

Espunya, M. C., Díaz, M., Moreno-Romero, J. and Martínez, M. C. (2006). Modification of intracellular levels of glutathione-dependent formaldehyde dehydrogenase alters glutathione homeostasis and root development. *Plant, Cell and Environment* **29**, 1002–1011.

Faske, M., Backhausen, J. E., Sendker, M., Singer-Bayrle, M., Scheibe, R. and von Schaewen, A. (1997). Transgenic tobacco plants expressing pea chloroplast *Nmdh* cDNA in sense and antisense orientation. *Plant Physiology* **115**, 705–715.

Fell, D. A. (1997). Understanding the Control of Metabolism. Portland Press, London and Miami.

Felsenstein, J. (1993). PHYLIP (Phylogeny Inference Package) Version 3.5c Department of Genetics, University of Washington, Seattle, USA.

Fernie, A. R., Swiedrych, A., Tauberger, E., Lytocchenko, A., Trethewey, R. N. and Willmitzer, L. (2002). Potato plants exhibiting combined antisense repression of cytosolic and plastidial isoforms of phosphoglucomutase surprisingly approximate wild type with respect to the rate of starch synthesis. *Plant Physiology and Biochemistry* **40**, 921–927.

Fey, V., Wagner, R., Bräutigam, K., Wirtz, M., Hell, R., Dietzmann, A., Leister, D., Oelmüller, R. and Pfannschmidt, T. (2005). Retrograde plastid redox signals in the expression of nuclear genes for chloroplast proteins of *Arabidopsis thaliana*. *The Journal of Biological Chemistry* **280**, 5318–5328.

Foyer, C. H. and Noctor, G. (2000). Oxygen processing in photosynthesis: Regulation and signalling. *New Phytologist* **146**, 359–388.

Foyer, C. H. and Noctor, G. (2005a). Oxidant and antioxidant signaling in plants: A re-evaluation of the concept of oxidative stress in a physiological context. *Plant, Cell and Environment* **28**, 1056–1071.

Foyer, C. H. and Noctor, G. (2005b). Redox homeostasis and antioxidant signaling: A metabolic interface between stress perception and physiological responses. *The Plant Cell* **17**, 1866–1875.

Foyer, C. H. and Noctor, G. (2009). Redox regulation in photosynthetic organisms: Signaling, acclimation, and practical implications. *Antioxidants and Redox Signaling* **11**, 861–905.

Foyer, C. H., Souriau, N., Perret, S., Lelandais, M., Kunert, K. J., Pruvost, C. and Jouanin, L. (1995). Overexpression of the glutathione reductase but not glutathione synthetase leads to increases in antioxidant capacity and resistance to photoinhibition in poplar trees. *Plant Physiology* **109**, 1047–1057.

Foyer, C. H., Bloom, A. J., Queval, G. and Noctor, G. (2009). Photorespiratory metabolism: Genes, mutants, energetic, and redox signaling. *Annual Review of Plant Biology* **60**, 455–484.

Fridlyand, L. E. and Scheibe, R. (1999). Controlled distribution of electrons between acceptors in chloroplasts: A theoretical consideration. *Biochimica et Biophysica Acta* **1413**, 31–42.

Fridlyand, L. E. and Scheibe, R. (2000). Regulation in metabolic systems under homeostatic flux control. *Archives of Biochemistry and Biophysics* **374**, 198–206.

Fridlyand, L. E., Backhausen, J. E. and Scheibe, R. (1999). Homeostatic regulation upon changes of enzyme activities in the Calvin cycle as an example for general mechanisms of flux control. What can we expect from transgenic plants? *Photosynthesis Research* **61**, 227–239.

Frugoli, J. A., Zhong, H. H., Nuccio, M. L., McCourt, P., McPeek, M. A., Thomas, T. L. and McClung, C. R. (1996). Catalase is encoded by a multigene family in *Arabidopsis thaliana* (L.) Heynh. *Plant Physiology* **112**, 327–336.

Gadjev, I., Stone, J. M. and Gechev, T. S. (2008). Programmed cell death in plants: New insights into redox regulation and the role of hydrogen peroxide. *International Review of Cell and Molecular Biology* **270**, 87–143.

Gelhaye, E., Rouhier, N., Navrot, N. and Jacquot, J.-P. (2005). The plant thioredoxin system. *Cellular and Molecular Life Sciences* **62**, 24–35.

Giacomelli, L., Masi, E. A., Ripoll, D. R., Lee, M. J. and van Wijk, K. J. (2007). *Arabidopsis thaliana* deficient in two chloroplast ascorbate peroxidases shows accelerated light-induced necrosis when levels of cellular ascorbate are low. *Plant Molecular Biology* **65**, 627–644.

Giraud, E., Ho, L. H. M., Clifton, R., Carroll, A., Estavillo, G., Tan, Y.-F., Howell, K. A., Ivanova, A., Pogson, B. J., Millar, A. H. and Whelan, J. (2008). The absence of ALTERNATIVE OXIDASE1a in *Arabidopsis* results in acute sensitivity to combined light and drought stress. *Plant Physiology* **147**, 595–610.

Giustarini, D., Rossi, R., Milzani, A., Colombo, R. and Dalle-Donne, I. (2004). *S*-Glutathionylation: From redox regulation of protein functions to human diseases. *Journal of Cellular and Molecular Medicine* **8**, 201–212.

Haake, V., Zrenner, R., Sonnewald, U. and Stitt, M. (1998). A moderate decrease of plastid aldolase activity inhibits photosynthesis, alters the levels of sugars and starch, and inhibits growth of potato plants. *The Plant Journal* **14**, 147–157.

Hajirezaei, M. R., Peisker, M., Tschiersch, H., Palatnik, J. F., Valle, E. M., Carrillo, N. and Sonnewald, U. (2002). Small changes in the activity of chloroplastic $NADP^+$-dependent ferredoxin oxidoreductase lead to impaired plant growth and restrict photosynthetic activity of transgenic tobacco plants. *The Plant Journal* **29**, 281–293.

Haldrup, A., Lunde, C. and Scheller, H. V. (2003). *Arabidopsis thaliana* plants lacking the PSI-D subunit of photosystem I suffer severe photoinhibition, have unstable photosystem I complexes, and altered redox homeostasis in the chloroplast stroma. *The Journal of Biological Chemistry* **278**, 33276–33283.

Halliwell, B. (2006). Reactive species and antioxidants: Redox biology is a fundamental theme of aerobic life. *Plant Physiology* **141**, 312–322.

Hanke, G. T. and Hase, T. (2008). Variable photosynthetic roles of two leaf-type ferredoxins in *Arabidopsis*, as revealed by RNA interference. *Photochemistry and Photobiology* **84**, 1302–1309.

Hanke, G. T., Endo, T., Satoh, F. and Hase, T. (2008). Altered photosynthetic electron channelling into cyclic electron flow and nitrite assimilation in a mutant of ferredoxin:NADP(H) reductase. *Plant, Cell and Environment* **31**, 1017–1028.

Hansson, A., Amann, K., Zygadlo, A., Meurer, J., Scheller, H. V. and Jensen, P. E. (2007). Knock-out of the chloroplast-encoded PSI-J subunit of photosystem I in *Nicotiana tabacum*. *FEBS Journal* **274**, 1734–1746.

Hashimoto, M., Endo, T., Peltier, G., Tasaka, M. and Shikanai, T. (2003). A nucleus-encoded factor, CRR2, is essential for the expression of chloroplast ndhB in *Arabidopsis*. *The Plant Journal* **36**, 541–549.

Häusler, R. E., Geimer, S., Kunz, H. H., Schmitz, J., Dörmann, P., Bell, K., Hetfeld, S., Guballa, A. and Flügge, U.-I. (2009). Chlororespiration and grana hyperstacking: How an *Arabidopsis* double mutant can survive despite defects in starch biosynthesis and daily carbon export from chloroplasts. *Plant Physiology* **149**, 515–533.

Heiber, I., Ströher, E., Raatz, B., Busse, I., Kahmann, U., Bevan, M. W., Dietz, K.-J. and Baier, M. (2007). The *redox imbalanced* mutants of *Arabidopsis* differentiate signaling pathways for redox regulation of chloroplast antioxidant enzymes. *Plant Physiology* **143**, 1774–1788.

Heineke, D., Bykova, N., Gardeström, P. and Bauwe, H. (2001). Metabolic response of potato plants to an antisense reduction of the P-protein of glycine decarboxylase. *Planta* **212**, 880–887.

Henkes, S., Sonnewald, U., Badur, R., Flachmann, R. and Stitt, M. (2001). A small decrease of plastid transketolase activity in antisense tobacco transformants has dramatic effects on photosynthesis and phenylpropanoid metabolism. *The Plant Cell* **13**, 535–551.

Henmi, K., Demura, T., Tsuboi, S., Fukuda, H., Iwabuchi, M. and Ogawa, K. (2005). Change in the redox state of glutathione regulates differentiation of tracheary elements in Zinnia cells and *Arabidopsis* roots. *Plant and Cell Physiology* **46**, 1757–1765.

Higuchi, M., Ozaki, H., Matsui, M. and Sonoike, K. (2009). A T-DNA insertion mutant of AtHMA1 gene encoding a Cu transporting ATPase in *Arabidopsis thaliana* has a defect in the water-water cycle of photosynthesis. *Journal of Photochemistry and Photobiology B* **94**, 205–213.

Holtgrefe, S., Bader, K. P., Horton, P., Scheibe, R., von Schaewen, A. and Backhausen, J. E. (2003). Decreased content of leaf ferredoxin changes electron distribution and limits photosynthesis in transgenic potato plants. *Plant Physiology* **133**, 1768–1778.

Hosoya-Matsuda, N., Motohashi, K., Yoshimura, H., Nozaki, A., Inoue, K., Ohmori, M. and Hisabori, T. (2005). Anti-oxidative stress system in cyanobacteria. Significance of type II peroxiredoxin and the role of 1-Cys peroxiredoxin in *Synechocystis* sp. strain PCC 6803. *The Journal Biological Chemistry* **280**, 840–846.

Igamberdiev, A. U., Bykova, N. V., Lea, P. J. and Gardeström, P. (2001). The role of photorespiration in redox and energy balance of photosynthetic plant cells:

A study with a barley mutant deficient in glycine decarboxylase. *Physiologia Plantarum* **111**, 427–438.

Ihalainen, J. A., Klimmek, F., Ganeteg, U., van Stokkum, I. H., van Grondelle, R., Jansson, S. and Dekker, J. P. (2005). Excitation energy trapping in photosystem I complexes depleted in Lhca1 and Lhca4. *FEBS Letters* **579**, 4787–4791.

Ihnatowicz, A., Pesaresi, P., Varotto, C., Richly, E., Schneider, A., Jahns, P., Salamini, F. and Leister, D. (2004). Mutants for photosystem I subunit D of *Arabidopsis thaliana*: Effects on photosynthesis, photosystem I stability and expression of nuclear genes for chloroplast functions. *The Plant Journal* **37**, 839–852.

Ihnatowicz, A., Pesaresi, P., Lohrig, K., Wolters, D., Müller, B. and Leister, D. (2008). Impaired photosystem I oxidation induces STN7-dependent phosphorylation of the light-harvesting complex I protein Lhca4 in *Arabidopsis thaliana*. *Planta* **227**, 717–722.

Ishihara, S., Takabayashi, A., Ido, K., Endo, T., Ifuku, K. and Sato, F. (2007). Distinct functions for the two PsbP-like proteins PPL1 and PPL2 in the chloroplast thylakoid lumen of *Arabidopsis*. *Plant Physiology* **145**, 668–679.

Jensen, P. E., Rosgaard, L., Knoetzel, J. and Scheller, H. V. (2002). Photosystem I activity is increased in the absence of the PSI-G subunit. *The Journal of Biological Chemistry* **277**, 2798–2803.

Jespersen, H. M., Kjaersgård, I. V., Ostergaard, L. and Welinder, K. G. (1997). From sequence analysis of three novel ascorbate peroxidases from *Arabidopsis thaliana* to structure, function and evolution of seven types of ascorbate peroxidase. *Biochemical Journal* **326**, 305–310.

Jing, H.-C., Hebeler, R., Oeljeklaus, S., Sitek, B., Stühler, K., Meyer, H. E., Sturre, M. J. G., Hille, J., Warscheid, B. and Dijkwel, P. P. (2008). Early leaf senescence is associated with an altered cellular redox balance in *Arabidopsis cpr5/old1* mutants. *Plant Biology* **10**, 85–98.

Joet, T., Genty, B., Josse, E. M., Kuntz, M., Cournac, L. and Peltier, G. (2002). Involvement of a plastid terminal oxidase in plastoquinone oxidation as evidenced by expression of the *Arabidopsis thaliana* enzyme in tobacco. *The Journal of Chemistry* **277**, 31623–31630.

Joliot, P. and Joliot, A. (2006). Cyclic electron flow in C3 plants. *Biochimica et Biophysica Acta* **1757**, 362–368.

Jones, D. P. (2006). Redefining oxidative stress. *Antioxidants and Redox Signaling* **8**, 1865–1879.

Kanervo, E., Suorsa, M. and Aro, E.-M. (2005). Functional flexibility and acclimation of the thylakoid membrane. *Photochemical and Photobiological Sciences* **4**, 1072–1080.

Kangasjärvi, S., Lepistö, A., Hännikäinen, K., Piippo, M., Luomala, E.-M., Aro, E.-M. and Rintamäki, E. (2008). Diverse roles for chloroplast stromal and thylakoid-bound ascorbate peroxidases in plant stress responses. *Biochemical Journal* **412**, 275–285.

Khandelwal, A., Elvitigala, T., Ghosh, B. and Quatrano, R. S. (2008). *Arabidopsis* transcriptome reveals control circuits regulating redox homeostasis and the role of an AP2 transcription factor. *Plant Physiology* **148**, 2050–2058.

Laisk, A., Eichelmann, H., Oja, V., Talts, E. and Scheibe, R. (2007). Rates and roles of cyclic and alternative electron flow in potato leaves. *Plant and Cell Physiology* **48**, 1575–1588.

Lauerer, M., Saftic, D., Quick, W. P., Labate, C., Fichtner, K., Schulze, E.-D., Rodermel, S. R., Bogorad, L. and Stitt, M. (1993). Decreased ribulose-1,

5-bisphosphate carboxylase-oxygenase in transgenic tobacco transformed with 'antisense' rbcS. *Planta* **190**, 332–345.

Lawson, T., Bryant, B., Lefebvre, S., Lloyd, J. C. and Raines, C. A. (2006). Decreased SBPase activity alters growth and development in transgenic tobacco plants. *Plant, Cell and Environment* **29**, 48–58.

Lemaire, S. D., Guillon, B., Le Maréchal, P., Keryer, E., Miginiac-Maslow, M. and Decottignies, P. (2004). New thioredoxin targets in the unicellular photosynthetic eukaryote *Chlamydomonas reinhardtii*. *Proceedings of the National Academy of Sciences of the United States of America* **19**, 7475–7480.

Lennartz, K., Plücken, H., Seidler, A., Westhoff, P., Bechtold, N. and Meierhoff, K. (2001). *HCF164* encodes a thioredoxin-like protein involved in the biogenesis of the cytochrome b_6f complex in *Arabidopsis*. *The Plant Cell* **13**, 2539–2551.

Lepistö, A., Kangasjärvi, S., Luomala, E.-M., Brader, G., Sipari, N., Keränen, M., Keinänen, M. and Rintamäki, E. (2009). Chloroplast NADPH-thioredoxin reductase interacts with photoperiodic development in *Arabidopsis thaliana*. *Plant Physiology* **149**, 1261–1276.

Li, S., Lauri, A., Ziemann, M., Busch, A., Bhave, M. and Zachgo, S. (2009). Nuclear activity of ROXY1, a glutaredoxin interacting with TGA factors, is required for petal development in *Arabidopsis thaliana*. *The Plant Cell* **21**, 429–441.

Lin, T.-P., Caspar, T., Somerville, C. and Preiss, J. (1988). Isolation and characterization of a starchless mutant of *Arabidopsis thaliana* (L.) Heynh. lacking ADPglucose pyrophosphorylase activity. *Plant Physiology* **86**, 1131–1135.

Lintala, M., Allahverdiyeva, Y., Kidron, H., Piippo, M., Battchikova, N., Suorsa, M., Rintamäki, E., Salminen, T. A., Aro, E.-M. and Mulo, P. (2007). Structural and functional characterization of ferredoxin-NADP$^+$-oxidoreductase using knock-out mutants of *Arabidopsis*. *The Plant Journal* **49**, 1041–1052.

Lintala, M., Allahverdiyeva, Y., Kangasjarvi, S., Lehtimaki, N., Keranen, M., Rintamäki, E., Aro, E.-M. and Mulo, P. (2009). Comparative analysis of leaf-type ferredoxin-NADP oxidoreductase isoforms in *Arabidopsis thaliana*. *The Plant Journal* **57**, 1103–1115.

Lunde, C., Jensen, P. E., Haldrup, A., Knoetzel, J. and Scheller, H. V. (2000). The PSI-H subunit of photosystem I is essential for state transitions in plant photosynthesis. *Nature* **408**, 613–615.

Lundin, B., Hansson, M., Schoefs, B., Vener, A. V. and Spetea, C. (2007). The *Arabidopsis* PsbO2 protein regulates dephosphorylation and turnover of the photosystem II reaction centre D1 protein. *The Plant Journal* **49**, 528–539.

Luo, R., Wei, H., Ye, L., Wang, K., Chen, F., Luo, L., Liu, L., Li, Y., Crabbe, M. J. C., Jin, L., Li, Y. and Zhong, Y. (2009). Photosynthetic metabolism of C_3 plants show highly cooperative regulation under changing environments: A systems biological analysis. *Proceedings of the National Academy of Sciences of the United States of America* **106**, 847–852.

Maiwald, D., Dietzmann, A., Jahns, P., Pesaresi, P., Joliot, P., Joliot, A., Levin, J. Z., Salamini, F. and Leister, D. (2003). Knock-out of the genes coding for the Rieske protein and the ATP-synthase delta-subunit of *Arabidopsis*. Effects on photosynthesis, thylakoid protein composition, and nuclear chloroplast gene expression. *Plant Physiology* **133**, 191–202.

Marty, L., Siala, W., Schwarzländer, M., Fricker, M. D., Wirtz, M., Sweetlove, L. J., Meyer, Y., Meyer, A. J., Reichheld, J.-P. and Hell, R. (2009). The NADPH-dependent thioredoxin system constitutes a functional backup for cytosolic glutathione reductase in *Arabidopsis. Proceedings of the National Academy of Sciences of the United States of America* **106**, 9109–9114.

Maughan, S. and Foyer, C. H. (2006). Engineering and genetic approaches to modulating the glutathione network in plants. *Physiologia Plantarum* **126**, 382–397.

May, M. J., Vernoux, T., Leaver, C. Van, Montagu, M. and Inzé, D. (1998). Glutathione homeostasis in plants: Implications for environmental sensing and plant development. *Journal of Experimental Botany* **49**, 649–667.

Meinke, D., Muralla, R., Sweeney, C. and Dickerman, A. (2008). Identifying essential genes in *Arabidopsis thaliana. Trends in Plant Science* **13**, 483–491.

Meyer, Y., Reichheld, J. P. and Vignols, F. (2005). Thioredoxins in *Arabidopsis* and other plants. *Photosynthesis Research* **86**, 419–433.

Meyer, Y., Siala, W., Bashandy, T., Riondet, C., Vignols, F. and Reichheld, J. P. (2008). Glutaredoxins and thioredoxins in plants. *Biochimica et Biophysica Acta* **1783**, 589–600.

Miao, Y., Lv, D., Wang, P., Wang, X.-C., Chen, J., Miao, C. and Song, C.-P. (2006). An *Arabidopsis* glutathione peroxidase functions as both a redox transducer and a scavenger in abscissic acid and drought stress responses. *The Plant Cell* **18**, 2749–2766.

Miles, D. (1994). The use of high chlorophyll fluorescence photosynthetic mutants in the analysis of thylakoid membrane assembly and function. *Maydica* **39**, 35–45.

Millar, H., Considine, M. J., Day, D. A. and Whelan, J. (2001). Unraveling the role of mitochondria during oxidative stress in plants. *IUBMB Life* **51**, 201–205.

Mittler, R., Vanderauwera, S., Gollery, M. and Van Breusegem, F. (2004). Reactive oxygen gene network of plants. *Trends in Plant Science* **9**, 490–498.

Miyake, C. and Asada, K. (1994). Ferredoxin-dependent photoreduction of monodehydroascorbate radicals in spinach thylakoids. *Plant and Cell Physiology* **35**, 539–549.

Morgan, M. J., Lehmann, M., Schwarzländer, M., Baxter, C. J., Sienkiewicz-Porzucek, A., Williams, T. C. R., Schauer, N., Fernie, A. R., Fricker, M. D., Ratcliffe, R. G., Sweetlove, L. J. and Finkemeier, I. (2008). Decrease in manganese superoxide dismutase leads to reduced root growth and affects tricarboxylic acid cycle flux and mitochondrial redox homeostasis. *Plant Physiology* **147**, 101–114.

Motohashi, K. and Hisabori, T. (2009). HCF 164 receives reducing equivalents from stromal thioredoxin across the thylakoid membrane and mediates reduction of target proteins in the thylakoid lumen. *The Journal of Biological Chemistry* **281**, 35039–35047.

Munekage, Y., Hojo, M., Meurer, J., Endo, T., Tasaka, M. and Shikanai, T. (2002). PGR5 is involved in cyclic electron flow around photosystem I and is essential for photoprotection in *Arabidopsis. Cell* **110**, 361–371.

Munekage, Y., Hashimoto, M., Miyake, C., Tomizawa, K., Endo, T., Tasaka, M. and Shikanai, T. (2004). Cyclic electron flow around photosystem I is essential for photosynthesis. *Nature* **429**, 579–582.

Nandha, B., Finazzi, G., Joliot, P., Hald, S. and Johnson, G. N. (2007). The role of PGR5 in the redox poising of photosynthetic electron transport. *Biochimica et Biophysica Acta* **1767**, 1252–1259.

Narendra, S., Venkataramani, S., Shen, G., Wang, J., Pasapula, V., Lin, Y., Kornyeyev, D., Holaday, A. S. and Zhang, H. (2006). The *Arabidopsis* ascorbate peroxidase 3 is a peroxisomal membrane-bound antioxidant enzyme and is dispensable for *Arabidopsis* growth and development. *Journal of Experimental Botany* **57**, 3033–3042.

Niyogi, K. K. (2000). Safety valves for photosynthesis. *Current Opinion in Plant Biology* **3**, 455–460.

Noctor, G., Dutilleul, C., De Paepe, R. and Foyer, C. H. (2004). Use of mitochondrial electron transport mutants to evaluate the effects of redox state on photosynthesis, stress tolerance and the integration of carbon/nitrogen metabolism. *Journal of Experimental Botany* **55**, 49–57.

Noctor, G., De Paepe, R. and Foyer, C. H. (2007). Mitochondrial redox biology and homeostasis in plants. *Trends in Plant Science* **12**, 125–134.

Oelze, M.-L., Kandlbinder, A. and Dietz, K.-J. (2008). Redox regulation and over-reduction control in the photosynthesizing cell: Complexity in redox regulatory networks. *Biochimica et Biophysica Acta* **1780**, 1261–1272.

Palatnik, J. F., Valle, E. M. and Carrillo, N. (1997). Oxidative stress causes ferredoxin-$NADP^+$ reductase solubilization from the thylakoid membranes in methyl viologen-treated plants. *Plant Physiology* **115**, 1721–1727.

Palatnik, J. F., Tognetti, V. B., Poli, H. O., Rodriguez, R. E., Blanco, N., Gattuso, M., Hajirezaei, M. R., Sonnewald, U., Valle, E. M. and Carrillo, N. (2003). Transgenic tobacco plants expressing antisense ferredoxin-NADP(H) reductase transcripts display increased susceptibility to photo-oxidative damage. *The Plant Journal* **35**, 332–341.

Pérez-Ruiz, J. M., Spínola, M. C., Kirchsteiger, K., Moreno, J., Sahrawy, M. and Cejudo, F. J. (2006). Rice NTRC is a high-efficiency redox system for chloroplast protection against oxidative damage. *The Plant Cell* **18**, 2356–2368.

Pesaresi, P., Varotto, C., Richly, E., Kurth, J., Salamini, F. and Leister, D. (2001). Functional genomics of *Arabidopsis* photosynthesis. *Plant Physiology and Biochemistry* **39**, 285–294.

Potters, G., Pasternak, T. P., Guisez, Y., Palme, K. J. and Jansen, M. A. K. (2007). Stress-induced morphogenic responses: Growing out of trouble? *Trends in Plant Science* **12**, 98–105.

Price, D. G., Evans, J. R., Van Caemmerer, S., Yu, J.-W. and Badger, M. R. (1995). Specific reduction of chloroplast glyceraldehyde-3-phosphate dehydrogenase activity by antisense RNA reduces CO_2 assimilation via a reduction in ribulose bisphosphate regeneration in transgenic tobacco plants. *Planta* **195**, 369–378.

Queval, G., Issakidis-Bourguet, E., Hoeberichts, F. A., Vandorpe, M., Gakière, B., Vanacker, H., Miginiac-Maslow, M., Van Breusegem, F. and Noctor, G. (2007). Conditional oxidative stress responses in the *Arabidopsis* photorespiratory mutant *cat2* demonstrate that redox state is a key modulator of daylength-dependent gene expression, and define photoperiod as a crucial factor in the regulation of H_2O_2-induced cell death. *The Plant Journal* **52**, 640–657.

Quick, W. P., Schurr, U., Fichtner, K., Schulze, E.-D., Rodermel, S. R., Bogorad, L. and Stitt, M. (1991a). The impact of decreased Rubisco on photosynthesis, growth, allocation and storage in tobacco plants which have been transformed with antisense *rbcS*. *The Plant Journal* **1**, 51–58.

Quick, W. P., Schurr, U., Scheibe, R., Schulze, E.-D., Rodermel, S. R., Bogorad, L. and Stitt, M. (1991b). Decreased ribulose-1,5-bisphosphate carboxylase-oxygenase in transgenic tobacco transformed with "antisense" rbcS. I. Impact on photosynthesis in ambient growth conditions. *Planta* **183**, 542–554.

Radhamony, R. N., Prasad, A. M. and Srinivasan, R. (2005). T-DNA insertional mutagenesis in *Arabidopsis*: A tool for functional genomics. *Electronic Journal of Biotechnology* **8**, 82–106.

Raghavendra, A. S. and Padmasree, K. (2003). Beneficial interactions of mitochondrial metabolism with photosynthetic carbon assimilation. *Trends in Plant Science* **8**, 546–563.

Raines, C. A. (2003). The Calvin cycle revisited. *Photosynthesis Research* **75**, 1–10.

Raines, C. A. and Paul, M. J. (2006). Products of leaf primary carbon metabolism modulate the developmental programme determining plant morphology. *Journal of Experimental Botany* **57**, 1857–1862.

Reichheld, J.-P., Khafif, M., Riondet, C., Droux, M., Bonnard, G. and Meyer, Y. (2007). Inactivation of thioredoxin reductases reveals a complex interplay between thioredoxin and glutathione pathways in *Arabidopsis* development. *The Plant Cell* **19**, 1851–1865.

Reiser, J., Linka, N., Lemke, L., Jeblick, W. and Neuhaus, H. E. (2004). Molecular physiological analysis of the two plastidic ATP/ADP transporters from *Arabidopsis*. *Plant Physiology* **136**, 3524–3536.

Rey, P., Bécuwe, N., Barrault, M.-B., Rumeau, D., Havaux, M., Biteau, B. and Toledano, M. B. (2007). The *Arabidopsis thaliana* sulfiredoxin is a plastidic cysteine-sulfinic acid reductase involved in the photooxidative stress response. *The Plant Journal* **49**, 505–514.

Rius, S. P., Casati, P., Iglesias, A. A. and Gomez-Casati, D. F. (2006). Characterization of an *Arabidopsis thaliana* mutant lacking a cytosolic non-phosphorylating glyceraldehyde-3-phosphate dehydrogenase. *Plant Molecular Biology* **61**, 945–957.

Rius, S. P., Casati, P., Iglesias, A. A. and Gomez-Casati, D. F. (2008). Characterization of *Arabidopsis* lines deficient in GAPC-1, a cytosolic NAD-dependent glyceraldehyde-3-phosphate dehydrogenase. *Plant Physiology* **148**, 1655–1667.

Rizhsky, L., Hallak-Herr, E., Van Breusegem, F., Rachmilevitch, S., Barr, J. E., Rodermel, S., Inze, D. and Mittler, R. (2002). Double antisense plants lacking ascorbate peroxidase and catalase are less sensitive to oxidative stress than single antisense plants lacking ascorbate peroxidase or catalase. *The Plant Journal* **32**, 329–342.

Rizhsky, L., Liang, H. and Mittler, R. (2003). The water–water cycle is essential for chloroplast protection in the absence of stress. *The Journal of Biological Chemistry* **278**, 38921–38925.

Robson, C. A. and Vanlerberghe, G. C. (2002). Transgenic plant cells lacking mitochondrial alternative oxidase have increased susceptibility to mitochondria-dependent and -independent pathways of programmed cell death. *Plant Physiology* **129**, 1908–1920.

Rodriguez, R. E., Lodeyro, A., Poli, H. O., Zurbriggen, M., Peisker, M., Palatnik, J. F., Tognetti, V. B., Tschiersch, H., Hajirezaei, M. R., Valle, E. M. and Carrillo, N. (2007). Transgenic tobacco plants overexpressing chloroplastic ferredoxin-NADP(H) reductase display normal rates of photosynthesis and increased tolerance to oxidative stress. *Plant Physiology* **143**, 639–649.

Rosgaard, L., Zygadlo, A., Scheller, H. V., Mant, A. and Jensen, P. E. (2005). Insertion of the plant photosystem I subunit G into the thylakoid membrane. *FEBS Journal* **272**, 4002–4010.

Rouhier, N. and Jacquot, J.-P. (2002). Plant peroxiredoxins: Alternative hydroperoxide scavenging enzymes. *Photosynthesis Research* **74**, 259–268.

Rouhier, N., Gelhaye, E. and Jacquot, J.-P. (2004). Plant glutaredoxins: Still mysterious reducing systems. *Cellular and Molecular Life Sciences* **61**, 1266–1277.

Sano, S., Tao, S., Endo, Y., Inaba, T., Hossain, M. A., Miyake, C., Matsuo, M., Aoki, H., Asada, K. and Saito, K. (2005). Purification and cDNA cloning of chloroplastic monodehydroascorbate reductase from spinach. *Bioscience, Biotechnology, and Biochemistry* **69**, 762–772.

Santos, M., Gousseau, H., Lister, C., Foyer, C., Creissen, G. and Mullineaux, P. (1996). Cytosolic ascorbate peroxidase from *Arabidopsis thaliana* L. is encoded by a small multigene family. *Planta* **198**, 64–69.

Sarvikas, P., Hakala, M., Patsikka, E., Tyystjarvi, T. and Tyystjarvi, E. (2006). Action spectrum of photoinhibition in leaves of wild type and npq1–2 and npq4–1 mutants of *Arabidopsis thaliana*. *Plant and Cell Physiology* **47**, 391–400.

Sazanov, L. A., Burrows, P. A. and Nixon, P. J. (1998). The chloroplast Ndh complex mediates the dark reduction of the plastoquinone pool in response to heat stress in tobacco leaves. *FEBS Letters* **429**, 115–118.

Scheibe, R. (1991). Redox-modulation of chloroplast enzymes. A common principle for individual control. *Plant Physiology* **96**, 1–3.

Scheibe, R. (2004). Malate valves to balance cellular energy supply. *Physiologia Plantarum* **120**, 21–26.

Scheibe, R., Backhausen, J. E., Emmerlich, V. and Holtgrefe, S. (2005). Strategies to maintain redox homeostasis during photosynthesis under changing conditions. *Journal of Experimental Botany* **56**, 1481–1489.

Schjoerring, J. K., Mäck, G., Nielsen, K. H., Husted, S., Suzuki, A., Driscoll, S., Boldt, R. and Bauwe, H. (2006). Antisense reduction of serine hydroxymethyltransferase results in diurnal displacement of NH_4^+ assimilation in leaves of *Solanum tuberosum*. *The Plant Journal* **45**, 71–82.

Schneider, A., Häusler, R. E., Kolukisaoglu, Ü., Kunze, R., van der Graaff, E., Schwacke, R., Catoni, E., Desimone, M. and Flügge, U.-I. (2002). An *Arabidopsis thaliana* knock-out mutant of the chloroplast triose phosphate/phosphate translocator is severely compromised only when starch synthesis, but not starch mobilisation is abolished. *The Plant Journal* **32**, 685–699.

Schütze, K., Steiner, S. and Pfannschmidt, T. (2008). Photosynthetic redox regulation of the plastocyanin promoter in tobacco. *Physiologia Plantarum* **133**, 557–565.

Schwenkert, S., Umate, P., Dal Bosco, C., Volz, S., Mlcochova, L., Zoryan, M., Eichacker, L. A., Ohad, I., Herrmann, R. G. and Meurer, J. (2006). PsbI affects the stability, function, and phosphorylation patterns of photosystem II assemblies in tobacco. *The Journal of Biological Chemistry* **281**, 34227–34238.

Serrato, A. J., Pérez-Ruiz, J. M., Spínola, M. C. and Cejudo, F. J. (2004). A novel HADPH thioredoxin reductase, localized in the chloroplast, which deficiency causes hypersensitivity to abiotic stress in *Arabidopsis thaliana*. *The Journal of Biological Chemistry* **42**, 43821–43827.

Sharma, S. S. and Dietz, K.-J. (2009). The relationship between metal toxicity and cellular redox imbalance. *Trends in Plant Science* **14**, 43–50.

Shikanai, T. (2007). Cyclic electron transport around photosystem I: Genetic approaches. *Annual Reviews of Plant Biology* **58**, 199–217.

Shikanai, T., Endo, T., Hashimoto, T., Yamada, Y., Asada, K. and Yokota, A. (1998). Directed disruption of the tobacco *ndhB* gene impairs cyclic electron flow around photosystem I. *Proceedings of the National Academy of Sciences of the United States of America* **95**, 9705–9709.

Somerville, C. R. and Ogren, W. L. (1982). Genetic modification of photorespiration. *Trends in Biochemical Sciences* **7**, 171–174.

Somerville, S. C. and Somerville, C. R. (1983). Effect of oxygen and carbon dioxide on photorespiration flux determined from glycine accumulation in a mutant of *Arabidopsis thaliana*. *Journal of Experimental Botany* **34**, 415–424.

Sommer, F., Hippler, M., Biehler, K., Fischer, N. and Rochaix, J.-D. (2003). Comparative analysis of photosensitivity in photosystem I donor and acceptor side mutants of *Chlamydomonas reinhardtii*. *Plant Cell and Environment* **26**, 1881–1892.

Stitt, M. and Sonnewald, U. (1995). Regulation of metabolism in transgenic plants. *Annual Reviews of Plant Physiology and Plant Molecular Biology* **46**, 341–368.

Stockel, J. and Oelmüller, R. (2004). A novel protein for photosystem I biogenesis. *The Journal of Biological Chemistry* **279**, 10243–10251.

Strodtkötter, I., Padmasree, K., Dinakar, C., Speth, B., Niazi, P. S., Wojtera, J., Voss, I., Do, P. T., Nunes-Nesi, A., Fernie, A. R., Linke, V. Raghavendra, A. S. *et al.* (2009). Induction of the AOX1D isoform of alternative oxidase in *A. thaliana* T-DNA insertion lines lacking isoform AOX1A is insufficient to optimize photosynthesis when treated with antimycin A. *Molecular Plant* **2**, 284–297.

Sweetlove, L. J., Lytovchenko, A., Morgan, M., Nunes-Nesi, A., Taylor, N. L., Baxter, C. J., Eickmeier, I. and Fernie, A. R. (2006). Mitochondrial uncoupling protein is required for efficient photosynthesis. *Proceedings of the National Academy of Sciences of the United States of America* **103**, 19587–19592.

Takabayashi, A., Endo, T., Shikanai, T. and Sato, F. (2002). Post-illumination reduction of the plastoquinone pool in chloroplast transformants in which chloroplastic NAD(P)H dehydrogenase was inactivated. *Biosciences, Biotechnology and Biochemistry* **66**, 2107–2111.

Takabayashi, A., Ishikawa, N., Obayashi, T., Ishida, S., Obokata, J., Endo, T. and Sato, F. (2009). Three novel subunits of *Arabidopsis* chloroplastic NAD(P)H dehydrogenase identified by bioinformatic and reverse genetic approaches. *The Plant Journal* **57**, 207–219.

Takahashi, H., Watanabe, A., Tanaka, A., Hashida, S. N., Kawai-Yamada, M., Sonoike, K. and Uchimiya, H. (2006). Chloroplast NAD kinase is essential for energy transduction through the xanthophyll cycle in photosynthesis. *Plant and Cell Physiology* **47**, 1678–1682.

Takahashi, S., Bauwe, H. and Badger, M. (2007). Impairment of the photorespiratory pathway accelerates photoinhibition of photosystem II by suppression of repair but not acceleration of damage processes in *Arabidopsis*. *Plant Physiology* **144**, 487–494.

Thorneycroft, D., Sherson, S. M. and Smith, S. M. (2001). Using gene knockouts to investigate plant metabolism. *Journal of Experimental Botany* **52**, 1593–1601.

Timm, S., Nunes-Nesi, A., Pärnik, T., Morgenthal, K., Wienkoop, S., Keerberg, O., Weckwerth, W., Kleczkowski, L. A., Fernie, A. R. and Bauwe, H. (2008). A cytosolic pathway for the conversion of hydroxypyruvate to glycerate during photorespiration in *Arabidopsis*. *The Plant Cell* **20**, 2848–2859.

Tzafrir, I., Pena-Muralla, R., Dickerman, A., Berg, M., Rogers, R., Hutchens, S., Sweeney, T. C., McElver, J., Aux, G., Patton, D. and Meinke, D. (2004). Identification of genes required for embryo development in *Arabidopsis*. *Plant Physiology* **135**, 1206–1220.

Umate, P., Schwenkert, S., Karbat, I., Dal Bosco, C., Mlcochova, L., Volz, S., Zer, H., Herrmann, R. G., Ohad, I. and Meurer, J. (2007). Deletion of PsbM in tobacco alters the Q_B site properties and the electron flow within photosystem II. *The Journal of Biological Chemistry* **282**, 9758–9767.

Urbanczyk-Wochniak, E., Usadel, B., Thimm, O., Nunes-Nesi, A., Carrari, F., Davy, M., Bläsing, O., Kowalczyk, M., Weicht, D., Polinceusz, A., Meyer, S. Stitt, M. *et al.* (2006). Conversion of MapMan to allow the analysis of transcript data from solanaceous species: Effects of genetic and environmental alterations in energy metabolism in the leaf. *Plant Molecular Biology* **60**, 773–792.

Van der Kroll, A. R., Lenting, P. E., Veenstra, J., van der Meer, I. M., Koes, R. E., Gerats, A. G. M., Mol, J. N. M. and Stuitje, A. R. (1988). An anti-sense chalcone synthase gene in transgenic plants inhibits flower pigmentation. *Nature* **333**, 866–869.

Varotto, C., Pesaresi, P., Jahns, P., Lessnick, A., Tizzano, M., Schiavon, F., Salamini, F. and Leister, D. (2002). Single and double knockouts of the genes for photosystem I subunits G, K, and H of *Arabidopsis*. Effects on photosystem I composition, photosynthetic electron flow, and state transitions. *Plant Physiology* **129**, 616–624.

Voss, I., Koelmann, M., Wojtera, J., Holtgrefe, S., Kitzmann, C., Backhausen, J. E. and Scheibe, R. (2008). Knockout of major leaf ferredoxin reveals new redox-regulatory adaptations in *Arabidopsis thaliana*. *Physiologia Plantarum* **133**, 584–598.

Wan, X., Tan, J., Lu, S., Lin, C., Hu, Y. and Guo, Z. (2009). Increased tolerance of oxidative stress in transgenic tobacco expressing a wheat oxalate oxidase gene via induction of antioxidant enzymes is mediated by H_2O_2. *Physiologia Plantarum* **136**, 30–44.

Wang, Z., Xing, S., Birkenbihl, R. P. and Zachgo, S. (2009). Conserved functions of *Arabidopsis* and rice CC-type glutaredoxins in flower development and pathogen response. *Molecular Plant* **2**, 323–335.

Watanabe, C. K., Hachiya, T., Terashima, I. and Noguchi, K. (2008). The lack of alternative oxidase at low temperature leads to a disruption of the balance in carbon and nitrogen metabolism, and to an up-regulation of antioxidant defence systems in *Arabidopsis thaliana* leaves. *Plant, Cell and Environment* **31**, 1190–1202.

Westerhoff, H. V., Kholodenko, B. N., Cascante, M. and Van Dam, K. (1995). Elusive control. *Journal of Bioenergetics and Biomembranes* **27**, 491–497.

Wingler, A., Lea, P. J. and Leegood, R. C. (1997). Control of photosynthesis in barley plants with reduced activities of glycine decarboxylase. *Planta* **202**, 171–178.

Wu, D., Wright, D. A., Wetzel, C., Voytas, D. F. and Rodermel, S. (1999). The IMMUTANS variegation locus of *Arabidopsis* defines a mitochondrial alternative oxidase homolog that functions during early chloroplast biogenesis. *The Plant Cell* **11**, 43–55.

Xing, S., Rosso, M. G. and Zachgo, S. (2005). *ROXY1*, a member of the plant glutaredoxin family, is required for petal development in *Arabidopsis thaliana*. *Development* **132**, 1555–1565.

Zer, H. and Ohad, I. (2003). Light, redox state, thylakoid-protein, phosphorylation and signaling gene expression. *Trends in Biochemical Sciences* **28**, 467–470.

Redundancy and Crosstalk Within the Thioredoxin and Glutathione Pathways: A New Development in Plants

J.-P. REICHHELD,*,[1] T. BASHANDY,* W. SIALA,* C. RIONDET,* V. DELORME,* A. MEYER[†] AND Y. MEYER*

**Genome et Developement des Plantes, Universite de Perpignan, CNRS-UP-IRD UMR 5096, Av P. Alduy, F 66860 Perpignan Cedex, France*

[†]Heidelberg Institute for Plant Science, Heidelberg University, Im Neuenheimer Feld 360, 69120 Heidelberg, Germany

[1]Corresponding author: Email: jpr@univ-perp.fr

Advances in Botanical Research, Vol. 52

0065-2296/09 $35.00
DOI: 10.1016/S0065-2296(09)52009-5

ABSTRACT

Thioredoxins (Trx) and glutaredoxins (Grx) are major disulfide reduction enzymes occurring in all living organisms that regulate the redox state of thiol groups of proteins. Initially discovered as a reductant of ribonucleotide reductase (RNR), an enzyme necessary for DNA synthesis, it is now established that they are involved in various biological processes. Trx and Grx have their own reduction system: typically, in most organisms and in the cytosol and mitochondria of plants the Trx pathway (NTS) comprises a redox cascade including NADPH, Trx reductase (NTR), and Trx, while the Grx pathway (NGS) is composed of NADPH, glutathione reductase (GR), glutathione (GSH), and Grx. These two systems act in parallel and have several common target proteins as shown by biochemical and genetic studies. Recent genetic studies in *Arabidopsis* show that the cytosolic Trx and Grx reduction systems are in fact more complex. In the cytosol, in absence of NTR, Trxs are reduced by a GSH-dependent pathway, while in the absence of GR oxidized glutathione (GSSG) is reduced by the NTR Trx pathway. By contrast in the chloroplast, Trxs have evolved a specific function of control of the light dark metabolism.

ABBREVIATIONS

FTS	ferredoxin thioredoxin system
GR	glutathione reductase
Grx	glutaredoxin
GSH	reduced glutathione
GSSG	oxidized glutathione
NGS	NADPH glutaredoxin system
NTS	NADPH thioredoxin system
NTR	Trx reductase
PAPS	phosphoadenylyl sulfate
RNR	ribonucleotide reductase
Trx	thioredoxin

I. INTRODUCTION

Reversible dithiol disulfide exchange is now well acknowledged as an important biological mechanism regulating various aspects of the metabolism, development, and adaptation to changing environmental situations. Several reductases, including ribonucleotide reductases (RNR), peroxiredoxins, and methionine sulfoxide reductases, gain their reduction power via a pair of cysteines switching from the disulfide to the dithiol state. In numerous proteins which do not need a redox flux for activity, including transcription factors, the disulfide to dithiol switch results in a conformational modification allowing or restricting activity. The reduction of disulfide bridges is performed by Trxs or Grxs, two phylogenetically unrelated proteins but which present a similar three-dimensional conformation, including at their surface a redox-active site CxxC with two cysteines separated by two amino acids. Almost all organisms encode several Trxs and Grxs. Trx and Grx have they own reduction system: in most organisms and in the cytosol and mitochondria of plants the Trx pathway comprises a redox cascade including NADPH, Trx reductase (NTR), and Trx, while the Grx pathway is composed of NADPH, glutathione reductase (GR), glutathione (GSH), and Grx. These two systems act in parallel and have several common target proteins as shown by biochemical and genetic studies. Two important redoxin targets are the subject of an abundant literature: (i) there are several types of RNR, which provide cells with the deoxyribonucleotides necessary for DNA. All organisms growing aerobically use Trx- or Grx-dependent RNR. (ii) Sulfate assimilation features a reduction step performed by a phosphoadenylyl sulfate (PAPS) reductase in bacteria and yeast and by an adenylyl sulfate (APS) reductase in plants, and these enzymes are also Trx dependent.

Recent genetic experiments in *Arabidopsis* show that the cytosolic Trx and Grx have certain common target proteins as is the case in most organisms, but, in addition, their reduction systems are in fact more complex. The reduction of cytosolic Trx and Grx can occur via alternative pathways. In the absence of NTR, cytosolic Trxs are reduced by a GSH-dependent pathway, while in the absence of cytosolic GR oxidized glutathione (GSSG) is reduced by the NTR Trx pathway. By contrast, sulfate reduction, which in plants is carried out in chloroplasts, is performed by a two-domain APS reductase with the APS domain associated to a Trx homolog reduced by Glutathione. Sulfate reduction in plants is thus independent from external Trx and Grx and their reductases.

II. NTS AND NGS OVERLAP IN BACTERIA AND YEAST

A. BACTERIA

In *Escherichia coli*, the NADPH Trx system (NTS) is composed by two Trxs, Trx1 and Trx3, encoded by *TrxA* and *TrxC* and one Trx reductase TrxR encoded by the *TrxB* gene (Laurent *et al.*, 1964; Miranda-Vizuete *et al.*, 1997). The NADPH Grx system (NGS) is composed of glutathione, synthesized by gamma-glutamylcysteine synthetase, a rate-limiting enzyme encoded by the GshA gene, glutathione synthase encoded by GshB

(Rietsch and Beckwith, 1998), GR encoded by Gor, three typical dicysteinic Grx (Grx1, Grx2, Grx3) encoded by *grxA*, *grxB*, and *grxC* (Holmgren, 1989) and two Grx-like proteins (Grx4 and NrdH) encoded by GrxD and NrdH (Fernandes *et al.*, 2005; Jordan *et al.*, 1997). Overlaps between NTS and NGS were initially discovered *E. coli* where neither NTS nor NGS is required for normal aerobic growth (Prinz *et al.*, 1997). Unlike other organisms, *E. coli* is able to grow in absence of GSH, as *gsha* and *gora* mutants are aphenotypic. However, mutations in either GshA or Gor in combination with TrxB or TrxA and TrxC are nonviable under aerobic conditions (Stewart *et al.*, 1998). The *trx1*, *trx2*, and *trx3* single or multiple mutants are aphenotypic as well as *grxa*, *grxb*, and *grxC* mutants. However, the *trxa trxc* mutant associated with *grxa* mutations is nonviable aerobically due to defects in the RNR reduction, a key enzyme in DNA synthesis. Under anaerobiosis ribonucleotide reduction is performed by an alternative RNR that uses formate as electron donor, instead of the thiol redox systems (Mulliez *et al.*, 1995). Further work combining mutations in different NTS or NGS components has implicated Trx1, Trx3, Grx1, or NrdH but not Grx2 and Grx3 in RNR reduction and shown that Trx1 or Grx1 but not Trx3, Grx2, or Grx3 are required for reduction of the PAPS reductase and sulfate assimilation (Russel *et al.*, 1990; Toledano *et al.*, 2007).

The sole redoxin essential in *E. coli* is the monocysteinic Grx4 but its function is still unclear: Grx4 forms a mixed disulfide with GSH and is most probably implicated in FeS cluster formation (Fernandes *et al.*, 2005; Fladvad *et al.*, 2005). The three canonical Grxs are reduced by Gor/GSH. The reduction of the atypical Grx-like NrdH which is an alternative reducer of the RNR was shown to be performed by TrxR (Gon *et al.*, 2006).

In short, *E. coli* can grow (very slowly) in anaerobiosis without NTS or NGS due to the presence of a formate-dependent RNR if a source of reduced sulfate is present in the culture medium. In aerobiosis the formate-dependent RNR is inactivated, and an active NTS or an active NGS is necessary for ribonucleotide reduction. In addition, under aerobiosis the antioxidative peroxiredoxins and methionine sulfoxide reductases which are also dependent of an active NTS or NGS may play a central role for the survival of the bacteria. NTS or NGS is also necessary for sulfate reduction. Thus, under laboratory conditions the NTS and NGS appear to play vital functions but are highly redundant. Nevertheless, most bacteria have maintained the two systems with several Trx and Grx genes during the evolution suggesting that in the natural environment the simultaneous presence of both systems provides an advantage. There is no evidence in *E. coli* that the NTR reduces Grx or GSSG neither that GSH and or Grx can reduce Trx. Thus the two pathways are redundant in terms of target proteins but no crosstalk seems to occur between them.

B. YEAST

Like all eukaryotes, *Saccharomyces cerevisiae* possesses a cytosolic and mitochondrial NTS and NGS. The genome encodes two cytosolic Trxs, a mitochondrial Trx, a cytosolic and a mitochondrial Trx reductase, one cytosolic and one mitochondrial dicystenic Grx, one dual-targeted (cytosolic and mitochondrial) GR. In addition, several monocysteinic Grxs with a redox-type CxxS have been identified. They are implicated in FeS cluster formation and do not present disulfide reductase activity. They will not be further considered in this review. The first genetic report was the demonstration that the inactivation of one cytosolic Trx is asymptomatic. Inactivation of both genes is not lethal, but this strain is unable to assimilate sulfate and grows only in the presence of reduced sulfur. In addition the double mutant presents a distorted cell cycle with a very long S phase and consequently no G1 (Muller, 1992, 1995). During the past decade, genetic studies in yeast have highlighted overlapping function of NTS and NGS in redox regulation. The first evidence came from the identification of GLR1, encoding a dual-targeted cytosolic and mitochondrial GR, in a genetic screen for mutations which confer a requirement for Trxs (Muller, 1996). The *glr1* mutant grows with a normal cell cycle, but in this mutant either one of the two cytosolic Trxs is essential for growth. Cells lacking both Trxs and GR are not viable under aerobic conditions and grow poorly in anaerobic conditions. To further test the requirement for components of the Trx and GSH/Grx systems, different combinations of mutants in the NTS and NGS pathways were performed. A quadruple *trx1 trx2 grx1 grx2* mutant is not viable but a single cytosolic disulfide reductase (Trx or Grx) is sufficient for viability if reduced sulfur is available (Draculic *et al.*, 2000). However, strains completely lacking the cytosolic NTS (*trx1 trx2 trr1*), the mitochondrial NTS (*trr2 and trx3*), both cytosolic/mitochondrial NTS (*trr1 trr2 and trx1 trx2 trx3*) or NGS (*gsh1 glr1*) systems are viable, but strains simultaneously deleted for components of both systems (*gsh1 trx1 trx2 and glr1 trr1*) are not viable (Trotter and Grant, 2002, 2003). The lethality of the *glr1 trr1* mutant is somewhat surprising because the *grl1 trr1 trx1 trx2* is viable. This is most probably due to the accumulation of oxidized Trx (Toledano *et al.*, 2007).

The overlap between the NTS and NGS has also been studied by measuring the redox state of Trxs and Grxs in mutants lacking GR and Trx reductase (Trotter and Grant, 2003). *Trr1* is required for normal cell growth, whereas no requirement for *Glr1* is apparent. Analysis of the redox state of Trxs and Grxs in *glr1* and *trr1* mutants reveals that Trxs are maintained reduced independent of the glutathione system. In contrast, there is a strong correlation between the redox state of Grxs and the oxidation state of the

GSSG/2GSH redox couple. Trotter and Grant (2003) have proposed that independent redox regulation of Trxs enables cells to survive under conditions where the GSH/Grx system is oxidized. In contrast to the *glr1* mutant, loss of TRR1 not only results in oxidation of cytosolic Trxs, but also shifts the redox state of Grxs to a more reduced form. Surprisingly, inactivation of the mitochondrial TRR2 has no effect on the redox state of the mitochondrial Trxh3. However, deletion of both TRR2 and GLR results in a partial oxidation of Trx3, indicating that both TRR2 and GLR1 have an overlapping function in maintaining the redox state of mitochondrial Trx (Trotter and Grant, 2005). Simultaneous defects in both disulfide reduction systems may therefore account for the growth defect observed in the *trr1* mutant (Trotter and Grant, 2003). In addition, Trx mutants contain elevated glutathione levels indicating a link between the Trx system and glutathione metabolism in the cell (Garrido and Grant, 2002; Muller, 1996). Mutants *trx1trx2* and *trr1* show an increase in both oxidized GSSG and reduced GSH levels with the redox state of the GSSG/2GSH similar to that in the wild-type strain (Trotter and Grant, 2003). It is unclear whether Trxs is involved directly or indirectly in the reduction of glutathione.

In short, in *S. cerevisiae* cytosolic Trxs are the sole hydrogen donors of PAPS reductase, while cytosolic Trxs or Grxs are able to reduce RNR, but Trxs are more efficient. The high content in oxidized glutathione (GSSG) in the *trx1 trx2* mutant suggests the implication of Trx in glutathione reduction even in the presence of an active GR. The mitochondrial Trx3 is apparently reduced by the TRR2 and GLR reactions. An additional difference in comparison to *E. coli* is the inability of *S. cerevisiae* to grow in the absence of glutathione, but this is clearly not related to its function in reducing Grxs.

III. OVERLAPS AND CROSSTALKS IN ANIMALS

The human and mice genomes each encode a cytosolic Trx1, a mitochondrial Trx2, a cytosolic TrxR1, and a mitochondrial TrxR2. Both reductases differ from the bacterial and plant NTRs by their size due to a C-terminal extension which contains a selenocysteine implicated in the reductase activity. The NGS is composed of a cytosolic Grx1, a mitochondrial Grx2, a cytosolic and a mitochondrial GR. In addition, the human genome encodes Trx-like and Grx-like proteins mostly expressed during spermatogenesis and other Trx- or Grx-like proteins. Inactivation of Trx1 (Matsui *et al.*, 1996) or TrxR1 (Jakupoglu *et al.*, 2005) in mice is embryo lethal at an early stage showing that neither the mitochondrial NTS nor the cytosolic NGS are substitutes of the cytosolic NTS. Nevertheless, both Trx1 and Grx1 are able to reduce human RNR *in vitro*, although by two different mechanisms (Avval and Holmgren, 2009). *Trx2* (Nonn *et al.*, 2003) or *TrxR2* (Conrad *et al.*, 2004) inactivation in mice leads also to embryo lethality, but at a later stage. Heart malformations are obvious in the KO mice suggesting an important role of the mitochondrial NTS in heart development or protection against oxidative stresses. Thus, in mammals NGS is not a backup of the cytosolic or mitochondrial NTS. Nevertheless the human Trx1 can be inactivated by glutathionylation of Cys73, a third cysteine which in not located in the redox site. Because deglutathionylation is generally performed by Grx, it is possible that the NGS controls the activity of the NTS under certain circumstances (Casagrande *et al.*, 2002). Inactivation of the cytosolic Grx1 (Ho *et al.*, 2007) does not have profound consequences on the development. It is possible that Grx1 performs functions which are redundant with the other Trx, Grx or Trx-like and Grx-like proteins.

IV. CROSSTALKS IN PLANTS

Plants are distinguishable from other eukaryotes by the presence of the chloroplast. In addition, they have a complex redoxin system as revealed by sequencing the *Arabidopsis thaliana* genome (Alonso *et al.*, 2000). About 40 genes coding for Trx and Trx-related proteins have been identified (Meyer *et al.*, 2006). Among them, at least eleven belong to the cytosolic Trxh group, but additional Trx-like proteins are predicted to have a cytosolic localization. A functional mitochondrial NTS system has been identified in *Arabidopsis* (Laloi *et al.*, 2001). Moreover, the chloroplast Trx system has been extensively studied in plants (Buchanan *et al.*, 2002; Schürmann and Buchanan, 2008; Schürmann and Jacquot, 2000). The genome of *A. thaliana* encodes 40 genes of Grx and Grx-like proteins.

A. GLUTATHIONE SYNTHESIS AND REDUCTION IN PLANTS

The metabolism of glutathione has been extensively characterized in plants as in other organisms. Glutathione is synthesized in two steps catalyzed by the gamma-glutamyl cysteine synthase (GSH1) and the glutathione synthase (GSH2) (Fig. 1). In higher plants, GSH1 is exclusively located in plastids, while GSH2 is dually targeted to plastids and cytosol (Pasternak *et al.*, 2008; Wachter *et al.*, 2005). As in other eukaryotes, GSH is required for plant development as shown by the embryo lethality of *gsh1* knockout lines (Cairns *et al.*, 2006; Tzafrir *et al.*, 2004). In *Arabidopsis*, forward genetic screens allowed the isolation of several mutants which after cloning and sequencing

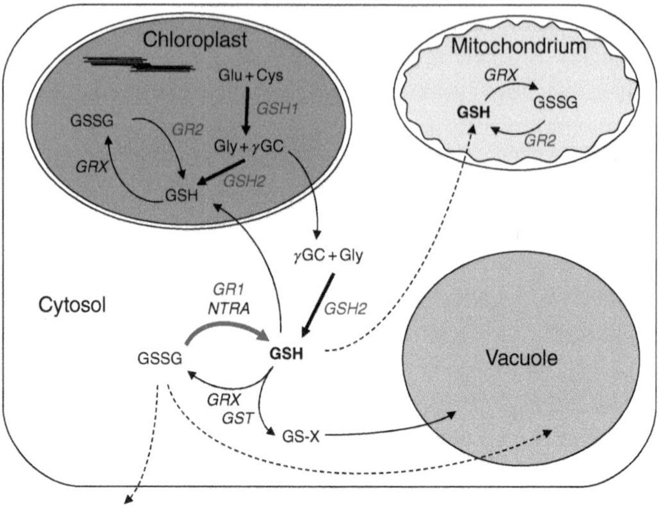

Fig. 1. Glutathione synthesis and reduction in plants. The GSH synthesis enzymes GSH1 and GSH2 are indicated in grey (brown in the web version). The GSSG reduction enzymes GR1 and GR2 are indicated in dark grey (blue in the web version). The alternative reduction pathway is represented by the grey arrow (red in the web version). The dotted arrows indicate putative GSH fluxes.

of the mutated genes appears to be weak alleles of GSH1. The first discovered mutant allele was *rml1* (*rootmeristemless1*) (Cheng *et al.*, 1995) which encodes an inefficient GSH1, allowing approximately only 5% GSH in comparison to the wild type (Vernoux *et al.*, 2000). Homozygote seeds are formed with a normal embryo, but the root meristem fails to grow during germination. In contrast the shoot meristem develops at least in the initial stage. Two other GSH1 alleles are available: *pad2-1* with about 16% GSH and *cad2* with about 20% GSH present a normal development under standard conditions and are fertile (Cobbett *et al.*, 1998). The *pad2-1* (*Phytoalexin-deficient2-1*) mutant was isolated on the basis of its sensitivity to several pathogens, possibly due to inefficient production of phytoalexins and glucosinolates (Parisy *et al.*, 2007). *cad2* (*cadmiumsensitive2*) is hypersensitive to Cd possibly due to a limited synthesis of phytochelatins (Howden *et al.*, 1995).

In contrast to the gsh1 KO mutant, a null mutant in GSH2 shows normal embryogenesis, but the seeds do not germinate. The *gsh2* mutant accumulates high levels of the substrate of GSH2, gamma-GC, suggesting that this low molecular weight thiol only very partially compensates GSH in the early stage of plant development (Pasternak *et al.*, 2008). This contrasts with *E. coli* and yeast in which elevated gamma-GC in the *gsh2* mutant was shown to compensate the absence of GSH (Faulkner *et al.*, 2008; Grant *et al.*, 1996). Interestingly, Pasternak *et al.* (2008) have shown by complementation of a *gsh2* mutant that the cytosolic GSH2 is sufficient for GSH synthesis, showing that gamma-GC is exported from the plastids to supply the cytosol with the precursor of GSH biosynthesis and that GSH is efficiently reimported into the plastids and mitochondria (Fig. 1).

As in most organisms, oxidized glutathione is reduced by GRs which are present in distinct cellular compartments. In *Arabidopsis* and other higher plants, two genes encode GRs. *GR2* was found to encode a dual-targeted chloroplast and mitochondrial GR, while *GR1* encodes a cytosolic protein. Although both GRs have similar reduction activities with GSSG, only GR2 is essential for plant development (Tzafrir *et al.*, 2004). It is still unknown whether the embryo lethality of the *gr2* mutant is due to inactivation of the chloroplast and/or mitochondrial isoform of GR2. Nevertheless, it clearly indicates that cytosolic GR1 is not able to compensate for the lack of organellar glutathione reduction. As previously stated, GSH synthesis takes place both in the chloroplast and the cytosol, but engineered *Arabidopsis* with only a cytosolic GSH2 are fully viable. Thus GSH import from the cytosol to organelles is sufficient, but GSSG reduction should take place in the organelles.

Biochemical characterization of *gr1* mutant plants has established that GR1 shows 65% of total GR activity of leaf extracts. Mutant gr1 accumulates high levels of GSSG. However, in contrast to *gr2* mutants, insertion mutants in *gr1* are aphenotypic even under stress conditions, indicating that cytosolic GR is dispensable for plant development (Marty *et al.*, 2009). This finding suggests that accumulation of GSSG in the cytosol is either not toxic for plants or that it is exported out of the cytosol. Export to the organelles seems unlikely because chloroplast and mitochondrial redox state of glutathione was shown to be unchanged in the *gr1* mutant (Marty *et al.*, 2009). However, vacuolar or apoplastic export cannot be excluded.

B. CYTOSOLIC NTS AND NGS

First genetic evidence of functional redundancies between the cytosolic NTS and NGS came from analyses of knockout mutants in several Trxs. The absence of phenotypes of single mutants in cytosolic *trxh* was presumably due to functional redundancies among members of this multigene family and or with cytosolic Grx. Presently in *Arabidopsis*, the only genetic evidence for a function of a cytosolic Trx is the activation of NPR1 by Trxh5 (Tada *et al.*, 2008), the only cytosolic Trxh induced by stress (Laloi *et al.*, 2004) and its role in victorin sensitivity (Sweat and Wolpert, 2007). In transgenic barley, overexpression of a Trx h in the endosperm enhanced the activity of starch-debranching enzyme (Cho *et al.*, 1999) and accelerated the appearance of alfa-amylase as well as the rate of germination (Wong *et al.*, 2002). In another report, a VIGS-mediated silencing genetic screen has identified the cytosolic Trx-like protein CITrx as implicated in the Cf9/Avr9-triggered hypersensitive response in tomato (Rivas *et al.*, 2004). Further analysis showed that CITrx acts in a redox-independent way (Nekrasov *et al.*, 2006). Nevertheless CITrx in tomato and in other plants presents a classical CGPC redox site suggesting that it has additional functions which are redox dependent. Other functions for Trxh can be supposed on the basis of the presence in the cytosol of antioxidant proteins like PRX and Msr, and by the necessity to reduce RNR. In addition several putative cytosolic Trx target proteins were identified by proteomics (Marchand *et al.*, 2004; Montrichard *et al.*, 2009), and some of them are reducible by Trx *in vitro* (Bréhélin *et al.*, 2003; Dietz, 2003; Rouhier *et al.*, 2006b). But which Trx is actually the reductant *in vivo* is presently not known. Cytosolic Trx are reduced by NTRA and NTRB which were identified in the *Arabidopsis* genome (Jacquot *et al.*, 1994). Each gene encodes a cytosolic and a mitochondrial isoform (Laloi *et al.*, 2001). Inactivation of NTRA or NTRB in *Arabidopsis* plants is aphenotypic under standard growth conditions, showing that both proteins have redundant functions (Reichheld *et al.*, 2005). Trxh3 is mainly found in the reduced state in the wild type and in the single *ntra* and *ntrb* mutants suggesting that a single NTR is sufficient to reduce cytosolic Trx. It remains to be determined whether the redox state of other types of cytosolic and mitochondrial Trx is affected in single mutants. Nevertheless, it has been shown more recently that both NTRs are not redundant in terms of pollen viability and SAR signaling (Marty *et al.*, 2009; Tada *et al.*, 2008).

The cytosolic NGS is composed of several Grx types including four typical Grx with a CPYC redox site and a Grx with a particular site CCxC or CCxS present only in plants. In the cytosol GSSG is reduced by GR1. Insertion mutants in typical Grx are available but are aphenotypic. Up to now a phenotype was associated only with a mutation in Roxy1 which induces petal modifications (Xing *et al.*, 2005). Roxy1 belongs to the CCxC Grx family but the wild-type phenotype can be restored by complementation with a CCxS variant suggesting that its function is related to a glutathionylation / deglutathionylation process rather than by disulfide reduction. Inactivation of Roxy2, an ortholog of Roxy1, is aphenotypic, but the double mutant *roxy1 roxy2* does not form anthers and is sterile (Xing and Zachgo, 2008). Furthermore other members of the family expressed ectopically under the control the Roxy1 promoter fully or partially reverse the phenotype of the *roxy1* mutant suggesting that the tissular localization or the level of expression determine the function *in planta* (Li *et al.*, 2009).

Inactivation of the NTR and of the GR1 genes has revealed further unexpected information on the cytosolic NTS and NGS. Inactivation of both NTRs in *A. thaliana* is not lethal: the *ntra ntrb* mutant has a slightly reduced size but is fertile and does not present a particular sensitivity to oxidants. This was rather surprising because the cytosolic Prxs are reduced by Trx, so far not by Grx (Bréhélin *et al.*, 2003; Navrot *et al.*, 2006). In fact the ability of the *ntra ntrb* mutant to show an almost normal growth pattern and resist oxidants results from the fact that the Trxhs are not completely oxidized: cytosolic Trxh3 was found to be only partially oxidized in the *ntra ntrb* mutant, suggesting that another reduction pathway is occurring in the absence of NTR. The implication of GSH in the alternative pathway was demonstrated by pharmacological and genetic approaches (Reichheld *et al.*, 2007). The *ntra ntrb* mutant is hypersensitive to partial inhibition of GSH biosynthesis by the buthionine sulfoximine (BSO) inhibitor. Second, the *ntra ntrb* mutant was crossed with *rml1* (*rootmeristemless1*), a mutant allele of the first enzyme of glutathione synthesis with a glutathione content of only 5% of the WT. Root growth was completely blocked in *rml1,* but the apical meristem developed like WT, at least at early germination stages. In the triple mutant both meristems were blocked early in germination. In this triple mutant, Trx *h3* is fully oxidized (i.e., inactive). This demonstrates genetically (i) the existence of an alternative, glutathione-dependent reduction system for at least some members of the Trx *h* family and (ii) the involvement of Trx *h* in apical meristem growth (Reichheld *et al.*, 2007) (Fig. 2). In contrast to *S. cerevisiae*, NTR inactivation in *Arabidopsis* does not lead to accumulation of GSSG, neosynthesis of GSH, or modification of the glutathione redox state. In addition, transcriptomic analyses have not shown any modification of the steady-state levels of GR or Grx gene expression, showing that the alternative reduction of Trx in the *ntra ntrb* mutant does not need strengthening of the NGS pathway and that this alternative pathway is likely occurring in wild-type plants (Bashandy *et al.*, 2009). Biochemical tests have shown that Trxh3 cannot be reduced by GSH alone but is reduced by Grxs (GrxC1 or GrxC2) (Fig. 3). Which Grx actually performs Trx reduction in planta is unknown as is the reduction mechanism.

Another type of crosstalk between NTS and NGS was described in poplar. Atypical members of plant Trx h (popTrxh4 and popCXXS3) were shown to be reduced exclusively by Grx (Gelhaye *et al.*, 2003). A four-step disulfide cascade mechanism involving the transient glutathionylation of an N-terminal Cys by Grx was proposed to convert popTrxh4 back to its active reduced form (Koh *et al.*, 2008). In *Arabidopsis*, it was shown that At1g11530, the ortholog of popCXXS3, has a high disulfide isomerase activity *in vitro* which is dependent on GSH but independent of Grx (Serrato *et al.*, 2008).

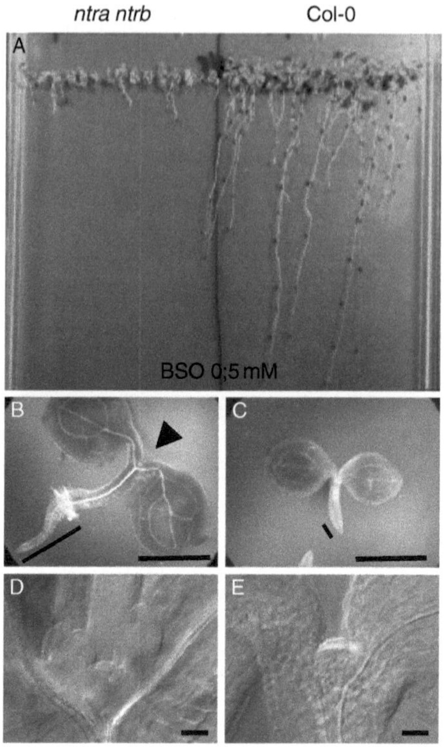

Fig. 2. Hypersensitivity of the *ntra ntrb* mutant to glutathione biosynthesis inhibition. (A) Growth inhibition of the *ntra ntrb* mutant under the specific glutathione biosynthesis inhibitor BSO. Seeds were soaked on vertical MS/2 agar plates supplemented by 0.5 mM BSO and grown under a 16-h light/8-h dark regime. (B, C) Phenotypes of *rml1* (B) and *ntra ntrb rml1* (C) homozygous mutants at 8 days after germination on MS/2 medium. Leaf primordia are indicated by an arrow in (B). Approximate root length is indicated by black lines. Note that root hairs are emerging in *rml1* plants, in contrast with ntra ntrb rml1 plants. (D, E) Magnification of the shoot meristem of *rml1* (D) and *ntra ntrb rml1* (E) homozygous mutants at 8 days after germination on MS/2 medium. Note the absence of leaf primordia in *ntra ntrb rml1* (E) in contrast with *rml1* (D). Figure 2 is from Reichheld *et al.* (2007) with the permission of Plant Cell "Copyright American Society of Plant Biologists." (See color plate section at the back of the book.)

Further, new data were recently obtained in consideration of the curious situation of cytosolic GR1, which represent the major GR activity and is highly regulated in the wild type under different conditions, but is apparently totally dispensable as suggested by the phenotype of the *grx1* mutant. One possibility is that, similar to Drosophila or Plasmodium falciparum which have no GR, the cytosolic GSSG could be reduced by an alternative reduction pathway involving the NTS (Kanzok *et al.*, 2001). The first indication was that *in vitro*, NADPH/NTRA, recombinant Trxh3 (as well as other Trxh)

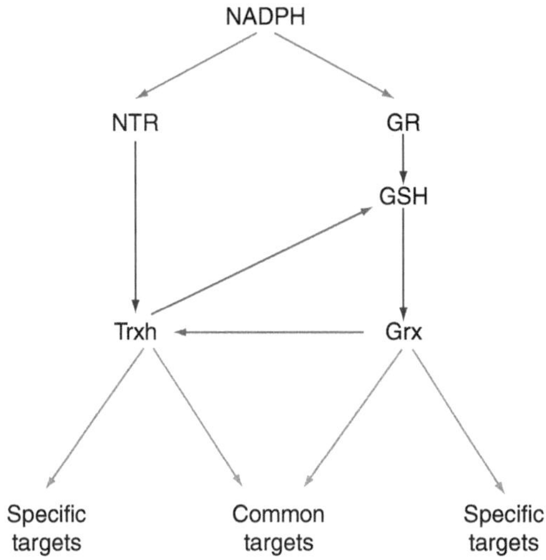

Fig. 3. NTS and NGS in the cytosol: the reduction pathways are represented by arrows (black, blue, red and green in the web version). Trxh are reduced by NTR or optionally by Grx. GSSG is reduced by GR or optionally by Trxh. Several cytosolic targets including RNR, antioxidant enzymes, and others were identified by biochemistry and proteomics (reviewed by Buchanan and Balmer, 2005; Montrichard *et al.*, 2009) but little is known on the specificity in planta.

were able to reduce GSSG but with a low efficiency rate ~200 lower than GR1. Nevertheless, this finding suggested that the NTS could constitute a functional backup for cytosolic GR in *Arabidopsis* (Fig. 1). The *ntra ntrb* mutant was then crossed with a *gr1* mutant but no triple homozygote was isolated. Analysis of the heterozygote *ntra/ntrantrb/ntrbgsh1/GSH1* revealed that only *ntra ntrb GSH1* pollen was produced but no *ntra ntrb gsh1*. In contrast *ntra ntrb gsh1* and *ntra ntrb GSH1* ovules were still produced in the 50%:50% Mendelian ratios. It was further shown that NTRA but not NTRB is able to allow fertile pollen formation in GR1 inactivated mutants. This demonstrates that in contrast to vegetative cells, NTRA and NTRB are not redundant in the pollen (Marty *et al.*, 2009). Whether overlap between NTR and GR1 is also occurring in the diploid stage of plant development is still to be determined, but the viability of the triple homozygote ovules suggests that this is not the case. One possibility is that GSSG reduction in the organelles could be sufficient to allow normal development in most tissues, but the low content of plastids in the pollen could be the reason for its sterility. Alternatively, mitochondria which play a major role in pollen development may be strongly perturbed in the triple homozygote state. Indeed, the picture is made more complex by the fact that GSH and other low molecular weight thiols may be trafficking between the different compartments, including mitochondria. Very little knowledge is available regarding the compartmentation of redox systems in mitochondria. High concentrations of glutathione were found in this organelle (Zechmann *et al.*, 2008), and GSSG reduction is also likely occurring due to the presence of GR2 in this compartment. Whether GSSG reduction is necessary in mitochondria and how GSH is synthesized or transported in mitochondria are still unclear (Fig. 1). These points should be analyzed in engineered mutants with the NTR limited to the cytosol or to the mitochondria and with GR2 limited to chloroplasts or mitochondria.

C. FTS, NTS, AND NGS IN THE ORGANELLES

Mitochondrial target proteins for NTS or NGS have been identified and have putatively essential functions (Balmer *et al.*, 2004; Winger *et al.*, 2007). Some of them can be reduced by the AtTrxo1 and PtTrxh2 which are yet the only Trx identified as mitochondrial proteins (Laloi *et al.*, 2001; Gelhaye *et al.*, 2004). Among all genes encoding Grx only S15 in poplar and *Arabidopsis* show a typical mitochondrial transit peptide (Herald *et al.*, 2003; Rouhier *et al.*, 2006a,b). The phenotypic perturbations of the *ntra ntrb* double mutant are all dependent on the cytosolic isoform of NTR (Reichheld *et al.*, 2007) in *Arabidopsis* KO mutant of the mitochondrial Trxo1 is aphenotypic (our unpublished data). The only genetic evidence concerning the function of NTS or NGS in mitochondria is given by the phenotype of *Atprx2f*. This mutant of the only mitochondrial PRX in *Arabidopsis* shows reduced root growth under various stresses. AtPrx2f can be reduced by Trx or Grx *in vitro*, but the actual reductant *in vivo* is not known (Finkemeier *et al.*, 2005).

The first Trx identified in plants was discovered as a component of a light-dependent chloroplast system in which it was reduced by a ferredoxin-dependent Trx reductase (FTR) (Ferredoxin/Trx system or FTS) (Fig. 4). Chloroplast FTR is itself reduced by ferredoxin which obtains its reducing power from electron transfer in photosynthesis. FTS in the chloroplast activates several enzymes of the Calvin cycle and the NADPH malate dehydrogenase allowing the use of NADPH and ATP for CO_2 fixation. In addition, the Trxs reduced by light inhibit the pentose phosphate pathway which is reactivated in the dark and reduces NADP to NADPH at the expense of sucrose. More recently it was shown that the chloroplast FTS also plays the role of antioxidant due to the presence of antioxidant Trx target proteins, a vital function because oxygenic photosynthesis is particularly prone to produce reactive oxygen species (ROS) and H_2O_2. The number of chloroplast Trxs is particularly high and some specialization of the

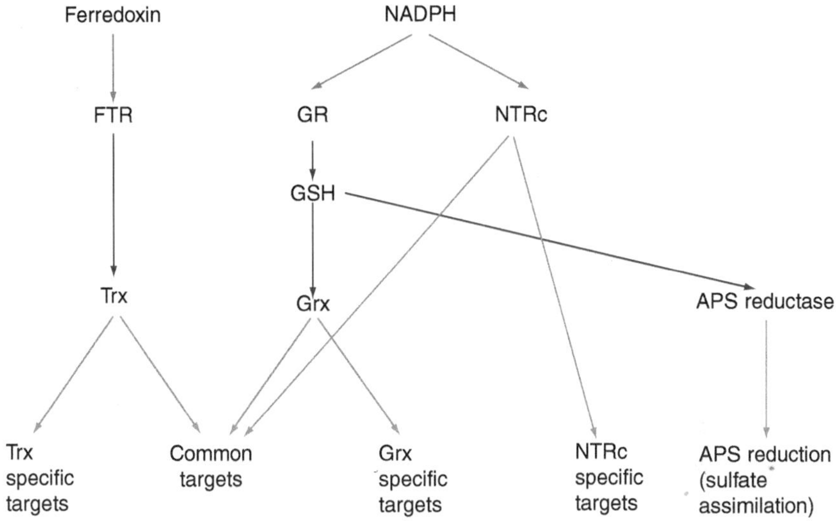

Fig. 4. Redox regulation in the plastid: the reduction pathways are represented by arrows (black, blue, red and green in the web version). Trx specific targets are the light-regulated proteins in the chloroplast (Calvin Cycle enzymes, NADP MDH, etc.), shared targets are mostly antioxidant enzymes, Grx specific are some antioxidant enzymes (Tarrago *et al.*, 2009), NTRc-specific targets are ADP-glucose pyrophosphorylase (AGPase) (Michalska *et al.*, 2009).

different types of Trx is suggested by biochemical studies (Vieira Dos Santos *et al.*, 2007). Nevertheless, while known in some cases, we are far from identifying which Trx reduces a particular target. Genetic results are limited to a partial inactivation of the FTR. FTR is a heterodimer composed of a redox-active peptide encoded by a single gene associated to a variable subunit encoded by two genes. A KO mutant of one variable subunit is available and shows some hypersensitivity to oxidants as well as a low reduction of the NADPH-dependent Malate dehydrogenase (Keryer *et al.*, 2004). Thus the chloroplast FTS system is generally considered as strictly light dependent: its function is the regulation of shift of chloroplast metabolism under light or darkness. Nevertheless, the chloroplast is only one of the differentiation states of the plastid. In nongreen tissues, roots, or seeds, for example, leucoplasts and amyloplasts are present and some plastidial Trxs are expressed. In addition, ferredoxins with higher redox potentials and reductible via NADPH and ferredoxin NADP reductase have been described, and they are likely to allow the functioning of a light-independent FTS in nongreen plastids. A recent paper has shown that Trxm3 a plastidial Trx performs a vital function during seed development, probably related to cell–cell communication (Benitez-Alfonso *et al.*, 2009).

The chloroplast NGS is composed of GR associated with several Grxs. Some monocysteinic Grxs are implicated in FeS cluster formation and transfer. They are not discussed in this review (for a discussion of these proteins, see Rouhier *et al.* (2008). The NGS is NADPH dependent and consequently is able to reduce its target proteins in light as well as under darkness. Several antioxidants targets are described (Tarrago *et al.*, 2009) suggesting that it plays an important function during the night when the FTS is inactive.

The FTS and NGS systems in plastids are complemented by a protein, NTRc, composed of two domains: NTRc is composed of an NADPH-dependent Trx reductase associated with a Trx domain (Fig. 4). It is encoded by a single gene in *Arabidopsis*. Inactivation of the NTRc demonstrates its role in the ROS detoxification at night via a Trx-dependent peroxidase (Pérez-Ruiz *et al.*, 2006; Serrato *et al.*, 2004). In addition, it has recently been implicated in the photoperiodic switch (Lepistö *et al.*, 2009) and in coordinating starch synthesis in tuber with sucrose synthesis in leaves (Michalska *et al.*, 2009). In contrast to *E. coli* and *S. cerevisiae*, reduction of sulfate in chloroplast is performed by an APS reductase with an APS reduction domain associated with a Trx-like domain, but, nevertheless, reduced by glutathione (Bick *et al.*, 1998; Martin *et al.*, 2005). Thus, sulfate reduction in plants is independent of the FTS, NTS, and of Grx.

V. CONCLUSIONS

A general conclusion for all organisms is the presence of Trxs and Grxs that in most cases perform functions that are, in part, redundant. Such observations obviously pose the question of redundant genes in the genome. An obvious advantage of redundant genes is to avoid deleterious effects of mutations in unique genes, but this cannot explain the maintenance of redundancy during evolution. A likely hypothesis is that multigene families are maintained because members are evolving specific characteristics even when they maintain an overlapping core function. Some specificities have been observed within redoxin families, in terms of expression patterns, biochemical activities, or yeast complementation activities. For example, within the multigene family of Trx h, the Trxh5 gene was the only one found to be transcriptionally induced by pathogen attack. These specific characteristics and their phenotypic consequences cannot be easily detected under laboratory conditions, but may give selective advantage under the complex and changing natural conditions. For example, Trxh5 which is highly similar to Trxh3 has recently been identified in a screen for resistance to victorin, a toxin secreted by *Cochliobolus victoriae* (Sweat and Wolpert, 2007) and in the SAR pathway (Tada *et al.*, 2008).

Turning to plants, it is interesting to consider the various evolutionary paths of redoxin redundancy in the cytosol and in the chloroplast. In the cytosol, Trxs and Grxs show redundant functions, for example, for the reduction of certain antioxidant enzymes likewise in *E. coli* and *S. cerevisiae*, but plants have evolved a further redundancy between their NTR- and GR-reducing enzymes, a mechanism not described in the microbes. In contrast, inactivation of mammalian cytosolic or mitochondrial Trx is lethal, clearly showing that the Grx are not Trx backups in animals. Extended redundancy in plants and absence of redundancy in mammals may be related to the complex environment of plants, including exposure to pathogens, while animals have developed different ways to escape changing environmental conditions and immunological systems to protect against pathogens. In contrast to the cytosol, redundancy is reduced in the chloroplast. The two-domain APS reductase renders sulfate assimilation independent of Trx or Grx, unlike most microbes, including cyanobacteria (contemporary organisms most closely related to the chloroplast ancestor) which have free APS reductases that are dependent on Trx or Grx. The regulation of photosynthesis and the related switch from carbon source to carbon sink in the dark could not be performed by Trx if it were redundant with Grx for the Calvin cycle and NADPH malate dehydrogenase regulation. The existence of the two-domain NTRc is a further example of a specialized redox regulator which performs functions independent of the free Trx, Grx, and their reductases. This specialization in independent blocks allows a fine regulation of the plastidial metabolism under various light situations and differentiation states. Nevertheless, within each block redundancy is obvious due to the high number of Trx and Grx members in each subfamily. Similarly four genes in Arabidopsis encode APS reductases responsible for sulfate reduction.

ACKNOWLEDGMENTS

The authors thank R. Cooke (Perpignan) and B. Buchanan (Berkeley) for reading and correcting the manuscript.

REFERENCES

Alonso, J. M., *et al.* (2000). Genome-wide insertional mutagenesis of Arabidopsis thaliana. *Science*. **301,** 653–657.

Avval, F. Z. and Holmgren, A. (2009). Molecular mechanisms of thioredoxin and glutaredoxin as hydrogen donors for Mammalian s phase ribonucleotide reductase. *The Journal of Biological Chemistry* **284**, 8233–8240.

Balmer, Y., Vensel, W. H., Tanaka, C. K., Hurkman, W. J., Gelhaye, E., Rouhier, N., Jacquot, J. -P., Manieri, W., Schurmann, P., Droux, M., and Buchanan, B. B. (2004). Thioredoxin links redox to the regulation of fundamental processes of plant mitochondria. *Proceedings of the National Academy of Sciences of the United States of America* **101**, 2642–2647.

Bashandy, T., Taconnat, L., Renou, J., Meyer, Y. and Reichheld, J. (2009). Accumulation of flavonoids in an ntra ntrb mutant leads to tolerance to UV-C. *Molecular Plant* **2**, 249–258.

Benitez-Alfonso, Y., Cilia, M., Roman, A. S., Thomas, C., Maule, A., Hearn, S. and Jackson, D. (2009). Control of *Arabidopsis* meristem development by thioredoxin-dependent regulation of intercellular transport. *Proceedings of the National Academy of Sciences of the United States of America* **106**, 3615–3620.

Bick, J. A., Aslund, F., Chen, Y. and Leustek, T. (1998). Glutaredoxin function for the carboxyl-terminal domain of the plant-type 5'-adenylylsulfate reductase. *Proceedings of the National Academy of Sciences of the United States of America* **95**, 8404–8409.

Bréhélin, C., Meyer, E. H., de Souris, J., Bonnard, G. and Meyer, Y. (2003). Resemblance and dissemblance of *Arabidopsis* type II peroxiredoxins: Similar sequences for divergent gene expression, protein localization, and activity. *Plant Physiology* **132**, 2045–2057.

Buchanan, B. B. and Balmer, Y. (2005). Redox regulation: A broadening horizon. *Annual Review of Plant Biology* **56**, 187–220.

Buchanan, B. B., Schürmann, P., Wolosiuk, R. A. and Jacquot, J. P. (2002). The ferredoxin/thioredoxin system: From discovery to molecular structures and beyond. *Photosynthesis Research* **73**, 215–222.

Cairns, N. G., Pasternak, M., Wachter, A., Cobbett, C. S., and Meyer, A. J. (2006). Maturation of Arabidopsis seeds is dependent on glutathione biosynthesis within the embryo. *Plant Physiology* **141**, 446–455.

Casagrande, S., Bonetto, V., Fratelli, M., Gianazza, E., Eberini, I., Massignan, T., Salmona, M., Chang, G., Holmgren, A. and Ghezzi, P. (2002). Glutathionylation of human thioredoxin: A possible crosstalk between the glutathione and thioredoxin systems. *Proceedings of the National Academy of Sciences of the United States of America* **99**, 9745–9749.

Cheng, J. C., Seeley, K. A. and Sung, Z. R. (1995). RML1 and RML2, *Arabidopsis* genes required for cell proliferation at the root tip. *Plant Physiology* **107**, 365–376.

Cho, M. J., Wong, J. H., Marx, C., Jiang, W., Lemaux, P. G. and Buchanan, B. B. (1999). Overexpression of thioredoxin h leads to enhanced activity of starch debranching enzyme (pullulanase) in barley grain. *Proceedings of the National Academy of Sciences of the United States of America* **96**, 14641–14646.

Cobbett, C. S., May, M. J., Howden, R. and Rolls, B. (1998). The glutathione-deficient, cadmium-sensitive mutant, cad2–1, of *Arabidopsis thaliana* is deficient in gamma-glutamylcysteine synthetase. *The Plant Journal: For Cell and Molecular Biology* **16**, 73–78.

Conrad, M., Jakupoglu, C., Moreno, S. G., Lippl, S., Banjac, A., Schneider, M., Beck, H., Hatzopoulos, A. K., Just, U., Sinowatz, F., Schmahl, W. Chien, K. R. *et al.* (2004). Essential role for mitochondrial thioredoxin reductase in hematopoiesis, heart development, and heart function. *Molecular and Cellular Biology* **24**, 9414–9423.

Dietz, K. (2003). Plant peroxiredoxins. *Annual Review of Plant Biology* **54**, 93–107.

Draculic, T., Dawes, I. W. and Grant, C. M. (2000). A single glutaredoxin or thioredoxin gene is essential for viability in the yeast *Saccharomyces cerevisiae*. *Molecular Microbiology* **36**, 1167–1174.

Faulkner, M. J., Veeravalli, K., Gon, S., Georgiou, G. and Beckwith, J. (2008). Functional plasticity of a peroxidase allows evolution of diverse disulfide-reducing pathways. *Proceedings of the National Academy of Sciences of the United States of America* **105**, 6735–6740.

Fernandes, A. P., Fladvad, M., Berndt, C., Andrésen, C., Lillig, C. H., Neubauer, P., Sunnerhagen, M., Holmgren, A. and Vlamis-Gardikas, A. (2005). A novel monothiol glutaredoxin (Grx4) from *Escherichia coli* can serve as a substrate for thioredoxin reductase. *The Journal of Biological Chemistry* **280**, 24544–24552.

Finkemeier, I., Goodman, M., Lamkemeyer, P., Kandlbinder, A., Sweetlove, L. J. and Dietz, K. (2005). The mitochondrial type II peroxiredoxin F is essential for redox homeostasis and root growth of *Arabidopsis thaliana* under stress. *The Journal of Biological Chemistry* **280**, 12168–12180.

Fladvad, M., Bellanda, M., Fernandes, A. P., Mammi, S., Vlamis-Gardikas, A., Holmgren, A. and Sunnerhagen, M. (2005). Molecular mapping of functionalities in the solution structure of reduced Grx4, a monothiol glutaredoxin from *Escherichia coli*. *The Journal of Biological Chemistry* **280**, 24553–24561.

Garrido, E. O. and Grant, C. M. (2002). Role of thioredoxins in the response of *Saccharomyces cerevisiae* to oxidative stress induced by hydroperoxides. *Molecular Microbiology* **43**, 993–1003.

Gelhaye, E., Rouhier, N. and Jacquot, J. P. (2003). Evidence for a subgroup of thioredoxin h that requires GSH/Grx for its reduction. *FEBS Letters* **555**, 443–448.

Gelhaye, E., Rouhier, N., Gérard, J., Jolivet, Y., Gualberto, J., Navrot, N., Ohlsson, P., Wingsle, G., Hirasawa, M., Knaff, D. B., Wang, H. Dizengremel, P. *et al.* (2004). A specific form of thioredoxin h occurs in plant mitochondria and regulates the alternative oxidase. *Proceedings of the National Academy of Sciences of the United States of America* **101**, 14545–14550.

Gon, S., Faulkner, M. J. and Beckwith, J. (2006). *In vivo* requirement for glutaredoxins and thioredoxins in the reduction of the ribonucleotide reductases of *Escherichia coli*. *Antioxidants & Redox Signaling* **8**, 735–742.

Grant, C. M., MacIver, F. H. and Dawes, I. W. (1996). Glutathione is an essential metabolite required for resistance to oxidative stress in the yeast *Saccharomyces cerevisiae*. *Current Genetics* **29**, 511–515.

Herald, V. L., Heazlewood, J. L., Day, D. A. and Millar, A. H. (2003). Proteomic identification of divalent metal cation binding proteins in plant mitochondria. *FEBS Letters* **537**, 96–100.

Ho, Y., Xiong, Y., Ho, D. S., Gao, J., Chua, B. H. L., Pai, H. and Mieyal, J. J. (2007). Targeted disruption of the glutaredoxin 1 gene does not sensitize adult mice to tissue injury induced by ischemia/reperfusion and hyperoxia. *Free Radical Biology & Medicine* **43**, 1299–1312.

Holmgren, A. (1989). Thioredoxin and glutaredoxin systems. *Journal of Biological Chemistry* **264**, 13963–13966.

Howden, R., Goldsbrough, P. B., Andersen, C. R. and Cobbett, C. S. (1995). Cadmium-sensitive, cad1 mutants of *Arabidopsis thaliana* are phytochelatin deficient. *Plant Physiology* **107**, 1059–1066.

Jacquot, J. P., Rivera-Madrid, R., Marinho, P., Kollarova, M., Le Maréchal, P., Miginiac-Maslow, M. and Meyer, Y. (1994). *Arabidopsis thaliana* NAPHP thioredoxin reductase. cDNA characterization and expression of the recombinant protein in *Escherichia coli. Journal of Molecular Biology* **235**, 1357–1363.

Jakupoglu, C., Przemeck, G. K. H., Schneider, M., Moreno, S. G., Mayr, N., Hatzopoulos, A. K., de Angelis, M. H., Wurst, W., Bornkamm, G. W., Brielmeier, M. and Conrad, M. (2005). Cytoplasmic thioredoxin reductase is essential for embryogenesis but dispensable for cardiac development. *Molecular and Cellular Biology* **25**, 1980–1988.

Jordan, A., Aslund, F., Pontis, E., Reichard, P. and Holmgren, A. (1997). Characterization of *Escherichia coli* NrdH. A glutaredoxin-like protein with a thioredoxin-like activity profile. *The Journal of Biological Chemistry* **272**, 18044–18050.

Kanzok, S. M., Fechner, A., Bauer, H., Ulschmid, J. K., Müller, H. M., Botella-Munoz, J., Schneuwly, S., Schirmer, R. and Becker, K. (2001). Substitution of the thioredoxin system for glutathione reductase in *Drosophila melanogaster. Science (New York, N.Y.)* **291**, 643–646.

Keryer, E., Collin, V., Lavergne, D., Lemaire, S. and Issakidis-Bourguet, E. (2004). *Characterization of* Arabidopsis *mutants for the variable subunit of ferredoxin:thioredoxin reductase photosynthesis research* **79**, 265–274.

Koh, C. S., Navrot, N., Didierjean, C., Rouhier, N., Hirasawa, M., Knaff, D. B., Wingsle, G., Samian, R., Jacquot, J., Corbier, C. and Gelhaye, E. (2008). An atypical catalytic mechanism involving three cysteines of thioredoxin. *The Journal of Biological Chemistry* **283**, 23062–23072.

Laloi, C., Rayapuram, N., Chartier, Y., Grienenberger, J. M., Bonnard, G. and Meyer, Y. (2001). Identification and characterization of a mitochondrial thioredoxin system in plants. *Proceedings of the National Academy of Sciences of the United States of America* **98**, 14144–14149.

Laloi, C., Mestres-Ortega, D., Marco, Y., Meyer, Y., and Reichheld, J. P. (2004). The Arabidopsis cytosolic thioredoxin h5 gene induction by oxidative stress and its W-box-mediated response to pathogen elicitor. *Plant Physiology* **134**, 1006–1016.

Laurent, T. C., Moore, E. C. and Reichard, P. (1964). Enzymatic synthesis of deoxyribonucleotides. IV. Isolation and characterization of thioredoxin, the hydrogen donor from *Escherichia coli* B. *The Journal of Biological Chemistry* **239**, 3436–3444.

Lepistö, A., Kangasjärvi, S., Luomala, E., Brader, G., Sipari, N., Keränen, M., Keinänen, M. and Rintamäki, E. (2009).*Chloroplast NADPH-thioredoxin reductase interacts with photoperiodic development in* Arabidopsis *plant physiology* **149**, 1261–1276.

Li, S., Lauri, A., Ziemann, M., Busch, A., Bhave, M. and Zachgo, S. (2009). Nuclear activity of ROXY1, a glutaredoxin interacting with TGA factors, is required for petal development in *Arabidopsis thaliana. The Plant Cell* **21**, 429–441.

Marchand, C., Le Maréchal, P., Meyer, Y., Miginiac-Maslow, M., Issakidis-Bourguet, E. and Decottignies, P. (2004). New targets of *Arabidopsis* thioredoxins revealed by proteomic analysis. *Proteomics* **4**, 2696–2706.

Martin, M. N., Tarczynski, M. C., Shen, B. and Leustek, T. (2005). The role of 5'-adenylylsulfate reductase in controlling sulfate reduction in plants. *Photosynthesis Research* **86**, 309–323.

Marty, L., Siala, W., Schwarzländer, M., Fricker, M. D., Wirtz, M., Sweetlove, L. J., Meyer, Y., Meyer, A. J., Reichheld, J. and Hell, R. (2009). The NADPH-dependent thioredoxin system constitutes a functional backup for cytosolic glutathione reductase in *Arabidopsis. Proceedings of the National Academy of Sciences of the United States of America* **106**, 9109–9114.

Matsui, M., Oshima, M., Oshima, H., Takaku, K., Maruyama, T., Yodoi, J. and Taketo, M. M. (1996). Early embryonic lethality caused by targeted disruption of the mouse thioredoxin gene. *Developmental Biology* **178**, 179–185.

Meyer, E. H., Giegé, P., Gelhaye, E., Rayapuram, N., Ahuja, U., Thöny-Meyer, L., Grienenberger, J. and Bonnard, G. (2005). AtCCMH, an essential component of the c-type cytochrome maturation pathway in *Arabidopsis* mitochondria, interacts with apocytochrome *c*. *Proceedings of the National Academy of Sciences of the United States of America* **102**, 16113–16118.

Meyer, Y., Riondet, C., Constans, L., Abdelgawwad, M. R., Reichheld, J. P. and Vignols, F. (2006). Evolution of redoxin genes in the green lineage. *Photosynthesis Research* **89**, 179–192.

Michalska, J., Zauber, H., Buchanan, B. B., Cejudo, F. J. and Geigenberger, P. (2009). NTRC links built-in thioredoxin to light and sucrose in regulating starch synthesis in chloroplasts and amyloplasts. *Proceedings of the National Academy of Sciences of the United States of America* **106**, 9908–9913.

Miranda-Vizuete, A., Damdimopoulos, A. E., Gustafsson, J. and Spyrou, G. (1997). Cloning, expression, and characterization of a novel *Escherichia coli* thioredoxin. *The Journal of Biological Chemistry* **272**, 30841–30847.

Montrichard, F., Alkhalfioui, F., Yano, H., Vensel, W. H., Hurkman, W. J. and Buchanan, B. B. (2009). Thioredoxin targets in plants: The first 30 years. *Journal of Proteomics* **72**, 452–474.

Muller, E. G. (1992). Thioredoxin genes in *Saccharomyces cerevisiae*: Map positions of TRX1 and TRX2. *Yeast (Chichester, England)* **8**, 117–120.

Muller, E. G. (1995). A redox-dependent function of thioredoxin is necessary to sustain a rapid rate of DNA synthesis in yeast. *Archives of Biochemistry and Biophysics* **318**, 356–361.

Muller, E. G. (1996). A glutathione reductase mutant of yeast accumulates high levels of oxidized glutathione and requires thioredoxin for growth. *Molecular Biology of the Cell* **7**, 1805–1813.

Mulliez, E., Ollagnier, S., Fontecave, M., Eliasson, R. and Reichard, P. (1995). Formate is the hydrogen donor for the anaerobic ribonucleotide reductase from *Escherichia coli*. *Proceedings of the National Academy of Sciences of the United States of America* **92**, 8759–8762.

Navrot, N., Collin, V., Gualberto, J., Gelhaye, E., Hirasawa, M., Rey, P., Knaff, D. B., Issakidis, E., Jacquot, J. and Rouhier, N. (2006). Plant glutathione peroxidases are functional peroxiredoxins distributed in several subcellular compartments and regulated during biotic and abiotic stresses. *Plant Physiology* **142**, 1364–1379.

Nekrasov, V., Ludwig, A. A. and Jones, J. D. G. (2006). CITRX thioredoxin is a putative adaptor protein connecting Cf-9 and the ACIK1 protein kinase during the Cf-9/Avr9- induced defence response. *FEBS Letters* **580**, 4236–4241.

Nonn, L., Williams, R. R., Erickson, R. P. and Powis, G. (2003). The absence of mitochondrial thioredoxin 2 causes massive apoptosis, exencephaly, and early embryonic lethality in homozygous mice. *Molecular and Cellular Biology* **23**, 916–922.

Parisy, V., Poinssot, B., Owsianowski, L., Buchala, A., Glazebrook, J. and Mauch, F. (2007). Identification of PAD2 as a gamma-glutamylcysteine synthetase highlights the importance of glutathione in disease resistance of *Arabidopsis*. *The Plant Journal: For Cell and Molecular Biology* **49**, 159–172.

Pasternak, M., Lim, B., Wirtz, M., Hell, R., Cobbett, C. S. and Meyer, A. J. (2008). Restricting glutathione biosynthesis to the cytosol is sufficient for normal plant development. *The Plant Journal: For Cell and Molecular Biology* **53**, 999–1012.

Pérez-Ruiz, J. M., Spínola, M. C., Kirchsteiger, K., Moreno, J., Sahrawy, M. and Cejudo, F. J. (2006). Rice NTRC is a high-efficiency redox system for chloroplast protection against oxidative damage. *The Plant Cell* **18**, 2356–2368.

Prinz, W. A., Aslund, F., Holmgren, A. and Beckwith, J. (1997). The role of the thioredoxin and glutaredoxin pathways in reducing protein disulfide bonds in the *Escherichia coli* cytoplasm. *The Journal of Biological Chemistry* **272**, 15661–15667.

Reichheld, J., Meyer, E., Khafif, M., Bonnard, G. and Meyer, Y. (2005). AtNTRB is the major mitochondrial thioredoxin reductase in *Arabidopsis thaliana*. *FEBS Letters* **579**, 337–342.

Reichheld, J., Khafif, M., Riondet, C., Droux, M., Bonnard, G. and Meyer, Y. (2007). Inactivation of thioredoxin reductases reveals a complex interplay between thioredoxin and glutathione pathways in *Arabidopsis* development. *The Plant Cell* **19**, 1851–1865.

Rietsch, A. and Beckwith, J. (1998). The genetics of disulfide bond metabolism. *Annual Review of Genetics* **32**, 163–184.

Rivas, S., Rougon-Cardoso, A., Smoker, M., Schauser, L., Yoshioka, H. and Jones, J. D. G. (2004). CITRX thioredoxin interacts with the tomato Cf-9 resistance protein and negatively regulates defence. *The EMBO Journal* **23**, 2156–2165.

Rouhier, N., Couturier, J. and Jacquot, J. P. (2006a). Genome-wide analysis of plant glutaredoxin systems. *Journal of Experimental Botany* **57**, 1685–1696.

Rouhier, N., Vieira Dos Santos, C., Tarrago, L. and Rey, P. (2006b). Plant methionine sulfoxide reductase A and B multigenic families. *Photosynthesis Research* **89**, 247–262.

Rouhier, N., Lemaire, S. D. and Jacquot, J. P. (2008). The role of glutathione in photosynthetic organisms: Emerging functions for glutaredoxins and glutathionylation. *Annual Review of Plant Biology* **59**, 143–166.

Russel, M., Model, P. and Holmgren, A. (1990). Thioredoxin or glutaredoxin in *Escherichia coli* is essential for sulfate reduction but not for deoxyribonucleotide synthesis. *Journal of Bacteriology* **172**, 1923–1929.

Schürmann, P. and Buchanan, B. B. (2008). The ferredoxin/thioredoxin system of oxygenic photosynthesis. *Antioxidants & Redox Signaling* **10**, 1235–1274.

Schürmann, P. and Jacquot, J. P. (2000). Plant thioredoxin systems revisited. *Annual Review of Plant Physiology and Plant Molecular Biology* **51**, 371–400.

Serrato, A. J., Pérez-Ruiz, J. M., Spínola, M. C. and Cejudo, F. J. (2004). A novel NADPH thioredoxin reductase, localized in the chloroplast, which deficiency causes hypersensitivity to abiotic stress in *Arabidopsis thaliana*. *The Journal of Biological Chemistry* **279**, 43821–43827.

Serrato, A. J., Guilleminot, J., Meyer, Y. and Vignols, F. (2008). AtCXXS: Atypical members of the *Arabidopsis thaliana* thioredoxin h family with a remarkably high disulfide isomerase activity. *Physiologia Plantarum* **133**, 611–622.

Stewart, E. J., Aslund, F. and Beckwith, J. (1998). Disulfide bond formation in the *Escherichia coli* cytoplasm: An *in vivo* role reversal for the thioredoxins. *The EMBO Journal* **17**, 5543–5550.

Sweat, T. A. and Wolpert, T. J. (2007). Thioredoxin h5 is required for victorin sensitivity mediated by a CC-NBS-LRR gene in *Arabidopsis*. *The Plant Cell* **19**, 673–687.

Tada, Y., Spoel, S. H., Pajerowska-Mukhtar, K., Mou, Z., Song, J., Wang, C., Zuo, J. and Dong, X. (2008). Plant immunity requires conformational charges of NPR1 via *S*-nitrosylation and thioredoxins. *Science (New York, N.Y.)* **321**, 952–956.

Tarrago, L., Laugier, E., Zaffagnini, M., Marchand, C., Le Maréchal, P., Rouhier, N., Lemaire, S. D. and Rey, P. (2009). Regeneration mechanisms of *Arabidopsis thaliana* methionine sulfoxide reductases B by glutaredoxins and thioredoxins. *The Journal of Biological Chemistry* **284**, 18963–18971.

Toledano, M. B., Kumar, C., Le Moan, N., Spector, D. and Tacnet, F. (2007). The system biology of thiol redox system in *Escherichia coli* and yeast: Differential functions in oxidative stress, iron metabolism and DNA synthesis. *FEBS Letters* **581**, 3598–3607.

Trotter, E. W. and Grant, C. M. (2002). Thioredoxins are required for protection against a reductive stress in the yeast *Saccharomyces cerevisiae. Molecular Microbiology* **46**, 869–878.

Trotter, E. W. and Grant, C. M. (2003). Non-reciprocal regulation of the redox state of the glutathione-glutaredoxin and thioredoxin systems. *EMBO Reports* **4**, 184–188.

Trotter, E. W. and Grant, C. M. (2005). Overlapping roles of the cytoplasmic and mitochondrial redox regulatory systems in the yeast *Saccharomyces cerevisiae. Eukaryotic Cell* **4**, 392–400.

Tzafrir, I., Pena-Muralla, R., Dickerman, A., Berg, M., Rogers, R., Hutchens, S., Sweeney, T. C., McElver, J., Aux, G., Patton, D. and Meinke, D. (2004). Identification of genes required for embryo development in *Arabidopsis. Plant Physiology* **135**, 1206–1220.

Vernoux, T., Wilson, R. C., Seeley, K. A., Reichheld, J. P., Muroy, S., Brown, S., Maughan, S. C., Cobbett, C. S., Van Montagu, M., Inzé, D., May, M. J. and Sung, Z. R. (2000). The ROOT MERISTEMLESS1/CADMIUM SENSITIVE2 gene defines a glutathione-dependent pathway involved in initiation and maintenance of cell division during postembryonic root development. *The Plant Cell* **12**, 97–110.

Vieira Dos Santos, C., Laugier, E., Tarrago, L., Massot, V., Issakidis-Bourguet, E., Rouhier, N. and Rey, P. (2007). Specificity of thioredoxins and glutaredoxins as electron donors to two distinct classes of *Arabidopsis* plastidial methionine sulfoxide reductases B. *FEBS Letters* **581**, 4371–4376.

Wachter, A., Wolf, S., Steininger, H., Bogs, J. and Rausch, T. (2005). Differential targeting of GSH1 and GSH2 is achieved by multiple transcription initiation: Implications for the compartmentation of glutathione biosynthesis in the Brassicaceae. *The Plant Journal: For Cell and Molecular Biology* **41**, 15–30.

Winger, A. M., Taylor, N. L., Heazlewood, J. L., Day, D. A. and Millar, A. H. (2007). Identification of intra- and intermolecular disulphide bonding in the plant mitochondrial proteome by diagonal gel electrophoresis. *Proteomics* **7**, 4158–4170.

Wong, J. H., Kim, Y., Ren, P., Cai, N., Cho, M., Hedden, P., Lemaux, P. G. and Buchanan, B. B. (2002). Transgenic barley grain overexpressing thioredoxin shows evidence that the starchy endosperm communicates with the embryo and the aleurone. *Proceedings of the National Academy of Sciences of the United States of America* **99**, 16325–16330.

Xing, S. and Zachgo, S. (2008). ROXY1 and ROXY2, two *Arabidopsis* glutaredoxin genes, are required for anther development. *The Plant Journal: For Cell and Molecular Biology* **53**, 790–801.

Xing, S., Rosso, M. G. and Zachgo, S. (2005). ROXY1, a member of the plant glutaredoxin family, is required for petal development in *Arabidopsis thaliana*. *Development (Cambridge, England)* **132**, 1555–1565.

Zechmann, B., Mauch, F., Sticher, L. and Muller, M. (2008). Subcellular immunocytochemical analysis detects the highest concentrations of glutathione in mitochondria and not in plastids. *Journal of Experimental Botany* **59**, 4017–4027.

Protein Import in Chloroplasts: An Emerging Regulatory Role for Redox

MÓNICA BALSERA,*[,†] JÜRGEN SOLL[‡] AND BOB B. BUCHANAN[§,1]

**Department of Structural Biology, Paul Scherrer Institute, Villigen, Switzerland*
[†]Departamento de Estrés Abiótico, Instituto de Recursos Naturales y Agrobiología de Salamanca (IRNASA-CSIC), Salamanca, Spain
[‡]Department of Biology I, Botany, Ludwig Maximilians University, Martinsried, Germany
[§]Department of Plant and Microbial Biology, University of California, Berkeley, USA

[1]Corresponding author: Email: view@nature.berkeley.edu

Advances in Botanical Research, Vol. 52
0065-2296/09 $35.00

DOI: 10.1016/S0065-2296(09)52010-1

ABSTRACT

Chloroplasts as well as mitochondria are surrounded by two envelope membranes that contain protein machineries to ensure well-organized communication and substrate distribution between the organelle and the rest of the cell. Protein import into organelles must be tightly coordinated with the internal protein synthesis machinery to ensure assembly of functional complexes of dual genetic origin. Import must also be coupled to the cellular metabolic state to guarantee requirements of the organelle in response to developmental changes and environmental conditions.

Several recently published findings point to a role for redox in regulating chloroplast protein import: light and cysteine-modifying reagents affect import, certain subunits of the protein import machinery contain redox-sensing components, and others are potential thioredoxin targets. Herein we review the recent structural, computational, genetic, and biochemical studies that have begun to identify key characteristics and properties underlying protein import in chloroplasts. The predicted topology of some components is discussed, pointing out conserved cysteines in the protein families that may play a role in linking redox to oxidative stress and changes in the metabolic state of chloroplasts.

I. INTRODUCTION

Chloroplasts, photosynthetic organelles in green plant cells, have developed different machineries during evolution to ensure a specific and efficient transport network for proteins that, while synthesized in the cytosol, exert their function within the plastid (Balsera *et al.*, 2009b; Cline and Dabney-Smith, 2008; Inaba and Schnell, 2008; Jarvis, 2008). Biogenesis of most chloroplast proteins implies, therefore, translocation through two lipid membranes (the outer and the inner envelope membranes) until they reach eventually the stroma of chloroplasts. In addition, some proteins are further transported from the stroma to thylakoids, with a final destination to either the thylakoidal membrane or lumen. Hence, these proteins have to face the insertion or complete translocation through a third lipid bilayer membrane.

The chloroplast proteome is, however, composed not only of nuclear-encoded proteins, but also a number of subunits that are encoded in the chloroplast genome (cpDNA), reminiscent of the ancient cyanobacterial ancestors according to the endosymbiotic theory (Gould *et al.*, 2008; Gray, 1992). Some essential multimeric protein complexes in chloroplasts are composed of proteins of dual genetic origin, and their assembly into functional units obviously depends on the availability of all subunits. For instance, photosystem II (PSII) in plant chloroplasts is made up of more than 25 extrinsic and intrinsic different subunits of dual genetic origin (Barber, 2006). Three of these subunits (OE33, OE23, and OE17) function coordinately at the oxygen-evolving complex. They are extrinsically associated with PSII in the interior lumen, yet synthesized in the cytosol of the plant cell. The three subunits use the general import pathway at the envelope level, involving the TOC and the TIC complexes (Balsera *et al.*, 2009b; Inaba and Schnell, 2008; Jarvis, 2008). At least two different protein translocation machineries participate in their transport into thylakoids: cpTat (twin-arginine translocation complex, with prefix "cp" indicating pathway in chloroplasts) for the translocation of folded OE23 and OE17 (Robinson and Bolhuis, 2004), and cpSec (secretory pathway in chloroplast) for the translocation of unfolded OE33 (Schuenemann *et al.*, 1999b). A third translocation machinery in thylakoids, Alb3/cpSRP (single recognition particle), assures correct insertion and folding of the light-harvesting complex within the thylakoid membrane posttranslationally (Schuenemann *et al.*, 1998). On the other hand, insertion of the chloroplast-encoded reaction center protein PsbA (or D1) in thylakoids and its assembly with other PSII subunits occurs cotranslationally in an Alb3/SRP-dependent pathway in cooperation with cpSec, following a mechanism similar to that mediating the transport of proteins to the bacterial plasma membrane (Luirink *et al.*, 2001; Zhang *et al.*, 2001). Membrane insertion of the thylakoidal proteins cytochrome *f* and *PsaF* is, however, independent of Alb3 and SRP and seems to require only cpSec (Karnauchov *et al.*, 1994; Nohara *et al.*, 1996). Finally, small intrinsic subunits like PsbW and PsbX are spontaneously inserted into the thylakoid (Tissier *et al.*, 2002).

Coordination and regulation of protein import is, therefore, central to chloroplast biogenesis as both developmental changes and environmental signals are believed to influence the capability for transport of proteins into chloroplasts as well as their internal routing within the plastid. In this respect, organelles have developed efficient means of communication with other parts of the plant cell (Woodson and Chory, 2008). Present knowledge indicates that chloroplast redox signals influence not only translation and translocation processes outside the chloroplast via retrograde signaling, but also the import of nuclear-encoded proteins. Light, temperature, and Cys-modifying reagents affect import (Bartsch *et al.*, 2008; Dutta *et al.*, 2009; Friedman and Keegstra, 1989; Hirohashi *et al.*, 2001; Pilon *et al.*, 1992; Row and Gray, 2001; Seedorf and Soll, 1995; Stengel *et al.*, 2009). The translocation machinery at the inner envelope membrane contains putative redox sensor components (Caliebe *et al.*, 1997; Hormann *et al.*, 2004; Kuchler *et al.*, 2002), and several members of the different protein transport complexes are linked to thioredoxin (Trx) (Balsera *et al.*, 2009a; Bartsch *et al.*, 2008; Mata-Cabana *et al.*, 2007)—findings that strongly suggest a role for redox in regulating and coordinating protein import with the rest of the plant cell.

II. PATHWAYS OF PROTEIN IMPORT IN CHLOROPLASTS

Most proteins targeted to chloroplasts are synthesized as preproteins or precursor proteins with a presequence (or transit peptide) at the N-terminus (Dobberstein *et al.*, 1977). The transit peptide is necessary for proper localization at the chloroplast surface (Bruce, 2000). Soluble proteins that are targeted to the thylakoid lumen contain a removable bipartite transit peptide, with the most N-proximal part analogous to the transit peptide of stromal proteins and the C-proximal counterpart specific for thylakoid transfer (Smeekens *et al.*, 1986). Thylakoid membrane proteins have a transit peptide for stromal targeting and contain internal motifs for insertion into the thylakoids. Preproteins are typically translocated in an unfolded state and after (or during) translocation, the transit peptide is removed by a specific protease. The mature protein is then folded in a functionally competent conformation and/or targeted to the proper chloroplast subcompartment (Smeekens *et al.*, 1990).

An interesting question deals with the location of chloroplast protein synthesis in the cytosol and how preproteins are targeted or delivered to the outer envelope membrane surface of chloroplasts. Different experiments, including electron microscopy studies, have demonstrated that the process of protein import into chloroplasts is posttranslational (Carde *et al.*, 1982; Chua and Schmidt, 1979) and that a cytosolic guidance complex, formed either by chaperones (for proteins with a transit peptide) or the ankyrin repeat containing protein AKR2 (for certain outer membrane proteins), accompanies the precursor protein to the chloroplast surface (Bae *et al.*, 2008; May and Soll, 2000; Qbadou *et al.*, 2006).

A. PROTEINS DESTINED TO THE STROMA

A general import pathway has been described for precursor proteins that contain a cleavable transit peptide and either reside in the stroma or have a stromal import intermediate.

The coordinate activities of the multimeric TOC and TIC complexes of the outer and the inner envelope membranes, respectively, facilitate the translocation of these substrates through these barriers (Agne and Kessler, 2009; Benz *et al.*, 2009) (Fig. 1A). Receptors at the outer envelope specifically recognize the chloroplast transit peptide and mediate precursor transfer to the pore and translocation at expense of GTP and ATP (Keegstra *et al.*, 1989; Olsen and Keegstra, 1992). Protein components in the intermembrane space assure the formation of supercomplexes between both translocases and the correct transfer of the precursor from TOC to TIC. Further transport across the inner envelope requires ATP for import motor at the *trans* side (Jackson-Constan *et al.*, 2001) until the preprotein finally reaches the stroma and the mature protein is folded and properly distributed.

B. PROTEINS DESTINED TO THE THYLAKOID MEMBRANE OR LUMEN

Following translocation to the stroma by the general import pathway, four different pathways mediate the insertion/translocation of proteins destined to the thylakoids and internal lumen compartment, three of which are dependent on proteinaceous machineries at the membrane, namely Alb3, cpSec, and cpTat. A fourth group, that includes certain small membrane proteins, is believed to be inserted within the thylakoids spontaneously (Aldridge *et al.*, 2009; Mori and Cline, 2001; Schunemann, 2004, 2007; Tissier *et al.*, 2002) (Fig. 1B).

Pathways that mediate thylakoidal protein biogenesis involve either the Alb3 insertase, or the cpSec translocase, or both. Alb3, a homolog of the bacterial YidC and mitochondrial Oxa1 membrane insertases, functions together with the cpSRP system in the posttranslational targeting and insertion of the light-harvesting chlorophyll-binding proteins (LHCPs) into the thylakoids (Luirink *et al.*, 2001; Sundberg *et al.*, 1997). The cpSRP/Alb3 and cpSec systems also cooperate in the cotranslational integration of a number of chloroplast-encoded subunits (Nilsson and van Wijk, 2002; Nilsson *et al.*, 1999; Zhang *et al.*, 2001). In other situations, the cpSec system can function independently of Alb3/cpSRP posttranslationally for the insertion of

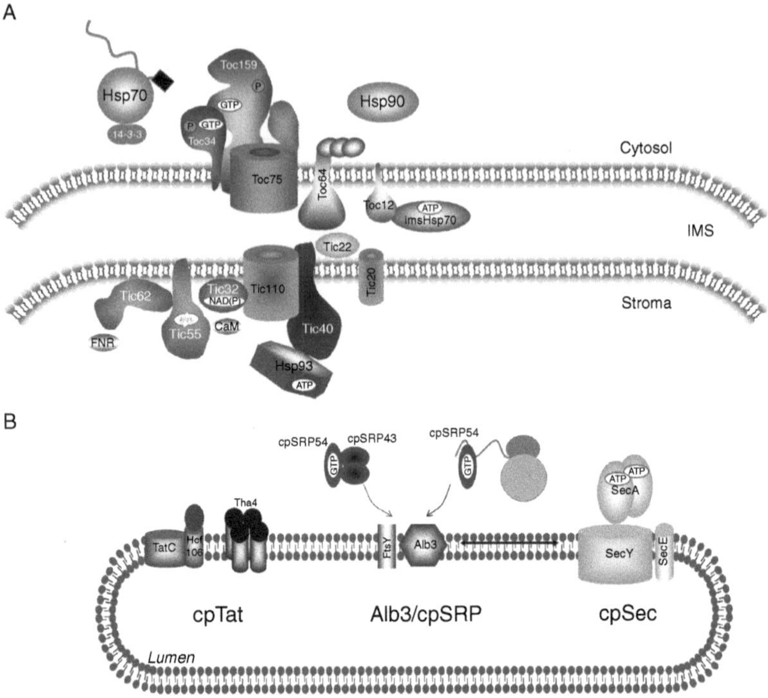

Fig. 1. Overview of protein machineries involved in protein import and routing in chloroplasts. (A) For translocation from the cytosol to the stroma, proteins that carry an N-terminal chloroplast transit peptide follow the general import pathway, involving the TOC and TIC complexes, respectively, at the outer and inner envelopes. (B) Proteins that are further transported to the thylakoid membrane or lumen follow different targeting pathways. Lumenal proteins are translocated through the thylakoids either by the cpSec or the cpTat machineries. Nuclear-encoded thylakoidal proteins are either inserted into the membrane posttranslationally by the cpSRP machinery or follow a mechanism of spontaneous insertion. Chloroplast-encoded thylakoidal proteins are cotranslationally inserted and folded into the membrane by the cpSRP system in cooperation with the cpSec machinery. The insertion of other thylakoidal proteins into the membrane is SRP-independent and seems to require only the cpSec machinery. (See color plate section at the back of the book.)

nuclear- and chloroplast-encoded thylakoidal proteins (Karnauchov *et al.*, 1994; Mori and Cline, 2001; Nohara *et al.*, 1996; Schuenemann *et al.*, 1999a).

The transport of soluble proteins into the lumen is mediated either by cpTat or cpSec pathways, each specialized for a subset of substrates (Braun and Theg, 2008; Braun *et al.*, 2007; Cline *et al.*, 1992; Mori and Cline, 2001; Schuenemann *et al.*, 1999a). Whereas the former transports prefolded proteins, the latter translocates unfolded substrates in an ATP-dependent manner.

C. Proteins Destined to the Intermembrane Space

Information on the biogenesis of proteins in the intermembrane space between the two envelopes is beginning to emerge. Two nuclear-encoded protein residents of the space, Tic22 and monogalactosyldiacylglycerol synthase isoform 1 (MGD1), have been studied (Kouranov *et al.*, 1999; Vojta *et al.*, 2007a). Whereas both seem to enter via the TOC complex (an observation still controversial for Tic22), it has been proposed that MGD1 reaches the stroma to the point at which its transit peptide is cleaved by the stromal processing peptidase and then retracts from TIC to the intermembrane space (Vojta *et al.*, 2007a). By contrast, Tic22 likely does not cross the inner envelope at any point of translocation.

D. Proteins Destined to the Inner Envelope Membrane

Two different insertion pathways have been described for proteins residing in the inner envelope membrane (Brink *et al.*, 1995; Firlej-Kwoka *et al.*, 2008) (Fig. 1A). The "stop-transfer" pathway is mediated by built-in hydrophobic signals that arrest them at the TIC level in the membrane and release them laterally into the lipid bilayer (Brink *et al.*, 1995; Firlej-Kwoka *et al.*, 2008). The "conservative sorting" pathway involves a soluble stromal intermediate that is retargeted to the envelope by an unknown mechanism (Li and Schnell, 2006; Lubeck *et al.*, 1996; Tripp *et al.*, 2007; Vojta *et al.*, 2007b).

E. Deviations from the General TOC/TIC Import Pathway

Proteomic studies have revealed many chloroplast proteins that are apparently transported without a canonical cleavage transit peptide (Kleffmann *et al.*, 2004). In these cases, the targeting information is contained in the mature protein sequence. Knowledge of the chloroplast import mechanism of these proteins is scanty, but novel pathways are being proposed. The import properties of two of these proteins, Tic32 and cQORH, that localize in the inner envelope membrane have been analyzed and, although energy in the form of ATP is required for import, no TOC/TIC components were found to be involved in their import (Miras *et al.*, 2002; Nada and Soll, 2004).

Distinct membrane insertion pathways are active for proteins residing at the outer envelope chloroplast membrane (Hofmann and Theg, 2005a; Inoue, 2007). The topological orientation for α-helical proteins anchored in the membrane by an N-terminal hydrophobic tail, represented by OEP7, is determined by the charged distribution flanking the transmembrane domain (Lee *et al.*, 2001; Salomon *et al.*, 1990; Schleiff *et al.*, 2001). These proteins might make use of the TOC channel for integration, without the need for the receptor components (Tsai *et al.*, 1999; Tu *et al.*, 2004). However, the insertion of proteins with an N_{out}–C_{in} topology, as Toc34, likely does not depend on protein machinery, but rather on the lipid composition of the membrane (Qbadou *et al.*, 2003). A mechanism for spontaneous insertion into the membrane is proposed for certain β-barrel outer membrane proteins, like OEP21 and OEP24, because neither proteinaceous machinery nor an obvious energy source seems to be required for their integration (Bolter *et al.*, 1999; Pohlmeyer *et al.*, 1998). The β-barrel protein Toc75 is unique among outer envelope proteins: it contains a cleavable bipartite transit peptide necessary for targeting to the outer envelope as well as for transport via an intricate transport pathway that likely involves stromal components (Tranel and Keegstra, 1996).

Another group of chloroplast proteins with secretory-pathway-targeting signal peptides has been identified and its relevance to their transport has been experimentally verified for a number of representatives (Chen *et al.*, 2004; Nanjo *et al.*, 2006; Villarejo *et al.*, 2005). How these proteins cross the envelope membranes and whether their transport involves vesicle fusion and the participation of TOC/TIC components is currently under investigation.

III. MOLECULAR MACHINERIES INVOLVED IN PROTEIN TRANSLOCATION THROUGH THE CHLOROPLAST ENVELOPE MEMBRANES: THE GENERAL IMPORT PATHWAY

Two multimeric protein complexes facilitate the translocation of most nuclear-encoded proteins into or across the two envelope membranes—the so-called TOC and TIC complexes (Fig. 1A). The subunits that comprise these complexes are designated by the molecular weight of the first subunit identified, usually in pea where much of the biochemical characterization has been done (Schnell *et al.*, 1997).

Protein import into chloroplasts has been divided into three distinct stages based on energy requirements (Perry and Keegstra, 1994): (a) Energy-independent stage during which proteins reversibly bind to the outer chloroplast surface (Kouranov and Schnell, 1997); (b) Early import intermediate stage, an irreversible step that corresponds to the stable binding/insertion and requires $\leq$20 μM ATP in the presence of GTP (Inoue and Akita, 2008; Kessler *et al.*, 1994); and (c) Translocation and processing, that requires $\geq$100 μM ATP (Pain and Blobel, 1987; Theg *et al.*, 1989). Upon emerging from the TIC complex, proteolytic cleavage of the transit peptide results in the formation of the mature, functional protein.

Toc75, the protein-conducting channel that is evolutionarily related to the family of Omp85 transporters in bacteria, is the central component of the TOC complex (Bolter *et al.*, 1998; Hinnah *et al.*, 1997, 2002; Schnell *et al.*, 1994). Toc75 associates with Toc159 and Toc34, two GTPases that act as import receptors at the outer membrane surface (Kessler *et al.*, 1994). These receptors receive the precursors from the cytosolic chaperones, preferentially the pair composed of Hsp70/14-3-3 that recognizes specific phosphorylated residues in the transit peptide of certain substrates (May and Soll, 2000). Intrinsic GTPase activity gives the receptors the capability to function not only in preprotein recognition, but also as molecular switches and/or molecular motors in transport itself (Kessler and Schnell, 2002; Schleiff *et al.*, 2003). There is evidence that the GTPase cycle of these receptors is modulated by precursor binding (Reddick *et al.*, 2007). The mechanistic model of protein import holds that Toc GTPases (in their GTP-bound state) have high affinity for precursor proteins. GTP hydrolysis (and/or nucleotide exchange) is stimulated upon precursor binding (Becker *et al.*, 2004b; Jelic *et al.*, 2002), releasing the protein to the Toc75 channel due to a loss of affinity of the GDP-loaded form (Jelic *et al.*, 2002; Schleiff *et al.*, 2002). The GTPase receptors are also susceptible to modification by phosphorylation (Jelic *et al.*, 2003; Oreb *et al.*, 2008; Sveshnikova *et al.*, 2000b). Functional analyses have demonstrated, moreover, that phosphorylation and GTPase activity are mutually exclusive, prompting the idea that cycles of phosphorylation/dephosphorylation act as additional regulatory points in protein import. The functional significance of phosphorylation *in vivo* is, however, unknown (Aronsson *et al.*, 2006).

In addition to the GTPase receptors, the TOC complex contains a third receptor subunit, termed Toc64, composed of two distinct functional domains (Fig. 1A). Toc64 is a receptor for the cytosolic Hsp90 chaperone that delivers a subset of precursors to Toc34 (Qbadou *et al.*, 2006, 2007; Sohrt and Soll, 2000). According to *in vivo* studies, its presence may not be strictly required for import, but is essential for highly efficient translocation (Aronsson *et al.*, 2007; Hofmann and Theg, 2005b). On the other hand, Toc64 forms a ternary complex of unknown properties with Toc12 (Becker *et al.*, 2004a) and a chaperone of the Hsp70 family in the intermembrane space (imsHsp70) (Ratnayake *et al.*, 2008). Moreover, Toc64 also recruits Tic22, the only soluble subunit of the TIC complex in the intermembrane space (Qbadou *et al.*, 2007). It has been demonstrated that these four subunits (Hsp90, Toc64, imsHsp70, Tic22) form a complex in an ATP-dependent manner, leading to the proposal of a putative role of imsHsp70 in pulling the precursor from the TOC complex in cooperation with the Toc64/Toc12 pair. An association with Tic22 would assure the continuity of import to the Tic110 inner envelope translocon channel (Balsera *et al.*, 2009a; Gross and Bhattacharya, 2009; Heins *et al.*, 2002; Kessler and Blobel, 1996; Kouranov *et al.*, 1998). During TIC translocation, a putative import motor formed by stromal Hsp93 (or ClpC) chaperone develops in coordination with the cochaperone Tic40 (Chou *et al.*, 2006; Kovacheva *et al.*, 2005, 2007; Nielsen *et al.*, 1997; Stahl *et al.*, 1999). In the current mechanistic model, precursor interaction with Tic110 promotes the association of the Tic40–Hsp93 complex with Tic110 that then supports transfer of the transit peptide to Hsp93 and stimulation of its ATPase activity, followed by delivery of the preprotein to the stroma.

Three redox sensor proteins, Tic62, Tic32, and Tic55, appear to regulate import according to the chloroplast metabolic redox state. Tic62 and Tic32 are classified as short-chain dehydrogenases (SDR)/reductases that associate with TIC in a regulatory manner dependent on NADPH, ferredoxin-NADP-oxidoreductase (FNR), and Ca^{2+} (Chigri *et al.*, 2006; Hormann *et al.*, 2004; Kuchler *et al.*, 2002; Stengel *et al.*, 2008). Tic55 contains a Rieske iron–sulfur center and a mononuclear iron-binding site (Caliebe *et al.*, 1997). It remains to be seen how these TIC-associated redox-active proteins function during import. The identification of their substrates and response to different cellular conditions also awaits further work.

IV. STRUCTURE–FUNCTION RELATIONS OF TOC AND TIC COMPONENTS: POTENTIAL FOR REDOX REGULATION

A. THE PORE-FORMING PROTEIN Toc75: POTRA MOTIFS AND CHANNEL PROPERTIES

Toc75, the protein-import channel in the TOC complex, is known as Toc75-III in *Arabidopsis thaliana*, referring to the chromosome where the protein is encoded (At3g46740). Two structurally and functionally different domains are recognized in the Toc75 protein sequence (Fig. 2A). The N-terminal soluble component, which functions as a receptor-binding site, harbors polypeptide-transport-associated (POTRA) domains involved in protein–protein interaction (Sanchez-Pulido *et al.*, 2003). The C-terminal membrane domain consists of antiparallel β-strands that form a β-barrel hydrophilic channel within the membrane through which precursor proteins are transported (Reddick *et al.*, 2008; Sveshnikova *et al.*, 2000a). Reconstitution

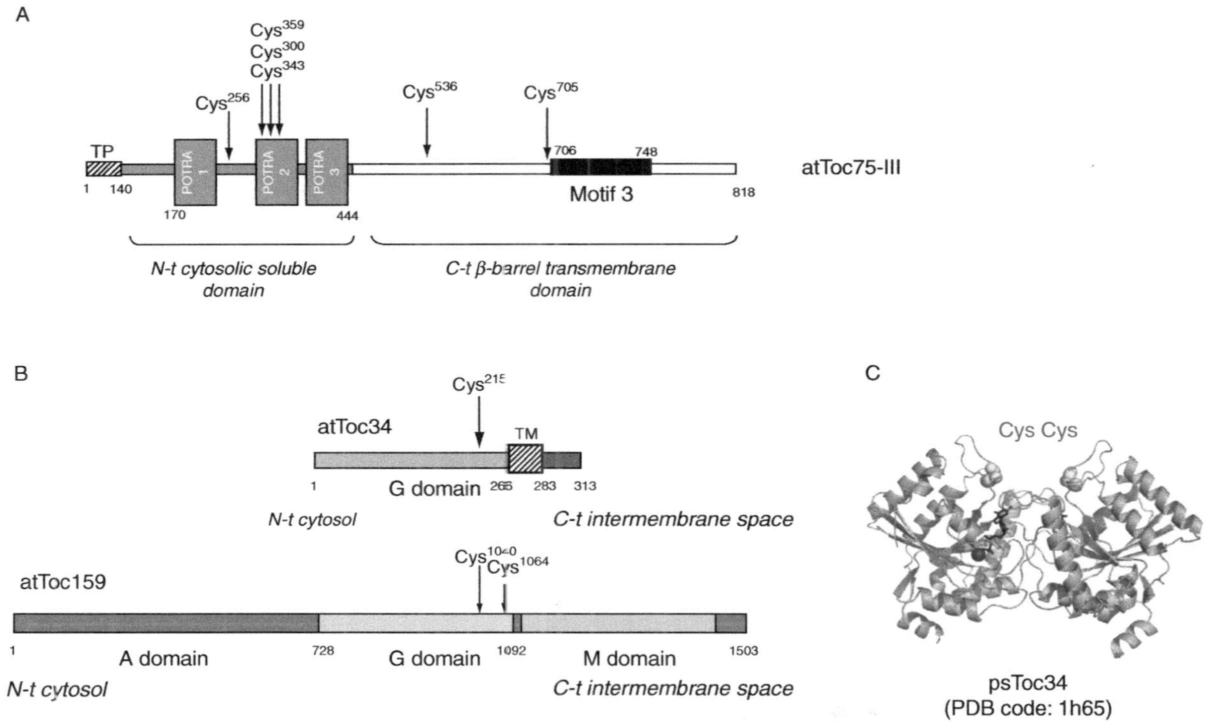

Fig. 2. The TOC core complex. (A) Toc75 has two structural and functional domains. The N-terminal cytosolic soluble domain is composed of three POTRA subdomains. In the C-terminal transmembrane β-barrel domain, the position of highly conserved motif 3 is shown. This motif resides inside the channel and, together with the POTRA domains, is proposed to be an active element in transport.

experiments with recombinant Toc75 in planar lipid bilayer membranes allowed *in vitro* characterization of the channel (Hinnah *et al.*, 2002) and demonstrated that the N-terminal region has the capability to modulate channel properties (Bredemeier *et al.*, 2007; Ertel *et al.*, 2005).

Recently, high-resolution structures of the N-terminal POTRA domain and the C-terminal β-barrel pore for Toc75 bacterial homologs have been obtained by X-ray crystallography (Clantin *et al.*, 2007; Gatzeva-Topalova *et al.*, 2008; Kim *et al.*, 2007), providing a general insight into the structural rearrangement and mechanism of transport. These studies, together with solution structures of the POTRA domain determined by small X-ray scattering and NMR (Knowles *et al.*, 2008), demonstrated the conformational flexibility of the POTRA domains that is essential for substrate binding. Functional analyses demonstrated that the POTRA domains differ in functional importance, consistent with the variation in the number of POTRA domains among species (Kim *et al.*, 2007). Toc75-III contains three POTRA domains (Fig. 2A), each with a $\beta1$–$\alpha1$–$\alpha2$–$\beta2$–$\beta3$ topology (Kim *et al.*, 2007). The structure of the membrane-embedded C-terminal β-barrel domain revealed the presence of 16 β-strand modules in FhaC from *Bordetella pertussi*, a bacterial Toc75 homolog (Clantin *et al.*, 2007). Based on this structure, homology models have been built for POTRA2 and 3 as well as the β-barrel for pea Toc75 (Reddick *et al.*, 2008). The crystallographic structure also contributed to understanding the so-called motif 3, a highly conserved region within the β-barrel domain identified previously in a protein sequence analysis of the Toc75 protein family (Moslavac *et al.*, 2005). This motif is made up of a long loop, termed L6, and the first half of strand B12 (Clantin *et al.*, 2007). In the conformational state with which the structure was solved, motif 3 is inside the transmembrane β-barrel (Clantin *et al.*, 2007). The authors proposed that motif 3 is an active element in transport and that conformational changes would expel L6 from the inside, opening the channel for translocation. Interestingly, a number of conserved Cys residues are uniquely found in the second POTRA domain and at the

The conserved cysteine residues in Toc75 from land plants, not present in green algae, are marked. The first 140 residues represent the transit peptide (TP); (B) Toc34 and Toc159 modular organization. The sequence numbers correspond to the sequences in *Arabidopsis*. Toc34 and Toc159 have a common GTPase domain, which accommodates the conserved cysteine in Toc34 implicated in interaction with other TOC subunits. This cysteine residue is conserved in all other isoforms of the TOC GTPase family and is present in all species (data not shown). The Toc159 subfamily contains an additional conserved cysteine in the G domain (no. 1064); (C) High resolution structure of the G-domain of pea Toc34 in its GDP-bound state. GDP and Mg^{2+} are represented as sticks and a ball, respectively. The conserved cysteine is marked in spacefill. The modular organization shown is not scaled.

beginning of motif 3 in the plant sequences (Fig. 2A). The functional or structural importance of these residues is unknown. However, according to the homology model, they are not sufficiently close to participate in an intramolecular interaction, but could interact with other TOC subunits in regulating protein import.

B. THE Toc159 AND Toc34 GTPase RECEPTORS: A MULTIGENE FAMILY

Toc159 and Toc34 operate as receptors in the outer membrane and provide the import process with specificity and efficiency (Agne and Kessler, 2009; Kessler and Schnell, 2002). Sequence analysis revealed that the TOC GTPases, Toc159 and Toc34, are evolutionarily related and similarly organized (Hiltbrunner *et al.*, 2001a; Hofmann and Theg, 2003; Reumann *et al.*, 2005). Toc34 is anchored to the membrane by a short transmembrane helix at the C-terminal tail and exposes the GTP-binding domain (G domain) to the cytosol (Fig. 2B). Toc159 is characterized by a large N-terminal acidic domain (A domain), followed by the Toc34-related GTP-binding domain (G domain) and a membrane-protected C-terminal domain (M domain) (Fig. 2B), exposing the A and G domains to the cytosol (Becker *et al.*, 2004b; Chen *et al.*, 2000; Hiltbrunner *et al.*, 2001a; Hirsch *et al.*, 1994; Kessler *et al.*, 1994; Schnell *et al.*, 1994; Seedorf *et al.*, 1995). Studies on truncated versions of Toc159 have demonstrated that the A domain is dispensable for activity, and its role in protein import remains evasive (Agne *et al.*, 2009; Chen *et al.*, 2000). On the other hand, even though it has been shown that precursor proteins interact directly with the M domain of Toc159 (Kouranov and Schnell, 1997), its function, structure, and membrane topology remain unknown. Secondary structure predictions indicate that the M domain is composed mainly of β-sheets (data not shown), while topological prediction programs (TBBpred and TMBpro (Natt *et al.*, 2004; Randall *et al.*, 2008)) indicate the M-domain is likely forming a transmembrane β-barrel module.

High-resolution structures of the soluble G-domain of pea and *Arabidopsis* Toc34 in the GDP- or GTP-bound form have contributed useful information not only about the functional mechanism and recognition of peptide-binding pocket, but also about the quaternary structure and dimerization mode of the protein (Koenig *et al.*, 2008a; Sun *et al.*, 2002) (Fig. 2C). Independent experiments have demonstrated the ability of Toc34 and Toc159 to undergo homo- and heterodimerization via their GTPase domains (Bauer *et al.*, 2002; Becker *et al.*, 2004b; Hiltbrunner *et al.*, 2001a; Koenig *et al.*, 2008b; Smith *et al.*, 2002; Weibel *et al.*, 2003; Yeh *et al.*, 2007). Recently, an interaction between Toc34 and Toc159 demonstrated *in vivo* in *Arabidopsis* (Rahim *et al.*, 2009) prompted the proposal that interchangeable homodimeric and heterodimeric states may serve as a regulatory mechanism. The two subunits of the dimer may function as GTPase-activating proteins that reciprocally stimulate or coordinate the hydrolysis of GTP. Apropos this point, the transit peptide likely plays a prominent role in stimulating the GTPase activity of the receptors and influencing their interaction (Becker *et al.*, 2004b).

Even though extensive work has been done on TOC GTPase receptors, the actual sequence of events in preprotein recognition and transfer of the precursor to the channel is still under debate. Two models are being considered. Both agree with GTP-mediated regulation of the Toc34/Toc159 receptors and the function of the central channel Toc75, but they differ in the mode of preprotein recognition. The first model (targeting model) favors Toc159 as the initial preprotein receptor (Bauer *et al.*, 2002; Hiltbrunner *et al.*, 2001b; Smith *et al.*, 2004). Binding of the preprotein to Toc159 promotes oligomerization of both GTPases, Toc34 and Toc159, in their GTP-bound state. The interaction with Toc75 would stimulate their GTPase activity, resulting in transfer of the cargo protein to the channel. A second model (motor model) proposes Toc34 to be the initial preprotein receptor and Toc159 to function as a GTP-driven motor that moves preproteins through the channel (Becker *et al.*, 2004b; Schleiff *et al.*, 2003; Sveshnikova *et al.*, 2000b). The mechanistic models proposed to date undoubtedly need further improvement. Structural analysis of the functional complex at high resolution is essential to understanding the mechanism of preprotein recognition and translocation involving the GTPases receptors and central channel.

Plant Toc159 and Toc34 are encoded by a multigene family (Bauer *et al.*, 2000; Gutensohn *et al.*, 2000; Hiltbrunner *et al.*, 2001a; Jackson-Constan and Keegstra, 2001; Jarvis *et al.*, 1998; Kubis *et al.*, 2004). Only one protein is known for each Toc159 and Toc34 component in pea (psToc159 and psToc34, respectively), where most of the biochemical studies have been performed. However, other isoforms should not be ignored until the genome sequence is completed.

Several studies have demonstrated functional specialization among the Toc159 and Toc34 isoforms. Further, genetic analysis of knockout mutants of Toc159 homologs in *Arabidopsis* (atToc159, atToc132, atToc120, and atToc90) indicated functional redundancy between Toc132 and Toc120 and minimal functional overlap among Toc132/Toc120, Toc159, and Toc90 (Bauer *et al.*, 2000; Ivanova *et al.*, 2004; Kubis *et al.*, 2004). Characterization of the expression and accumulation of photosynthetic proteins in *Arabidopsis* mutants led to the conclusion that Toc159 is a receptor with specificity for highly abundant, photosynthetic proteins and that Toc132 and Toc120 are preferentially involved in the import of nonphotosynthetic proteins, especially in nongreen tissues (Bauer *et al.*, 2000; Kubis *et al.*, 2004). Toc90, the shortest member of the Toc159 family that lacks the yet uncharacterized A domain, may support the accumulation of photosynthetic proteins in plastids jointly with Toc159 (Hiltbrunner *et al.*, 2004). Further evidence for the existence of an import pathway with preference for photosynthetic precursors was gained by detailed studies on Toc34 mutant plants. In *Arabidopsis*, the Toc34 family is composed of atToc34 and atToc33, isoforms with different biochemical properties and tissue distribution (Jarvis *et al.*, 1998; Jelic *et al.*, 2003; Kubis *et al.*, 2003). atToc33 is mainly expressed in photosynthetic tissues and can be specifically phosphorylated, whereas atToc34 is the dominant isoform in roots and is not modified by phosphorylation. These differences are consistent with different modes of regulation—a proposal supported by genetic studies with *Arabidopsis* revealing that these two isoforms may function as receptors for different subsets of substrates. Thus, the *toc33* knockout mutant is deficient in expression, import, and accumulation of photosynthetic proteins. Specific and preferential association between the different Toc34 and Toc159 isoforms meets functional needs: atToc34 with atToc132 and atToc120, and atToc159 with atToc33 (Bauer *et al.*, 2000; Gutensohn *et al.*, 2000; Jarvis *et al.*, 1998; Jelic *et al.*, 2003).

C. ACCESSORY TOC COMPONENTS: Toc64, Toc12, AND ImHsp70

A third preprotein receptor, termed Toc64, has been implicated in the protein import process in chloroplasts (Qbadou *et al.*, 2007; Sohrt and Soll, 2000). Toc64 has been described as having bimodular architecture and dual functionality, acting at both sides of the outer envelope membrane (Fig. 3A). Topological studies indicate that Toc64 is anchored to the membrane by three transmembrane helices in an N_{in}–C_{out} topology (Lee *et al.*, 2004b; Qbadou *et al.*, 2007). The C-terminal domain, composed of three tetratricopeptide repeats (TPR), is exposed to the cytosol and is involved in the recognition and interaction with cytosolic Hsp90 chaperones preloaded with precursor proteins (Mirus *et al.*, 2009; Qbadou *et al.*, 2006, 2007). The N-terminal amidase-like domain, oriented to the intermembrane space, seems to be involved in recruiting chaperones via interaction with a cochaperone, Toc12, and likely participates in coordination events at the intermembrane space side after preprotein translocation through Toc75 (Qbadou *et al.*, 2007). Functional analyses demonstrated that Toc64 lacks the capability to act as an amidase. Moreover, it seems like its amidase conformation has been dramatically disrupted by hydrophobic helices traversing the outer

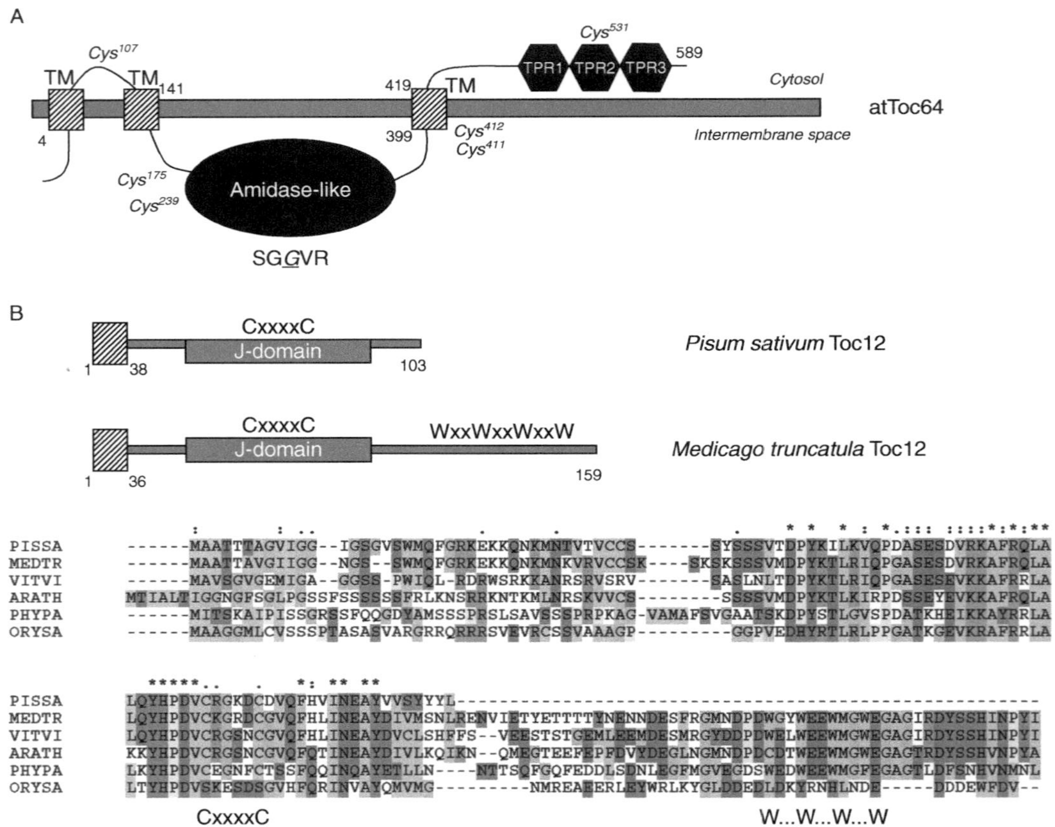

Fig. 3. Accessory components of the TIC complex. (A) Toc64 has been described as bi-functional protein with three transmembrane regions (TM). An amidase-like domain is predicted to be located in the intermembrane space. The protein is, however, not functional as an amidase due to a point mutation in the catalytic center (depicted in the figure). This region is likely involved in the interaction with partners in

envelope membrane (Qbadou *et al.*, 2007; Sohrt and Soll, 2000) (Fig. 3A). Although perhaps essential for high-efficiency protein import, genetic studies have assigned a secondary role to Toc64 because it may not be required for protein import per se (Aronsson *et al.*, 2007; Hofmann and Theg, 2005b). Furthermore, a detailed analysis of the *Arabidopsis* genome indicates the presence of a second Toc64 homolog in chloroplasts (J. Soll, unpublished data). Alternatively, it might have a very specific function for the import of just a subset of substrates, as demonstrated for its homolog in mitochondria (Lister *et al.*, 2007).

As the most recently described member of the TOC complex, little is known about Toc12 (Becker *et al.*, 2004a). It is anchored to the outer envelope membrane by a short motif at its N-terminus. The rest of the protein is exposed to the intermembrane space and shows sequence similarity to J-domains (Fig. 3B). In fact, biochemical studies have identified an interaction with Hsp70-like proteins in the intermembrane space (Becker *et al.*, 2004a). The exact nature and properties of the imsHsp70 that interacts with Toc12 in chloroplasts remain to be defined (Becker *et al.*, 2004a; Marshall *et al.*, 1990; Qbadou *et al.*, 2007; Ratnayake *et al.*, 2008; Schnell *et al.*, 1994; Waegemann and Soll, 1991).

A three-dimensional model of pea Toc12 has been built from an NMR structure and a disulfide bridge involving the Cys of its characteristic CXGXXC motif has been proposed (Becker *et al.*, 2004a) (Fig. 3B). The disulfide bridge may be essential for its cochaperone activity because the mutation of the N-terminal Cys into serine dramatically affects Toc12 functionality in pea (Becker *et al.*, 2004a). Taking into account that reversible disulfide bridge formation is a common mechanism to alter chloroplast enzyme activity, the authors proposed that the bridge maybe important for the regulation of the translocation event. Although there is no direct evidence

---

the intermembrane space such as Toc12 and Tic22. Three TPR motifs in the cytosol account for its function as a receptor for a subset of substrates. Toc64 has been implicated in an intermolecular interaction with other TOC components via disulfide bridges. Conserved cysteines are indicated. According to the present topological model, two cysteines are oriented to the cytosol, one of each is situated in the second TPR motif. Two cysteines (Cys411 and Cys412) formed part of the third transmembrane helix. Two other cysteines are located in the amidase-like domain in the intermembrane space; (B) Modular organization and multiple sequence alignment of several Toc12 proteins. The sequence from pea is shorter than in other family members and lacks a C-terminal domain characterized by tryptophan-rich region (except in rice). The indicated cysteines, localized in the J-domain, form part of the intramolecular disulfide bridge in Toc12. Abbreviations used: ARATH, *Arabidopsis thaliana*; MEDTR, *Medicago truncatula*; ORYSA, *Oryza sativa*; PHYPA, *Physcomitrella patens*; PISSA, *Pisum sativum*; VITVI, *Vitis vinifera*. The modular organization shown is not scaled.

for this to date, a potential disulfide bridge constitutes an interesting statement on the basis of redox properties of the chloroplast intermembrane space. In pea, Toc12 is a protein of 103 amino acids, of which the first 38 residues may function as noncleavage targeting signal to chloroplasts (as reported in ARAMEMNON database; http://aramemnon.botanik.uni-koeln.de/) as well as membrane anchor (Becker *et al.*, 2004a). Homologs of Toc12 are found in *Arabidopsis*, *Zea*, *Medicago*, and *Physcomitrella*, albeit longer in sequence (160 residues). A multiple sequence alignment shows that pea lacks the last 52–54 amino acid residues found in other members of the family (Fig. 3B). Actually the *toc12* gene is composed of three exons, the third of which corresponds to the missing component in the Toc12 pea sequence (Fig. 3B). This structure suggests that either pea Toc12 sequence is incomplete or that the pea protein, indeed, lacks this domain possibly due to alternative splicing. Interestingly, a highly conserved tryptophan-rich motif (WxxWxxWxxW) is found in this domain. The functional significance of the absence of the C-terminal extension in pea Toc12 remains to be explored. The protein sequence from rice constitutes an exception as the Cys residues in the CXGXXC motif are replaced by serine and lacks the conserved tryptophans (Fig. 3B). Additional studies are necessary to analyze the role of the C-terminal region absent in pea and not conserved in rice as well as the significance of the Cys mutation in rice and possibly other cereals.

D. Tic22, A PUTATIVE LINKER IN THE INTERMEMBRANE SPACE

Little is known about the structure and function of Tic22—a soluble protein in the intermembrane space, peripherally associated with the outer face of the inner membrane (Fig. 1A) (Kouranov *et al.*, 1998). It has been found closely associated to Tic20, Tic110, and, as mentioned earlier, in a complex with a number of TOC components (Becker *et al.*, 2004a; Qbadou *et al.*, 2007). These findings have led to the proposal that Tic22 may participate as a molecular linker in the interaction between TOC and TIC during protein import. Phylogenetic analyses revealed that Tic22 is a protein of cyanobacterial origin. Its homolog in *Synechocystis* (slr0924) has been experimentally located in the thylakoid lumen and, to a minor extent, in the periplasmic space (Fulda *et al.*, 2002). Tic22 is an essential protein in cyanobacteria and, interestingly, its content seems to be regulated by the redox state of the cell. Two putative roles were ascribed to slr0924 protein in *Synechocystis*, one in the transport of protein subunits and the other linked (directly or indirectly) to electron transfer (Fulda *et al.*, 2002). The relevance of these ideas awaits further experiments.

Two Tic22 isoforms are recognized in the genome of *Arabidopsis* (Fig. 4A): Tic22-IV (At4g33350) has two conserved Cys in two different motifs (Gly-X-X-Cys-Phe and Cys-Pro) and represents the ortholog of Tic22 in pea. The other isoform, Tic22-III (At3g23710), has one conserved Cys. Secondary structure predictions indicate that Tic22 is a mixed α/β protein (Fig. 4A). According to PHYRE, a protein structure prediction program (Kelley and Sternberg, 2009), templates for structural modeling with highest likelihood to be correct (45% estimated precision) are related to a number of proteins that contain a Trx-like fold and belong to the Trx superfamily. Two groups of proteins are recognized: one group is constituted of proteins that have evolved to bind metals to act as chaperones (metal-binding motif: Cys-Pro-X-X-Cys-Pro; PDB codes: 2cx4, 2ggt, 2b7k); the others are members of the DsbD-DsbC family (catalytic motif: Cys-X-X-Cys; PDB code: 1eej). The results from PHYRE are in accord with those obtained by the BIOINFO server (Ginalski *et al.*, 2003). Although the presence of redox-active Cys cannot be disregarded as both Tic22 isoforms have conserved Cys, their function as a Trx is not expected due to the absence of the canonical Trx catalytic motif, Cys-X-X-Cys. Apropos this point, it is noted that recent phylogenetic analyses have identified a number of Trx- and glutaredoxin-related sequences containing a single Cys at the putative active site (Fig. 4A) and the conserved metal-binding motif is not present. Moreover, due to the absence of Cys in the cyanobacterial homolog of Tic22, a relation with bacterial DsbD-DsbC is not appropriate (Fig. 4A). A putative function could be envisaged as a chaperone that involves a noncovalent interaction with substrates—like either protein disulfide isomerase-related ERp29 in the endoplasmic reticulum that contains an inactive Trx fold and assists in the maturation and transport of many secretory proteins (Barak *et al.*, 2009), or the Q8ZP25_SALTY and HYAE_ECOLI proteins that interact with the Tat signal peptide and function as molecular chaperones (Parish *et al.*, 2008). Structural and functional characterizations of Tic22 would give useful insight into the function of this putative chaperone in the chloroplast intermembrane space.

E. THE PROTEIN-CONDUCTING CHANNEL OF THE INNER ENVELOPE: Tic110 AND Tic20

Tic110 is an essential and specific protein in chloroplast-containing organisms (Inaba *et al.*, 2005; Schnell *et al.*, 1994; Wu *et al.*, 1994). It is found as a single copy in all plants, except for *Physcomitrella patens* in which a second isoform has been detected by genome sequence analysis (Kalanon and McFadden, 2008). There is extensive evidence that Tic110 forms the protein

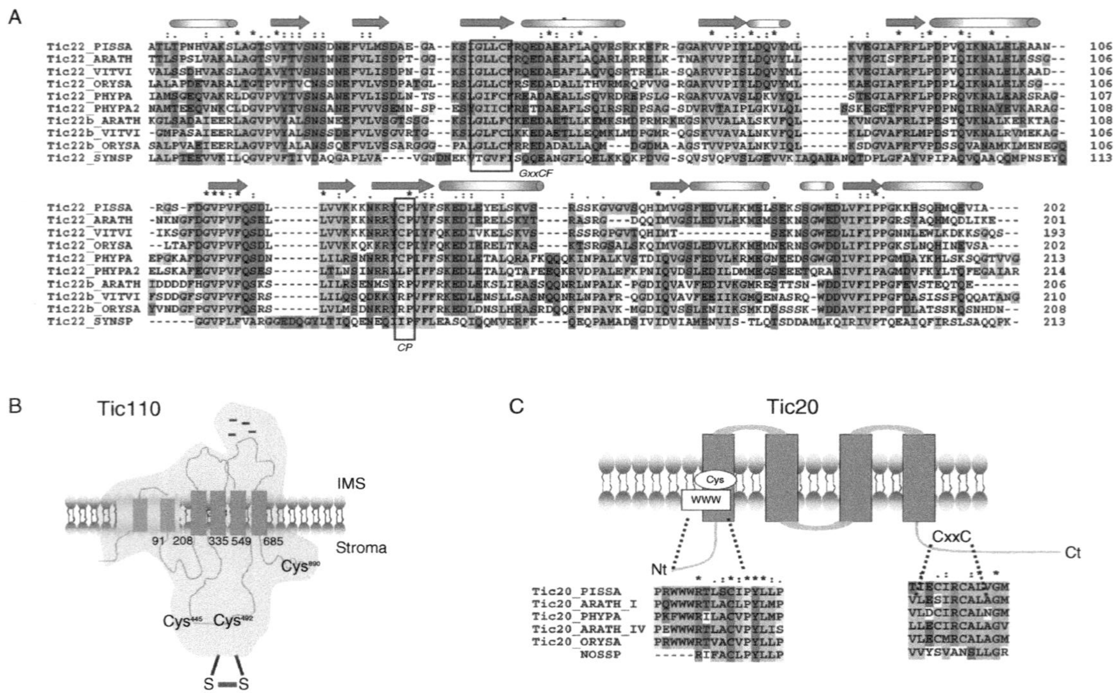

Fig. 4. Tic22, Tic110 and Tic20 components of the TIC machinery. (A) Multiple sequence alignment of selected Tic22 proteins. In the alignment, a cyanobacterial homolog (SYNSP, slr0924 from *Synechocystis*) has been included for comparison. Two subgroups of plant Tic22, are identified as Tic22 and Tic22b. All of the proteins, except the one from bacteria, have a conserved cysteine in a motif depicted as GxxCF. A second conserved cysteine is, however, found only in the first group of Tic22 sequences (CP motif). Predicted secondary structure

translocation channel at the inner envelope membrane of chloroplasts (Fig. 1A), although other specialized import pathways appear to be present, for example, Tic20 (Kouranov *et al.*, 1998).

Tic110 comprises two hydrophobic membrane-spanning α-helices at the extreme N-terminus, both necessary for anchoring into the inner envelope (Lubeck *et al.*, 1996). A new topological model of Tic110 incorporates four additional transmembrane helices, with amphipathic features, within the region of residues 92–959 in the pea sequence (Balsera *et al.*, 2009a) (Fig. 4B). This new model for Tic110 reconciles the results of numerous studies during the past several years. The model contains regions in the intermembrane space suitable to form supercomplexes with the TOC machinery and to receive the precursor as well as a large region in the stroma that could interact with molecular chaperones (Inaba *et al.*, 2003, 2005). However, other studies suggest that Tic110 forms a large globular domain on the stromal side, without channel activity, to recruit chaperones (Inaba *et al.*, 2003; Jackson *et al.*, 1998). Electrophysiological characterization of reconstituted planar lipid bilayer membranes of native Tic110 as well as recombinant mutants of the protein lacking the two hydrophobic transmembrane helices at the N-terminus favor the former topological model (Balsera *et al.*, 2009a; Heins *et al.*, 2002) (Fig. 4B). Experiments have shown that Tic110 has properties of a cation-selective channel that is sensitive to chloroplast transit peptides. These studies have assigned a prominent role to the amphipathic transmembrane helices in the channel structure of Tic110. Interestingly, Ca^{2+} seems to act as an effector of gating and selectivity in the Tic110 channel.

Tic110 is composed mainly of α-helices (Balsera *et al.*, 2009a; Inaba *et al.*, 2003). Computational analyses have detected the presence of two helical unit

elements are shown above the alignment as cylinders (α helices) and arrows (β-strand); (B) Tic110 topological model in which two hydrophobic helices at the N-terminus and four amphipathic helices create a channel at the inner envelope (represented by rectangles within the membrane). Regions in the intermembrane space (IMS) are likely involved in contacts with TOC components. A highly disordered negatively charged motif is a feature of Tic110 sequences in this region. At the stromal side, domains of Tic110 dock with the motor complex. Note that the lines connecting putative transmembrane regions do not represent secondary structural features. Tic110 is predicted to have a high content of α-helices and to form a reversible disulfide bridge regulated by thioredoxin. Although the participating cysteines are not known, several conserved cysteines shown in the figure are candidates for formation of such a disulfide bridge in land plants; (C) Tic20 topological predictions show that four hydrophobic α-helices traverse the envelope membrane. Three conserved cysteines are represented in the topological model. Multiple sequence alignment around these residues is shown for a number of Tic20 proteins. A cyanobacterial homolog is shown for comparison (NOSSP; all4804 from *Nostoc sp. PCC 7120*). A list of names of the organisms is shown in Figure 3. The modular organization shown is not scaled.

repetitions in the stroma that resemble those of the so-called HEAT motifs (Andrade *et al.*, 2001; Balsera *et al.*, 2009a). Interestingly, a large number of the HEAT repeats are involved in cytoplasmic transport, where the two helical units appear to function as flexible joints that can wrap around target substrates and act as scaffolding on which other molecular components assemble (Andrade and Bork, 1995; Andrade *et al.*, 2001). A highly disordered, negatively charged region may be important for protein interaction, likely in the intermembrane space (Balsera *et al.*, 2009a). In addition, a number of fully conserved Cys are spread over the sequence and, according to the proposed topological model (Balsera *et al.*, 2009a), they are likely localized in the stromal region of the protein (Fig. 4B). Biochemical experiments uncovered a redox-active disulfide bridge potentially functional in regulating Tic110 via stromal Trx. Owing to its presence in the reduced state in chloroplasts isolated from darkened plants, Tic110 was assigned a role in allowing the channel to adapt to environmental change (Balsera *et al.*, 2009a). Thus, when channel activity is disturbed by oxidation, Trx could restore the system to normal activity. The validity of this interpretation awaits further experiments.

Tic20 (Kouranov *et al.*, 1998) and Tic21 (Teng *et al.*, 2006) are two other candidates proposed to constitute translocation channels at the inner envelope membrane of chloroplasts. Participation of Tic20 in the import of preproteins has been assessed by crosslinking experiments in which the protein was found to form part of an active import supercomplex in the chloroplast envelope together with other TOC and TIC components (Kouranov and Schnell, 1997; Kouranov *et al.*, 1998). Although Tic20 knockout plants are not viable, *in vivo* analyses have been carried out in transgenic *Arabidopsis* plants in which levels of expression of the protein were altered (Chen *et al.*, 2002). It was concluded that, while the import of certain nuclear-encoded plastid proteins was impaired, others were not affected. More recently, analyses of a conditional mutant generated in *Toxoplasma gondii* showed that Tic20 is essential for protein import albeit likely not as the general import pore, but as an accessory or regulatory component of the import complex (van Dooren *et al.*, 2008). Topological studies have not yet been addressed, but four α-helical transmembrane regions are predicted by hydrophobicity analyses (Fig. 4C). GFP fusion at the Tic20 C-terminus in *Toxoplasma gondii* demonstrated that this region is oriented toward the stroma, suggesting an N_{in}–C_{in} topology (van Dooren *et al.*, 2008).

Databases reveal two isoforms of Tic20 in *A. thaliana*: Tic20-I (At1g04940) is the ortholog of psTic20 and Tic20-IV (At4g03320). Two other *Arabidopsis* proteins distantly related to Tic20 (At2g47840 and At5g55710) represent the closest orthologs to cyanobacterial counterparts (Kalanon and McFadden,

2008). A chloroplast location is predicted for all four proteins by the ARAMMENON database (http://aramemnon.botanik.uni-koeln.de/). Given that the function of these isoforms is not known, further characterizations of the cyanobacterial counterparts may help understanding the role of Tic20 in plants (Gross and Bhattacharya, 2009). Tic20 may form a channel independent of Tic110.

Analysis of protein sequences of Tic20-I family members reveals a highly conserved C-terminal CXXC motif (except in *Arabidopsis* where it is replaced by an SXXC motif) and a highly conserved CXP motif close to a WWW motif within the first putative transmembrane helix. Assuming four transmembrane helices in Tic20 and topology similar to *Toxoplasma gondii* (van Dooren *et al.*, 2008), the CXXC motif would face the chloroplast stroma (Fig. 4C). However, whether these Cys act in a regulatory capacity remains to be experimentally verified.

The other inner membrane component proposed to function as channel for protein import, Tic21 (At5g15290), has four predicted transmembrane helical segments (Teng *et al.*, 2006). However, its affiliation to the TIC complex is a matter of debate as the protein has been identified as a metal permease (Duy *et al.*, 2007).

F. ENERGY DRIVING FORCE IN THE TIC COMPLEX: Tic40 AND Hsp93/ClpC

One of the functional aspects of Tic110 is its capability to recruit the stromal motor complex composed of the membrane-attached Tic40 cochaperone and the stromal Hsp93 (or ClpC) chaperone at the inner envelope membrane (Fig. 1A) (Nielsen *et al.*, 1997; Stahl *et al.*, 1999). Tic40 is anchored to the inner membrane by a single transmembrane helix at its N-terminus, exposing a large C-terminal domain to the stroma that contains a putative TPR domain followed by a Sti1 domain that is characteristic of cochaperones Hsp70-interacting protein (Hip) and Hsp70/Hsp90-organizing protein (Hop) (Bedard *et al.*, 2007; Chou *et al.*, 2003; Stahl *et al.*, 1999) (Fig. 5A). The TPR motif is a degenerate 34-amino acid sequence that structurally consists of two antiparallel α-helices often found in tandem repetitions in a protein sequence. This motif is found in certain cochaperones, especially for interaction with Hsp90 (Odunuga *et al.*, 2003). It has been demonstrated that the Tic40 Sti1 domain is functionally equivalent to the Hip/Hop family of cochaperones (Bedard *et al.*, 2007; Chou *et al.*, 2006). The mode of action currently envisaged involves the ternary interaction of Tic40 with Hsp93 and Tic110 during a late stage of translocation. A region that includes the putative TPR domain of Tic40 has been implicated in interaction with Tic110, while the Sti1 domain seems to stimulate the ATPase activity of the Hsp93 chaperone.

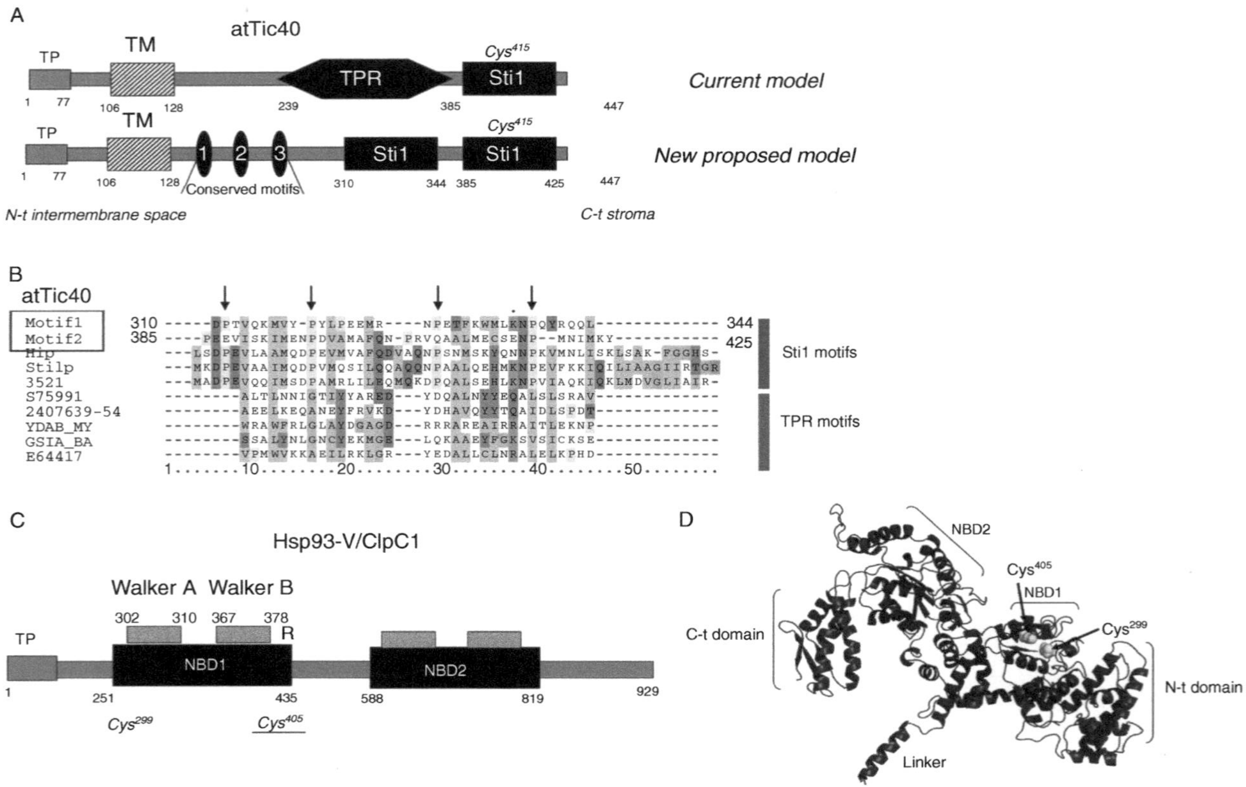

Fig. 5. The molecular motor of the TIC complex. (A) Schemes of the current and a new proposed Tic40 models are shown. Whereas the former model held that the C-terminal region of Tic40 is composed of TPR and Sti1 domains, new sequence analysis suggests a possible repetition of the Sti1 motifs as shown in (B). The new model shows the three conserved motifs recognized in Tic40 family of sequences

Analyses of mutant plants suggest that Tic40 might not essential for protein import, but is important for its efficiency (Chou *et al.*, 2003). Refinements of the current model are awaited with interest (Jarvis, 2008).

An analysis of its amino acid sequence with the SMART search engine (Letunic *et al.*, 2006) revealed that Tic40 likely consists of two Sti1 modules (Karpenahalli *et al.*, 2007). Surprisingly, the TPR motifs were not recognized. A further search with the TPRpred and REP programs (Andrade *et al.*, 2000) yielded negative results, suggesting that, if TPR motifs are present in Tic40, they are largely degenerated. In fact, the protein sequence region where putative TPR motifs were identified in Tic40 is composed of strategically situated proline residues that fit better with Sti1 motifs than with TPR motifs (Fig. 5B). Based on this finding, an alternative module organization is proposed here in which two Sti1 motifs are arranged in tandem at the C-terminus of Tic40 (Fig. 5A). Considering that Sti1 modules are also composed of α-helices, the secondary structure of Tic40 would fit with the second alternative model presented here. Structural analyses are, however, necessary to confirm a given model. Bedard and collaborators pointed out that the Tic40 family of proteins contains only one conserved Cys within their Sti1 domain (Bedard *et al.*, 2007). This observation may explain how Tic40 covalently interacts with Tic110 in the presence of copper chloride (II)—that is, by forming a crosslink between this Cys and a counterpart in Tic110 (Stahl *et al.*, 1999). The Tic40 Cys residue is, however, in the region proposed to interact with Hsp93, not Tic110. This point should be further studied with other interaction partners under controlled redox conditions.

Hsp93 (or ClpC) belongs to class I of the Hsp100/Clp superfamily—molecular chaperones that use ATP to promote changes in the folding of proteins and belong to the functionally heterogeneous AAA+ family of

(Bedard *et al.*, 2007). The location of the conserved cysteine residue is shown. TP and TM stand for transit peptide and transmembrane regions, respectively; (B) Motif1 (regions 310–344) and motif2 (regions 385–425) in *Arabidopsis* Tic40 are compared with sequences that contain either Sti1 or TPR motifs (a complete list can be found at http://smart.embl-heidelberg.de/smart/do_annotation.pl?DOMAIN=SM00727 and http://coot.embl.de/~andrade/rep/get_info.pl?TPR). The arrows mark conserved proline residues within the Sti1 motifs. These two motifs (motif1 and motif2) are shown in (A) as Sti1 boxes; (C) Hsp93 structural organization. The two non-identical nucleotide binding domains are shown in the figure with the Walker A and Walker B motifs indicated in grey boxes. The two conserved cysteines in the plant protein are indicated. Cys405 (underlined in the figure) is also conserved in the cyanobacterial homologs (data not shown); (D) Homology model of Hsp93 based on ClpB as template (PDB code: 1QVR). The different functional domains are identified and the conserved cysteines residues of the NBD1 motif are represented in spacefill. Note that the central linker region cannot be modelled due to sequence differences between Hsp93 and ClpB, where there is a long coiled-coil structure that is shorten in Hsp93. Cys299 is not present in cyanobacterial Hsp93. The modular organization shown is not scaled.

proteins (ATPases associated with various cellular activities). They can function as independent molecular chaperones or as regulatory components of the Clp protease. The Hsp93 population associated with the inner envelope in chloroplasts is recognized as a component of the TIC machinery that acts as the protein motor in the final stages of import (Akita *et al.*, 1997; Kovacheva *et al.*, 2005, 2007; Nielsen *et al.*, 1997). At the protein sequence level, Hsp93 in chloroplasts is very similar to the counterparts in cyanobacteria (data not shown), and no obvious motif can be identified at first sight as a determinant for a specific function in protein import transport. *Arabidopsis* has two isoforms of the protein: Hsp93-V or ClpC1 (At5g50920) and Hsp93-III or ClpC2 (At3g48870) (Constan *et al.*, 2004; Kovacheva *et al.*, 2007). Dimers of ClpC1, but not ClpC2, were detected in the *Arabidopsis* stroma by proteomics (Peltier *et al.*, 2004). On the other hand, in maize mesophyll chloroplasts, a hexameric ClpC was observed (Majeran *et al.*, 2008) in accord with the expected functional assembly (Andersson *et al.*, 2006).

Recently, Hsp93 was identified as a potential Trx target in cyanobacteria, although the physiological significance of this observation is not known (Mata-Cabana *et al.*, 2007). Moreover, the expression of *clpC* seems to respond to redox state in *Synechocystis* (Hihara *et al.*, 2003). Interestingly, knockout mutants plants of chloroplast NADPH-Trx reductase (NTRC), a protein involved in the response to environmental stress, showed increased levels of transcripts for Hsp93-V in short-day *ntrc* rosette leaves (Lepistö *et al.*, 2009).

The Hsp93 sequence has two ATP-binding domains (NBD1 and NBD2), each containing two conserved motifs (known as Walker A and Walker B) that participate in the binding of nucleotide and the Mg^{2+} cation (Schirmer *et al.*, 1996). NBD1 and NBD2 are linked by a central spacer that provides flexibility to the molecule. Further, the so-called arginine finger (R-finger) senses ATP binding and hydrolysis and transmits conformational changes essential for Hsp93 activity (Fig. 5C). Cyanobacterial Hsp93 contains a single conserved Cys, suggesting that a regulatory intermolecular disulfide bridge links activity to Trx. The conserved Cys residue is located at the beginning of the Walker A motif in the first nucleotide-binding domain (Fig. 5C) and thus may have a direct effect on ATPase activity. In plants, the situation is different: a second conserved Cys is present in the NBD1 domain (Fig. 5C). According to a theoretical structure built by homology modeling using a high-resolution structure of ClpB as template (PDB code: 1QVR; 49% identity), these two Cys residues (Cys299 and Cys405 in Hsp93-V from *Arabidopsis*) are sufficiently close to form an intramolecular disulfide bridge (Fig. 5D). If so, the disulfide bridge could either be structural or regulate direct ATPase activity. The applicability of either of these views awaits future experiments.

G. REGULATORY COMPONENTS OF THE TIC COMPLEX: Tic55, Tic62, AND Tic32

Three modular proteins, Tic55, Tic62, and Tic32, have been classified as members of a redox regulatory network at the TIC machinery (Caliebe *et al.*, 1997; Hormann *et al.*, 2004; Kuchler *et al.*, 2002). Tic55 is a multi-domain integral membrane protein that consists of: (i) a Rieske domain that binds a [2Fe-2S] cluster via the motif Cys-X-His-X16-Cys-X2-His; (ii) a catalytic domain that contains a mononuclear nonheme mononuclear Fe(II) ion active site, likely coordinated by two histidine residues and one carboxylate ligand from an aspartate residue, a type of structural motif known as "2-His-1-carboxylate facial triad"; (iii) a hydrophilic CXXC domain of unknown function; (iv) a C-terminal transmembrane domain predicted to contain two transmembrane hydrophobic helices, TM (Caliebe *et al.*, 1997) (Fig. 6A). *In vitro* protein import experiments in chloroplasts showed that the translocation of precursor proteins into chloroplasts was diminished by diethylpyrocarbonate, possibly via modification of the Rieske iron–sulfur cluster. However, the role for Tic55 in protein import is not known. It is of cyanobacterial origin evolutionarily and is related to the LLS1(*Lethal-leaf spot 1*)-family of oxygenases in plants (Gray *et al.*, 2004).

This type of Rieske nonheme protein, which is widespread among electron transfer proteins, participates in widely divergent enzyme reactions. Whether Tic55 participates in an electron transfer reaction or as a redox sensor element that controls TIC activity, or both, remains to be investigated (Caliebe *et al.*, 1997; Hidalgo *et al.*, 1997; JagerVottero *et al.*, 1997). There is evidence that the oxidation state of the Rieske center acts as a regulatory element based on findings with similar proteins (Martins *et al.*, 2005). Interestingly, Tic55 has recently been identified as a potential Trx target (Bartsch *et al.*, 2008). While this identification needs to be supported by biochemical or genetic studies, it is in accord with the presence of a number of conserved Cys not involved in iron binding in the chloroplast protein (Fig. 6A). Of particular interest is a fully conserved CxxC motif located in a domain of Tic55, LLS1, and PTC52 and not found in other Rieske proteins (Gray *et al.*, 2004). Relevant possibilities also come from initial homology modeling of the Rieske domain which pointed to a thiol pair (Cys 176 and 193 in the pea Tic55 sequence, Fig. 6A) as possible participants in disulfide bridge formation and, therefore, in interacting with Trx. These Cys residues are located in the so-called proline loop characteristic of Rieske proteins that connects the cluster-binding domain with the basal domain (Colbert *et al.*, 2000). How the Rieske cluster regulates the active site, whose regulatory redox motif is recognized by

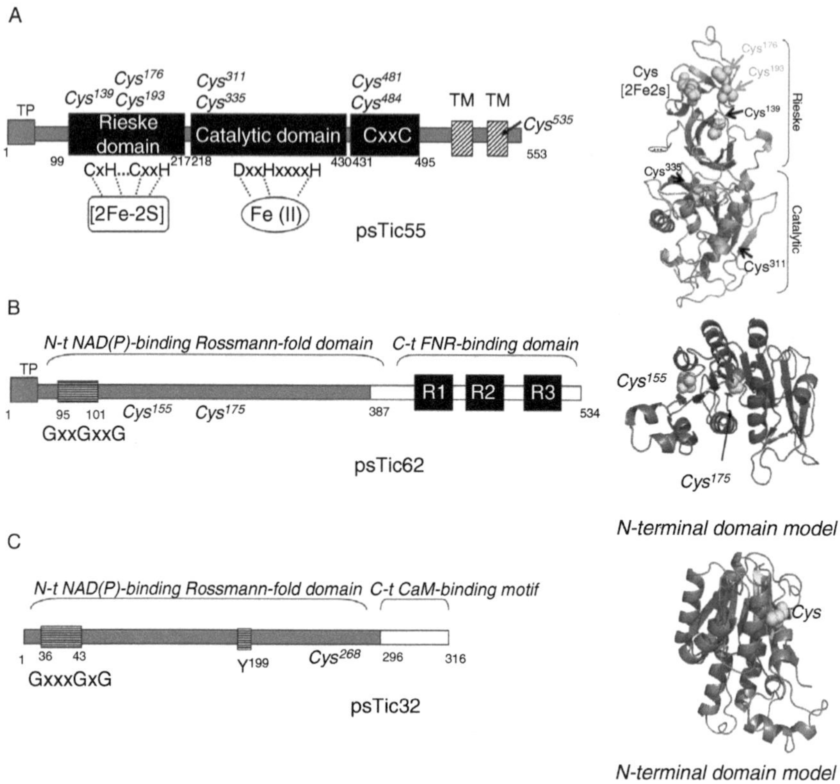

Fig. 6. Redox sensor proteins in the TIC complex. (A) Tic55 structural and functional module organization. TP and TM stand for transit peptide and transmembrane regions, respectively. Residues implicated in coordinating iron are indicated below the functional modules. The Tic55 sequence is composed of which Cys144 and Cys163 (not shown) with His146 and His166 participate in the coordination of the Rieske center. Two cysteines (Cys481-Cys484), which may form a disulfide bridge, are part of a soluble domain, the so-called CxxC domain, not found in other Rieske nonheme Fe proteins. The other cysteines are shown in the scheme. Tic55 has cyanobacterial homologs that present a similar modular organization (data not shown). Cys535 (in the second TM helix), Cys311 and Cys335 are the only conserved cysteine residues in plant Tic55 proteins. On the right, a homology model (using the protein structure with PDB code 1Z01 as template) is shown for the Rieske and catalytic domains. A circle with dots indicates where the protein sequence continues (CxxC domain and TM regions). Cysteines are represented in spacefill; (B) Tic62 bimodular organization and homology model prediction (Balsera *et al.*, 2007) of the N-terminal NAD(P)-binding domain. Tic62 sequence feature is a fully conserved GxxGxxG motif, involved in nucleotide coordination, and belongs to the SDR superfamily. The N-terminal catalytic domain is predicted to have a Rossmann-fold. Three repetitions (R1, R2 and R3) in pea Tic62, enriched in proline and serine residues, are involved in interaction with FNR in the stroma. Tic62 proteins have two conserved cysteines, shown in spacefill in the model on the right. Cys175 is in the protein core where the catalytic residues cluster and may participate in electron transfer reactions but likely not in a disulfide bridge formation; (C) Tic32 belongs

Trx and elucidation of the role of Tic55 in regulating the complete TIC machinery are very interesting questions to pursue in the future.

Tic62 and Tic32 activities are linked to NADPH (Hormann *et al.*, 2004; Kuchler *et al.*, 2002). They belong to the extended and classical families of the SDR superfamily, respectively, and are predicted to contain a Rossmann-type fold. Their reductase activity has been demonstrated experimentally by *in vitro* assays using NBT as substrate, but endogenous substrates are still unknown. To date, no direct functional link or interaction has been detected between them, and they may possibly function independently. Both are extrinsically attached to the inner envelope membrane at the stromal side and show similar modes of dynamic binding to the TIC complex: more oxidizing conditions (or lower NADPH/NADP ratio) favor their association with TIC, whereas more reducing conditions (or higher NADPH/NADP ratio) dissociate the subunits from the complex (Chigri *et al.*, 2006; Stengel *et al.*, 2008). Despite their similarities and common cyanobacterial origin, they are not evolutionarily related and interact with different partners: Tic62 interacts with FNR, and Tic32 binds to calmodulin (CaM)—a primary transducer of intracellular Ca^{2+} signals in eukaryotes (Chigri *et al.*, 2006; Kuchler *et al.*, 2002).

Tic62 is characterized by a bimodular structure: the N-terminal domain contains the NADPH-binding motif, while the C-terminal domain is predicted to be highly disordered in solution and to contain several repetitive proline-rich modules (Kuchler *et al.*, 2002) (Fig. 6B). These repetitive modules have been shown to interact with FNR. Evolutionarily, only vascular plants possess the "full-length Tic62" protein, that is, they contain the C-terminal, FNR-binding domain that is missing in other plants like *Physcomitrella patens* (Balsera *et al.*, 2007). This leads to the question of what defines Tic62 as a TIC component: is it reductase capacity or FNR-binding capacity? As noted earlier, the redox environment, reflected by the stromal NADPH/NADP ratio and FNR availability, strongly influences Tic62 functionality and its binding to the TIC complex (Stengel *et al.*, 2008). However, not only these factors affect the reversible binding of Tic62 to the TIC complex: its distribution between the soluble stromal compartment and the inner envelope membrane: more reducing conditions

also to the SDR superfamily, specifically to the tyrosine-dependent reductase family (tyrosine residue shown in the figure), and contains a fully conserved GxxxGxG motif. Tic32 contains a C-terminal extension involved in CaM interaction. A homology model has been obtained for the N-terminal reductase domain (using the protein structure with PDB code 2C07 as template). Just one conserved cysteine is found in Tic32 family of sequences. (See color plate section at the back of the book.)

released Tic62 from the membrane to the stroma, thereby possibly increasing its affinity for FNR. The significance of this stromal effect is not known.

Tic32 consists of a catalytic (NADP-binding) N-terminal domain and a C-terminal CaM-binding extension (Fig. 6C) (Chigri *et al.*, 2006). It is not known whether CaM regulates Tic32 oxidoreductase activity or provokes a conformational rearrangement that perturbs interaction with its partners. The second option seems plausible because coimmunoprecipitation experiments showed that NADPH abolished the interaction of Tic32 with the TIC machinery. It was assumed, therefore, that CaM would favor binding to the translocon complex. Protein sequence analyses showed the presence of a number of Tic32 homologs in plants (Kalanon and McFadden, 2008). However, because no functional data are available for these homologs, a relation with the TIC complex cannot be ensured. Further, due to lack of a transit peptide (Nada and Soll, 2004), a chloroplast location cannot be predicted for the Tic32 homologs. Thus, much remains to be done to understand the role of the redox-sensitive Tic62 and Tic32 molecules that are switches closely associated with the TIC complex. Interestingly, both Tic62 and Tic32 contain conserved Cys in their Rossmann-fold domain (Fig. 6B and C). Based on homology modeling, they seem unfit for an intramolecular interaction (Fig. 6B and C). It is of interest to learn whether they participate in intermolecular binding.

V. REGULATION OF CHLOROPLAST PROTEIN IMPORT BY METABOLIC AND ENVIRONMENTAL REDOX STATE

A. THE OUTER ENVELOPE, THE FIRST MEMBRANE BARRIER IN PROTEIN IMPORT

Biochemical characterization of protein import in chloroplasts with *in vitro* systems has shown that chloroplast protein translocation is affected by Cys-modifying reagents such as *N*-ethyl-maleimide (NEM) and dithiothreitol (DTT). Experiments with the precursor of the small subunit of RuBisCO (pSSu) as a model protein for import showed that the affinity for the receptor-binding sites was dramatically reduced after NEM treatment, while the number of binding sites was not affected (Friedman and Keegstra, 1989). These results were later confirmed not just for pSSu, but also for the precursor of ferredoxin-NADP-oxidoreductase (pFNR) (Row and Gray, 2001). Significantly, mutating the Cys residues of the transit peptides did not affect import. These findings thus let us to the view that at least one component of the chloroplast protein import machinery participating in the formation of early import intermediates contains one or more functionally important Cys residues.

Subsequent work showed that import yield was stimulated after chloroplasts were treated with DTT or reduced glutathione (Pilon *et al.*, 1992; Stengel *et al.*, 2009). In another set of experiments, the oxidizing agent copper chloride (II) was shown to inhibit the formation of an early import intermediate and subsequent import of pSSu (Seedorf and Soll, 1995). Inhibition was relieved by DTT. In this case, the effect was ascribed to the formation of a crosslinked complex consisting of Toc159, Toc34, and Toc75. The opposite effects of DTT and $CuCl_2$ treatments point to the presence of a reversible disulfide bridge at the level of outer envelope membrane in land plant chloroplasts. By means of diagonal 2D redox SDS-PAGE, different redox treatments on outer envelope vesicles isolated from pea chloroplasts confirmed the formation of intermolecular disulfide bridges among the TOC components. It was concluded that the redox state of Cys in the TOC complex affects the efficiency of protein import (Stengel *et al.*, 2009). The longstanding question of whether DTT can influence the chloroplast protein import machinery has thus at long last been answered.

The only subunit of the import machinery in chloroplasts whose structure is known to date at high resolution is Toc34 (Koenig *et al.*, 2008b; Sun *et al.*, 2002). X-ray crystallography structural studies revealed that the Toc34 monomer is divided into a globular GTP-binding domain (G domain) and a small C-terminal α-helical part (Figs. 2C and 7). The core is defined by six β-sheets flanked by six α- and two 3_{10}-helices. A feature of the Toc34 structure is a long loop connecting $\beta 6$ and $\alpha 6$ secondary elements. Many conserved residues in the Toc34 sequence are localized in this loop, and a role in the nucleotide exchange process is proposed for this region (Sun *et al.*, 2002). As noted earlier, this cytosolic loop contains the fully conserved Cys residue in the Toc34 family that was shown to interact with other TOC components in the presence of copper chloride using pea as a model plant (Seedorf and Soll, 1995; Sun *et al.*, 2002). Interestingly, the long loop and the conserved Cys residue are also predicted in the Toc159 family (Figs. 2C and 7).

In view of these findings, determination of the structure of the outer envelope channel protein, Toc75, becomes even more timely. A high-resolution structure of a bacterial homolog of Toc75 (Gatzeva-Topalova *et al.*, 2008) highlights the need to analyze the conformation of the complete N-terminal POTRA-containing domain and its organization with respect to the pore domain and the membrane. As seen earlier, a number of conserved Cys residues are found in the second POTRA domain—in the cytosol and at

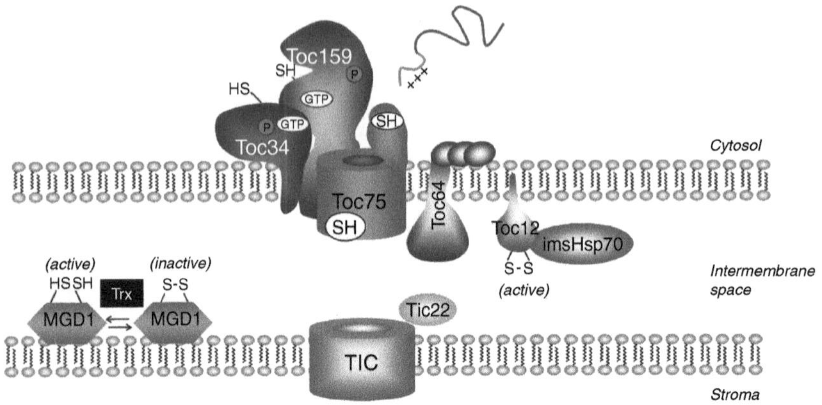

Fig. 7. Regulatory components in the outer membrane barrier and intermembrane space of chloroplasts. Different modes of regulation have been identified in the TOC complex: (i) modulation of the GTPase activities of the receptors; (ii) phosphorylation/dephosphorylation of the receptors; (iii) characteristic protein composition, according to the isoform of the multigene family present; (iv) intermolecular disulfide bridge formation involving cysteines of the TOC components. In the intermembrane space, oxidative folding machinery may exist for proteins that require the formation of disulfide bridges, like Toc12 (see text). Proteins, such as MGD1, contain disulfide bridges sensitive to thioredoxin, although no thioredoxin or thioredoxin-like protein has been described in this compartment. (See color plate section at the back of the book.)

the beginning of motif 3 within the channel (Fig. 2A). As noted earlier, these residues may participate in the interaction with Toc34.

Finally, the Toc64 receptor has also been implicated in covalent interactions with other TOC components using copper chloride (II) as a specific Cys crosslinker (Sohrt and Soll, 2000). Of the six Cys conserved in land plants (Fig. 3A), only two are predicted to be located to the cytosolic side (Qbadou *et al.*, 2007), one in the loop between the first two putative transmembrane helices and the second in the second TPR domain, the latter also conserved in the mitochondrial isoform (Kalanon and McFadden, 2008). These residues may be responsible for interacting with other TOC components. Mutation studies would be useful to assess the role that these Cys play at the outer envelope interface.

B. BETWEEN TWO BARRIERS, THE INTERMEMBRANE SPACE

It has been known for some time that at least two chloroplast precursor proteins, nonphotosynthetic ferredoxin (FdIII) and FNR isoform II (FNRII), are differentially imported into chloroplasts under light and dark conditions (Hirohashi *et al.*, 2001). High light resulted in mistargeting and accumulation of these preproteins in the intermembrane space, whereas the import of the precursor forms of photosynthetic counterparts (FdI and FNRI) was not affected. This effect of protein arrest in the intermembrane space under high light was ascribed to a folding event, likely due to the sensitivity of the folding machinery to redox state in the intermembrane space (Becker *et al.*, 2004a).

Furthermore, two residents of the intermembrane space, Toc12 and MGD1, potentially contain disulfide bridges (Fig. 7). The redox state of the Cys residues in MGD1 has been linked to function, showing that the enzyme is active only in its reduced form. Further, Trx has been reported to modulate the activity of MGD1, resulting in the proposal that redox regulation may be essential to coordinate galactolipid biosynthesis and chloroplast development (Benning and Ohta, 2005; Yamaryo *et al.*, 2006). By contrast, the oxidized form of Toc12 is believed to be the form active in coordinating the chaperone system of the intermembrane space with protein import (Becker *et al.*, 2004a). However, the presence of a disulfide bridge in Toc12 has yet to be confirmed. The intermembrane studies raise a number of interesting questions. First, how does MGD1 sense the redox state of the chloroplast? One could think of the presence of Trxs, and a proteomic study has indeed detected a form of m-type Trx associated with the *Arabidopsis* envelope membrane. It is likely, however, that it is present at the stromal face of the inner envelope membrane (Ferro *et al.*, 2003). Even so, one could envisage a mechanism similar to the one reported for thylakoids for which a transmembrane reducing equivalent transfer system has been proposed to link the stroma to the lumen (Motohashi and Hisabori, 2006). There is, however, no evidence for such a system in the envelope. Finally, machineries are known to catalyze the oxidation of proteins in the intermembrane space of mitochondria and the periplasmic space of bacteria (Herrmann *et al.*, 2009). Here again, although the presence of semiquinone radicals, flavoproteins, and iron–sulfur proteins has been experimentally detected at the inner envelope membrane of chloroplasts and an electron transfer chain employing these carriers might be present (JagerVottero *et al.*, 1997), no equivalent systems have been detected so far in this region.

At this point it would be necessary to identify if a disulfide bridge is present in Toc12, to get additional insight into the regulatory disulfide bridge in MGD1 and look for associated elements in the intermembrane space. Further experiments are also needed to understand how light affects the transport of a subset of proteins resulting in their folding and arrest in the intermembrane space. It would be interesting to know whether Cys are implicated and the identity of neighboring proteins that participate in this process.

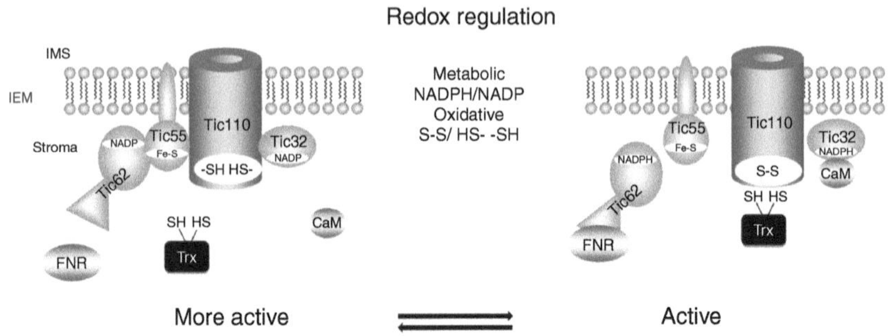

Fig. 8. Environmental and metabolic regulation of the TIC machinery. Different components of the TIC machinery seem to sense the chloroplast redox environment. On one side, the redox sensors Tic62 and Tic32 show a dynamic and reversible interaction with TIC core components and other interaction partners (FNR and CaM, respectively) affected by the NADPH/NADP ratio. Independent experiments have shown that the import efficiency for a subset of substrates is, indeed, altered by this ratio, indicating that the composition of TIC adapts to the redox metabolic state of the chloroplast. On the other side, disulfide bridges on Tic110 (and Tic55) are potential targets of thioredoxin regulation in the stroma, potentially linking function to environmental adaptations.

C. CHLOROPLAST REDOX STATE AND THE INNER ENVELOPE BARRIER

It is now accepted that the activity of the TIC complex is likely linked to both metabolic and oxidative regulation (Fig. 8). Recently, three TIC components have been functionally linked to stromal thioreodoxins. The finding that Tic110, the putative translocation channel, contains a regulatory disulfide bridge linked to specific stromal Trxs has opened a new line of investigation in the regulation of protein import. It has been proposed that Tic110 may respond to environmental factors via a change in redox and that Trx then reactivates Tic110 by reduction (Balsera *et al.*, 2009a). Future experiments will determine whether a particular Trx acts *in vivo* in this capacity as well as the functional consequences of the effect of change in redox state. Another TIC subunit, the redox sensor Tic55, has been also classified as a Trx target based on its interaction with a mutant Trx protein (Bartsch *et al.*, 2008). Although sequence analysis demonstrated the presence of well-conserved Cys in strategic positions (Fig. 6A), further experiments are needed to confirm the presence of regulatory disulfide bridge(s) in Tic55 and, eventually, its functional significance.

Evidence for a final component of the TIC machinery as a putative Trx target, Hsp93, was obtained with a cyanobacterial homolog (Mata-Cabana *et al.*, 2007). Although the TIC machinery is not present in cyanobacteria, work is justified to determine whether plant and cyanobacterial Hsp93 chaperone functions are comparable. As mentioned earlier, a proteomic study has identified a Trx isoform closely associated with the inner envelope membrane, raising the possibility of a functional link between these groups of proteins (Ferro *et al.*, 2003).

On the other hand, a dynamic interaction between Tic62 and Tic32 and the TIC machinery is dependent on the ratio of another redox pair, NADPH/NADP (Chigri *et al.*, 2006; Stengel *et al.*, 2008). This ratio appears to play a central role due to their functionality both as redox carriers and signaling components (Fig. 8). Although variations in the NADPH/NADP ratio in chloroplasts are not extensive due to the photosynthetic machinery operating during the day and the oxidative pentose phosphate pathway at night, the equilibrium may be significantly perturbed during metabolism, particularly under stress conditions. It thus seems feasible that by a dynamic association, the Tic62 and Tic32 subunits could influence the import properties of, at least, a subset of proteins by altering the composition of the TIC machinery in response to changing conditions of the chloroplast. Indeed, recent results showed that the protein import efficiency of a subset of protein substrates is affected after altering the NADPH/NADP ratio in isolated chloroplasts by the addition of different metabolites—a decrease in the NADPH/NADP ratio enhanced protein import (Stengel *et al.*, 2009).

The metabolic factors, FNR and CaM, also affect Tic62 and Tic32 binding, respectively, as mentioned earlier (Chigri *et al.*, 2006; Kuchler *et al.*, 2002). Again, the molecular mechanisms of these interactions remain largely unknown. Thus, it remains to be seen whether FNR transmits redox information to Tic62 when thylakoid bound and part of the photosynthetic machinery or when dissociated and soluble in the stroma (Forti and Bracale, 1984). Alternatively, a signal could be generated when the enzyme undergoes binding and induces a conformational change in Tic62.

On the other hand, the functional interpretation of the Tic32–CaM interaction is also elusive. CaM responds to Ca^{2+} signals and transmits information to a receptor protein whose activity is then affected (Zielinski, 1998). It is known that light/dark conditions influence the level of Ca^{2+} in the chloroplast (Sai and Johnson, 2002) and that the import of a subset of proteins that contain a cleavable transit peptide is inhibited either by calmodulin inhibitors or calcium ionophores. Further, the addition of external calmodulin or calcium restores import activity (Chigri *et al.*, 2005) and calcium acts as an effector in control of gating and selectivity of the Tic110 protein channel (Balsera *et al.*, 2009a). It becomes interesting to know if Ca^{2+}, CaM, Tic32, and Tic110 are functionally connected.

A role for redox in controlling the composition of the TIC machinery has been strengthened by a recent proteomic study on envelope membranes of C3 (pea) chloroplasts and C4 (maize) mesophyll chloroplasts (Brautigam *et al.*, 2008). Whereas the channel-forming proteins Tic110 and Toc75 showed a relatively high spectral abundance in both samples, the study showed that the relative abundance of Tic55, Hsp93, and FNR is markedly decreased in C4 mesophyll envelopes. On the other hand, Tic32 and Tic62 were not identified in the envelope of C4 mesophyll chloroplasts. The authors proposed that, due to a spatial separation of reduction equivalents produced in mesophyll and bundle sheath chloroplasts, C3 and C4 mesophyll chloroplasts have different modes of redox-dependent import. Independent proteomic experiments with total chloroplast membranes from maize mesophyll and bundle sheath cells revealed high levels of Tic110, Tic40 (and Tic21) in mesophyll chloroplasts under conditions attributed to increased protein influx (Majeran and van Wijk, 2009; Majeran *et al.*, 2008). In sharp contrast to the Brautigam study (Brautigam *et al.*, 2008), subunits of the translocation machinery other than Hsp93 were not detected specifically in mesophyll cells. At this point, no further conclusions can be drawn regarding the distribution of the redox regulatory subunits.

VI. FURTHER POSSIBLE REDOX TARGETS IN CHLOROPLAST PROTEIN IMPORT

A. THE DRIVEN MOTOR FORCE FOR THE SEC MACHINERY

The cpSec system in thylakoids consists of a membrane-embedded protein-conducting channel, formed by the SecE and SecY subunits, and the peripherally associated ATPase motor SecA (Laidler *et al.*, 1995; Nakai *et al.*, 1994; Schuenemann *et al.*, 1999a; Yuan *et al.*, 1994) (Fig. 1B). Upon complex formation with the precursor protein in the stroma, SecA associates with the thylakoidal SecE/Y channel. Cycles of ATP hydrolysis together with proton motive force energy provide the driving force for the complete translocation of the preprotein through the channel (Economou and Wickner, 1994; Karamanou *et al.*, 1999; Nakai *et al.*, 1994).Out of the two SecA-related sequences in *Arabidopsis* (At4g01800 and At1g21650), the former represents the true homolog to cyanobacterial SecA and its localization in chloroplasts has been confirmed by proteomics (Kleffmann *et al.*, 2004).

Little is known specifically about SecA in chloroplasts but its functional equivalence and high sequence similarity to bacterial homologs allows direct comparisons with many functional and structural studies performed on the bacterial system (Yuan *et al.*, 1994). Under physiological conditions, SecA forms homodimers in solution. However, the oligomerization state of the protein during translocation is a matter of debate since certain groups have proposed that the dimer dissociates during translocation. This view is not shared by others, and it has been proposed that a dynamic equilibrium between different oligomerization states exists in SecA that depends on environmental conditions, lipids, or even the presence of transit peptides (Benach *et al.*, 2003; Driessen, 1993; Duong, 2003; Or and Rapoport, 2007; Osborne *et al.*, 2004; Shin *et al.*, 2006; Woodbury *et al.*, 2002).

High-resolution structures have been obtained by X-ray crystallography for several bacterial SecA proteins in different nucleotide-bound states (Hunt *et al.*, 2002; Osborne *et al.*, 2004; Papanikolau *et al.*, 2007; Sharma *et al.*, 2003; Vassylyev *et al.*, 2006; Zimmer *et al.*, 2008). SecA is composed of two domains: an N-terminal motor domain and a C-terminal translocation domain. The motor domain shares similarities with the DEAD motor domains of ATPase helicases. With this type of protein, the binding of ATP and release of ADP generate a force that is translated into large protein conformational changes and movement of the protein.

Based on studies with SecA homologs, several structural elements can be identified in the chloroplast SecA protein from *Arabidopsis* (Fig. 9A). Two nucleotide-interacting regions, NBD1 and NBD2 (or IRA2), have been identified in the motor domain (Kull *et al.*, 1996) as forming a mononucleotide cleft (Sharma *et al.*, 2003). Whereas NBD1 contains the minimal ATPase catalytic machinery, the highly flexible NBD2 (or IRA2) subdomain acts as intramolecular receptor for ATPase hydrolysis (Sianidis *et al.*, 2001). The motor region also contains a family-specific variable region (VAR) that shows significant length and sequence variability in the different SecA proteins. Cyanobacterial and plant SecA present one of the longest VAR regions in the SecA superfamily (Papanikolau *et al.*, 2007) that, interestingly, is the most divergent region between cyanobacterial and plant SecA. Although a function has not been identified, this region may be involved in SecA oligomerization (Papanikolau *et al.*, 2007). Another region identified in the motor domain of chloroplast SecA is the preprotein-binding domain (PBD). On the other hand, the C-terminal translocation domain functions as a regulatory domain that physically associates with NBD2 (or IRA2) and restricts its activator function (Sianidis *et al.*, 2001).

Recently, a proteomic study carried out with membrane preparations from a cyanobacterium has identified SecA as a potential Trx target (Mata-Cabana *et al.*, 2007). The high sequence similarity between the bacterial SecA homologs and *Synechocystis* and *Arabidopsis* SecA proteins (42–49% identity) prompted us to apply homology modeling to build a three-dimensional structure for cpSecA to obtain insight into the conformational organization of the conserved Cys (Fig. 9B). Independent of the

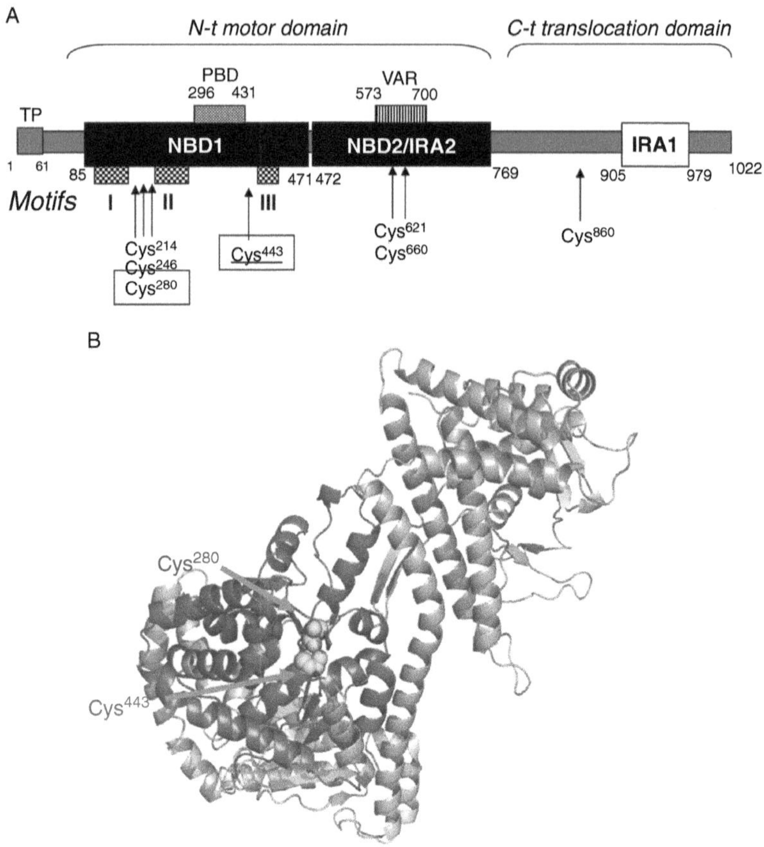

Fig. 9. The SecA import motor of the cpSec machinery. (A) Simplified scheme showing SecA structural elements. The sequence motifs are defined by comparison with the *E. coli* SecA homolog (Papinikolau *et al.*, 2007). SecA is composed of two domains: an N-terminal motor domain and a C-terminal translocation or modulator domain. The motor domain is composed of NBD1 and NBD2 that contains two nonequivalent nucleotide binding domains. NBD2 is also known as IRA2 (intramolecular regulator of ATP hydrolysis). Motif I and II, respectively, represent the Walker A and B motifs in NBD1; motif III denotes a conserved TGT triad in NBD1. The protein binding domain (PBD) forms part of the NBD1 region. The variable region (VAR) in the NBD2 characterizes each SecA subfamily. A number of conserved cysteines are found in the sequence of plant SecA proteins as depicted in the diagram. Based on a structural model (B) the residues Cys280 and Cys443 are candidates for an intramolecular disulfide bridge. Of these two residues, only Cys443 (underlined) is found in cyanobacterial SecA, constituting a potential thioredoxin target; (B) SecA homology model based on *E. coli* SecA as template (PDB code: 2FSF). The two conserved Cys280 and Cys443 residues are shown in spacefill. The modular organization shown is not scaled.

template used for the modeling (PDB codes: 2fsf, 1nkt, 2vda, 2ibm, 1tf5), two conserved Cys (Cys280 and Cys443) in the *Arabidopsis* model were found to be located in a favorable position for forming a disulfide bridge. Interestingly, the first of these Cys precedes the catalytic motif II or Walker B DEAD-motif, whereas the second is positioned closed to the so-called motif III (Hunt *et al.*, 2002; Sharma *et al.*, 2003) (Fig. 9A). It was found that motif III is intertwined with the catalytic motif II and IRA2. The formation of a putative disulfide bridge close to the catalytic site places it in strategic position for regulation. However, due to the lack of the first Cys, a similar disulfide bridge is not expected for the cyanobacterial SecA counterpart (see also the paper by Nishiyama and Hisabori in this volume concerning the distribution and specificity of Cys residues distribution in cyanobacterial enzymes) (Fig. 9A). Considering that SecA from *Synechocystis* was experimentally identified as a potential Trx target (Mata-Cabana *et al.*, 2007), further analyses are necessary to determine whether the cyanobacterial Cys targeted by Trx are linked in an intermolecular disulfide bond.

B. Transport and Folding of Redox-Active Proteins in the Thylakoidal Lumen

Since the turn of the current century, a number of groups have contributed to the elucidation of the composition, mechanisms, and metabolic activities of proteins in the thylakoidal lumen—a long-neglected compartment of thylakoids. To date, more than 70 different proteins have been detected in the lumen, all nuclear encoded, that function not only in the light reactions, but also in folding, proteolysis, and protection against oxidative stress (Peltier *et al.*, 2002; Schubert *et al.*, 2002). Recent observations provide evidence that a number of these proteins contain disulfide bridges and have the potential to participate in redox signaling (Gopalan *et al.*, 2004; Hall *et al.*, 2008; Marchand *et al.*, 2006). However, in contrast to their stromal counterparts that are usually active in the reduced state, lumenal proteins are active when oxidized (Buchanan and Luan, 2005).

Import behavior has been analyzed for two lumenal proteins that contain disulfide bridges essential for activity—the immunophilin FKBP13 and oxygen-evolving complex extrinsic subunit OE33 (or OEE1/PsbO) (Fig. 10A). Both proteins have been identified as Trx targets and are thus candidates for redox regulation (Gopalan *et al.*, 2004; Lee *et al.*, 2004a; Lemaire *et al.*, 2004; Marchand *et al.*, 2006). *In vitro* import experiments suggested that FKBP13 uses cpTat for import (Gupta *et al.*, 2002)—a pathway that employs folded substrates, thereby raising the question of whether FKBP13 is transported in the oxidized or reduced state. Both scenarios are plausible since biochemical

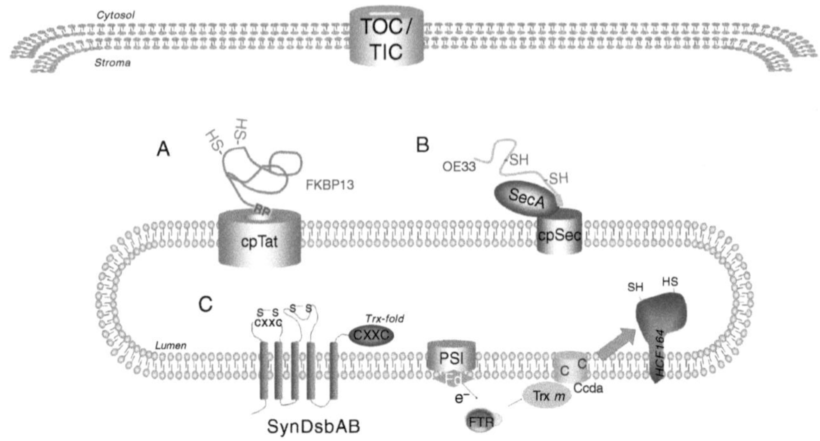

Fig. 10. Schematic view of transport of lumenal redox-active proteins through chloroplast thylakoids and enzymes possibly involved in redox regulation in the thylakoid lumen. Two lumenal redox-active proteins follow divergent pathways of translocation through the thylakoid membrane: (A) The immunophilin FKBP13 is transported folded via the cpTat pathway; (B) The extrinsic subunit of the oxygen evolving complex OE33 is kept unfolded during translocation by the cpSec machinery. Once in the lumen, both proteins require oxidation to achieve an active conformation; (C) Recent studies demonstrate the existence of a number of enzymes that may function to transport reducing equivalents from the stroma to the thylakoid lumen. HCF164, a protein bound to thylakoids, contains a lumenal thioredoxin-like domain, that has been proposed to accept reducing equivalents from stromal thioredoxin *m* via the polytopic membrane protein Ccda (cytochrome *c* defective A) (Motohashi and Hisabari, 2006). Further, SynDsbBA, a protein required for disulfide bond formation in the periplasmic space of bacteria, has been recently localized in thylakoids, adding a possible new participant in regulating thiol/disulfide formation of proteins within the lumen. (See color plate section at the back of the book.)

and structural studies revealed that the disulfide bridge formation is essential not for folding but for enzymatic activity. By contrast, OE33 follows a cpSec-dependent pathway of translocation (Cline *et al.*, 1992). OE33 contains disulfide bridges that are essential both for activity and incorporation as part of PSII structure (Tanaka and Wada, 1988). OE33 has thus to be kept in an unfolded, competent state in the stroma as well as when crossing the thylakoidal membrane. Once in the lumen, the protein is oxidized and folded (Fig. 10B).

The evolutionary and molecular basis for the acquisition of a transit peptide specific for the cpTat or cpSec import pathway is not clear. The situation is complicated by the requirement of particular proteins for transport in either the folded or unfolded state. The disulfide bond-containing aprotinin, a therapeutic protease inhibitor, is an interesting case in point.

By fusing the protein to transit peptides specific for each pathway, aprotinin has been successfully translocated into the lumen of tobacco plastid transformants by using either the cpSec or cpTat pathways (Tissot *et al.*, 2008). In each case, the protein was fully functional after translocation—an indication that disulfide bridges were correctly formed, independent of import pathway. At least for this example, the efficiency of transport was comparable and a functional structure was similarly acquired in the stroma and the lumen. A requirement for specific lumenal chaperones may be an important factor in defining the path of translocation. On the other hand, keeping in mind that SecA is a putative Trx target, selection of a particular pathway may be linked to regulation of import for a subset of substrates (Fig. 10). This is an attractive idea that needs in-depth experiments. First, SecA should be confirmed as a Trx target biochemically or genetically and its regulatory disulfide bridge(s) should be identified. These studies should include demonstration that Trx alters SecA activity and protein import. Different situations can be envisaged for an effect on import: the redox state of SecA could influence its oligomerization state (a subject highly debated in the literature), its interaction with the transit peptide or its catalytic activity, the most common mechanism of action of Trx in the stroma (Schürmann and Buchanan, 2008).

Little is known about how newly imported proteins are correctly folded for disulfide bridge formation once in the lumen. This could be either enzyme mediated or achieved spontaneously in an environment that is highly oxidizing, particularly in the light (Fig. 10C). The mystery is heightened by the apparent absence of Trx in a compartment that, so far, appears to house a number of Trx targets. Recent work, however, offers relevant possibilities.

HCF164, a protein residing in the thylakoid, that contains a soluble Trx-like domain exposed to the lumen, has been proposed to act as transmembrane transducer of reducing equivalents from photoreduced stromal Trx (Motohashi and Hisabori, 2006) (Fig. 10C). In a separate report, structural analysis prompted the suggestion that cytochrome *c6A*, a redox-active lumenal component, could act as catalyst for the oxidation of protein dithiols by molecular oxygen (Marcaida *et al.*, 2006). Finally, more recently, a protein required for disulfide bond formation in the extracytosolic space has been identified in the inner membrane of the cyanobacterium *Synechocystis,* SynDsbBA (Singh *et al.*, 2008) (Fig. 10C). The full-length protein is the result of the evolutionary fusion of two homologs from Gram-negative bacteria: (i) DsbA, a soluble periplasmic protein with a catalytic Cys-X-X-Cys motif that acts as a disulfide bond carrier and rapidly oxidizes Cys residues in protein substrates, and (ii) DsbB, a membrane-localized thiol oxidoreductase protein that transfers redox equivalents to DsbA. Genome analyses revealed that this fused system is conserved in all oxygenic photosynthetic organisms (Singh *et al.*, 2008). Further, its ortholog in *Arabidopsis* (At4g35760) seems to be localized in the thylakoidal membrane according to the plastid proteome database (http://ppdb.tc.cornell.edu). Future studies on the location, function, and substrate specificity of this fused protein will surely give new insight into redox regulation. Establishing the system in chloroplasts would give also support to the believed prokaryotic evolutionary origin of this compartment (Herrmann *et al.*, 2009). A tangential area awaiting exploration concerns the role of thiol/disulfide exchange in regulating the activity of lumenal proteins and the mechanism(s) by which these changes are achieved.

ACKNOWLEDGMENTS

B. B. B. gratefully acknowledges receipt of a Research Award from the Alexander von Humboldt Foundation. Work in the laboratory of J. S. was supported by Deutsche Forschungsgemeinschaft Grant SFB594.

REFERENCES

Agne, B. and Kessler, F. (2009). Protein transport in organelles: The Toc complex way of preprotein import. *FEBS Journal* **276**, 1156–1165.

Agne, B., Infanger, S., Wang, F., Hofstetter, V., Rahim, G., Martin, M., Lee, D. W., Hwang, I., Schnell, D. and Kessler, F. (2009). A Toc159 import receptor mutant, defective in hydrolysis of GTP, supports preprotein import into chloroplasts. *Journal of Biological Chemistry* **284**, 8661–8670.

Akita, M., Nielsen, E. and Keegstra, K. (1997). Identification of protein transport complexes in the chloroplastic envelope membranes via chemical cross-linking. *Journal of Cell Biology* **136**, 983–994.

Aldridge, C., Cain, P. and Robinson, C. (2009). Protein transport in organelles: Protein transport into and across the thylakoid membrane. *FEBS Journal* **276**, 1177–1186.

Andersson, F. I., Blakytny, R., Kirstein, J., Turgay, K., Bukau, B., Mogk, A. and Clarke, A. K. (2006). Cyanobacterial ClpC/HSP100 protein displays intrinsic chaperone activity. *Journal of Biological Chemistry* **281**, 5468–5475.

Andrade, M. A. and Bork, P. (1995). Heat repeats in the Huntingtons-disease protein. *Nature Genetics* **11**, 115–116.

Andrade, M. A., Ponting, C. P., Gibson, T. J. and Bork, P. (2000). Homology-based method for identification of protein repeats using statistical significance estimates. *Journal of Molecular Biology* **298**, 521–537.

Andrade, M. A., Petosa, C., O'Donoghue, S. I., Muller, C. W. and Bork, P. (2001). Comparison of ARM and HEAT protein repeats. *Journal of Molecular Biology* **309**, 1–18.

Aronsson, H., Combe, J., Patel, R. and Jarvis, P. (2006). *In vivo* assessment of the significance of phosphorylation of the *Arabidopsis* chloroplast protein import receptor, atToc33. *FEBS Letters* **580**, 649–655.

Aronsson, H., Boij, P., Patel, R., Wardle, A., Topel, M. and Jarvis, P. (2007). Toc64/OEP64 is not essential for the efficient import of proteins into chloroplasts in *Arabidopsis thaliana*. *Plant Journal* **52**, 53–68.

Bae, W., Lee, Y. J., Kim, D. H., Lee, J., Kim, S., Sohn, E. J. and Hwang, I. (2008). AKr2A-mediated import of chloroplast outer membrane proteins is essential for chloroplast biogenesis. *Nature Cell Biology* **10**, 220–227.

Balsera, M., Stengel, A., Soll, J. and Bolter, B. (2007). Tic62: A protein family from metabolism to protein translocation. *BMC Evolutionary Biology* **7**, 43.

Balsera, M., Goetze, T., Kovács-Bogdán, E., Schürmann, P., Wagner, R., Buchanan, B., Soll, J. and Bölter, B. (2009a). Characterization of TIC110, a channel-forming protein at the inner envelope membrane of chloroplasts, unveils a response to Ca^{2+} and a stromal regulatory disulfide bridge. *Journal of Biological Chemistry* **284**, 2603–2616.

Balsera, M., Soll, J. and Bölter, B. (2009b). Protein import machineries in endosymbiotic organelles. *Cellular and Molecular Life Sciences* **66**, 1903–1923.

Barak, N. N., Neumann, P., Sevvana, M., Schutkowski, M., Naumann, K., Malesevic, M., Reichardt, H., Fischer, G., Stubbs, M. T. and Ferrari, D. M. (2009). Crystal structure and functional analysis of the protein disulfide isomerase-related protein ERp29. *Journal of Molecular Biology* **385**, 1630–1642.

Barber, J. (2006). Photosystem II: An enzyme of global significance. *Biochemical Society Transactions* **34**, 619–631.

Bartsch, S., Monnet, J., Selbach, K., Quigley, F., Gray, J., von Wettstein, D., Reinbothe, S. and Reinbothe, C. (2008). Three thioredoxin targets in the inner envelope membrane of chloroplasts function in protein import and chlorophyll metabolism. *Proceedings of the National Academy of Sciences of the United States of America* **105**, 4933–4938.

Bauer, J., Chen, K. H., Hiltbunner, A., Wehrli, E., Eugster, M., Schnell, D. and Kessler, F. (2000). The major protein import receptor of plastids is essential for chloroplast biogenesis. *Nature* **403**, 203–207.

Bauer, J., Hiltbrunner, A., Weibel, P., Vidi, P. A., Alvarez-Huerta, M., Smith, M. D., Schnell, D. J. and Kessler, F. (2002). Essential role of the G-domain in targeting of the protein import receptor atToc159 to the chloroplast outer membrane. *Journal of Cell Biology* **159**, 845–854.

Becker, T., Hritz, J., Vogel, M., Caliebe, A., Bukau, B., Soll, J. and Schleiff, E. (2004a). Toc12, a novel subunit of the intermembrane space preprotein translocon of chloroplasts. *Molecular Biology of the Cell* **15**, 5130–5144.

Becker, T., Jelic, M., Vojta, A., Radunz, A., Soll, J. and Schleiff, E. (2004b). Preprotein recognition by the Toc complex. *EMBO Journal* **23**, 520–530.

Bedard, J., Kubis, S., Bimanadham, S. and Jarvis, P. (2007). Functional similarity between the chloroplast translocon component, Tic40, and the human co-chaperone, Hsp70-interacting protein (Hip). *Journal of Biological Chemistry* **282**, 21404–21414.

Benach, J., Chou, Y. T., Fak, J. J., Itkin, A., Nicolae, D. D., Smith, P. C., Wittrock, G., Floyd, D. L., Golsaz, C. M., Gierasch, L. M. and Hunt, J. F. (2003). Phospholipid-induced monomerization and signal-peptide-induced oligomerization of SecA. *Journal of Biological Chemistry* **278**, 3628–3638.

Benning, C. and Ohta, H. (2005). Three enzyme systems for galactoglycerolipid biosynthesis are coordinately regulated in plants. *Journal of Biological Chemistry* **280**, 2397–2400.

Benz, J. P., Soll, J. and Bolter, B. (2009). Protein transport in organelles: The composition, function and regulation of the Tic complex in chloroplast protein import. *FEBS Journal* **276**, 1166–1176.

Bolter, B., Soll, J., Schulz, A., Hinnah, S. and Wagner, R. (1998). Origin of a chloroplast protein importer. *Proceedings of the National Academy of Sciences of the United States of America* **95**, 15831–15836.

Bolter, B., Soll, J., Hill, K., Hemmler, R. and Wagner, R. (1999). A rectifying ATP-regulated solute channel in the chloroplastic outer envelope from pea. *EMBO Journal* **18**, 5505–5516.

Braun, N. A. and Theg, S. M. (2008). The chloroplast Tat pathway transports substrates in the dark. *Journal of Biological Chemistry* **283**, 8822–8828.

Braun, N. A., Davis, A. W. and Theg, S. M. (2007). The chloroplast Tat pathway utilizes the transmembrane electric potential as an energy source. *Biophysical Journal* **93**, 1993–1998.

Brautigam, A., Hofmann-Benning, S. and Weber, A. P. M. (2008). Comparative proteomics of chloroplast envelopes from C-3 and C-4 plants reveals specific adaptations of the plastid envelope to C-4 photosynthesis and candidate proteins required for maintaining C-4 metabolite fluxes. *Plant Physiology* **148**, 568–579.

Bredemeier, R., Schlegel, T., Ertel, F., Vojta, A., Borissenko, L., Bohnsack, M. T., Groll, M., Von Haeseler, A. and Schleiff, E. (2007). Functional and phylogenetic properties of the pore-forming beta-barrel transporters of the Omp85 family. *Journal of Biological Chemistry* **282**, 1882–1890.

Brink, S., Fischer, K., Klosgen, R. B. and Flugge, U. I. (1995). Sorting of nuclear-encoded chloroplast membrane-proteins to the envelope and the thylakoid membrane. *Journal of Biological Chemistry* **270**, 20808–20815.

Bruce, B. D. (2000). Chloroplast transit peptides: Structure, function and evolution. *Trends in Cell Biology* **10**, 440–447.

Buchanan, B. B. and Luan, S. (2005). Redox regulation in the chloroplast thylakoid lumen: A new frontier in photosynthesis research. *Journal of Experimental Botany* **56**, 1439–1447.

Caliebe, A., Grimm, R., Kaiser, G., Lubeck, J., Soll, J. and Heins, L. (1997). The chloroplastic protein import machinery contains a Rieske-type iron-sulfur cluster and a mononuclear iron-binding protein. *EMBO Journal* **16**, 7342–7350.

Carde, J. P., Joyard, J. and Douce, R. (1982). Electron-microscopic studies of envelope membranes from spinach plastids. *Biology of the Cell* **44**, 315–324.

Chen, K. H., Chen, X. J. and Schnell, D. J. (2000). Initial binding of preproteins involving the Toc159 receptor can be bypassed during protein import into chloroplasts. *Plant Physiology* **122**, 813–822.

Chen, X. J., Smith, M. D., Fitzpatrick, L. and Schnell, D. J. (2002). *In vivo* analysis of the role of atTic20 in protein import into chloroplasts. *Plant Cell* **14**, 641–654.

Chen, M. H., Huang, L. F., Li, H. M., Chen, Y. R. and Yu, S. M. (2004). Signal peptide-dependent targeting of a rice a-amylase and cargo proteins to plastids and extracellular compartments of plant cells. *Plant Physiology* **135**, 1367–1377.

Chigri, F., Soll, J. and Vothknecht, U. C. (2005). Calcium regulation of chloroplast protein import. *Plant Journal* **42**, 821–831.

Chigri, F., Hormann, F., Stamp, A., Stammers, D. K., Bolter, B., Soll, J. and Vothknecht, U. C. (2006). Calcium regulation of chloroplast protein translocation is mediated by calmodulin binding to Tic32. *Proceedings of the National Academy of Sciences of the United States of America* **103**, 16051–16056.

Chou, M. L., Fitzpatrick, L. M., Tu, S. L., Budziszewski, G., Potter-Lewis, S., Akita, M., Levin, J. Z., Keegstra, K. and Li, H. M. (2003). Tic40, a membrane-anchored co-chaperone homolog in the chloroplast protein translocon. *EMBO Journal* **22**, 2970–2980.

Chou, M. L., Chu, C. C., Chen, L. J., Akita, M. and Li, H. M. (2006). Stimulation of transit-peptide release and ATP hydrolysis by a cochaperone during protein import into chloroplasts. *Journal of Cell Biology* **175**, 893–900.

Chua, N. H. and Schmidt, G. W. (1979). Transport of proteins into mitochondria and chloroplasts. *Journal of Cell Biology* **81**, 461–483.

Clantin, B., Delattre, A. S., Rucktooa, P., Saint, N., Meli, A. C., Locht, C., Jacob-Dubuisson, F. and Villeret, V. (2007). Structure of the membrane protein FhaC: A member of the Omp85-TpsB transporter superfamily. *Science* **317**, 957–961.

Cline, K. and Dabney-Smith, C. (2008). Plastid protein import and sorting: Different paths to the same compartments. *Current Opinion in Plant Biology* **11**, 585–592.

Cline, K., Ettinger, W. F. and Theg, S. M. (1992). Protein-specific energy-requirements for protein-transport across or into thylakoid membranes—2 lumenal proteins are transported in the absence of ATP. *Journal of Biological Chemistry* **267**, 2688–2696.

Colbert, C. L., Couture, M. M. J., Eltis, L. D. and Bolin, J. T. (2000). A cluster exposed: Structure of the Rieske ferredoxin from biphenyl dioxygenase and the redox properties of Rieske Fe-S proteins. *Structure* **8**, 1267–1278.

Constan, D., Froehlich, J. E., Rangarajan, S. and Keegstra, K. (2004). A stromal Hsp100 protein is required for normal chloroplast development and function in *Arabidopsis*. *Plant Physiology* **136**, 3605–3615.

Dobberstein, B., Blobel, G. and Chua, N. H. (1977). *In vitro* synthesis and processing of a putative precursor for small subunit of ribulose-1,5-bisphosphate carboxylase of *Chlamydomonas reinhardtii*. *Proceedings of the National Academy of Sciences of the United States of America* **74**, 1082–1085.

Driessen, A. J. M. (1993). SecA, the peripheral subunit of the *Escherichia coli* precursor protein pranslocase, is functional as a dimer. *Biochemistry* **32**, 13190–13197.

Duong, F. (2003). Binding, activation and dissociation of the dimeric SecA ATPase at the dimeric SecYEG translocase. *EMBO Journal* **22**, 4375–4384.

Dutta, S., Mohanty, S. and Tripathy, B. C. (2009). Role of temperature stress on chloroplast biogenesis and protein import in pea (*Pisum sativum* L). *Plant Physiology* **150**, 1050–1061.

Duy, D., Wanner, G., Meda, A. R., von Wiren, N., Soll, J. and Philippar, K. (2007). PIC1, an ancient permease in *Arabidopsis* chloroplasts, mediates iron transport. *Plant Cell* **19**, 986–1006.

Economou, A. and Wickner, W. (1994). SecA promotes preprotein translocation by undergoing ATP-driven cycles of membrane insertion and deinsertion. *Cell* **78**, 835–843.

Ertel, F., Mirus, O., Bredemeier, R., Moslavac, S., Becker, T. and Schleiff, E. (2005). The evolutionarily related beta-barrel polypeptide transporters from *Pisum sativum* and *Nostoc PCC7120* contain two distinct functional domains. *Journal of Biological Chemistry* **280**, 28281–28289.

Ferro, M., Salvi, D., Brugiere, S., Miras, S., Kowalski, S., Louwagie, M., Garin, J., Joyard, J. and Rolland, N. (2003). Proteomics of the chloroplast envelope membranes from *Arabidopsis thaliana*. *Molecular & Cellular Proteomics* **2**, 325–345.

Firlej-Kwoka, E., Strittmatter, P., Soll, J. and Bölter, B. (2008). Import of preproteins into the chloroplast inner envelope membrane. *Plant Molecular Biology* **68**, 505–519.

Forti, G. and Bracale, M. (1984). Ferredoxin Ferredoxin NADP Reductase interaction—Catalytic differences between the soluble and thylakoid-bound complex. *FEBS Letters* **166**, 81–84.

Friedman, A. L. and Keegstra, K. (1989). Chloroplast protein import—Quantitative-analysis of precursor binding. *Plant Physiology* **89**, 993–999.

Fulda, S., Norling, B., Schoor, A. and Hagemann, M. (2002). The Slr0924 protein of *Synechocystis* sp strain PCC 6803 resembles a subunit of the chloroplast protein import complex and is mainly localized in the thylakoid lumen. *Plant Molecular Biology* **49**, 107–118.

Gatzeva-Topalova, P. Z., Walton, T. A. and Sousa, M. C. (2008). Crystal structure of YaeT: Conformational flexibility and substrate recognition. *Structure* **16**, 1873–1881.

Ginalski, K., Elofsson, A., Fischer, D. and Rychlewski, L. (2003). 3D-jury: A simple approach to improve protein structure predictions. *Bioinformatics* **19**, 1015–1018.

Gopalan, G., He, Z., Balmer, Y., Romano, P., Gupta, R., Heroux, A., Buchanan, B. B., Swaminathan, K. and Luan, S. (2004). Structural analysis uncovers a role for redox in regulating FKBP13, an immunophilin of the chloroplast thylakoid lumen. *Proceedings of the National Academy of Sciences of the United States of America* **101**, 13945–13950.

Gould, S. B., Waller, R. R. and McFadden, G. I. (2008). Plastid evolution. *Annual Review of Plant Biology* **59**, 491–517.

Gray, M. W. (1992). The endosymbiont hypothesis revisited. *International Review of Cytology—A Survey of Cell Biology* **141**, 233–357.

Gray, J., Wardzala, E., Yang, M. L., Reinbothe, S., Haller, S. and Pauli, F. (2004). A small family of LLS1-related non-heme oxygenases in plants with an origin amongst oxygenic photosynthesizers. *Plant Molecular Biology* **54**, 39–54.

Gross, J. and Bhattacharya, D. (2009). Revaluating the evolution of the Toc and Tic protein translocons. *Trends in Plant Science* **14**, 13–20.

Gupta, R., Mould, R. M., He, Z. Y. and Luan, S. (2002). A chloroplast FKBP interacts with and affects the accumulation of Rieske subunit of cytochrome bf complex. *Proceedings of the National Academy of Sciences of the United States of America* **99**, 15806–15811.

Gutensohn, M., Schulz, B., Nicolay, P. and Flugge, U. I. (2000). Functional analysis of the two *Arabidopsis* homologues of Toc34, a component of the chloroplast protein import apparatus. *Plant Journal* **23**, 771–783.

Hall, M., Schröder, W. P. and Kieselbach, T. (2008). Thioredoxin interactions of the chloroplast lumen of *Arabidopsis thaliana* indicate a redox regulation of the xanthophyll cycle. *In* "Photosynthesis. Energy from the Sun", (J. F. Allen, E. Gantt, J. H. Golbeck and B. Osmond, eds.), pp. 1099–1102. Springer, The Netherlands.

Heins, L., Mehrle, A., Hemmler, R., Wagner, R., Kuchler, M., Hormann, F., Sveshnikov, D. and Soll, J. (2002). The preprotein conducting channel at the inner envelope membrane of plastids. *EMBO Journal* **21**, 2616–2625.

Herrmann, J. M., Kauff, F. and Neuhaus, H. E. (2009). Thiol oxidation in bacteria, mitochondria and chloroplasts: Common principles but three unrelated machineries? *Biochimica et Biophysica Acta—Molecular Cell Research* **1793**, 71–77.

Hidalgo, E., Ding, H. G. and Demple, B. (1997). Redox signal transduction via iron-sulfur clusters in the SoxR transcription activator. *Trends in Biochemical Sciences* **22**, 207–210.

Hihara, Y., Sonoike, K., Kanehisa, M. and Ikeuchi, M. (2003). DNA microarray analysis of redox-responsive genes in the genome of the cyanobacterium *Synechocystis* sp. strain PCC 6803. *Journal of Bacteriology* **185**, 1719–1725.

Hiltbrunner, A., Bauer, J., Alvarez-Huerta, M. and Kessler, F. (2001a). Protein translocon at the *Arabidopsis* outer chloroplast membrane. *Biochemistry and Cell Biology—Biochimie et Biologie Cellulaire* **79**, 629–635.

Hiltbrunner, A., Bauer, J., Vidi, P. A., Infanger, S., Weibel, P., Hohwy, M. and Kessler, F. (2001b). Targeting of an abundant cytosolic form of the protein import receptor at Toc159 to the outer chloroplast membrane. *Journal of Cell Biology* **154**, 309–316.

Hiltbrunner, A., Grunig, K., Alvarez-Huerta, M., Infanger, S., Bauer, J. and Kessler, F. (2004). AtToc90, a new GTP-binding component of the *Arabidopsis* chloroplast protein import machinery. *Plant Molecular Biology* **54**, 427–440.

Hinnah, S. C., Hill, K., Wagner, R., Schlicher, T. and Soll, J. (1997). Reconstitution of a chloroplast protein import channel. *EMBO Journal* **16**, 7351–7360.

Hinnah, S. C., Wagner, R., Sveshnikova, N., Harrer, R. and Soll, J. (2002). The chloroplast protein import channel Toc75: Pore properties and interaction with transit peptides. *Biophysical Journal* **83**, 899–911.

Hirohashi, T., Hase, T. and Nakai, M. (2001). Maize non-photosynthetic ferredoxin precursor is mis-sorted to the intermembrane space of chloroplasts in the presence of light. *Plant Physiology* **125**, 2154–2163.

Hirsch, S., Muckel, E., Heemeyer, F., Vonheijne, G. and Soll, J. (1994). A receptor component of the chloroplast protein translocation machinery. *Science* **266**, 1989–1992.

Hofmann, N. R. and Theg, S. M. (2003). *Physcomitrella patens* as a model for the study of chloroplast protein transport: Conserved machineries between vascular and non-vascular plants. *Plant Molecular Biology* **53**, 643–654.

Hofmann, N. R. and Theg, S. M. (2005a). Chloroplast outer membrane protein targeting and insertion. *Trends in Plant Science* **10**, 450–457.

Hofmann, N. R. and Theg, S. M. (2005b). Toc64 is not required for import of proteins into chloroplasts in the moss *Physcomitrella patens. Plant Journal* **43**, 675–687.

Hormann, F., Kuchler, M., Sveshnikov, D., Oppermann, U., Li, Y. and Soll, J. (2004). Tic32, an essential component in chloroplast biogenesis. *Journal of Biological Chemistry* **279**, 34756–34762.

Hunt, J. F., Weinkauf, S., Henry, L., Fak, J. J., McNicholas, P., Oliver, D. B. and Deisenhofer, J. (2002). Nucleotide control of interdomain interactions in the conformational reaction cycle of SecA. *Science* **297**, 2018–2026.

Inaba, T. and Schnell, D. J. (2008). Protein trafficking to plastids: One theme, many variations. *Biochemical Journal* **413**, 15–28.

Inaba, T., Li, M., Alvarez-Huerta, M., Kessler, F. and Schnell, D. J. (2003). AtTic110 functions as a scaffold for coordinating the stromal events of protein import into chloroplasts. *Journal of Biological Chemistry* **278**, 38617–38627.

Inaba, T., Alvarez-Huerta, M., Li, M., Bauer, J., Ewers, C., Kessler, F. and Schnell, D. J. (2005). Arabidopsis Tic110 is essential for the assembly and function of the protein import machinery of plastids. *Plant Cell* **17**, 1482–1496.

Inoue, K. (2007). The chloroplast outer envelope membrane: The edge of light and excitement. *Journal of Integrative Plant Biology* **49**, 1100–1111.

Inoue, H. and Akita, M. (2008). Three sets of translocation intermediates are formed during the early stage of protein import into chloroplasts. *Journal of Biological Chemistry* **283**, 7491–7502.

Ivanova, Y., Smith, M. D., Chen, K. H. and Schnell, D. J. (2004). Members of the Toc159 import receptor family represent distinct pathways for protein targeting to plastids. *Molecular Biology of the Cell* **15**, 3379–3392.

Jackson, D. T., Froehlich, J. E. and Keegstra, K. (1998). The hydrophilic domain of Tic110, an inner envelope membrane component of the chloroplastic protein translocation apparatus, faces the stromal compartment. *Journal of Biological Chemistry* **273**, 16583–16588.

Jackson-Constan, D. and Keegstra, K. (2001). Arabidopsis genes encoding components of the chloroplastic protein import apparatus. *Plant Physiology* **125**, 1567–1576.

Jackson-Constan, D., Akita, M. and Keegstra, K. (2001). Molecular chaperones involved in chloroplast protein import. *Biochimica et Biophysica Acta—Molecular Cell Research* **1541**, 102–113.

JagerVottero, P., Dorne, A. J., Jordanov, J., Douce, R. and Joyard, J. (1997). Redox chains in chloroplast envelope membranes: Spectroscopic evidence for the presence of electron carriers, including iron-sulfur centers. *Proceedings of the National Academy of Sciences of the United States of America* **94**, 1597–1602.

Jarvis, P. (2008). Targeting of nucleus-encoded proteins to chloroplasts in plants. *New Phytologist* **179**, 257–285.

Jarvis, P., Chen, L. J., Li, H. M., Pete, C. A., Fankhauser, C. and Chory, J. (1998). An *Arabidopsis* mutant defective in the plastid general protein import apparatus. *Science* **282**, 100–103.

Jelic, M., Sveshnikova, N., Motzkus, M., Horth, P., Soll, J. and Schleiff, E. (2002). The chloroplast import receptor Toc34 functions as preprotein-regulated GTPase. *Biological Chemistry* **383**, 1875–1883.

Jelic, M., Soll, J. and Schleiff, E. (2003). Two Toc34 homologues with different properties. *Biochemistry* **42**, 5906–5916.

Kalanon, M. and McFadden, G. I. (2008). The chloroplast protein translocation complexes of *Chlamydomonas reinhardtii*: A bioinformatic comparison of Toc and Tic components in plants, green algae and red algae. *Genetics* **179**, 95–112.

Karamanou, S., Vrontou, E., Sianidis, G., Baud, C., Roos, T., Kuhn, A., Politou, A. S. and Economou, A. (1999). A molecular switch in SecA protein couples ATP hydrolysis to protein translocation. *Molecular Microbiology* **34**, 1133–1145.

Karnauchov, I., Cai, D. G., Schmidt, I., Herrmann, R. G. and Klosgen, R. B. (1994). The thylakoid translocation of subunit-3 of photosystem-I, the PsaF Gene-Product, depends on a bipartite transit peptide and proceeds along an azide-sensitive pathway. *Journal of Biological Chemistry* **269**, 32871–32878.

Karpenahalli, M. R., Lupas, A. N. and Soding, J. (2007). TPRpred: A tool for prediction of TPR-, PPR- and SELI-like repeats from protein sequences. *BMC Bioinformatics* **8**, 2.

Keegstra, K., Olsen, L. J. and Theg, S. M. (1989). Chloroplastic precursors and their transport across the envelope membranes. *Annual Review of Plant Physiology and Plant Molecular Biology* **40**, 471–501.

Kelley, L. A. and Sternberg, M. J. (2009). Protein structure prediction on the Web: A case study using the Phyre server. *Nature Protocols* **4**, 363–371.

Kessler, F. and Blobel, G. (1996). Interaction of the protein import and folding machineries in the chloroplast. *Proceedings of the National Academy of Sciences of the United States of America* **93**, 7684–7689.

Kessler, F. and Schnell, D. J. (2002). A GTPase gate for protein import into chloroplasts. *Nature Structural Biology* **9**, 81–83.

Kessler, F., Blobel, G., Patel, H. A. and Schnell, D. J. (1994). Identification of 2 GTP-binding proteins in the chloroplast protein import machinery. *Science* **266**, 1035–1039.

Kim, S., Malinverni, J. C., Sliz, P., Silhavy, T. J., Harrison, S. C. and Kahne, D. (2007). Structure and function of an essential component of the outer membrane protein assembly machine. *Science* **317**, 961–964.

Kleffmann, T., Russenberger, D., von Zychlinski, A., Christopher, W., Sjolander, K., Gruissem, W. and Baginsky, S. (2004). The *Arabidopsis thaliana* chloroplast proteome reveals pathway abundance and novel protein functions. *Current Biology* **14**, 354–362.

Knowles, T. J., Jeeves, M., Bobat, S., Dancea, F., McClelland, D., Palmer, T., Overduin, M. and Henderson, I. R. (2008). Fold and function of polypeptide transport-associated domains responsible for delivering unfolded proteins to membranes. *Molecular Microbiology* **68**, 1216–1227.

Koenig, P., Oreb, M., Hofle, A., Kaltofen, S., Rippe, K., Sinning, I., Schleiff, E. and Tews, I. (2008a). The GTPase cycle of the chloroplast import receptors Toc33/Toc34: Implications from monomeric and dimeric structures. *Structure* **16**, 585–596.

Koenig, P., Oreb, M., Rippe, K., Muhle-Goll, C., Sinning, I., Schleiff, E. and Tews, I. (2008b). On the significance of Toc-GTPase homodimers. *Journal of Biological Chemistry* **283**, 23104–23112.

Kouranov, A. and Schnell, D. J. (1997). Analysis of the interactions of preproteins with the import machinery over the course of protein import into chloroplasts. *Journal of Cell Biology* **139**, 1677–1685.

Kouranov, A., Chen, X. J., Fuks, B. and Schnell, D. J. (1998). Tic20 and Tic22 are new components of the protein import apparatus at the chloroplast inner envelope membrane. *Journal of Cell Biology* **143**, 991–1002.

Kouranov, A., Wang, H. A. and Schnell, D. J. (1999). Tic22 is targeted to the intermembrane space of chloroplasts by a novel pathway. *Journal of Biological Chemistry* **274**, 25181–25186.

Kovacheva, S., Bedard, J., Patel, R., Dudley, P., Twell, D., Rios, G., Koncz, C. and Jarvis, P. (2005). *In vivo* studies on the roles of Tic110, Tic40 and Hsp93 during chloroplast protein import. *Plant Journal* **41**, 412–428.

Kovacheva, S., Bedard, J., Wardle, A., Patel, R. and Jarvis, P. (2007). Further *in vivo* studies on the role of the molecular chaperone, Hsp93, in plastid protein import. *Plant Journal* **50**, 364–379.

Kubis, S., Baldwin, A., Patel, R., Razzaq, A., Dupree, P., Lilley, K., Kurth, J., Leister, D. and Jarvis, P. (2003). The Arabidopsis pp i1 mutant is specifically defective in the expression, chloroplast import, and accumulation of photosynthetic proteins. *Plant Cell* **15**, 1859–1871.

Kubis, S., Patel, R., Combe, J., Bedard, J., Kovacheva, S., Lilley, K., Biehl, A., Leister, D., Rios, G., Koncz, C. and Jarvis, P. (2004). Functional specialization amongst the *Arabidopsis* Toc159 family of chloroplast protein import receptors. *Plant Cell* **16**, 2059–2077.

Kuchler, M., Decker, S., Hormann, F., Soll, J. and Heins, L. (2002). Protein import into chloroplasts involves redox-regulated proteins. *EMBO Journal* **21**, 6136–6145.

Kull, F. J., Sablin, E. P., Lau, R., Fletterick, R. J. and Vale, R. D. (1996). Crystal structure of the kinesin motor domain reveals a structural similarity to myosin. *Nature* **380**, 550–555.

Laidler, V., Chaddock, A. M., Knott, T. G., Walker, D. and Robinson, C. (1995). A SecY homolog in *Arabidopsis thaliana* sequence of a full-length cDNA clone and import of the precursor protein into chloroplasts. *Journal of Biological Chemistry* **270**, 17664–17667.
Lee, Y. J., Kim, D. H., Kim, Y. W. and Hwang, I. (2001). Identification of a signal that distinguishes between the chloroplast outer envelope membrane and the endomembrane system *in vivo*. *Plant Cell* **13**, 2175–2190.
Lee, K., Lee, J., Kim, Y., Bae, D., Kang, K. Y., Yoon, S. C. and Lim, D. (2004a). Defining the plant disulfide proteome. *Electrophoresis* **25**, 532–541.
Lee, Y. J., Sohn, E. J., Lee, K. H., Lee, D. W. and Hwang, I. (2004b). The transmembrane domain of AtTco64 and its C-terminal lysine-rich flanking region are targeting signals to the chloroplast outer envelope membrane. *Molecules and Cells* **17**, 281–291.
Lemaire, S. D., Guillon, B., Le Marechal, P., Keryer, E., Miginiac-Maslow, M. and Decottignies, P. (2004). New thioredoxin targets in the unicellular photosynthetic eukaryote *Chlamydomonas reinhardtii*. *Proceedings of the National Academy of Sciences of the United States of America* **101**, 7475–7480.
Lepistö, A., Kangasjärvi, S., Luomala, E., Brader, G., Sipari, N., Keränen, M., Keinänen, M. and Rintamäki, E. (2009). Chloroplast NADPH-thioredoxin reductase interacts with photoperiodic development in *Arabidopsis*. *Plant Physiology* **149**, 1261–1276.
Letunic, I., Copley, R. R., Pils, B., Pinkert, S., Schultz, J. and Bork, P. (2006). SMART 5: Domains in the context of genomes and networks. *Nucleic Acids Research* **34**, D257–D260.
Li, M. and Schnell, D. J. (2006). Reconstitution of protein targeting to the inner envelope membrane of chloroplasts. *Journal of Cell Biology* **175**, 249–259.
Lister, R., Carrie, C., Duncan, O., Ho, L. H. M., Howell, K. A., Murcha, M. W. and Whelan, J. (2007). Functional definition of outer membrane proteins involved in preprotein import into mitochondria. *Plant Cell* **19**, 3739–3759.
Lubeck, J., Soll, J., Akita, M., Nielsen, E. and Keegstra, K. (1996). Topology of IEP110, a component of the chloroplastic protein import machinery present in the inner envelope membrane. *EMBO Journal* **15**, 4230–4238.
Luirink, J., Samuelsson, T. and de Gier, J. W. (2001). YidC/Oxa1p/Alb3: Evolutionarily conserved mediators of membrane protein assembly. *FEBS Letters* **501**, 1–5.
Majeran, W. and van Wijk, K. J. (2009). Cell-type-specific differentiation of chloroplasts in C4 plants. *Trends in Plant Science* **14**, 100–109.
Majeran, W., Zybailov, B., Ytterberg, A. J., Dunsmore, J., Sun, Q. and van Wijk, K. J. (2008). Consequences of C4 differentiation for chloroplast membrane proteomes in maize mesophyll and bundle sheath cells. *Molecular & Cellular Proteomics* **7**, 1609–1638.
Marcaida, M. J., Schlarb-Ridley, B. G., Worrall, J. A. R., Wastl, J., Evans, T. J., Bendall, D. S., Luisi, B. F. and Howe, C. J. (2006). Structure of cytochrome *c*(6A), a novel dithio-cytochrome of *Arabidopsis thaliana*, and its reactivity with plastocyanin: Implications for function. *Journal of Molecular Biology* **360**, 968–977.
Marchand, C., Marechal, P., Meyer, Y. and Decottignies, P. (2006). Comparative proteomic approaches for the isolation of proteins interacting with thioredoxin. *Proteomics* **6**, 6528–6537.
Marshall, J. S., Derocher, A. E., Keegstra, K. and Vierling, E. (1990). Identification of heat-shock protein Hsp70 Homologs in chloroplasts. *Proceedings of the National Academy of Sciences of the United States of America* **87**, 374–378.
Martins, B. M., Svetlitchnaia, T. and Dobbek, H. (2005). 2-Oxoquinoline 8-monooxygenase oxygenase component: Active site modulation by Rieske-[2Fe-2S] center oxidation/reduction. *Structure* **13**, 817–824.

Mata-Cabana, A., Florencio, F. J. and Lindahl, M. (2007). Membrane proteins from the cyanobacterium *Synechocystis* sp. PCC 6803 interacting with thioredoxin. *Proteomics* **7**, 3953–3963.

May, T. and Soll, J. (2000). 14-3-3 proteins form a guidance complex with chloroplast precursor proteins in plants. *Plant Cell* **12**, 53–63.

Miras, S., Salvi, D., Ferro, M., Grunwald, D., Garin, J., Joyard, J. and Rolland, N. (2002). Non-canonical transit peptide for import into the chloroplast. *Journal of Biological Chemistry* **277**, 47770–47778.

Mirus, O., Bionda, T., von Haeseler, A. and Schleiff, E. (2009). Evolutionarily evolved discriminators in the 3-TPR domain of the Toc64 family involved in protein translocation at the outer membrane of chloroplasts and mitochondria. *Journal of Molecular Modeling* **15**, 971–982.

Mori, H. and Cline, K. (2001). Post-translational protein translocation into thylakoids by the Sec and Delta pH-dependent pathways. *Biochimica et Biophysica Acta—Molecular Cell Research* **1541**, 80–90.

Moslavac, S., Mirus, O., Bredemeier, R., Soll, J., von Haeseler, A. and Schleiff, E. (2005). Conserved pore-forming regions in polypeptide-transporting proteins. *FEBS Journal* **272**, 1367–1378.

Motohashi, K. and Hisabori, T. (2006). HCF164 receives reducing equivalents from stromal thioredoxin across the thylakoid membrane and mediates reduction of target proteins in the thylakoid lumen. *Journal of Biological Chemistry* **281**, 35039–35047.

Nada, A. and Soll, J. (2004). Inner envelope protein 32 is imported into chloroplasts by a novel pathway. *Journal of Cell Science* **117**, 3975–3982.

Nakai, M., Goto, A., Nohara, T., Sugita, D. and Endo, T. (1994). Identification of the SecA protein homolog in pea-chloroplasts and its possible involvement in thylakoidal protein-transport. *Journal of Biological Chemistry* **269**, 31338–31341.

Nanjo, Y., Oka, H., Ikarashi, N., Kaneko, K., Kitajima, A., Mitsui, T., Munoz, F. J., Rodriguez-Lopez, M., Baroja-Fernandez, E. and Pozueta-Romero, J. (2006). Rice plastidial *N*-glycosylated nucleotide pyrophosphatase/phosphodiesterase is transported from the ER-Golgi to the chloroplast through the secretory pathway. *Plant Cell* **18**, 2582–2592.

Natt, N. K., Kaur, H. and Raghava, G. P. S. (2004). Prediction of transmembrane regions of beta-barrel proteins using ANN- and SVM-based methods. *Proteins—Structure Function and Bioinformatics* **56**, 11–18.

Nielsen, E., Akita, M., DavilaAponte, J. and Keegstra, K. (1997). Stable association of chloroplastic precursors with protein translocation complexes that contain proteins from both envelope membranes and a stromal Hsp 100 molecular chaperone. *EMBO Journal* **16**, 935–946.

Nilsson, R. and van Wijk, K. J. (2002). Transient interaction of cpSRP54 with elongating nascent chains of the chloroplast-encoded D1 protein; 'cpSRP54 caught in the act'. *FEBS Letters* **524**, 127–133.

Nilsson, R., Brunner, J., Hoffman, N. E. and van Wijk, K. J. (1999). Interactions of ribosome nascent chain complexes of the chloroplast-encoded D1 thylakoid membrane protein with cpSRP54. *EMBO Journal* **18**, 733–742.

Nohara, T., Asai, T., Nakai, M., Sugiura, M. and Endo, T. (1996). Cytochrome *f* encoded by the chloroplast genome is imported into thylakoids via the SecA-dependent pathway. *Biochemical and Biophysical Research Communications* **224**, 474–478.

Odunuga, O. O., Hornby, J. A., Bies, C., Zimmermann, R., Pugh, D. J. and Blatch, G. L. (2003). Tetratricopeptide repeat motif-mediated Hsc70-mSTI1 interaction—Molecular characterization of the critical contacts for successful binding and specificity. *Journal of Biological Chemistry* **278**, 6896–6904.

Olsen, L. J. and Keegstra, K. (1992). The binding of precursor proteins to chloroplasts requires nucleoside triphosphates in the intermembrane space. *Journal of Biological Chemistry* **267**, 433–439.

Or, E. and Rapoport, T. (2007). Cross-linked SecA dimers are not functional in protein translocation. *FEBS Letters* **581**, 2616–2620.

Oreb, M., Hofle, A., Mirus, O. and Schleiff, E. (2008). Phosphorylation regulates the assembly of chloroplast import machinery. *Journal of Experimental Botany* **59**, 2309–2316.

Osborne, A. R., Clemons, W. M. and Rapoport, T. A. (2004). A large conformational change of the translocation ATPase SecA. *Proceedings of the National Academy of Sciences of the United States of America* **101**, 10937–10942.

Pain, D. and Blobel, G. (1987). Protein import into chloroplasts requires a chloroplast ATPase. *Proceedings of the National Academy of Sciences of the United States of America* **84**, 3288–3292.

Papanikolau, Y., Papadovasilaki, M., Ravelli, R. B. G., McCarthy, A. A., Cusack, S., Economou, A. and Petratos, K. (2007). Structure of dimeric SecA, the *Escherichia coli* preprotein translocase motor. *Journal of Molecular Biology* **366**, 1545–1557.

Parish, D., Benach, J., Liu, G., Singarapu, K. K., Xiao, R., Acton, T., Su, M., Bansal, S., Prestegard, J. H., Hunt, J., Montelione, G. T. and Szyperski, T. (2008). Protein chaperones Q8ZP25_SALTY from *Salmonella typhimurium* and HYAE_ECOLI from *Escherichia coli* exhibit thioredoxin-like structures despite lack of canonical thioredoxin active site sequence motif. *Journal of Structural and Functional Genomics* **9**, 41–49.

Peltier, J. B., Emanuelsson, O., Kalume, D. E., Ytterberg, J., Friso, G., Rudella, A., Liberles, D. A., Soderberg, L., Roepstorff, P., von Heijne, G. and van Wijk, K. J. (2002). Central functions of the lumenal and peripheral thylakoid proteome of *Arabidopsis* determined by experimentation and genome-wide prediction. *Plant Cell* **14**, 211–236.

Peltier, J. B., Ytterberg, A. J., Sun, Q. and van Wijk, K. J. (2004). New functions of the thylakoid membrane proteome of *Arabidopsis thaliana* revealed by a simple, fast, and versatile fractionation strategy. *Journal of Biological Chemistry* **279**, 49367–49383.

Perry, S. E. and Keegstra, K. (1994). Envelope membrane-proteins that interact with chloroplastic precursor proteins. *Plant Cell* **6**, 93–105.

Pilon, M., Dekruijff, B. and Weisbeek, P. J. (1992). New insights into the import mechanism of the ferredoxin precursor into chloroplasts. *Journal of Biological Chemistry* **267**, 2548–2556.

Pohlmeyer, K., Soll, J., Grimm, R., Hill, K. and Wagner, R. (1998). A high-conductance solute channel in the chloroplastic outer envelope from pea. *Plant Cell* **10**, 1207–1216.

Qbadou, S., Tien, R., Soll, J. and Schleiff, E. (2003). Membrane insertion of the chloroplast outer envelope protein, Toc34: Constrains for insertion and topology. *Journal of Cell Science* **116**, 837–846.

Qbadou, S., Becker, T., Mirus, O., Tews, I., Soll, J. and Schleiff, E. (2006). The molecular chaperone Hsp90 delivers precursor proteins to the chloroplast import receptor Toc64. *EMBO Journal* **25**, 1836–1847.

Qbadou, S., Becker, T., Bionda, T., Reger, K., Ruprecht, M., Soll, J. and Schleiff, E. (2007). Toc64—A preprotein-receptor at the outer membrane with bipartide function. *Journal of Molecular Biology* **367**, 1330–1346.

Rahim, G., Bischof, S., Kessler, F. and Agne, B. (2009). *In vivo* interaction between *at*Toc33 and *at*Toc159 GTP-binding domains demonstrated in a plant split-ubiquitin system. *Journal of Experimental Botany* **60**, 257–267.

Randall, A., Cheng, J. L., Sweredoski, M. and Baldi, P. (2008). TMBpro: Secondary structure, beta-contact and tertiary structure prediction of transmembrane beta-barrel proteins. *Bioinformatics* **24**, 513–520.

Ratnayake, R. M. U., Inoue, H., Nonami, H. and Akita, M. (2008). Alternative processing of *Arabidopsis* Hsp70 precursors during protein import into chloroplasts. *Bioscience, Biotechnology and Biochemistry* **72**, 2926–2935.

Reddick, L. E., Vaughn, M. D., Wright, S. J., Campbell, I. M. and Bruce, B. D. (2007). *In vitro* comparative kinetic analysis of the chloroplast Toc GTPases. *Journal of Biological Chemistry* **282**, 11410–11426.

Reddick, L. E., Chotewutmontri, P., Crenshaw, W., Dave, A., Vaughn, M. and Bruce, B. D. (2008). Nano-scale characterization of the dynamics of the chloroplast Toc translocon. *Methods in Nano Cell Biology* **90**, 365–398.

Reumann, S., Inoue, K. and Keegstra, K. (2005). Evolution of the general protein import pathway of plastids (Review). *Molecular Membrane Biology* **22**, 73–86.

Robinson, C. and Bolhuis, A. (2004). Tat-dependent protein targeting in prokaryotes and chloroplasts. *Biochimica et Biophysica Acta—Molecular Cell Research* **1694**, 135–147.

Row, P. E. and Gray, J. C. (2001). The effect of amino acid-modifying reagents on chloroplast protein import and the formation of early import intermediates. *Journal of Experimental Botany* **52**, 57–66.

Sai, J. Q. and Johnson, C. H. (2002). Dark-stimulated calcium ion fluxes in the chloroplast stroma and cytosol. *Plant Cell* **14**, 1279–1291.

Salomon, M., Fischer, K., Flugge, U. I. and Soll, J. (1990). Sequence-analysis and protein import studies of an outer chloroplast envelope polypeptide. *Proceedings of the National Academy of Sciences of the United States of America* **87**, 5778–5782.

Sanchez-Pulido, L., Devos, D., Genevrois, S., Vicente, M. and Valencia, A. (2003). POTRA: A conserved domain in the FtsQ family and a class of beta-barrel outer membrane proteins. *Trends in Biochemical Sciences* **28**, 523–526.

Schirmer, E. C., Glover, J. R., Singer, M. A. and Lindquist, S. (1996). HSP100/Clp proteins: A common mechanism explains diverse functions. *Trends in Biochemical Sciences* **21**, 289–296.

Schleiff, E., Tien, R., Salomon, M. and Soll, J. (2001). Lipid composition of outer leaflet of chloroplast outer envelope determines topology of OEP7. *Molecular Biology of the Cell* **12**, 4090–4102.

Schleiff, E., Soll, J., Sveshnikova, N., Tien, R., Wright, S., Dabney-Smith, C., Subramanian, C. and Bruce, B. D. (2002). Structural and guanosine triphosphate/diphosphate requirements for transit peptide recognition by the cytosolic domain of the chloroplast outer envelope receptor, Toc34. *Biochemistry* **41**, 1934–1946.

Schleiff, E., Jelic, M. and Soll, J. (2003). A GTP-driven motor moves proteins across the outer envelope of chloroplasts. *Proceedings of the National Academy of Sciences of the United States of America* **100**, 4604–4609.

Schnell, D. J., Kessler, F. and Blobel, G. (1994). Isolation of components of the chloroplast protein import machinery. *Science* **266**, 1007–1012.

Schnell, D. J., Blobel, G., Keegstra, K., Kessler, F., Ko, K. and Soll, J. (1997). A consensus nomenclature for the protein-import components of the chloroplast envelope. *Trends in Cell Biology* **7**, 303–304.

Schubert, M., Petersson, U. A., Haas, B. J., Funk, C., Schroder, W. P. and Kieselbach, T. (2002). Proteome map of the chloroplast lumen of *Arabidopsis thaliana*. *Journal of Biological Chemistry* **277**, 8354–8365.

Schuenemann, D., Gupta, S., Persello-Cartieaux, F., Klimyuk, V. I., Jones, J. D. G., Nussaume, L. and Hoffman, N. E. (1998). A novel signal recognition particle targets light-harvesting proteins to the thylakoid membranes. *Proceedings of the National Academy of Sciences of the United States of America* **95**, 10312–10316.

Schuenemann, D., Amin, P., Hartmann, E. and Hoffman, N. E. (1999a). Chloroplast SecY is complexed to SecE and involved in the translocation of the 33-kDa but not the 23-kDa subunit of the oxygen-evolving complex. *Journal of Biological Chemistry* **274**, 12177–12182.

Schuenemann, D., Amin, P. and Hoffman, N. E. (1999b). Functional divergence of the plastid and cytosolic forms of the 54-kDa subunit of signal recognition particle. *Biochemical and Biophysical Research Communications* **254**, 253–258.

Schunemann, D. (2004). Structure and function of the chloroplast signal recognition particle. *Current Genetics* **44**, 295–304.

Schunemann, D. (2007). Mechanisms of protein import into thylakoids of chloroplasts. *Biological Chemistry* **388**, 907–915.

Schürmann, P. and Buchanan, B. B. (2008). The ferredoxin/thioredoxin system of oxygenic photosynthesis. *Antioxidants & Redox Signaling* **10**, 1235–1273.

Seedorf, M. and Soll, J. (1995). Copper chloride, an inhibitor of protein import into chloroplasts. *FEBS Letters* **367**, 19–22.

Seedorf, M., Waegemann, K. and Soll, J. (1995). A constituent of the chloroplast import complex represents a new-type of GTP-binding protein. *Plant Journal* **7**, 401–411.

Sharma, V., Arockiasamy, A., Ronning, D. R., Savva, C. G., Holzenburg, A., Braunstein, M., Jacobs, W. R. and Sacchettini, J. C. (2003). Crystal structure of *Mycobacterium tuberculosis* SecA, a preprotein translocating ATPase. *Proceedings of the National Academy of Sciences of the United States of America* **100**, 2243–2248.

Shin, J. Y., Kim, M. and Ahn, T. (2006). Effects of signal peptide and adenylate on the oligomerization and membrane binding of soluble SecA. *Journal of Biochemistry and Molecular Biology* **39**, 319–328.

Sianidis, G., Karamanou, S., Vrontou, E., Boulias, K., Repanas, K., Kyrpides, N., Politou, A. S. and Economou, A. (2001). Cross-talk between catalytic and regulatory elements in a DEAD motor domain is essential for SecA function. *EMBO Journal* **20**, 961–970.

Singh, A. K., Bhattacharyya-Pakrasi, M. and Pakrasi, H. B. (2008). Identification of an atypical membrane protein involved in the formation of protein disulfide bonds in oxygenic photosynthetic organisms. *Journal of Biological Chemistry* **283**, 15762–15770.

Smeekens, S., Bauerle, C., Hageman, J., Keegstra, K. and Weisbeek, P. (1986). The role of the transit peptide in the routing of precursors toward different chloroplast compartments. *Cell* **46**, 365–375.

Smeekens, S., Weisbeek, P. and Robinson, C. (1990). Protein-transport into and within chloroplasts. *Trends in Biochemical Sciences* **15**, 73–76.

Smith, M. D., Hiltbrunner, A., Kessler, F. and Schnell, D. J. (2002). The targeting of the atToc159 preprotein receptor to the chloroplast outer membrane is mediated by its GTPase domain and is regulated by GTP. *Journal of Cell Biology* **159**, 833–843.

Smith, M. D., Rounds, C. M., Wang, F., Chen, K. H., Afitlhile, M. and Schnell, D. J. (2004). AtToc159 is a selective transit peptide receptor for the import of nucleus-encoded chloroplast proteins. *Journal of Cell Biology* **165**, 323–334.

Sohrt, K. and Soll, J. (2000). Toc64, a new component of the protein translocon of chloroplasts. *Journal of Cell Biology* **148**, 1213–1221.

Stahl, T., Glockmann, C., Soll, J. and Heins, L. (1999). Tic40, a new "old" subunit of the chloroplast protein import translocon. *Journal of Biological Chemistry* **274**, 37467–37472.

Stengel, A., Benz, P., Balsera, M., Soll, J. and Bolter, B. (2008). TIC62 redox-regulated translocon composition and dynamics. *Journal of Biological Chemistry* **283**, 6656–6667.

Stengel, A., Benz, J. P., Buchanan, B. B., Soll, J. and Bölter, B. (2009). Preprotein import into chloroplasts via the Toc and the Tic complexes is regulated by redox signals in *Pisum sativum*. *Molecular Plant*. Online published on July 6, 2009; doi:10.1093/mp/ssp043.

Sun, Y. J., Forouhar, F., Li, H. M., Tu, S. L., Yeh, Y. H., Kao, S., Shr, H. L., Chou, C. C., Chen, C. P. and Hsiao, C. D. (2002). Crystal structure of pea Toc34, a novel GTPase of the chloroplast protein translocon. *Nature Structural Biology* **9**, 95–100.

Sundberg, E., Slagter, J. G., Fridborg, I., Cleary, S. P., Robinson, C. and Coupland, G. (1997). *ALBINO3*, an *Arabidopsis* nuclear gene essential for chloroplast differentiation, encodes a chloroplast protein that shows homology to proteins present in bacterial membranes and yeast mitochondria. *Plant Cell* **9**, 717–730.

Sveshnikova, N., Grimm, R., Soll, J. and Schleiff, E. (2000a). Topology studies of the chloroplast protein import channel Toc75. *Biological Chemistry* **381**, 687–693.

Sveshnikova, N., Soll, J. and Schleiff, E. (2000b). Toc34 is a preprotein receptor regulated by GTP and phosphorylation. *Proceedings of the National Academy of Sciences of the United States of America* **97**, 4973–4978.

Tanaka, S. and Wada, K. (1988). The status of cysteine residues in the extrinsic 33kDa protein of spinach Photosystem-II complexes. *Photosynthesis Research* **17**, 255–266.

Teng, Y. S., Su, Y. S., Chen, L. J., Lee, Y. J., Hwang, I. and Li, H. M. (2006). Tic21 is an essential translocon component for protein translocation across the chloroplast inner envelope membrane. *Plant Cell* **18**, 2247–2257.

Theg, S. M., Bauerles, C., Olsen, L. J., Selman, B. R. and Keegstra, K. (1989). Internal ATP is the only energy requirement for the translocation of precursor proteins across chloroplastic membranes. *Journal of Biological Chemistry* **264**, 6730–6736.

Tissier, C., Woolhead, C. A. and Robinson, C. (2002). Unique structural determinants in the signal peptides of 'spontaneously' inserting thylakoid membrane proteins. *European Journal of Biochemistry* **269**, 3131–3141.

Tissot, G., Canard, H., Nadai, M., Martone, A., Botterman, J. and Dubald, M. (2008). Translocation of aprotinin, a therapeutic protease inhibitor, into the thylakoid lumen of genetically engineered tobacco chloroplasts. *Plant Biotechnology Journal* **6**, 309–320.

Tranel, P. J. and Keegstra, K. (1996). A novel, bipartite transit peptide targets OEP75 to the outer membrane of the chloroplastic envelope. *Plant Cell* **8**, 2093–2104.

Tripp, J., Inoue, K., Keegstra, K. and Froehlich, J. E. (2007). A novel serine/proline-rich domain in combination with a transmembrane domain is required for the insertion of AtTic40 into the inner envelope membrane of chloroplasts. *Plant Journal* **52**, 824–838.

Tsai, L. Y., Tu, S. L. and Li, H. M. (1999). Insertion of atToc34 into the chloroplastic outer membrane is assisted by at least two proteinaceous components in the import system. *Journal of Biological Chemistry* **274**, 18735–18740.

Tu, S. L., Chen, L. J., Smith, M. D., Su, Y. S., Schnell, D. J. and Li, H. M. (2004). Import pathways of chloroplast interior proteins and the outer-membrane protein OEP14 converge at Toc75. *Plant Cell* **16**, 2078–2088.

van Dooren, G. G., Tomova, C., Agrawal, S., Humbel, B. M. and Striepen, B. (2008). *Toxoplasma gondii* Tic20 is essential for apicoplast protein import. *Proceedings of the National Academy of Sciences of the United States of America* **105**, 13574–13579.

Vassylyev, D. G., Mori, H., Vassylyeva, M. N., Tsukazaki, T., Kimura, Y., Tahirov, T. H. and Ito, K. (2006). Crystal structure of the translocation ATPase SecA from *Thermus thermophilus* reveals a parallel, head-to-head dimer. *Journal of Molecular Biology* **364**, 248–258.

Villarejo, A., Buren, S., Larsson, S., Dejardin, A., Monne, M., Rudhe, C., Karlsson, J., Jansson, S., Lerouge, P., Rolland, N., von Heijne, G. Grebe, M. *et al.* (2005). Evidence for a protein transported through the secretory pathway en route to the higher plant chloroplast. *Nature Cell Biology* **7**, 1224–1231.

Vojta, L., Soll, J. and Bolter, B. (2007a). Protein transport in chloroplasts—Targeting to the intermembrane space. *FEBS Journal* **274**, 5043–5054.

Vojta, L., Soll, J. and Bolter, B. (2007b). Requirements for a conservative protein translocation pathway in chloroplasts. *FEBS Letters* **581**, 2621–2624.

Waegemann, K. and Soll, J. (1991). Characterization of the protein import apparatus in isolated outer envelopes of chloroplasts. *Plant Journal* **1**, 149–158.

Weibel, P., Hiltbrunner, A., Brand, L. and Kessler, F. (2003). Dimerization of Toc-GTPases at the chloroplast protein import machinery. *Journal of Biological Chemistry* **278**, 37321–37329.

Woodbury, R. L., Hardy, S. J. S. and Randall, L. L. (2002). Complex behavior in solution of homodimeric SecA. *Protein Science* **11**, 875–882.

Woodson, J. D. and Chory, J. (2008). Coordination of gene expression between organellar and nuclear genomes. *Nature Reviews Genetics* **9**, 383–395.

Wu, C. B., Seibert, F. S. and Ko, K. (1994). Identification of chloroplast envelope proteins in close physical proximity to a partially translocated chimeric precursor protein. *Journal of Biological Chemistry* **269**, 32264–32271.

Yamaryo, Y., Motohashi, K., Takamiya, K., Hisabori, T. and Ohta, H. (2006). *In vitro* reconstitution of monogalactosyldiacylglycerol (MGDG) synthase regulation by thioredoxin. *FEBS Letters* **580**, 4086–4090.

Yeh, Y. H., Kesavulu, M. M., Li, H. M., Wu, S. Z., Sun, Y. J., Konozy, E. H. E. and Hsiao, C. D. (2007). Dimerization is important for the GTPase activity of chloroplast translocon components at Toc33 and psToc159. *Journal of Biological Chemistry* **282**, 13845–13853.

Yuan, J. G., Henry, R., McCaffery, M. and Cline, K. (1994). SecA homolog in protein-transport within chloroplasts—Evidence for endosymbiont-derived sorting. *Science* **266**, 796–798.

Zhang, L. X., Paakkarinen, V., Suorsa, M. and Aro, E. M. (2001). A SecY homologue is involved in chloroplast-encoded D1 protein biogenesis. *Journal of Biological Chemistry* **276**, 37809–37814.

Zielinski, R. E. (1998). Calmodulin and calmodulin-binding proteins in plants. *Annual Review of Plant Physiology and Plant Molecular Biology* **49**, 697–725.

Zimmer, J., Nam, Y. S. and Rapoport, T. A. (2008). Structure of a complex of the ATPase SecA and the protein-translocation channel. *Nature* **455**, 936–943.

Glutaredoxins in Development and Stress Responses of Plants

SHUTIAN LI AND SABINE ZACHGO[1]

Department of Botany, Osnabrück University, 49076 Osnabrück, Germany

[1]Corresponding author: Email: zachgo@biologie.uni-osnabrueck.de

Advances in Botanical Research, Vol. 52 0065-2296/09 $35.00

DOI: 10.1016/S0065-2296(09)52011-3

ABSTRACT

Glutaredoxins (GRXs) are small ubiquitous glutathione (GSH)-dependent oxidoreductases and are known to exert a crucial function in the response to oxidative stresses in a range of pro- and eukaryotes, such as *Escherichia coli*, yeast, and human. Plant genomes encode three GRX classes, the CPYC, the CGFS, and the CC-type; however, up till now little has been known about their functions. Whereas the CPYC and CGFS classes occur ubiquitously in all thus far analyzed species, the CC-type GRXs exist only in land plants. Moreover, only the CC-type class expanded markedly during the evolution of land plants, providing clues for their involvement in the formation of more complex plants adapted to life on land. Accumulating evidence indicates that GRXs participate in various cellular processes in plants. In this review, focus is given to the recently emerging functions of land plant-specific GRXs in flower development and pathogen resistance. Strikingly, comparisons of the involved signaling pathways that have been thus far considered to be unrelated with one another reveal that similar protein molecules, CC-type GRXs as well as TGA transcription factors, seem to have been recruited to participate in these different pathways. The small CC-type oxidoreductases likely have the potential to alter TGA activities in a redox-sensitive manner, thereby causing differential gene expression and consequently affecting associated downstream biological processes. The activities of plant CPYC and CGFS GRXs in the assembly and delivery of Fe–S clusters and arsenic resistance indicate a broad functional spectrum for the other two classes of GRXs. Furthermore, *S*-nitrosylation of protein thiols adds a new layer of complexity to redox regulation in plant cells. As GRXs require GSH to reduce their target proteins, GSH-associated developmental processes, which affect flowering time, root and shoot development, are also discussed. The identification of more plant GRX targets, albeit challenging, will help to uncover further roles of plant GRXs and facilitate the investigation of functional redundancies and crosstalk between GRXs and TRXs.

ABBREVIATIONS

GRX	glutaredoxin
TRX	thioredoxin
GSH	reduced glutathione
GSSG	oxidized glutathione
PRX	peroxiredoxin
BiFC	bimolecular fluorescence complementation
GA	gibberellin
SAR	systemic acquired resistance
SA	salicylic acid
JA	jasmonic acid
GSNO	S-nitrosoglutathione
ROS	reactive oxygen species
RNS	reactive nitrogen species
EST	expressed sequence tag
H_2O_2	hydrogen peroxide

SRX	sulfiredoxin
SOH	sulfenic acid
SO_2H	sulfinic acid
SO_3H	sulfonic acid
GR	glutathione reductase
TR	thioredoxin reductase
NTR	NADPH thioredoxin reductase
γ-GCS	γ-glutamylcysteine synthase
QC	quiescent center
roGFP	redox-sensitive green fluorescent protein

I. INTRODUCTION

Glutaredoxins (GRXs) are small ubiquitous glutathione (GSH)-dependent oxidoreductases that catalyze the reversible reduction of disulfide bonds and thus regulate protein activities in a large variety of cellular processes (Buchanan and Balmer, 2005; Fernandes and Holmgren, 2004). Together with thioredoxins (TRXs), protein disulfide isomerase (PDI), glutathione-*S*-transferase (GST), and glutathione peroxidase, GRXs belong to the TRX superfamily sharing a common thioredoxin fold ($\beta1$-$\alpha1$-$\beta2$-$\alpha2$-$\beta3$-$\beta4$-$\alpha3$; Martin, 1995). Since the first GRX was identified as a hydrogen donor for ribonucleotide reductase in an *Escherichia coli* mutant lacking TRX1 (Holmgren *et al.*, 1975), GRXs that are currently known to exist in all living organisms have been extensively investigated in *E. coli*, yeast, and human, where they play crucial roles in the response to oxidative stress (Fernandes and Holmgren, 2004). GRXs function as redox proteins through two different mechanisms. In the dithiol mechanism, electrons are transferred from NADPH to GSH reductase (GR), then to GSH, and from there to GRXs. Finally, GRXs reduce target proteins by dithiol-disulfide exchange reactions in a manner similar to TRXs. In contrast to TRXs, GRXs can also efficiently and specifically mediate reversible reduction of protein GSH-mixed disulfides via the monothiol mechanism, a process known as glutathionylation/deglutathionylation (Fernandes and Holmgren, 2004; Rouhier *et al.*, 2004). Similar to TRXs, GRXs are a multigenic family of proteins, represented by various isoforms and targeted to multiple subcellular compartments (Lemaire, 2004). Based on their primary structures and active site motif compositions, GRXs are divided into three classes, the CPYC, CGFS, and CC-type, respectively. The dithiol CPYC and monothiol CGFS GRXs are common to all pro- and eukaryotes, whereas CC-type GRXs occur only in land plants (Lemaire, 2004; Rouhier *et al.*, 2004, 2006; Xing *et al.*, 2006).

The first plant GRX sequence was identified in rice (Minakuchi *et al.*, 1994) and the first report on a plant GRX protein localization was published slightly later (Morell *et al.*, 1995). The roles of plant TRXs in chloroplast and mitochondrial processes, seed development and germination, as well as self-incompatibility are well established (Buchanan and Balmer, 2005), but less is known about the biochemical and biological functions of plant GRXs. The concept of plant GRXs was pronouncedly modified by sequencing projects of several plant model species, which have revealed an unexpected number of genes coding for GRXs. Proteomic approaches were developed, allowing the identification of numerous GRX target proteins implicated in different aspects of plant life, including development and adaptation to environmental alterations and stresses (Ito *et al.*, 2003; Rouhier *et al.*, 2005). Several timely updates describing impressive progress on plant GRXs have been published in recent years. These reviews focused on major recent achievements at that time, such as GRX subfamily compositions of evolutionarily informative plant species, identification of target proteins, established roles and putative functions (Lemaire, 2004; Rouhier *et al.*, 2004, 2006; Xing *et al.*, 2006). Recently, analysis of two Arabidopsis CC-type GRX mutants together with protein interaction studies unraveled novel functions of plant GRXs in floral organ development (Li *et al.*, 2009; Wang *et al.*, 2009; Xing and Zachgo, 2008; Xing *et al.*, 2005). Strikingly, another closely related CC-type GRX, which also interacts with TGA transcription factors, participates in defense responses (Ndamukong *et al.*, 2007). Similarly, our understanding of the CPYC and CGFS GRXs is expanding, highlighting their function in assembly and delivery of iron–sulfur clusters (Bandyopadhyay *et al.*, 2008; Cheng *et al.*, 2006; Rouhier *et al.*, 2007), and arsenic resistance (Sundaram *et al.*, 2008). Here, we provide a synoptic overview of GRXs that participate in known cellular processes of plants, with an emphasis on recent findings indicating an intriguing involvement of the CC-type GRXs in stress responses and developmental processes.

II. EVOLUTIONARY IMPLICATIONS OF LAND PLANT-SPECIFIC CC-TYPE GRXs

The availability of whole genome sequences and EST data allows the identification and comparative analysis of GRX classes in oxygenic photosynthetic organisms representing different stages of plant evolution (Lemaire, 2004; Rouhier *et al.*, 2004; Xing *et al.*, 2006). Table I shows the sizes of three GRX classes present in seven plant species. Two single-celled green algae, considered to be primitive plant life forms, *Chlamydomonas reinhardtii* and

TABLE I
Comparisons of sizes of three GRX classes in seven species representing different stages of plant evolution

	CPYC	CGFS	CC-type	Total
Chlamydomonas reinhardtii	2	4	0	6
Synechocystis sp. *PCC6803*	2	1	0	3
Physcomitrella patens	4	6	2	12
Pinus taeda	4	8	5	17
Oryza sativa	5	5	17	27
Populus trichocarpa	5	6	22	33
Arabidopsis thaliana	6	4	21	31

Synechocystis sp. *PCC6803*, only contain CPYC and CGFS GRXs but no CC-type GRXs could be identified in all investigated algae thus far. In the bryophyte *Physcomitrella patens*, a representative of basal nonvascular land plants, two CC-type GRXs exist (Xing *et al.*, 2006). The number of the CC-type GRXs progressively increases from the bryophytes (two in *Physcomitrella patens*) to the gymnosperms (five in *Pinus taeda*) and finally to the angiosperms comprising monocots (17 in *Oryza sativa*) and dicots (22 and 21 in *Populus trichocarpa* and *Arabidopsis thaliana*, respectively). Contrarily, sizes of the other two classes remain relatively stable during the evolution of higher land plants, ranging from 4 to 6 for the CPYC and 4 to 8 for the CGFS classes. The striking difference in the overall number of GRXs among these analyzed species is thus essentially attributable to the presence and strong expansion of land plant-specific CC-type GRXs. Given these observations, comparisons of the GRX class compositions in evolutionarily informative plant species thus indicate an intriguing correlation between the expansion of the land plant-specific CC-type GRXs and plant adaptations to life on land accompanied by the formation of more complex land plants.

III. *ROXY1* AND *ROXY2*, TWO CC-TYPE GRX GENES, REGULATE FLOWER DEVELOPMENT

Flower development of higher plants is a complex process involving genetic interactions among several well-characterized floral homeotic genes. Genetic analysis of floral homeotic mutants led to the formulation of the classic ABC model for specifying Arabidopsis floral organ identity (Coen and Meyerowitz, 1991). This model proposes that class A genes specify sepals in the first whorl, a combination of class A and B genes controls petal organogenesis in the second whorl, class B and C genes together govern stamen development in the third whorl, and class C genes alone specify carpel organogenesis in the fourth whorl. In Arabidopsis, *APETALA1/2* (*AP1/2*) represents class A genes, *PISTILLATA* (*PI*) and *APETALA3* (*AP3*) are class B genes, and *AGAMOUS* (*AG*) is classified as a class C gene. Further genetic and biochemical studies identified *SEPALLATA1/2/3/4* (*SEP/1/2/3/4*) as class E floral homeotic genes encoding related proteins that form ternary or quaternary complexes with ABC proteins and act redundantly to specify floral organ development (Ditta *et al.*, 2004; Honma and Goto, 2001; Pelaz *et al.*, 2000). Furthermore, antagonistic interactions are present between class A (*AP1* and *AP2*) and C genes (*AG*): *AP2* represses *AG* in the first two whorls and *AG* excludes *AP1* in whorls 3 and 4 (Drews *et al.*, 1991; Gustafson-Brown *et al.*, 1994).

Whereas ABCE genes specify floral organ identity, the initiation of floral organ primordia is determined before the onset of the class ABCE activity and much less is known about the regulatory processes that control the positioning and initiation of floral organ primordia. Recently, two CC-type GRX mutants were characterized, demonstrating a novel, unexpected function for these proteins during Arabidopsis flower development (Xing and Zachgo, 2008; Xing *et al.*, 2005). The *roxy1* mutant initiates a reduced number of petal primordia and forms 2.5 petals on average instead of 4.0 petals as observed in wild-type flowers. During further petal differentiation, abnormalities such as lack of blade expansion and abnormal bending occur in the *roxy1* mutant (Xing *et al.*, 2005). *ROXY1* mRNA expression is highly dynamic during flower development and only transiently detectable in young floral buds as well as in all young floral organs. As soon as floral organs start to differentiate, *ROXY1* expression ceases and residual weak expression remains in anthers where the lobes are formed (Xing *et al.*, 2005). Intrigued by this novel function of GRXs in flower development, loss-of-function mutants were analyzed from the closest *ROXY1* homolog, *ROXY2*, sharing over 80% amino acid similarity with *ROXY1*. The *roxy2* single mutant does not reveal any flower defects; however, construction of the *roxy1 roxy2* double mutant unraveled redundant activities of these CC-type GRXs. Differentiation of anther lobes and thus microspore formation are disrupted in *roxy1 roxy2* double mutant flowers that consequently fail to set seeds (Xing and Zachgo, 2008). The male-sterile phenotype of the *roxy1 roxy2* double mutant, which is observed in neither of the two single mutants, indicates that these two small proteins act redundantly to secure crucial, possibly redox-balanced processes during the production of male germline cells. Furthermore, double mutant analysis of *roxy1* with the class A mutant *ap2* revealed that *ROXY1* also participates in the negative regulation of expression of the class C gene *AG* in the first and second whorl (Xing *et al.*, 2005).

IV. CC-TYPE GRXs WITH A CONSERVED C-TERMINUS CAN MODIFY THE SAME TARGET PROTEINS IF EXPRESSED PROPERLY

Complementation experiments of the *roxy1* mutant using different Arabidopsis CC-type GRX genes driven under the control of endogenous *ROXY1* promoter sequences have shown that all complementing GRXs share a short motif, A[L/I]WL, located at the end of the C-terminus (Li *et al.*, 2009). These findings highlight the importance of the ROXY1 C-terminus, particularly the presence of an ALWL motif for its functionality. *ROXY4* (At3g62950), one of the CC-type GRXs possessing the ALWL motif and complementing the *roxy1* mutant, has been recently shown to be a target gene that is upregulated by DELLA and to participate in gibberellin (GA) signaling and floral organ development (Hou *et al.*, 2008). Surprisingly, *ROXY19*, an even more distantly related CC-type GRX, initially called *GRX480* and known to participate in pathogen defense (Ndamukong *et al.*, 2007), shares this C-terminal short motif and can replace *ROXY1* and act in petal development if appropriately expressed (Li *et al.*, 2009). Together, these data emphasize the importance of expression regulation as a crucial mechanism by which specificity is conferred to the function of CC-type GRXs sharing the conserved C-terminal extension (Li *et al.*, 2009).

Furthermore, *ROXY1* homologs from the distantly related monocot species rice (*Oryza sativa*) were isolated and characterized (Wang *et al.*, 2009). Similar to *ROXY1* and *ROXY2* from Arabidopsis, *OsROXY1* and *OsROXY2* show highly dynamic floral expression patterns and are transiently expressed in young floral organs. Additionally, these GRXs also possess the C-terminal ALWL motif. However, the rice floral architecture deviates strongly from that of eudicot flowers. The small second whorl organs, two lodicules, contribute to the opening of rice buds and to pollen dispersal. Thus, rice lodicules differ from typical eudicot petals in their morphology and function. Nevertheless, if *OsROXY1/2* are expressed under the regulatory elements of *ROXY1*, they can function in Arabidopsis petal development and restore the *roxy1* mutant phenotype. This implies that the monocot *ROXY* homologs, if appropriately expressed in the second whorl of an Arabidopsis flower, are indeed able to recognize and posttranslationally modify the same target proteins as ROXY1/2 in the eudicot Arabidopsis. Thus, the biochemical functions of the ROXY1 and OsROXY1/2 proteins seem to have been conserved for more than 120 million years when the last common ancestor of the mono- and eudicot lineages existed (Schmidt and Ambrose, 1998). During the evolution of higher land plants, duplication of *ROXY* genes likely led to the strong enlargement of the CC-type class.

Gene duplication events are known to be followed by sub- and neofunctionalization processes, which contribute to diversifying gene functions (Lynch and Conery, 2000). The exchangeability of the *ROXY* homologs from Arabidopsis and rice indicates that changes in *cis*-regulatory elements of these CC-type genes seem to have been a major mechanism for the realization of the sub- and neofunctionalization of the expanding CC-type GRXs. Different, tightly regulated spatial and temporal expression patterns of CC-type GRX genes seemed to have evolved that allowed their participation in different flower developmental processes. Crucial pathways, such as microspore formation, even involve redundant CC-type GRX activities to secure germline cell production. Contrarily, protein alterations, such as single or multiple exchanges of amino acids in the ROXY protein, which could lead to altered biochemical activities, seem to have contributed less to the sub- and neofunctionalization of the CC-type GRXs.

Although *ROXY1* is required for petal development, double mutant studies with a floral homeotic class B mutant revealed that its function is specific to the second whorl organogenesis, rather than being specifically required for petal development (Xing *et al.*, 2005). Lack of the B function in the *apetala3* (*ap3*) mutant causes a failure to produce petals in the second whorl where sepals develop instead (Jack *et al.*, 1992). However, in *ap3 roxy1* double mutants, sepals are formed in the second whorl that develop further defects due to a lack of the ROXY function. Given this second whorl-specific *ROXY1* activity in Arabidopsis, it will thus be interesting to investigate if the development of rice lodicules with a different morphology and function would also be disturbed if the *OsROXY1/2* function is affected.

V. ROXY1 INTERACTS WITH TGA TRANSCRIPTION FACTORS IN THE NUCLEUS

The conserved biochemical activity of several ROXY homologs and altered, differential expression patterns emphasize the importance of understanding which target proteins are modified by these CC-type GRXs. Several TGA transcription factors were recently isolated as ROXY1-interacting proteins by a yeast two-hybrid screen. The interactions were further investigated with independent techniques such as bimolecular fluorescence complementation (BiFC) using transient expression assays in tobacco leaf epidermal cells (Li *et al.*, 2009). BiFC analysis confirmed a nuclear interaction of ROXY1 with TGA transcription factors in planta. Overlapping expression patterns of *ROXY1* and TGA genes during flower development allow ROXY1/TGA protein interactions to occur *in vivo* and support their biological relevance in petal development (Li *et al.*, 2009). Intracellular localization studies showed a nucleocytoplasmic expression of a functional YFP–ROXY1 fusion protein. The availability of the *roxy1* mutant allowed the determination of an *in vivo* relevance of a nuclear versus a cytoplasmic *ROXY1* localization to the *ROXY1* function. The fusion of an NLS (nuclear localization signal) to ROXY1 (NLS–ROXY1) led to an exclusive nuclear protein expression, whereas a successive fusion of three YFP fragments to ROXY1 (3× YFP–ROXY1) retained the fusion protein in the cytoplasm. Complementation experiments verified that only nuclear ROXY1 protein localization is capable of restoring the development of four normal petals in the *roxy1* mutant. The functionality of the 3× YFP–ROXY1 fusion protein was proven by N-terminally fusing an NLS (NLS–3×YFP–ROXY1), as this fusion protein can still complement the *roxy1* mutant (Li *et al.*, 2009). These experiments delivered proof for a novel, nuclear activity of land plant-specific GRXs and indicate the existence of a nuclear, likely redox-dependent mechanism, which modulates the activity of plant transcription factors. Posttranslational modifications of transcription factors by GRXs have also been demonstrated for other organisms. In yeast, the CGFS GRX3 and GRX4 interact with the transcription factor ScAFT1, which induces the transcription of iron regulon genes in iron-deficient yeast (Ojeda *et al.*, 2006). Overexpression of PICOT, a human GRX, inhibits transactivation activities of AP-1 and NF-κB (Witte *et al.*, 2000).

VI. GENETIC INTERACTION OF *ROXY1* WITH TGA GENES

The TGA transcription factor gene family comprises 10 members in Arabidopsis. *PERIANTHIA (PAN)* is thus far the only TGA gene with a described function in flower development (Chuang *et al.*, 1999), whereas seven other TGAs were shown to be associated with stress-related functions and seem to act redundantly (Kesarwani *et al.*, 2007; Zhang *et al.*, 2003). Genetic analysis of *roxy1* and *pan* single mutants revealed opposite effects on petal development, which is depicted in Fig. 1A. Whereas *roxy1* produces only 2.5 petals on average, *pan* mutants form a pentamerous second whorl that is composed of five petals. In *roxy1 pan* plants, pentamerous flowers develop, indicating that *PAN* is epistatic to *ROXY1* with respect to their function in the regulation of petal primordia initiation (Li *et al.*, 2009). Moreover, pentamerous petals of double mutants also exhibit an abnormal later petal morphogenesis, such as bending of petals, a feature resembling petal abnormalities observed in the *roxy1* single mutant (Li *et al.*, 2009).

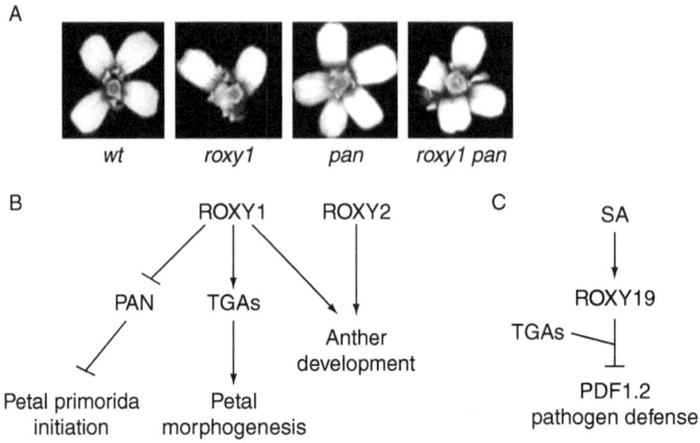

Fig. 1. CC-type GRXs implicated in floral development and pathogen defense. (A) Comparison of petal phenotypes. In Arabidopsis wild-type (wt) flowers, four equally sized petals are formed. Petal number is reduced in the *roxy1* mutant and later petal morphogenesis is also disturbed. Floral organ numbers in the *pan* mutant are increased in whorls 1 and 2, such that a pentamerous pattern is formed. In *roxy1 pan* double mutants, similar to the *pan* single mutant, flowers develop a pentamerous phenotype in the outer whorls. Thus, *PAN* is epistatic to *ROXY1* with respect to its function of regulating petal primorida initiation. However, the pentamerous double mutant also displays an abnormal later petal morphogenesis, such as the formation of smaller or folded petals as observed in the *roxy1* mutant. (B) Regulatory interactions between ROXYs and TGAs in flower development. As revealed by the *pan* mutant phenotype, *PAN* represses petal primordia initiation. At early stages of petal development, ROXY1 controls petal primordia initiation probably by negatively regulating PAN activity. Additionally, acting independently from PAN, ROXY1 also affects later petal morphogenesis. ROXY1 likely positively modulates the activity of further TGA factors acting together in later petal differentiation. Furthermore, *ROXY1* and its closest homolog *ROXY2* act redundantly during anther development. (C) Regulatory interactions between ROXY19 and TGAs in pathogen defense. SA-inducible ROXY19 interacts with TGA factors and suppresses JA-responsive *PDF1.2* transcription. Arrows and hatchets denote positive and negative regulation, respectively. (See color plate section at the back of the book.)

This phenotype implies that an additional function for *ROXY1*, which is independent of the *PAN* activity, likely modifies proteins involved in later petal morphogenesis (Fig. 1B).

The chimeric repressor gene-silencing technology (CRES-T) has been reported as a useful technique for overcoming functional redundancies of transcription activators (Heyl *et al.*, 2008; Hiratsu *et al.*, 2003; Koyama *et al.*, 2007). Functionality of the CRES-T was proven by the observation that T1 transgenic *pan* mutants harboring a *PAN* transgene fused to the EAR-motif repression domain (SRDX) developed about 20% of flowers with a *pan*-like pentamerous flower morphology, most likely due to successful silencing of *PAN* target gene expression (Li *et al.*, 2009). This indicates that *PAN* probably activates genes that prevent petal primordia initiation. Surprisingly, over 50% of T1 transgenic flowers resembled those of the *roxy1 pan* double mutant, where, in addition to the formation of pentamerous flowers, further abnormalities occurred during later petal morphogenesis (Li *et al.*, 2009). These later petal defects were also observed in about 20% of examined T1 flowers that did not form a pentamerous second whorl. Thus, in these plants, besides *PAN*, further TGA genes, acting redundantly during later petal morphogenesis, were most likely silenced. The chimeric repressor approach thus seemed to have successfully uncoupled a dual function of TGA factors, namely the negative effect of PAN on petal primordia initiation and the positive effect of further TGA factors on later petal morphogenesis (Li *et al.*, 2009; Fig. 1B).

VII. CC-TYPE GRXs AND DISEASE RESISTANCE

Plants have evolved both innate and induced immune responses to protect themselves against attacks from herbivorous insects and microbial pathogens (Jones and Dangl, 2006; Koornneef and Pieterse, 2008). Besides pre-existing defense mechanisms directed against specific invaders, plants are able to activate systemic acquired resistance (SAR) that functions against a broad spectrum of attackers systemically and effectively (Sticher *et al.*, 1997). The roles of plant hormones and their crosstalk in the regulation of induced SAR are well understood. The best characterized cross communication between phytohormones is an antagonistic interaction between salicylic acid (SA) and jasmonic acid (JA) defense signaling pathways, which involves many key players, such as NON-EXPRESSOR OF PR GENES1 (NPR1), WRKY factors, and MPK4 (Koornneef and Pieterse, 2008). Defense responses are often associated with changes in cellular redox states (Mou *et al.*, 2003; Tada *et al.*, 2008). Conformational changes of NPR1 via *S*-nitrosylation and thioredoxin (TRX) activity in the response to altered intracellular redox potential are required to initiate plant immunity (Tada *et al.*, 2008). In unchallenged plant cells, NPR1 is sequestered in the cytoplasm as an oligomer through intermolecular disulfide bonds, and *S*-nitrosylation of NPR1 by *S*-nitrosoglutathione (GSNO) at cysteine 156 facilitates its oligomerization, which maintains protein homeostasis upon SA induction (Mou *et al.*, 2003; Tada *et al.*, 2008). Upon pathogen infection and SA accumulation, intermolecular disulfide bonds holding NPR1 proteins together are reduced by cytosolic TRXs. Subsequently, monomeric NPR1 is translocated from the cytoplasm into the nucleus (Mou *et al.*, 2003). There, NPR1 interacts with TGA factors and serves as a coactivator, mediating

DNA binding and thus initiating transcription of *PR* genes. The NPR1/TGA1 protein interaction depends on the reduced state of two cysteines in the TGA1 protein as an intramolecular disulfide bridge formed between them prevents this interaction (Després *et al.*, 2003). SA-induced *ROXY19* is likely involved in the reduction of TGA1, as this glutaredoxin physically interacts with several TGA factors and negatively regulates JA-responsive *PDF1.2* transcription in Arabidopsis (Ndamukong *et al.*, 2007). The suppressive effect of *ROXY19* on *PDF1.2* induction is abolished in *tga2 tga5 tga6* triple mutants, further supporting the observation that the interaction between ROXY19 and TGA factors is a prerequisite for the *ROXY19*-dependent crosstalk (Fig. 1C). Overproduction of *OsWRKY13*, a positive regulator of disease resistance, induces the expression of two CC-type GRXs possibly linked to redox homeostasis in rice (Qiu *et al.*, 2008). Arabidopsis and rice *ROXY1/2* have also been demonstrated to participate in defense response (Wang *et al.*, 2009). Ectopic expression of these CC-type GRXs in Arabidopsis leads to increased accumulation of hydrogen peroxide (H_2O_2) and renders transgenic plants highly susceptible to pathogen infection. Furthermore, developmental abnormalities can be observed, altogether revealing the importance of ROXY homologs participating in balanced redox processes likely affecting defense responses as well as floral organ development.

VIII. COMPARISONS OF SIGNALING MECHANISMS INVOLVED IN DISEASE RESISTANCE AND FLOWER DEVELOPMENT

NPR1 is a member of a small gene family in Arabidopsis. The five members all share structural motifs comprising a BTB/POZ domain and four ankyrin repeats known to mediate interactions with TGA factors (Mou *et al.*, 2003). Two members of this family, *BLADE-ON-PETIOLE1* (*BOP1*) and *BOP2*, control floral patterning and development as well as leaf patterning. This, along with the *NPR1* function in pathogen response, reveals a broad function for *NPR1*-like genes (Hepworth *et al.*, 2005). Likewise, members of the TGA family and the CC-type GRX class were shown to be able to participate in flower development as well as pathogen responses. Pathogen-related GRXs such as *ROXY19* can even replace the ROXY function in flower development, if expressed appropriately (Li *et al.*, 2009). Similar to the NPR1/TGA protein interactions, biochemical studies revealed that BOP1 and BOP2 also interact with PAN, a floral TGA factor. As their expression patterns overlap in younger flowers, these interactions are likely biologically relevant. Phenotypic analysis of *bop1 bop2 pan* triple mutants provided further evidence for the involvement of BOP and PAN proteins in the same genetic pathway (Hepworth *et al.*, 2005). As mentioned earlier, genetic analysis of the *roxy1 pan* double mutant indicates that *PAN* is epistatic to *ROXY1*. Thus, the same protein molecules, TGA factors, NPR1 homologs, and CC-type GRXs, participate in signal transduction processes causing differential gene expression with two different outcomes, namely regulating target genes participating in pathogen defense as well as in flower development, which is summarized in a simplified scheme in Fig. 2. This raises an intriguing question of whether the same mechanisms, particularly identical or similar posttranslational modifications, contribute to altering gene expression patterns in these processes. Mutagenesis of either of the two conserved crucial cysteines (Cys82 or Cys216) in NPR1 is sufficient to cause nuclear localization of mutant proteins and thereby nuclear interaction with TGA1 leading to an activation of defense gene expression (Mou *et al.*, 2003). Sequence alignment

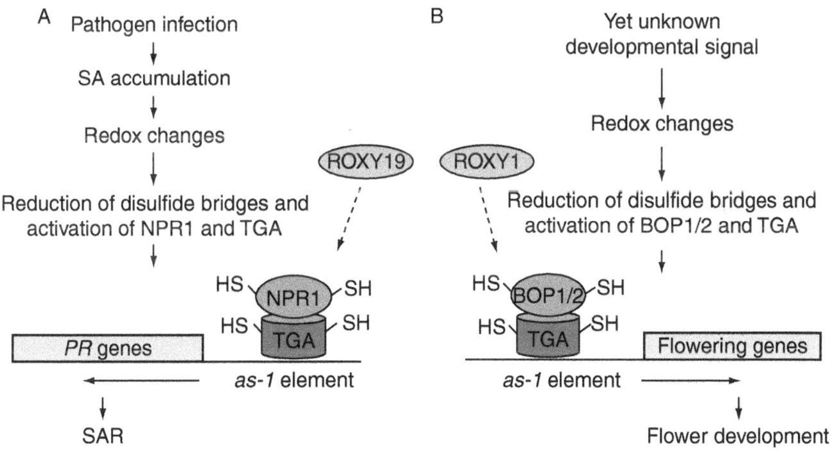

Fig. 2. Signaling mechanisms involved in disease resistance and flower development employ similar protein molecules. (A) Upon pathogen attack, cellular redox changes cause reduction of inter- and intramolecular disulfide bonds of NPR1 and TGA proteins, respectively. Subsequently, active NPR1/TGA protein complexes can bind to *as-1* elements of target genes, thereby inducing expression of *PR* genes and establishment of SAR. ROXY19 interacts with and thus likely posttranslationally modifies TGA factors, as indicated by the dashed arrow. (B) Similarly, redox changes caused by a yet unknown developmental signal might lead to the formation of active BOP/TGA protein complexes, which might in turn bind to *as-1* elements of flowering genes and activate transcription of flowering genes. Nuclear interactions of ROXY1 with TGA factors were shown to be crucial and indispensable for normal petal development (dashed arrow). Interestingly, ROXY19 can replace ROXY1 activity if appropriately expressed. Thus, ROXY19 can exert the same nuclear activity as ROXY1 in flower development. (See color plate section at the back of the book.)

```
A                C82
 NPR1 77 VSFHRCVLSARSSFFKSALAAAKKEKD-----------SNNTAAVKLELKEI
 BOP1 38 VHAHRCILAARSLFFRKFFCESDPSQP--GAEPANQ---TGSGAR-AAAVGGV
 BOP2 37 VHAHRCILAARSLFFRKFFCGTDSPQPVTGIDPTQHGSVPASPTRGSTAPAGI

 NPR1    AKDYEVGFDSVVTVLAYVYSSRVRPPPKGVS---ECADENCCHVACRPAVDFM
 BOP1    IPVNSVGYEVFLLLLQFLYSGQVSIVPHKHEPRSNCGDRGCWHTHCTAAVDLS
 BOP2    IPVNSVGYEVFLLLLQFLYSGQVSIVPQKHEPRPNCGERGCWHTHCSAAVDLA
                                                      C216
 NPR1    LEVLYLAFIFKIPELITLYQRHLLDVVDKVVIEDTLVILKLANICGKACMKLL 220
 BOP1    LDILAAARYFGVEQLALLTQKHLTSMVEKASIEDVMKVLIAS---RKQDMHQL 187
 BOP2    LDTLAASRYFGVEQLALLTQKQLASMVEKASIEDVMKVLIAS---RKQDMHQL 192

B                  C260  C266
 TGA1 249 DVLTDQQLLDVCNLKQSCQQAEDALTQGMEKLQHTL 284
 TGA4 245 DPLTDQQLLDVCNLRQSCQQAEDALSQGMEKLQHTL 280
 PAN  329 DPLTDQQLIGICNLQQSSQQAEDALSQGMEALQQSL 364
```

Fig. 3. Redox-regulated crucial cysteines in members of the TGA and NPR1 families. (A) Alignment of the predicted BTB/POZ domain for NPR1 and BOP1/2. C82 and C216 denote conserved cysteines that mediate redox control of NPR1 oligomerization and nuclear localization. Amino acids corresponding to C82 and C216 of NPR1 are highlighted with gray and white boxes, respectively. (B) Alignment of an N-terminal conserved Gln-rich region for TGA1, TGA4, and PAN. C260 and C266 in TGA1 indicate conserved cysteines that mediate the formation of an intramolecular disulfide bridge, which prevents the interaction of NPR1 and TGA1. In contrast, there exists only one conserved cysteine in PAN, namely C340, which corresponds to C260 in TGA1 and has been shown to be crucial for PAN function. Given the differences in the degree of conservation of crucial cysteine residues, NPR1 and BOP1/2 as well as TGA1 and PAN, might be modulated posttranslationally via different mechanisms. Amino acids corresponding to C260 and C266 of TGA1 are highlighted with gray and white boxes, respectively.

reveals that both BOP proteins contain only one conserved cysteine corresponding to the N-terminal Cys82 in NPR1 (Fig. 3A). However, as mutagenesis of one cysteine in NPR1 is sufficient to cause its nuclear localization, this might also be the case for BOP1/2. Similarly, amino acid comparisons of TGA1, TGA4, and PAN (Fig. 3B) also show deviations in the conservation degree of cysteines known to be crucial for posttranslational modifications. Whereas Cys260 and Cys266 in TGA1 are involved in the intramolecular disulfide bridge formation impeding the interaction with NPR1, PAN contains only one conserved cysteine, Cys340, in the position corresponding to the TGA1 Cys260. Mutagenesis of PAN Cys340 into serine caused a failure of mutagenized cDNA to complement the *pan* mutant, demonstrating the importance of this single cysteine in PAN (Li *et al.*, 2009). Therefore, the mechanisms by which TGA1 and PAN are posttranslationally modulated might differ. Further mutagenesis experiments can shed light on the contribution of the conserved cysteines to the intracellular localization of BOP1/2 and how this relates to their function in flower development.

To summarize, similar protein molecules from the TGA family of transcription factors, the CC-type GRX class and the NPR1 family, were recruited during land plant evolution to participate in these two different processes. However, slight differences might exist as to how they are modified and act in redox-related processes that could have contributed to the evolution of specific functions in these superficially unrelated pathways. This is further supported by the observation that *ROXY19* transcription requires both NPR1 and TGA factors (Ndamukong *et al.*, 2007), whereas *ROXY1* expression is unlikely to rely on TGA factors and BOP proteins, as binding sites of TGA factors are found in the promoter region of *ROXY19* but are not present in a 3.6-kb promoter fragment that confers endogenous *ROXY1* expression (Xing *et al.*, 2005; Li and Zachgo, unpublished data). Whether and how redox changes act as a signal to regulate expression and activities of CC-type GRXs and thereby alter activities of transcription factors functioning in flower development remain to be further elucidated.

IX. CPYC AND CGFS GRXs ACT IN IRON–SULFUR CLUSTER FORMATION AND ARSENIC RESISTANCE

Iron–sulfur (Fe–S) clusters are found in a variety of metalloproteins. Proteins with Fe–S cofactors characterized by the presence of sulfide-linked di-, tri-, and tetrairon centers are ubiquitous in all living organisms and play central roles in fundamental cellular processes, including redox reactions, metabolic catalysis, and iron sensing (Beinert and Kiley, 1999; Beinert *et al.*, 1997). Biosynthetic processes of Fe–S clusters are highly conserved across all biological systems and involve a complex assemblage of proteins (Frazzon and Dean, 2003; Lill and Kispal, 2000). Mitochondria and/or chloroplasts contain many Fe–S proteins, such as [2Fe–2S] ferredoxin and [4Fe–4S] ferredoxin-thioredoxin reductase, and are two subcellular compartments involved in the biosynthesis of organellar or cytosolic Fe–S proteins (Mühlenhoff *et al.*, 2003). In plants, Fe–S cluster biosynthesis primarily occurs in mitochondria using the ISC machinery with Isu, IscA, and Nfu as potential scaffold proteins, and in chloroplasts employing the SUF machinery with SufA, SufB, and Nfu proteins as potential scaffold proteins (Balk and Lobreaux, 2005; Ye *et al.*, 2006).

In recent years, GRXs with CPYC and CGFS motifs have unexpectedly been identified as crucial players in the assembly and delivery of Fe–S clusters. Yeast GRX5 is a mitochondrial CGFS GRX, and loss of its function results in iron enrichment in the cell, constitutive oxidative damage, inactivation of enzymes requiring Fe–S clusters for their activity, and accumulation of Fe–S clusters on the scaffold protein Isu1p (Mühlenhoff *et al.*, 2003; Rodriguez-Manzaneque *et al.*, 2002). These defects suggest that GRX5 performs a direct function confined to a step after assembly of a Fe–S cluster on Isu1p in Fe–S protein biogenesis. Human mitochondrial GRX2 with a CSYC active site motif exists as an enzymatically active monomer or a [2Fe–2S]-bridged quiescent dimer coordinated by four cysteine residues, indicating that [2Fe–2S] clusters likely serve as a redox sensor for activating GRX2 under oxidative stresses (Johansson *et al.*, 2007; Lillig *et al.*, 2005). Like human GRX2, the CGFS SyGRX3 from *Synechocystis PCC6803* and the cytosolic poplar GRXC1 with a CGYC active site are also present as a monomeric apoprotein or a dimeric holoprotein with a subunit-bridging [2Fe–2S] cluster that is ligated by the catalytic cysteines of two GRXs and the cysteines of two GSH molecules (Feng *et al.*, 2006; Picciocchi *et al.*, 2007; Rouhier *et al.*, 2007; Fig. 4A). Biochemical characterization of representative CGFS monothiol GRXs from *Gloeobacter violaceus* (GvGRX3), *Thermosynechococcus elongates* (TeGRX3), *Saccharomyces cerevisiae* (ScGRX3-5), *A. thaliana* (AtGRXS14,16), and *Populus trichocarpa* (PtGRX14,16) has revealed that the incorporation of a GSH-ligated [2Fe–2S] center is a common characteristic of prokaryotic and eukaryotic CGFS GRXs (Picciocchi *et al.*, 2007), implicating a possible involvement of these enzymes in iron sensing and/or biogenesis and transfer of Fe–S cofactors. Several chloroplast-localized CGFS GRXs, such as AtGRXS14/16 and PtGRXS14/16, have been recently shown to function potentially as scaffold proteins for the assembly of [2Fe–2S] clusters that can be transferred to physiologically relevant acceptor proteins (Fig. 4B). Alternatively, they may act in the storage and/or delivery of preformed Fe–S clusters or in the regulation of the chloroplastic Fe–S cluster assembly machinery (Bandyopadhyay *et al.*, 2008, Fig. 4B). In plants, the CGFS GRX class comprises four members, but only Arabidopsis AtGRXS14 is characterized functionally. Loss of function of AtGRXS14 displayed increased protein carbonylation within chloroplasts and defects in early seedling growth under oxidative stresses, supporting a possible role of AtGRXS14 in protection against oxidative protein damage via maintaining iron homeostasis and modulating iron-generated oxygen radicals within chloroplasts (Cheng *et al.*, 2006, Fig. 4B).

Besides being involved in iron–sulfur cluster formation, GRXs with CPYC and CGFS motifs have also been shown to function in arsenic resistance. Arsenic is a notoriously poisonous metalloid, and inorganic arsenic, including the oxidized form arsenate [As(V)] and the reduced form arsenite [As(III)], is widely distributed in the environment. Arsenic resistance in *E. coli* is conferred by the arsenic resistance (*ars*) operon. The *arsC* gene of this operon encodes an arsenate reductase that catalyzes the reduction of

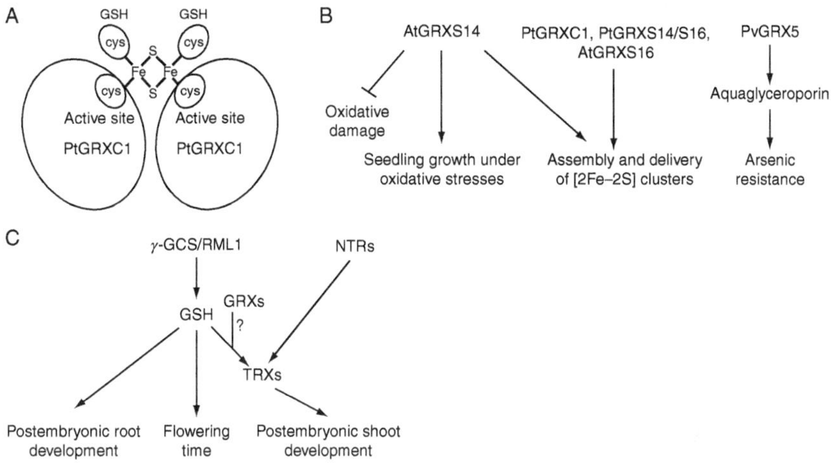

Fig. 4. Other redox-sensitive cellular processes in plants. (A) Coordination patterns of a [2Fe–2S] cluster for poplar dimeric holo PtGRXC1. Cytosolic poplar PtGRXC1 with a CGYC active site is present as either a monomeric apoprotein or a dimeric holoprotein with a subunit-bridging [2Fe–2S] cluster that is ligated by catalytic cysteines of two PtGRXC1 molecules and cysteines of two GSH molecules. (B) AtGRXS14 with a CGFS active site protects cells against oxidative damage and is necessary for early seedling growth under oxidative stresses. As PtGRXC1, AtGRXS14/S16 and PtGRXS14/S16 also play a role in the assembly and delivery of [2Fe–2S] clusters. PvGRX5 with a CGFS active site likely regulates intracellular arsenite levels, either by directly or indirectly modulating aquaglyceroporins. (C) *RML1* codes for γ-GCS, the first enzyme of GSH biosynthesis. Analyses of Arabidopsis *rml1* mutants have revealed that GSH promotes flowering and positively regulates postembryonic root development. Inactivation of two NTRs and further analyses of *ntra ntrb rml1* triple mutants allowed characterization of a complex interplay between TRXs and a GSH-dependent pathway in Arabidopsis postembryonic shoot development. Biochemical data indicate that GRXs might participate in alternative reduction TRXs by the GSH pathway. Arrows and hatchets denote positive and negative regulation, respectively. Abbreviations: At, *Arabidopsis thaliana*; Pt, *Populus trichocarpa*; Pv, *Pteris vittata*; SA, salicylic acid; GRX, glutaredoxin; γ-GCS, γ-glutamylcysteine synthase; GSH, glutathione; NTR, NADPH thioredoxin reductase; TRX, thioredoxin.

arsenate to arsenite using the CPYC GRX2 and GSH as hydrogen donors. The reduced arsenite is subsequently exported from *E. coli* cells into the extracellular milieu, thereby conferring arsenic resistance (Rossen, 1999). Chinese brake fern *Pteris vittata* is resistant to arsenic and is able to hyperaccumulate a larger quantity of arsenic in fronds (Ma *et al.*, 2001). Arsenate absorbed by this fern is transported via the xylem, then reduced to arsenite in fronds, and finally likely stored in vacuoles. To explore the molecular mechanism of arsenic resistance, Sundaram *et al.* (2008) isolated a CGFS GRX (PvGRX5) involved in the regulation of intracellular arsenite from *P. vittata* fronds based on its capacity to increase arsenic resistance in *E. coli*. Expression of this unusual GRX increased arsenic tolerance in an *E. coli* mutant in which the *ars* operon was removed. Contrarily, production of PvGRX5 in another mutant strain in which both the *ars* operon and the gene coding for the GlpF aquaglyceroporin were knocked out, did not affect arsenic resistance significantly. These findings suggest that PvGRX5 confers cellular arsenic resistance independent of the *ars* operon genes but dependent on GlpF. Aquaglyceroporins have been shown to be arsenite channels that move arsenite into and out of cells (Liu *et al.*, 2002). Therefore, it is most likely that PvGRX5 regulates intracellular arsenite levels, either by directly or indirectly modulating aquaglyceroporins. In *P. vittata*, PvGRX5 might possibly regulate a vacuolar GlpF homolog to alter arsenite transport into vacuoles. Given the ubiquity to pro- and eukaryotes of CPYC and CGFS GRXs and the specificity to land plants of CC-type GRXs, the involvement of GRXs in assembly and delivery of iron–sulfur clusters and arsenic resistance seems to represent ancestral GRX functions. It will thus be interesting to determine if the mechanisms by which CC-type GRXs exert their activity have evolved independently from ancestral GRXs.

X. GSH-ASSOCIATED DEVELOPMENTAL PROCESSES

GSH is a key component of plant antioxidant networks and is required for both the dithiol and monothiol reduction mechanism mediated by GRXs. In the dithiol mechanism, GSH is necessary for the regeneration of GRX. In contrast, in the monothiol mechanism, GSH is directly attached to a cysteine and thereby reversibly modifies target proteins posttranslationally. GSH is synthesized in two ATP-dependent steps that are sequentially catalyzed by γ-glutamylcysteine synthase (γ-GCS) and glutathione synthase. GSH is present at millimolar concentrations in plant cells and participates in stress resistance and adaptation, such as detoxification of heavy metals and scavenging of reactive oxygen species (ROS) generated during normal cell metabolism and induced by abiotic factors such as UV light, drought, and chilling as well as by biotic factors such as pathogens (Cobbert, 2000; Noctor and Foyer, 1998). However, besides playing critical roles in stress responses, evidence is currently emerging that GSH is also crucial for many plant developmental processes (Fig. 4C).

Recent data indicate that GSH plays an important role in determining flowering time. Ogawa *et al.* (2001, 2004) examined the relationship between

GSH levels, photosynthesis, and flowering in Arabidopsis and found that flowering is regulated by the key reaction of GSH biosynthesis catalyzed by γ-GCS. Arabidopsis mutants, defective in glutathione biosynthesis or possessing nonfunctional light-harvesting antenna of photosystem II, produce reduced GSH levels and hence develop a late flowering phenotype (Cobbett *et al.*, 1998; Ogawa *et al.*, 2001, 2004).

Embedded within root meristems is a population of slowly dividing cells collectively designated as the quiescent center (QC). Postembryonic root development entails activation of cell division in this center. Analysis of Arabidopsis plants homozygous for a mutation in *ROOT MERISTEMLESS1* (*RML1*)/*CADMIUM SENSITIVE2* (*CAD2*), which encodes the γ-GCS, identified a GSH-dependent pathway involved in establishing the postembryonic root meristem (Cobbett *et al.*, 1998; Vernoux *et al.*, 2000). The *rml1/cad2* mutant is almost devoid of GSH due to strongly reduced GSH synthesis, thereby causing an arrest in root development due to abolished cell division after germination in the root but not in the shoot (Vernoux *et al.*, 2000). In maize roots, quiescent center formation is associated with an auxin-regulated oxidizing environment. High levels of auxin and oxidized GSH occur in the QC, contrasting with reduced GSH and low levels of auxin favored in adjacent, rapidly dividing cells in the root meristem. Decreasing auxin levels in the QC by perturbing polar auxin transport leads to a reduced environment and thereby activates the QC (Jiang *et al.*, 2003). These experiments corroborate earlier observations that GSH participates in the regulation of cell division in the apical meristem of Arabidopsis roots (Sánchez-Fernández *et al.*, 1997).

NADPH thioredoxin reductases (NTRs) are key regulatory enzymes determining the redox state of the TRX system. The Arabidopsis genome contains two genes encoding NTRs (*NTRA* and *NTRB*) and one gene coding NTRC, a chloroplastic hybrid enzyme containing both an NTR and a TRX modules. The *ntra ntrb* double mutant is viable and fertile, and the cytosolic TRXh3 in this mutant is only partially oxidized. Crossing *ntra ntrb* with *rml1/cad2* leads to complete inhibition of both shoot and root growth, indicating that a GSH-dependent pathway is implicated in the alternative reduction of TRXh3 and is indispensable for postembryonic activity in the shoot and root apical meristems (Reichheld *et al.*, 2007). However, a direct reduction of type-h TRXs by GSH is excluded by biochemical data, suggesting that GRXs may be responsible for the alternative reduction of TRXs in the *ntra ntrb* double mutant (Reichheld *et al.*, 2007), as observed *in vitro* for poplar TRX h4 (Gelhaye *et al.*, 2003; Koh *et al.*, 2008).

XI. OXIDATIVE STRESS RESPONSES

Oxidative stress from environmental sources and plant developmental processes generates various oxidants, such as reactive oxygen, sulfur, and nitrogen, which can cause serious damage to biological macromolecules. In higher plants, chloroplasts and mitochondria are two major organelles that contribute to the production of ROS during photosynthesis and carbon metabolism. To prevent oxidative damage, plants have evolved effective mechanisms comprising nonenzymatic antioxidant molecules and enzymes (Michelet *et al.*, 2006). Oxidants have been traditionally considered as poisonous molecules with deleterious effects on plant cells. However, recent studies suggest that these reactive species, particularly ROS and reactive nitrogen species (RNS), are involved in redox signaling through reversible posttranslational modifications of protein thiols (Foyer and Noctor, 2005; Michelet *et al.*, 2006). Free protein thiols can be oxidized into different reversible states, such as *S*-glutathionylation, *S*-nitrosylation, sulfenic (SOH) or sulfinic acids (SO_2H), and intra- or interprotein disulfide bonds (Fig. 5). *S*-Glutathionylation of protein sulfhydryl groups protects cysteinyl residues from irreversible oxidation to sulfonic acids. Apart from sulfinic acids that are reduced by sulfiredoxins (SRXs), these oxidized cysteines can be efficiently reduced by GRXs, providing evidence that GRXs play an important role in oxidative stress signaling (Foyer and Noctor, 2005; Michelet *et al.*, 2006). Antioxidant defense involves the activation of redox-responsive transcription factors in *E. coli*, yeast, and mammals (Delaunay *et al.*, 2000; Rahman and MacNee, 2000). In apoptosis signaling of mammalian cells, TRXs are known to catalyze denitrosylation of cysteine residues of caspase-3, a process required for caspase-3 activation (Benhar *et al.*, 2008). Higher plants are also thought to acclimate to oxidative stress through coordinate modulation of a battery of antioxidant genes. Tsukamoto *et al.* (2005) identified a novel conserved 28-bp long *cis*-element on the promoter regions of three antioxidant rice genes, one of which codes for a rice GRX. This short regulatory sequence is responsive to oxidative stress induced by methyl viologen treatment and regulates expression of these antioxidant genes. In Arabidopsis, a network of at least 152 genes is involved in producing and scavenging ROS, among which 27 encode GRXs functioning as ROS-scavenging enzymes (Mittler *et al.*, 2004). As mentioned earlier, overexpression of *ROXY1* and rice *OsROXY1/2* in Arabidopsis leads to accumulation of H_2O_2 and increased susceptibility of transgenic plants to pathogen infection, indicating a link between oxidative stress, plant development, and pathogen responses (Wang *et al.*, 2009). The use of ROS as signaling molecules by plant cells suggests that plants are able to achieve a high degree of control over ROS toxicity during the course of evolution. The first piece of

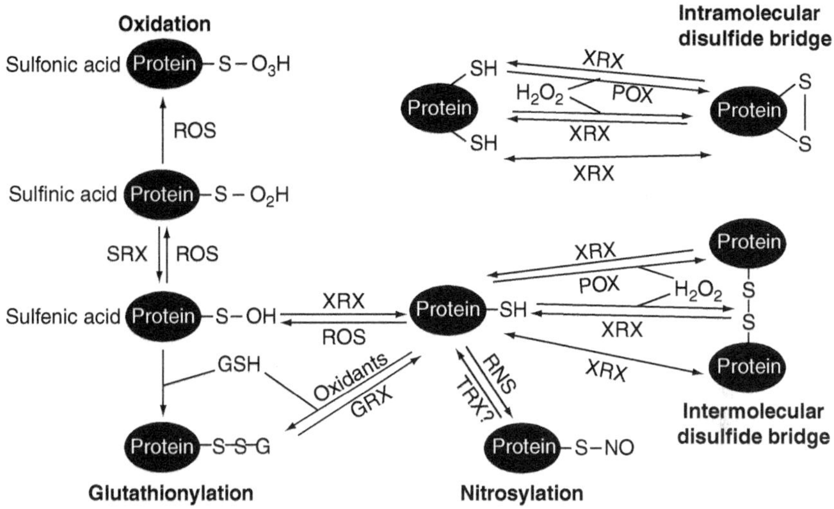

Fig. 5. Oxidation and reduction of protein thiols. Under oxidizing conditions, free and accessible protein thiols undergo several posttranslational modifications, which can be reversible or not. Protein cysteines can be oxidized by ROS into sulfenic acid (SOH), which can be reduced by XRX (GRX or TRX). Further oxidation of sulfenic acid by ROS can result in the formation of sulfinic acid (SO_2H), which can be reversed by sulfiredoxins (SRX), or irreversibly oxidized to sulfonic acid (SO_3H). The presence of oxidants and/or GSH allows glutathionylation of protein cysteines to occur via different mechanisms. Deglutathionylation can be catalyzed by GRX. Reversible formation of intra- and intermolecular disulfide bridges is mediated by XRX. Direct H_2O_2 dependent oxidation of cysteines to intra- and intermolecular disulfides and peroxidase (POX)-catalyzed H_2O_2 sensing can be reversed by XRX. In addition, protein cysteines also undergo nitrosylation in the presence of reactive nitrogen species (RNS), a reversible process that could be catalyzed by TRX.

biochemical evidence indicating the possible involvement of plant GRXs in oxidative stress responses was obtained from a poplar CPYC GRX that is able to reduce a type II peroxiredoxin (PRX; Rouhier *et al.*, 2001, 2002). Further independent observations have shown that GRXs and PRXs are abundantly colocalized in the phloem sap, allowing in planta reduction reactions to occur and preventing oxidative damage of sieve-tube proteins (Balachandran *et al.*, 1997; Ishiwatari *et al.*, 1995; Szederkenyi *et al.*, 1997). A redox-sensitive green fluorescent protein (roGFP) has been constructed and is now available for nondestructive, real-time measurement of the redox potential in both animal and plant cells (Hanson *et al.*, 2004; Jiang *et al.*, 2006). The CPYC GRXC1 from Arabidopsis is able to reduce disulfide bonds of roGFPs (Meyer *et al.*, 2007), demonstrating the potential of this novel tool to monitor plant redox changes and indicating a role for GRXC1 in maintaining the cellular redox homeostasis.

XII. IDENTIFICATION OF GRX TARGETS

Systemic dissection of the earlier described diverse cellular processes associated with GRXs entails identification and functional analysis of target proteins. In plants, target proteins of TRXs have been extensively explored and most of TRX targets were identified using affinity chromatography approaches (Buchanan and Balmer, 2005). In contrast, data on target proteins of plant GRXs have been scarce, whereas the numbers of genes encoding GRXs suggests a high representation of this type of proteins in plant species. Cytosolic poplar PRX II was identified as the first GRX target (Rouhier *et al.*, 2001). Another target protein is a vacuolar CAX1 H^+/Ca^{2+} antiporter in Arabidopsis (Cheng and Hirschi, 2003). The development of proteomics-based approaches and the availability of complete protein databases allow large-scale identification of plant GRX targets. Lee *et al.* (2004) applied thiol affinity chromatography to identify proteins with disulfide bonds in Arabidopsis and isolated 65 putative disulfide proteins that are likely targets of GRXs and TRXs. Using liquid chromatography coupled to tandem mass spectrometry, Rouhier *et al.* (2005) isolated over 90 targets from the poplar CPYC GRXC4, of which many are involved in cellular processes including stress responses, photorespiration, translation, nitrogen, sulfur, carbon metabolisms, and protein folding. Most of these proteins are encoded by housekeeping genes and likely expressed at higher levels. However, proteins involved in signaling or transcription factors, such as TGA factors, which are normally expressed at lower levels, might escape detection by this method and can thus be better isolated by yeast two-hybrid screens (Li *et al.*, 2009). As GRXs can reduce protein-GSH mixed disulfides, analysis of glutathionylated proteins represents another efficient strategy for the determination of potential targets of GRXs. Ito *et al.* (2003) labeled proteins using biotinylated GSH in Arabidopsis cells and identified two key enzymes for sugar metabolism as putative GRX targets. To identify proteins capable of undergoing glutathionylation under oxidative conditions, biotinylated, oxidized GSH (GSSG-biotin) was utilized to label proteins in Arabidopsis suspension cultures treated with oxidants (Dixon *et al.*, 2005). Following 2D-PAGE and MALDI-TOF MS, 79 distinct proteins were isolated and most of them are involved in protein turnover and metabolism. Proteins identified by this proteomic approach include not only putative targets undergoing glutathionylation but also those interacting with glutathionylated proteins rather than undergoing this modification themselves. Thus far, proteomic approaches, along with yeast two-hybrid screens, have identified a large array of GRX targets implicated in many cellular processes. However, an *in vivo* physical interaction of GRXs and their respective targets needs to be confirmed using additional independent techniques. Particularly, genetic experiments, such as the analysis of double mutants, can reveal if GRXs and their putative targets act in the same genetic pathway. With more new GRX targets identified by a broad range of currently available techniques, each having its inherent advantages and disadvantages, it will be a challenging but promising task to define the complete proteome redox-regulated by GRXs.

XIII. CROSSTALKS BETWEEN GRXs AND TRXs

TRXs and GRXs belong to the thioredoxin superfamily. TRXs are involved in chloroplast and mitochondrial processes, seed development, and germination (Buchanan and Balmer, 2005). In plants, cytosolic and mitochondrial TRXs are reduced by NADPH and NADPH thioredoxin reductase (NTR), whereas chloroplast TRXs are reduced by two proteins containing Fe–S clusters, known as ferredoxin and ferredoxin thioredoxin reductase (Rouhier *et al.*, 2004). Despite different modes of reduction between GRXs and TRXs, emerging evidence suggests that these two systems likely constitute a complex network of redox signaling by crosstalk. Many target proteins of plant GRXs arising from proteomic approaches have also been identified as partners of TRXs, underlying connections and redundancy between GRXs and TRXs (Rouhier *et al.*, 2005). A poplar TRX is reduced by a poplar GRX and GSH instead of NTR, indicating that these two redox systems are dependent on one another (Gelhaye *et al.*, 2003; Koh *et al.*, 2008). Chloroplastic f type thioredoxins present in several plant species as well as a poplar mitochondrial TRX are shown to undergo reversible glutathionylation. Glutathionylation of these TRXs affects enzymatic activity toward their targets and thereby leads to metabolic changes (Gelhaye *et al.*, 2004; Michelet *et al.*, 2006). As GRXs are able to catalyze deglutathionylation reactions, it is most likely that GRXs regulate enzymatic activity of these specific TRXs by removing GSH attached to them. Uncovering functional redundancies between GRXs and TRXs and crosstalk among GRXs, TRXs and GSH are key points of interest for further studies on cellular redox signaling.

XIV. CONCLUDING REMARKS

Overall, our understanding of the biological functions of plant GRXs is currently expanding. Mutant analysis has recently revealed exciting insights into their contribution to diverse plant processes. Increasing evidence indicates the importance of the CC-type GRXs in exerting posttranslational modifications of transcription factors, such as TGA factors, and thereby affecting their transcriptional activity. Interestingly, TGA transcription factors are known to function in two processes that seem at first glimpse not to be connected, namely during floral development and in disease resistance. Similarly, the TGA coactivators belonging to the small NPR1 family in Arabidopsis act in both pathways. Therefore, CC-type GRXs, TGA factors, and NPR1 homologs were likely all co-opted and integrated during land plant evolution into these distinctive processes. Further analysis of more CC-type GRXs and their target proteins will unravel their contribution to different processes. This will also shed light on the evolutionary mechanisms by which GRXs have been recruited and integrated into a diverse, broad spectrum of biological processes that seem to be regulated in a redox-sensitive manner, turning GRX activities into a prime mechanism to modulate protein functions.

REFERENCES

Balachandran, S., Xiang, Y., Schobert, C., Thompson, G. A. and Lucas, W. J. (1997). Phloem sap proteins from cucurbita maxima and ricinus communis have the capacity to traffic cell to cell through plasmodesmata. *Proceedings of the National Academy of Sciences of the United States of America* **94**, 14150–14155.

Balk, J. and Lobreaux, S. (2005). Biogenesis of iron–sulfur proteins in plants. *Trends in Plant Science* **10**, 324–331.

Bandyopadhyay, S., Gama, F., Molina-Navarro, M. M., Gualberto, J. M., Claxton, R., Natik, S. G., Huynh, B. H., Herrero, E., Jacquot, J. P., Johnson, M. K. and Rouhier, N. (2008). Chloroplast monothiol glutaredoxins as scaffold proteins for the assembly and delivery of [2Fe–2S] clusters. *EMBO Journal* **27**, 1122–1133.

Beinert, H. and Kiley, P. J. (1999). Fe–S proteins in sensing and regulatory functions. *Current Opinion in Chemical Biology* **3**, 152–157.

Beinert, H., Holm, R. H. and Münck, E. (1997). Iron–sulfur clusters: Nature's modular, multipurpose structures. *Science* **277**, 653–659.

Benhar, M., Forrester, M. T., Hess, D. T. and Stamler, J. S. (2008). Regulated protein denitrosylation by cytosolic and mitochondrial thiredoxins. *Science* **320**, 1050–1054.

Buchanan, B. B. and Balmer, Y. (2005). Redox regulation: A broadening horizon. *Annual Review of Plant Biology* **56**, 187–220.

Cheng, N. H. and Hirschi, K. D. (2003). Cloning and characterization of CXIP1, a novel PICOT domain-containing Arabidopsis protein that associates with CAX1. *The Journal of Biological Chemistry* **278**, 6503–6509.

Cheng, N. H., Liu, J. Z., Brock, A., Nelson, R. S. and Hirschi, K. D. (2006). AtGRXcp, an Arabidopsis chloroplastic glutaredoxin, is critical for protection against protein oxidative damage. *The Journal of Biological Chemistry* **281**, 26280–26288.

Chuang, C. F., Running, M. P., Williams, R. W. and Meyerowitz, E. M. (1999). The *PERIANTHIA* gene encodes a bZIP protein involved in the determination of floral organ number in *Arabidopsis thaliana*. *Genes and Development* **13**, 334–344.

Cobbert, C. S. (2000). Phytochelatin biosynthesis and function in heavy-metal detoxification. *Current Opinion in Plant Biology* **3**, 211–216.

Cobbett, C. S., May, M. J., Howden, R. and Rolls, B. (1998). The glutathione-deficient, cadmium-sensitive mutant, cad2-1, of *Arabidopsis thaliana* is deficient in gamma-glutamylcysteine synthetase. *Plant Journal* **16**, 73–78.

Coen, E. S. and Meyerowitz, E. M. (1991). The war of the whorls: Genetic interactions controlling flower development. *Nature* **353**, 31–37.

Delaunay, A., Isnard, A. D. and Toledano, M. B. (2000). H_2O_2 sensing through oxidation of the Yap1 transcription factor. *EMBO Journal* **19**, 5157–5166.

Després, C., Chubak, C., Rochon, A., Clark, R., Bethune, T., Desveaux, D. and Fobert, P. R. (2003). The Arabidopsis NPR1 disease resistance protein is a novel cofactor that confers redox regulation of DNA binding activity to the basic domain/leucine zipper transcription factor TGA1. *Plant Cell* **15**, 2181–2191.

Ditta, G., Pinyopich, A., Robles, P., Pelaz, S. and Yanofsky, M. F. (2004). The *SEP4* gene of *Arabidopsis thaliana* functions in floral organ and meristem identity. *Current Biology* **14**, 1935–1940.

Dixon, D. P., Skipsey, M., Grundy, N. M. and Edwards, R. (2005). Stress-induced protein S-glutathionylation in Arabidopsis. *Plant Physiology* **138**, 2233–2244.

Drews, G. N., Bowman, J. L. and Meyerowitz, E. M. (1991). Negative regulation of the Arabidopsis homeotic gene *AGAMOUS* by the *APETALA2* product. *Cell* **14**, 991–1002.

Feng, Y., Zhong, N., Rouhier, N., Hase, T., Kusunoki, M., Jacquot, J. P., Jin, C. and Xia, B. (2006). Structural insight into poplar glutaredoxin C1 with a bridging iron-sulfur cluster at the active site. *Biochem* **45**, 7998–8008.

Fernandes, A. P. and Holmgren, A. (2004). Glutaredoxins: Glutathione-dependent redox enzymes with functions far beyond a simple thioredoxin backup system. *Antioxidants Redox Signaling* **6**, 63–74.

Foyer, C. H. and Noctor, G. (2005). Redox homeostasis and antioxidant signaling: A metabolic interface between stress perception and physiological responses. *Plant Cell* **17**, 1866–1875.

Frazzon, J. and Dean, D. R. (2003). Formation of iron–sulfur clusters in bacteria-an emerging field in bioinorganic chemistry. *Current Opinion in Chemical Biology* **7**, 166–173.

Gelhaye, E., Rounier, N. and Jacquot, J. P. (2003). Evidence for a subgroup of thioredoxin h that requires GSH/GRX for its reduction. *FEBS Letters* **555**, 443–448.

Gelhaye, E., Rounier, N., Gerard, J., Jolivet, Y., Gualberto, J., Navrot, N., Ohlsson, P. I., Wingsle, G., Hirasawa, M., Knaff, D. B., Wang, H. Dizengremel, P. *et al.* (2004). A specific form of thioredoxin h occurs in plant mitochondria and regulates the alternative oxidase. *Proceedings of the National Academy of Sciences of the United States of America* **101**, 14545–14550.

Gustafson-Brown, C., Savidge, B. and Yanofsky, M. F. (1994). Regulation of the Arabidopsis floral homeotic gene *APETALA1*. *Cell* **76**, 131–143.

Hanson, G. T., Aggeler, R., Oglesbee, D., Cannon, M., Capaldi, R. A., Tsien, R. Y. and Remington, S. J. (2004). Investigating mitochondrial redox potential with redox-sensitive green fluorescent protein. *The Journal of Biological Chemistry* **279**, 13044–13053.

Hepworth, S. R., Zhang, Y., McKim, S., Li, X. and Haughn, G. W. (2005). *BLADE-ON-PETIOLE*-dependent signaling controls leaf and floral patterning in Arabidopsis. *Plant Cell* **17**, 1434–1448.

Heyl, A., Ramireddy, E., Brenner, W. G., Riefler, M., Allemeersch, J. and Schmulling, T. (2008). The transcriptional repressor ARR1-SRDX suppresses pleiotropic cytokinin activities in *Arabidopsis. Plant Physiology* **147**, 13801395.

Hiratsu, K., Matsui, K., Koyama, T. and Ohme-Takagi, M. (2003). Dominant repression of target genes by chimeric repressors that include the EAR motif, a repression domain, in *Arabidopsis. Plant Journal* **34**, 733739.

Holmgren, A., Soberberg, B. O., Eklund, H. and Branden, C. I. (1975). Three-dimensional structure of *Escherichia coli* thioredoxin-S2 to 2.8 angstrom resolution. *Proceedings of the National Academy of Sciences of the United States of America* **72**, 2305–2309.

Honma, T. and Goto, K. (2001). Complexes of MADS-box proteins are sufficient to convert leaves into floral organs. *Nature* **409**, 525–529.

Hou, X. L., Hu, W. W., Shen, L. S., Lee, L. Y. C., Tao, Z., Han, J. H. and Hao, Yu. (2008). Global identification of DELLA target genes during Arabidopsis flower development. *Plant Physiology* **147**, 1126–1142.

Ishiwatari, Y., Honda, C., Kawashima, I., Nakamura, S., Hirano, H., Mori, S., Fujiwara, T., Hayashi, H. and Chino, M. (1995). Thioredoxin h is one of the major proteins in rice phloem sap. *Planta* **195**, 456–463.

Ito, H., Iwabuchi, M. and Ogawa, K. (2003). The sugar-metabolic enzymes aldolase and triose-phosphate isomerase are targets of glutathionylation in *Arabidopsis thaliana*: Detection using biotinylated glutathione. *Plant and Cell Physiology* **44**, 655–660.

Jack, T., Brockmann, L. L. and Meyerowitz, E. M. (1992). The homeotic gene APETALA3 of *Arabidopsis thaliana* encodes a MADS box and is expressed in petals and stamens. *Cell* **68**, 683–697.

Jiang, K., Meng, Y. L. and Feldman, L. J. (2003). Quiescent center formation in maize roots is associated with an auxin-regulated oxidizing environment. *Development* **130**, 1429–1438.

Jiang, K., Schwarzer, C., Lally, E., Zhang, S., Ruzin, S., Machen, T., Remington, S. J. and Feldman, L. (2006). Expression and characterization of a redox-sensing green fluorescent protein (reduction-oxidation-sensitive green fluorescent protein) in Arabidopsis. *Plant Physiology* **141**, 397–403.

Johansson, C., Kavanagh, K. L., Gileadi, O. and Oppermann, U. (2007). Reversible sequestration of active site cysteines in a 2Fe–2S-bridged dimmer provides a mechanism for glutaredoxin 2 regulation in human mitochondria. *The Journal of Biological Chemistry* **282**, 3077–3082.

Jones, J. D. G. and Dangl, J. L. (2006). The plant immune system. *Nature* **444**, 323–329.

Kesarwani, M., Yoo, J. and Dong, X. (2007). Genetic interactions of TGA transcription factors in the regulation of pathogenesis-related genes and disease resistance in Arabidopsis. *Plant Physiology* **144**, 336–346.

Koh, C. S., Navrot, N., Didierjean, C., Rouhier, N., Hirasawa, M., Knaff, D. B., Wingsle, G., Samian, R., Jacquot, J. P., Corbier, C. and Gelhaye, E. (2008). An atypical catalytic mechanism involving three cysteines of thioredoxin. *The Journal of Biological Chemistry* **283**, 23062–23072.

Koornneef, A. and Pieterse, C. M. J. (2008). Cross talk in defense signaling. *Plant Physiology* **146**, 839–844.

Koyama, T., Furutani, M., Masao, T. and Ohme-Takagi, M. (2007). TCP transcription factors control the morphology of shoot lateral organs via negative regulation of the expression of boundary-specific genes in *Arabidopsis*. *Plant Cell* **19**, 473–484.

Lee, K., Lee, J., Kim, Y., Bae, D., Kang, K. Y., Yoon, S. C. and Lim, D. B. (2004). Defining the plant disulfide proteome. *Electrophoresis* **25**, 532–541.

Lemaire, S. D. (2004). The glutaredoxin family in oxygenic photosynthetic organisms. *Photosynthesis Research* **79**, 305–318.

Lill, R. and Kispal, G. (2000). Mutation of cellular Fe/S proteins: The essential function of mitochondria. *Trends in Biochemical Sciences* **25**, 352–356.

Li, S., Lauri, A., Ziemann, M., Busch, A., Bhave, M. and Zachgo, S. (2009). Nuclear activity of ROXY1, a glutaredoxin interacting with TGA factors, is required for petal development in *Arabidopsis thaliana*. *Plant Cell* **21**, 429–441.

Liu, Z., Shen, J., Carbrey, J. M., Mukhopadhyay, R., Agre, P. and Rosen, B. P. (2002). Arsenite transport by mammalian aquaglyceroporins AQP7 and AQP9. *Proceedings of the National Academy of Sciences of the United States of America* **99**, 6053–6058.

Lynch, M. and Conery, J. S. (2000). The evolutionary fate and consequences of duplicate genes. *Science* **290**, 1151–1155.

Ma, L. Q., Komar, K. M., Tu, C., Zhang, W., Cai, Y. and Kennelley, E. D. (2001). *A fern that hyperaccumulates arsenic. Nature* **409**, 579.

Martin, J. L. (1995). Thioredoxin—A fold for all reasons. *Structure* **3**, 245–250.

Meyer, A. J., Brach, T., Marty, L., Kreye, S., Rouhier, N., Jacquot, J. P. and Hell, R. (2007). Redox-sensitive GFP in *Arabidopsis thaliana* is a quantitative biosensor for the redox potential of the cellular glutathione redox buffer. *Plant Journal* **52**, 973–986.

Michelet, L., Zaffagnini, M., Massot, V., Keryer, E., Vanacker, H., Miginiac-Maslow, M., Issakidis-Bourguet, E. and Lemaire, S. D. (2006). Thioredoxins, glutaredoxins, and glutathionylation: New crosstalk's to explore. *Photosynthesis Research* **89**, 225–245.

Minakuchi, K., Yahushita, T., Masumura, T., Ichihara, K. and Tanaka, K. (1994). Cloning and sequence analysis of a cDNA encoding rice glutaredoxin. *FEBS Letters* **337**, 157–160.

Mittler, R., Vanderauwera, S., Gollery, M. and Breusegem, F. V. (2004). Reactive oxygen gene network of plants. *Trends in Plant Science* **10**, 1360–1385.

Morell, S., Follmann, H. and Häberlein, I. (1995). Identification and localization of the first glutaredoxin in leaves of a higher plant. *FEBS Letters* **369**, 149–152.

Mou, Z., Fan, W. H. and Dong, X. N. (2003). Inducers of plant systemic acquired resistance regulate NPR1 function through redox changes. *Cell* **113**, 935–944.

Mühlenhoff, U., Gerber, J., Richhardt, N. and Lill, R. (2003). Components involved in assembly and dislocation of iron-sulfur clusters on the scaffold protein Isu1p. *EMBO Journal* **22**, 4815–4825.

Ndamukong, I., Abdallat, A. A., Thurow, C., Fode, B., Zander, M., Weigel, R. and Gatz, C. (2007). SA-inducible Arabidopsis glutaredoxin interacts with TGA factors and suppresses JA-responsive *PDF1.2* transcription. *Plant Journal* **50**, 128–139.

Noctor, G. and Foyer, C. H. (1998). Ascorbate and glutathione: Keeping active oxygen under control. *Annual Review of Plant Physiology and Plant Molecular Biology* **49**, 249–279.

Ogawa, K., Tasaka, Y., Mino, M., Tanaka, Y. and Iwabuchi, M. (2001). Association of glutathione with flowering in *Arabidopsis thaliana*. *Plant and Cell Physiology* **42**, 524–530.

Ogawa, K., Hatano-Iwasaki, A., Yanagida, M. and Iwabuchi, M. (2004). Level of glutathione is regulated by ATP-dependent Ligation of glutamateand cysteine through photosynthesis in *Arabidopsis thaliana*: Mechanism of strong interaction of light intensity with flowering. *Plant and Cell Physiology* **45**, 1–8.

Ojeda, L., Keller, G., Muhlenhoff, U., Rutherford, J. C., Lill, R. and Winge, D. R. (2006). Role of glutaredoxin-3 and glutaredoxin-4 in the iron regulation of the Aft1 transcriptional activator in *Saccharomyces cerevisiae*. *The Journal of Biological Chemistry* **281**, 17661–17669.

Pelaz, S., Ditta, G. S., Baumann, E., Wisman, E. and Yanofsky, M. F. (2000). B and C floral organ identify functions require *SEPALLATA* MADS-box genes. *Nature* **405**, 200–203.

Picciocchi, A., Saguez, C., Boussac, A., Cassier-Chauvat, C. and Chauvat, F. (2007). CGFS type monothiol glutaredoxins from the cyanobacterium *Synechocystis PCC6803* and other evolutionary distant model organisms possess a glutathione-ligated [2Fe–2S] cluster. *Biochemistry* **46**, 15018–15026.

Qiu, D., Xiao, J., Xie, W., Liu, H., Li, X., Li, Xiong and Wang, S. (2008). Rice gene network inferred from expression profiling of plants overexpressing *OsWRKY13*, a positive regulator of disease resistance. *Molecular Plant* **1**, 538–551.

Rahman, I. and MacNee, W. (2000). Regulation of redox glutathione levels and gene transcription in lung inflammation: Therapeutic approaches. *Free Radical Biology and Medicine* **28**, 1405–1420.

Reichheld, J. P., Khafif, M., Riondet, C., Droux, M., Bonnard, G. and Meyer, Y. (2007). Inactivation of thioredoxin reductases reveals a complex interplay between thioredoxin and glutathione pathways in Arabidopsis development. *Plant Cell* **19**, 1851–1865.

Rodriguez-Manzaneque, M. T., Tamarit, J., Belli, G., Ros, J. and Herrero, E. (2002). Grx5 is a mitochondrial glutaredoxin required for the activity of iron/sulfur enzymes. *Molecular Biology of the Cell* **13**, 1109–1121.

Rossen, B. P. (1999). Families of arsenic transporters. *Trends in Microbiology* **7**, 207–212.

Rouhier, N., Gelhaye, E., Sautiere, P. E., Brun, A., Laurent, P., Tagu, D., Gérard, J., de Faÿ, E., Meyer, Y. and Jacquot, J. P. (2001). Isolation and characterization of a new peroxiredoxin from poplar sieve tubes that uses either glutaredoxin or thioredoxin as a proton donor. *Plant Physiology* **127**, 1299–1309.

Rouhier, N., Gelhaye, E. and Jacquot, J. P. (2002). Glutaredoxin-dependent peroxiredoxin from poplar: Protein–protein interaction and catalytic mechanism. *The Journal of Biological Chemistry* **277**, 13609–13614.

Rouhier, N., Gelhaye, E. and Jacquot, J. P. (2004). Plant glutaredoxins: Still mysterious reducing systems. *Cellular and Molecular Life Sciences* **61**, 1266–1277.

Rouhier, N., Villarejo, A., Srivastava, M., Gelhaye, E., Keech, O., Droux, M., Finkemeier, I., Samuelsson, G., Dietz, K. J., Jacquot, J. P. and Wingsle, G. (2005). Identification of plant glutaredoxin targets. *Antioxidants Redox Signaling* **7**, 919–929.

Rouhier, N., Couturier, J. and Jacquot, J. P. (2006). Genome-wide analysis of plant glutaredoxin systems. *Journal of Experimental Botany* **57**, 1685–1696.

Rouhier, N., Unno, H., Bandyopadhyay, S., Masip, L., Kim, S. K., Hirasawa, M., Gualberto, J. M., Lattard, V., Knaff, D. B., Georgiou, G., Hase, T. Johnson, M. K. *et al.* (2007). Functional, structural, and spectroscopic characterization of a glutathione-ligated [2Fe–2S] cluster in poplar glutaredoxin C1. *Proceedings of the National Academy of Sciences of the United States of America* **104**, 7379–7384.

Sánchez-Fernández, R., Fricker, M., Corben, L. B., White, N. S., Sheard, N., Leaver, C. J., Montagu, M. V., Inzé, D. and May, M. J. (1997). Cell proliferation and hair tip growth in the Arabidopsis root are under mechanistically different forms of redox control. *Proceedings of the National Academy of Sciences of the United States of America* **94**, 2745–2750.

Schmidt, R. J. and Ambrose, B. A. (1998). The blooming of grass flower development. *Current Opinion in Plant Biology* **1**, 60–67.

Sticher, L., Mauch-Mani, B. and Metraux, J. P. (1997). Systemic acquired resistance. *Annual review of Phytopathology* **35**, 235–270.

Sundaram, S., Rathinasabapathi, B., Ma, L. Q. and Rosen, B. P. (2008). An arsenate-activated glutaredoxin from the arsenic hyperaccumulator fern *Pteris vittata* L. regulates intracellular arsenite. *The Journal of Biological Chemistry* **283**, 6095–6101.

Szederkenyi, J., Komor, E. and Schobert, C. (1997). Cloning of the cDNA for glutaredoxin, an abundant sieve-tube exudate protein from *Ricinus communis* L. and characterisation of the glutathione-dependent thiol-reduction system in sieve tubes. *Planta* **202**, 349–356.

Tada, Y., Spoel, S. H., Pajerowska-Mukhtar, K., Mou, Z., Song, J., Wang, C., Zuo, J. and Dong, X. (2008). Plant immunity requires conformational changes of NPR1 via *S*-nitrosylation and thioredoxins. *Science* **321**, 952–956.

Tsukamoto, S., Morita, S., Hirano, E., Yokoi, H., Masumura, T. and Tanaka, K. (2005). A novel *cis*-element that is responsive to oxidative stress regulates three antioxidant defense genes in rice. *Plant Physiology* **137**, 317–327.

Vernoux, T., Wilson, R. C., Seeley, K. A., Reichheld, J. P., Muroy, S., Brown, S., Maughan, S. C., Cobbett, C. S., Montagu, M. V., Inzé, D., May, M. J. and Sung, Z. R. (2000). The *ROOT MERISTEMLESS1/CADMIUM SENSITIVE2* gene defines a glutathione-dependent pathway involved in initiation and maintenance of cell division during postembryonic root development. *Plant Cell* **12**, 97–109.

Wang, Z., Xing, S., Birkenbihl, R. P. and Zachgo, S. (2009). Conserved functions of Arabidopsis and rice CC-type glutaredoxins in flower development and pathogen response. *Molecular Plant* **2**, 323–335.

Witte, S., Villalba, M., Bi, K., Liu, Y., Isakov, N. and Altman, A. (2000). Inhibition of the c-Jun N-terminal kinase/AP-1 and NF-kappaB pathways by PICOT, a novel protein kinase C-interacting protein with a thioredoxin homology domain. *The Journal of Biological Chemistry* **275**, 1902–1909.

Xing, S. and Zachgo, S. (2008). *ROXY1* and *ROXY2*, two Arabidopsis glutaredoxin genes, are required for anther development. *Plant Journal* **53**, 790–801.

Xing, S., Rosso, M. G. and Zachgo, S. (2005). ROXY1, a member of the plant glutaredoxin family, is required for petal development in *Arabidopsis thaliana*. *Development* **132**, 1555–1565.

Xing, S., Lauri, A. and Zachgo, S. (2006). Redox regulation and flower development: A novel function for glutaredoxins. *Plant Biology* **8**, 547–555.

Ye, H., Abdel-Ghany, S. E., Anderson, T. D., Pilon-Smits, E. A. and Pilon, M. (2006). CpSufE activates the cysteine desulfurase CpNifS for chloroplastic Fe–S cluster formation. *The Journal of Biological Chemistry* **281**, 8958–8969.

Zhang, Y. L., Tessaro, M. J., Lassner, M. and Li, X. (2003). Knockout analysis of Arabidopsis transcription factors TGA2, TGA5, and TGA6 reveals their redundant and essential roles in systemic acquired resistance. *Plant Cell* **15**, 2647–2653.

Glutathionylation in Photosynthetic Organisms

XING-HUANG GAO, MARIETTE BEDHOMME,
LAURE MICHELET, MIRKO ZAFFAGNINI
AND STÉPHANE D. LEMAIRE[1]

*Institut de Biotechnologie des Plantes, UMR 8618, CNRS,
Univ Paris-Sud, Bâtiment 630, 91405 Orsay Cedex, France*

[1]Corresponding author: Email: stephane.lemaire@u-psud.fr

Advances in Botanical Research, Vol. 52

0065-2296/09 $35.00
DOI: 10.1016/S0065-2296(09)52012-5

ABSTRACT

Protein glutathionylation is a reversible posttranslational modification promoted by oxidative and nitrosative stresses and consisting of the formation of a mixed disulfide between glutathione and a protein cysteine residue. This modification can protect specific cysteines from irreversible oxidation but can also modulate protein activities, either positively or negatively, and thereby play a role in many cellular processes including signaling. While the mechanism of glutathionylation prevailing *in vivo* remains unclear, the reverse reaction, called deglutathionylation, is mainly catalyzed by small disulfide oxidoreductases of the thioredoxin family named glutaredoxins (GRXs). This chapter will provide an overview of our current knowledge of the underlying molecular mechanisms, and especially the functions of GRXs, but will also review the targets and the possible physiological functions of protein glutathionylation.

I. INTRODUCTION

Living organisms continually need to adapt to fluctuating environmental conditions and have thus developed several systems for sensing modifications of the environment which, coupled to complex signaling pathways, allow adaptation to changing life conditions. Different types of biotic and abiotic stresses result directly or indirectly in the production of reactive oxygen species (ROS) or reactive nitrogen species (RNS). This production activates several signaling pathways resulting in transcriptional, posttranscriptional, and posttranslational responses which will, *in fine*, allow adaptation to new environmental conditions (Fig. 1). These past decades, redox modifications have emerged as central mechanisms in these processes, at the interface between ROS/RNS and the signaling pathways triggering adaptative responses.

It is generally observed that ROS/RNS production and ROS/RNS-induced damage increase during abiotic and biotic stress. In this view, ROS and RNS, considered as toxic molecules, have to be scavenged and detoxified efficiently and continuously. Plant cells exhibit a remarkable ability to cope with high rates of ROS/RNS production as a result of a complex scavenging system that includes either low molecular weight antioxidant molecules such as ascorbic acid, glutathione, tocopherol, and carotenoids or several enzymes, for example, SODs, catalases, and peroxidases (Foyer *et al.*, 2009).

However, emerging evidence suggests that ROS/RNS can play an important role in redox signaling, mainly through a set of reversible posttranslational modifications of cysteine residues on proteins. Indeed, cysteine residues can undergo different states of oxidation such as sulfenic, sulfinic, and sulfonic acid but also protein disulfide bridges (intra- or intermolecular), *S*-thiolation (mainly glutathionylation), or nitrosylation. These posttranslational redox

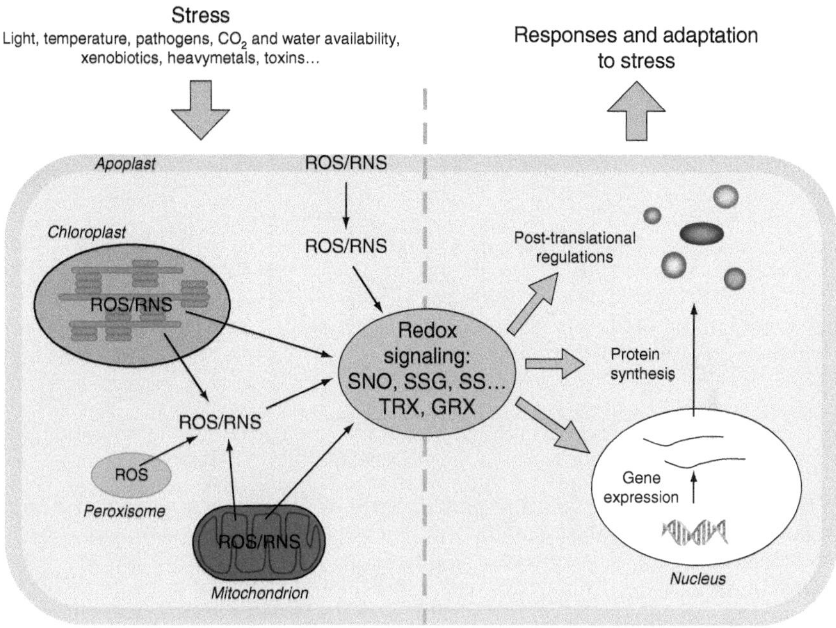

Fig. 1. Redox signaling and plant adaptation. Redox signaling plays a central role in the response of plants to growth, developmental, and environmental stimuli. Various stresses result in the production of reactive oxygen or nitrogen species (ROS, RNS) in different cell compartments, which leads to the induction of specific signaling pathways resulting in modulation of gene expression, protein synthesis and protein activity and allowing adaptation to new environmental conditions. Redox signaling is at the heart of these adaptive responses, at the interface between ROS/RNS and signaling pathways. Redox signaling is mediated by several types of post-translational modifications including glutathionylation (SSG), oxidoreduction of disulfide bridges (SS), and nitrosylation (SNO) and by different enzymes such as thioredoxin (TRX) and glutaredoxin (GRX).

modifications are mainly under the control of two types of ubiquitous disulfide oxidoreductases: thioredoxins (TRXs) and glutaredoxins (GRXs).

Oxidoreduction of intra- or interprotein disulfide bridges is probably the most extensively studied redox modification. It is mainly controlled by TRXs, which play a major role in redox signaling and oxidative stress responses. In plants, different approaches led to the identification of ca. 400 putative TRX targets implicated in nearly all cell processes (Michelet *et al.*, 2006; Montrichard *et al.*, 2009).

Besides TRX-dependent regulation of the redox state of protein disulfide bonds, glutathionylation has recently emerged, among other thiol-based posttranslational modifications, as an important redox-based signaling mechanism under stress conditions. This modification consists of the

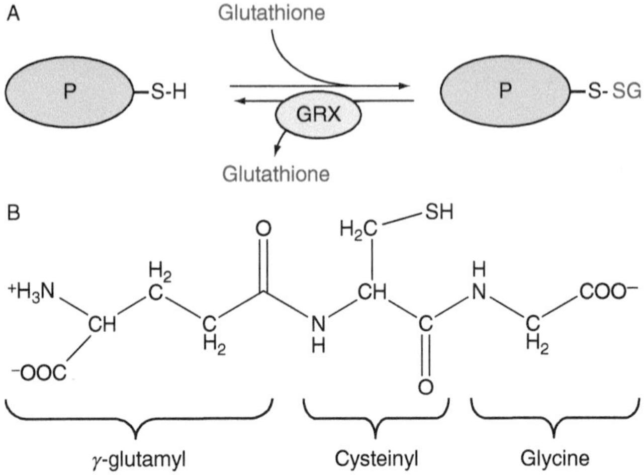

Fig. 2. Glutathionylation and glutathione. (A) Protein glutathionylation and deglutathionylation. Protein glutathionylation can occur under oxidative or nitrosative stress conditions, in the presence of glutathione. This modification consists of the formation of a mixed disulfide between a free thiol of a protein and a molecule of glutathione. Glutathionylation is reversible and glutaredoxin (GRX) can efficiently catalyze protein deglutathionylation. (B) Chemical structure of reduced glutathione. Glutathione is a 307-Da tripeptide (γ-L-glutamyl-L-cysteinyl-L-glycine).

formation of a mixed disulfide between an accessible free thiol on a protein and a molecule of glutathione (protein-SSG) (Fig. 2A).

Glutathione is a highly abundant (1–5 mM) tripeptide (γ-L-glutamyl-L-cysteinyl-L-glycine) of 307 Da found in almost all living organisms (Fig. 2B). As the main free soluble thiol of low molecular weight, glutathione is classically considered to constitute a redox buffer which maintains the intracellular environment in a reduced state. Indeed, it exists in two different forms: the reduced form (GSH), which is the major form, and the oxidized form (GSSG). GSSG is continuously regenerated into GSH by glutathione reductase using NADPH as a reductant. Glutathione is involved in various cellular processes in photosynthetic organisms including detoxification of ROS, xenobiotics and heavy metals, control of the G1/S cell cycle transition, cell differentiation, flowering, anthocyanin accumulation, programmed cell death but also resistance toward pathogens (reviewed in Foyer and Noctor, 2005; Meyer and Hell, 2005; Noctor, 2006; Ogawa, 2005; Rouhier *et al.*, 2008).

In addition to all these functions, recent studies have suggested that glutathione could also play a major role in redox signaling through modification of cysteine residues by glutathionylation. Protein glutathionylation can act as a mechanism of protection of protein cysteine residues from irreversible oxidation but can also modulate, either positively or negatively, protein activities. The exact mechanism(s) leading to protein glutathionylation remain(s) to be determined, but this modification seems to occur under oxidative or nitrosative stress conditions. By contrast, the reverse reaction, named deglutathionylation, is likely catalyzed by GRXs, proteins belonging to the TRX family. To date, this modification has been mainly studied in animals where proteomic studies, essentially based on ^{35}S-cysteine labeling, allowed identification of ca. 150 targets involved in several processes, such as glycolysis, signal transduction, protein degradation, or protein folding (Dalle-Donne *et al.*, 2008; Ghezzi and Di Simplicio, 2007; Michelet *et al.*, 2006; Rouhier *et al.*, 2008; Shelton and Mieyal, 2008). Concerning photosynthetic organisms, much less is known about glutathionylation, despite the importance of these organisms on Earth. Indeed, photosynthetic organisms have a great ecological importance as major actors in global oceanic function, in the carbon cycle and consequently in the evolution of climate. Moreover, photosynthesis is the basis of our food and energy supply, and innovative utilization of photosynthetic organisms is likely to be of increasingly critical importance in the future. Nevertheless, recent studies indicate that glutathionylation could also constitute an important type of redox signaling in plants (Rouhier *et al.*, 2008). This chapter will provide an overview of our current knowledge of the underlying molecular mechanisms, and especially the functions of GRXs, but will also review the targets and the possible physiological functions of protein glutathionylation.

II. GLUTATHIONYLATION REACTIONS

Protein glutathionylation is a dynamic process. Theoretically, several mechanisms can lead to protein glutathionylation but the precise mechanisms occurring *in vivo* are still far from being well understood. Glutathionylation can occur spontaneously but several enzymes have also been suggested to catalyze this reaction under specific conditions or with specific substrates, though the relevance of these potential mechanisms of catalyzed glutathionylation reactions remains to be established under physiological conditions (Gallogly and Mieyal, 2007).

A. THIOL/DISULFIDE EXCHANGE

One of the most widely studied mechanisms of glutathionylation involves spontaneous thiol/disulfide exchange between GSSG and a protein free thiol (Fig. 3, reaction 1). However, GSSG is poorly efficient to trigger glutathionylation and high concentrations (1–5 mM) and long incubation times

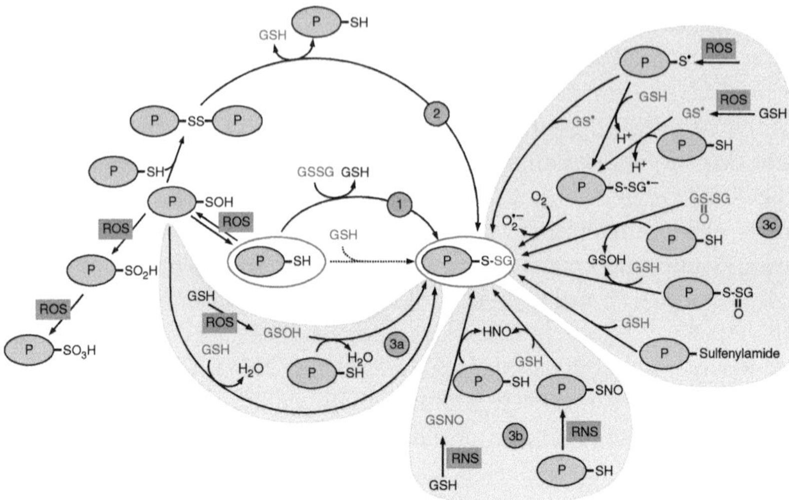

Fig. 3. Potential mechanisms of protein glutathionylation. Several mechanisms can theoretically trigger protein glutathionylation. Glutathionylation can occur via a thiol/disulfide exchange between the protein and oxidized glutathione (GSSG) (1); by direct oxidation of both protein and reduced glutathione (GSH) (low probability, dotted line) or via reaction with intermolecular disulfide bridges (2). Glutathionylation can also involve activated thiol derivatives such as sulfenic acids (–SOH) (3a), *S*-nitrosylated thiols (–SNO) (3b) or thiyl radicals (–S•), thiosulfinates (–S(O)SG), and sulfenylamides (3c). See Section II. for further details. (See color plate section at the back of the book.)

(usually several hours) are usually required to yield substantial amounts of the glutathionylated form of the target protein *in vitro*. This mechanism implies that the glutathionylation rate is directly related to the redox state of the glutathione pool which is linked to the GSH:GSSG ratio and to the total glutathione concentration. However, in most cells and subcellular compartments the glutathione pool is maintained highly reduced by glutathione reductase with GSH:GSSG ratios around 100:1 in nonstressed cells (Gilbert, 1995; Gutscher *et al.*, 2008; Jiang *et al.*, 2006; Meyer *et al.*, 2007). For most protein cysteines, even a dramatic decline of the intracellular GSH:GSSG ratio to 1:1, very unlikely even under artificial oxidative stress, would only lead to 50% conversion of protein-SH to protein-SSG through spontaneous thiol/disulfide exchange between GSSG and a protein thiol (Gallogly and Mieyal, 2007; Starke *et al.*, 1997; Wang *et al.*, 2001). Therefore, although glutathionylation by thiol/disulfide exchange can lead to glutathionylation *in vitro* and is a useful method to study the biochemical and structural consequences of protein glutathionylation, this mechanism is, for most proteins, not likely to contribute significantly to glutathionylation *in vivo*. For example, during the oxidative burst in human neutrophils, a massive increase in protein glutathionylation is observed without any increase in GSSG concentration (Chai *et al.*, 1994). Interestingly, human actin, an established glutathionylated protein, does not seem to undergo glutathionylation through thiol/disulfide exchange *in vitro* as well as *in vivo* (Dalle-Donne *et al.*, 2003, 2005). Some exceptions to these general considerations may however exist. Indeed, some proteins, such as the c-Jun transcription factor (Klatt *et al.*, 1999a) or caspase 3 (Huang *et al.*, 2008), appear to undergo GSSG-mediated glutathionylation at physiological GSH:GSSG ratios and GSSG concentrations.

B. DIRECT OXIDATION AND DISULFIDE BONDS

Glutathionylation can also occur through direct oxidation. However, a direct oxidation of two thiols in the presence of oxygen for instance is unlikely to occur because this reaction would implicate a ternary collision between the two thiols and the oxidant.

Disulfide bonds with compatible redox potentials can be reduced by GSH, leading to glutathionylation of one of the two thiols. When the disulfide bond is intramolecular, the glutathionylated form constitutes a transient form which will either lead to the fully reduced form after reaction with a second GSH molecule or return back to the disulfide form after nucleophilic attack of the second cysteine on the glutathionylated thiol. However, when the disulfide bond is intermolecular and contributes significantly to the stability of the multimer, glutathionylation of one subunit by reaction of GSH with the disulfide might lead to dissociation of the subunits and lead to formation of a stable glutathionylated form (Fig. 3, reaction 2).

C. ACTIVATED THIOLS

The mechanisms of glutathionylation that are more likely to occur *in vivo* involve activated thiol derivatives such as sulfenic acids (–SOH), *S*-nitrosylated thiols (–SNO), or thiyl radicals ($-S^{\bullet}$).

1. Mechanisms involving sulfenic acids

Protein glutathionylation can be mediated by a sulfenic acid formed by two-electron oxidation of one of the two thiols involved in the reaction. Reaction of cysteines with ROS or RNS (e.g., H_2O_2) can lead to the formation of protein sulfenates or glutathione sulfenate which can trigger glutathionylation by reaction with GSH or a protein thiol, respectively (Fig. 3, reactions 3a). With the exception of several ROS-scavenging enzymes such as peroxiredoxins (PRXs), glutathione peroxidases, NADH peroxidases, or methionine sulfoxide reductases which are able to stabilize sulfenic acids within their active site (Poole *et al.*, 2004), sulfenic acids are unstable molecules which are either rapidly converted by further oxidations to sulfinic (SO_2H) or sulfonic (SO_3H) acids or react with vicinal thiols to form disulfides. In most cases, sulfinic and sulfonic acid forms are irreversible and lead to the degradation of the overoxidized proteins (Reddie and Carroll, 2008). One known exception is 2-cys PRXs whose sulfinic form can be reduced back to the sulfenic form by sulfiredoxin (SRX) or some sestrins (Biteau *et al.*, 2003; Budanov *et al.*, 2004; Rhee *et al.*, 2008; Woo *et al.*, 2005). Glutathionylation being reversible, it could constitute a mechanism of protection of protein cysteine residues from irreversible oxidation, as demonstrated *in vitro* for photosynthetic glyceraldehyde-3-phosphate dehydrogenase (GAPDH) (Zaffagnini *et al.*, 2007). Many proteins have been suggested to be regulated by sulfenic acid formation (Poole and Nelson, 2008; Poole *et al.*, 2004), but in most cases the experiments were performed in the absence of GSH. Albeit in some cases the sulfenic and glutathionylated forms may both be present *in vivo* and may have distinct consequences on protein activities, as shown for the OxyR transcription factor (Kim *et al.*, 2002), in the cell, most sulfenic acids would be converted to glutathionylated thiols (Gallogly *et al.*, 2009). This was recently shown for protein tyrosine phosphatase 1B (PTP-1B), initially suggested to be regulated by hydrogen peroxide through sulfenic acid formation (Denu and Tanner, 1998) and later shown to undergo glutathionylation after reaction of GSH with the sulfenic form (Barrett *et al.*, 1999a,b). Similarly, the *Bacillus subtilis* organic peroxide sensor OhrR was initially suggested to be inactivated by sulfenic acid formation (Fuangthong and Helmann, 2002), but it was recently demonstrated that the sulfenic forms are converted to *S*-thiolated forms after reaction with low molecular weight thiols (Lee *et al.*, 2007). Therefore, it has been suggested that although protein-SOH may be the initial form of oxidative modification for many proteins, in most cases protein-SSG is expected to serve as the more stable intermediate in redox signaling (Gallogly and Mieyal, 2007).

2. Mechanisms involving nitrosylated thiols

S-Nitrosylation is a ubiquitous reversible posttranslational modification playing a major role in cell signaling in many organisms (Hess *et al.*, 2005; Janssen-Heininger *et al.*, 2008). RNS can trigger the formation of nitrosylated thiols on proteins *in vivo* and can also generate nitrosoglutathione (GSNO), a relatively stable molecule considered as a NO reservoir which can trigger both nitrosylation and glutathionylation (Hogg, 2002). GSNO can promote glutathionylation by reaction with a protein thiol (Fig. 3, reactions 3b) as reported, either *in vivo* or *in vitro*, for many proteins such as c-Jun (Klatt *et al.*, 1999b), GAPDH (Chandra *et al.*, 1997; Giustarini *et al.*, 2005; Mohr *et al.*, 1999), papain and creatine phosphokinase (Giustarini *et al.*, 2005), carbonic anhydrase and H-ras (Ji *et al.*, 1999), or mitochondrial and cytosolic branched-chain aminotransferases (Coles *et al.*, 2009). Altogether, these data suggest that GSNO could mediate protein glutathionylation in cells. However, the protein properties or conditions favoring nitrosylation versus glutathionylation upon reaction of protein thiols with GSNO remain largely unclear (Giustarini *et al.*, 2005). Alternatively, glutathionylation could also be mediated by reaction of GSH with a nitrosylated protein thiol (Fig. 3, reaction 3b), but little information is available on this mechanism.

3. Mechanisms involving thiyl radicals, thiosulfinates, and sulfenylamides
One-electron oxidation of a protein thiol or GSH, for example, by the hydroxyl radical ($HO^\bullet$), leads to the formation of a protein thiyl radical ($PS^\bullet$) or a glutathione thiyl radical ($GS^\bullet$) which can form a radical mixed disulfide by reaction with GSH or a protein thiol, respectively. This radical mixed disulfide will then transfer an electron to oxygen to form the superoxide anion ($O_2^{\bullet -}$), leaving a glutathionylated protein (Fig. 3, reactions 3c). Thiyl radicals are likely to be formed under oxidative or nitrosative stress conditions (Karoui *et al.*, 1996; Kwak *et al.*, 1995; Maples *et al.*, 1990) and are among the shortest-lived activated thiols (Wardman and von Sonntag, 1995). Several proteins have been shown to undergo thiyl-mediated glutathionylation *in vitro*, and the reaction was suggested to be catalyzed by GRXs (Gallogly *et al.*, 2008; Starke *et al.*, 2003).

The thiosulfinate derivative of GSH, GS(O)SG is a product of GSNO decomposition which is highly reactive with protein thiols and can lead to the formation of protein-SSG (Fig. 3, reactions 3c). Thiosulfinates were proposed to mediate the glutathionylation of rat brain neurogranin (Huang and Huang, 2002; Li *et al.*, 2001), tyrosine hydroxylase (Sadidi *et al.*, 2005), and matrix metalloproteinases (Okamoto *et al.*, 2001). Sulfenylamides, such as the one formed on PTP-1B under extreme oxidative conditions (Salmeen *et al.*, 2003; Sarma and Mugesh, 2007; van Montfort *et al.*, 2003), could also lead to protein glutathionylation (Fig. 3, reactions 3c).

Finally, even if thiyl radicals, thiosulfinates, and sulfenylamides can trigger glutathionylation *in vitro*, it remains unclear whether these short-lived and highly reactive species significantly contribute to protein glutathionylation *in vivo*.

D. CATALYSIS OF PROTEIN GLUTATHIONYLATION

Despite the fact that glutathionylation can occur spontaneously, especially in the presence of ROS/RNS, several enzymes could also catalyze protein glutathionylation. Glutathione-*S*-transferase π (GSTπ) was shown to interact with the sulfenic form of 1-Cys-PRX (1-Cys-PRX-SOH) and to convert it into 1-Cys-PRX-SSG (Manevich *et al.*, 2004; Ralat *et al.*, 2006). In addition, GSTπ may have a more general role in the control of protein glutathionylation under oxidative and nitrosative stress (Tew, 2007). Protein glutathionylation was found to be significantly decreased in GSTπ deficient mouse cell lines treated with GSSG or PABA/NO (Townsend *et al.*, 2009). GSTπ could facilitate the conversion of protein-SOH to protein-SSG. However, it is not known whether this mechanism of glutathionylation is widespread or only affects specific targets able to interact with GSTπ such as 1-Cys-PRX.

Although GRXs are the main enzymes contributing to deglutathionylation reactions in cells (detailed below), they have been proposed to catalyze glutathionylation under specific conditions. Human GRX1 and GRX2 were shown to enhance the rate of glutathionylation of several proteins, such as actin, GAPDH, or PTP-1B, in the presence of thiyl radicals (Gallogly *et al.*, 2008; Starke *et al.*, 2003). GRXs were proposed to stabilize $GS^{\bullet}$ as an enzyme disulfide anion radical intermediate ($GRX1\text{-}SSG^{\bullet -}$), thereby facilitating GS-radical recombination with a protein thiyl radical (Gallogly and Mieyal, 2007; Starke *et al.*, 2003).

Finally, the flavoprotein sulfhydryl oxidase (Qsox) was also proposed as a potential catalyst of protein glutathionylation (Gallogly and Mieyal, 2007). However, a direct demonstration and detailed biochemical analysis supporting this function is still required.

Overall, all the mechanisms described are possible and most of them have been demonstrated *in vitro*, but the one prevalent *in vivo* remains unknown. With regard to specificity, it is unclear what features contribute to the sensitivity of a given cysteinyl residue to protein glutathionylation though the accessibility, the reactivity, and the microenvironment of the cysteine are likely to play a major role (Dalle-Donne *et al.*, 2009).

III. DEGLUTATHIONYLATION REACTIONS

A. GLUTAREDOXINS

GRXs are small disulfide oxidoreductases belonging to the TRX superfamily. They are considered to represent the major deglutathionylating enzymes in cells. Historically, GRX was first discovered as an enzyme-catalyzing GSH-dependent reduction of oxidized ribonucleotide reductase (RNR) in a mutant of *Escherichia coli* lacking TRX (Holmgren, 1976). Recently, mammalian RNR was shown to be reduced by GRX through a mechanism involving a deglutathionylation step rather than by reduction of RNR disulfide bond by dithiol/disulfide exchange (Avval and Holmgren, 2009). Two decades after its discovery, GRX was described as a specific glutathionyl-mixed disulfide oxidoreductase (Gravina and Mieyal, 1993). GRXs exhibit a very high selectivity for protein-SSG substrates compared to other protein mixed disulfides such as protein-SSCys and catalyze deglutathionylation much more efficiently than other disulfide oxidoreductases such as TRX or protein disulfide isomerase (PDI) (Chrestensen *et al.*, 2000; Jung and Thomas, 1996; Nulton-Persson *et al.*, 2003; Peltoniemi *et al.*, 2006; Zaffagnini *et al.*, 2008).

GRXs belong to a multigenic family with around 30 members in higher plants, while the genomes of lower photosynthetic organisms contain much less GRX genes (six in the green alga *Chlamydomonas reinhardtii* and three in the cyanobacteria *Synechocystis* sp. *PCC6803*) (Lemaire, 2004; Rouhier *et al.*, 2004, 2006). In plants, GRXs have been classified into three subgroups based on phylogenetic analyses and on their active-site sequence. Classical GRXs, such as *E. coli* GRX1 or human GRX1, harbor a CPYC active-site sequence with two vicinal cysteines forming a glutathione reducible disulfide. All GRXs from bacteria, mammals, yeast, and plants containing a CPYC active site apparently catalyze protein disulfide reduction and deglutathionylation (Fernandes and Holmgren, 2004; Gallogly *et al.*, 2009; Herrero *et al.*, 2008; Rouhier *et al.*, 2008; Zaffagnini *et al.*, 2008). A second group, identified more recently, corresponds to proteins with a CGFS active site (Herrero and de la Torre-Ruiz, 2007). Despite the presence of only one active-site cysteine, some CGFS-type GRXs have been shown to contain an intramolecular disulfide involving a partially conserved C-terminal cysteine (Johansson *et al.*, 2004; Tamarit *et al.*, 2003; Zaffagnini *et al.*, 2008). These proteins do not apparently catalyze disulfide reduction but were reported to catalyze protein deglutathionylation. CGFS-type GRXs are not reduced by GSH but rather by NADPH thioredoxin reductase (NTR) or ferredoxin-thioredoxin reductase (FTR). In *Chlamydomonas*, the reduction of a chloroplastic CGFS GRX (GRX3) by FTR was suggested to be linked to the low redox potential of the active-site disulfide (Zaffagnini *et al.*, 2008). CPYC and CGFS GRXs are ubiquitous since they are present in almost all prokaryotic and eukaryotic genomes (Alves *et al.*, 2009; Lemaire and Miginiac-Maslow, 2004; Lillig *et al.*, 2008; Rouhier *et al.*, 2004, 2006; Sagemark *et al.*, 2007). By contrast, the third type of GRX, called the CC type and corresponding to proteins with a CCXC or CCXS active site, is specific to higher plants, since no counterparts are present in nonphotosynthetic organisms or lower photosynthetic eukaryotes (Alves *et al.*, 2009; Li and Zachgo, this issue; Lemaire and Miginiac-Maslow, 2004). With approximately 20 members in most plant genomes, the CC-type gene family represents the most extended subgroup. Two CC-type GRXs from *Arabidopsis*, designated ROXY1 and ROXY2, were recently shown to play a role in petal development (Li *et al.*, 2009; Wang *et al.*, 2009; Xing and Zachgo, 2008; Xing *et al.*, 2005). ROXY1 interacts with TGA transcription factors, an interaction also shown for other CC-type GRXs that were suggested to participate in jasmonic acid/ salicylic acid responses (Ndamukong *et al.*, 2007). Nevertheless, the biochemical properties of CC-type GRXs have not yet been investigated and, therefore, the molecular basis of their functions remains unknown, as well as their ability to catalyze deglutathionylation.

GRXs have been much less studied than TRXs. Most of these studies were focused on the functions of GRXs in nonphotosynthetic organisms, especially bacteria, yeast, and mammals, while much less is known on GRXs from photosynthetic organisms. One of the most documented functions of GRXs in plants concerns their role in oxidative stress responses. For example, they are directly involved in the reduction of peroxides, PRXs, and methionine sulfoxide reductases (Finkemeier *et al.*, 2005; Gama *et al.*, 2007; Lee *et al.*, 2002; Rouhier *et al.*, 2001, 2002b; Vieira Dos Santos *et al.*, 2007). An *Arabidopsis* GRX mutant was also shown to be sensitive to oxidative damage (Cheng *et al.*, 2006). Overexpression of a poplar GRX in *E. coli* was also reported to confer resistance to oxidative stress (Rouhier *et al.*, 2003).

Many studies indicate that GRXs from diverse organisms including yeast, mammals, poplar, and Synechocystis are able to assemble a [2Fe-2S] cluster (Bandyopadhyay *et al.*, 2008a,b; Comini *et al.*, 2008; Feng *et al.*, 2006; Hudemann *et al.*, 2009; Johansson *et al.*, 2007; Lonn *et al.*, 2008; Mesecke *et al.*, 2008; Picciocchi *et al.*, 2007; Rouhier *et al.*, 2007). In yeast, mitochondrial CGFS GRX5 appears required for the assembly/biogenesis of iron sulfur clusters (Herrero and de la Torre-Ruiz, 2007; Rodriguez-Manzaneque *et al.*, 2002). Several plant GRXs are able to complement the yeast GRX5 mutant and restore iron sulfur cluster assembly/biogenesis (Bandyopadhyay *et al.*, 2008b; Cheng, 2008; Cheng *et al.*, 2006). A poplar GRX was shown to transfer rapidly and quantitatively its bound [2Fe–2S] to apoferredoxin (Bandyopadhyay *et al.*, 2008b). Alternatively, it was proposed for human mitochondrial GRX2 that the binding of the cluster inactivates the protein by sequestering active-site residues and that loss of the cluster through changes in subcellular redox status would create a catalytically active protein (Johansson *et al.*, 2007). Clearly, further work is required to determine the function of iron–sulfur bound GRX.

The classification and functions of GRXs have been described in detail in several recent reviews (Berndt *et al.*, 2008; Couturier *et al.*, 2009a; Gallogly *et al.*, 2009; Herrero and de la Torre-Ruiz, 2007; Lemaire, 2004; Lillig *et al.*, 2008; Meyer *et al.*, 2008; Michelet *et al.*, 2006; Rouhier *et al.*, 2004, 2006, 2008).

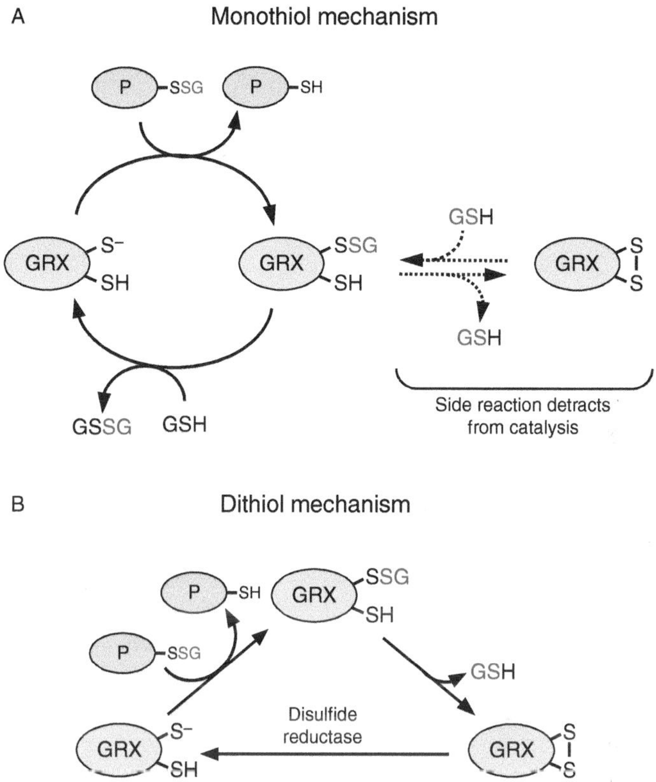

Fig. 4. Protein deglutathionylation by glutaredoxins. The first step of protein deglutathionylation is common to both monothiol and dithiol mechanisms. The most reactive cysteine of GRX (the N-terminal active-site cysteine) performs a nucleophilic attack on the glutathione-mixed disulfide, resulting in the release of a deglutathionylated substrate (protein-SH) and the formation of a glutathionylated GRX intermediate (GRX-SSG). (A) Monothiol mechanism. In the monothiol mechanism, the second step consists of the reduction of GRX-SSG by a second molecule of GSH to form GSSG and reduced GRX. CPYC-type GRXs were shown to catalyze deglutathionylation by this monothiol mechanism. (B) Dithiol mechanism. The dithiol mechanism requires the presence of a second vicinal cysteine of GRX that will perform a second nucleophilic attack and leads to the formation of oxidized GRX and GSH. GRX is subsequently reduced by a disulfide reductase (NTR or FTR). This mechanism was suggested to be used by CGFS-type GRXs which are not reduced by GSH.

B. Mechanisms of Deglutathionylation by GRXs

Reduced GRXs can catalyze deglutathionylation either through a monothiol or a dithiol mechanism (Fig. 4). The first step is common to both mechanisms and consists of a nucleophilic attack of the glutathione-mixed disulfide by the most reactive cysteine of GRX (the N-terminal active-site cysteine) and results in the release of a deglutathionylated substrate (protein-SH) and the formation of a glutathionylated GRX intermediate (GRX-SSG). In the monothiol mechanism, the second step consists of the reduction of GRX-SSG by a second molecule of GSH to form GSSG and reduced GRX (Fig. 4A). In the dithiol mechanism, the second nucleophilic attack is performed by a vicinal cysteine of GRX and leads to the formation of oxidized GRX and GSH (Fig. 4B).

Classical dithiol CPYC-type GRXs are able to catalyze deglutathionylation of both artificial (glutathionylated mercaptoethanol synthesized from β-hydroxyethyl disulfide by reaction with GSH) and protein substrates through a monothiol mechanism requiring only the most N-terminal cysteine of the active site of GRX. Indeed, GRX monocysteinic mutants, where the C-terminal active-site cysteine has been replaced by serine (CPYS active site), retain their ability to catalyze deglutathionylation (Bushweller *et al.*, 1992; Gallogly *et al.*, 2008; Johansson *et al.*, 2004; Peltoniemi *et al.*, 2006; Yang *et al.*, 1998). Natural monocysteinic variants, such as the recently characterized chloroplastic GRX-S12 from poplar (CSYS active site), are also efficient catalysts of protein deglutathionylation (Couturier *et al.*, 2009b). GRX-catalyzed protein deglutathionylation proceeds by a nucleophilic, double-displacement (ping-pong) mechanism in which rate enhancement is attributed to the special reactivity of the low pK_a cysteine at its active site, and to increased nucleophilicity of the second substrate, GSH (Discola *et al.*, 2009; Gallogly *et al.*, 2008, 2009; Srinivasan *et al.*, 1997). For example, the pK_as of the N-terminal active-site cysteines of human GRX1 and GRX2 are 3.5 and 4.6, respectively (Gallogly *et al.*, 2008; Mieyal *et al.*, 1991), and we recently measured values within the same range for poplar GRX-S12 (Couturier *et al.*, 2009b) and for *Chlamydomonas* GRX1 and GRX2 (Gao *et al.*, manuscript in preparation). However, due to the presence of the C-terminal active-site cysteine of CPYC GRXs, a side reaction involving the formation of the intramolecular disulfide form of GRX decreases the turnover rate of the monothiol mechanism (Fig. 4A). This intramolecular disulfide has to react with two GSH molecules to regenerate the reduced enzyme for another cycle of catalysis. Therefore, removal of the second active-site cysteine in mammalian GRXs, yeast GRX1, and *Chlamydomonas* GRX1 resulted in a two- to fivefold increase of specific activity (Discola *et al.*, 2009; Gallogly *et al.*, 2008; Gao and Lemaire, unpublished results; Johansson *et al.*, 2004; Yang and Wells, 1991; Yang *et al.*, 1998). However, for *E. coli* GRX1 and GRX3, yeast GRX2 and poplar GRX C4, the same mutation resulted in a 75% decrease of specific activity, suggesting that the presence of the second active-site cysteine is playing some role in the deglutathionylation reaction (Bushweller *et al.*, 1992; Discola *et al.*, 2009; Nordstrand *et al.*, 1999; Peltoniemi *et al.*, 2006; Rouhier *et al.*, 2002a). Recently, the second active-site cysteine of *E. coli* GRX1 was suggested to determine the specificity for glutathione in the second step of the deglutathionylation reaction (Saaranen *et al.*, 2009).

Among CGFS-type GRXs, yeast GRX5, *E. coli* GRX4 and *Chlamydomonas* GRX3 have been reported to catalyze deglutathionylation, most likely through a dithiol mechanism (Tamarit *et al.*, 2003; Zaffagnini *et al.*, 2008). These GRXs contain a disulfide bond formed between the active-site cysteine and a partially conserved C-terminal cysteine. The use of a dithiol mechanism might be linked to the fact that this disulfide is not reduced by GSH but by thioredoxin reductases (NTR or FTR).

C. OTHER ENZYMES

Besides GRXs, several other enzymes have been suggested to catalyze protein deglutathionylation. SRX was proposed to participate in deglutathionylation reactions based on the observation that overexpression of SRX in human cells diminished protein glutathionylation after PABA/NO treatment and that the glutathionylation of actin and PTP-1B by PABA/NO and GSH could be reversed by SRX *in vitro* (Findlay *et al.*, 2006). However, the underlying mechanisms have not been established. Further studies are required to clarify whether SRX catalyzes deglutathionylation or rather indirectly affects glutathionylated proteins, for example, by interfering with ROS or activated thiol intermediates involved in the glutathionylation mechanisms. Several GST isoforms have also been suggested to catalyze deglutathionylation, but the significance of these results has been questioned (Gallogly *et al.*, 2009).

IV. METHODS FOR IDENTIFICATION AND ANALYSIS OF GLUTATHIONYLATED PROTEINS

A number of methods have been developed to identify and analyze proteins undergoing glutathionylation. These methods, which range from the analysis of single purified proteins *in vitro* to large-scale proteomic studies, allowed identification of nearly 200 targets, mainly in mammalian cells (Dalle-Donne *et al.*, 2007; Fratelli *et al.*, 2004; Ghezzi and Di Simplicio, 2007; Giustarini *et al.*, 2004; Klatt and Lamas, 2000; Rouhier *et al.*, 2008). These approaches have been extensively described in several recent reviews (Dalle-Donne *et al.*, 2008; Fratelli *et al.*, 2004; Gao *et al.*, 2009). Here, we will rapidly present these techniques, compare their advantages and drawbacks, and present their application to the identification and analysis of glutathionylated proteins in photosynthetic organisms.

A. ^{35}S RADIOLABELING

The most popular technique for large-scale proteomic analysis of glutathionylated proteins is based on the use of radiolabeled ^{35}S-cysteine to label the glutathione pool. Cell cultures are (1) incubated in the presence of protein synthesis inhibitors, (2) supplemented with ^{35}S-cysteine to label the glutathione pool, and (3) submitted to an oxidative stress treatment, usually by addition of diamide or hydrogen peroxide, which will induce protein glutathionylation (Fig. 5). Then, total proteins are extracted and separated on 2D gels; radiolabeled proteins are visualized by fluorography and identified by mass spectrometry. *S*-Thiolation is easily reversed by strong

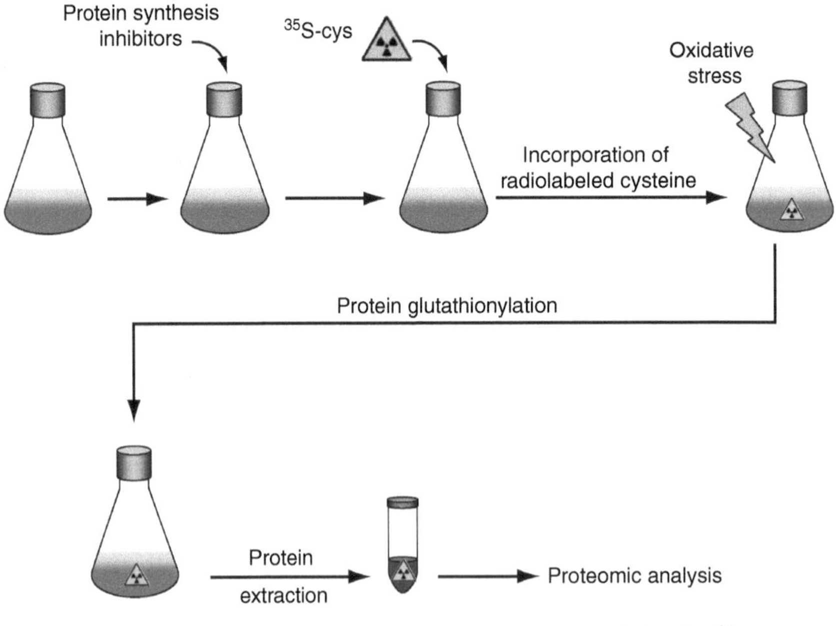

Fig. 5. Identification of glutathionylated proteins using radiolabeled ^{35}S-cysteine. Cell cultures are first treated with protein synthesis inhibitors so as to avoid protein synthesis throughout the experiment. Radiolabeled ^{35}S-cysteine is then added to the culture, enters the cells, and can be incorporated into glutathione, leading to the radiolabeling of the glutathione pool. An oxidative stress is subsequently applied to cells to trigger protein glutathionylation. Proteins are then extracted and separated on two-dimensional gels. Radiolabeled proteins can be identified by proteomic analysis.

reductants such as dithiothreitol (DTT). A control gel in the presence of DTT is therefore performed to check that the radioactive spots are lost, thereby confirming that the signal is indeed linked to *S*-thiolation. The first large-scale proteomic analysis based on this method allowed identification of 38 *S*-thiolated proteins in human T lymphocytes (Fratelli *et al.*, 2002). Subsequently, similar studies allowed identification of a number of *S*-thiolated proteins in several cell types and organisms (reviewed in Gao *et al.*, 2009). We recently employed this strategy to identify *S*-thiolated proteins in the unicellular green alga *C. reinhardtii* (Michelet *et al.*, 2008). Oxidative treatment with diamide led to the identification of 25 targets, mainly chloroplastic and involved in diverse metabolic processes (detailed below). On the other hand, attempts to employ this method with *Arabidopsis* cell cultures proved unsuccessful due to low levels of labeling that precluded identification of *S*-thiolated proteins (Dixon *et al.*, 2005b).

This method has several drawbacks. The major one is the requirement of a pretreatment with protein synthesis inhibitors which will unavoidably perturb cell physiology. Moreover, this method cannot distinguish protein-SSG from other possible types of *S*-thiolation such as *S*-cysteinylation, the formation of a mixed disulfide between a protein thiol and a molecule of cysteine. Nevertheless, glutathione being by far the most abundant low molecular weight thiol, most detected proteins likely correspond to glutathionylated proteins as suggested by treatments with the glutathione biosynthesis inhibitor buthionine sulfoximine (Fratelli *et al.*, 2002). The method is also limited by the necessity to perform 2D gels to visualize the *S*-thiolated proteins, which will favor the identification of abundant proteins and limit the identification of proteins with extreme pI and molecular weights. In addition, this method can only be used with cell cultures, thereby preventing studies on whole plants under physiological conditions. Finally, this approach can only detect proteins undergoing glutathionylation during treatment while some proteins might be already glutathionylated under basal conditions.

B. BIOTINYLATED GLUTATHIONE

Methods based on biotinylated glutathione (BioGSH/BioGSSG) or the membrane-permeant biotinylated glutathione ethyl ester (BioGEE) have also been developed. The presence of the biotin tag allows detection of glutathionylated proteins by multiple methods such as immunoblotting with or without prior immunoprecipitation using biotin antibodies or HRP-avidin (Reynaert *et al.*, 2006a), batch or column-based avidin affinity purifications, or cellular localization by fluorescence microscopy

(Brennan *et al.*, 2006). Compared to ^{35}S radiolabeling, the methods based on biotinylated glutathione present several advantages for proteomic analyses. First, the affinity purification steps can partially overcome the limitations of 2D gels and allow identification of less abundant proteins. In addition, the affinity purified proteins can also be analyzed directly by highly sensitive and high-throughput proteomic methods such as nanoLC–MS/MS. Another major advantage is that these methods are specific for glutathionylated proteins. Originally used in mammalian systems (Sullivan *et al.*, 2000), these methods have also been applied to the analysis of protein glutathionylation in photosynthetic organisms.

BioGEE allowed visualization of 22 glutathionylated proteins after H_2O_2 treatment of *Arabidopsis* cell cultures, two of which, cytosolic triose phosphate isomerase (TPI) and chloroplastic fructose-1,6-bisphosphate aldolase (FBA), were identified (Ito *et al.*, 2003). In the presence of BioGSSG, nine other proteins which underwent glutathionylation *in vivo* in *Arabidopsis* dark-grown cell cultures after treatment with *tert*-butyl hydroperoxide (tBOOH) were identified (Dixon *et al.*, 2005b). In addition, in the same study, *in vitro* treatment of total cell extracts from *Arabidopsis* cell cultures with BioGSSG followed by streptavidin agarose affinity purification and 2D gels led to the identification of 72 proteins. However, only 22 proteins could be eluted by DTT from the column under denaturing conditions, suggesting that some proteins might not be glutathionylated but rather interact with glutathionylated proteins. Recently, BioGEE was used for *in vitro* analysis of the glutathionylation of cytosolic GAPDH isoforms from *Arabidopsis* (Holtgrefe *et al.*, 2008).

The major drawback of the methods based on biotinylated glutathione is the fact that proteins are not glutathionylated by the cellular GSH itself but by an exogenous molecule that is chemically different and is also characterized by a greater steric hindrance. The presence of the biotin tag on the glutathione molecule might perturb the function of proteins interacting with glutathione molecule and especially GRXs (Zaffagnini and Lemaire, unpublished data). Another drawback, shared by the ^{35}S labeling method, is that proteins glutathionylated under basal conditions are not detected.

C. ANTIGLUTATHIONE ANTIBODIES

Commercially available antiglutathione antibodies have also been employed to detect glutathionylated proteins. This strategy could overcome most problems encountered with the two methods described earlier. Indeed, with this method, glutathionylated proteins could be analyzed under more physiological conditions and easily detected by Western blots with 1D or 2D gels, by immunoprecipitation, or even by immunocytolocalization. However, due to a lack of specificity and sensitivity, antiglutathione antibodies are not adapted for large-scale proteomic analyses although they recently proved useful to study the glutathionylation of individual proteins such as the 20S proteasome (Silva *et al.*, 2008), the interferon regulatory factor 3 (IRF3) (Prinarakis *et al.*, 2008), or Complex I (Hurd *et al.*, 2008). The specificity issues may be linked to the flexibility of the glutathione molecule which can potentially exhibit hundreds of conformations, either in solution or bound to proteins (Lampela *et al.*, 2003). Furthermore, we observed that several purified proteins containing a glutathionylated thiol, as confirmed by mass spectrometry, were not detected by antiglutathione antibodies in Western blots experiments (unpublished data). In addition, it has been reported that some proteins containing a Glu-Cys-Gly sequence in their primary structure could react with antiglutathione antibodies (Demasi *et al.*, 2008).

D. IDENTIFICATION OF GLUTATHIONYLATED PROTEINS THROUGH REDUCTION BY GRX

CPYC-type GRXs can catalyze both disulfide reduction and deglutathionylation, while monocysteinic GRX mutants with a CPYS active site only catalyze deglutathionylation (see Section B). Based on these properties, an interesting method has been developed to identify glutathionylated proteins by their ability to be deglutathionylated by GRXs (Lind *et al.*, 2002). Protein extracts are initially heavily alkylated with *N*-ethyl maleimide (NEM) to block all free thiols and, after reduction of protein-SSG by a monocysteinic GRX, the newly accessible thiols are derivatized by biotin maleimide (NEM-Biotin) to tag proteins that were initially glutathionylated in the extract (Fig. 6). These tagged proteins are then purified by avidin-based affinity chromatography and identified by proteomic analysis.

This approach initially allowed identification of 22 proteins undergoing glutathionylation after diamide treatment of human ECV304 endothelial cells but also 21 proteins glutathionylated in untreated cells (Lind *et al.*, 2002). It was also used to analyze the glutathionylation of γ-actin, hsp60, elongation factor 1-α-1 (Hamnell-Pamment *et al.*, 2005), and α-ketoglutarate dehydrogenase (Applegate *et al.*, 2008).

This approach has several major advantages including the detection of proteins already glutathionylated under basal conditions and the identification of glutathionylated proteins in more physiological conditions than the labeling methods, since no pretreatment is required. Moreover, this method is not restricted to cell cultures and could therefore allow analysis of protein glutathionylation in extracts from different tissues and organs of

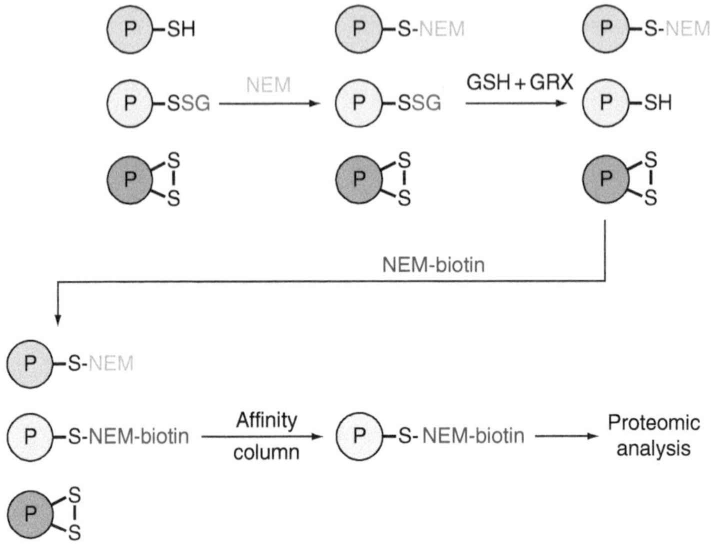

Fig. 6. Identification of glutathionylated proteins through reduction by GRX. The method is based on the property of glutaredoxins to catalyze protein deglutathionylation. In a first step, free thiols present in the protein extract are blocked by alkylation using *N*-ethylmaleimide (NEM). Then a monocysteinic mutant of glutaredoxin, which is no longer able to reduce protein disulfides, specifically catalyzes protein deglutathionylation. The new accessible thiols are derivatized using NEM-biotin. The biotinylated proteins are then purified by avidin/streptavidin pull-down and identified by proteomic analyses.

higher plants. One limitation of this approach is linked to the possible specificity and selectivity of the monocysteinic mutant GRX for some protein substrates. However, this limitation could also be useful to explore the specificity of GRXs. Indeed, it could be determined whether different proteins are identified after reduction of protein extracts by different GRX isoforms.

E. OTHER METHODS

Several alternative strategies aimed at identifying glutathionylated proteins have been described, but they were generally only used in a single study and therefore require further validation. They include the GST overlay approach based on the ability of a GST from *Schistosoma japonicum* to specifically bind to the glutathione moiety of glutathionylated proteins (Cheng *et al.*, 2005) and methods based on immobilized glutathione and its analogs (Niture *et al.*, 2005). These methods have been discussed elsewhere (Gao *et al.*, 2009).

All the methods for proteomic analysis of glutathionylated proteins described are complementary, since they exhibit different advantages and drawbacks. In future studies, a need of all these methods as well as newly developed approaches will be required to get further insights into the diversity of proteins undergoing glutathionylation *in vivo* in different organisms, different cells, and different physiological conditions.

V. GLUTATHIONYLATION IN NONPHOTOSYNTHETIC ORGANISMS

Most studies on glutathionylation have been performed in mammals where approximately 200 glutathionylated proteins have been identified, thus implicating this modification in the regulation of various processes and diverse metabolic and signaling pathways. For numerous proteins, the biochemical and/or functional consequences of glutathionylation have been investigated. These aspects have been detailed in several recent reviews (Dalle-Donne *et al.*, 2007, 2009; Fratelli *et al.*, 2004; Gao *et al.*, 2009; Michelet *et al.*, 2006; Mieyal *et al.*, 2008; Rouhier *et al.*, 2008; Shelton and Mieyal, 2008). Here we will present examples of cellular processes, including metabolic and signaling pathways, whose regulation by glutathionylation has been well described.

Many metabolic enzymes have been reported to be susceptible to redox regulation by glutathionylation. The most extensively studied of these enzymes is cytosolic glyceraldehyde 3-phosphate dehydrogenase (GAPDH), a glycolytic enzyme playing a central role in energy production but also involved in many other functions including regulation of apoptosis, DNA repair, or nuclear RNA export (reviewed in Chuang *et al.*, 2005; Sirover, 1999, 2005). GAPDH possesses a very reactive active-site cysteine which can be specifically glutathionylated thereby inhibiting the glycolytic activity of the enzyme (Grant *et al.*, 1999; Mohr *et al.*, 1999; Ravichandran *et al.*, 1994; Schuppe-Koistinen *et al.*, 1994). Recently, glutathionylation of GAPDH has been shown to regulate endothelin-1 (ET-1) expression by altering the binding of GAPDH to the 3′ untranslated region of ET-1 mRNA thereby increasing its stability (Rodriguez-Pascual *et al.*, 2008). In yeast cells, exposure to H_2O_2 leads to glutathionylation of GAPDH, enolase, and alcohol dehydrogenase (Shenton and Grant, 2003). Other glycolytic enzymes (pyruvate kinase, aldolase, phosphoglycerate kinase (PGK), and TPI) can undergo glutathionylation in human cells exposed to artificial oxidant conditions (Fratelli *et al.*, 2003). These results suggest that glutathionylation could coordinate cellular metabolism in response to oxidative stress by modulating glycolysis.

Several other enzymes involved in energy metabolism are regulated by glutathionylation. Mitochondrial and cytosolic NAD^+-dependent isocitrate dehydrogenases have a glutathionylated catalytic cysteine, and the activity of both enzymes can be restored by GRX (Kil and Park, 2005; Shin *et al.*, 2009). In mitochondria, complexes I and II were shown to be glutathionylated, this modification resulting in activation in the latter case (Chen *et al.*, 2007; Hurd *et al.*, 2008).

Glutathionylation also plays an important role in the regulation of many cell signaling pathways and transcription factors regulating cell growth, differentiation, and apoptosis (Dalle-Donne *et al.*, 2007, 2009; England and Cotter, 2005; Klatt and Lamas, 2000). Numerous protein kinases and phosphatases are regulated, either positively or negatively, by glutathionylation such as protein kinase C (Ward *et al.*, 2000, 2002), MEKK1 (Cross and Templeton, 2004), and PTP-1B (Barrett *et al.*, 1999b; Rinna *et al.*, 2006). Very recently, Fas, a receptor important for regulation of programmed cell death, was shown to be glutathionylated after caspase-dependent degradation of GRX1, thereby increasing subsequent caspase activation and apoptosis (Anathy *et al.*, 2009).

In addition, by perturbing their DNA binding sites, glutathionylation inhibits the DNA binding activity of several redox-sensitive transcription factors including c-Jun (Klatt *et al.*, 1999a), Pax-8 (Cao *et al.*, 2005), IRF3 (Prinarakis *et al.*, 2008), and p53 (Velu *et al.*, 2007). Glutathionylation has also been reported for up to 13 proteins within the NF-KappaB pathway (reviewed in Mieyal *et al.*, 2008), including the p50 and p65 subunits of NF-KappaB (Pineda-Molina *et al.*, 2001; Qanungo *et al.*, 2007) and the Inhibitory KappaB kinase (Reynaert *et al.*, 2006b).

However, protein glutathionylation does not always have an inhibitory effect. For example, the glutathionylation of p21ras, a small GTPase, increases its activity and mediates downstream signaling (Adachi *et al.*, 2004a; Clavreul *et al.*, 2006). The sarco(endo)plasmic reticulum calcium ATPase (SERCA) is also activated by glutathionylation, whereas this kind of regulation inhibits the chloride channel CFTR (Adachi *et al.*, 2004b; Wang *et al.*, 2005).

Cytoskeletal arrangements and intracellular trafficking can also be regulated by glutathionylation. Indeed, this modification can regulate actin polymerization (Dalle-Donne *et al.*, 2003, 2005; Wang *et al.*, 2001, 2003). Moreover, annexin A2, which interacts with actin, is also glutathionylated (Caplan *et al.*, 2004).

Finally, *S*-glutathionylation has also been implicated in the regulation of protein folding and stability. The activity of the 20S proteasome is inhibited by glutathionylation and reactivated by GRX (Demasi *et al.*, 2003;

Silva *et al.*, 2008). Moreover, several protein chaperones such as HSP70, HSP60, or PDI have also been identified among glutathionylated proteins by proteomic studies (reviewed in Michelet *et al.*, 2006). The multimeric aggregate size of HSP27 seems to be regulated by glutathionylation (Eaton *et al.*, 2002), and the chaperonin activity of HSP70 was suggested to be increased by this modification (Hoppe *et al.*, 2004).

VI. GLUTATHIONYLATION IN PHOTOSYNTHETIC ORGANISMS

Most studies on protein glutathionylation have been performed in mammals but this posttranslational modification has also emerged recently as a mechanism of redox signaling in photosynthetic organisms. To our knowledge, the first publications reporting the glutathionylation of plant proteins appeared in 2002. The first one identified the birch PR-10c protein, which belongs to the family of intracellular pathogenesis-related (PR) proteins, as a target of glutathionylation (Koistinen *et al.*, 2002). Mass spectrometry analyses revealed that PR-10c purified from birch roots or expressed recombinantly in *E. coli* was glutathionylated on a nonconserved cysteine residue. In addition, the ribonuclease activity of PR-10c was not affected by glutathionylation of this residue. Therefore, the functional significance of this modification remains unclear. The second early study reported that four GST isoforms from *Arabidopsis thaliana* could undergo glutathionylation (Dixon *et al.*, 2002). DHAR1 and DHAR2, two cytosolic GSTs possessing dehydroascorbate (DHA) reductase and glutathione-dependent thioltransferase activities, were shown to be glutathionylated *in vitro* in the presence of GSSG. The mixed disulfide form was proposed to constitute a key intermediate in the catalytic mechanism of DHA reduction. In addition, two isoforms belonging to the Lambda GST family (GSTL) were also characterized. Cytosolic GSTL1 and chloroplastic GSTL2 only possess thioltransferase activity and can be glutathionylated *in vitro* upon GSSG treatments. Proteomic analyses have also suggested that multiple isoforms of GSTs could be glutathionylated *in vitro* (Dixon *et al.*, 2005b). In this study, the glutathionylation of the catalytic cysteine of a zeta-class GST (GSTZ1) was confirmed *in vitro*.

As described earlier, PTP-1B is a well-characterized target of glutathionylation in human cells (Barrett *et al.*, 1999b; Rinna *et al.*, 2006). A soybean PTP was also found to be glutathionylated *in vitro* in the presence of GSSG (Dixon *et al.*, 2005a). However, compared to its mammalian counterpart that readily reacts with H_2O_2 to form a sulfenic acid on the catalytic cysteine, the plant PTP was insensitive to H_2O_2 but hypersensitive to GSSG.

This inactivation of soybean PTP by GSSG was proposed to constitute a mechanism of protection of the enzyme under highly oxidizing conditions.

PRXs are thiol-dependent peroxidases which can be regenerated by different electron donors, including TRXs and GRXs (Dayer *et al.*, 2008; Rouhier and Jacquot, 2005). Many PRXs have been identified among glutathionylated proteins in mammals (Fratelli *et al.*, 2002, 2003, 2004; Lind *et al.*, 2002; Manevich *et al.*, 2004; Sullivan *et al.*, 2000) but also in photosynthetic organisms. For example, the *in vitro* glutathionylation of the peroxidatic cysteine of a poplar type II PRX was shown to trigger the dissociation of the noncovalent homodimers into monomers (Noguera-Mazon *et al.*, 2006). *In vitro* treatments with BioGSSG also suggested that *Arabidopsis* 2-cys PRXs could undergo glutathionylation (Dixon *et al.*, 2005b). In *Chlamydomonas*, ^{35}S-cysteine labeling experiments revealed that both cytosolic and chloroplastic 2-cys PRXs are *S*-thiolated *in vivo* (Michelet *et al.*, 2008). *In vitro*, the glutathionylation of *Chlamydomonas* chloroplastic PRX led to a dimer/monomer switch. Glutathionylation could thus constitute a mechanism of regulation of the oligomerization/activity of PRXs and might be involved in a redox-dependent dimer/monomer switch within the PRX family (Michelet *et al.*, 2008; Noguera-Mazon *et al.*, 2006). Alternatively, glutathionylation might be an intermediate in the mechanism of reactivation of some PRXs, as previously observed for mammalian 1-cys PRX which is converted to the glutathionylated form by interaction with GSTπ (Manevich *et al.*, 2004; Ralat *et al.*, 2006).

Human TRX was shown to undergo glutathionylation *in vivo*, on an extra cysteine, distinct from the two active-site cysteines (Casagrande *et al.*, 2002; Fratelli *et al.*, 2002). The glutathionylation of human TRX appears to partially inhibit its oxidoreductase activity, probably by decreasing its reduction by NTR. These results on human TRX have prompted several groups to examine the ability of plant TRXs to undergo glutathionylation *in vitro*. In photosynthetic organisms, TRXs are encoded by a multigenic family and have been classified into six different types (Gelhaye *et al.*, 2005; Lemaire *et al.*, 2007; Meyer and Hell, 2005; Meyer *et al.*, 2008). TRX h and TRX o are localized in the cytosol and/or mitochondria where they are reduced by NADPH via NTRs. In addition, 4 canonical TRX types, named f, m, x, and y are chloroplast localized and reduced in the light, by way of photoreduced ferredoxin and FTR. Once reduced, TRXs are able to reduce key disulfides on their target proteins, with different specificities for their targets. TRX x and TRX y are preferential electron donors to several antioxidant enzymes such as PRXs, glutathione peroxidases, or methionine sulfoxide reductases, while f-type TRXs are more specifically involved in the light-dependent regulation of carbon metabolism enzymes, including several

Calvin cycle enzymes (Dayer *et al.*, 2008; Lemaire *et al.*, 2007; Schurmann and Buchanan, 2008). An extra active-site cysteine of poplar mitochondrial TRX h was shown to undergo glutathionylation *in vitro* with a concomitant increase of the active-site disulfide redox potential suggesting that the modification might perturb the function of this TRX (Gelhaye *et al.*, 2004). The analysis of all chloroplastic TRXs from *Arabidopsis* and *Chlamydomonas* revealed that only f-type TRXs could undergo glutathionylation *in vitro* (Michelet *et al.*, 2005). The glutathionylated residue is an extra cysteine, distinct from the two active-site cysteines, which is conserved in all f-type TRXs and is localized in the three-dimensional structure of the enzyme, close to the active site. The glutathionylation of TRX f strongly decreased its ability to activate A_2B_2-GAPDH and NADP-malate dehydrogenase likely by perturbing the interaction with FTR since glutathionylated TRX f is less efficiently reduced in the light. This suggests that glutathionylation could affect all TRX f targets. Several Calvin cycle enzymes were also reported to undergo glutathionylation *in vivo*: FBA was identified in *Arabidopsis* (Ito *et al.*, 2003) but also PGK and ribose-5-phosphate isomerase (RPI) in *Chlamydomonas*. However, the effect of glutathionylation on the activity of these enzymes is not yet known. We have also recently analyzed the glutathionylation of the two higher plant GAPDH isoforms, the heterotetrameric A_2B_2-GAPDH and the homotetrameric A_4-GAPDH, that participate in the Calvin cycle (Zaffagnini *et al.*, 2007). The NADPH-dependent activity of the major isoform, A_2B_2-GAPDH, is specifically light regulated by TRX f while the A_4 isoform is not (Marri *et al.*, 2009). *Arabidopsis* A_4-GAPDH is glutathionylated *in vitro* on its catalytic cysteine with a concomitant loss of enzyme activity. The enzyme is very sensitive to oxidants and is rapidly and irreversibly inactivated by H_2O_2. However, incubation of the enzyme with H_2O_2 in the presence of GSH leads to glutathionylation, most likely through a mechanism involving a sulfenic acid intermediate. Therefore, glutathionylation efficiently protects A_4-GAPDH from irreversible oxidation. Moreover, glutathionylated A_4-GAPDH was reported to be efficiently reactivated by GRXs (Zaffagnini *et al.*, 2008). By contrast, the A_2B_2-GAPDH isoform or its higher oligomeric state A_8B_8-GAPDH was not found to undergo glutathionylation *in vitro*. Nevertheless, the activity of A_2B_2-GAPDH is under the control of TRX f while A_4-GAPDH is not. Consequently, under conditions leading to protein glutathionylation in chloroplasts, the activity of both types of GAPDH is likely to be decreased. More generally, the glutathionylation of TRX f and of several Calvin cycle enzymes suggests that this posttranslational modification could constitute a new mechanism of regulation of photosynthetic metabolism. We have recently proposed that such a mechanism could allow a fine tuning of the Calvin cycle allowing redistribution of the reducing power within chloroplasts under oxidative stress thereby favoring ROS scavenging (Lemaire *et al.*, 2007; Michelet *et al.*, 2005).

In *Arabidopsis*, cytosolic GAPDH isoforms were also identified among glutathionylated proteins *in vivo* (Dixon *et al.*, 2005b) and shown to be inactivated by glutathionylation of their catalytic cysteine *in vitro* (Holtgrefe *et al.*, 2008). Similarly, TPI was found to be glutathionylated *in vivo* and this modification inactivates purified TPI *in vitro* (Ito *et al.*, 2003). These results suggest that, as in nonphotosynthetic organisms, glycolysis might be regulated by glutathionylation in plants.

Several protein chaperones were reported to be glutathionylated *in vivo* and *in vitro* in mammals but also in *Arabidopsis* and *Chlamydomonas* (Dixon *et al.*, 2005b; Michelet *et al.*, 2008). Among these, *Chlamydomonas* chloroplastic HSP70B could be glutathionylated *in vitro* on a partially conserved cysteine residue located in the ATPase domain of the protein. This suggests that glutathionylation could affect the activity of this HSP70 isoform which has been implicated in the folding of chloroplast proteins, the assembly/disassembly of VIPP1 oligomers, and the protection of photosystem II under high light illumination (Liu *et al.*, 2007; Schroda, 2004).

In photosynthetic organisms, the glutathionylation of a few additional enzymes involved in diverse pathways was also reported. Glutathionylation was shown to reversibly inactivate *Chlamydomonas* isocitrate lyase, a key enzyme of acetate assimilation (Michelet *et al.*, 2008). A recent *in vitro* study also indicated that galactonolactone dehydrogenase, which catalyzes the terminal step of L-ascorbate biosynthesis, is sensitive to H_2O_2 and can be protected from irreversible oxidation by glutathionylation (Leferink *et al.*, 2009). Moreover, the glutathionylation of *Arabidopsis* methionine synthase and alcohol dehydrogenase was reported, but a protein factor present in *Arabidopsis* extracts was required to trigger glutathionylation of these enzymes *in vitro* (Dixon *et al.*, 2005b).

Proteomic analyses allowed identification, either *in vivo* or *in vitro*, of many additional potential targets of glutathionylation in *Arabidopsis* and *Chlamydomonas* (Dixon *et al.*, 2005b; Michelet *et al.*, 2008). These proteins are involved in diverse processes and metabolic pathways including photosynthesis, oxidative stress responses, protein folding, amino acid biosynthesis, lipid metabolism, translation, ATP metabolism, cytoskeletal arrangements, etc. (reviewed in Gao *et al.*, 2009). Several of these proteins and processes were previously suggested to be regulated by glutathionylation in mammals. However, all these candidates await further characterization to determine the targeted residues, the effect of glutathionylation on their activity, and the functional importance of glutathionylation for the regulation of the corresponding cellular processes.

VII. MULTIPLE INTERCONNECTIONS

Emerging evidence suggests the existence of multiple interconnections between different types of posttranslational redox modifications under the control of TRXs and GRXs. GRXs appear specifically involved in the control of glutathionylation while TRXs, which are well established to regulate the oxidoreduction state of protein disulfide bonds, were recently suggested to catalyze denitrosylation reactions (Benhar *et al.*, 2008). Moreover, TRXs and GRXs were reported to undergo both nitrosylation and glutathionylation (Hashemy and Holmgren, 2008; Hashemy *et al.*, 2007). Many pathways have been shown to be regulated by multiple posttranslational redox modifications. One example in photosynthetic organisms

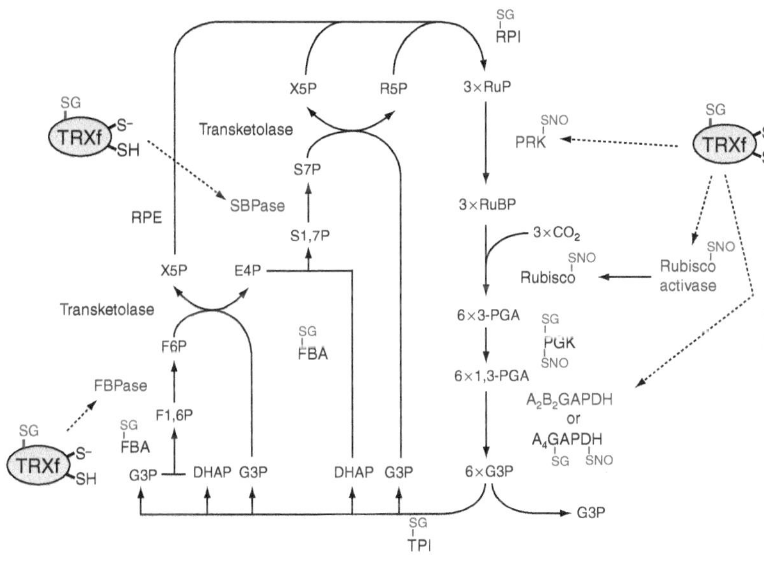

Fig. 7. Redox regulation of Calvin cycle enzymes. Most Calvin cycle enzymes have been shown to undergo one or several post-translational redox modifications: oxidoreduction of disulfide bridges by thioredoxin f (dotted arrows pointed at established TRX targets indicated in blue), glutathionylation (–SG in red), and/or nitrosylation (–SNO in green). Enzymes (green): Rubisco, ribulose-1,5-bisphosphate carboxylase/oxygenase; PGK, phosphoglycerate kinase; GAPDH, glyceraldehyde-3-phosphate dehydrogenase; TPI, triose phosphate isomerase; FBA, fructose-1, 6-bisphosphate aldolase; FBPase, fructose-1,6-bisphosphatase; SBPase, sedoheptulose-1,7-bisphosphatase; RPE, ribulose-5-phosphate epimerase; RPI, ribose-5-phosphate isomerase; PRK, phosphoribulokinase. Metabolites: RuBP, ribulose-1, 5-bisphosphate; 3-PGA, 3-phosphoglycerate; 1,3-BPGA, 1,3-bisphosphoglycerate; G3P, glyceraldehyde-3-phosphate; DHAP, dihydroxyacetone phosphate; F1,6P, fructose-1,6-bisphosphate; F6P, fructose-6-phosphate; X5P, xylulose-5-phosphate; E4P, erythrose-4-phosphate; S1,7BP, sedoheptulose-1,7-bisphosphate; S7P, sedoheptulose-7-phosphate; R5P, ribose-5-phosphate; RuP, ribulose-5-phosphate. (See color plate section at the back of the book.)

is the Calvin cycle (Fig. 7). Five enzymes of this pathway are established targets of TRXs but all Calvin cycle enzymes were identified among putative TRX targets eluted from monocysteinic TRX affinity columns (Lemaire *et al.*, 2007; Montrichard *et al.*, 2009; Schurmann and Buchanan, 2008). In addition, as detailed earlier, glutathionylation is also likely regulating the activity of Calvin cycle enzymes, either directly or indirectly through glutathionylation of TRX f. Finally, several Calvin cycle enzymes where also reported to undergo nitrosylation in *Arabidopsis* (Lindermayr *et al.*, 2005; Romero-Puertas *et al.*, 2008). In addition, the analysis of an *Arabidopsis* mutant deficient in mitochondrial and cytosolic TRX reductases indicated that some cytosolic TRXs are still partially reduced through an unknown glutathione-dependent pathway that might involve some GRXs (Reichheld *et al.*, 2007).

Overall, the complex interplay between TRXs, GRXs, and the posttranslational redox modifications under their control has only emerged recently. Efforts aimed at understanding these complex interconnections will have to be pursued to clarify the respective specific functions and redundancies of the different components of the redox signaling network. Unraveling the importance of this crosstalk will certainly constitute a major challenge for future studies.

ACKNOWLEDGMENTS

The authors would like to thank Myroslawa Miginiac-Maslow for critical reading of the manuscript and helpful suggestions. This work was supported by Agence Nationale de la Recherche Grant JC-45651.

REFERENCES

Adachi, T., Pimentel, D. R., Heibeck, T., Hou, X., Lee, Y. J., Jiang, B., Ido, Y. and Cohen, R. A. (2004a). *S*-Glutathiolation of Ras mediates redox-sensitive signaling by angiotensin II in vascular smooth muscle cells. *The Journal of Biological Chemistry* **279**, 29857–29862.

Adachi, T., Weisbrod, R. M., Pimentel, D. R., Ying, J., Sharov, V. S., Schoneich, C. and Cohen, R. A. (2004b). *S*-Glutathiolation by peroxynitrite activates SERCA during arterial relaxation by nitric oxide. *Nature Medicine* **10**, 1200–1207.

Alves, R., Vilaprinyo, E., Sorribas, A. and Herrero, E. (2009). Evolution based on domain combinations: The case of glutaredoxins. *BMC Evolutionary Biology* **9**, 66.

Anathy, V., Aesif, S. W., Guala, A. S., Havermans, M., Reynaert, N. L., Ho, Y. S., Budd, R. C. and Janssen-Heininger, Y. M. (2009). Redox amplification of apoptosis by caspase-dependent cleavage of glutaredoxin 1 and *S*-glutathionylation of Fas. *The Journal of Cell Biology* **184**, 241–252.

Applegate, M. A., Humphries, K. M. and Szweda, L. I. (2008). Reversible inhibition of alpha-ketoglutarate dehydrogenase by hydrogen peroxide: Glutathionylation and protection of lipoic acid. *Biochemistry* **47**, 473–478.

Avval, F. Z. and Holmgren, A. (2009). Molecular mechanisms of thioredoxin and glutaredoxin as hydrogen donors for Mammalian s phase ribonucleotide reductase. *The Journal of Biological Chemistry* **284**, 8233–8240.

Bandyopadhyay, S., Chandramouli, K. and Johnson, M. K. (2008a). Iron–sulfur cluster biosynthesis. *Biochemical Society Transactions* **36**, 1112–1119.

Bandyopadhyay, S., Gama, F., Molina-Navarro, M. M., Gualberto, J. M., Claxton, R., Naik, S. G., Huynh, B. H., Herrero, E., Jacquot, J. P., Johnson, M. K. and Rouhier, N. (2008b). Chloroplast monothiol glutaredoxins as scaffold proteins for the assembly and delivery of [2Fe-2S] clusters. *EMBO Journal* **27**, 1122–1133.

Barrett, W. C., DeGnore, J. P., Keng, Y. F., Zhang, Z. Y., Yim, M. B. and Chock, P. B. (1999a). Roles of superoxide radical anion in signal transduction mediated by reversible regulation of protein-tyrosine phosphatase 1B. *The Journal of Biological Chemistry* **274**, 34543–34546.

Barrett, W. C., DeGnore, J. P., Konig, S., Fales, H. M., Keng, Y. F., Zhang, Z. Y., Yim, M. B. and Chock, P. B. (1999b). Regulation of PTP1B via glutathionylation of the active site cysteine 215. *Biochemistry* **38**, 6699–6705.

Benhar, M., Forrester, M. T., Hess, D. T. and Stamler, J. S. (2008). Regulated protein denitrosylation by cytosolic and mitochondrial thioredoxins. *Science* **320**, 1050–1054.

Berndt, C., Lillig, C. H. and Holmgren, A. (2008). Thioredoxins and glutaredoxins as facilitators of protein folding. *Biochimica et Biophysica Acta* **1783**, 641–650.

Biteau, B., Labarre, J. and Toledano, M. B. (2003). ATP-dependent reduction of cysteine-sulphinic acid by *S. cerevisiae* sulphiredoxin. *Nature* **425**, 980–984.

Brennan, J. P., Miller, J. I., Fuller, W., Wait, R., Begum, S., Dunn, M. J. and Eaton, P. (2006). The utility of *N,N*-biotinyl glutathione disulfide in the study of protein *S*-glutathiolation. *Molecular and Cellular Proteomics* **5**, 215–225.

Budanov, A. V., Sablina, A. A., Feinstein, E., Koonin, E. V. and Chumakov, P. M. (2004). Regeneration of peroxiredoxins by p53-regulated sestrins, homologs of bacterial AhpD. *Science* **304**, 596–600.

Bushweller, J. H., Aslund, F., Wuthrich, K. and Holmgren, A. (1992). Structural and functional characterization of the mutant *Escherichia coli* glutaredoxin (C14—S) and its mixed disulfide with glutathione. *Biochemistry* **31**, 9288–9293.

Cao, X., Kambe, F., Lu, X., Kobayashi, N., Ohmori, S. and Seo, H. (2005). Glutathionylation of two cysteine residues in paired domain regulates DNA binding activity of Pax-8. *The Journal of Biological Chemistry* **280**, 25901–25906.

Caplan, J. F., Filipenko, N. R., Fitzpatrick, S. L. and Waisman, D. M. (2004). Regulation of annexin A2 by reversible glutathionylation. *The Journal of Biological Chemistry* **279**, 7740–7750.

Casagrande, S., Bonetto, V., Fratelli, M., Gianazza, E., Eberini, I., Massignan, T., Salmona, M., Chang, G., Holmgren, A. and Ghezzi, P. (2002). Glutathionylation of human thioredoxin: A possible crosstalk between the glutathione and thioredoxin systems. *Proceedings of the National Academy of Sciences of the United States of America* **99**, 9745–9749.

Chai, Y. C., Ashraf, S. S., Rokutan, K., Johnston, R. B., Jr. and Thomas, J. A. (1994). *S*-Thiolation of individual human neutrophil proteins including actin by stimulation of the respiratory burst: evidence against a role for glutathione disulfide. *Archives of Biochemistry and Biophysics* **310**, 273–281.

Chandra, A., Srivastava, S., Petrash, J. M., Bhatnagar, A. and Srivastava, S. K. (1997). Modification of aldose reductase by *S*-nitrosoglutathione. *Biochemistry* **36**, 15801–15809.

Chen, Y. R., Chen, C. L., Pfeiffer, D. R. and Zweier, J. L. (2007). Mitochondrial complex II in the post-ischemic heart: Oxidative injury and the role of protein *S*-glutathionylation. *The Journal of Biological Chemistry* **282**, 32640–32654.

Cheng, N. H. (2008). AtGRX4, an *Arabidopsis* chloroplastic monothiol glutaredoxin, is able to suppress yeast grx5 mutant phenotypes and respond to oxidative stress. *FEBS Letters* **582**, 848–854.

Cheng, G., Ikeda, Y., Iuchi, Y. and Fujii, J. (2005). Detection of *S*-glutathionylated proteins by glutathione *S*-transferase overlay. *Archives of Biochemistry and Biophysics* **435**, 42–49.

Cheng, N. H., Liu, J. Z., Brock, A., Nelson, R. S. and Hirschi, K. D. (2006). AtGRXcp, an *Arabidopsis* chloroplastic glutaredoxin, is critical for protection against protein oxidative damage. *The Journal of Biological Chemistry* **281**, 26280–26288.

Chrestensen, C. A., Starke, D. W. and Mieyal, J. J. (2000). Acute cadmium exposure inactivates thioltransferase (Glutaredoxin), inhibits intracellular reduction of protein-glutathionyl-mixed disulfides, and initiates apoptosis. *The Journal of Biological Chemistry* **275**, 26556–26565.

Chuang, D. M., Hough, C. and Senatorov, V. V. (2005). Glyceraldehyde-3-phosphate dehydrogenase, apoptosis, and neurodegenerative diseases. *Annual Review of Pharmacology and Toxicology* **45**, 269–290.

Clavreul, N., Adachi, T., Pimental, D. R., Ido, Y., Schoneich, C. and Cohen, R. A. (2006). *S*-Glutathiolation by peroxynitrite of p21ras at cysteine-118 mediates its direct activation and downstream signaling in endothelial cells. *FASEB Journal* **20**, 518–520.

Coles, S. J., Easton, P., Sharrod, H., Hutson, S. M., Hancock, J., Patel, V. B. and Conway, M. E. (2009). *S*-Nitrosoglutathione inactivation of the mitochondrial and cytosolic BCAT proteins: *S*-nitrosation and *S*-thiolation. *Biochemistry* **48**, 645–656.

Comini, M. A., Rettig, J., Dirdjaja, N., Hanschmann, E. M., Berndt, C. and Krauth-Siegel, R. L. (2008). Monothiol glutaredoxin-1 is an essential iron-sulfur protein in the mitochondrion of African trypanosomes. *The Journal of Biological Chemistry* **283**, 27785–27798.

Couturier, J., Koh, C. S., Zaffagnini, M., Winger, A. M., Gualberto, J. M., Corbier, C., Decottignies, P., Jacquot, J. P., Lemaire, S. D., Didierjean, C. and Rouhier, N. (2009a). Structure–function relationship of the chloroplastic glutaredoxin S12 with an atypical WCSYS active site. *The Journal of Biological Chemistry* **284**, 9299–9310.

Couturier, J., Jacquot, J. P. and Rouhier, N. (2009a). Evolution and diversity of glutaredoxins in photosynthetic organisms. *Cellular and Molecular Life Sciences* (In press).

Cross, J. V. and Templeton, D. J. (2004). Oxidative stress inhibits MEKK1 by site-specific glutathionylation in the ATP-binding domain. *The Biochemical Journal* **381**, 675–683.

Dalle-Donne, I., Rossi, R., Giustarini, D., Colombo, R. and Milzani, A. (2003). Actin *S*-glutathionylation: Evidence against a thiol-disulphide exchange mechanism. *Free Radical Biology and Medicine* **35**, 1185–1193.

Dalle-Donne, I., Giustarini, D., Colombo, R., Milzani, A. and Rossi, R. (2005). *S*-Glutathionylation in human platelets by a thiol-disulfide exchange-independent mechanism. *Free Radical Biology and Medicine* **38**, 1501–1510.

Dalle-Donne, I., Rossi, R., Giustarini, D., Colombo, R. and Milzani, A. (2007). *S*-Glutathionylation in protein redox regulation. *Free Radical Biology and Medicine* **43**, 883–898.

Dalle-Donne, I., Milzani, A., Gagliano, N., Colombo, R., Giustarini, D. and Rossi, R. (2008). Molecular mechanisms and potential clinical significance of *S*-glutathionylation. *Antioxidants and Redox Signaling* **10**, 445–473.

Dalle-Donne, I., Rossi, R., Colombo, G., Giustarini, D. and Milzani, A. (2009). Protein *S*-glutathionylation: A regulatory device from bacteria to humans. *Trends in Biochemical Sciences* **34**, 85–96.

Dayer, R., Fischer, B. B., Eggen, R. I. and Lemaire, S. D. (2008). The peroxiredoxin and glutathione peroxidase families in *Chlamydomonas reinhardtii*. *Genetics* **179**, 41–57.

Demasi, M., Silva, G. M. and Netto, L. E. (2003). 20 S proteasome from *Saccharomyces cerevisiae* is responsive to redox modifications and is *S*-glutathionylated. *The Journal of Biological Chemistry* **278**, 679–685.

Demasi, M., Piassa Filho, G. M., Castro, L. M., Ferreira, J. C., Rioli, V. and Ferro, E. S. (2008). Oligomerization of the cysteinyl-rich oligopeptidase EP24.15 is triggered by *S*-glutathionylation. *Free Radical Biology and Medicine* **44**, 1180–1190.

Denu, J. M. and Tanner, K. G. (1998). Specific and reversible inactivation of protein tyrosine phosphatases by hydrogen peroxide: Evidence for a sulfenic acid intermediate and implications for redox regulation. *Biochemistry* **37**, 5633–5642.

Discola, K. F., de Oliveira, M. A., Rosa Cussiol, J. R., Monteiro, G., Barcena, J. A., Porras, P., Padilla, C. A., Guimaraes, B. G. and Netto, L. E. (2009). Structural aspects of the distinct biochemical properties of glutaredoxin 1 and glutaredoxin 2 from *Saccharomyces cerevisiae*. *Journal of Molecular Biology* **385**, 889–901.

Dixon, D. P., Davis, B. G. and Edwards, R. (2002). Functional divergence in the glutathione transferase superfamily in plants. Identification of two classes with putative functions in redox homeostasis in *Arabidopsis thaliana*. *The Journal of Biological Chemistry* **277**, 30859–30869.

Dixon, D. P., Fordham-Skelton, A. P. and Edwards, R. (2005a). Redox regulation of a soybean tyrosine-specific protein phosphatase. *Biochemistry* **44**, 7696–7703.

Dixon, D. P., Skipsey, M., Grundy, N. M. and Edwards, R. (2005b). Stress-induced protein *S*-glutathionylation in *Arabidopsis*. *Plant Physiology* **138**, 2233–2244.

Eaton, P., Fuller, W. and Shattock, M. J. (2002). *S*-Thiolation of HSP27 regulates its multimeric aggregate size independently of phosphorylation. *The Journal of Biological Chemistry* **277**, 21189–21196.

England, K. and Cotter, T. G. (2005). Direct oxidative modifications of signalling proteins in mammalian cells and their effects on apoptosis. *Redox Report: Communications in free radical research* **10**, 237–245.

Feng, Y., Zhong, N., Rouhier, N., Hase, T., Kusunoki, M., Jacquot, J. P., Jin, C. and Xia, B. (2006). Structural insight into poplar glutaredoxin C1 with a bridging iron-sulfur cluster at the active site. *Biochemistry* **45**, 7998–8008.

Fernandes, A. P. and Holmgren, A. (2004). Glutaredoxins: Glutathione-dependent redox enzymes with functions far beyond a simple thioredoxin backup system. *Antioxidants and Redox Signaling* **6**, 63–74.

Findlay, V. J., Townsend, D. M., Morris, T. E., Fraser, J. P., He, L. and Tew, K. D. (2006). A novel role for human sulfiredoxin in the reversal of glutathionylation. *Cancer Research* **66**, 6800–6806.

Finkemeier, I., Goodman, M., Lamkemeyer, P., Kandlbinder, A., Sweetlove, L. J. and Dietz, K. J. (2005). The mitochondrial type II peroxiredoxin F is essential for redox homeostasis and root growth of *Arabidopsis thaliana* under stress. *The Journal of Biological Chemistry* **280**, 12168–12180.

Foyer, C. H. and Noctor, G. (2005). Redox homeostasis and antioxidant signaling: A metabolic interface between stress perception and physiological responses. *The Plant Cell* **17**, 1866–1875.

Foyer, C. H., Noctor, G., Buchanan, B., Dietz, K. J. and Pfannschmidt, T. (2009). Redox regulation in photosynthetic organisms: Signaling, acclimation, and practical implications. *Antioxidants and Redox Signaling* **11**, 861–905.

Fratelli, M., Demol, H., Puype, M., Casagrande, S., Eberini, I., Salmona, M., Bonetto, V., Mengozzi, M., Duffieux, F., Miclet, E., Bachi, A. Vandekerckhove, J. *et al.* (2002). Identification by redox proteomics of glutathionylated proteins in oxidatively stressed human T lymphocytes. *Proceedings of the National Academy of Sciences of the United States of America* **99**, 3505–3510.

Fratelli, M., Demol, H., Puype, M., Casagrande, S., Villa, P., Eberini, I., Vandekerckhove, J., Gianazza, E. and Ghezzi, P. (2003). Identification of proteins undergoing glutathionylation in oxidatively stressed hepatocytes and hepatoma cells. *Proteomics* **3**, 1154–1161.

Fratelli, M., Gianazza, E. and Ghezzi, P. (2004). Redox proteomics: Identification and functional role of glutathionylated proteins. *Expert Review of Proteomics* **1**, 365–376.

Fuangthong, M. and Helmann, J. D. (2002). The OhrR repressor senses organic hydroperoxides by reversible formation of a cysteine-sulfenic acid derivative. *Proceedings of the National Academy of Sciences of the United States of America* **99**, 6690–6695.

Gallogly, M. M. and Mieyal, J. J. (2007). Mechanisms of reversible protein glutathionylation in redox signaling and oxidative stress. *Current Opinion in Pharmacology* **7**, 381–391.

Gallogly, M. M., Starke, D. W., Leonberg, A. K., Ospina, S. M. and Mieyal, J. J. (2008). Kinetic and mechanistic characterization and versatile catalytic properties of mammalian glutaredoxin 2: Implications for intracellular roles. *Biochemistry* **47**, 11144–11157.

Gallogly, M. M., Starke, D. W. and Mieyal, J. J. (2009). Mechanistic and kinetic details of thiol-disulfide exchange by glutaredoxins and potential mechanisms of regulation. *Antioxidants and Redox Signaling* **11**(5), 1059–1081.

Gama, F., Keech, O., Eymery, F., Finkemeier, I., Gelhaye, E., Gardestrom, P., Dietz, K. J., Rey, P., Jacquot, J.-P. and Rouhier, N. (2007). The mitochondrial type II peroxiredoxin from poplar. *Physiologia Plantarum* **129**, 196–206.

Gao, X.-H., Bedhomme, M., Veyel, D., Zaffagnini, M. and Lemaire, S. D. (2009). Methods for analysis of protein glutathionylation and their application to photosynthetic organisms. *Molecular Plant* **2**, 218–235.

Gelhaye, E., Rouhier, N., Gérard, J., Jolivet, Y., Gualberto, J., Navrot, N., Ohlsson, P. I., Wingsle, G., Hirasawa, M., Knaff, D. B., Wang, H. Dizengremel, P. *et al.* (2004). A specific form of thioredoxin h occurs in plant mitochondria and regulates the alternative oxidase. *Proceedings of the National Academy of Sciences of the United States of America* **101**, 14545–14550.

Gelhaye, E., Rouhier, N., Navrot, N. and Jacquot, J. P. (2005). The plant thioredoxin system. *Cellular and Molecular Life Sciences* **62**, 24–35.

Ghezzi, P. and Di Simplicio, P. (2007). Glutathionylation pathways in drug response. *Current Opinion in Pharmacology* **7**, 398–403.

Gilbert, H. F. (1995). Thiol/disulfide exchange equilibria and disulfide bond stability. *Methods in Enzymology* **251**, 8–28.

Giustarini, D., Rossi, R., Milzani, A., Colombo, R. and Dalle-Donne, I. (2004). *S*-Glutathionylation: From redox regulation of protein functions to human diseases. *Journal of Cellular and Molecular Medicine* **8**, 201–212.

Giustarini, D., Milzani, A., Aldini, G., Carini, M., Rossi, R. and Dalle-Donne, I. (2005). *S*-Nitrosation versus *S*-glutathionylation of protein sulfhydryl groups by *S*-nitrosoglutathione. *Antioxidants and Redox Signaling* **7**, 930–939.

Grant, C. M., Quinn, K. A. and Dawes, I. W. (1999). Differential protein *S*-thiolation of glyceraldehyde-3-phosphate dehydrogenase isoenzymes influences sensitivity to oxidative stress. *Molecular and Cellular Biology* **19**, 2650–2656.

Gravina, S. A. and Mieyal, J. J. (1993). Thioltransferase is a specific glutathionyl mixed disulfide oxidoreductase. *Biochemistry* **32**, 3368–3376.

Gutscher, M., Pauleau, A. L., Marty, L., Brach, T., Wabnitz, G. H., Samstag, Y., Meyer, A. J. and Dick, T. P. (2008). Real-time imaging of the intracellular glutathione redox potential. *Nature Methods* **5**, 553–559.

Hamnell-Pamment, Y., Lind, C., Palmberg, C., Bergman, T. and Cotgreave, I. A. (2005). Determination of site-specificity of *S*-glutathionylated cellular proteins. *Biochemical and Biophysical Research Communications* **332**, 362–369.

Hashemy, S. I. and Holmgren, A. (2008). Regulation of the catalytic activity and structure of human thioredoxin 1 via oxidation and *S*-nitrosylation of cysteine residues. *The Journal of Biological Chemistry* **283**, 21890–21898.

Hashemy, S. I., Johansson, C., Berndt, C., Lillig, C. H. and Holmgren, A. (2007). Oxidation and *S*-nitrosylation of cysteines in human cytosolic and mitochondrial glutaredoxins: Effects on structure and activity. *The Journal of Biological Chemistry* **282**, 14428–14436.

Herrero, E. and de la Torre-Ruiz, M. A. (2007). Monothiol glutaredoxins: A common domain for multiple functions. *Cellular and Molecular Life Sciences* **64**, 1518–1530.

Herrero, E., Ros, J., Belli, G. and Cabiscol, E. (2008). Redox control and oxidative stress in yeast cells. *Biochimica et Biophysica Acta* **1780**, 1217–1235.

Hess, D. T., Matsumoto, A., Kim, S. O., Marshall, H. E. and Stamler, J. S. (2005). Protein *S*-nitrosylation: Purview and parameters. *Nature Reviews. Molecular Cell Biology* **6**, 150–166.

Hogg, N. (2002). The biochemistry and physiology of *S*-nitrosothiols. *Annual Review of Pharmacology and Toxicology* **42**, 585–600.

Holmgren, A. (1976). Hydrogen donor system for *Escherichia coli* ribonucleoside-diphosphate reductase dependent upon glutathione. *Proceedings of the National Academy of Sciences of the United States of America* **73**, 2275–2279.

Holtgrefe, S., Gohlke, J., Starmann, J., Druce, S., Klocke, S., Altmann, B., Wojtera, J., Lindermayr, C. and Scheibe, R. (2008). Regulation of plant cytosolic glyceraldehyde 3-phosphate dehydrogenase isoforms by thiol modifications. *Physiologia Plantarum* **133**, 211–228.

Hoppe, G., Chai, Y. C., Crabb, J. W. and Sears, J. (2004). Protein *S*-glutathionylation in retinal pigment epithelium converts heat shock protein 70 to an active chaperone. *Experimental Eye Research* **78**, 1085–1092.

Huang, K. P. and Huang, F. L. (2002). Glutathionylation of proteins by glutathione disulfide S-oxide. *Biochemical Pharmacology* **64**, 1049–1056.

Huang, Z., Pinto, J. T., Deng, H. and Richie, J. P., Jr. (2008). Inhibition of caspase-3 activity and activation by protein glutathionylation. *Biochemical Pharmacology* **75**, 2234–2244.

Hudemann, C., Lonn, M. E., Godoy, J. R., Zahedi Avval, F., Capani, F., Holmgren, A. and Lillig, C. H. (2009). Identification, expression pattern, and characterization of mouse glutaredoxin 2 isoforms. *Antioxidants and Redox Signaling* **11**, 1–14.

Hurd, T. R., Requejo, R., Filipovska, A., Brown, S., Prime, T. A., Robinson, A. J., Fearnley, I. M. and Murphy, M. P. (2008). Complex I within oxidatively stressed bovine heart mitochondria is glutathionylated on Cys-531 and Cys-704 of the 75-kDa subunit: Potential role of CYS residues in decreasing oxidative damage. *The Journal of Biological Chemistry* **283**, 24801–24815.

Ito, H., Iwabuchi, M. and Ogawa, K. (2003). The sugar-metabolic enzymes aldolase and triose-phosphate isomerase are targets of glutathionylation in *Arabidopsis thaliana*: Detection using biotinylated glutathione. *Plant and Cell Physiology* **44**, 655–660.

Janssen-Heininger, Y. M., Mossman, B. T., Heintz, N. H., Forman, H. J., Kalyanaraman, B., Finkel, T., Stamler, J. S., Rhee, S. G. and van der Vliet, A. (2008). Redox-based regulation of signal transduction: Principles, pitfalls, and promises. *Free Radical Biology and Medicine* **45**, 1–17.

Ji, Y., Akerboom, T. P., Sies, H. and Thomas, J. A. (1999). *S*-Nitrosylation and *S*-glutathiolation of protein sulfhydryls by *S*-nitroso glutathione. *Archives of Biochemistry and Biophysics* **362**, 67–78.

Jiang, K., Schwarzer, C., Lally, E., Zhang, S., Ruzin, S., Machen, T., Remington, S. J. and Feldman, L. (2006). Expression and characterization of a redox-sensing green fluorescent protein (reduction-oxidation-sensitive green fluorescent protein) in *Arabidopsis*. *Plant Physiology* **141**, 397–403.

Johansson, C., Lillig, C. H. and Holmgren, A. (2004). Human mitochondrial glutaredoxin reduces *S*-glutathionylated proteins with high affinity accepting electrons from either glutathione or thioredoxin reductase. *The Journal of Biological Chemistry* **279**, 7537–7543.

Johansson, C., Kavanagh, K. L., Gileadi, O. and Oppermann, U. (2007). Reversible sequestration of active site cysteines in a 2Fe-2S-bridged dimer provides a mechanism for glutaredoxin 2 regulation in human mitochondria. *The Journal of Biological Chemistry* **282**, 3077–3082.

Jung, C. H. and Thomas, J. A. (1996). *S*-Glutathiolated hepatocyte proteins and insulin disulfides as substrates for reduction by glutaredoxin, thioredoxin, protein disulfide isomerase, and glutathione. *Archives of Biochemistry and Biophysics* **335**, 61–72.

Karoui, H., Hogg, N., Frejaville, C., Tordo, P. and Kalyanaraman, B. (1996). Characterization of sulfur-centered radical intermediates formed during the oxidation of thiols and sulfite by peroxynitrite. ESR-spin trapping and oxygen uptake studies. *The Journal of Biological Chemistry* **271**, 6000–6009.

Kil, I. S. and Park, J. W. (2005). Regulation of mitochondrial NADP + -dependent isocitrate dehydrogenase activity by glutathionylation. *The Journal of Biological Chemistry* **280**, 10846–10854.

Kim, S. O., Merchant, K., Nudelman, R., Beyer, W. F., Jr., Keng, T., DeAngelo, J., Hausladen, A. and Stamler, J. S. (2002). OxyR: A molecular code for redox-related signaling. *Cell* **109**, 383–396.

Klatt, P. and Lamas, S. (2000). Regulation of protein function by *S*-glutathiolation in response to oxidative and nitrosative stress. *European Journal of Biochemistry* **267**, 4928–4944.

Klatt, P., Molina, E. P., De Lacoba, M. G., Padilla, C. A., Martinez-Galesteo, E., Barcena, J. A. and Lamas, S. (1999a). Redox regulation of c-Jun DNA binding by reversible *S*-glutathiolation. *FASEB Journal* **13**, 1481–1490.

Klatt, P., Molina, E. P. and Lamas, S. (1999b). Nitric oxide inhibits c-Jun DNA binding by specifically targeted *S*-glutathionylation. *The Journal of Biological Chemistry* **274**, 15857–15864.

Koistinen, K., Kokko, H., Hassinen, V., Tervahauta, I., Auriola, S. and Kärenlampi, S. O. (2002). Stress-related RNase PR-10c is post-translationally modified by glutathione in birch. *Plant, Cell and Environment* **25**, 707–715.

Kwak, H. S., Yim, H. S., Chock, P. B. and Yim, M. B. (1995). Endogenous intracellular glutathionyl radicals are generated in neuroblastoma cells under hydrogen peroxide oxidative stress. *Proceedings of the National Academy of Sciences of the United States of America* **92**, 4582–4586.

Lampela, O., Juffer, A. H. and Rauk, A. (2003). Conformational analysis of glutathione in aqueous solution with molecular dynamics. *The Journal of Physical Chemistry A* **107**, 9208–9220.

Lee, K. O., Lee, J. R., Yoo, J. Y., Jang, H. H., Moon, J. C., Jung, B. G., Chi, Y. H., Park, S. K., Lee, S. S., Lim, C. O., Yun, D. J. Cho, M. J. *et al.* (2002). GSH-dependent peroxidase activity of the rice (*Oryza sativa*) glutaredoxin, a thioltransferase. *Biochemical and Biophysical Research Communications* **296**, 1152–1156.

Lee, J. W., Soonsanga, S. and Helmann, J. D. (2007). A complex thiolate switch regulates the *Bacillus subtilis* organic peroxide sensor OhrR. *Proceedings of the National Academy of Sciences of the United States of America* **104**, 8743–8748.

Leferink, N. G., van Duijn, E., Barendregt, A., Heck, A. J. and van Berkel, W. J. (2009). Galactonolactone dehydrogenase requires a redox-sensitive thiol for optimal production of vitamin C. *Plant Physiology*.

Lemaire, S. D. (2004). The glutaredoxin family in oxygenic photosynthetic organisms. *Photosynthesis Research* **79**, 305–318.

Lemaire, S. D. and Miginiac-Maslow, M. (2004). The thioredoxin superfamily in *Chlamydomonas reinhardtii. Photosynthesis Research* **82**, 203–220.

Lemaire, S. D., Michelet, L., Zaffagnini, M., Massot, V. and Issakidis-Bourguet, E. (2007). Thioredoxins in chloroplasts. *Current Genetics* **51**, 343–365.

Li, J., Huang, F. L. and Huang, K. P. (2001). Glutathiolation of proteins by glutathione disulfide S-oxide derived from *S*-nitrosoglutathione. Modifications of rat brain neurogranin/RC3 and neuromodulin/GAP-43. *The Journal of Biological Chemistry* **276**, 3098–3105.

Li, S., Lauri, A., Ziemann, M., Busch, A., Bhave, M. and Zachgo, S. (2009). Nuclear activity of ROXY1, a glutaredoxin interacting with TGA factors, is required for petal development in *Arabidopsis thaliana. The Plant Cell* **21**, 429–441.

Lillig, C. H., Berndt, C. and Holmgren, A. (2008). Glutaredoxin systems. *Biochimica et Biophysica Acta* **1780**, 1304–1317.

Lind, C., Gerdes, R., Hamnell, Y., Schuppe-Koistinen, I., von Lowenhielm, H. B., Holmgren, A. and Cotgreave, I. A. (2002). Identification of *S*-glutathionylated cellular proteins during oxidative stress and constitutive metabolism by affinity purification and proteomic analysis. *Archives of Biochemistry and Biophysics* **406**, 229–240.

Lindermayr, C., Saalbach, G. and Durner, J. (2005). Proteomic identification of *S*-nitrosylated proteins in *Arabidopsis. Plant Physiology* **137**, 921–930.

Liu, C., Willmund, F., Golecki, J. R., Cacace, S., Hess, B., Markert, C. and Schroda, M. (2007). The chloroplast HSP70B-CDJ2-CGE1 chaperones catalyse assembly and disassembly of VIPP1 oligomers in *Chlamydomonas*. *The Plant Journal: For Cell and Molecular Biology* **50**, 265–277.

Lonn, M. E., Hudemann, C., Berndt, C., Cherkasov, V., Capani, F., Holmgren, A. and Lillig, C. H. (2008). Expression pattern of human glutaredoxin 2 isoforms: Identification and characterization of two testis/cancer cell-specific isoforms. *Antioxidants and Redox Signaling* **10**, 547–557.

Manevich, Y., Feinstein, S. I. and Fisher, A. B. (2004). Activation of the antioxidant enzyme 1-CYS peroxiredoxin requires glutathionylation mediated by heterodimerization with pi GST. *Proceedings of the National Academy of Sciences of the United States of America* **101**, 3780–3785.

Maples, K. R., Kennedy, C. H., Jordan, S. J. and Mason, R. P. (1990). *In vivo* thiyl free radical formation from hemoglobin following administration of hydroperoxides. *Archives of Biochemistry and Biophysics* **277**, 402–409.

Marri, L., Zaffagnini, M., Collin, V., Issakidis-Bourguet, E., Lemaire, S. D., Pupillo, P., Sparla, F., Miginiac-Maslow, M. and Trost, P. (2009). Prompt and easy activation by specific thioredoxins of calvin cycle enzymes of *Arabidopsis thaliana* associated in the GAPDH/CP12/PRK supramolecular complex. *Molecular Plant* **2**, 259–269.

Mesecke, N., Mittler, S., Eckers, E., Herrmann, J. M. and Deponte, M. (2008). Two novel monothiol glutaredoxins from *Saccharomyces cerevisiae* provide further insight into iron-sulfur cluster binding, oligomerization, and enzymatic activity of glutaredoxins. *Biochemistry* **47**, 1452–1463.

Meyer, A. J. and Hell, R. (2005). Glutathione homeostasis and redox-regulation by sulfhydryl groups. *Photosynthesis Research* **86**, 435–457.

Meyer, A. J., Brach, T., Marty, L., Kreye, S., Rouhier, N., Jacquot, J. P. and Hell, R. (2007). Redox-sensitive GFP in *Arabidopsis thaliana* is a quantitative biosensor for the redox potential of the cellular glutathione redox buffer. *The Plant Journal : For Cell and Molecular Biology* **52**, 973–986.

Meyer, Y., Siala, W., Bashandy, T., Riondet, C., Vignols, F. and Reichheld, J. P. (2008). Glutaredoxins and thioredoxins in plants. *Biochimica et Biophysica Acta* **1783**, 589–600.

Michelet, L., Zaffagnini, M., Marchand, C., Collin, V., Decottignies, P., Tsan, P., Lancelin, J. M., Trost, P., Miginiac-Maslow, M., Noctor, G. and Lemaire, S. D. (2005). Glutathionylation of chloroplast thioredoxin f is a redox signaling mechanism in plants. *Proceedings of the National Academy of Sciences of the United States of America* **102**, 16478–16483.

Michelet, L., Zaffagnini, M., Massot, V., Keryer, E., Vanacker, H., Miginiac-Maslow, M., Issakidis-Bourguet, E. and Lemaire, S. D. (2006). Thioredoxins, glutaredoxins, and glutathionylation: New crosstalks to explore. *Photosynthesis Research* **89**, 225–245.

Michelet, L., Zaffagnini, M., Vanacker, H., Le Marechal, P., Marchand, C., Schroda, M., Lemaire, S. D. and Decottignies, P. (2008). *In vivo* targets of *S*-thiolation in *Chlamydomonas reinhardtii*. *The Journal of Biological Chemistry* **283**, 21571–21578.

Mieyal, J. J., Starke, D. W., Gravina, S. A., Dothey, C. and Chung, J. S. (1991). Thioltransferase in human red blood cells: Purification and properties. *Biochemistry* **30**, 6088–6097.

Mieyal, J. J., Gallogly, M. M., Qanungo, S., Sabens, E. A. and Shelton, M. D. (2008). Molecular mechanisms and clinical implications of reversible protein *S*-glutathionylation. *Antioxidants and Redox Signaling* **10**, 1941–1988.

Mohr, S., Hallak, H., de Boitte, A., Lapetina, E. G. and Brune, B. (1999). Nitric oxide-induced *S*-glutathionylation and inactivation of glyceraldehyde-3-phosphate dehydrogenase. *The Journal of Biological Chemistry* **274**, 9427–9430.

Montrichard, F., Alkhalfioui, F., Yano, H., Vensel, W. H., Hurkman, W. J. and Buchanan, B. B. (2009). Thioredoxin targets in plants: The first 30 years. *Journal of Proteomics* **72**, 452–474.

Ndamukong, I., Abdallat, A. A., Thurow, C., Fode, B., Zander, M., Weigel, R. and Gatz, C. (2007). SA-inducible *Arabidopsis* glutaredoxin interacts with TGA factors and suppresses JA-responsive PDF1.2 transcription. *The Plant Journal : For Cell and Molecular Biology* **50**, 128–139.

Niture, S. K., Velu, C. S., Bailey, N. I. and Srivenugopal, K. S. (2005). *S*-Thiolation mimicry: Quantitative and kinetic analysis of redox status of protein cysteines by glutathione-affinity chromatography. *Archives of Biochemistry and Biophysics* **444**, 174–184.

Noctor, G. (2006). Metabolic signalling in defence and stress: The central roles of soluble redox couples. *Plant, Cell and Environment* **29**, 409–425.

Noguera-Mazon, V., Lemoine, J., Walker, O., Rouhier, N., Salvador, A., Jacquot, J. P., Lancelin, J. M. and Krimm, I. (2006). Glutathionylation induces the dissociation of 1-Cys D-peroxiredoxin non-covalent homodimer. *The Journal of Biological Chemistry* **281**, 31736–31742.

Nordstrand, K., Aslund, F., Meunier, S., Holmgren, A., Otting, G. and Berndt, K. D. (1999). Direct NMR observation of the Cys-14 thiol proton of reduced Escherichia coli glutaredoxin-3 supports the presence of an active site thiol-thiolate hydrogen bond. *FEBS Letters* **449**, 196–200.

Nulton-Persson, A. C., Starke, D. W., Mieyal, J. J. and Szweda, L. I. (2003). Reversible inactivation of alpha-ketoglutarate dehydrogenase in response to alterations in the mitochondrial glutathione status. *Biochemistry* **42**, 4235–4242.

Ogawa, K. (2005). Glutathione-associated regulation of plant growth and stress responses. *Antioxidants and Redox Signaling* **7**, 973–981.

Okamoto, T., Akaike, T., Sawa, T., Miyamoto, Y., van der Vliet, A. and Maeda, H. (2001). Activation of matrix metalloproteinases by peroxynitrite-induced protein *S*-glutathiolation via disulfide S-oxide formation. *The Journal of Biological Chemistry* **276**, 29596–29602.

Peltoniemi, M. J., Karala, A. R., Jurvansuu, J. K., Kinnula, V. L. and Ruddock, L. W. (2006). Insights into deglutathionylation reactions. Different intermediates in the glutaredoxin and protein disulfide isomerase catalyzed reactions are defined by the gamma-linkage present in glutathione. *The Journal of Biological Chemistry* **281**, 33107–33114.

Picciocchi, A., Saguez, C., Boussac, A., Cassier-Chauvat, C. and Chauvat, F. (2007). CGFS-type monothiol glutaredoxins from the cyanobacterium *Synechocystis* PCC6803 and other evolutionary distant model organisms possess a glutathione-ligated [2Fe-2S] cluster. *Biochemistry* **46**, 15018–15026.

Pineda-Molina, E., Klatt, P., Vazquez, J., Marina, A., Garcia de Lacoba, M., Perez-Sala, D. and Lamas, S. (2001). Glutathionylation of the p50 subunit of NF-kappaB: A mechanism for redox-induced inhibition of DNA binding. *Biochemistry* **40**, 14134–14142.

Poole, L. B. and Nelson, K. J. (2008). Discovering mechanisms of signaling-mediated cysteine oxidation. *Current Opinion in Chemical Biology* **12**, 18–24.

Poole, L. B., Karplus, P. A. and Claiborne, A. (2004). Protein sulfenic acids in redox signaling. *Annual Review of Pharmacology and Toxicology* **44**, 325–347.

Prinarakis, E., Chantzoura, E., Thanos, D. and Spyrou, G. (2008). *S*-Glutathionylation of IRF3 regulates IRF3-CBP interaction and activation of the IFN beta pathway. *EMBO Journal* **27**, 865–875.

Qanungo, S., Starke, D. W., Pai, H. V., Mieyal, J. J. and Nieminen, A. L. (2007). Glutathione supplementation potentiates hypoxic apoptosis by *S*-glutathionylation of p65-NFkappaB. *The Journal of Biological Chemistry* **282**, 18427–18436.

Ralat, L. A., Manevich, Y., Fisher, A. B. and Colman, R. F. (2006). Direct evidence for the formation of a complex between 1-cysteine peroxiredoxin and glutathione *S*-transferase pi with activity changes in both enzymes. *Biochemistry* **45**, 360–372.

Ravichandran, V., Seres, T., Moriguchi, T., Thomas, J. A. and Johnston, R. B., Jr. (1994). *S*-Thiolation of glyceraldehyde-3-phosphate dehydrogenase induced by the phagocytosis-associated respiratory burst in blood monocytes. *The Journal of Biological Chemistry* **269**, 25010–25015.

Reddie, K. G. and Carroll, K. S. (2008). Expanding the functional diversity of proteins through cysteine oxidation. *Current Opinion in Chemical Biology* **12**, 746–754.

Reichheld, J. P., Khafif, M., Riondet, C., Droux, M., Bonnard, G. and Meyer, Y. (2007). Inactivation of thioredoxin reductases reveals a complex interplay between thioredoxin and glutathione pathways in *Arabidopsis* development. *The Plant Cell* **19**, 1851–1865.

Reynaert, N. L., Ckless, K., Guala, A. S., Wouters, E. F., van der Vliet, A. and Janssen-Heininger, Y. M. (2006a). *In situ* detection of *S*-glutathionylated proteins following glutaredoxin-1 catalyzed cysteine derivatization. *Biochimica et Biophysica Acta* **1760**, 380–387.

Reynaert, N. L., van der Vliet, A., Guala, A. S., McGovern, T., Hristova, M., Pantano, C., Heintz, N. H., Heim, J., Ho, Y. S., Matthews, D. E., Wouters, E. F. and Janssen-Heininger, Y. M. (2006b). Dynamic redox control of NF-kappaB through glutaredoxin-regulated *S*-glutathionylation of inhibitory kappaB kinase beta. *Proceedings of the National Academy of Sciences of the United States of America* **103**, 13086–13091.

Rhee, S. G., Woo, H. A., Bae, S. H. and Park, S. (2008). Sestrin 2 is not a reductase for cysteine sulfinic acid of peroxiredoxins. *Antioxidants and Redox Signaling* **11**(4), 739–745.

Rinna, A., Torres, M. and Forman, H. J. (2006). Stimulation of the alveolar macrophage respiratory burst by ADP causes selective glutathionylation of protein tyrosine phosphatase 1B. *Free Radical Biology and Medicine* **41**, 86–91.

Rodriguez-Manzaneque, M. T., Tamarit, J., Belli, G., Ros, J. and Herrero, E. (2002). Grx5 is a mitochondrial glutaredoxin required for the activity of iron/sulfur enzymes. *Molecular Biology of the Cell* **13**, 1109–1121.

Rodriguez-Pascual, F., Redondo-Horcajo, M., Magan-Marchal, N., Lagares, D., Martinez-Ruiz, A., Kleinert, H. and Lamas, S. (2008). Glyceraldehyde-3-phosphate dehydrogenase regulates endothelin-1 expression by a novel, redox-sensitive mechanism involving mRNA stability. *Molecular and Cellular Biology* **28**, 7139–7155.

Romero-Puertas, M. C., Campostrini, N., Matte, A., Righetti, P. G., Perazzolli, M., Zolla, L., Roepstorff, P. and Delledonne, M. (2008). Proteomic analysis of *S*-nitrosylated proteins in *Arabidopsis thaliana* undergoing hypersensitive response. *Proteomics* **8**, 1459–1469.

Rouhier, N. and Jacquot, J. P. (2005). The plant multigenic family of thiol peroxidases. *Free Radical Biology and Medicine* **38**, 1413–1421.

Rouhier, N., Gelhaye, E., Sautiere, P. E., Brun, A., Laurent, P., Tagu, D., Gérard, J., de Faÿ, E., Meyer, Y. and Jacquot, J. P. (2001). Isolation and characterization of a new peroxiredoxin from poplar sieve tubes that uses either glutaredoxin or thioredoxin as a proton donor. *Plant Physiology* **127**, 1299–1309.

Rouhier, N., Gelhaye, E. and Jacquot, J. P. (2002a). Exploring the active site of plant glutaredoxin by site-directed mutagenesis. *FEBS Letters* **511**, 145–149.

Rouhier, N., Gelhaye, E. and Jacquot, J. P. (2002b). Glutaredoxin-dependent peroxiredoxin from poplar: Protein–protein interaction and catalytic mechanism. *The Journal of Biological Chemistry* **277**, 13609–13614.

Rouhier, N., Vlamis-Gardikas, A., Lillig, C. H., Berndt, C., Schwenn, J. D., Holmgren, A. and Jacquot, J. P. (2003). Characterization of the redox properties of poplar glutaredoxin. *Antioxidants and Redox Signaling* **5**, 15–22.

Rouhier, N., Gelhaye, E. and Jacquot, J. P. (2004). Plant glutaredoxins: Still mysterious reducing systems. *Cellular and Molecular Life Sciences* **61**, 1266–1277.

Rouhier, N., Couturier, J. and Jacquot, J. P. (2006). Genome-wide analysis of plant glutaredoxin systems. *Journal of Experimental Botany* **57**, 1685–1696.

Rouhier, N., Unno, H., Bandyopadhyay, S., Masip, L., Kim, S. K., Hirasawa, M., Gualberto, J. M., Lattard, V., Kusunoki, M., Knaff, D. B., Georgiou, G. Hase, T. *et al.* (2007). Functional, structural, and spectroscopic characterization of a glutathione-ligated [2Fe-2S] cluster in poplar glutaredoxin C1. *Proceedings of the National Academy of Sciences of the United States of America* **104**, 7379–7384.

Rouhier, N., Lemaire, S. D. and Jacquot, J. P. (2008). The role of glutathione in photosynthetic organisms: Emerging functions for glutaredoxins and glutathionylation. *Annual Review of Plant Biology* **59**, 143–166.

Saaranen, M., Salo, K., Latva-Ranta, M., Kinnula, V. and Ruddock, L. W. (2009). The C-terminal active site cysteine of *Escherichia coli* glutaredoxin 1 determines the glutathione specificity of the second step of peptide deglutathionylation. *Antioxidants and Redox Signaling* **11**(8), 1819–1828.

Sadidi, M., Geddes, T. J. and Kuhn, D. M. (2005). *S*-Thiolation of tyrosine hydroxylase by reactive nitrogen species in the presence of cysteine or glutathione. *Antioxidants and Redox Signaling* **7**, 863–869.

Sagemark, J., Elgan, T. H., Burglin, T. R., Johansson, C., Holmgren, A. and Berndt, K. D. (2007). Redox properties and evolution of human glutaredoxins. *Proteins* **68**, 879–892.

Salmeen, A., Andersen, J. N., Myers, M. P., Meng, T. C., Hinks, J. A., Tonks, N. K. and Barford, D. (2003). Redox regulation of protein tyrosine phosphatase 1B involves a sulphenyl-amide intermediate. *Nature* **423**, 769–773.

Sarma, B. K. and Mugesh, G. (2007). Redox regulation of protein tyrosine phosphatase 1B (PTP1B): A biomimetic study on the unexpected formation of a sulfenyl amide intermediate. *Journal of the American Chemical Society* **129**, 8872–8881.

Schroda, M. (2004). The *Chlamydomonas* genome reveals its secrets: Chaperone genes and the potential roles of their gene products in the chloroplast. *Photosynthesis Research* **82**, 221–240.

Schuppe-Koistinen, I., Moldeus, P., Bergman, T. and Cotgreave, I. A. (1994). *S*-Thiolation of human endothelial cell glyceraldehyde-3-phosphate dehydrogenase after hydrogen peroxide treatment. *European Journal of Biochemistry* **221**, 1033–1037.

Schurmann, P. and Buchanan, B. B. (2008). The ferredoxin/thioredoxin system of oxygenic photosynthesis. *Antioxidants and Redox Signaling* **10**, 1235–1274.

Shelton, M. D. and Mieyal, J. J. (2008). Regulation by reversible *S*-glutathionylation: Molecular targets implicated in inflammatory diseases. *Molecules and Cells* **25**, 332–346.

Shenton, D. and Grant, C. M. (2003). Protein *S*-thiolation targets glycolysis and protein synthesis in response to oxidative stress in the yeast *Saccharomyces cerevisiae*. *The Biochemical Journal* **374**, 513–519.

Shin, S. W., Oh, C. J., Kil, I. S. and Park, J. W. (2009). Glutathionylation regulates cytosolic NADP+-dependent isocitrate dehydrogenase activity. *Free Radical Research* **43**, 409–416.

Silva, G. M., Netto, L. E., Discola, K. F., Piassa-Filho, G. M., Pimenta, D. C., Barcena, J. A. and Demasi, M. (2008). Role of glutaredoxin 2 and cytosolic thioredoxins in cysteinyl-based redox modification of the 20S proteasome. *FEBS Journal* **275**, 2942–2955.

Sirover, M. A. (1999). New insights into an old protein: the functional diversity of mammalian glyceraldehyde-3-phosphate dehydrogenase. *Biochimica et Biophysica Acta* **1432**, 159–184.

Sirover, M. A. (2005). New nuclear functions of the glycolytic protein, glyceraldehyde-3-phosphate dehydrogenase, in mammalian cells. *Journal of Cellular Biochemistry* **95**, 45–52.

Srinivasan, U., Mieyal, P. A. and Mieyal, J. J. (1997). pH profiles indicative of rate-limiting nucleophilic displacement in thioltransferase catalysis. *Biochemistry* **36**, 3199–3206.

Starke, D. W., Chen, Y., Bapna, C. P., Lesnefsky, E. J. and Mieyal, J. J. (1997). Sensitivity of protein sulfhydryl repair enzymes to oxidative stress. *Free Radical Biology and Medicine* **23**, 373–384.

Starke, D. W., Chock, P. B. and Mieyal, J. J. (2003). Glutathione-thiyl radical scavenging and transferase properties of human glutaredoxin (thioltransferase). Potential role in redox signal transduction. *The Journal of Biological Chemistry* **278**, 14607–14613.

Sullivan, D. M., Wehr, N. B., Fergusson, M. M., Levine, R. L. and Finkel, T. (2000). Identification of oxidant-sensitive proteins: TNF-alpha induces protein glutathiolation. *Biochemistry* **39**, 11121–11128.

Tamarit, J., Belli, G., Cabiscol, E., Herrero, E. and Ros, J. (2003). Biochemical characterization of yeast mitochondrial Grx5 monothiol glutaredoxin. *The Journal of Biological Chemistry* **278**, 25745–25751.

Tew, K. D. (2007). Redox in redux: Emergent roles for glutathione *S*-transferase P (GSTP) in regulation of cell signaling and *S*-glutathionylation. *Biochemical Pharmacology* **73**, 1257–1269.

Townsend, D. M., Manevich, Y., He, L., Hutchens, S., Pazoles, C. J. and Tew, K. D. (2009). Novel role for glutathione *S*-transferase pi. Regulator of protein *S*-glutathionylation following oxidative and nitrosative stress. *The Journal of Biological Chemistry* **284**, 436–445.

van Montfort, R. L., Congreve, M., Tisi, D., Carr, R. and Jhoti, H. (2003). Oxidation state of the active-site cysteine in protein tyrosine phosphatase 1B. *Nature* **423**, 773–777.

Velu, C. S., Niture, S. K., Doneanu, C. E., Pattabiraman, N. and Srivenugopal, K. S. (2007). Human p53 is inhibited by glutathionylation of cysteines present in the proximal DNA-binding domain during oxidative stress. *Biochemistry* **46**, 7765–7780.

Vieira Dos Santos, C., Laugier, E., Tarrago, L., Massot, V., Issakidis-Bourguet, E., Rouhier, N. and Rey, P. (2007). Specificity of thioredoxins and glutaredoxins as electron donors to two distinct classes of *Arabidopsis* plastidial methionine sulfoxide reductases B. *FEBS Letters* **581**, 4371–4376.

Wang, J., Boja, E. S., Tan, W., Tekle, E., Fales, H. M., English, S., Mieyal, J. J. and Chock, P. B. (2001). Reversible glutathionylation regulates actin polymerization in A431 cells. *The Journal of Biological Chemistry* **276**, 47763–47766.

Wang, J., Tekle, E., Oubrahim, H., Mieyal, J. J., Stadtman, E. R. and Chock, P. B. (2003). Stable and controllable RNA interference: Investigating the physiological function of glutathionylated actin. *Proceedings of the National Academy of Sciences of the United States of America* **100**, 5103–5106.

Wang, W., Oliva, C., Li, G., Holmgren, A., Lillig, C. H. and Kirk, K. L. (2005). Reversible silencing of CFTR chloride channels by glutathionylation. *The Journal of General Physiology* **125**, 127–141.

Wang, Z., Xing, S., Birkenbihl, R. P. and Zachgo, S. (2009). Conserved functions of *Arabidopsis* and rice CC-Type glutaredoxins in flower development and pathogen response. *Molecular Plant* **2**, 323–335.

Ward, N. E., Stewart, J. R., Ioannides, C. G. and O'Brian, C. A. (2000). Oxidant-induced *S*-glutathiolation inactivates protein kinase C-alpha (PKC-alpha): A potential mechanism of PKC isozyme regulation. *Biochemistry* **39**, 10319–10329.

Ward, N. E., Chu, F. and O'Brian, C. A. (2002). Regulation of protein kinase C isozyme activity by *S*-glutathiolation. *Methods in Enzymology* **353**, 89–100.

Wardman, P. and von Sonntag, C. (1995). Kinetic factors that control the fate of thiyl radicals in cells. *Methods in Enzymology* **251**, 31–45.

Woo, H. A., Jeong, W., Chang, T. S., Park, K. J., Park, S. J., Yang, J. S. and Rhee, S. G. (2005). Reduction of cysteine sulfinic acid by sulfiredoxin is specific to 2-cys peroxiredoxins. *The Journal of Biological Chemistry* **280**, 3125–3128.

Xing, S. and Zachgo, S. (2008). ROXY1 and ROXY2, two *Arabidopsis* glutaredoxin genes, are required for anther development. *The Plant Journal: For Cell and Molecular Biology* **53**, 790–801.

Xing, S., Rosso, M. G. and Zachgo, S. (2005). ROXY1, a member of the plant glutaredoxin family, is required for petal development in *Arabidopsis thaliana*. *Development* **132**, 1555–1565.

Yang, Y. F. and Wells, W. W. (1991). Identification and characterization of the functional amino acids at the active center of pig liver thioltransferase by site-directed mutagenesis. *The Journal of Biological Chemistry* **266**, 12759–12765.

Yang, Y., Jao, S., Nanduri, S., Starke, D. W., Mieyal, J. J. and Qin, J. (1998). Reactivity of the human thioltransferase (glutaredoxin) C7S, C25S, C78S, C82S mutant and NMR solution structure of its glutathionyl mixed disulfide intermediate reflect catalytic specificity. *Biochemistry* **37**, 17145–17156.

Zaffagnini, M., Michelet, L., Marchand, C., Sparla, F., Decottignies, P., Le Marechal, P., Miginiac-Maslow, M., Noctor, G., Trost, P. and Lemaire, S. D. (2007). The thioredoxin-independent isoform of chloroplastic glyceraldehyde-3-phosphate dehydrogenase is selectively regulated by glutathionylation. *FEBS Journal* **274**, 212–226.

Zaffagnini, M., Michelet, L., Massot, V., Trost, P. and Lemaire, S. D. (2008). Biochemical characterization of glutaredoxins from *Chlamydomonas reinhardtii* reveals the unique properties of a chloroplastic CGFS-type glutaredoxin. *The Journal of Biological Chemistry* **283**, 8868–8876.

Glutaredoxin: The Missing Link Between Thiol-Disulfide Oxidoreductases and Iron Sulfur Enzymes

BENJAMIN SELLES, NICOLAS ROUHIER,
KAMEL CHIBANI, JEREMY COUTURIER,
FILIPE GAMA AND JEAN-PIERRE JACQUOT[1]

Unité Mixte de Recherches 1136 INRA UHP Interactions Arbre-Microorganismes, IFR 110 EFABA, Ecosystèmes Forestiers, Agroressources, Bioprocédés et Alimentation, Faculté des Sciences, Nancy Université, BP 239, 54506 Vandoeuvre Cedex, France

[1]Corresponding author: Email: j2p@scbiol.uhp-nancy.fr

Advances in Botanical Research, Vol. 52

0065-2296/09 $35.00
DOI: 10.1016/S0065-2296(09)52013-7

ABSTRACT

The CXXC motif is present in many disulfide oxidoreductases as thioredoxins, glutaredoxins, and protein disulfide isomerases. It is also present in several metal-binding structures including hemoproteins and iron sulfur proteins. Although the 3D structure of ferredoxins and thioredoxins is radically different, the presence of this motif in both proteins suggests that thioredoxins and their derivatives might be able to accommodate iron sulfur centers (ISCs) as well. Several studies have indeed proven the presence of metals, such as iron or zinc, in thioredoxin-like structures either as natural products or after mutagenesis as in *Escherichia coli* thioredoxin 1. Moreover, it was recently demonstrated that some glutaredoxin species with CGYC or CGFS active sites can assemble a [2Fe–2S] ISC in a homodimer. Quite surprisingly, the ligands are the glutaredoxin catalytic cysteine and an external glutathione molecule. As a yeast CGFS glutaredoxin is thought to be involved in the transfer of preassembled ISCs from scaffold to acceptor apoproteins, this suggests that glutaredoxins are involved in these pathways through their own capacity to assemble such centers and transfer them efficiently. Altogether, these data provide firm evidence that glutaredoxins are a link between the world of thiol-disulfide reductases and iron sulfur enzymes.

I. INTRODUCTION

A large number of biochemical reactions require the participation of metals, including the mitochondrial and plastidial electron transfer chains, the metabolic pathways leading to the assimilation of nitrogen or sulfur, the synthesis of ribosomes, or the process of DNA repair, to cite only a few of these processes. Overwhelmingly, the metal ligands in proteins are histidine (through its nitrogen atom) and methionine and cysteine residues via their sulfur atom. Occasionally, acidic residues as aspartate or glutamate are also encountered. Metals as diverse as copper, molybdenum, iron, manganese, zinc, and nickel or even vanadium and cadmium (Lane *et al.*, 2005; Messerschmidt and Wever, 1996) are assembled into apoproteins via post-translational reactions. With a few exceptions, most metalloenzymes play a redox role, transferring electrons to neighboring proteins or chemical compounds by shifting back and forth between reduced and oxidized states. The microenvironment of the metal is extremely important in determining the redox potential of the metallic structures and hence, their reactivity. Among the many metalloenzymes, iron-containing enzymes play a prominent role in biology, and thus we discuss thereafter more extensively the organization of these metallic centers in macromolecules.

II. IRON-CONTAINING ENZYMES

Several types of iron-containing structures are present in proteins, the more frequently encountered being the di-iron centers, the hemes, and the iron sulfur centers (ISCs). It has been proposed that ISCs are the most primitive structures, as the primitive earth atmosphere is thought to have contained reduced sulfur and iron, the necessary components for ISC assembly (Milner-White and Russell, 2005). Interestingly, the Nest theory assumes the importance of glycine residues in early iron sulfur-binding peptides (see later the glutaredoxin (Grx) requirements for iron sulfur binding). It is postulated that hemes appeared later in evolution.

A. DI-IRON CENTERS

Di-ferric iron centers are present in a number of proteins, most notably ribonucleotide reductase (RNR), a key enzyme necessary for the biosynthesis of deoxyribonucleotides. In aerobic type RNR, the two irons are μ-oxo-linked and coordinated to the protein via ligands that include four carboxylic residues and two histidines. The di-iron site is close to the Tyr radical of the enzyme and serves for the generation and stabilization of the radical. Unlike the situation of cytochromes of the *c* type and of iron sulfur proteins, there is no direct link between this di-iron site in RNR and thioredoxin (Trx) except that RNR is an enzyme that requires thioredoxin or glutaredoxin for its catalytic activity (Avval and Holmgren, 2009). The regeneration of the active form of RNR involves the successive reduction of disulfide bonds, one situated at the C-terminus with the sequence CESGAC is more specifically the thioredoxin or glutaredoxin target (Persson *et al.*, 1997). Interestingly, some model compounds have been created where the two iron atoms are not oxo-linked but rather dithiolate bridged (Borg *et al.*, 2004).

B. HEMES

Several types of hemes are present in proteins, including the a, b, and c classes. All hemes are inserted posttranslationally into the apoproteins but their binding differs depending on the class considered. The a and b hemes are noncovalently attached, and are bound mostly via hydrophobic bonds. On the other hand, the c-type hemes are linked via thioether bonds that include two vicinal cysteine residues separated by two amino acids with the general formula CXXCH. The two X residues are variable, for example, in horse heart cytochrome *c*, the attachment sequence is CAQC (Abriata *et al.*, 2009). It is CAAC in *Chlamydomonas reinhardtii* cytochrome *c*6 (Merchant and Bogorad, 1987), CSQC in chicken cytochrome *c* (Chan and Margoliash, 1966), and CANC in cytochrome *f* either from *Mastigocladus laminosus* or from *C. reinhardtii* (Baniulis *et al.*, 2008). In all c-type cytochromes, the iron atom in the center of the heme possesses two axial ligands, a histidine, and a methionine. It is well known that spacing two cysteines with two amino acids results in having the two sulfur atoms in close proximity in the 3D space, and thus in the possibility of creating a disulfide bond between the two cysteines following translation of the polypeptide. This is probably the reason why heme assembly requires the participation of thioredoxin-like molecules, for example, CCMH for *Arabidopsis thaliana* mitochondrial cytochrome *c* (Meyer *et al.*, 2005). In this case, the thioredoxin-like protein is required for the reduction of the disulfide prior to heme assembly.

C. ISC AND IRON SULFUR PROTEINS

1. Nature of the center

There are many types of ISCs in proteins, from the most simple type containing a single iron and no heterosulfur (rubredoxin-like) to more complex structures as [2Fe–2S] centers (chloroplastic ferredoxin, chloroplastic, and mitochondrial Rieske protein, mitochondrial adrenodoxin), [3Fe–4S] centers (e.g., in aconitase), and [4Fe–4S] centers (e.g., in chloroplastic photosystem I, nitrite reductase, sulfite reductase, glutamate oxoglutarate aminotransferase, and also in mitochondrial complex I) (Sazanov and Hinchliffe, 2006).

2. Nature of the ligands, position in the primary structure, and the CXXC motif

Except for the Rieske protein in which the ISC is ligated via two cysteines and two histidines, the amino acid ligands for ISCs are overwhelmingly cysteines and the ligation involves the sulfur atom of the side chain. In the rubredoxin mono iron type, the ligands coordinate with the iron atom in a tetrahedral organization. Interestingly for the *Clostridium pasteurianum* rubredoxin, the four cysteines are organized in two pairs separated by two amino acids with the sequences CTVC and CPLC (Mathieu *et al.*, 1992). In [2Fe–2S] ISCs, the four coordinating sulfur atoms are in the same plane than the two iron atoms, but in addition there are two inorganic sulfur atoms (also called labile sulfur) linking the two irons above and below the above defined plane. In [3Fe–4S] and [4Fe–4S] centers, the iron atoms and the hetero sulfur atoms are alternating in a cubane-like structure, the iron atoms being themselves covalently bound via the sulfur atoms of the ligand cysteines. In chloroplastic ferredoxins, the position of the cysteines is conserved across species (in position 38, 43, 46, and 76 in *C. reinhardtii*) with a pair of cysteines in a CXXC motif which is absolutely conserved (Stein *et al.*, 1993). In general, the positions of the cysteine ligands are quite variable depending on which enzyme is considered, but examples abound where a CXXC motif is necessary for the binding of either [2Fe–2S] or [4Fe–4S] ISCs (Amman *et al.*, 2004; Brandt, 2006; Bych *et al.*, 2008; Raux-Deery *et al.*, 2005; Song and Lee, 2008; Yabe *et al.*, 2008; Zhang *et al.*, 2008). The remark concerning the cysteine spacing in cytochrome *c* and their potential oxidation into a disulfide after protein synthesis and before heme insertion obviously also applies to the many ISC-containing proteins with the CXXC motifs. Overall, nearly all of the ISC-containing proteins contain multiple iron atoms (except for the rubredoxin-type) and sometimes even multiple ISCs are present in a single poplypeptide, as in PsaC of photosystem I (Takahashi *et al.*, 1991). Nevertheless, although they contain multiple iron atoms, the ISCs are able to transfer only one electron at a time, this being possibly related to the position of the iron atoms in the 3D structure. For example, for a spinach chloroplastic ferredoxin (pdb accession number 1A70), it is indeed clear that one of the iron atoms is located closer to the surface of the protein and the other more deeply buried and thus unable to participate in electron transfer reactions (Binda *et al.*, 1998).

3. Pathways where iron sulfur enzymes are required

It has been briefly mentioned that nitrite reductase and glutamate synthase contain an ISC, and that nitrogen assimilation requires the participation of iron sulfur enzymes (Swamy *et al.*, 2005). Likewise, the iron sulfur-containing sulfite reductase, an enzyme with high analogy to nitrite reductase, and APS reductase are involved in sulphate assimilation (Hirasawa *et al.*, 2004; Kim *et al.*, 2006). As noted above the electron transfer chains of mitochondria and chloroplasts contain multiple ISCs (Complex I, II, and III in mitochondria, and cytochrome *b6f* and PSI in chloroplasts). The energetic metabolism and formation of ATP are thus dependent on these enzymes and so is the carbon assimilation in plants. In the cytosol, one enzyme involved in leucine biosynthesis, isopropylmalate isomerase (Leu1p) also contains an ISC (Sipos *et al.*, 2002). Scaffold proteins such as CnfU, a key iron sulfur cluster biosynthetic scaffold that is required for biogenesis of ferredoxin and photosystem I in chloroplasts also contain ISCs (Yabe *et al.*, 2008). This is also the case for the proteins Dre2, Nar1, and SirB which are involved in cytosolic iron sulfur biogenesis and siroheme biosynthesis, respectively (Raux-Deery *et al.*, 2005; Song and Lee, 2008; Zhang *et al.*, 2008). These examples are certainly not exhaustive but are demonstrative of how essential these structures are for central metabolism, be it carbon, nitrogen, or sulfur metabolisms (and autotrophy in plants) or amino acid biosynthesis and iron sulfur assembly.

4. 3D structure of [2Fe–2S]-containing ferredoxins

Experiments described subsequently in this chapter have indicated that the thioredoxin and glutaredoxin molecules can harbor ISCs of the [2Fe–2S] type, either in "natural" proteins or in engineered mutagenized versions of the proteins and thus become "ferredoxin-like," so we describe here the 3D structure of most well-known ferredoxins from cyanobacteria or chloroplasts. These proteins are extremely well conserved with 96–99 amino acids in their mature form and the ligand cysteines in invariant positions. The protein is held together by several β-strands forming a β-sheet at the back of the molecule (Fig. 1). To the front of the molecule, three short α-helices surround the iron sulfur cluster, with one of the iron atoms positioned closer to the surface and thus better suited for transferring electrons. In the three helices lie key acidic residues that are required for protein–protein interaction (Binda *et al.*, 1998; Jacquot *et al.*, 1997). Other types of structures harboring a low potential [2Fe–2S] center are thioredoxin-like proteins present in bacteria such as *Azotobacter vinelandii, C. pasteurianum*, and *Chlorobium tepidum* (Meyer, 2001). Interestingly, these proteins form dimers, each

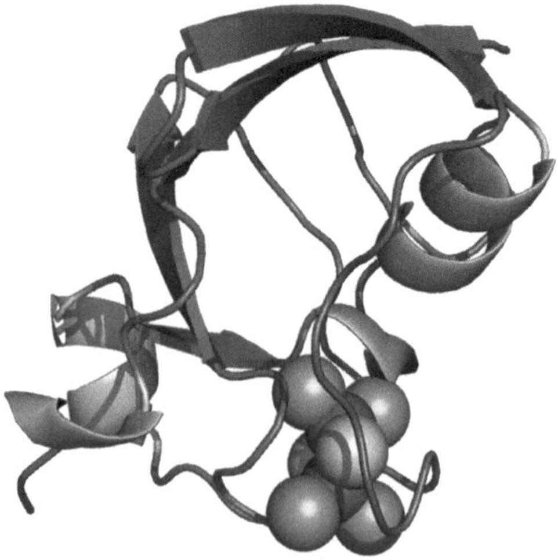

Fig. 1. Crystal structure of the E92K mutant of [2Fe–2S] ferredoxin I from *Spinacia oleracea* (Binda *et al.*, 1998; pdb accession number 1A70). Sulfur atoms belonging to cysteine residues 38, 43, 46, and 76 are represented by green spheres, iron atoms by red, and labile sulfur atoms by yellow. (See color plate section at the back of the book.)

monomer having a half "thioredoxin-like" architecture with a central pleated β-sheet surrounded by α-helices (Fig. 2) (Yeh *et al.*, 2000). In the dimer, the two β-sheets face one another and two of the helices surrounding them in more traditional thioredoxins are missing.

D. CURRENT MECHANISM OF ISC ASSEMBLY IN PLANT PLASTIDS AND MITOCHONDRIA

Two of the three bacterial ISC assembly systems, nif, suf, and isc have been retained in photosynthetic eukaryote organisms, the suf machinery being present in plastids and the isc machinery in mitochondria. In addition, nonplant eukaryotes possess an ISC export machinery and a cytosolic

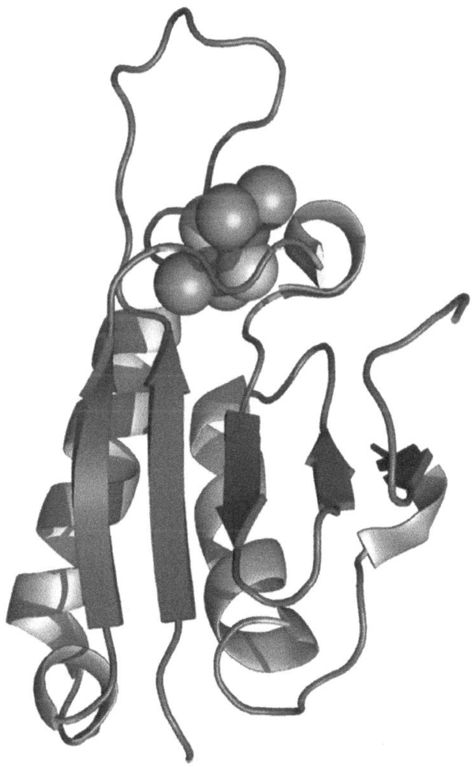

Fig. 2. Crystal structure of a thioredoxin-like [2Fe–2S] ferredoxin from *Aquifex aeolicus* (Yeh *et al.*, 2000; pdb 1F37). Sulfur atoms belonging to cysteine residues 9, 22, 55, and 59 are represented by green spheres, iron atoms by red and labile sulfur atoms by yellow. (See color plate section at the back of the book.)

assembly machinery for cytosolic and nuclear FeS proteins (Lill and Mühlenhoff, 2008). Based on the conservation of the proteins involved in the two latter systems, it is likely that these pathways also exist in plants (Balk and Lobreaux, 2005). It has been suggested from genetic studies in yeast that Grx5 is involved in the transfer of preassembled clusters from Isu proteins to acceptor proteins (Muhlenhoff *et al.*, 2003). As most Grxs from mammals, plants, and cyanobacteria can complement the defects associated with the deletion of Grx5 in yeast, this suggests that they fulfil similar functions in their respective organelles (Molina-Navarro *et al.*, 2006; Picciocchi *et al.*, 2007; Rouhier *et al.*, 2008). The latest developments concerning the involvement of Grxs in plant ISC biogenesis will be described in a subsequent section.

III. THIOL-DISULFIDE OXIDOREDUCTASES

The family of oxidoreductases includes disulfide reductases called thioredoxins, glutaredoxins, disulfide oxidases, or isomerases belonging to the sulfhydryl oxidase (SOX) and protein disulfide isomerase (PDI) families. Grxs, Trxs, and PDIs belonging to the large Trx superfamily possess many features in common, the active site, the 3D structure, the reaction mechanism. They are all ubiquitous or nearly ubiquitous redox proteins with conserved redox-active sites CXXC/S possessing generally oxidoreductase activity in dithiol-disulfide exchange reactions (Holmgren, 1985). The major difference lies in the redox potential of their active pair of cysteines, Trxs having a more electronegative redox potential than Grxs and even more than PDI. Furthermore, Chivers and collaborators have shown that the modification of the CXXC motif and consequently of the redox potential can modify the oxidoreductase properties of the protein. For example, a *Sc*Trx mutant with a higher redox potential can efficiently replace *Sc*PDI1 null mutant (Chivers *et al.*, 1996, 1997).

Photosynthetic and nonphotosynthetic organisms contain a large number of genes encoding these three classes of enzymes, for example, in poplar, around 40 Trx- and Grx-encoding genes and 13 PDI encoding genes (Chibani *et al.*, 2009; Couturier *et al.*, 2009b; Houston *et al.*, 2005; Morel *et al.*, 2008). All these proteins have variable subcellular localizations. In eukaryotic cells, Trxs and Grxs are rather present in compartments with a reducing environment, whereas PDI and SOX are present in compartments supposed to have an oxidizing environment. For example, PDIs are mostly found in the endoplasmic reticulum and their function is to fold properly other proteins via cysteine rearrangement.

A. THE THIOREDOXIN MODEL

1. Active-site sequence

Most frequently, Trxs have a highly conserved classical dicysteinic CXXC active-site sequence (overwhelmingly WC[G/P]PC), comprising two vicinal cysteines separated by two variable amino acids. Besides, in plants, several genomic studies have highlighted the presence of thioredoxin-like proteins with dicysteinic or monocysteinic active sites (CXXS), but little is known about their biochemical properties (Chibani *et al.*, 2009; Gelhaye *et al.*, 2004; Meyer *et al.*, 2007). In fact, Serrato and colleagues have demonstrated that some *Arabidopsis* CXXS Trxs have a disulfide reductase activity (Serrato *et al.*, 2008).

2. Redox potential and reaction mechanism

In Trxs containing a pair of cysteine residues, both cysteines play different roles. The first cysteine is involved in the nucleophilic attack on disulfide bonds present in target proteins, leading to the formation of a disulfide bond between the target protein and Trx. The second cysteine, called the backup/resolving or recycling cysteine, subsequently cleaves the disulfide formed between Trx and its target. In general, Trxs with a conventional active site WCGPC, have a low redox potential comprised between − 270 and − 330 mV compared to the other redoxins (Bréhélin *et al.*, 2004; Collin *et al.*, 2003). The two residues located between the two cysteines in the active-site motif are important in controlling the redox properties of the proteins. Changing one residue of the active site by site-directed mutagenesis affects the redox potential of the protein. For example, swapping the proline to histidine in the *Escherichia coli* Trx active-site sequence induces a higher redox potential (− 235 mV) than the wild type (Krause *et al.*, 1991). This mutation confers the ability to function as a disulfide isomerase and also has an impact on its interaction with folding protein substrates (Eklund *et al.*, 1991; Holmgren, 1995). Recently, it has also been demonstrated that the residue preceding the *cis* Pro conserved in all Trx superfamily members is crucial in determining the redox potential of the oxidoreductases (Ren *et al.*, 2009). A recent study showed that atypical chloroplastic Trxs called Trx-lilium from *Arabidopsis* with CGSC, CGGC, or CASC active sites display higher redox potentials between −237 and −240 mV, suggesting they cannot reduce all the usual Trx partners but that they could instead have specific functions (Dangoor *et al.*, 2009).

3. Subcellular localization and physiological roles

Thioredoxins are involved in a wide variety of fundamental biological functions including dithiol hydrogen donation to RNR, regulation of the activity of photosynthetic enzymes including fructose-1,6-bisphosphatase and

NADP malate dehydrogenase and of some eukaryotic transcription factors (Schürmann and Buchanan, 2008; Schürmann and Jacquot, 2000). In addition, it has been recently well documented that thioredoxins serve as regenerating systems for peroxiredoxins (Prx) and methionine sulfoxide reductases, enzymes in which the catalytic cysteine becomes oxidized into a sulfenic acid residue upon catalysis (Rouhier *et al.*, 2001, 2007a, 2008). In addition to these well-described interactions, the list of Trx targets involved in many metabolic pathways and cellular processes is growing with their identification by proteomic studies (for a review see Montrichard *et al.*, 2009 and also the paper by Nishiyama and Hisabori in this issue). There are multiple Trx isoforms localized in different compartments such as the cytosol (h), chloroplast (f, m, x, y, CDSP32, lilium), mitochondria (o, h2), nucleus (nucleoredoxin), apoplasm (h), endoplasmic reticulum (s), and Trxs with unknown localization (Trx-like and clot) (Alkhalfioui *et al.*, 2008; Chibani *et al.*, 2009; Meyer *et al.*, 2006). In chloroplasts, Trxs are essentially reduced by ferredoxin via ferredoxin-thioredoxin reductase (FTR), whereas cytosolic or mitochondrial Trxs are reduced by NADPH via a NADPH-Trx reductase (NTR) (Jacquot *et al.*, 2009). An alternative chloroplastic Trx reduction pathway involves NADPH and the hybrid enzyme NTRC (it contains a built-in Trx module in the C-terminal part of a NTR module). This protein might be specifically devoted to the reduction of some Prx types. A recent review summarizes the evolution and properties of thioredoxin reductases in photosynthetic organisms (Jacquot *et al.*, 2009).

4. 3D structures of thioredoxins

The first 3D Trx structure is the one of the *E. coli* Trx1 and it has been solved by X-ray crystallography (Holmgren *et al.*, 1975). Subsequently, several structures of Trxs from different organisms have been determined, and they display a high degree of homology, most of these proteins being rather homogeneous in length with ca. 110–120 amino acid residues (Fig. 3). Trx structures are also described in this volume in a chapter by Hägglund and colleagues. All Trx structures have a well-conserved hydrophobic core and most amino acid variants are located on the surface of the protein, affecting surface patches only locally. Thioredoxins and glutaredoxins are characterized by a common fold, the thioredoxin fold, which is a central five stranded β-sheet flanked by three or four α-helices and a CXXC active-site motif. The secondary elements are in the order $\beta1$, $\alpha1$, $\beta2$, $\alpha2$, $\beta3$, $\alpha3$, $\beta4$, $\beta5$, $\alpha4$. The conserved redox-active site forms the link between the second β-strand and the subsequent $\alpha2$ helix, and the cysteines are located at the N-terminus of the helix and rather exposed at the surface of the molecule, especially the catalytic one (Dai *et al.*, 2000; Mössners *et al.*, 1998). Concerning plant Trxs, the

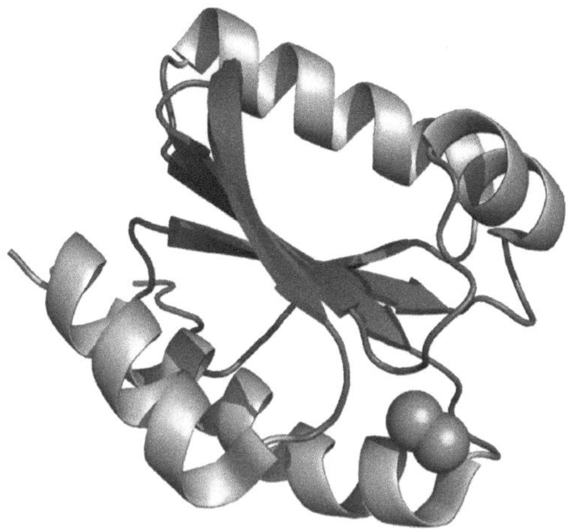

Fig. 3. Crystal structure of thioredoxin h1 from *Arabidopsis thaliana* (Peterson *et al.*, 2005; pdb 1XFI). Sulfur atoms belonging to cysteine residues 40 and 43 are represented in spheres (green in the web version).

structures of some Trxs f, m, and h have been determined in several organisms and solved by X-ray crystallography and NMR. The crystal structure of recombinant spinach Trx m has been solved in the oxidized and reduced state at 2.1 and 2.3 Å resolution, respectively (Capitani *et al.*, 2000). The structure of *C. reinhardtii* thioredoxin *m* has been solved by NMR and it is very similar to the spinach protein (Lancelin *et al.*, 2000). The spinach Trx f structure shares much similarity with the m, however, the f protein is more positively charged with some of these charges surrounding the active site where they must be instrumental in orientating the protein correctly upon interaction with its targets. Despite their structural similarities, a striking difference is the presence of a conserved third Cys73 in the C-terminal part of the f sequence. This cysteine is structurally exposed on the surface of the structure, 9.7 Å away from the first Cys of the active site (Brandes *et al.*, 1993; del Val *et al.*, 1999). Thus, the overall structure of the spinach Trx f and m does not differ markedly from the *E. coli* model (Schürmann and Buchanan, 2008). The structures of h-type Trxs have been determined for *C. reinhardtii,* barley, *A. thaliana*, and poplar enzymes (Coudevylle *et al.*, 2005; Koh *et al.*, 2008; Maeda *et al.*, 2008; Menchise *et al.*, 2001; Peterson *et al.*, 2005). In general, the h structure presents one major difference compared to other thioredoxins, namely an elongated α1 helix. Analysis of the 3D-structure of the *C. reinhardtii h* together with calorimetric studies showed that thioredoxin h has a much reduced thermal stability compared to thioredoxin m and has more similarity to the mammalian protein-type (Lemaire *et al.*, 2000;

Richardson *et al.*, 2000; Stein *et al.*, 1995). This is in agreement with the phylogenetic analyses that proposed an eukaryotic origin for thioredoxin f and h and a prokaryotic origin for thioredoxin m (Meyer *et al.*, 2002). These hypotheses were recently strengthened in a study reporting the effect of applying force to individual thioredoxin molecules from various sources (Perez-Jimenez *et al.*, 2009). A recent study has allowed the comparison of structures of two barley Trx h isoforms (Trxh1 and Trxh2). Barley Trx h1 and h2 have been solved in the oxidized state at 1.7 Å and Trxh2 in the reduced state at 2 Å resolution. The Trxh1 Arg101 can play a particularly crucial role in the association with target proteins by forming electrostatic interactions with a protein motif bound in the substrate-binding loop motif. The presence of Arg101 in Trxh1 and the uncharged Ile107 in Trxh2 may therefore give rise to differential isoform interactions with some redox patterns (Hagglund *et al.*, in this issue; Maeda *et al.*, 2008).

B. PDIs DERIVE FROM THIOREDOXINS

PDIs are generally multidomain proteins sharing structural and amino acid sequence similarities with thioredoxins (Ferrari and Söling, 1999). "Classical" PDI are constituted by five independent domains (a-b-b′-a′-c). The a, a′, b, and b′ domains have homologies to thioredoxins, while the a and a′ domains generally possess the CXXC active site, the b and b′ domains do not. The c domain is a short, acidic amino acid sequence (Edman *et al.*, 1985; Freedman *et al.*, 1994; Kemmink *et al.*, 1999). Most PDI also share an endoplasmic reticulum retention signal (K/HDEL) at the C-terminal end (Freedman *et al.*, 1994). Other classes of PDI differ from the "classical" representative member by the number and distribution of active and inactive TRX modules and the presence of additional domains sharing no similarity with Trxs (Appenzeller-Herzog and Ellgaard, 2008). Concerning plants, PDIs were first detected in higher plants at the end of the 1970s (Grynberg *et al.*, 1977) and then in the photosynthetic alga *C. reinhardii* (Myllylä *et al.*, 1989). Some comparative genomic studies performed on different photosynthetic organisms (*C. reinhardii*, *A. thaliana*, *Oryza sativa*, *Zea mays*) highlighted the complexity of the plant PDI family, similar to the situation in mammals and yeast (Houston *et al.*, 2005; Lemaire and Miginiac-Maslow, 2004).

1. Active site

The plant PDIs characterized so far generally possess one or two TRX active domains with a catalytic WCGHC motif that can be extended to the EFYAPW*CGHC*K/Q sequence, based on amino acid sequence comparisons

(Houston *et al.*, 2005; Lemaire and Miginiac-Maslow, 2004; Xu *et al.*, 2002). However, as for the thioredoxin and glutaredoxin families, a large variability exists concerning the amino acids present between the cysteines and even the number of cysteines in the catalytic modules leading to CxxS or SxxC active-site motifs (Maattanen *et al.*, 2006). Although plants have such isoforms, none of them has been characterized yet. As mentioned before, an *Arabidopsis* atypical thioredoxin h with a CXXS active site has been shown to catalyze isomerization of scrambled RNAse (Serrato *et al.*, 2008). This study suggests the existence of plants atypical PDI with efficient isomerase activity.

2. Redox potential

The global redox potential of PDI is around − 175 mV, with values ranging from − 188 mV for the a domain to − 155 mV for the a′ domain. These data are coherent with the fact that the redox state of the two modules can be different, the active site of the a domain being natively oxidized while the one of the a′ domain is reduced (Tian *et al.*, 2006). The prokaryotic periplasmic oxidase, DsbA, which has a function similar to eukaryotic PDI has a redox potential of around − 130 mV. This could explain why PDIs, which have a redox potential similar to Grxs, can also be efficient reductases (Hawkins *et al.*, 1991; Wunderlich and Glockshuber, 1993).

3. Subcellular localization and physiological role

As mentioned previously, PDIs are generally residents of the endoplasmic reticulum and they present all features required and sufficient for its targeting and retention into the ER. Nevertheless it was shown that RB60, a *C. reinhardtii* PDI member with a KDEL retention signal was partitioned between the stroma and thylakoids of chloroplasts (Trebitsh *et al.*, 2001). Concerning their physiological role, PDI is probably one of the most versatile members of the thioredoxin superfamily as it is able to catalyze *in vitro* disulfide oxidation, reduction, and isomerization (Kaska *et al.*, 1990; Myllylä *et al.*, 1989; Shorrosh *et al.*, 1993; Wadahama *et al.*, 2007). Some data obtained *in vivo* with mammalian and yeast PDI confirmed oxidase and isomerase activities (Frand and Kaiser, 1999; Laboissiere *et al.*, 1995), and also chaperone activities in the absence of an active Trx domain (Song and Wang, 1995).

4. 3D structures of PDIs

The 3D structures of some nonplant eukaryotic PDI members have been solved, essentially from human and yeast (Fig. 4). Only a few of them represent whole proteins (*Hs*ERp29, *Hs*ERp18, *Sc*PDI1p, and *Sc*Mpd1 with respective pdb accession numbers: 2QC7, 1SEN, 2B5E, 3ED3) while some others only describe isolated protein domains (Barak *et al.*, 2009; Rowe

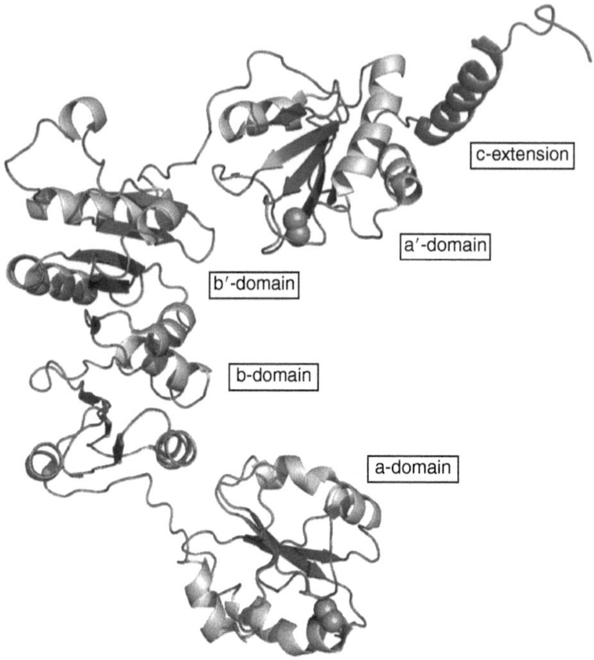

Fig. 4. Crystal structure of protein disulfide isomerase *Sc*PDI1p from Baker's yeast *Saccharomyces cerevisiae* (Tian *et al.*, 2006; pdb accession number 3BOA). Active Trx domains are named a and a′, inactive Trx domain b and b′ and acidic C-terminal extension c. Sulfur atoms belonging to cysteines 43, 46, 388, and 391 are represented by green spheres. (See color plate section at the back of the book.)

et al., 2009; Tian *et al.*, 2006; Vitu *et al.*, 2008). These four proteins vary by the domain composition, the number of "active" Trx modules, and exemplify various PDI classes. Here, we will focus our attention on the 3D structure of *Saccharomyces cerevisiae* PDI1p, a "classical" PDI. This protein is composed of the four domains a, a′, b, and b′ with thioredoxin fold. The four Trx domains and the acidic C-terminal tail of *Sc*PDI form a twisted "U" shape with active sites (CGHC) of active TRX domain (a-a′) facing each other at the end of the U branch. As in thioredoxins, each active site is located at the N-terminal side of the second α-helix. The inactive Trx domain (b-b′), forming the base of the "U," is present at the same relative position as the hydrophobic patch forming a continuous hydrophobic surface at the inside face of the "U." This hydrophobic "pocket" seems to be important for substrate recognition, particularly when PDI acts as a folding and chaperone protein (Kozlov *et al.*, 2006; Zheng and Gilbert, 2001). The connections between the Trx modules are different. The noncatalytic b and b′ modules are connected by a sequence of 17 amino acid residues in an extended conformation susceptible to rigidity. The link between a-b and b′-a′ domains are shorter and presumably more flexible, suggesting that PDI can accommodate diverse substrates (Tian *et al.*, 2006).

C. GLUTAREDOXINS ARE GLUTATHIONE-DEPENDENT PROTEINS DERIVED FROM THIOREDOXINS

1. Active-site sequence

Glutaredoxins are ubiquitous proteins present in most prokaryotes and eukaryotes (Couturier *et al.*, 2009b; Fernandes and Holmgren, 2004). They were initially categorized, based on the active-site sequence, into two groups, a dithiol (CPY/FC motif) and a monothiol (CGFS motif) subgroup (Rodríguez-Manzaneque *et al.*, 1999). In plants, these two types of Grxs represent what we defined as class I and II, respectively. Classes I and II have been further divided into several subgroups called Grx C1, C2, C3, C4, and C5/S12 for class I and GrxS14, S15, S16, and S17 for class II. The specificity of these subgroups has been detailed previously (Couturier *et al.*, 2009b; Rouhier *et al.*, 2004a, 2006). A third Grx class, which contains proteins with a CCxx active-site motif, was also identified specifically in terrestrial plants (Lemaire, 2004; Rouhier *et al.*, 2004a), see the chapter by Li and Zachgo in this volume concerning the Roxy protein of this group. More recently, it has been demonstrated, notably in eukaryotic photosynthetic organisms, that there is a fourth class comprising multidomain proteins containing in their N-terminal part one Grx module with atypical active sites (essentially, CRDC in higher plants and CPHC in algae) (Couturier *et al.*, 2009b). Moreover, genome analysis of cyanobacterial genomes identified two new Grx classes composed of multidomain proteins containing a Grx domain with atypical active site, either CP[W/Y]G in class V, or CPWC/S in class VI (Couturier *et al.*, 2009b). With some exceptions, Grxs from other kingdoms fit well with the above-defined classification.

2. Redox potential and mechanism

Basically, whatever the organism considered, half of the Grxs possess a CxxC active site whereas the remaining sequences have a CxxS active site. Hence, this sequence property makes Grxs very efficient in reducing glutathionylated proteins (the topic of the chapter by Gao and colleagues in this volume). Only in a few cases, has it been shown that Grxs reduce intra- or intermolecular disulfide bonds on target proteins. For the reduction of protein-glutathione adducts, two distinct mechanisms have been elucidated, a monothiol or a dithiol one. In both cases, the N-terminal cysteine of the active site is employed for reducing mixed disulfides between glutathione and the target protein.

The striking difference between the two mechanisms is the involvement of a second cysteine for solving the intermediate disulfide bridge in the case of dithiol mechanism, whereas a glutathione molecule is used for solving the intermediate disulfide bridge in the case of monothiol mechanism. Until recently, the Grxs were considered as glutathione (GSH)-dependent oxidoreductases, but the identification of some Grxs in organisms lacking glutathione, especially in Gram negative bacteria, the identification of fusion proteins between FTR and Grx modules and the biochemical evidence that some Grxs from various sources are reduced by NADPH- or ferredoxin-dependent thioredoxin reductases have moderated this assumption (Jacquot *et al.*, 2009; Johansson *et al.*, 2004; Jordan *et al.*, 1997; Reynolds *et al.*, 2002; Zaffagnini *et al.*, 2008). These observations also suggest that Grxs derived from a Trx ancestor by the acquisition of a glutathione binding site, but some of these Grxs, although possessing the necessary residues, kept the possibility to be regenerated by thioredoxin reductases. This is exemplified by the human Grx2, which is reduced both by GSH and NTR, depending on the protein oxidation state that has been generated during its catalytic cycle (Johansson *et al.*, 2004). A redox potential can be measured for Grxs with two cysteine residues involved in the formation of a disulfide bridge. This led to the conclusion that Grxs are less efficient reductants than Trxs as they possess redox potential values around − 170 mV at pH 7 for poplar GrxC4 and GrxC1, compared to values around − 300 mV for thioredoxins (Rouhier *et al.*, 2007b). In the case of a class II Grx (*C. reinhardtii* Grx3 possessing a CGFS active site), a disulfide bond is formed during its catalytic cycle between the active-site cysteine and a cysteine positioned in the C-terminal part. A lower redox potential has been measured for this protein (around − 323 mV at pH 7.9), explaining that this protein is reduced by FTR but not by glutathione (Zaffagnini *et al.*, 2008). Assuming this is a two-electron process, the value at pH 7, extrapolated from the one at pH 7.9, would be − 269 mV, still much more negative than those determined for classical Grxs with regular dithiol active sites.

3. Subcellular localization and function

There are multiple Grx isoforms localized in different compartments, such as the cytosol, chloroplast, mitochondria, nucleus, and probably apoplasm (Rouhier *et al.*, 2008). The functions of Grxs in plants have been reviewed recently and are detailed in the chapter by Li and Zachgo. Basically, these proteins emerged as stress response proteins involving either their specific deglutathionylation activity or not. They are involved in some developmental processes such as petal development and most likely participate in iron sensing and/or iron sulfur biogenesis (reviewed in Rouhier *et al.*, 2008). Indeed, they are able to reduce dehydroascorbate and are involved in the regeneration of thiol-dependent antioxidant enzymes, especially type II peroxiredoxins and methionine sulfoxide reductases (Msr) (Rouhier *et al.*, 2001, 2002, 2004b, 2007a; ; Tarrago *et al.*, 2009; Vieira dos Santos *et al.*, 2007). Normally, they do not reduce the same Msr and Prx classes than Trxs.

4. 3D structure of Grxs

Grx structures from various organisms have been solved both by NMR and X-ray crystallography. In general, most X-ray determined structures have been obtained with mutated proteins on the second active-site cysteine which allowed the irreversible binding of a glutathione molecule on the first active-site cysteine, most likely stabilizing the protein. The solved Grx structures (Grx1, 2, 3, and 4 from *E. coli*, pig, and human Grxs, Grxs from viruses) generally indicate a monomeric organization and the arrangement of secondary structures is quite similar to Trx structures, with β-strands forming a β-sheet flanked by several α-helices. Interestingly, some eukaryotic Grxs possess two supplementary α-helices in the N- and C-terminal regions compared to most prokaryotic Grxs. Several of the nonphotosynthetic 3D structures of Grx are discussed in the following reviews (Fernandes and Holmgren, 2004; Qin *et al.*, 2000). To date, three structures of plant Grxs (poplar GrxC1, C4, and S12) have been solved, GrxC1 and GrxC4 existing in a monomeric and a dimeric form, whereas GrxS12 exists only in the monomeric form (Figs. 4 and 5) (Couturier *et al.*, 2009a; Feng *et al.*, 2006; Noguera *et al.*, 2005; Rouhier *et al.*, 2007b). These different studies have notably demonstrated that dimeric GrxC1 bridges an iron sulfur cluster, whereas dimeric GrxC4 cannot ligate such a cluster due to the presence of an active-site proline residue (Glycine in GrxC1). The negative impact of another conserved *cis* proline residue for ISC incorporation into disulfide oxidoreductases has also been described by Su and colleagues (Su *et al.*, 2007). Some differences exist between the dimers of GrxC1 and GrxC4, since in GrxC4 the monomers are arranged in a head-to-tail orientation, while in GrxC1 the monomers are in a mirrored conformation. We discuss in more detail in the next section the arrangement of the iron sulfur cluster insertion in glutaredoxins as it is relevant to the function in iron sulfur assembly and to the CXXC motif described throughout this chapter.

IV. EARLY EXPERIMENTS SUGGESTING A LINK BETWEEN IRON SULFUR ENZYMES AND REDOXINS

Early experimental evidence for a connection between the thioredoxin world and proteins of the iron sulfur world originated from the laboratory of Jacques Meyer. As explained above Meyer and colleagues have isolated an

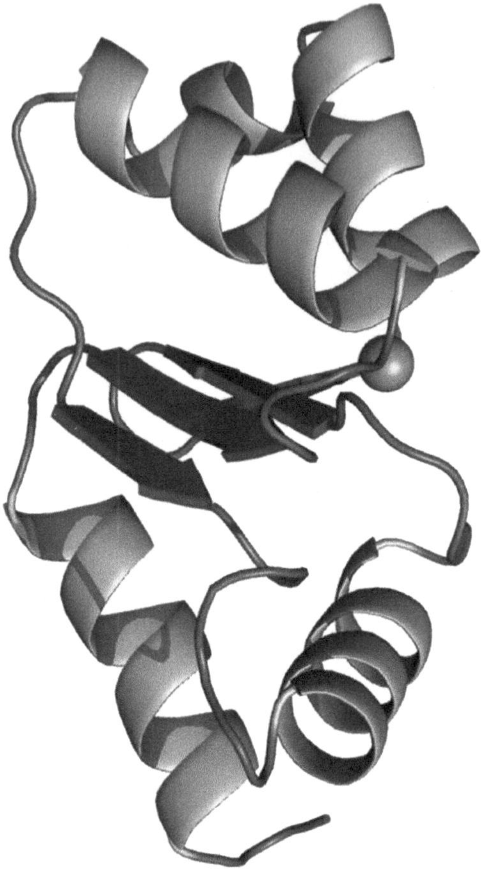

Fig. 5. RMN structure of *Populus tremula* GrxC4 exhibiting a CPYS active site (Noguera *et al.*, 2005). The sulfur atom belonging to cysteine 27 is shown as a sphere (green in the web version).

unusual ferredoxin from *Aquiflex aeolicus* that they have named ferredoxin of the third kind. Instead of having an architecture similar to traditional ferredoxins, this protein incorporates a [2Fe–2S] center in a thioredoxin-like molecule which forms dimers, each monomer containing an ISC (Fig. 2; Yeh *et al.*, 2000, 2002). Similar proteins have been identified in *C. pasteurianum*, *A. vinelandii*, and *C. tepidum* (Meyer, 2001). These studies have been instrumental in indicating that a thioredoxin fold could incorporate an ISC.

That thioredoxin was able to incorporate metals was also recognized in several other instances. In 2003, Collet and colleagues have observed that *E. coli* thioredoxin 2, an elongated form of thioredoxin, can bind zinc with a high affinity. Compared to traditional thioredoxins, Trx2 contains an

N-terminal extension with two CxxC motifs (CTHC and CGRC) that provide the zinc binding site (Collet *et al.*, 2003). The presence of this zinc center influences the reactivity and redox properties of the protein (El Hajjaji *et al.*, 2009).

Another approach that has led to the incorporation of metals in thioredoxin was through rational design and site-directed mutagenesis. In a pioneering work, Hellinga and colleagues have introduced a series of mutations in *E. coli* thioredoxin allowing the insertion of metals as copper or mercury (Hellinga *et al.*, 1991). Likewise, the substitution of W28 and I75 in *E. coli* thioredoxin into cysteines has led to the creation of a rubredoxin-like center with one iron atom linked to four cysteines (Benson *et al.*, 1998). Similar strategies have led to the creation of a mononuclear Cys2His2Zinc binding site in *E. coli* thioredoxin (Wisz *et al.*, 1998) and to the incorporation of a variety of transition metals with possible gain of enzymatic function (e.g., superoxide dismutase activity linked to zinc insertion) (Benson *et al.*, 2000).

In the present context of discussing the CXXC motif in redoxins and iron sulfur proteins, the most significant achievement has been the introduction of a [2Fe–2S] into *E. coli* thioredoxin by manipulating the WCGPC active-site sequence and altering it into a WCACA active site (Fig. 6.) (Masip *et al.*, 2004). This has been achieved by random mutagenesis and the structure of the CACA mutant has suggested that an exposure of a second cysteine (it is generally buried) allows the bridging of a 2Fe–2S cluster into a homodimer, the sulfur atoms of the two cysteines of each subunit serving as ligands to the iron atoms (Collet *et al.*, 2005).

Except for the bacterial ferredoxins of the third kind, all of the experiments described above create *de novo* artificial metal-containing thioredoxins by rational design and site-directed mutagenesis and it was unclear whether these manipulations were related to the ability of molecules with a thioredoxin fold to bind ISCs *in vivo*.

V. Glutaredoxins Bind ISCs

The simultaneous observation that two glutaredoxins, one human with a SCSYC active-site sequence and the other from poplar with a YCGYC active site sequence, were able to bind [2Fe–2S] clusters *in vitro* and *in vivo* was a remarkable discovery (Feng *et al.*, 2006, Lillig *et al.*, 2005; Rouhier *et al.*, 2007b). Site-directed mutagenesis, together with the elucidation of the 3D structure of the two proteins, has indicated that the ISC is inserted at the interface of dimers with only the first Cys of the active site being a ligand. Remarkably, in the two structures, the additional ligands are two external

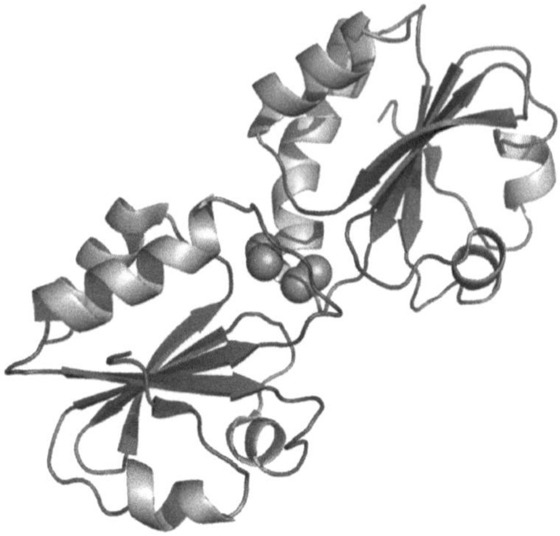

Fig. 6. Crystal structure of the *E. coli* Trx A CACA mutant dimer (Collet *et al.*, 2005; pdb 1ZCP). Sulfur atoms belonging to protein cysteine residues 32 and 34 are shown as green spheres. (See color plate section at the back of the book.)

glutathione molecules situated in *trans* with respect to the ISC (Fig. 7.; Johansson *et al.*, 2007; Rouhier *et al.*, 2007b). Further investigations with poplar Grxs (GrxC1, YCGYC active site; GrxC2, YCPFC active site; GrxC3, YCPYC active site; GrxC4, YCPYC active site; and GrxS12, WCSYS active site) have established that the ISC assembly in these structures requires a Tyr or another small amino acid residue before the first Cys residue and a small amino acid (Gly or Ser but not Pro) after the first Cys of the active site. On the other hand, the second Cys of the active site is dispensable, and as a consequence, Grxs with a CGFS active site sequence could accommodate an ISC in a dimer as poplar GrxC1 (Rouhier *et al.*, 2007b). This finding is remarkable in many respects, (i) it shows that glutaredoxins which possess a thioredoxin fold are able to incorporate an ISC, thereby providing a link between disulfide oxidoreductases and iron sulfur proteins; (ii) it indicates that the prediction of the binding of an ISC to a given polypeptide is complex. So far, it has essentially been based on sequence comparisons and the presence of CXXC motifs in proteins or conserved Cys and His for the Rieske type ISC. The possibility for CGFS Grxs to incorporate ISC in a homodimer with external glutathione molecules as ligands would have been impossible to predict in this context; (iii) it brings forth a hypothesis that glutaredoxins of the CGFS type can actually be involved in reactions related to iron homeostasis rather than in redox reactions. Many CGFS Grxs (*E. coli* Grx4, yeast Grx5 and, *A. thaliana* and polar GrxS14 and S16) characterized so far are indeed able to bind a labile

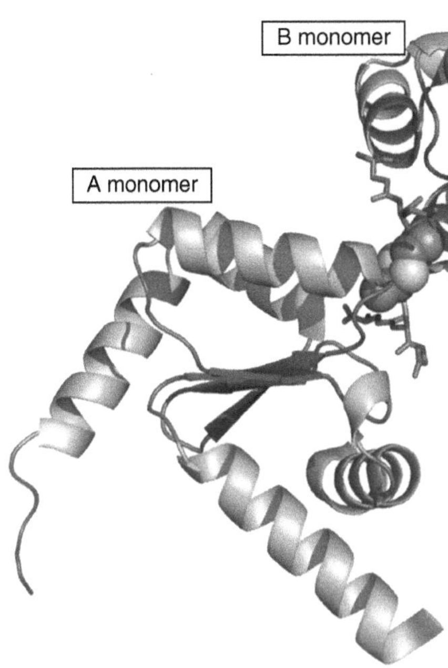

Fig. 7. Crystal structure of the holo form of glutaredoxin C1 from *Populus Tremula X Tremuloides* (Rouhier *et al.*, 2007b; pdb 2E7P). Sulfur atoms belonging to protein cysteine residues 30 and 33 of each monomer (A and B) are shown as green spheres, iron atoms as red, labile sulfur atoms as yellow and glutathione in stick. (See color plate section at the back of the book.)

[2Fe–2S] cluster when overexpressed in *E. coli*, presumably, as in GrxC1, with the catalytic cysteines and external glutathione molecules, but some differences in the spectroscopic signature of these proteins suggest that the cluster environment might be different (Bandyopadhyay *et al.*, 2008; Picciocchi *et al.*, 2007). No structure for ISC-containing CGFS Grxs has been solved yet. The recent biochemical and structural characterization of the *A. thaliana* CNFU scaffold protein indicated that the protein, similar to Grxs, bridged an all cysteinyl-ligated and labile [2Fe–2S] cluster into a dimer but that the ligands are two cysteines of each monomer comprised in a CXXC motif (Yabe *et al.*, 2008).

The study of Grxs from various sources led to the characterization of other ISC-containing Grxs, as the yeast Grx6 or the *Trypanosoma brucei* Grx1 (Comini *et al.*, 2008; Mesecke *et al.*, 2008). Considering that different Grxs, located in various subcellular compartments, can incorporate ISCs with variable stabilities, different putative roles have been proposed. First, based on the observation that GSH can stabilize the cluster in human Grx2 and poplar GrxC1, it has been proposed that the cluster might, similar to some bacterial ISC-containing transcription factors, serve as a redox sensor, which would respond to cellular variation in the GSH/GSSG ratio (Lillig *et al.*, 2005; Rouhier *et al.*, 2007b). Alternatively, these ISC-containing Grxs could sense the iron status of a given compartment or of the cells. It has been shown in yeast that Grx3 and 4 can regulate the iron regulon in response to the cellular iron status through their interaction with the transcriptional activator Aft1 (Kumánovics *et al.*, 2008; Ojeda *et al.*, 2006). Last but not least, additional evidence indicates that CGFS Grxs participate in the ISC assembly (see Section VI).

VI. GLUTAREDOXINS HELP TRANSFER ISCs IN APOPROTEINS

The first evidence for a role of Grxs in ISC assembly came from the study of the yeast null *grx5* mutant strain (Rodríguez-Manzaneque *et al.*, 2002). In this mutant, the activity of two mitochondrial enzymes requiring iron sulfur clusters, aconitase and succinate dehydrogenase, was considerably decreased, whereas the content and activity of cytochrome *c*, a heme-containing protein is not affected. Other deficiencies include inability to grow in respiratory conditions and iron accumulation in their mitochondria. The latter observation might be responsible for the hypersensitivity of this Δ*grx5* mutant to external oxidants. It has been proposed later, still in yeast, based on the accumulation in this mutant of high amount of Fe/S clusters bound to the Isu1p scaffold protein that this mitochondrial Grx5 would help in the transfer of preassembled clusters from scaffold to acceptor proteins (Muhlenhoff *et al.*, 2003). With a few exceptions, almost all CGFS Grxs, either from prokaryotes or eukaryotes, are able to complement the yeast Grx5 deleted strain, suggesting that they could play similar roles in their own cells (Bandyopadhyay *et al.*, 2008; Molina *et al.*, 2004; Molina-Navarro *et al.*, 2006). The presence of a labile ISC in many CGFS Grxs and the possibility of the poplar chloroplastic monothiol Grx, GrxS14, to efficiently transfer intact [2Fe–2S] cluster to chloroplastic *Synechocystis* apoferredoxin, then supported the view that CGFS Grxs could also act directly as scaffold proteins (Bandyopadhyay *et al.*, 2008). The fact that GrxC1, which contains a more stable cluster and a different active site, was not able to transfer its cluster suggests that only Grxs with CGFS active sites and containing a labile cluster could be involved in ISC biogenesis (Bandyopadhyay *et al.*, 2008). The presence of CGFS Grxs in all cellular compartments (chloroplast, mitochondria, cytosol), where an ISC assembly machinery exists, supports the hypothesis that the function of these Grxs is conserved. Nevertheless, as many scaffold proteins of the SUF, NFU, or ISC types have already been characterized in chloroplasts and mitochondria, one may wonder what the roles of these Grxs would be (for an exhaustive list, see Balk and Lobreaux, 2005). In addition, it is not known whether other Grxs with less classical active sites can also incorporate a cluster, whether all Grxs containing a labile cluster can transfer it to various acceptor proteins. Clearly, there is a pressing need to investigate if Grxs can transfer their cluster to other acceptor proteins and if some specificity exists for these reactions, both *in vitro* and *in vivo*.

VII. CONCLUDING REMARKS

At first sight, disulfide oxidoreductases and iron sulfur proteins have little in common, for example, the 3D architecture of ferredoxin is radically different from those of thioredoxin, both having a pleated β-sheet, central in thioredoxin but in the back of the molecule for ferredoxin and the arrangement of the α-helices is completely different. Nevertheless, they have in common to possess CxxC motifs that participate either in disulfide/dithiol exchange (thioredoxin) or in ISC binding (ferredoxin). The recent advances made with glutaredoxins in animal, plant, and bacterial systems have finally firmly connected these two separate worlds. The observation that some glutaredoxins bind ISCs in a way that has never been observed before for other ISC-containing proteins, together with the capacity to transfer those centers with high efficiency to apoferredoxin have absolutely transformed our vision of this research domain. Much remains to be done, however, for example, it is not known if Grxs are able to bind ISCs different from the [2Fe–2S] type. Can they bind [4Fe–4S] centers for example? Would they be able to transfer those? What about more complex proteins such as nitrogenase that contains both iron sulfur and iron molybdenum centers? All these questions are still unanswered and much remains to be done to further clarify the role that glutaredoxins play in ISC synthesis and assembly.

REFERENCES

Abriata, L. A., Cassina, A., Tórtora, V., Marín, M., Souza, J. M., Castro, L., Vila, A. J. and Radi, R. (2009). Nitration of solvent-exposed tyrosine 74 on cytochrome *c* triggers heme iron-methionine 80 bond disruption. Nuclear magnetic resonance and optical spectroscopy studies. *The Journal of Biological Chemistry* **284**, 17–26.

Alkhalfioui, F., Renard, M., Frendo, P., Keichinger, C., Meyer, Y., Gelhaye, E., Hirasawa, M., Knaff, D. B., Ritzenthaler, C. and Montrichard, F. (2008).

A novel type of thioredoxin dedicated to symbiosis in legumes. *Plant Physiology* **148**, 424–435.

Amman, K., Lezhneva, L., Wanner, G., Herrmann, R. G. and Meurer, J. (2004). ACCUMULATION OF PHOTOSYSTEM ONE1, a member of a novel gene family, is required for accumulation of [4Fe-4S] cluster-containing chloroplast complexes and antenna proteins. *The Plant Cell* **16**, 3084–3097.

Appenzeller-Herzog, C. and Ellgaard, L. (2008). The human PDI family: Versatility packed into a single fold. *Biochimica et Biophysica Acta* **4**, 535–548.

Avval, F. Z. and Holmgren, A. (2009). Molecular mechanisms of thioredoxin and glutaredoxin as hydrogen donors for Mammalian s phase ribonucleotide reductase. *The Journal of Biological Chemistry* **284**, 8233–8240.

Balk, J. and Lobreaux, S. (2005). Biogenesis of iron sulfur proteins in plants. *Trends in Plant Science* **10**, 324–331.

Bandyopadhyay, S., Gama, F., Molina-Navarro, M. M., Gualberto, J. M., Claxton, R., Naik, S. G., Huynh, B. H., Herrero, E., Jacquot, J. P., Johnson, M. K. and Rouhier, N. (2008). Chloroplast monothiol glutaredoxins as scaffold proteins for the assembly and delivery of [2Fe-2S] clusters. *EMBO Journal* **27**, 1122–1133.

Baniulis, D., Yamashita, E., Zhang, H., Hasan, S. S. and Cramer, W. A. (2008). Structure–function of the cytochrome *b6f* complex. *Photochemistry and Photobiology* **84**, 1349–1358.

Barak, N. N., Neumann, P., Sevvana, M., Schutkowski, M., Naumann, K., Malešević, M., Reichardt, H., Fischer, G., Stubbs, M. T. and Ferrari, D. M. (2009). Crystal structure and functional analysis of the protein disulfide isomerase-related protein ERp29. *Journal of Molecular Biology* **385**, 1630–1642.

Benson, D. E., Wisz, M. S., Liu, W. and Hellinga, H. W. (1998). Construction of a novel redox protein by rational design: Conversion of a disulfide bridge into a mononuclear iron sulfur center. *Biochemistry* **37**, 7070–7076.

Benson, D. E., Wisz, M. S. and Hellinga, H. W. (2000). Rational design of nascent metalloenzymes. *Proceedings of the National Academy of Sciences of the United States of America* **97**, 6292–6297.

Binda, C., Coda, A., Aliverti, A., Zanetti, G. and Mattevi, A. (1998). Structure of the mutant E92K of [2Fe-2S] ferredoxin I from *Spinacia oleracea* at 1.7 A resolution. *Acta Crystallographica. Section D, Biological Crystallography* **54**, 1353–1558.

Borg, S. J., Behrsing, T., Best, S. P., Razavet, M., Liu, X. and Pickett, C. J. (2004). Electron transfer at a dithiolate-bridged diiron assembly: Electrocatalytic hydrogen evolution. *Journal of the American Chemical Society* **126**, 16988–16999.

Brandes, H. K., Larimer, F. W., Geck, M. K., Stringer, C. D., Schürmann, P. and Hartman, F. C. (1993). Direct identification of the primary nucleophile of thioredoxin f. *Journal of Biological Chemistry* **268**, 18411–18414.

Brandt, U. (2006). Energy converting NADH:quinone oxidoreductase (complex I). *Annual Review of Biochemistry* **75**, 69–92.

Bréhélin, C., Laloi, C., Setterdahl, A. T., Knaff, D. B. and Meyer, Y. (2004). Cytosolic, mitochondrial thioredoxins and thioredoxin reductases in *Arabidopsis thaliana*. *Photosynthesis Research* **79**, 295–304.

Bych, K., Kerscher, S., Netz, D. J., Pierik, A. J., Zwicker, K., Huynen, M. A., Lill, R., Brandt, U. and Balk, J. (2008). The iron sulfur protein Ind1 is required for effective complex I assembly. *EMBO Journal* **27**, 1736–1746.

Capitani, G., Markovic-Housley, Z., del Val, G., Morris, M., Jansonius, J. N. and Schürmann, P. (2000). Crystal structures of two functionally different thioredoxins in spinach chloroplasts. *Journal of Molecular Biology* **302**, 135–154.

Chan, S. K. and Margoliash, E. (1966). Amino acid sequence of chicken heart cytochrome *c*. *The Journal of Biological Chemistry* **241**, 507–515.

Chibani, K., Wingsle, G., Jacquot, J. P., Gelhaye, E. and Rouhier, N. (2009). Comparative genomic study of the thioredoxin family in photosynthetic organisms with emphasis on *Populus trichocarpa*. *Molecular Plant* **2**, 308–322.

Chivers, P. T., Laboissière, M. C. and Raines, R. T. (1996). The CXXC motif: Imperatives for the formation of native disulfide bonds in the cell. *EMBO Journal* **15**, 2659–2667.

Chivers, P. T., Prehoda, K. E. and Raines, R. T. (1997). The CXXC motif: A rheostat in the active site. *Biochemistry* **36**, 4061–4066.

Collet, J. F., D'Souza, J. C., Jakob, U. and Bardwell, J. C. (2003). Thioredoxin 2, an oxidative stress-induced protein, contains a high affinity zinc binding site. *The Journal of Biological Chemistry* **14**, 45325–45332.

Collet, J. F., Peisach, D., Bardwell, J. C. and Xu, Z. (2005). The crystal structure of TrxA (CACA): Insights into the formation of a [2Fe-2S] iron sulfur cluster in an *Escherichia coli* thioredoxin mutant. *Protein Science* **14**, 1863–1869.

Collin, V., Issakidis-Bourguet, E., Marchand, C., Hirasawa, M., Lancelin, J. M., Knaff, D. B. and Miginiac-Maslow, M. (2003). The *Arabidopsis* plastidial thioredoxins: New functions and new insights into specificity. *The Journal of Biological Chemistry* **278**, 23747–23752.

Comini, M. A., Rettig, J., Dirdjaja, N., Hanschmann, E. M., Berndt, C. and Krauth-Siegel, R. L. (2008). Monothiol glutaredoxin-1 is an essential iron sulfur protein in the mitochondrion of African trypanosomes. *The Journal of Biological Chemistry* **283**, 27785–27798.

Coudevylle, N., Thureau, A., Hemmerlin, C., Gelhaye, E., Jacquot, J. P. and Cung, M. T. (2005). Solution structure of a natural CPPC active site variant, the reduced form of thioredoxin h1 from poplar. *Biochemistry* **44**, 2001–2008.

Couturier, J., Koh, C. S., Zaffagnini, M., Winger, A. M., Gualberto, J. M., Corbier, C., Decottignies, P., Jacquot, J. P., Lemaire, S. D., Didierjean, C. and Rouhier, N. (2009a). Structure–function relationship of the chloroplastic glutaredoxin S12 with an atypical WCSYS active site. *The Journal of Biological Chemistry* **284**, 9299–9310.

Couturier, J., Jacquot, J. P. and Rouhier, N. (2009b). Evolution and diversity of glutaedoxins in photosynthetic organisms. *Cellular and Molecular Life Sciences* **66**, 2539–2557.

Dai, S., Schwendtmayer, C., Johansson, K., Ramaswamy, S., Schürmann, P. and Eklund, H. (2000). How does light regulate chloroplast enzymes? Structure–function studies of the ferredoxin/thioredoxin system. *Quarterly Reviews of Biophysics* **33**, 67–108.

Dangoor, I., Peled-Zehavi, H., Levitan, A., Passand, O. and Danon, A. (2009). A small family of chloroplast atypical thioredoxins. *Plant Physiology* **149**, 1240–1250.

del Val, G., Maurer, F., Stutz, E. and Schürmann, P. (1999). Modification of the reactivity of spinach chloroplast thioredoxin *f* by site-directed mutagenesis. *Plant Science* **149**, 183–190.

Edman, J. C., Ellis, L., Blacher, R. W., Roth, R. A. and Rutter, W. J. (1985). Sequence of protein disulfide isomerase and implications of its relationship to thioredoxin. *Nature* **317**, 267–270.

Eklund, H., Gleason, F. K. and Holmgren, A. (1991). Structural and functional relations among thioredoxins of different species. *Proteins Structure Function and Genetics* **11**, 13–28.

El Hajjaji, H., Dumoulin, M., Matagne, A., Colau, D., Roos, G., Messens, J. and Collet, J. F. (2009). The zinc center influences the redox and thermodynamic properties of *Escherichia coli* thioredoxin 2. *Journal of Molecular Biology* **386**, 60–71.

Feng, Y., Zhong, N., Rouhier, N., Hase, T., Kusunoki, M., Jacquot, J. P., Jin, C. and Xia, B. (2006). Structural insight into poplar glutaredoxin C1 with a bridging iron sulfur cluster at the active site. *Biochemistry* **45**, 7998–8008.

Fernandes, A. P. and Holmgren, A. (2004). Glutaredoxins: Glutathione-dependent redox enzymes with functions far beyond a simple thioredoxin backup system. *Antioxidants and Redox Signaling* **6**, 63–74.

Ferrari, D. M. and Söling, H.-D. (1999). The protein disulfide isomerase family: Unravelling a string of folds. *Biochemical Journal* **339**, 1–10.

Frand, A. R. and Kaiser, C. A. (1999). ERO1p oxidizes protein disulfide isomerase in a pathway for disulfide bond formation in the endoplasmic reticulum. *Molecular Cell* **4**, 469–477.

Freedman, R. B., Hirst, T. R. and Tuite, M. F. (1994). Protein disulfide isomerase: Building bridges in protein folding. *Trends in Biochemical Sciences* **19**, 331–336.

Gelhaye, E., Rouhier, N. and Jacquot, J.-P. (2004). The thioredoxin h system of higher plants. *Plant Physiology and Biochemistry* **42**, 265–271.

Grynberg, A., Nicolas, J. and Drapron, R. (1977). Presence of a protein disulfide isomerase (EC 5.3.4.1) in wheat germ. *Comptes Rendus de l'Académie des Sciences Série D Sciences Naturelles* **284**, 235–238.

Hawkins, H. C., de Nardi, M. and Freedman, R. B. (1991). Redox properties and cross-linking of the dithiol/disulfide active sites of mammalian protein disulfide isomerase. *The Biochemical journal* **275**, 341–348.

Hellinga, H. W., Caradonna, J. P. and Richards, F. M. (1991). Construction of new ligand binding sites in proteins of known structure. II. Grafting of a buried transition metal binding site into Escherichia coli thioredoxin. *Journal of Molecular Biology* **222**, 787–803.

Hirasawa, M., Nakayama, M., Hase, T. and Knaff, D. B. (2004). Oxidation-reduction properties of maize ferredoxin: Sulfite oxidoreductase. *Biochimica et Biophysica Acta* **1608**, 140–148.

Holmgren, A. (1985). Thioredoxin. *Annual Review of Biochemistry* **54**, 237–271.

Holmgren, A. (1995). Thioredoxin structure and mechanism: Conformational changes on oxidation of the active-site sulfhydryls to a disulfide. *Structure* **3**, 239–243.

Holmgren, A., Söderberg, B. O., Eklund, H. and Brändén, C. I. (1975). Three-dimensional structure of *Escherichia coli* thioredoxin-S2 to 2.8 A resolution. *Proceedings of the National Academy of Sceinces of the United States of America* **72**, 2305–2309.

Houston, N. L., Fan, C., Xiang, Q.-Y., Schulze, J. M., Jung, R. and Boston, R. S. (2005). Phylogenetic analyses identify 10 classes of the protein disulfide isomerase family in plants, including single-domain protein disulfide isomerase-related proteins. *Plant Physiology* **137**, 762–778.

Jacquot, J. P., Stein, M., Suzuki, A., Liottet, S., Sandoz, G. and Miginiac-Maslow, M. (1997). Residue Glu-91 of *Chlamydomonas reinhardtii* ferredoxin is essential for electron transfer to ferredoxin-thioredoxin reductase. *FEBS Letters* **400**, 293–296.

Jacquot, J. P., Eklund, H., Rouhier, N. and Schürmann, P. (2009). Thioredoxin reductases in photosynthetic organisms. *Trends in Plant Science* **14**, 336–343.

Johansson, C., Lillig, C. H. and Holmgren, A. (2004). Human mitochondrial glutaredoxin reduces *S*-glutathionylated proteins with high affinity accepting electrons from either glutathione or thioredoxin reductase. *The Journal of Biological Chemistry* **279**, 7537–7543.

Johansson, C., Kavanagh, K. L., Gileadi, O. and Oppermann, U. (2007). Reversible sequestration of active site cysteines in a 2Fe-2S-bridged dimer provides a mechanism for glutaredoxin 2 regulation in human mitochondria. *The Journal of Biological Chemistry* **282**, 3077–3082.

Jordan, A., Aslund, F., Pontis, E., Reichard, P. and Holmgren, A. (1997). Characterization of *Escherichia coli* NrdH. A glutaredoxin-like protein with a thioredoxin-like activity profile. *The Journal of Biological Chemistry* **272**, 18044–18050.

Kaska, D. D., Kivirikko, K. I. and Myllylä, R. (1990). Purification and characterization of protein disulfide-isomerase from unicellular alga *Chlamidomonas reinhardii*. *The Biochemical journal* **268**, 63–68.

Kemmink, J., Dijkstra, K., Mariani, M., Scheek, R. M., Penka, E., Nilges, M. and Darby, N. J. (1999). The structure in solution of the b domain of protein disulfide isomerase. *Journal of Biomolecular NMR* **13**, 357–368.

Kim, S. K., Rahman, A., Conover, R. C., Johnson, M. K., Mason, J. T., Gomes, V., Hirasawa, M., Moore, M. L., Leustek, T. and Knaff, D. B. (2006). Properties of the cysteine residues and the iron sulfur cluster of the assimilatory 5'-adenylyl sulfate reductase from *Enteromorpha intestinalis*. *Biochemistry* **45**, 5010–5018.

Koh, C. S., Navrot, N., Didierjean, C., Rouhier, N., Hirasawa, M., Knaff, D. B., Wingsle, G., Samian, R., Jacquot, J. P., Corbier, C. and Gelhaye, E. (2008). An atypical catalytic mechanism involving three cysteines of thioredoxin. *The Journal of Biological Chemistry* **283**, 23062–23072.

Kozlov, G., Maattanen, P., Schrag, J. D., Pollock, S., Cygler, M., Nagar, B., Thomas, D. Y. and Gehring, Kalle (2006). Crystal structure of the bb domains of the protein disulfide isomerase ERp57. *Structure* **14**, 1331–1339.

Krause, G., Lundstrom, J., Barea, J. L., Pueyo de la Cuesta, C. and Holmgren, A. (1991). Mimicking the active site of protein disulfide-isomerase by substitution of proline 34 in *Escherichia coli* thioredoxin. *The Journal of Biological Chemistry* **266**, 9494–9500.

Kumánovics, A., Chen, O. S., Li, L., Bagley, D., Adkins, E. M., Lin, H., Dingra, N. N., Outten, C. E., Keller, G., Winge, D., Ward, D. M. and Kaplan, J. (2008). Identification of *FRA1* and *FRA2* as genes involved in regulating the yeast iron regulon in response to decreased mitochondrial iron sulfur cluster synthesis. *The Journal of Biological Chemistry* **16**, 10276–10286.

Laboissiere, M. C., Sturley, S. L. and Raines, R. T. (1995). The essential function of protein disulfide isomerase is to unscramble non-native disulfide bonds. *The Journal of Biological Chemistry* **270**, 28006–28009.

Lancelin, J. M., Guilhaudis, L., Stein, M., Krimm, I., Blackledge, M. J., Marion, D. and Jacquot, J. P. (2000). NMR Structures of thioredoxin *m* from the green alga *Chlamydomomas reinhardtii*. *Proteins Structure Function and Genetics* **41**, 334–349.

Lane, T. W., Saito, M. A., George, G. N., Pickering, I. J., Prince, R. C. and Morel, F. M. (2005). Biochemistry: A cadmium enzyme from a marine diatom. *Nature* **435**, 42.

Lemaire, S. D. (2004). The glutaredoxin family in oxygenic photosynthetic organisms. *Photosynthesis Research* **3**, 305–318.

Lemaire, S. D. and Miginiac-Maslow, M. (2004). The thioredoxin superfamily in *Chlamydomonas reinhardii*. *Photosynthesis Research* **82**, 203–220.

Lemaire, S. D., Richardson, J. M., Goyer, A., Keryer, E., Lancelin, J. M., Makhatadze, G. I. and Jacquot, J. P. (2000). Primary structure determinants of the pH- and temperature-dependent aggregation of thioredoxin. *Biochimica et Biophysica Acta (BBA)—Protein Structure and Molecular Enzymology* **2**, 311–323.

Lill, R. and Mühlenhoff, U. (2008). Maturation of iron sulfur proteins in eukaryotes: Mechanisms, connected processes, and diseases. *Annual Review of Biochemistry* **77**, 669–700.

Lillig, C. H., Berndt, C., Vergnolle, O., Lönn, M. E., Hudemann, C., Bill, E. and Holmgren, A. (2005). Characterization of human glutaredoxin 2 as iron sulfur protein: A possible role as redox sensor. *Proceedings of the National Academy of Sciences of the United States of America* **102**, 8168–8173.

Maattanen, P., Kozlov, G., Gehring, K. and Thomas, D. Y. (2006). Erp57 and PDI: Multifunctional protein disulfide isomerase with similar domain architectures but differing substrate-partner associations. *Biochemistry and Cell Biology* **84**, 881–889.

Maeda, K., Hägglund, P., Finnie, C., Svensson, B. and Henriksen, A. (2008). Crystal structures of barley thioredoxin h isoforms HvTrxh1 and HvTrxh2 reveal features involved in protein recognition and possibly in discriminating the isoform specificity. *Protein Science* **17**, 1015–1024.

Masip, L., Pan, J. L., Haldar, S., Penner-Hahn, J. E., DeLisa, M. P., Georgiou, G., Bardwell, J. C. and Collet, J. F. (2004). An engineered pathway for the formation of protein disulfide bonds. *Science* **303**, 1185–1189.

Mathieu, I., Meyer, J. and Moulis, J. M. (1992). Cloning, sequencing and expression in *Escherichia coli* of the rubredoxin gene from *Clostridium pasteurianum*. *Biochemical Journal* **285**, 255–262.

Menchise, V., Corbier, C., Didierjean, C., Saviano, M., Benedetti, E., Jacquot, J. P. and Aubry, A. (2001). Crystal structure of the wild-type and D30A mutant thioredoxin h of *Chlamydomonas reinhardtii* and implications for the catalytic mechanism. *Biochemical Journal* **359**, 65–75.

Merchant, S. and Bogorad, L. (1987). The Cu(II)-repressible plastidic cytochrome *c*. Cloning and sequence of a complementary DNA for the pre-apoprotein. *The Journal of Biological Chemistry* **262**, 9062–9067.

Mesecke, N., Mittler, S., Eckers, E., Herrmann, J. M. and Deponte, M. (2008). Two novel monothiol glutaredoxins from *Saccharomyces cerevisiae* provide further insight into iron sulfur cluster binding, oligomerization, and enzymatic activity of glutaredoxins. *Biochemistry* **47**, 1452–1463.

Messerschmidt, A. and Wever, R. (1996). X-ray structure of a vanadium-containing enzyme: Chloroperoxidase from the fungus *Curvularia inaequalis*. *Proceedings of the National Academy of Sciences of the United States of America* **93**, 392–396.

Meyer, J. (2001). Ferredoxins of the third kind. *FEBS Letters* **30**, 1–5.

Meyer, Y., Vignols, F. and Reichheld, J. P. (2002). Classification of plant thioredoxins by sequence similarity and intron position. *Methods in Enzymology* **347**, 394–402.

Meyer, E. H., Giegé, P., Gelhaye, E., Rayapuram, N., Ahuja, U., Thöny-Meyer, L., Grienenberger, J. M. and Bonnard, G. (2005). AtCCMH, an essential component of the c-type cytochrome maturation pathway in *Arabidopsis* mitochondria, interacts with apocytochrome *c*. *Proceedings of the National Academy of Sciences of the United States of America* **102**, 16113–16118.

Meyer, Y., Riondet, C., Constans, L., Abdelgawwad, M. R., Reichheld, J. P. and Vignols, F. (2006). Evolution of redoxin genes in the green lineage. *Photosynthesis Research* **89**, 179–192.

Meyer, Y., Siala, W., Bashandy, T., Riondet, C., Vignols, P. and Reichheld, J. P. (2007). Glutaredoxins and thioredoxins in plants. *Biochimica et Biophysica Acta* **1783**, 589–600.

Milner-White, E. J. and Russell, M. J. (2005). Sites for phosphates and iron sulfur thiolates in the first membranes: 3 to 6 residue anion-binding motifs (nests). *Origins of Life and Evolution of the Biosphere* **35**, 19–27.

Molina, M. M., Bellı, G., de la Torre, M. A., Rodriguez-Manzaneque, M. T. and Herrero, E. (2004). Nuclear monothiol glutaredoxins of *Saccharomyces cerevisiae* can function as mitochondrial glutaredoxins. *The Journal of Biochemistry* **50**, 51923–51930.

Molina-Navarro, M. M., Casas, C., Piedrafita, L., Bellí, G. and Herrero, E. (2006). Prokaryotic and eukaryotic monothiol glutaredoxins are able to perform the functions of Grx5 in the biogenesis of Fe/S clusters in yeast mitochondria. *FEBS Letters* **580**, 2273–2280.

Montrichard, F., Alkhalfioui, F., Yano, H., Venseld, W. H., Hurkmand, W. J. and Buchanan, B. B. (2009). Thioredoxin targets in plants: The first 30 years. *Journal of Proteomics* **72**, 452–474.

Morel, M., Kohler, A., Martin, F., Gelhaye, E. and Rouhier, N. (2008). Comparison of the thiol-dependent antioxidant systems in the ectomycorrhizal *Laccaria bicolor* and the saprotrophic *Phanerochaete chrysosporium*. *New Phytologist* **180**, 391–407.

Mössners, E., Huber-Wunderlich, M. and Glockshuber, R. (1998). Characterization of *Escherichia coli* thioredoxin variants mimicking the active-sites of other thiol/disulfide oxidoreductases. *Protein Science* **7**, 1233–1244.

Muhlenhoff, U., Gerber, J., Richhardt, N. and Lill, R. (2003). Components involved in assembly and dislocation of iron sulfur clusters on the scaffold protein Isu1p. *EMBO Journal* **22**, 4815–4825.

Myllylä, R., Kaska, D. D. and Kivirikko, K. I. (1989). The catalytic mechanism of the hydroxylation reaction of peptidylproline and lysine does not require protein disulfide-isomerase activity. *The Biochemical Journal* **263**, 609–611.

Noguera, V., Walker, O., Rouhier, N., Jacquot, J. P., Krimm, I. and Lancelin, J. M. (2005). NMR reveals a novel glutaredoxin–glutaredoxin interaction interface. *Journal of Molecular Biology* **353**, 629–641.

Ojeda, L., Keller, G., Muhlenhoff, U., Rutherford, J. C., Lill, R. and Winge, D. R. (2006). Role of glutaredoxin-3 and glutaredoxin-4 in the iron regulation of the Aft1 transcriptional activator in *Saccharomyces cerevisiae*. *The Journal of Biological Chemistry* **281**, 17661–17669.

Perez-Jimenez, R., Li, J., Kosuri, P., Sanchez-Romero, I., Wiita, A. P., Rodriguez-Larrea, D., Chueca, A., Holmgren, A., Miranda-Vizuete, A., Becker, K., Cho, S. H. Beckwith, J. *et al.* (2009). Diversity of chemical mechanisms in thioredoxin catalysis. *Nature Structural and Molecular Biology* **16**, 890–896.

Persson, A. L., Eriksson, M., Katterle, B., Pötsch, S., Sahlin, M. and Sjöberg, B. M. (1997). A new mechanism-based radical intermediate in a mutant R1 protein affecting the catalytically essential Glu441 in *Escherichia coli* ribonucleotide reductase. *The Journal of Biological Chemistry* **272**, 31533–31541.

Peterson, F. C., Lytle, B. L., Sampath, S., Vinarov, D., Tyler, E., Shahan, M., Markley, J. L. and Volkman, B. F. (2005). Solution structure of thioredoxin *h*1 from *Arabidopsis thaliana*. *Protein science* **8**, 2195–2200.

Picciocchi, A., Saguez, C., Boussac, A., Cassier-Chauvat, C. and Chauvat, F. (2007). CGFS-type monothiol glutaredoxins from the cyanobacterium *Synechocystis* PCC6803 and other evolutionary distant model organisms possess a glutathione-ligated [2Fe-2S] cluster. *Biochemistry* **46**, 15018–15026.

Qin, J., Yang, Y., Velyvis, A. and Gronenborn, A. (2000). Molecular views of redox regulation: Three-dimensional structures of redox regulatory proteins and protein complexes. *Antioxidants and Redox Signaling* **2**, 827–840.

Raux-Deery, E., Leech, H. K., Nakrieko, K. A., McLean, K. J., Munro, A. W., Heathcote, P., Rigb, S. E., Smith, A. G. and Warren, M. J. (2005). Identification and characterization of the terminal enzyme of siroheme biosynthesis from *Arabidopsis thaliana*: A plastid-located sirohydrochlorin ferrochelatase containing a 2FE-2S center. *The Journal of Biological Chemistry* **280**, 4713–4721.

Ren, G., Stephan, D., Xu, Z., Zheng, Y., Tang, D., Harrison, R. S., Kurz, M., Jarrott, R., Shouldice, S. R., Hiniker, A., Martin, J. L. Heras, B. *et al.* (2009). Properties of the thioredoxin fold superfamily are modulated by a single amino acid residue. *The Journal of Biological Chemistry* **284**, 10150–10159.

Reynolds, C. M., Meyer, J. and Poole, L. B. (2002). An NADH-dependent bacterial thioredoxin reductase-like protein in conjunction with a glutaredoxin homologue form a unique peroxiredoxin (AhpC) reducing system in *Clostridium pasteurianum*. *Biochemistry* **41**, 1990–2001.

Richardson, J. M., III, Lemaire, S. D., Jacquot, J. P. and Makhatadze, G. I. (2000). Difference in the mechanisms of the cold and heat induced unfolding of thioredoxin h from *Chlamydomonas reinhardtii*: Spectroscopic and calorimetric studies. *Biochemistry* **39**, 11154–11162.

Rodríguez-Manzaneque, M. T., Ros, J., Cabiscol, E., Sorribas, A. and Herrero, E. (1999). Grx5 glutaredoxin plays a central role in protection against protein oxidative damage in *Saccharomyces cerevisiae*. *Molecular and Cellular Biology* **19**, 8180–8190.

Rodríguez-Manzaneque, M. T., Tamarit, J., Bellí, G., Ros, J. and Herrero, E. (2002). Grx5 is a mitochondrial glutaredoxin required for the activity of iron/sulfur enzymes. *Molecular Biology of the Cell* **4**, 1109–1121.

Rouhier, N., Gelhaye, E., Sautiere, P. E., Brun, A., Laurent, P., Tagu, D., Gerard, J., de Fay, E., Meyer, Y. and Jacquot, J. P. (2001). Isolation and characterization of a new peroxiredoxin from poplar sieve tubes that uses either glutaredoxin or thioredoxin as a proton donor. *Plant Physiology* **127**, 1299–1309.

Rouhier, N., Gelhaye, E. and Jacquot, J. P. (2002). Glutaredoxin dependent peroxiredoxin from poplar: Protein–protein interaction and catalytic mechanism. *The Journal of Biological Chemistry* **277**, 13609–13614.

Rouhier, N., Gelhaye, E. and Jacquot, J. P. (2004a). Plant glutaredoxins: Still mysterious reducing systems. *Cellular and Molecular Life Science* **61**, 1266–1277.

Rouhier, N., Gelhaye, E., Gualberto, J., Jordy, M. N., de Fay, E., Hirasawa, M., Duplessis, S., Lemaire, S., Frey, P., Martin, F., Manieri, W. Knaff, D. B. *et al.* (2004b). Poplar peroxiredoxin Q: A thioredoxin-linked chloroplast antioxidant functional in pathogen defense. *Plant Physiology* **134**, 1027–1038.

Rouhier, N., Vieira Dos Santos, C., Tarrago, L. and Rey, P. (2006). Plant methionine sulfoxide reductase A and B multigenic families. *Photosynthesis Research* **89**, 247–262.

Rouhier, N., Kauffmann, B., Tete-Favier, F., Paladino, P., Gans, P., Branlant, G., Jacquot, J. P. and Boschi-Muller, S. (2007a). Functional and structural aspects of poplar cytosolic and plastidial type a methionine sulfoxide reductases. *The Journal of Biological Chemistry* **282**, 3367–3378.

Rouhier, N., Unno, H., Bandyopadhyay, S., Masip, L., Kim, S. K., Hirasawa, M., Gualberto, J. M., Lattard, V., Kusunoki, M., Knaff, D. B., Georgiou, G. Hase, T. *et al.* (2007b). Functional, structural, and spectroscopic characterization of a glutathione-ligated [2Fe-2S] cluster in poplar glutaredoxin C1. *Proceedings of the National Academy of Sciences of the United States of America* **104**, 7379–7384.

Rouhier, N., Lemaire, S. and Jacquot, J. P. (2008). The role of glutathione in photosynthetic organisms: Emerging functions for glutaredoxins and glutathionylation. *Annual Reviews of Plant Biology* **59**, 143–166.

Rowe, M. L., Ruddock, L. W., Jurgen, G. K., Schmidt, M., Williamson, R. A. and Howard, M. J. (2009). Solution structure and dynamics of ERp18, a small endoplasmic reticulum resident oxidoreductase. *Biochemistry* **2**, 4596–4606.

Sazanov, L. A. and Hinchliffe, P. (2006). Structure of the hydrophilic domain of respiratory complex I from *Thermus thermophilus*. *Science* **31**, 1430–1436.

Schürmann, P. and Buchanan, B. B. (2008). The ferredoxin/thioredoxin system of oxygenic photosynthesis. *Antioxidants and Redox Signaling* **10**, 1235–1274.

Schürmann, P. and Jacquot, J. P. (2000). Plant thioredoxin systems revisited. *Annual Review of Plant Physiology and Plant Molecular Biology* **51**, 371–400.

Serrato, A. J., Guilleminot, J., Meyer, Y. and Vignols, F. (2008). AtCXXS: Atypical members of the *Arabidopsis thaliana* thioredoxin h family with a remarkably high disulfide isomerase activity. *Physiologia Plantarum* **133**, 611–622.

Shorrosh, B. S., Subramaniam, J., Schubert, K. R. and Dixon, R. A. (1993). Expression and localisation of plant protein disulfide isomerase. *Plant Physiology* **103**, 719–726.

Sipos, K., Lange, H., Fekete, Z., Ullmann, P., Lill, R. and Kispal, G. (2002). Maturation of cytosolic iron sulfur proteins requires glutathione. *The Journal of Biological Chemistry* **277**, 26944–26949.

Song, D. and Lee, F. S. (2008). A role for IOP1 in mammalian cytosolic iron sulfur protein biogenesis. *The Journal of Biological Chemistry* **283**, 9231–9238.

Song, J.-L. and Wang, C. C. (1995). Chaperon-like activity of protein disulfide-isomerase in the refolding of rhodanese. *European Journal of Biochemistry/FEBS Journal* **231**, 312–316.

Stein, M., Jacquot, J. P. and Miginiac-Maslow, M. (1993). A cDNA clone encoding *Chlamydomonas reinhardtii* preferredoxin. *Plant Physiology* **102**, 1349–1350.

Stein, M., Jacquot, J. P., Jeannette, E., Decottignies, P., Hodges, M., Lancelin, J. M., Mittard, V., Schmitter, J. M. and Miginiac-Maslow, M. (1995). *Chlamydomonas reinhardtii* thioredoxins: Structure of the genes coding for the chloroplastic *m* and cytosolic *h* isoforms; expression in *Escherichia coli* of the recombinant proteins, purification and biochemical properties. *Plant Molecular Biology* **28**, 487–503.

Su, D., Berndt, C., Fomenko, D. E., Holmgren, A. and Gladyshev, V. N. (2007). A conserved cis-proline precludes metal binding by the active site thiolates in members of the thioredoxin family of proteins. *Biochemistry* **46**, 6903–6910.

Swamy, U., Wang, M., Tripathy, J. N., Kim, S. K., Hirasawa, M., Knaff, D. B. and Allen, J. P. (2005). Structure of spinach nitrite reductase: Implications for multi-electron reactions by the iron sulfur:siroheme cofactor. *Biochemistry* **44**, 16054–16063.

Takahashi, Y., Goldschmidt-Clermont, M., Soen, S. Y., Franzén, L. G. and Rochaix, J. D. (1991). Directed chloroplast transformation in *Chlamydomonas reinhardtii*: Insertional inactivation of the psaC gene encoding the iron sulfur protein destabilizes photosystem I. *EMBO Journal* **10**, 2033–2040.

Tarrago, L., Laugier, E., Zaffagnini, M., Marchand, C., Le Maréchal, P., Rouhier, N., Lemaire, S. D. and Rey, P. (2009). Regeneration mechanisms of *Arabidopsis thaliana* methionine sulfoxide reductases B by glutaredoxins and thioredoxins. *The Journal of Biological Chemistry* **284**, 18963–18971.

Tian, G. G., Xiang, S., Noiva, R., Lennarz, W. and Schindelin, H. (2006). The crystal structure of yeast protein disulfide isomerase suggest cooperativity between its active sites. *Cell* **124**, 61–73; Erratum in: Cell (2006). 124, 1085-1088.

Trebitsh, T., Meiri, E., Ostersetzeri, O., Adami, Z. and Danon, A. (2001). The protein disulfide isomerase-like RB60 is partitioned between stroma and thylakoids in *Chlamydomonas reinhardtii* chloroplasts. *The Journal of Biological Chemistry* **7**, 4564–4569.

Vieira Dos Santos, C., Laugier, E., Tarrago, L., Massot, V., Issakidis-Bourguet, E., Rouhier, N. and Rey, P. (2007). Specificity of thioredoxins and glutaredoxins as electron donors to two distinct classes of *Arabidopsis* plastidial methionine sulfoxide reductases B. *FEBS Letters* **581**, 4371–4376.

Vitu, E., Gross, E., Greenblatt, H. M., Sevier, C. S., Kaiser, C. A. and Fass, D. (2008). Yeast Mpd1p reveals the structural diversity of the protein disulfide isomerase family. *Journal of Molecular Biology* **384**, 631–640.

Wadahama, H., Kamauchi, S., Ishimoto, M., Kawada, T. and Urade, R. (2007). Protein disufide isomerase family proteins involved in soybean protein biogenesis. *The FEBS Journal* **274**, 687–703.

Wisz, M. S., Garrett, C. Z. and Hellinga, H. W. (1998). Construction of a family of Cys2His2 zinc binding sites in the hydrophobic core of thioredoxin by structure-based design. *Biochemistry* **37**, 8269–8277.

Wunderlich, M. and Glockshuber, R. (1993). *In vivo* control of redox potential during protein folding catalyzed by bacterial protein disulfide-isomerase (DsbA). *The Journal of Biological Chemistry* **268**, 24547–24550.

Xu, Z.-J., Ueda, K., Masuda, K., Ono, M. and Inoue, M. (2002). Molecular characterization of a novel protein disulfide isomerase in carrot. *Gene* **284**, 225–231.

Yabe, T., Yamashita, E., Kikuchi, A., Morimoto, K., Nakagawa, A., Tsukihara, T. and Nakai, M. (2008). Structural analysis of *Arabidopsis* CnfU protein: An iron sulfur cluster biosynthetic scaffold in chloroplasts. *Journal of Molecular Biology* **381**, 160–173.

Yeh, A. P., Chatelet, C., Soltis, S. M., Kuhn, P., Meyer, J. and Rees, D. C. (2000). Structure of a thioredoxin-like [2Fe-2S] ferredoxin from *Aquifex aeolicus*. *Journal of Molecular Biology* **300**, 587–595.

Yeh, A. P., Ambroggio, X. I., Andrade, S. L., Einsle, O., Chatelet, C., Meyer, J. and Rees, D. C. (2002). High resolution crystal structures of the wild type and Cys-55–>Ser and Cys-59–>Ser variants of the thioredoxin-like [2Fe-2S] ferredoxin from *Aquifex aeolicus*. *The Journal of Biological Chemistry* **13**, 34499–34507.

Zaffagnini, M., Michelet, L., Massot, V., Trost, P. and Lemaire, S. D. (2008). Biochemical characterization of glutaredoxins from *Chlamydomonas reinhardtii* reveals the unique properties of a chloroplastic CGFS-type glutaredoxin. *The Journal of Biological Chemistry* **283**, 8868–8876.

Zhang, Y., Lyver, E. R., Nakamaru-Ogiso, E., Yoon, H., Amutha, B., Lee, D. W., Bi, E., Ohnishi, T., Daldal, F., Pain, D. and Dancis, A. (2008). Dre2, a conserved eukaryotic Fe/S cluster protein, functions in cytosolic Fe/S protein biogenesis. *Molecular and Cellular Biology* **28**, 5569–5582.

Zheng, J. and Gilbert, H. F. (2001). Discrimination between native and non-native disulfides by protein disulfide isomerase. *The Journal of Biological Chemistry* **276**, 15747–15752.

Oxidative Stress and Thiol-Based Antioxidants in Cereal Seeds

PABLO PULIDO, FERNANDO DOMÍNGUEZ
AND FRANCISCO JAVIER CEJUDO[1]

Instituto de Bioquímica Vegetal y Fotosíntesis, Universidad de Sevilla and CSIC, Avda Américo Vespucio 49, 41092-Sevilla, Spain

[1]Corresponding author: Email: fjcejudo@us.es

Advances in Botanical Research, Vol. 52

0065-2296/09 $35.00
DOI: 10.1016/S0065-2296(09)52014-9

ABSTRACT

Cereals are among the most important cultivable crops in the world. Two phases in the life cycle of the cereal seed are distinguished: development and germination. Seed development is initiated upon fertilization and culminates in the stage of maturation, a process of massive loss of water that produces mature seed. The germination phase is initiated upon seed imbibition under appropriate environmental conditions. This process is activated by gibberellins, plant hormones synthesized in the embryo that activate the synthesis and secretion of hydrolytic enzymes by the aleurone cells, thereby allowing the mobilization of storage compounds from the starchy endosperm. During development and germination, different seed tissues suffer oxidative stress. Consequently, for the successful completion of the seed life cycle, antioxidant mechanisms are required. Here, we focus on a thiol-based antioxidant system formed by a 1-Cys peroxiredoxin (1-Cys Prx), thioredoxin *h* (Trx *h*) and NADPH-dependent thioredoxin reductase (NTR). We summarize the evidence supporting the classical function proposed for the NTR/Trx redox system in the activation of storage mobilization, thus facilitating seed germination. However, evidence showing the accumulation of these proteins in nuclei of seed cells suffering oxidative stress suggests additional functions for this system.

I. INTRODUCTION

Cereals may be considered as a group of cultivable plants with the highest agronomical importance in the world. Wheat, rice, maize or barley seeds are the most important source of carbon (starch) and proteins in the human diet, and constitute the only source of these nutrients for large populations in different regions of the world. While the world production of cereals has only been slowly increasing during the last years, its consumption has been increasing at a much higher rate, a problem that is now getting worse due to the alternative use of cereals, like biofuel production, in developed countries.

Due to its agronomical, economical, and cultural importance, cereals have been studied from very different points of view. These include the improvement and production of new varieties with higher yield and performance based on classical genetic approaches. Since 2002, the sequence of the rice genome is available (Goff *et al.*, 2002), a major achievement that allows a substantial increase of the genetic and molecular tools available for research in this important cereal. This year the genome sequence of another model cereal, *Brachypodium distachyon*, is expected to be completed and the sequencing of wheat is in progress. The short life cycle of this new cereal model as well as the facilities of growth and transformation will be of great help in improving cereal studies, providing information on other cereals with much more complex genomes far from being completely sequenced. There is no doubt that these new tools will have an important biotechnological impact on improving this essential source of food.

In this chapter, we describe the processes of cereal seed development and germination, and the tissues that suffer oxidative stress. Among the different antioxidant mechanisms available in the cereal seed, we describe an enzymatic system composed of thioredoxin, thioredoxin reductase and a 1-Cys peroxiredoxin, enzymes that are very abundant in cereal seeds and accumulate in the nucleus of cells suffering oxidative stress. We discuss the different functions of these antioxidant systems. In this volume, a chapter by Hägglund *et al.* also describes some structural aspects concerning thioredoxins from cereals like barley.

II. THE LIFE CYCLE OF CEREAL SEEDS: DEVELOPMENT AND GERMINATION

A. THE SEED ENDOSPERM IS A SINK TISSUE

The life cycle of the cereal seed may be divided into two phases—development, or formation of mature seed, and germination. Seed development is initiated by the process of fertilization, which in cereals is a double event. Fertilization of two nuclei of the embryo sac by one sperm cell produces a triploid endosperm, whereas a second sperm cell fertilizes the egg to produce a diploid embryo (Sabelli and Larkins, 2009). Both the triploid endosperm and the diploid embryo undergo different differentiation programmes during seed development. The process of embryogenesis leads to the formation of an embryo, which is contained in the scutellum of the mature seed occupying one of the seed axes. Most of the seed volume, however, is occupied by the endosperm, which becomes the seed storage organ. Due to the agronomical and economical importance of these crops, endosperm development has been studied in the most important cereals. In barley, Bosnes *et al.* (1992) identified four main stages of endosperm development—syncytial, cellularization, differentiation and maturation. In the mature seed, endosperm is formed by five cell types: central starchy endosperm, subaleurone layer, aleurone layer, basal endosperm transfer layer, and the embryo-surrounding region (Olsen *et al.*, 1999). The aleurone tissue, formed by one to three layers of cells depending on the cereal, differentiates from the outermost layer of the endosperm cells. This differentiation is an early event occurring 6–10 days after pollination (DAP) in maize (Sabelli and Larkins, 2009), and about 8 DAP in wheat (Morrison *et al.*, 1975). The aleurone cells have a peculiar morphology characterized by a large size, dense cytoplasm, and the presence of numerous inclusion bodies. The rest of the endosperm cells form the starchy endosperm, a tissue specialized in the accumulation of storage compounds like starch, which constitutes about 70% of the seed dry weight, and proteins that constitute about 15% of the dry weight, though this content is variable among the different cereals and varieties. The protein content is the most important component for the viscoelastic properties of dough obtained from the grain, and therefore, determines its quality for commercial use (bread, pasta, etc.) of the different cereals (Fitzgerald *et al.*, 2009).

The high rates of synthesis of storage compounds make the starchy endosperm an important sink tissue during seed development. The transfer of photosynthates to the endosperm occurs symplastically through the nucellar projection cells in the endosperm cavity (Wang *et al.*, 1995). The endosperm is surrounded by a maternal tissue, the nucellus, formed by parenchymal and epidermal cells, which suffers degeneration at early stages of seed development (Domínguez and Cejudo, 1998). Though the function of the nucellus is not yet well understood, it has been suggested that it participates in the nourishment of the growing endosperm at early stages, hence being important in the control of its growth and development (Thorne, 1985). Additional evidence of the relationship between both tissues, nucellus and endosperm, is provided by the effect of mutations affecting maternal tissues on the endosperm development (Colombo *et al.*, 1997).

B. THE MATURATION PHASE

Once the seed has reached its maximum volume, and all the tissues are differentiated, the seed undergoes a process termed maturation, which consists in a massive loss of water. This is the final stage of development, and it culminates with the production of the mature seed, which has a water content lower than 10%. This process of desiccation is unique among complex organisms, resulting in increased longevity of the seed as well as high resistance to adverse environmental conditions.

The mechanism that allows seed desiccation is still not completely understood. It is clear that the plant hormone, ABA, plays an essential role, as it is responsible for the regulation of genes encoding proteins that have a great importance both in the protection of cell structures and in the establishment of seed dormancy to avoid precocious germination (Bewley, 1997).

Because seed development involves growth and formation of new tissues, it is a process that requires a high rate of cell division; however, cell death is also an important component of seed development. Indeed, the nucellus, the maternal tissue described as degenerating at early stages of development, does suffer a process of programmed cell death (PCD) characterized by morphological features that include high vacuolarization, nuclear membrane instability, chromatin condensation, and the expression of genes encoding proteolytic enzymes (Domínguez and Cejudo, 1998; Domínguez *et al.*, 2001). Similarly, endosperm cells undergo PCD at later stages of development (Young and Gallie, 1999, 2000), such that the starchy endosperm is formed by dead cells in mature cereal seeds. Therefore, a mature seed is made up of different tissues that have undergone cell death, notably endosperm and maternal tissues, whereas the scutellum, embryo and aleurone cells remain alive, though in dormant state, ready to become metabolically active during germination.

C. SEED GERMINATION

The process of germination is started when the mature grain enters in contact with water under adequate environmental conditions. This process is tightly regulated by hormones. About 18 h after seed imbibition ent-kaurene, a precursor of gibberellins, and gibberellins accumulate in after-ripened barley embryos, whereas the level of ABA is decreased (Jacobsen *et al.*, 2002). Gibberellins diffuse into the starchy endosperm (Appleford and Lenton, 1997) and are perceived by the aleurone cells, which in response to the hormone induce the expression of genes encoding hydrolytic enzymes, including α-amylases, glucanases and proteases (Fincher, 1989). Among the proteases secreted from the aleurone layer, a significant component is constituted by thiol-proteases showing optimum activity at acidic pH (Domínguez and Cejudo, 1995), raising the question of whether an acidification of the starchy endosperm is required for the efficient mobilization of storage compounds. To answer this question Domínguez and Cejudo (1999) used an approach based on *in situ* detection of acidity by staining longitudinal sections of wheat seeds at different stages of post-germination with the pH indicator Bromocresol purple, and gene expression by *in situ* hybridization of GA-responsive genes. Following germination, a coordination of the pattern of gene expression and acidification in wheat seeds was observed, this coordination being achieved by gibberellins.

The effect of gibberellins on aleurone cells can be subdivided into two phases: in the short term, gibberellins promote the metabolic activation of aleurone cells inducing the synthesis and secretion of large amounts of hydrolytic enzymes that allow starchy endosperm mobilization. Once this process is completed, gibberellins activate aleurone cell death (Fath *et al.*, 2002). Indeed, aleurone cells have become a model system for the morphological and biochemical analysis of PCD in plants (Domínguez *et al.*, 2004, Domínguez and Cejudo, 2006). Therefore, the progression of the two phases in the life cycle of the cereal seed, development and germination, relies on an intense cell division activity to produce the tissues forming the mature seed as well as seedling growth after germination. However, cell death occurs in a co-ordinated manner and it is essential for the success of seed development and germination.

III. DEVELOPING AND GERMINATING SEEDS SUFFER OXIDATIVE STRESS

Aerobic metabolism produces oxygen derivatives including singlet oxygen ($^1O_2$), superoxide anion (O_2^-), hydrogen peroxide (H_2O_2) or hydroxyl radicals ($OH^·$), termed reactive oxygen species (ROS) (Apel and Hirt, 2004). ROS accumulation has a toxic effect due to the high reactivity of these species that cause damage to the lipids, proteins and nucleic acids. The diffusion rate and toxicity of the different ROS are variable. Singlet oxygen and superoxide anions are highly reactive and it has been considered that both interact with molecules at their site of production, as they have low diffusion rates. However, it has been shown in *Chlamydomonas* cells under high light-stress that singlet oxygen produced by the PSII reaction centre leaves the chloroplast and activates the expression of nuclear genes (Fischer *et al.*, 2007), thus indicating a higher diffusion rate than previously expected. By contrast, hydrogen peroxide has a lower reactivity but it diffuses readily through biological membranes. Hydrogen peroxide thus has a higher diffusion rate and may be perceived in cell locations far from the site of its production. However, hydrogen peroxide is spontaneously converted to hydroxyl radicals, a very reactive and toxic ROS, so that cells require antioxidant systems to avoid high accumulation of hydrogen peroxide.

Although ROS are produced as a consequence of aerobic metabolism, their levels are maintained relatively low under standard growth conditions. However, environmental stress like salinity, drought, high light intensity, low or high temperature, or pathogen attack disturb the cellular homeostasis and increase ROS production, thus causing oxidative stress (Mittler *et al.*, 2004). Besides its well-established toxic effect, ROS also have an important signalling function. The expression of a large number of genes is affected by hydrogen peroxide as shown by transcriptomic analysis in different plants (Desikan *et al.*, 2001; Vandenabeele *et al.*, 2003; Vanderauwera *et al.*, 2005). Similarly, singlet oxygen has an important signalling activity, transduction being achieved by a different pathway than hydrogen peroxide (Op den Camp *et al.*, 2003). Though the antioxidant systems were considered as mechanisms needed to protect cells from oxidative damage, our present vision is that these systems are able to balance ROS production and scavenging, and make possible the signalling activity of these species. In this regard, antioxidant mechanisms have an important signalling function, therefore, involved in the regulation of different processes of plant development and response to the environment.

ROS are key components of different aspects of seed biology (Bailly, 2004). In contrast to other plant organs, photosynthetic production of ROS is elevated at early stages of seed development, the only phase where photosynthesis is important. Respiration becomes a significant source of ROS production during the early stages of embryogenesis but decreases considerably during maturation, a phase with a low metabolic activity (Bailly, 2004). In this phase, however, the seed suffers oxidative stress due to the production of ROS that occurs as a consequence of massive loss of water (Leprince *et al.*, 1993, 1994). Finally, seed germination, a phase of metabolic reactivation that implies resumption of respiration, is associated with high production of ROS in the embryo and scutellum cells as shown in different cereal seeds (Caliskan and Cuming, 1998; Hite *et al.*, 1999).

As discussed above, cereal seeds possess tissues with a high rate of cell division and metabolic activity during embryogenesis and endosperm development. Similarly, in germinating and post-germinating seeds there are cells showing a high metabolic activity, as is the case with the aleurone layer. All of these tissues suffer oxidative stress, as shown in wheat seeds (Serrato and Cejudo, 2003). The growth and activity of these tissues is coordinated with tissues that undergo PCD during development and germination. Oxidative damage may indeed be important to accelerate the process of death as was shown for the aleurone cells in barley (Beligni *et al.*, 2002). Therefore, ROS play an essential role in seed biology, and hence the presence of different antioxidant systems, both enzymatic and non-enzymatic, able to cope with ROS production is not surprising. Among the non-enzymatic systems, the presence of molecules with antioxidant activity in seed tissues including ascorbate, tocopherols and polyphenols has been shown (Howitt and Pogson, 2006; Sattler *et al.*, 2004). The antioxidant activity of the phytic acid through iron complexation has also been shown in maize seed embryos (Doria *et al.*, 2009). The non-enzymatic antioxidant systems have a relevant role in ageing seeds (Galleschi *et al.*, 2002). Among the enzymatic antioxidant systems, seeds express superoxide dismutases responsible for the conversion of superoxide anion to hydrogen peroxide, which is reduced to water by different peroxidases, such as ascorbate peroxidase, catalases, and peroxiredoxins (Bailly *et al.*, 2004).

IV. SEED REDOX SYSTEMS

A. A 1-CYS PEROXIREDOXIN IS ALMOST EXCLUSIVELY EXPRESSED IN SEEDS

The antioxidant systems have an important function of ROS scavenging to protect seed tissues from the toxic effect of these species. However, as mentioned above, ROS also have a signalling function, and are probably involved in the expression of genes that form part of the tolerance mechanism to oxidative stress and also in the process of cell death. Among the different enzymes able to reduce peroxides, we focus in this chapter on peroxiredoxins (Prx). Prxs are thiol-based peroxidases with a molecular mass ranging from 17 to 22 kDa that show a typical thioredoxin fold. Although Prxs were the last peroxidases to be described, it is now well-established that these enzymes are present in all types of organisms from *Archae* and *Eubacteria* to animals and vascular plants. Peroxidase activity of Prxs is based on the action of one or two Cys residues that form the active site since, in contrast with other peroxidases, Prxs have no metals or prosthetic groups as redox cofactors. Based on the peculiarities of the reaction mechanism and enzyme properties, Prxs are classified into four types: 2-Cys Prx, Prx Q, type-II Prx and 1-Cys Prx (Dietz, 2003). This classification should be extended to five types when including the so-called glutathione peroxidases, which in plants are actually thioredoxin-dependent peroxidases as are peroxiredoxins (Koh *et al.*, 2007; Navrot *et al.*, 2006; Rouhier and Jacquot, 2005).

Typical 2-Cys Prxs are homodimeric and contain two Cys residues of the active site localized in different subunits, whereas atypical 2-Cys Prxs (Prx Q, Prx II) are monomeric with the two Cys residues of the active site localized in the same subunit (Wood *et al.*, 2003). Both typical and atypical Prxs have a similar reaction mechanism. The first Cys residue, termed peroxidatic, localized at the N-terminus of the enzyme, attacks the peroxide and becomes oxidized as sulphenic acid. The second Cys residue, termed resolving, localized at the C-terminus, reacts with this sulphenic intermediate so that a molecule of water is produced and the two catalytic Cys are oxidized forming an intramolecular disulfide bridge (in the case of atypical Prx II and Prx Q) or intermolecular disulfide bridge (in the case of typical 2-Cys Prxs), which needs to be reduced by the thioredoxin system for catalysis to carry on (Dietz, 2003; Rouhier *et al.*, 2004, Wood *et al.*, 2003). Atypical type-II Prx can nevertheless be reduced by oxidants other than thioredoxin (namely, glutathione and glutaredoxin), but in the latter case the second Cys of the Prx is not absolutely needed (Rouhier *et al.*, 2002). Less is known about the reaction mechanism of 1-Cys Prxs. These Prxs have only one conserved active site Cys, which is considered to be the peroxidatic residue attacking the peroxide, as in 2-Cys Prx, therefore forming the sulphenic intermediate. The mechanism of reduction of the sulphenic intermediate is not well known but it has been proposed to be mediated by Trxs (Pedrajas *et al.*, 2000) or by glutathyonilation (Manevich *et al.*, 2004; Ralat *et al.*, 2006).

The phylogenetic analysis of the different types of Prxs shows a higher relationship for 2-Cys Prx, Prx Q, and 1-Cys Prx, suggesting an independent origin for type-II Prxs (Dietz *et al.*, 2002). The Prx gene family is formed by 10 genes in *Arabidopsis* and eight in rice; however, both dicot and monocot plants have a similar distribution of Prxs in the different cell compartments: chloroplast, mitochondria, cytoplasm, and nucleus (Dietz *et al.*, 2006). Of these, chloroplasts, the organelles with higher rate of hydrogen peroxide production in plant cells (Pitzschke *et al.*, 2006), contain all types of 2-Cys Prxs, both typical and atypical.

Regarding the seed peroxide scavenging systems, it is worth mentioning the presence of a 1-Cys Prx, which is almost exclusively expressed in seed tissues. In a search of genes involved in seed dormancy, Stacy *et al.* (1996) described the first 1-Cys Prx reported in plants, termed PER1. Though initially identified as being related to seed dormancy, an antioxidant function of this Prx became soon evident as shown by DNA protection assays *in vitro*. This 1-Cys Prx is expressed in seed tissues, aleurone, and scutellum, expression being higher during the desiccation stage (Haleskas *et al.*, 2003; Stacy *et al.*, 1996). The orthologous wheat gene encoding a 1-Cys Prx was recently cloned and characterized by our group (Pulido *et al.*, 2009). As in barley, this 1-Cys Prx is highly expressed in developing wheat seeds, where it accumulates in the endosperm, scutellum, and aleurone cells. Surprisingly, in the starchy endosperm of germinated seeds, the 1-Cys Prx was detected almost exclusively as dimer (Fig. 1A) which probably reflects the highly oxidant environment of this tissue. Both aleurone (results not shown) and scutellum cells (Fig. 1B) of germinating seeds show as well a high content of 1-Cys Prx, which decreased following germination, though in these tissues it was detected in monomeric form.

The 1-Cys Prx, PER1, from barley has a nuclear localization signal (NLS) at the C-terminus also found in the wheat enzyme; thus, explaining the nuclear localization of the enzyme in aleurone and scutellum cells of developing wheat seeds (Fig. 2), as shown in barley seeds (Stacy *et al.*, 1999). The nuclear localization, in conjunction with the DNA protection activity shown by *in vitro* assays, suggested a major antioxidant function for PER1 during the desiccation phase (Haleskas *et al.*, 2003; Stacy *et al.*, 1996). Indeed, tobacco transgenic plants overexpressing PER1 show a reduced sensitivity to hydrogen peroxide as well as a lower level of protein carbonylation

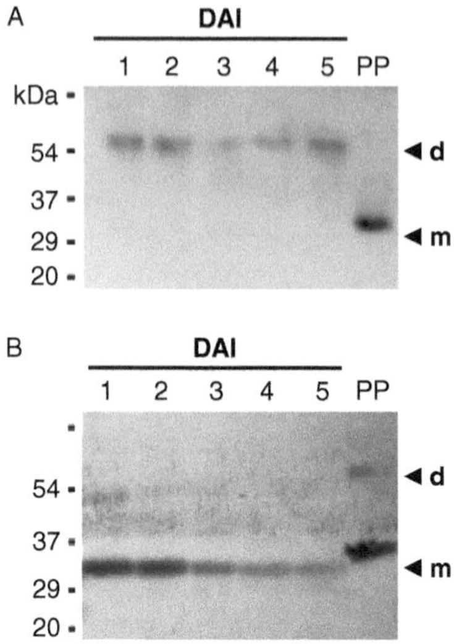

Fig. 1. Analysis of the content and redox state of 1-Cys Prx in wheat seeds following germination. Wheat seeds (*Triticum aestivum*, cv. Chinese Spring) were imbibed on filter paper soaked in water under sterile conditions at room temperature. Seeds were collected up to 5 days after imbibition (DAI) and the starchy endosperm (A) and scutellum cells (B) were dissected. Protein extracts (20 μg of protein) were subjected to SDS-PAGE under non-reducing conditions and Western blot analysis probing the filters with a polyclonal anti-1-Cys Prx antibody (Pulido *et al.*, 2009). Molecular weight markers were loaded and are indicated on the left, and purified recombinant protein (PP) from wheat (50 ng) was also loaded.

(Lee *et al.*, 2000). The pattern of expression and localization of this protein in wheat seeds (Fig. 2; Pulido *et al.*, 2009) fully supports this protective function.

B. THE NADP/THIOREDOXIN SYSTEM OF CEREAL SEEDS

The characteristic pattern of expression of a 1-Cys Prx in seed cells suffering oxidative stress, in conjunction with its nuclear localization and antioxidant activity, suggested a function for this enzyme in the control of the level of peroxides in the nucleus. This system may be relevant not only as a mechanism to protect nuclear DNA from oxidative damage, but also as a possible mechanism for redox regulation of nuclear processes, like transcription. Whatever the function of the nuclear-localized 1-Cys Prx, it is clear that it requires a source of reducing power in order to maintain activity, a function carried out in most organisms by Trxs.

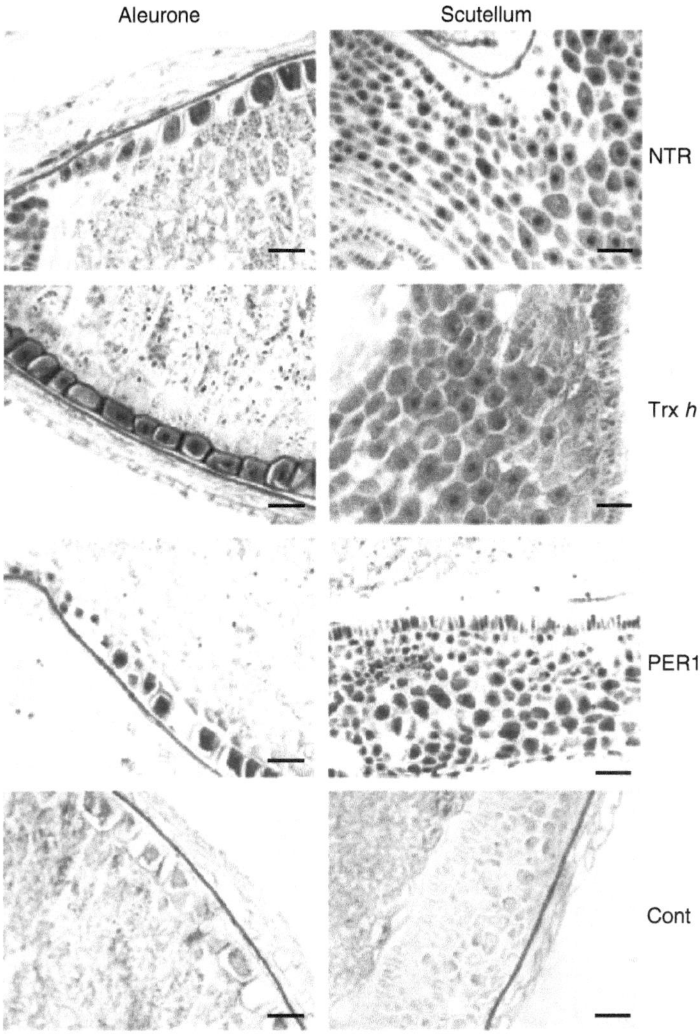

Fig. 2. Immunocytochemical analysis of 1-Cys Prx (PER1), NTR, and Trx *h* in aleurone and scutellum cells of developing wheat seeds. Sections (10 μm) of developing wheat seeds harvested at 25 days post anthesis were probed with anti-PER1, anti-NTR or anti-Trx *h* antibodies (as indicated) or pre-immune serum (Cont). Images show details of the aleurone layer and the scutellum cells. These three proteins show a similar pattern of localization in the two types of wheat seed cells that suffer oxidative stress, hence showing the presence of an antioxidant system in the nucleus of these cells. Bar, 50 μm. (See color plate section at the back of the book.)

Trxs are small proteins (12–14 kDa) with a conserved active site formed by the consensus sequence CGPC, in which the two Cys residues are very efficient disulfide reductants (Schürmann and Jacquot, 2000). Trxs and

Trx-like proteins have a characteristic folding organized in five β-strands surrounded by four α-helices, which is maintained even in members with residue conservation lower than 25% (Eklund *et al.*, 1991). The active site is exposed at the surface of the protein in a hydrophobic region, though charged groups in conserved positions in each type of Trx have been proposed to play a role in the specificity of the interaction with targets (Eklund *et al.*, 1984; Wangensteen *et al.*, 2001).

Trxs are universally distributed from bacteria to unicellular eukaryotes, animals, and plants. However, whereas all organisms contain one or two genes encoding Trx, plants constitute an exception because of the large gene family encoding Trxs or Trx-like proteins, formed by more than 30 genes in *Arabidopsis* (Meyer *et al.*, 2005). Of these genes, the more complex group encodes *h*-type Trxs, which is formed by eight members in *Arabidopsis*, additionally classified into three subgroups (Gelhaye *et al.*, 2004a). The genes encoding *h*-type Trxs have two introns placed at conserved positions (Sahrawy *et al.*, 1996), and the encoded proteins are essentially cytosolic, although a mitochondrial form of Trx *h* has also been described (Gelhaye *et al.*, 2004b).

Type-*h* Trxs are very abundant in the cereal seed. By methods based on antibody or degenerated oligonucleotide screening of cDNA libraries, up to three cDNAs encoding Trx *h* were isolated from wheat (Cazalis *et al.*, 2006; Gautier *et al.*, 1998; Serrato *et al.*, 2001). However, the availability of the rice genome sequence has allowed a more in-depth analysis of the Trx gene family of this important cereal. Up to 30 genes were identified encoding Trx in rice (Nuruzzaman *et al.*, 2008), thus confirming the high complexity of this gene family as previously reported in *Arabidopsis*. As in *Arabidopsis*, the type-*h* group of Trxs, formed by nine genes, is the most complex in the rice genome.

As the two Cys residues of the active site of Trxs become oxidized after catalysis, for a new catalytic cycle Trxs must be reduced in a reaction catalyzed by Trx reductase. In plants, there are up to four types of Trx reductases depending on their degree of evolution, the source of reducing power, and the subcellular localization (Jacquot *et al.*, 2009). The cytosol and mitochondria of land plant cells contain a NADPH-dependent Trx reductase (NTR), similar to the enzyme from bacteria and yeast. These enzymes belong to the short type of NTR, in contrast with the larger form of NTR found in animals, which contains an active site selenocysteine at the C-terminus (Jacquot *et al.*, 1994; Tamura and Stadtman, 1996). NTR is encoded by two genes in *Arabidopsis* and rice (Serrato *et al.*, 2004). In addition, land plants contain an additional type of Trx reductase specifically localized in plastids, which depends on ferredoxin reduced by the photosynthetic electron chain (Schürmann and Buchanan, 2008), thus involved in the reduction of plastidial Trxs. Plants, as organisms that perform oxygenic photosynthesis, have a chloroplast-localized NTR with a joint Trx domain at the C-terminus, called NTRC (Serrato *et al.*, 2004). This enzyme is localized in the chloroplast, uses NADPH as electron donor, and is able to conjugate both NTR and Trx activity to efficiently reduce disulfides in target proteins as 2-Cys Prxs (Alkhalfioui *et al.*, 2007; Moon *et al.*, 2006; Pérez-Ruiz *et al.*, 2006; Sueoka *et al.*, 2009) or ADP-glucose pyrophosphorylase (Michalska *et al.*, 2009).

The two NTR isoforms are as well abundant in cereal seeds as in barley (Shahpiri *et al.*, 2008), in agreement with its high Trx *h* content (Serrato *et al.*, 2002). These two isoforms show differential regulation of expression, though most probably overlapping roles. No description of NTRC in cereal seeds has been reported so far.

C. THE FUNCTION OF THE NTR/TRX SYSTEM DURING SEED GERMINATION

Extensive evidence has been reported suggesting that a major function of the NTR/Trx system in cereal seeds is to accelerate germination by facilitating the mobilization of the storage compounds of the starchy endosperm. Most of the proteins of the starchy endosperm are in the oxidized state, and reduction is required to facilitate the action of proteases involved in their degradation. Indeed, most of the proteolytic activity found in the starchy endosperm of the germinating seed is of the thiol-protease type (Domínguez and Cejudo, 1995), and thiocalsin, a Ser protease involved in this degradative process, which has been analysed in more detail, is activated in reduced state (Besse *et al.*, 1996). Similarly, Trx *h* is involved in the inactivation of α-amylase and trypsin inhibitors, thus facilitating starch and protein degradation (Kobrehel *et al.*, 1991, 1992). A proteomics-based approach allowed identification of Trx targets in wheat starchy endosperm that confirmed the important role of this redox system in seed germination (Wong *et al.*, 2003).

Besides the effect of the NTR/Trx *h* system in facilitating starch and protein degradation in germinating seeds, it was reported that transgenic barley seeds overexpressing Trx *h* show enhanced gibberellin synthesis in the embryo and that a higher Trx *h* content has an activating effect on the expression of α-amylase in aleurone cells, thus suggesting a function for this enzyme in the communication between the embryo and aleurone tissues (Wong *et al.*, 2002). Thus, these results raise a new aspect of Trx function as a signalling molecule in germinating seeds. Previously, the discovery of Trx *h* as one of the most abundant proteins of the phloem sap in rice

(Ishiwatari *et al.*, 1995, 2000) led to the proposal that Trx *h* might serve as a redox signal transducer in vascular tissues, thus connecting different parts of the plant.

Localization studies performed in developing wheat seeds showed the presence of Trx *h* in the vascular tissue, as well as in transfer cells (Serrato and Cejudo, 2003). Similarly, Trx *h* accumulates in maize pedicel and chalazal cells (Santandrea *et al.*, 2002), thus confirming the presence of these proteins in vascular tissue and lending support to the proposal of its function as redox signal. However, the immunocytochemical analysis carried out in wheat seeds revealed a previously unknown property of Trx *h*, which is its accumulation in nuclei of seed cells, aleurone, and scutellum, that suffer oxidative stress during development and germination (Serrato and Cejudo, 2003; Serrato *et al.*, 2001). The nuclear localization of Trx *h* raises several questions concerning the function of these enzymes, besides the degradation of storage compounds that take place in the starchy endosperm. As scutellum and aleurone cells suffer oxidative stress during desiccation at late stages of seed development, both types of cells need elaborate mechanisms to survive this stress. The presence of Trx *h* in the nucleus might be related to a function protecting nuclear structures. Moreover, scutellum and aleurone cells have an essential function in germinating seeds, which requires the expression of a large set of genes before they enter in a process of cell death (Domínguez *et al.*, 2004). So, the additional possibility of redox control of gene expression has to be taken into consideration.

D. AN ANTIOXIDANT REDOX SYSTEM IN THE NUCLEUS OF SEED CELLS SUFFERING OXIDATIVE STRESS

The first question raised by the finding of the high accumulation of Trx *h* in the nucleus of seed cells suffering oxidative stress was to determine whether or not NTR, which is an abundant protein in cereal seeds (Serrato *et al.*, 2002), shows nuclear localization. The immunocytochemical analysis revealed a pattern of localization of NTR very similar to the one previously reported for Trx *h* (Pulido *et al.*, 2009; Serrato and Cejudo, 2003). NTR is abundant in vascular tissues and transfer cells showing co-localization with Trx *h*, which supports the signalling function previously proposed for this system in rice (Ishiwatari *et al.*, 2000). In addition, NTR is highly expressed in cells with a high rate of division, such as developing embryo and root meristem, a pattern reflecting the involvement of these proteins in nucleotide metabolism and cell cycle progression as shown in bacteria and yeast (Muller, 1991; Nordlund and Reichard, 2006).

The immunocytochemical analysis also revealed the localization of NTR in the nucleus of seed cells, aleurone, and scutellum, thereby showing co-localization with Trx *h* (Fig. 2). It is worth mentioning that no NLS has been described in either NTR or Trx *h*. Whilst the size of Trx *h* is lower than the exclusion size through the nuclear pores, and therefore, Trx *h* may have free diffusion to the nucleus, the approximately 70 kDa size of the dimeric NTR (Dai *et al.*, 1996) would impede its diffusion through the nuclear pores, thus suggesting the existence of a mechanism to introduce NTR into the nucleus in these cells.

As the nuclear localization of the NTR/Trx redox system is specific to cells suffering oxidative stress, an initial possibility to be considered is that it has an antioxidant function. In this regard it should be noted that, as mentioned above, the 1-Cys Prx (PER1) initially identified in barley (Stacy *et al.*, 1996) shows seed-specific expression and accumulation in the nucleus of embryo and aleurone cells (Stacy *et al.*, 1999). We have confirmed this pattern of expression for the orthologous gene isolated from wheat, as well as the nuclear localization of the encoded protein (Fig. 2).

The co-localization of NTR, Trx *h*, and PER1 in the nucleus of seed cells suffering oxidative stress (Fig. 2) suggested the possibility of these enzymes forming an antioxidant system able to use reducing power, NADPH, to reduce hydrogen peroxide. This may be an important mechanism to maintain the redox homeostasis of the nucleus because it has been shown in tobacco cells that the nucleus has an elevated level of hydrogen peroxide (Ashtamker *et al.*, 2007), so that a highly oxidant environment in the nucleus of the seed cells would be expected. This possibility was addressed by *in vitro* reconstitution of the system using purified recombinant enzymes from wheat. The results showed that these enzymes effectively catalyzed the reduction of hydrogen peroxide (Pulido *et al.*, 2009). Unexpectedly, Trx *h* was not required, thus showing that NTR is able to directly transfer electrons from NADPH to the 1-Cys Prx. This result was intriguing because NTR is an efficient reductant of disulfide bridges, which were not expected in a 1-Cys Prx. As most of the 1-Cys Prx in the starchy endosperm of germinating seeds was identified in dimeric form, it could be expected that under oxidant conditions the 1-Cys Prx might be able to dimerize, this form being the substrate of NTR. The ability of 1-Cys Prx to dimerize under oxidant conditions was tested by incubation of the purified recombinant protein in the presence of increasing concentrations of hydrogen peroxide. Figure 3 shows that H_2O_2 concentration up to 5 mM promoted the dimerization of the 1-Cys Prx. At higher concentration (10–20 mM), the monomeric form of the enzyme was detected (Fig. 3) most probably due to the overoxidation of the peroxidatic Cys residue to sulphinic acid (Pulido *et al.*, 2009).

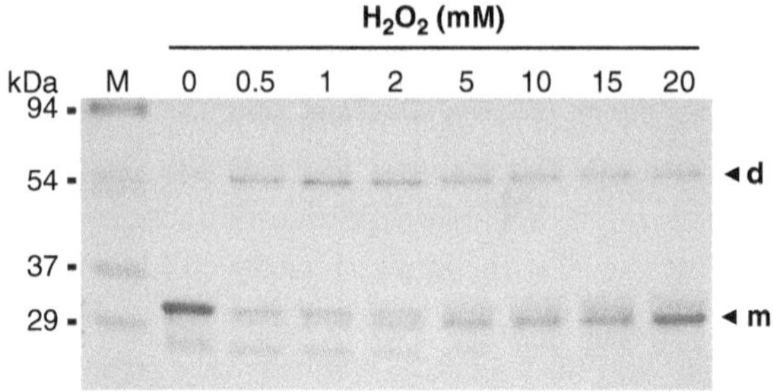

Fig. 3. Effect of hydrogen peroxide on the dimerization capacity of wheat 1-Cys Prx. Purified recombinant 1-Cys Prx from wheat (3 μg) was incubated in the presence of increasing concentrations of hydrogen peroxide as indicated for 30 min. Protein was then subjected to SDS-PAGE under non-reducing conditions and the gels were stained with Coomassie Brilliant Blue. Molecular weight markers (M) were loaded and are indicated on the left. d, dimer; m, monomer.

E. 1-CYS PRX OVEROXIDATION SUGGESTS A NUCLEAR SIGNALLING FUNCTION FOR HYDROGEN PEROXIDE

A more in-depth analysis of this system based on two-dimensional and mass spectrometry analysis revealed that the peroxidatic Cys residue of the 1-Cys Prx became overoxidized to sulphinic acid (-SO_2H) upon exposure to hydrogen peroxide. Moreover, the overoxidized form of the 1-Cys Prx could be detected *in vivo* as shown by two-dimensional and Western blot analysis of protein extracts from aleurone cells of developing and germinating wheat seeds (Pulido *et al.*, 2009). Studies performed with a different type of Prx, 2-Cys Prx, in animal and yeast cells has shown that overoxidation of the peroxidatic Cys residue leads to the inactivation of the enzyme (Rhee *et al.*, 2005), which is also the case for the 1-Cys Prx. Whereas the inactivation of the 2-Cys Prx is reversed in a process catalyzed by sulfiredoxins as shown in yeast (Biteau *et al.*, 2003) and most probably in plants (Rey *et al.*, 2007), it is very unlikely that the overoxidation, and inactivation, of the nuclear 1-Cys Prx is a reversible process since sulfiredoxin is a chloroplast-localized protein in plants, therefore not present in the nucleus of seed cells.

Consequently, if the inactivation of the 1-Cys Prx in the nucleus of seed cells is an irreversible process, it might serve to allow an increase of hydrogen peroxide (Fig. 4), which might cause the alteration of the redox homeostasis of the nucleus. These conditions may favour the process of PCD that both scutellum and aleurone cells undergo in post-germinative seeds (Domínguez *et al.*, 2004).

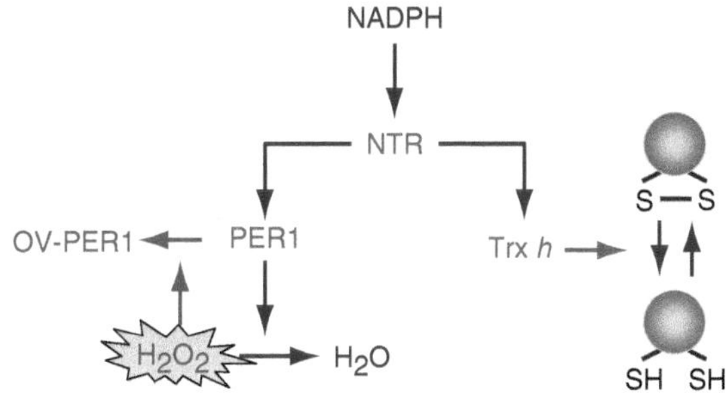

Fig. 4. An antioxidant redox system in the nucleus of wheat seed cells. The immunocytochemical analysis showed co-localization of NTR, Trx *h* and 1-Cys Prx in the nucleus of wheat seed cells, aleurone and scutellum, suffering oxidative stress. NTR is able to use NADPH to reduce Trx *h*, which in turn reduces disulfides of nuclear targets, thus controlling the redox status of these proteins. NTR is also able to support the peroxidase activity of the 1-Cys Prx (PER1) thus allowing the use of NADPH to reduce hydrogen peroxide and maintain the redox homeostasis of the nucleus. Under oxidant conditions, the peroxidatic Cys residue of PER1 is over-oxidized causing the inactivation of the enzyme (OV-PER1). The progressive over-oxidation of PER1 in post-germinative seeds may raise the level of hydrogen peroxide thus favouring programmed cell death in aleurone and scutellum cells.

V. CONCLUDING REMARKS AND FUTURE PROSPECTS

Type-*h* thioredoxins and NTR have been identified as abundant proteins in cereal seeds. Because the redox conversion of disulfide/dithiol is a process of major importance in germination, different studies have shown that the function of the NTR/Trx system is to activate seed germination by facilitating the mobilization of reserves stored in the starchy endosperm. Recent evidence showing the localization of Trx *h* and NTR in seed vascular tissues and transfer cells suggests a role in redox signalling linked to the coordination of the redox status of the different tissues of the seed. Lastly, the accumulation of these enzymes in the nucleus of seed cells suffering oxidative stress, and the finding that NTR is able to support the hydrogen peroxide reduction activity through a seed-specific nuclear-localized 1-Cys Prx, reveals the existence of a mechanism to control the level of hydrogen peroxide in the nucleus (Fig. 4). This system is probably important as an antioxidant mechanism to protect nuclear structures, notably DNA, from oxidative damage. However, as hydrogen peroxide has been shown to have an important capacity to exert redox regulation of transcription factors in yeast cells (Delaunay *et al.*, 2002), this finding opens the possibility that the control of the level of hydrogen peroxide in the nucleus of seed cells may have an additional effect of regulating nuclear processes like transcription, though the knowledge of redox regulation of transcription factors in plants is still scarce. We consider that identification of the nuclear targets of Trx *h* in cereal seeds will allow the identification of these new functions. It is expected that this knowledge has an important biotechnological impact in future prospects of these important crops.

ACKNOWLEDGMENTS

This work was supported by Grant BIO2007-60644 from the Ministerio de Educación y Ciencia, Spain, and Grants CVI-182 and P06-CVI-01578 from the Junta de Andalucía, Spain. P. P. was a recipient of a pre-doctoral fellowship from Universidad de Sevilla.

REFERENCES

Alkhalfioui, F., Renard, M. and Montrichard, F. (2007). Unique properties of NADP-thioredoxin reductase C in legumes. *Journal of Experimental Botany* **58**, 969–978.

Apel, K. and Hirt, H. (2004). Reactive oxygen species: Metabolism, oxidative stress, and signal transduction. *Annual Review of Plant Biology* **55**, 373–399.

Appleford, N. E. J. and Lenton, J. R. (1997). Hormonal regulation of α-amylase gene expression in germinating wheat (*Triticum aestivum*) grains. *Physiologia Plantarum* **100**, 534–542.

Ashtamker, C., Kiss, V., Sagi, M., Davydov, O. and Fluhr, R. (2007). Diverse subcellular locations of cryptogein-induced reactive oxygen species production in tobacco bright yellow-2 cells. *Plant Physiology* **143**, 1817–1826.

Bailly, C. (2004). Active oxygen species and antioxidants in seed biology. *Seed Science Research* **14**, 93–107.

Bailly, C., Leymarie, J., Lehner, A., Rousseau, S., Come, D. and Corbineau, F. (2004). Catalase activity and expression in developing sunflower seeds as related to drying. *Journal of Experimental Botany* **55**, 475–483.

Beligni, M. V., Fath, A., Bethke, P. C., Lamattina, L. and Jones, R. L. (2002). Nitric oxide acts as an antioxidant and delays programmed cell death in barley aleurone layers. *Plant Physiology* **129**, 1642–1650.

Besse, I., Wong, J. H., Kobrehel, K. and Buchanan, B. B. (1996). Thiocalsin: A thioredoxin-linked, substrate-specific protease dependent on calcium. *Proceedings of the National Academy of Science of the United States of America* **93**, 3169–3175.

Bewley, J. D. (1997). Seed dormancy and germination. *Plant Cell* **9**, 1055–1066.

Biteau, B., Labarre, J. and Toledano, M. (2003). ATP-dependent reduction of cysteine-sulphinic acid by *S. cerevisiae* sulphiredoxin. *Nature* **425**, 980–984.

Bosnes, M., Weideman, F. and Olsen, O. A. (1992). Endosperm differentiation in barley wild-type and sex mutants. *Plant Journal* **2**, 661–674.

Caliskan, M. and Cuming, A. C. (1998). Spatial specificity of H_2O_2-generating oxalate oxidase gene expression during wheat embryo germination. *Plant Journal* **15**, 165–171.

Cazalis, R., Pulido, P., Aussenac, T., Pérez-Ruiz, J. M. and Cejudo, F. J. (2006). Cloning and characterization of three thioredoxin *h* isoforms from wheat showing differential expression in seeds. *Journal of Experimental Botany* **57**, 2165–2172.

Colombo, L., Franken, J., van der Krol, A. R., Wittich, P. E., Dons, H. J. M. and Angenet, G. C. (1997). Downregulation of ovule-specific MADS box genes from petunia results in maternally controlled defects in seed development. *Plant Cell* **9**, 703–715.

Dai, S., Saarinen, M., Ramaswamy, S., Meyer, Y., Jacquot, J. P. and Eklund, H. (1996). Crystal structure of *Arabidopsis thaliana* NADPH dependent thioredoxin reductase at 2.5A resolution. *Journal of Molecular Biology* **264**, 1044–1057.

Delaunay, A., Pflieger, D., Barrault, M. B., Vinh, J. and Toledano, M. B. (2002). A thiolperoxidase is an H_2O_2 receptor and redox-transducer in gene activation. *Cell* **111**, 471–481.

Desikan, R., Mackerness, S. A. H., Hancock, J. T. and Neill, S. J. (2001). Regulation of the *Arabidopsis* transcriptome by oxidative stress. *Plant Physiology* **127**, 159–172.

Dietz, K.-J. (2003). Plant peroxiredoxins. *Annual Review of Plant Biology* **54**, 93–107.

Dietz, K.-J., Horling, F., Konig, J. and Baier, M. (2002). The function of the chloroplast 2-cysteine peroxiredoxin in peroxide detoxification and its regulation. *Journal of Experimental Botany* **53**, 1321–1329.

Dietz, K.-J., Jacob, S., Oelze, M. L., Laxa, M., Tognetti, V., de Miranda, S. M., Baier, M. and Finkemeier, I. (2006). The function of peroxiredoxins in plant organelle redox metabolism. *Journal of Experimental Botany* **57**, 1697–1709.

Domínguez, F. and Cejudo, F. J. (1995). Pattern of endoproteolysis following wheat grain germination. *Physiologia Plantarum* **95**, 253–259.

Domínguez, F. and Cejudo, F. J. (1998). Germination-related genes encoding proteolytic enzymes are expressed in the nucellus of developing wheat grains. *Plant Journal* **15**, 569–574.

Domínguez, F. and Cejudo, F. J. (1999). Patterns of starchy endosperm acidification and protease gene expression in wheat grains following germination. *Plant Physiology* **119**, 81–87.

Domínguez, F. and Cejudo, F. J. (2006). Identification of a nuclear-localized nuclease from wheat cells undergoing programmed cell death that is able to trigger DNA fragmentation and apoptotic morphology on nuclei from human cells. *Biochemical Journal* **397**, 529–536.

Domínguez, F., Moreno, J. and Cejudo, F. J. (2001). The nucellus degenerates by a process of programmed cell death during the early stages of wheat grain development. *Planta* **213**, 352–360.

Domínguez, F., Moreno, J. and Cejudo, F. J. (2004). A gibberellin-induced nuclease is localized in the nucleus of wheat aleurone cells undergoing programmed cell death. *Journal of Biological Chemistry* **279**, 11530–11536.

Doria, E., Galleschi, L., Calucci, L., Pinzino, C., Pilu, R., Cassani, E. and Nielsen, E. (2009). Phytic acid prevents oxidative stress in seeds: evidence from a maize (*Zea mays* L.) low phytic acid mutant. *Journal of Experimental Botany* **60**, 967–978.

Eklund, H., Cambillau, C., Sjoberg, B. M., Holmgren, A., Jornvall, H., Hoog, J. O. and Branden, C. I. (1984). Conformational and functional similarities between glutaredoxin and thioredoxins. *EMBO Journal* **3**, 1443–1449.

Eklund, H., Gleason, F. K. and Holmgren, A. (1991). Structural and functional relations among thioredoxins of different species. *Proteins* **11**, 13–28.

Fath, A., Bethke, P., Beligni, V. and Jones, R. L. (2002). Active oxygen and cell death in cereal aleurone layers. *Journal of Experimental Botany* **53**, 1273–1282.

Fincher, G. B. (1989). Molecular and cellular biology associated with endosperm mobilization in germinating cereal grains. *Annual Review of Plant Physiology and Plant Molecular Biology* **40**, 305–346.

Fischer, B. B., Krieger-Liszkay, A., Hideg, E., Šnyrychová, I., Wiesendanger, M. and Eggen, R. I. L. (2007). Role of singlet oxygen in chloroplast to nucleus retrograde signaling in *Chlamydomonas reinhardtii. FEBS Letters* **581**, 5555–5560.

Fitzgerald, M. A., McCouch, S. R. and Hall, R. D. (2009). Not just a grain of rice: The quest for quality. *Trends in Plant Science* **14**, 133–139.

Galleschi, L., Capocchi, A., Ghiringhelli, S., Saviozzi, F., Calucci, L., Pinzino, C. and Zandomeneghi, M. (2002). Antioxidants, free radicals, storage proteins, and proteolytic activities in wheat (*Triticum durum*) seeds during accelerated aging. *Journal of Agricultural and Food Chemistry* **50**, 5450–5457.

Gautier, M. F., Lullien-Pellerin, V., Lamotte-Guery, F., Guirao, A. and Joudrier, P. (1998). Characterization of wheat thioredoxin *h* cDNA and production of an active *Triticum aestivum* protein in *E. coli. European Journal of Biochemistry* **252**, 314–324.

Gelhaye, E., Rouhier, N., Gerard, J., Jolivet, Y., Gualberto, J., Navrot, N., Ohlsson, P. I., Wingsle, G., Hirasawa, M., Knaff, D. B., Wang, H. Dizengremel, P. *et al.* (2004a). A specific form of thioredoxin h occurs in plant mitochondria and regulates the alternative oxidase. *Proceedings of the National Academy of Science of the United States of America* **101**, 14545–14550.

Gelhaye, E., Rouhier, N. and Jacquot, J. P. (2004b). The thioredoxin h system of higher plants. *Plant Physiology and Biochemistry* **42**, 265–271.

Goff, S. A., Ricke, D., Lan, T. H., Presting, G., Wang, R., Dunn, M., Glazebrook, J., Sessions, A., Oeller, P., Varma, H., Hadley, D. Hutchison, D. *et al.* (2002). A draft sequence of the rice genome (*Oryza sativa* L. sp. *japonica*). *Science* **296**, 92–100.

Haleskas, C., Viken, M. K., Grini, P. E., Nygaard, V., Nordgard, S. H., Meza, T. J. and Aalen, R. B. (2003). Seed 1-Cysteine peroxiredoxin antioxidants are not involved in dormancy, but contribute to inhibition of germination during seed stress. *Plant Physiology* **133**, 1148–1157.

Hite, D. R. C., Auh, C. and Scandalios, J. G. (1999). Catalase activity and hydrogen peroxide levels are inversely correlated in maize scutella during seed germination. *Redox Reports* **4**, 29–34.

Howitt, C. A. and Pogson, B. J. (2006). Carotenoid accumulation and function in seeds and non-green tissues. *Plant Cell and Environment* **29**, 435–445.

Ishiwatari, Y., Honda, C., Kawashima, I., Nakamura, S., Hirano, H., Mori, S., Fujiwara, T., Hayashi, H. and Chino, M. (1995). Thioredoxin *h* is one of the major proteins in rice phloem sap. *Planta* **195**, 456–463.

Ishiwatari, Y., Remoto, K., Fujiwara, T., Chino, M. and Hayashi, H. (2000). *In situ* hybridization study of the rice phloem thioredoxin *h* mRNA accumulation—Possible involvement in the differentiation of vascular tissues. *Physiologia Plantarum* **109**, 90–96.

Jacobsen, J. V., Pearce, D. W., Poole, A. T., Pharis, R. P. and Mander, L. N. (2002). Abscisic acid, phaseic acid and gibberellin contents associated with dormancy and germination in barley. *Plant Physiology* **115**, 428–441.

Jacquot, J. P., Rivera-Madrid, R., Marinho, P., Kollarova, M., Le Maréchal, P., Miginiac-Maslow, M. and Meyer, Y. (1994). *Arabidopsis thaliana* NADPH thioredoxin reductase. *Journal of Molecular Biology* **235**, 1357–1363.

Jacquot, J. P., Eklund, H., Rouhier, N. and Schürmann, P. (2009). Thioredoxin reductases in photosynthetic organisms. *Trends in Plant Science.* **14**, 336–343.

Kobrehel, K., Yee, B. C. and Buchanan, B. B. (1991). Role of the NADP/thioredoxin system in the reduction of alpha-amylase and trypsin inhibitor proteins. *Journal of Biological Chemistry* **266**, 16135–16140.

Kobrehel, K., Wong, J. H., Balogh, A., Kiss, F., Yee, B. C. and Buchanan, B. B. (1992). Specific reduction of wheat storage proteins by thioredoxin *h*. *Plant Physiology* **99**, 919–924.

Koh, C. S., Didierjean, C., Navrot, N., Panjikar, S., Mulliert, G., Jacquot, J. P., Rouhier, N., Aubry, A., Shawakataly, O. and Corbier, C. (2007). Crystal structures of poplar thioredoxin peroxidase which exhibits the structure of glutathione peroxidases: Insights into redox driven conformational changes. *Journal of Molecular Biology* **370**, 512–529.

Lee, K. O., Jang, H. H., Jung, B. G., Chi, Y. H., Lee, J. Y., Choi, Y. O., Lee, J. R., Lim, C. O., Cho, M. J. and Lee, S. Y. (2000). Rice 1 Cys-peroxiredoxin over-expressed in transgenic tobacco does not maintain dormancy but enhances antioxidant activity. *FEBS Letters* **486**, 103–106.

Leprince, O., Hendry, G. A. F. and McKersie, B. D. (1993). The mechanisms of desiccation tolerance in developing seeds. *Seed Science Research* **3**, 231–246.

Leprince, O., Atherton, N. M., Deltour, R. and Hendry, G. A. F. (1994). The involvement of respiration in free radical processes during loss of desiccation tolerance in germinating *Zea mays* L.. *Plant Physiology* **104**, 1333–1339.

Manevich, Y., Feinstein, S. I. and Fisher, A. B. (2004). Activation of the antioxidant enzyme 1-Cys peroxiredoxin requires glutathionylation mediated by hetero-dimerization with πGST. *Proceedings of the National Academy of Science of the United States of America* **101**, 3780–3785.

Meyer, Y., Reichheld, J. and Vignols, F. (2005). Thioredoxins in *Arabidopsis* and other plants. *Photosynthesis Research* **86**, 419–433.

Michalska, J., Zauber, H., Buchanan, B. B., Cejudo, F. J. and Geigenberger, P. (2009). NTRC links built-in thioredoxin to light and sucrose in regulating starch synthesis in chloroplasts and amyloplasts. *Proceedings of the National Academy of Science USA* **106**, 9908–9913.

Mittler, R., Vanderauwera, S., Gollery, M. and Van Breusegem, F. (2004). Reactive oxygen gene network of plants. *Trends in Plant Science* **9**, 490–498.

Moon, J. C., Jang, H. H., Chae, H. B., Lee, J. R., Lee, S. Y., Jung, Y. J., Shin, M. R., Lim, H. S., Chung, W. S., Yun, D.-J., Lee, K. O. and Lee, S. Y. (2006). The C-type *Arabidopsis* thioredoxin reductase ANTR-C acts as an electron donor to 2-Cys peroxiredoxins in chloroplasts. *Biochemical and Biophysical Research Communications* **348**, 478–484.

Morrison, I. N., Kuo, J. and O'Brien, T. P. (1975). Histochemistry and fine structure of developing wheat aleurone cells. *Planta* **123**, 105–116.

Muller, E. G. D. (1991). Thioredoxin deficiency in yeast prolongs S phase and shortens the G1 interval of the cell cycle. *Journal of Biological Chemistry* **266**, 9194–9202.

Navrot, N., Collin, V., Gualberto, J., Gelhaye, E., Hirasawa, M., Rey, P., Knaff, D. B., Issakidis, E., Jacquot, J. P. and Rouhier, N. (2006). Plant glutathione peroxidases are functional peroxiredoxins distributed in several subcellular compartments and regulated during biotic and abiotic stresses. *Plant Physiology* **142**, 1364–1379.

Nordlund, P. and Reichard, P. (2006). Ribonucleotide reductases. *Annual Review of Biochemistry* **75**, 681–706.

Nuruzzaman, M., Gupta, M., Zhang, C., Wang, L., Xie, W., Xiong, L., Zhang, Q. and Lian, X. (2008). Sequence and expression analysis of the thioredoxin protein gene family in rice. *Molecular Genetics and Genomics* **280**, 139–151.

Olsen, O.-A., Linnestad, C. and Nichols, S. E. (1999). Developmental biology of the cereal endosperm. *Trends in Plant Science* **4**, 253–257.

Op den Camp, R. G., Przybyla, D., Ochsenbein, C., Laloi, C., Kim, C., Danon, A., Wagner, D., Hideg, E., Gobel, C., Feussner, I., Nater, M. and Apel, K. (2003). Rapid induction of distinct stress responses after the release of singlet oxygen in *Arabidopsis*. *Plant Cell* **15**, 2320–2332.

Pedrajas, J. R., Miranda-Vizuete, A., Javanmardy, N., Gustafsson, J.-K. and Spyrou, G. (2000). Mitochondria of *Saccharomyces cerevisiae* contain one-conserved cysteine type peroxiredoxin with thioredoxin peroxidase activity. *Journal of Biological Chemistry* **21**, 16296–16301.

Pérez-Ruiz, J. M., Spinola, M. C., Kirchsteiger, K., Moreno, J., Sahrawy, M. and Cejudo, F. J. (2006). Rice NTRC is a high-efficiency redox system for chloroplast protection against oxidative damage. *Plant Cell* **18**, 2356–2368.

Pitzschke, A. C., Forzani, C. and Hirt, H. (2006). Reactive oxygen species signalling in plants. *Antioxidant Redox Signaling* **8**, 1757–1764.

Pulido, P., Cazalis, R. and Cejudo, F. J. (2009). An antioxidant redox system in the nucleus of wheat seed cells suffering oxidative stress. *Plant Journal* **57**, 132–145.

Ralat, L. A., Manevich, Y., Fisher, A. B. and Colman, R. F. (2006). Direct evidence for the formation of a complex between 1-cysteine peroxiredoxin and glutathione *S*-transferase pi with activity changes in both enzymes. *Biochemistry* **45**, 360–372.

Rey, P., Bécuwe, N., Barrault, M.-B., Rumeau, D., Havaux, M., Biteau, B. and Toledano, M. (2007). The *Arabidopsis thaliana* sulfiredoxin is a plastidic cysteine-sulfinic acid reductase involved in the photooxidative stress response. *Plant Journal* **49**, 505–514.

Rhee, S. G., Chae, H. Z. and Kim, K. (2005). Peroxiredoxins: A historical overview and speculative preview of novel mechanisms and emerging concepts in cell signaling. *Free Radical Biology and Medicine* **38**, 1543–1552.

Rouhier, N. and Jacquot, J. P. (2005). Plant peroxiredoxins: A multigenic family of thiol peroxidases. *Free Radicals in Biology and Medicine* **38**, 1413–1421.

Rouhier, N., Gelhaye, E. and Jacquot, J. P. (2002). Glutaredoxin dependent peroxiredoxin from poplar: Protein–protein interaction and catalytic mechanism. *Journal of Biological Chemistry* **277**, 13609–13614.

Rouhier, N., Gelhaye, E., Gualberto, J., Jordy, M. N., de Fay, E., Hirasawa, M., Duplessis, S., Lemaire, S., Frey, P., Martin, F., Manieri, W. Knaff, D. B. *et al.* (2004). Poplar peroxiredoxin Q: A thioredoxin-linked chloroplast antioxidant functional in pathogen defense. *Plant Physiology* **134**, 1027–1038.

Sabelli, P. A. and Larkins, B. A. (2009). The development of endosperm in grasses. *Plant Physiology* **149**, 14–26.

Sahrawy, M., Hecht, V., López-Jaramillo, J., Chueca, A., Chartier, Y. and Meyer, Y. (1996). Intron position as an evolutionary marker of thioredoxins and thioredoxin domains. *Journal of Molecular Evolution* **42**, 422–431.

Santandrea, G., Guo, Y., O'Connell, T. and Thompson, R. D. (2002). Post-phloem protein trafficking in the maize caryopsis: zmTRXh1, a thioredoxin specifically expressed in the pedicel parenchyma of *Zea mays* L., is found predominantly in the placentochalaza. *Plant Molecular Biology* **50**, 743–756.

Sattler, S. E., Gilliland, L. U., Magallanes-Lundback, M., Pollard, M. and DellaPenna, D. (2004). Vitamin E is essential for seed longevity and for preventing lipid peroxidation during germination. *Plant Cell* **16**, 1419–1432.

Schürmann, P. and Buchanan, B. B. (2008). The ferredoxin/thioredoxin system of oxygenic photosynthesis. *Antioxidant Redox Signaling* **10**, 1235–1273.

Schürmann, P. and Jacquot, J. P. (2000). Plant thioredoxin systems revisited. *Annual Review of Plant Physiology and Plant Molecular Biology* **51**, 371–400.

Serrato, A. J. and Cejudo, F. J. (2003). Type-*h* thioredoxins accumulate in the nucleus of developing wheat seed tissues suffering oxidative stress. *Planta* **217**, 392–399.

Serrato, A. J., Crespo, J. L., Florencio, F. J. and Cejudo, F. J. (2001). Characterization of two thioredoxins *h* with predominant localization in the nucleus of aleurone and scutellum cells of germinating wheat seeds. *Plant Molecular Biology* **46**, 361–371.

Serrato, A. J., Pérez-Ruiz, J. M. and Cejudo, F. J. (2002). Cloning of thioredoxin *h* reductase and characterization of the thioredoxin reductase-thioredoxin *h* system from wheat. *Biochemical Journal* **367**, 491–497.

Serrato, A., Pérez-Ruiz, J. M., Spinola, M. C. and Cejudo, F. J. (2004). A novel NADPH thioredoxin reductase, localized in the chloroplast, which deficiency causes hypersensitivity to abiotic stress in *Arabidopsis thaliana. Journal of Biological Chemistry* **279**, 43821–43827.

Shahpiri, A., Svenson, B. and Finnie, C. (2008). The NADPH-dependent thioredoxin reductase/thioredoxin system in germinating barley seeds: gene expression, protein profiles, and interactions between isoforms of thioredoxin *h* and thioredoxin reductase. *Plant Physiology* **146**, 789–799.

Stacy, R. A. P., Munthe, E., Steinum, T., Sharma, B. and Aalen, R. (1996). A peroxiredoxin antioxidant is encoded by a dormancy-related gene, *Per1*, expressed during late development in the aleurone and embryo of barley grains. *Plant Molecular Biology* **31**, 1205–1216.

Stacy, R. A P., Nordeng, T. W., Culiáñez-Maciá, F. A. and Aalen, R. (1999). The dormancy-related peroxiredoxin anti-oxidant, PER1, is localized to the nucleus of barley embryo and aleurone cells. *Plant Journal* **19**, 1–8.

Sueoka, K., Yamazaki, T., Hiyama, T. and Nakamoto, H. (2009). The NADPH Thioredoxin reductase C as an electron donor to 2-Cys peroxiredoxin in a thermophylic cyanobacterium *Thermosynechococcus elongatus* BP-1. *Biochemical and Biophysical Research Communications* **380**, 520–524.

Tamura, T. and Stadtman, T. C. (1996). A new selenoprotein from human lung adenocarcinoma cells: Purification, properties, and thioredoxin reductase activity. *Proceedings of the National Academy of Science of the United States of America* **93**, 1006–1011.

Thorne, J. H. (1985). Phloem unloading of C and N assimilates in developing seeds. *Annual Review of Plant Physiology* **36**, 317–343.

Vandenabeele, S., Van der Kelen, K., Dat, J., Gadjev, I., Bonefaes, T., Morsa, S., Rttiers, P., Sloten, L., Van Montagu, M., Zabeau, M., Inzé, D. and Van Breusegem, F. (2003). A comprehensive analysis of hydrogen peroxide-induced gene expression in tobacco. *Proceedings of the National Academy of Science of the United States of America* **100**, 16113–16118.

Vanderauwera, S., Zimmermann, P., Rombauts, S., Vandenabeele, S., Langebartels, C., Gruissem, W., Inzé, D. and Van Breusegem, F. (2005). Genome-wide analysis of hydrogen peroxide-regulated gene expression in *Arabidopsis* reveals a high light-induced transcriptional cluster involved in anthocyanin biosynthesis. *Plant Physiology* **139**, 806–821.

Wang, H. L., Offler, C. E. and Patrick, J. W. (1995). The cellular pathway of photosynthate transfer in the developing wheat grain. III. A structural analysis and physiological studies of the pathway from the endosperm cavity to the starchy endosperm. *Plant Cell and Environment* **18**, 389–407.

Wangensteen, O. S., Chueca, A., Hirasawa, M., Sahrawy, M., Knaff, D. B. and López Gorgé, J. (2001). Binding features of chloroplast fructose-1,6-bisphosphatase-thioredoxin interaction. *Biochimica et Biophysica Acta* **1547**, 156–166.

Wong, J. H., Kim, Y.-B., Ren, P.-H., Cho, M.-J., Hedden, P., Lemaux, P. G. and Buchanan, B. B. (2002). Transgenic barley grain overexpressing thioredoxin shows evidence that the starchy endosperm communicates with the embryo and the aleurone. *Proceedings of the National Academy of Science of the United States of America* **99**, 16325–16330.

Wong, J. H., Balmer, Y., Cai, N., Tanaka, C. K., Vensel, W. H., Hurkman, W. J. and Buchanan, B. B. (2003). Unraveling thioredoxin-linked metabolic processes of cereal starchy endosperm using proteomics. *FEBS Letters* **547**, 151–156.

Wood, Z. A., Schröder, E., Harris, J. R. and Poole, L. B. (2003). Structure, mechanism and regulation of peroxiredoxins. *Trends Biochemical Science* **28**, 32–40.

Young, T. E. and Gallie, D. R. (1999). Analysis of programmed cell death in wheat endosperm reveals differences in endosperm development between cereals. *Plant Molecular Biology* **39**, 915–926.

Young, T. E. and Gallie, D. R. (2000). Regulation of programmed cell death in maize endosperm by abscisic acid. *Plant Molecular Biology* **42**, 397–414.

Molecular Recognition in NADPH-Dependent Plant Thioredoxin Systems—Catalytic Mechanisms, Structural Snapshots and Target Identifications

PER HÄGGLUND,*[,1] KRISTINE GROTH KIRKENSGAARD,* KENJI MAEDA,* CHRISTINE FINNIE,* ANETTE HENRIKSEN[†] AND BIRTE SVENSSON*

**Enzyme and Protein Chemistry, Department of Systems Biology, Technical University of Denmark, DK-2800 Kgs. Lyngby, Denmark*
[†]*Protein Chemistry Group, Carlsberg Laboratory, Gamle Carlsberg Vej 10, DK-2500 Valby, Denmark*

[1]Corresponding author: Email: ph@bio.dtu.dk

Advances in Botanical Research, Vol. 52

0065-2296/09 $35.00
DOI: 10.1016/S0065-2296(09)52015-0

ABSTRACT

NADPH-dependent thioredoxin systems (NTS) control enzymatic activities and provide reducing equivalents to metabolic pathways in all types of organisms, from bacteria to mammals. In these redox systems, thioredoxin reduces disulfide bonds in target proteins and receives electrons from NADPH via thioredoxin reductase (NTR). Plant NTS were first discovered in wheat seeds some 30 years ago and were demonstrated to play a key role in the seed germination process. Since then, NTS have been identified in a large variety of photosynthetic organisms, and an organelle-specific pattern for their cellular localization is established. The last decade has witnessed a remarkable expansion of the knowledge about these systems and novel molecular architectures, catalytic mechanisms and target proteins have been revealed. In general, these findings have provided a wealth of information about the physiological role and molecular mechanisms of plant NTS, and this chapter will highlight some of the recent developments in this area.

ABBREVIATIONS

BASI	barley α-amylase/subtilisin inhibitor
DHAR	dehydroascorbate reductase
FO	flavin-oxidizing
FR	flavin-reducing
Grx	glutaredoxin
GR	glutathione reductase
GSH	glutathione
IAM	iodoacetamide
ICAT	isotope-coded affinity tags
mBBr	monobromobimane
MSR	methionine sulfoxide reductase
NTR	NADPH-dependent thioredoxin reductase
NTS	NADPH-dependent thioredoxin systems
TCEP	tris(2-carboxyethyl)phosphine
TNB	2-nitro-5-thiobenzoate
Trx	thioredoxin
4-VP	4-vinylpyridine

I. INTRODUCTION

Protein thiol groups are highly reactive and may undergo oxidative modifications under stress conditions. Furthermore, reversible thiol modifications are utilized for electron transfer in a range of enzymatic processes and redox signalling pathways. Therefore, control of intracellular thiol oxidation is essential and in general maintained by the NADPH-dependent thioredoxin (Trx) and glutaredoxin (Grx) systems present in all kingdoms of life (Holmgren, 1989). Trx and Grx reduce disulfides in target proteins and receive reducing equivalents from NADPH via the flavoenzyme thioredoxin reductase (NTR) and glutathione reductase (GR), respectively. Trx is reduced directly by NTR, whereas GR donates electrons to the tripeptide glutathione (GSH), which in turn reduces Grx (Fig. 1). Examples of cross-talk between these NADPH-dependent systems include reduction of Trx by Grx (Gelhaye *et al.*, 2003; Koh *et al.*, 2008), Grx by NTR (Johansson *et al.*, 2004) and GSH by Trx (Kanzok *et al.*, 2001). The physiological significance of these overlapping interactions is underpinned by the non-viable phenotype of *Escherichia coli* and *Saccharomyces cerevisiae* double knock-out mutants lacking both systems, while single knock-out mutants are viable (Muller, 1996; Prinz *et al.*, 1997). Trx and Grx play key roles in many cellular processes, and targets can be divided into three categories: (i) enzymes, such as ribonucleotide reductase and methionine sulfoxide reductase (MSR) that depend on Trx/Grx as suppliers of reducing equivalents, (ii) redox-regulated transcription factors and enzymes (e.g. chloroplastic malate dehydrogenase) that are activated/deactivated through disulfide reduction and (iii) proteins such as T7 DNA polymerase that incorporate Trx/Grx as subunits of protein complexes (Arnér and Holmgren, 2000).

Plants have evolved advanced redox control systems that are remarkable in terms of complexity and diversity. For example, *Arabidopsis thaliana* contains more than 20 Trx genes that are grouped into categories (Trx-f,-m,-h,-o,-x,-y and -s) based on sequence similarity (Meyer *et al.*, 2002). A similar range of Trx genes has been found in other model organisms, such as rice (Nuruzzaman *et al.*, 2008). Trx-f and Trx-m are the most well-characterized categories of plant Trxs, and play important roles in the

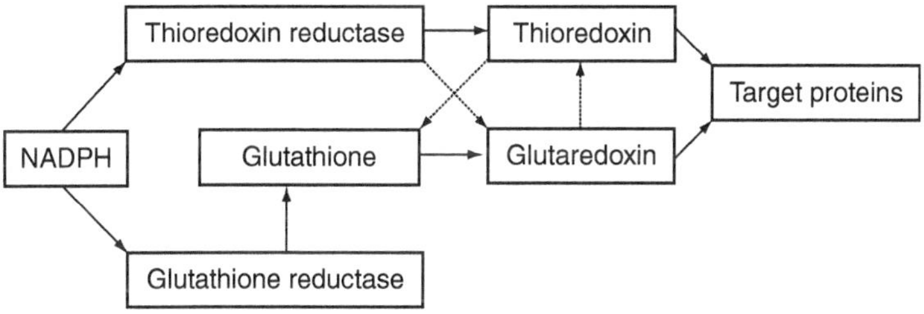

Fig. 1. NADPH-dependent redox systems. Solid arrows represent the general pathways for electron transfer and dotted arrows represent alternative pathways for reduction of Trx by Grx (Gelhaye *et al.*, 2003; Koh *et al.*, 2008), Grx by NTR (Johansson *et al.*, 2004) and GSH by Trx (Kanzok *et al.*, 2001).

regulation of key enzymes in the Calvin cycle and pentose phosphate pathway (Buchanan and Balmer, 2005). These chloroplastic Trxs are reduced by light via ferredoxin and ferredoxin:thioredoxin reductase in a unique thioredoxin system recently reviewed elsewhere (Schürmann and Buchanan, 2008). This chapter addresses the current knowledge of catalytic mechanisms and molecular recognition in NADPH-dependent thioredoxin systems (NTS) from plants with focus on recent data from proteome analyses and structural investigations. A very recent review has summarized data concerning NTR in plants (Jacquot *et al.*, 2009).

II. COMPONENTS OF NADPH-DEPENDENT Trx SYSTEMS IN PLANTS

A. Trx

1. Trx-h

Trx-h was the first NTR-dependent Trx to be discovered in plants and its name derives from its heterotrophic origin as opposed to the previously characterized chloroplastic Trxs (Johnson *et al.*, 1987; Suske *et al.*, 1979). Trx-h comprises the largest and most diverse group of plant Trxs. For example, *A. thaliana* contains at least eight Trx-h isoforms. The h isoforms show differential time and tissue-specific expression patterns (Cazalis *et al.*, 2006; Reichheld *et al.*, 2002). Although h-type Trxs are considered to be mainly cytosolic, some Trx-h are apparently translocated to other compartments (Fig. 2). For example, Trx-h has been isolated from the nucleus of

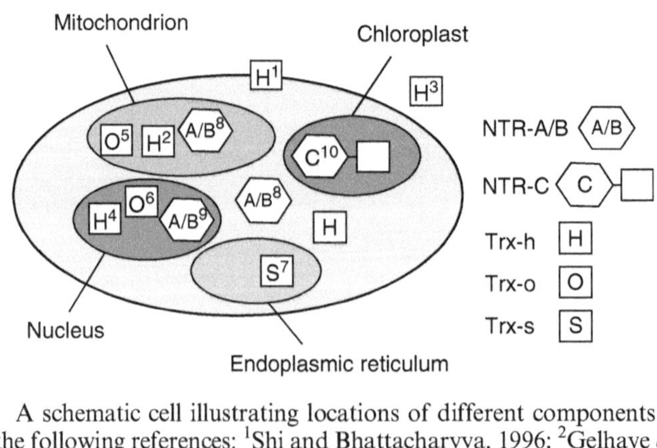

Fig. 2. A schematic cell illustrating locations of different components from NTS based on the following references: [1]Shi and Bhattacharyya, 1996; [2]Gelhaye *et al.*, 2004; [3]Juárez-Díaz *et al.*, 2006; [4]Serrato *et al.*, 2001; [5]Laloi *et al.*, 2001; [6]Marti *et al.*, 2009; [7]Alkhalfioui *et al.*, 2008; [8]Reichheld *et al.*, 2005; [9]Pulido *et al.*, 2009; [10]Serrato *et al.*, 2004.

cells in developing wheat seeds (Serrato and Cejudo, 2003; Serrato *et al.*, 2001). Furthermore, Trx-h transport to phloem sieve tubes is shown in many plant species, including rice and maize (Ishiwatari *et al.*, 1995, 1998; Santandrea *et al.*, 2002). Trx-h can be subdivided into three groups based on sequence similarity (Gelhaye *et al.*, 2005). In short, subgroup I contains several well-characterized cytosolic Trxs, and some members harbour a RKDD motif implicated in cell-to-cell transport (Fig. 3). Subgroup II contains proteins with N-terminal extensions that are translocated to mitochondria, the plasma membrane, or the extracellular matrix (Gelhaye *et al.*, 2002, 2004; Juárez-Díaz *et al.*, 2006; Shi and Bhattacharyya, 1996). Trx-h from subgroup III are exceptional as they are reduced by the GSH/Grx system and not by NTR (Gelhaye *et al.*, 2003; Juttner *et al.*, 2000; Koh *et al.*, 2008).

The first physiological functions ascribed to Trx-h were related to the germination of cereal seeds. Trx-mediated disulfide bond reduction was demonstrated to facilitate germination by (i) inactivating small proteinaceous inhibitors of proteolytic and amylolytic enzymes, (ii) activating hydrolytic enzymes, such as thiocalsin and pullulanase and (iii) enhancing the solubility of storage proteins (Besse *et al.*, 1996; Kobrehel *et al.*, 1991, 1992). Indeed, overexpression of Trx-h in barley seeds results in an accelerated germination rate, an increase in gibberellic acid concentration, and an increased α-amylase release from the aleurone layer (Cho *et al.*, 1999; Wong *et al.*, 2002). Trx-h isoforms display different temporal and spatial distributions in cereal seed tissues indicating that they may have different roles during seed germination (Cazalis *et al.*, 2006; Maeda *et al.*, 2003; Shahpiri *et al.*, 2008).

The role of Trx-h in oxidative stress resistance has been investigated in complementation studies with yeast mutants deprived of endogenous Trxs (Bréhélin *et al.*, 2000; Mouaheb *et al.*, 1998). It was concluded that Trx-h acts as an electron donor for MSR as judged by the ability of several *A. thaliana* Trx-h to restore growth on methionine sulfoxide as sole sulphur source. This was confirmed in a later study that described the mode of regeneration of plant MSR of the A type (Rouhier *et al.*, 2007). The *A. thaliana* Trx-h isoforms appear to have at least partially non-redundant functions as two isoforms, *At*Trxh2 and *At*Trxh3, confer a different extent of tolerance to hydrogen peroxide (Bréhélin *et al.*, 2000; Mouaheb *et al.*, 1998). In this context, it is interesting to note that *At*Trxh3 has been demonstrated to interact directly with peroxiredoxin *in vivo* (Verdoucq *et al.*, 1999) and also *in vitro* (Rouhier *et al.*, 2002). In addition, Trx-h plays a key role in pollen self-incompatibility by inhibiting an S-locus receptor kinase in *Brassica oleracea* (Bower *et al.*, 1996; Cabrillac *et al.*,

```
                                                                                                            .   :   .          :
h-typeI_AtTrxh1    -----------------------------------------------MA-------------------------------SEEGQVIACHTVETWNEQLQKANESKTLVV   32
h-typeI_AtTrxh3    -----------------------------------------------MAA--------------------------------EGEVIACHTVEDWTEKLKAANESKKLIV   31
h-typeI_AtTrxh4    -----------------------------------------------MA-------------------------------AEEGQVIGCHTNDVWTVQLDKAKESNKLIV   32
h-typeI_AtTrxh5    -----------------------------------------------MAG--------------------------------EGEVIACHTLEVWNEKVKDANESKKLIV   31
h-typeI_HvTrxh2    -----------------------------------------------MAASATA-------------------------AAVAAEVISVHSLEQWTMQIEEANTAKKLVV   38
h-typeI_OsTrxh1    -----------------------------------------------MA-------------------------------AEEGVVIACHNKDEFDAQMTKAKEAGKVVI   32
h-typeI_PsTrxh1    -----------------------------------------------MAGS-----------------------------SEEGQVISCHSVDAWNDILHRGNESKKLIV   34
h-typeI_PtTrxh1    -----------------------------------------------MA--------------------------------EEGQVIACHTVDVWKEQFEKGKGTQKLIV   31
h-typeI_TaTrxh1    -----------------------------------------------MA-------------------------------AEEGAVIACHTKQEFDTHMANGKETGKLVI   32
h-typeII_AtTrxh2   -----------------------------------------------MGGALST---------VFGSGEDATAAG---TESEPSRVLKFSSSARWQLHFNEIKESNKLLV   51
h-typeII_AtTrxh7   -----------------------------------------------MGSNVSS-------VHDVHSSMEITS--------NG-FVVEIESRRQWKSLFDSMKGSNKLLV   47
h-typeII_AtTrxh8   -----------------------------------------------MGANVSTPDQRFQVTHFRSTKPWTPRPEIYPFKVNSPCIVEIKNMNQWKSRLNALKDTNKLLV   63
h-typeII_PsTrxh3   -----------------------------------------------MA--------------------------------EEGQVIGVHTVDAWKEQLEKGKASKKLIV   31
h-typeIII_AtTrxh9  -----------------------------------------------MGSCVSK-------------GKGD-DDSVHNVEFSGGNVHLITTKESWDDKLAEADRDGKIVV   49
h-typeIII_PtTrxh4  -----------------------------------------------MGLCLAK-------------RNHDADDDEPHIELAGGNVHLITTKERWDQKLSEASRDGKIVL   50
h-typeIII_AtCXXS1  ----------------------------------------------------------------------------------MARVVKIDSAESWNFYVSQAKNQNCPIV   28
h-typeIII_PtCXXS3  --------------------------------------------------------------------------MESQEQQPKSRVIKVESVESWDFYITQATNQACPIV   36
o-type_AtTrxo1     MKGNWSIVRKVLHRQFSTLRSSTPSSRLSTSIRPLVLAPNSISSLIARNSLFTASNIGPSIDFNFSNTSLPHRRSLCSEAGGENGVVLVKSEEEFINAMSKAQDGSLPSV  110
o-type_AtTrxo2     MKSQWSNFHQ-----------------------------------IGRNSFLAASTVYVSNEFNFLNTSLLNRRSFCAEGDRSSFVVLKSEAEFNSALSKARDGSLPSV   75
s-type_MtTrxs1     -----------------------------------------MTTVTITMSSFIFIILFHTLSASMATAQLLN-----AATQDASNGVAYVTDENFGSLVLNSET---LVL   61
s-type_MtTrxs2     -----------------------------------------MATVTLAMSSFIILILFHTLSSSMATVQLESLPSYLLSAADASDGVAPVTDETFGSFVPMSKN---LVL   66
                   1.......10........20........30........40........50........60........70........80........90........100.......110

                     *    *   .  :       :           :: *                *  . :.: *     :.*
h-typeI_AtTrxh1    VDFTASWCGP-CRFIAPFFADLAKKLPN-VLFLKVDTDE--LKSVASDWAIQAMPTFMFLKEGKILDKVVGAKKDELQSTIAKHLA---------  114
h-typeI_AtTrxh3    IDFTATWCPP-CRFIAPVFADLAKKHLD-VVFFKVDVDE--LNTVAEEFKVQAMPTFIFMKEGEIKETVVGAAKEEIIANLEKHKTV---VAAA-  118
h-typeI_AtTrxh4    IDFTASWCPP-CRMIAPIFNDLAKKFMSSAIFFKVDVDE--LQSVAKEFGVEAMPTFVFIKAGEVVDKLVGANKEDLQAKIVKHTG----VTTA-  119
h-typeI_AtTrxh5    IDFTASWCPP-CRFIAPVFAEMAKKFTN-VVFFKIDVDE--LQAVAQEFKVEAMPTFVFMKEGNIIDRVVGAAKDEINEKLMKHGGL---VASA-  118
h-typeI_HvTrxh2    IDFTASWCGP-CRIMAPVFADLAKKFPN-AVFLKVDVDE--LKPIAEQFSVEAMPTFLFMKEGDVKDRVVGAIKEELTAKVGLH------AAAQ-  122
h-typeI_OsTrxh1    IDFTASWCGP-CRFIAPVFAEYAKKFPG-AVFLKVDVDE--LKEVAEKYNVEAMPTFLFIKDGAEADKVVGARKDDLQNTIVKHVGATAASASA-  122
h-typeI_PsTrxh1    VDFTASWCGP-CRFIAPFLGELAKKFTN-VIFLKVDVDE--LKSVAQDWAVEAMPTFVFVKEGTILGKVVGAKKEELQQTIERHVAS----SNA-  120
h-typeI_PtTrxh1    VDFTASWCPP-CKFIAPVFADLAKKFTN-VTFLKVDVDE--LKPVAAEWEVEAMPTFIFLKDGKLVDKIVGADKDGLPALVEKHSVY----TA--  116
h-typeI_TaTrxh1    IDFTASWCGP-CRVIAPVFAEYAKKFPG-AIFLKVDVDE--LKDVAEAYNVEAMPTFLFIKDGAKVDTVVGGRKDDIHTKIVALMG----SASA-  118
h-typeII_AtTrxh2   VDFSASWCGP-CRMIEPAIHAMADKFND-VDFVKLDVDE--LPDVAKEFNVTAMPTFVLVKRGKEIERIIGAKKDELEKKVSKLRA---------  133
h-typeII_AtTrxh7   IDFTAVWCGP-CKAMEPRVREIASKYSE-AVFARVDVDR--LMDVAGTYRAITLPAFVFVKRGEEIDRVVGAKPDELVKKIEQHRV---------  129
h-typeII_AtTrxh8   IEFTAKWCGP-CKTLEPKLEELAAKYTD-VEFVKIDVDV--LMSVWMEFNLSTLPAIVFMKRGREVDMVVGVKVDELERKLNKYTQSFF------  148
h-typeII_PsTrxh3   VDFTASWCGP-CRFIAPILAEIAKKLTH-VTFLKVDVDE--LKTVSEEWGIEAMPTFLFLKDGELVDKVVGAKKEELQLKIDKHAA---------  113
h-typeIII_AtTrxh9  ANFSATWCGP-CKIVAPFFIELSEKHSS-LMFLLVDVDE--LSDFSSSWDIKATPTFFFLKNGQQIGKLVGANKPELQKKVTSIIDSVPESPQRP  140
h-typeIII_PtTrxh4  ANFSARWCGP-CKQIAPYYIELSENYPS-LMFLVIDVDE--LSDFSASWEIKATPTFFFLRDGQQVDKLVGANKPELHKKITAILDSLPPSDK--  139
h-typeIII_AtCXXS1  AHFTALWCIP-SVFMNSFFEELAFNYKD-ALFLIVDVDE--VKEVASQLEVKAMPTFLFLKDGNAMDKLVGANPDEIKKRVDGFVQSSRVVHIA-  118
h-typeIII_PtCXXS3  VHFTALWCMP-SVAMNPVFEELASAYPD-GLFLIVDVDE--VKEVATKMEVKAMPTFVLMKDGAQVDKIVGANPEEIRKRIDGFVQSIRAYNA--  125
o-type_AtTrxo1     FYFTAAWCGP-CRFISPVIVELSKQYPD-VTTYKVDIDEGGISNTISKLNITAVPTLHFFKGGSKKGEVVGADVTKLKNLMEQLYK---------  194
o-type_AtTrxo2     FYFTAAWCGP-CRLISPVILELSNKYPD-VTTYKVDIDEGGLSNAIGKLNVSAVPTLQFFKGGVKKAEIVGVDVVRLKSVMEQLYK---------  159
s-type_MtTrxs1     VEFFAPLCSP-CKNVDFKIVELANEYAGEVEFFKLNVDD--NQLIPSKYGIKGIPNVLIFKNGEQRDTLFGNLPKATFIKRMEQNL---------  144
s-type_MtTrxs2     VEFYNPWCGQNCKNIHSIMVELANDYAGKVDFYKLNIDE--NPYITNRYVIQDLPTVVFIKYGMQRDRLVGDVPKATFIELIQLSI---------  150
                   .......120.......130.......140.......150.......160.......170.......180.......190.......200.....
```

Fig. 3. Multiple sequence alignment of Trx-h, Trx-o and Trx-s. The catalytic cysteines in the CXXC motif and a third cysteine implicated in the catalytic mechanism of class III h-type Trxs are highlighted in light and dark gray (yellow and red in the web version), respectively. Residues aligned with barley *Hv*Trxh2 M88 and A106 implicated in target recognition are in dark gray (green in the web version). The RKDD motif implicated in cell-to-cell transport is highlighted in mid gray (orange in the web version). *At*Trxh1 (P29448); *At*Trxh3 (Q42403); *At*Trxh4 (Q39239); *At*Trxh5 (Q39241); *Hv*Trxh2 (Q7XZK2); *Os*Trxh1(Q0D840); *Ps*Trxh1 (Q9AR82); *Pt*Trxh1 (B9GRI3)); *Ta*Trxh1 (Q8GVD3); *At*Trxh2 (Q38879); *At*Trxh7 (NP_176182); *At*Trxh8 (NP_177146); *Ps*Trxh3 (Q8GUR9); *At*Trxh9 (Q9C9Y6); *Pt*Trxh4 (P85801); *At*CXXS1 (NP_172620); *Pt*CXXS3 (EEE85494); *At*Trxo1 (AAC12840); *At*Trxo2 (AF396650); *Mt*Trxs1 (A9RAA5); *Mt*Trxs2 (A9RAA6). The sequences were aligned using CLUSTAL-W (Thompson *et al.*, 1994).

2001). Other processes associated with Trx-h include sulphate assimilation (Bréhélin *et al.*, 2000), cell cycle regulation (Mouaheb *et al.*, 1998), DNA damage repair (Sarkar *et al.*, 2005) and pathogen interactions (Sweat and Wolpert, 2007).

2. Trx-o and Trx-s

The o-type Trxs have N-terminal transit peptides (Fig. 3), and *At*Trxo1 from *A. thaliana* has been localized in mitochondria (Laloi *et al.*, 2001), while *Ps*Trxo1 from garden pea (*Pisum sativum*) was recently reported to be present in the nucleus (Marti *et al.*, 2009). So far the physiological importance of Trx-o is not well defined, but it has been shown to interact with the mitochondria-specific peroxiredoxin PrxIIF (Barranco-Medina *et al.*, 2008) and to activate the alternative oxidase that acts as an electron acceptor in the electron transport chain under oxidative stress conditions (Marti *et al.*, 2009).

Recently, two Trx sequences that did not match the hitherto described Trx types were isolated from *Medicago truncatula* (Alkhalfioui *et al.*, 2008). The genes contained N-terminal signal peptides (Fig. 3), and it was shown that the proteins were translocated to the endoplasmic reticulum (Fig. 2). The proteins appear to be specifically expressed in *M. truncatula* grown in symbiosis with the nitrogen-fixing bacterium *Sinorhizobium meliloti* and were assigned to a new Trx type, coined as Trx-s. These proteins seem to be unique to *M. truncatula* since no orthologs were identified in other plant species (Alkhalfioui *et al.*, 2008). Notably, only one of the two proteins (*Mt*Trxs1) is reduced by *M. truncatula* NTR and both proteins lack disulfide reductase activity in the insulin-based Trx assay (Holmgren, 1979).

B. NTR

1. NTR-A/B

Even though NTS in plants were described 30 years ago, it was not until 1994 that the first gene encoding a plant NTR was isolated from *A. thaliana* (Jacquot *et al.*, 1994). The protein was named NTR-B, and later the highly similar NTR-A was isolated from mitochondrial fractions (Laloi *et al.*, 2001). The *A. thaliana* genes encoding NTR-A and NTR-B are transcribed in short and long versions, and the products of the longer transcripts are translocated to mitochondria. It has been concluded that NTR-B is the major mitochondrial form, while NTR-A is most abundant in the cytoplasm (Reichheld *et al.*, 2005). Remarkably, *ntra ntrb* knock-outs in *A. thaliana* are viable and fertile but hypersensitive to buthionine sulfoximine, an inhibitor of glutathione biosynthesis (Reichheld *et al.*, 2007). Highly similar NTR-A and NTR-B orthologs have also been identified in several other plant species, for example, barley (Shahpiri *et al.*, 2008) and wheat (Serrato *et al.*, 2002). Wheat NTR has been demonstrated to co-localize with Trx-h in the nucleus (Pulido *et al.*, 2009).

Enzyme kinetics analyses suggest that both NTR-A and NTR-B are efficient electron donors for Trx-h and Trx-o (Laloi *et al.*, 2001). Furthermore, NTR-B shows similar affinity for four out of five tested *A. thaliana* Trx-h isoforms, and two forms of barley Trx-h (*Hv*Trxh1 and *Hv*Trxh2) are reduced at a comparable rate by the two endogenous NTR-A/B orthologs, *Hv*NTR1 and *Hv*NTR2 (Rivera-Madrid *et al.*, 1995; Shahpiri *et al.*, 2008). Taken together, these results could suggest that NTRs have rather broad specificities for Trxs or that the variations between NTR-A/B and Trx isoforms within a species are relatively conservative (Table I). However, cross-species comparisons reveal that interactions between NTRs and Trxs are highly species-dependent and that NTR–Trx interactions are specific (Table I). For example, barley NTR has a hundred-fold reduced affinity (increased K_M) for *E. coli* Trx when compared to endogenous Trx-h (Shahpiri *et al.*, 2008). Besides the capacity to transfer electrons to oxidized Trx, NTR has been demonstrated to activate 1-Cys peroxiredoxin in a Trx-independent manner (Pulido *et al.*, 2009).

2. *NTR-C*

NTR-C was recently isolated from rice and *A. thaliana* (Serrato *et al.*, 2004). This protein has a different architecture, with an N-terminal domain similar to NTR-A/B and a Trx-like domain in a C-terminal extension. Noticeably, this molecular architecture seems to be found mainly among plants, but recurs in the pathogenic bacterium *Mycobacterium leprae* (Wieles *et al.*, 1995a,b). NTR-C is localized in chloroplasts and acts as an efficient electron donor to 2-Cys peroxiredoxin (Alkhalfioui *et al.*, 2007a; Moon *et al.*, 2006). If either the NTR or Trx domain is removed, the capacity to reduce 2-Cys peroxiredoxin is lost indicating that NTR-C functions as a complete NTS in a single polypeptide chain. In *A. thaliana*, *ntrc* knock-outs show severe growth inhibition, oxidative stress sensitivity, perturbed chlorophyll biosynthesis and lower rate of photosynthesis (Pérez-Ruiz *et al.*, 2006; Serrato *et al.*, 2004; Stenbaek *et al.*, 2008).

TABLE I

Interactions between NTRs and Trxs from different sources measured as K_M in enzyme kinetics assays with Trxs in varying concentrations as substrates for NTRs

NTR	Trx	K_M (μM)
A. thaliana NTRA[a]	*A. thaliana* Trxo1	2.6
A. thaliana NTRA[a]	*A. thaliana* Trxo2	1.8
A. thaliana NTRA[a]	*A. thaliana* Trxh3	2.8
A. thaliana NTRB[a]	*A. thaliana* Trxo1	2.2
A. thaliana NTRB[a]	*A. thaliana* Trxo2	2.1
A. thaliana NTRB[a]	*A. thaliana* Trxh3	3.0
A. thaliana NTRB[b]	*A. thaliana* Trxh1	2.0
A. thaliana NTRB[b]	*A. thaliana* Trxh2	0.8
A. thaliana NTRB[b]	*A. thaliana* Trxh3	0.7
A. thaliana NTRB[b]	*A. thaliana* Trxh4	20
A. thaliana NTRB[b]	*A. thaliana* Trxh5	1.6
A. thaliana NTRB[c]	*C. reinhardtii* Trx-h	1.6
A. thaliana NTRB[c]	*C. reinhardtii* Trx-m	5.8
A. thaliana NTRB[c]	*E. coli* Trx	81
H. vulgare NTR1[d]	*H. vulgare* Trxh1	1.2
H. vulgare NTR1[d]	*H. vulgare* Trxh2	1.8
H. vulgare NTR2[d]	*H. vulgare* Trxh1	1.1
H. vulgare NTR2[d]	*H. vulgare* Trxh2	1.3
H. vulgare NTR2[d]	*E. coli* Trx	107
A. thaliana NTRB[d]	*H. vulgare* Trxh1	25
A. thaliana NTRB[d]	*H. vulgare* Trxh2	27
T. aestivum NTR[e]	*T. aestivum* Trx-h	7.6
T. aestivum NTR[e]	*E. coli* Trx	36

[a]Laloi *et al.* (2001)
[b]Rivera-Madrid *et al.* (1995)
[c]Jacquot *et al.* (1994)
[d]Shahpiri *et al.* (2008)
[e]Serrato *et al.* (2002)

III. STRUCTURAL SNAPSHOTS AND CATALYTIC MECHANISMS

A. Trx

1. *Catalytic mechanisms*

Trx reduces protein disulfides through a dithiol/disulfide exchange reaction involving two thiol groups in a conserved C_NXXC_C motif (Figs. 3 and 4). The cysteine residue at the N-terminal end of this motif (C_N) has a perturbed, low thiol-pK_a value and the deprotonated thiolate form acts as a nucleophile and attacks a protein disulfide to transiently form an intermolecular disulfide bond (Kallis and Holmgren, 1980). The thiol group of the other active-site

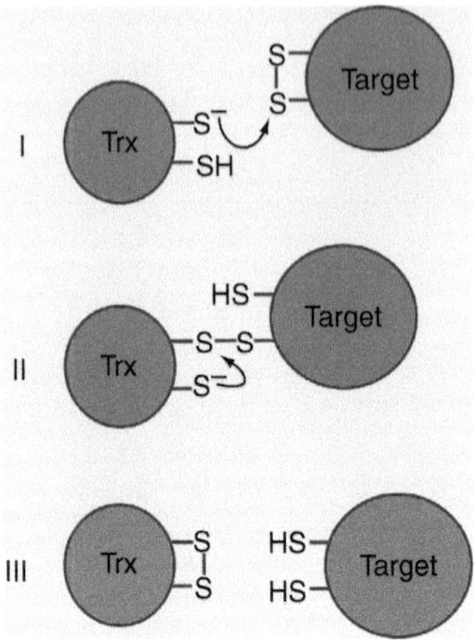

Fig. 4. Catalytic mechanism of Trx. (i) The N-terminal cysteine (C_N) in the Trx redox-active C_NXXC_C motif makes a nucleophilic attack on the target disulfide bond and forms an intermolecular Trx-target disulfide intermediate. (ii) The disulfide is attacked by the C-terminal cysteine (C_C). (iii) The reduced target protein and oxidized Trx is released.

cysteine residue (C_C) subsequently attacks this disulfide bond, resulting in the release of the reduced target protein and oxidized Trx. An aspartic acid residue (D26 in *E. coli* Trx) is proposed to act as a general acid/base catalyst for the protonation/deprotonation of the C_C thiol group during Trx oxidoreduction (Chivers and Raines, 1997; Menchise *et al.*, 2001). A tryptophan residue at the immediate N-terminal side of the CXXC motif (Fig. 3) is conserved, and also suggested to be important for efficient catalysis (Krause and Holmgren, 1991; Krimm *et al.*, 1998).

In most plant Trxs, the C_NXXC_C motif is CGPC, but in some h-type Trxs from subgroup I it is replaced by CPPC (Fig. 3). This substitution of glycine with proline appears not to change the redox potential or the overall protein disulfide reductase activity of Trx (Behm and Jacquot, 2000; Bréhélin *et al.*, 2004; Rivera-Madrid *et al.*, 1995), but it nevertheless influences certain biological functions, such as sulphate assimilation (Bréhélin *et al.*, 2000). In the recently described s-type Trx from *M. truncatula*, the active-site motif is either CSPC (*Mt*Trxs1) or CGQNC (*Mt*Trxs2). Finally, some h-type Trxs from subgroup III lack one of the catalytic cysteines (CXXS) and consequently are not to be considered as 'true' Trxs according to the mechanism described above (Fig. 4). Indeed, Trxs with this type of CXXS motif show Grx-like activity (Gelhaye *et al.*, 2003).

2. 3D structures

Trxs are small proteins of ~12 kDa that display highly conserved 3D structures with a central five-stranded β-sheet surrounded by four α-helices in a $\beta\alpha\beta\alpha\beta\alpha\beta\beta\alpha$ topology (Holmgren *et al.*, 1975). The 3D structure of Trx includes the so-called Trx-fold ($\beta\alpha\beta\alpha\beta\beta\alpha$) (Martin, 1995), which also appears in other redox proteins with different functions, including glutathione transferase, Grx and protein disulfide isomerase. The active-site C_NXXC_C motif is located at the N-terminus of $\alpha 2$ and the preceding loop (Fig. 5A). In accordance with the catalytic mechanism, the catalytic C_N is exposed to the solvent, while C_C is buried and inaccessible. The conserved aspartic acid residue proposed to act as a general acid/base during catalysis is located further toward the protein interior. The solvent accessible surface surrounding C_N is composed of hydrophobic and uncharged residues that form a shallow groove, suggested to act as a target-binding site (Eklund *et al.*, 1984).

3D structures have been determined for several plant Trx-h, including barley *Hv*Trxh1 and *Hv*Trxh2 (Fig. 5A), *A. thaliana At*Trxh1, poplar *Pt*Trxh1 and *Pt*Trxh4 (Fig. 5B), and *Chlamydomonas reinhardtii* Trxh1 (Coudevylle *et al.*, 2005; Koh *et al.*, 2008; Maeda *et al.*, 2008;

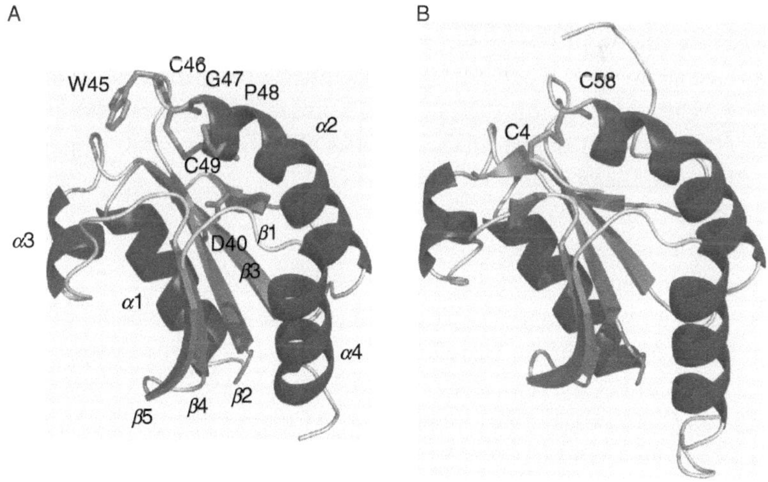

Fig. 5. (A) The structure of *Hv*Trxh2 from barley (h-type Trx from subclass I) in the oxidized form (Maeda *et al.*, 2008). The two conserved redox-active cysteines in the $_{45}$WCGPC$_{49}$ motif are connected by a disulfide bond. D40 corresponds to the aspartic acid residue proposed to act as a general acid/base during catalysis. β-Strands and α-helices are labelled. (B) The structure of a $C_C{\rightarrow}S$ mutant of *Pt*Trxh4 from poplar (h-type Trx from subgroup III) in the oxidized form (Koh *et al.*, 2008). A disulfide bond is formed between C58 (C_N) and C4 (C_{NT}) conserved in the N-terminal part of Trx-h from subgroup III. (See color plate section at the back of the book.)

Mittard *et al.*, 1997; Peterson *et al.*, 2005). In general, the overall molecular architectures and topologies of these proteins are similar to Trxs from non-photosynthetic organisms (Katti *et al.*, 1990; Weichsel *et al.*, 1996). The structures of *A. thaliana* *At*Trxh1 and barley *Hv*Trxh2 in the reduced and oxidized forms revealed no major redox-induced structural changes although few residues sequentially close to the CGPC motif do change conformation (Maeda *et al.*, 2008; Peterson *et al.*, 2005). Dimers of barley *Hv*Trxh1 are formed in the crystal lattice and the Trx–Trx interface is stabilized by three backbone–backbone hydrogen bonds in a pattern that resembles the intermolecular contacts observed in Trx-target complexes as described below (Maeda *et al.*, 2008).

The recently reported structure of the poplar *Pt*Trxh4 sheds light on the observed Grx-dependent activity of Trx-h from subgroup III (Gelhaye *et al.*, 2003; Koh *et al.*, 2008). In the oxidized wild-type protein, a disulfide is formed between the two cysteines in the $C_N XXC_C$ motif. In a $C_C \rightarrow S$ mutant, however, a disulfide is formed between C_N and a conserved cysteine in the N-terminal part of Trx-h from subgroup III (Figs. 3 and 5B). Based on these structures and biochemical data, a novel catalytic mechanism was suggested (Koh *et al.*, 2008). The proposed reaction pathway is initiated as described above for 'classical' Trx, that is, a target disulfide bond is reduced and a C_N–C_C disulfide is formed in Trx (Fig. 6). Then the C_N–C_C disulfide is attacked by the third N-terminal cysteine (C_{NT}) and the C_{NT}–C_N disulfide is formed (Fig. 5B). Reduction of this disulfide is proposed to proceed through glutathionylation of C_{NT} as supported by experimental evidence for a GSH-mixed disulfide of C_{NT} *in vitro* (Koh *et al.*, 2008). Finally, the glutathione is released from C_{NT} by Grx.

3. *Thioredoxin-target complexes*

As described above, Trx and its target proteins are linked with a disulfide bond as an intermediate in the catalytic cycle of disulfide reduction (Fig. 4). This intermediate can be trapped by an approach based on site-directed mutagenesis of CXXC that was originally developed for NTR–Trx complex formation (Veine *et al.*, 1998, Wang *et al.*, 1996). Briefly, the thiol group of C_C in the Trx $C_N XXC_C$ motif is removed through a $C \rightarrow S$ mutation to generate a mutant capable of performing the initial nucleophilic attack on the target disulfide but lacking the ability to reduce the intermolecular disulfide (Fig. 7). To facilitate the formation of a Trx-target complex, a $C \rightarrow S$ mutation is also introduced in one of the cysteines from the target disulfide bond, and the remaining single cysteine is activated by reaction with Ellman's reagent (5,5'-dithiobis-2-nitrobenzoic acid; DTNB) yielding a 2-nitro-5-thiobenzoate (TNB) conjugate that is subsequently replaced by the

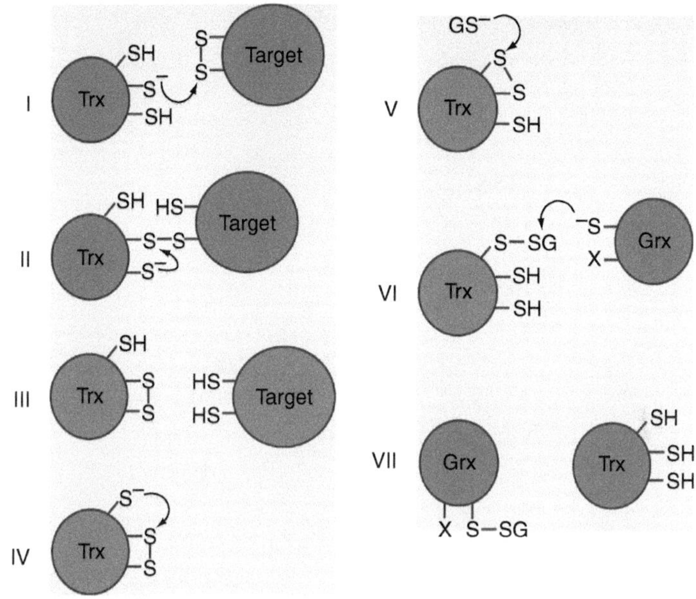

Fig. 6. Catalytic mechanism of h-type Trxs from subgroup III as proposed by Koh *et al.* (2008). (i–iii) As in 'classical' Trx, the pathway is initiated by the nucleophilic attack of C_N on the target disulfide bond followed by the release of the resulting Trx-target disulfide bond by C_C. (iv) The C_N–C_C disulfide is then reduced by the third N-terminal cysteine (C_{NT}) and a C_{NT}–C_N disulfide is formed. (v) C_{NT} is attacked by reduced glutathione (GS^-) and finally (vi) the glutathione is released from C_{NT} by Grx.

C_N thiol of the Trx mutant (Fig. 7). The release of TNB can be monitored at 412 nm. This approach has been successfully applied to obtain high yields of products for structure determination of Trx-target complexes from different biological systems (Chartron *et al.*, 2007; Li *et al.*, 2007; Maeda *et al.*, 2006; Qin *et al.*, 1995, 1996).

The only Trx-target structure from a plant source so far is the complex between barley h-type Trx (*Hv*Trxh2) and an α-amylase/subtilisin inhibitor (BASI) involved in the barley seed germination process (Maeda *et al.*, 2006). The structure reveals a relatively small interaction area of 762 Å^2. In the *Hv*Trxh2–BASI interface, C_N in *Hv*Trxh2 (C46) is conjugated to the BASI target cysteine (C148), and the neighbouring D146 and W147 are positioned into a shallow hydrophobic groove of *Hv*Trxh2 (Fig. 8). The residues in D146-C148 are exposed in an extended backbone conformation and form numerous van der Waals interactions and three hydrogen bonds to *Hv*Trxh2 in a manner resembling the interactions in an anti-parallel β-sheet. The hydrogen bond partners in *Hv*Trxh2 are M88 and A106 positioned in two

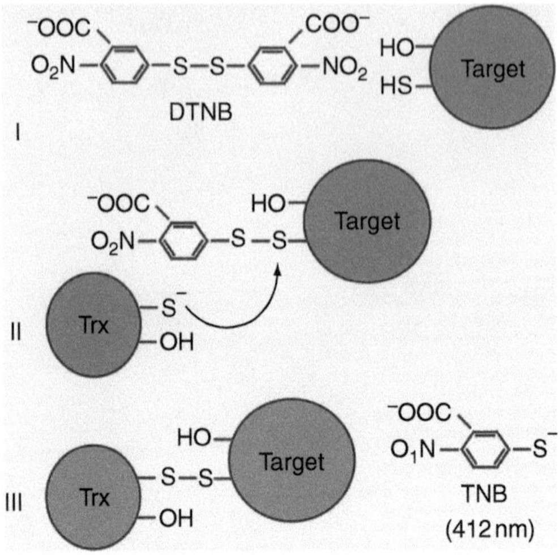

Fig. 7. A strategy to form Trx-target complexes adapted from Wang *et al.* (1996). (i) Single cysteine (C→S) mutants of target disulfides are activated by TNB conjugation. (ii) The activated target disulfide is attacked by C_N in a single cysteine (C→S) Trx mutant that lacks C_C. (iii) A kinetically stable intermolecular Trx-target disulfide in formed and the reaction can be monitored spectrophotometrically by following the release of TNB at 412 nm.

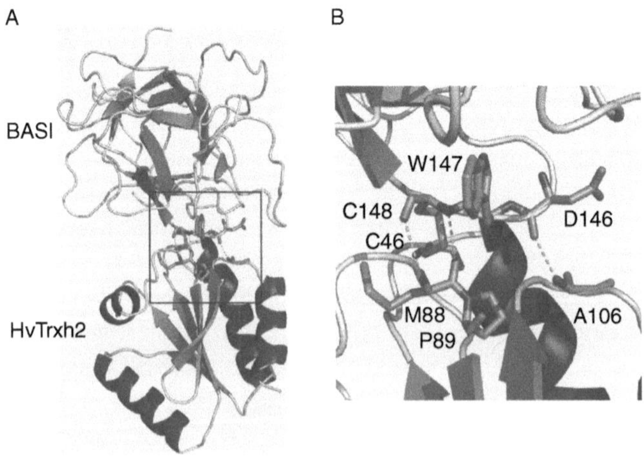

Fig. 8. The crystal structure of *Hv*Trxh2-S-S-BASI (Maeda *et al.*, 2006). (i) Overall structure of *Hv*Trxh2-S-S-BASI in cartoon display with β-strands and α-helices in green and blue, respectively. (ii) Close up view of the interaction between *Hv*Trxh2 and BASI. The segment $_{146}DWC_{148}$ of BASI and C_{46}, $_{88}MP_{89}$ and A_{106} of *Hv*Trxh2 are shown in stick representation. Carbon, oxygen, nitrogen and sulphur are shown in green, red, blue and yellow, respectively. Intermolecular hydrogen bonds are shown as dashed yellow lines. (See color plate section at the back of the book.)

loops that form the shallow groove surrounding the active site. M88 and A106 form backbone-backbone hydrogen bonds to C148 and D146 in BASI, respectively (Fig. 8). Remarkably, the hydrogen bond pattern involving M88 and the target cysteine C148 appears to be conserved in disulfide bonded protein–protein complexes involving Trxs and other related redox proteins, suggesting that Trx-like proteins share this common molecular feature involved in the recognition of target proteins (Chartron *et al.*, 2007; Li *et al.*, 2007; Qin *et al.*, 1995, 1996; Rozhkova *et al.*, 2004). Indeed the importance of the residue corresponding to M88 in several Trx-like proteins has recently been validated by biochemical data (Ren *et al.*, 2009). M88 is conserved among h-type Trxs from subgroup I and is replaced by similar large hydrophobic residues in other plant Trx isoforms (Fig. 3). A106 is also fairly well conserved among plant Trxs.

B. NTR

1. Catalytic mechanisms

NTR (E.C. 1.8.1.9) transfers electrons from NADPH to the oxidized CXXC motif in Trx via a bound FAD and a redox-active C_NXXC_C motif (Fig. 9). Briefly, the reduced C_C in the NTR C_NXXC_C motif attacks the oxidized CXXC motif in Trx resulting in a covalent NTR–Trx intermediate that is subsequently reduced by C_N from the C_NXXC_C motif in NTR (Veine *et al.*, 1998). The now oxidized CXXC motif in NTR is in turn reduced by FAD, which receives reducing equivalents from NADPH. In the 'flavin-oxidizing' state (FO), FAD is positioned for transfer of electrons to the oxidized NTR disulfide and separated by a distance of 17 Å from NADPH. In this state, the C_NXXC_C motif is buried and inaccessible to Trx (Fig. 9). A large domain rearrangement is needed to expose the active-site cysteines to Trx, and to position the FAD isoalloxazine rings in close proximity of the NADPH nicotinamide ring for electron transfer (Fig. 9). The latter conformation is referred to as the 'flavin-reducing' (FR) conformation (Fig. 9) and the FO to FR conformational change is most likely the rate-limiting step of the reaction (Lennon and Williams, 1997). It is not yet known if this catalytic mechanism also applies to the recently described NTR-C.

2. 3D structures

NTR is a member of the family of pyridine nucleotide-disulfide oxidoreductases (Pai, 1991) that contain two Rossman-type nucleotide-binding domains binding FAD and NADPH, respectively. NTR from higher eukaryotes are relatively rigid homodimeric proteins ($\geq$ 55 kDa) that have an additional C-terminal domain at the subunit interface (Manstein *et al.*, 1988;

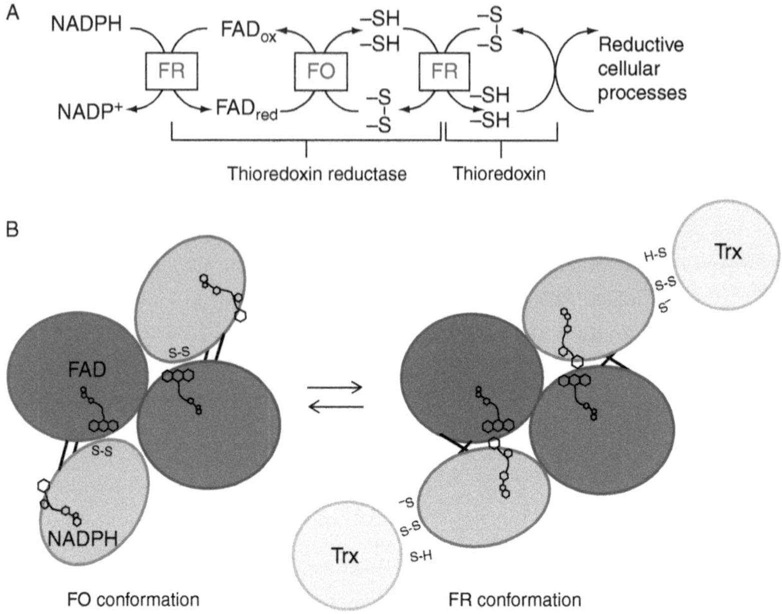

Fig. 9. (A) The catalytic mechanism of NTR. Reducing equivalents are transferred from NADPH to a tightly bound FAD in NTR. From FAD, the electrons are transferred to a CXXC in the enzyme and subsequently to the oxidized target CXXC disulfide in Trx. In order to catalyze the entire reaction, NTR needs to shift between two conformations, the FO and FR conformation. The electron transfers linked to each conformation are indicated. (B) The two conformations as proposed by Waksman *et al.* (1994) are shown schematically. The two subunits of the NTR dimer are shown in blue and green, respectively. The darker coloured ovals symbolize the FAD domains, while the lighter coloured show the NADPH domains. Disulfides and dithiols are indicated as S–S and S–H, respectively. The black lines connecting the two domains symbolize the anti-parallel β-sheet around which a 66° rotation occurs to bring NTR from the FO to the FR conformation. The rotation position of the nicotinamide ring in proximity of the flavin ring system and the CXXC dithiol is brought to the surface of the protein, where it can interact with thioredoxin (shown in yellow). Adapted from Waksman *et al.* (1994). (See color plate section at the back of the book.)

Waksman *et al.*, 1994; Williams *et al.*, 2000). NTRs from prokaryotes, yeast and plants ($\sim$ 35 kDa) are also homodimeric proteins, but each NTR subunit only contains the core of two Rossman-type nucleotide-binding domains. The structures of *E. coli* NTR in the FO and FR conformations revealed that the FAD domain needs to undergo a 66° rotation relative to the NADPH domain in order to switch from one conformation to the other (Lennon *et al.*, 2000).

The *A. thaliana* NTR-B in the FO conformation (Dai *et al.*, 1996) and recently the A/B-type barley *Hv*NTR2 (Kirkensgaard *et al.*, 2009) are the only crystal structures available for plant NTRs. These structures show the same overall homodimeric molecular organization as displayed in other low-molecular-weight NTR structures (Akif *et al.*, 2005; Gustafsson *et al.*, 2007; Kuriyan *et al.*, 1991; Lennon *et al.*, 1999, 2000; Ruggiero *et al.*, 2009; Waksman *et al.*, 1994; Zhang *et al.*, 2009; pdb: 2q7v, unpublished data). *At*NTR-B is shown superimposed on the structure of *E. coli* NTR (Waksman *et al.*, 1994) in Fig. 10. Each subunit of the homodimeric protein contains the composite FAD domain made from the C- and N-terminal regions of the polypeptide and the central NADPH domain carrying the redox-active CXXC motif. Each domain has a central five-stranded parallel β-sheet.

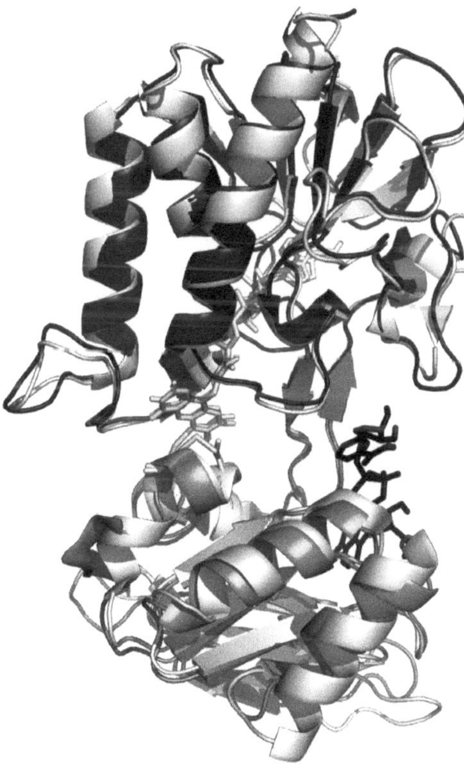

Fig. 10. Superposition of one subunit of the functionally dimeric NTRs from *A. thaliana* (pdb: 1vdc, shown in blue/green) and *E. coli* (shown in white, pdb: 1tdf). The enzymes consist of two domains with similar Rossman-folds. One domain (blue) binds FAD (yellow sticks) and the other domain (green) binds NADPH. The two domains are connected by two anti-parallel β-strands (pink). The structures are in the FO conformation, having FAD in close proximity of the redox-active CXXC motif (yellow sticks) from the NADPH domain. The *E. coli* structure was crystallized in the presence of $NADP^+$ (black). (See color plate section at the back of the book.)

The central sheet in the FAD domain is flanked by a four-stranded β-sheet on one side and three α-helices on the other. In the NADPH domain, the central sheet has a three-stranded β-sheet on one side and two α-helices plus an extra short helix containing the redox-active motif on the other. Two anti-parallel β-strands ($\beta9$ + $\beta18$) form a hinge between the two domains (Fig. 10), which are otherwise separated by a broad cleft with only few inter-domain contacts. In *At*NTR-B, there are hydrogen bonds from T53 in the FAD domain to W140 and N141 in the NADPH domain. Hydrogen bonds are found in the same regions in *Ec*NTR, but the residues involved in the inter-domain interactions are not conserved and the relative position of interacting residues are shifted by one or two residues in the sequence. Hence, G129 and R130 from the NADPH domain are hydrogen bonded with T47 and E48 in the FAD domain, respectively.

An alignment of plant NTR sequences shows that both T53 and W140 are strictly conserved within this group, whereas N141 is only conserved among the NTRs of the A/B type and replaced by a serine in the C-type (Fig. 11). The loop containing W140 and N141 is one of the variable loops in plant NTRs, and the consequences of the sequence variation for inter-domain interactions are hard to predict for C-type NTRs. The relative orientation of the FAD and NADPH domains is not the same among NTR structures. For example, the NADPH domains of *At*NTR-B and *S. cerevisiae* Trr1 (Zhang *et al.*, 2009) must be rotated by 8° to overlay the corresponding *Ec*NTR domain, when the respective FAD domains are superimposed. The *Mycobacterium tuberculosis* NTR domains have to be rotated by 11° for superposition with *Ec*NTR domains (Akif *et al.*, 2005). It is not clear if this variation in domain packing reflects a general flexibility of NTRs, or if it could be relevant for the reaction mechanism and the transformation between the FO and FR states. Two conserved arginines (R190 and R195 in *At*-B) impose NADPH over NADH specificity to short chain NTRs by donating hydrogen bonds to the 2′-phosphate of NADPH (Gustafsson *et al.*, 2007). The NADPH/NADH diphosphate is positioned by hydrogen bonds donated by conserved R, I and S/T amino acid residues.

As stated earlier, enzyme kinetics analyses suggest species-dependent interactions between plant NTRs and Trxs (Table I). Such specificity determinants are likely to reside in loop regions, where the major differences between plant NTR structures have been observed (Kirkensgaard *et al.*, 2009). For example, the group of monocot A/B type NTRs are characterized by an [H/Y]-F-[S/P/A]-G-S-D-[T/A] loop between strand $\beta9$ and $\beta10$, while the corresponding loop in dicot NTRs can be extended by four residues and have a [S/N/P]-F-[T/V/A]-G-S-[G/E]-[E/K/T/D]-[G/A]-[N/P/S]-[G/N]-G motif (Fig. 11). Another loop segregating the monocot and dicot NTRs of

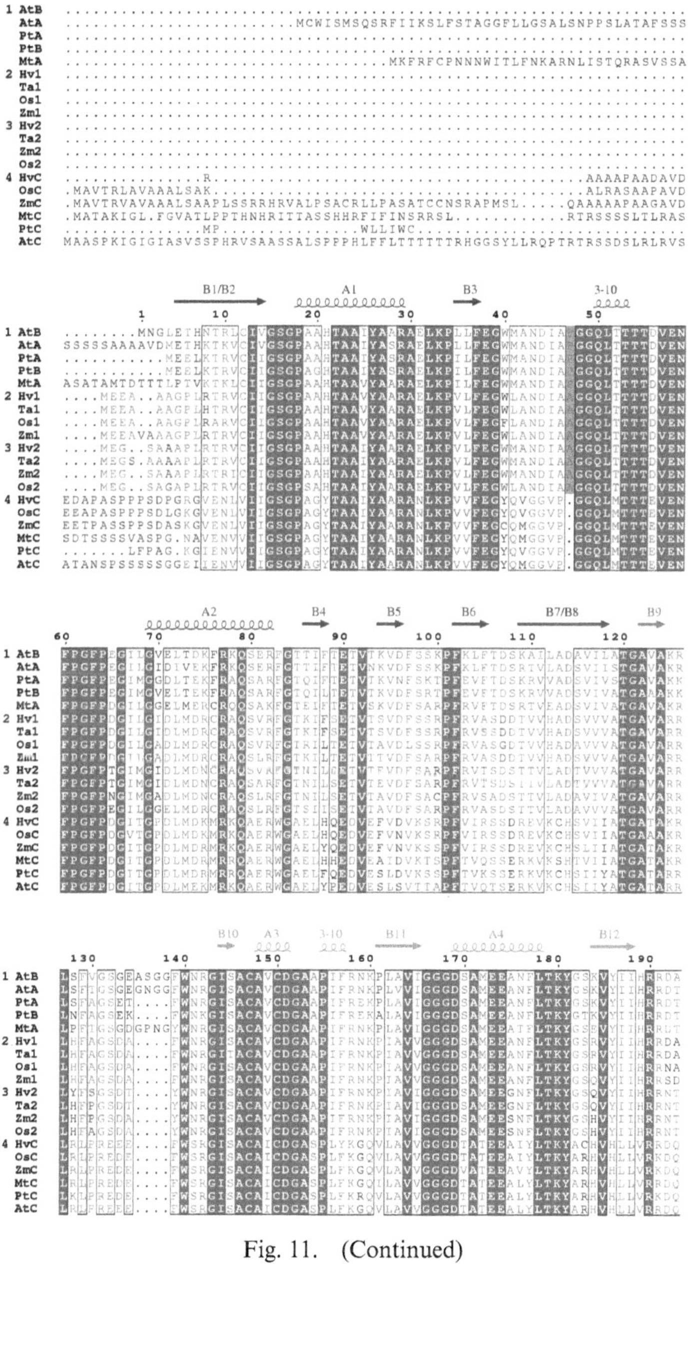

Fig. 11. (Continued)

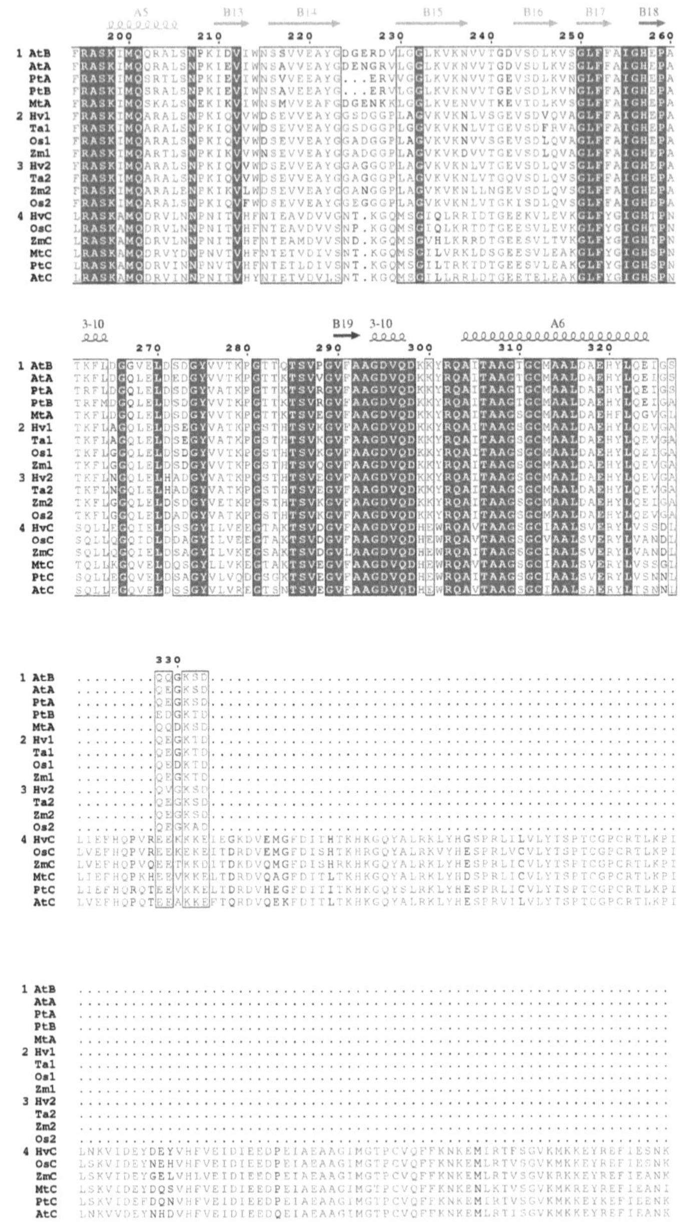

Fig. 11. (Continued)

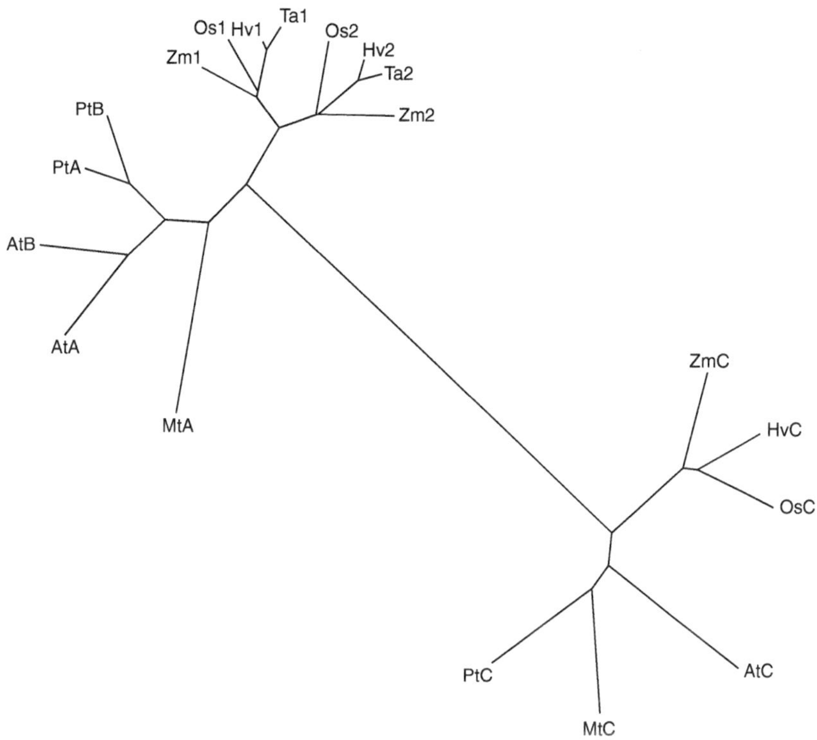

Fig. 11. Alignment of plant NTRs. The NTRs and their accession numbers in parentheses are *Hv*NTR1 (EU314717), *Hv*NTR2 (EU250021) and *Hv*NTRC from *Hordeum vulgare* (barley); *Ta*NTR1 (Q8VX47) and *Ta*NTR2 (TC297680) from *Triticum aestivum* (wheat); *Os*NTR1 (Q69PS6), *Os*NTR2 (Q6ZFU6) and *Os*NTRC (Q70G58) from *Oryza sativa* (rice); *Zm*NTR1 (EU966898), *Zm*NTR2 (BT054285) and *Zm*NTRC (BT037345) from *Zea mays* (maize); *At*NTRA (Q39242), *At*NTRB (Q39243) and *At*NTRC (O22229) from *Arabidopsis thaliana* (Mouse-ear cress); *Pt*NTRA (AC149479), *Pt*NTRB (XM_002317595) and *Pt*NTRC (XM_002308899) from *Populus trichocarpa* (western balsam poplar) and *Mt*NTRA and *Mt*NTRC from *Medicago truncatula* (Barrel Medic, legume). Residues strictly conserved have a dark gray background (red in the web version), residues well conserved within a group according to the Risler matrix (Risler *et al.*, 1988) are indicated by dark gray letters (red in the web version), residues conserved between groups are boxed and residues conserved within a group, but showing significant differences between groups, have a light gray background (orange in the web version). The secondary structure of *At*NTR-B was added using ESPript, and coloured according to domain; dark gray (blue in the web version) is the FAD domain (B1-B8 and B19-A6), (light gray) (green in the web version) the NADPH domain (B10-B17) and medium gray (pink in the web version) the β-sheets functioning as a link between the two. The sequences where aligned using CLUSTAL-W (Thompson *et al.*, 1994), and fall into three main groups, which can be characterized as dicotyledon A/B type (group 1), monocotyledon A/B type (group 2 and 3), and chloroplastic C-type (group 4). Monocotyledons appear to always have two NTRs of the A/B-type, which can be phylogenetically subgrouped (group 2 and 3). The alignment is followed by the phylogentic tree produced by the same CLUSTAL-W analysis and is illustrated in TreeView.

the A/B type is the loop between strand β14 and β15. This loop appears more flexible in monocots, where the G-G-[A/E/S]-[N/G/D]-G-G-P-L-[A/G] motif is found. The corresponding loop is variable in both length and sequence in dicots with no clear sequence motif. Both of these loops are likely to face and interact with the Trx substrate (Kirkensgaard *et al.*, 2009). A third loop placed between strand β3 and a short 3_{10}-helix, which is only present in plant and yeast NTR structures, is predicted to bind to Trx (Zhang *et al.*, 2009). This loop is strictly conserved in dicots with the sequence E-G-W-M-A-N-D-I-A-P-G-G. In monocots, the proline in this loop is exchanged with an alanine and the loop sequence is more variable: E-G-[W/F]-[M/L]-A-N-D-I-A-A-G-G. The C-type NTRs have a quite different sequence motif: E-G-[Y/C]-Q-[M/V]-G-G-V-P-G-G.

IV. IDENTIFICATION OF Trx TARGETS BY PROTEOMICS APPROACHES

A. PROTEOMICS TECHNIQUES APPLICABLE AT THE PROTEIN IDENTIFICATION LEVEL

Proteome analysis involves polyacrylamide gel-based or chromatographic multidimensional separation of protein mixtures, combined with highly sensitive mass spectrometry for protein identification and characterization. As such, proteomics enable parallel analysis and identification of many proteins in complex mixtures and has been applied with success for *in vitro* identification of Trx target proteins. Broadly, two approaches have been applied: (i) labelling of thiol groups released by Trx treatment of protein extracts followed by detection of labelled proteins on 2D gels and (ii) affinity isolation of proteins binding to immobilized active-site mutants of Trx.

1. *Labelling approaches*

A method was developed using the thiol-specific fluorescent probe monobromobimane (mBBr) to label free thiol groups in protein extracts treated with recombinant Trx prior to their separation by 2D gel electrophoresis (Yano *et al.*, 2001). Labelled proteins were visualized on 2D gels by UV-illumination, and comparison with protein extracts without Trx treatment demonstrated the appearance of additional fluorescent 2D gel spots containing putative Trx target proteins. Using this procedure, five potential target proteins were identified by Edman degradation in peanut seed extracts (Yano *et al.*, 2001). Similar procedures have also been applied together with mass spectrometry to identify five targets in barley embryo and 23 targets in the starchy endosperm of mature wheat seeds (Marx *et al.*, 2003; Wong *et al.*, 2003). The same strategy applied in parallel with a method based on a C→S mutated-Trx affinity column led to the identification of 111 putative Trx target proteins in germinating *M. truncatula* seeds (Alkhalfioui *et al.*, 2007b). In a related procedure, mBBR was replaced with a more sensitive Cy5 maleimide dye, which resulted in the identification of 16 putative *Hv*Trxh1 and *Hv*Trxh2 target proteins in mature barley seeds or seeds after 72 h germination (Maeda *et al.*, 2004). Other thiol-specific reagents have also been used, for example target proteins have been radio-labelled by alkylation with ^{14}C-iodoacetamide or biotinylated by reaction with thiol-reactive biotin derivatives (Marchand *et al.*, 2006). The latter approach enables affinity isolation of labelled proteins on immobilized avidin.

2. *Affinity isolation approaches*

The two-step catalytic mechanism of Trx (Fig. 4) has been exploited to trap target proteins in an intermolecular disulfide complex with an immobilized Trx, where the active-site CXXC has been replaced by CXXS through site-directed mutagenesis. Non-targets are washed away, and the target proteins can subsequently be released from the mutant Trx by addition of a strong reducing agent, such as DTT. This methodology has been used for the identification of putative target proteins from mitochondrial (Balmer *et al.*, 2004) and cytosolic extracts (Alkhalfioui *et al.*, 2007b; Marchand *et al.*, 2006; Wong *et al.*, 2004; Yamazaki *et al.*, 2004).

B. TECHNIQUES FOR IDENTIFICATION OF TARGET DISULFIDE BONDS

Information of specific target sites in target proteins can provide insight into molecular details of target recognition. Determination of specific target disulfides can be achieved by analysing mass spectra of peptides from proteolytically digested target proteins conjugated with different thiol-reactive reagents. For example, iodoacetamide (IAM) and 4-vinylpyridine (4-VP) were used to identify target disulfides in mature barley seed extract (Maeda *et al.*, 2005). Samples were incubated ± barley Trx-h followed by labelling of accessible thiol groups with IAM. Prior to the second dimension in 2D gel electrophoresis, protein thiol groups were reduced and the remaining free cysteines labelled with 4-VP. Thus, cysteines from target disulfides are labelled with IAM in the Trx-treated sample and with 4-VP in a control without Trx. Cysteines labelled with the two forms of modifications have different masses and are thus are distinguished in mass spectra. This procedure was used to analyse proteins in 2D-gel spots that were indicated to contain target proteins on the basis of Cy5 maleimide labelling patterns (Maeda *et al.*, 2004). By comparing samples incubated in the presence or absence of Trx, nine disulfides in eight proteins were identified as Trx-h targets.

A quantitative proteomics approach for site-specific identification of target disulfides was recently developed (Hägglund *et al.*, 2008). In this method, IAM-based isotope-coded affinity tag (ICAT) reagents were used. The ICAT reagents react with free thiol groups and are available with nine either 'light' (^{12}C; $ICAT_L$) or 'heavy' (^{13}C; $ICAT_H$) carbon atoms. The isotopically labelled ICAT reagents are identical except for a mass difference of 9 Da, and are therefore compatible with quantitative mass spectrometric detection. Basically, two identical samples of protein extract were incubated in the presence (+Trx) or absence (−Trx) of Trx (Fig. 12). Subsequently, IAM was added to quench the Trx activity and block cysteine thiols released from the target disulfides, followed by complete reduction of remaining protein disulfides by tris(2-carboxyethyl)phosphine (TCEP) under denaturing conditions in SDS. The samples were ICAT-labelled (+Trx, $ICAT_L$; −Trx, $ICAT_H$), mixed in a 1:1 ratio and digested by trypsin. Cysteine-containing peptides are isolated by avidin affinity chromatography and analyzed by mass spectrometry. Because non-target disulfide bonds remain intact until the TCEP reduction, they give rise to identical amounts of thiol groups available for ICAT labelling in the ±Trx samples and $ICAT_H/ICAT_L$ peptide ratios of 1 are expected when the samples are mixed 1:1. In contrast, cysteine residues released from Trx target disulfide bonds are blocked by IAM and hence do not undergo $ICAT_L$-labelling in the +Trx sample. Peptides carrying Trx-targeted residues will thus exhibit $ICAT_H/ICAT_L$ ratios >1. The method was applied to identify targets of barley *Hv*Trxh1 in protein extracts of 48 h germinated barley seed embryo.

C. EXAMPLES OF TARGET PROTEINS

Several proteomics studies of Trxs from different organisms, tissues, and organelle extracts have provided over 400 putative Trx-target proteins that were recently compiled into a comprehensive list (Montrichard *et al.*, 2009). Some examples of targets from NTS are described below.

1. Seed germination

As previously stated, Trx-h plays an important role in the germination of cereal seeds by facilitating release of storage reserves through reduction of disulfide bonds in storage proteins, hydrolytic enzymes, and their inhibitors (Besse *et al.*, 1996; Kobrehel *et al.*, 1991, 1992). Storage proteins identified as target proteins include hordeins and globulins from barley and globulins, glutenins, and avenins in wheat (Hägglund *et al.*, 2008; Marx *et al.*, 2003;

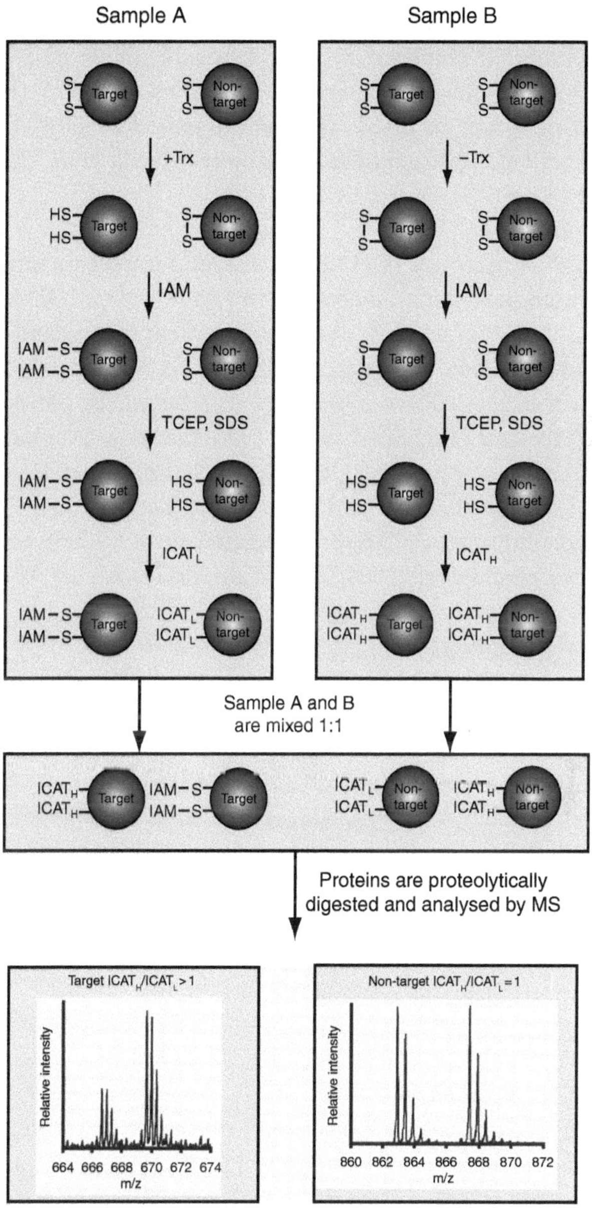

Fig. 12. An approach for the identification of target protein disulfides using ICAT-labelling. Two identical samples of protein extract were incubated in the presence (+ Trx) or absence (− Trx) of Trx and IAM was added to quench Trx activity and block cysteine thiols. Proteins were denatured by SDS and remaining disulfides were reduced by TCEP. The samples were ICAT-labelled (+ Trx, $ICAT_L$; − Trx, $ICAT_H$), mixed in a 1:1 ratio, digested with trypsin and analysed by mass spectrometry. $ICAT_H$/$ICAT_L$ peptide ratios of 1 are expected for non-targets and peptides carrying Trx-targeted residues are expected to show $ICAT_H$/$ICAT_L$ ratios >1.

Wong *et al.*, 2004). Several inhibitors of α-amylases and proteases have been identified, including the barley α-amylase/subtilisin inhibitor (BASI) that has been structure-determined in complex with barley *Hv*Trxh2 (Fig. 8). The specific disulfide target in BASI was determined to be C144-C148 by the IAM/4-VP labelling approach described above (Maeda *et al.*, 2005).

2. Redox control

Dehydroascorbate reductase (DHAR) participates in redox control mediated through the ascorbate–glutathione cycle in higher plants. DHAR has been identified as a Trx-h target in *A. thaliana* and wheat (Marchand *et al.*, 2004, 2006; Wong *et al.*, 2004). The cysteine proposed to be the catalytic residue in DHAR (Dixon *et al.*, 2002) was identified as a target for barley h-type Trx using ICAT labelling (Hägglund *et al.*, 2008). Other examples of putative Trx-targeted redox proteins include 1-Cys peroxiredoxin (Hägglund *et al.*, 2008; Maeda *et al.*, 2004, Marx *et al.*, 2003; Wong *et al.*, 2004), 2-Cys peroxiredoxin (Alkhalfioui *et al.*, 2007b; Marchand *et al.*, 2006) and ascorbate peroxidase (Marchand *et al.*, 2004, 2006; Yamazaki *et al.*, 2004).

3. Amino acid metabolism

Several enzymes involved in amino acid metabolism have been identified to be targets, including adenosylhomocysteinase (Hägglund *et al.*, 2008), alanine aminotransferase (Balmer *et al.*, 2004; Marchand *et al.*, 2004, 2006; Wong *et al.*, 2004), aspartate aminotransferase (Balmer *et al.*, 2004), aspartate-semialdehyde dehydrogenase (Hägglund *et al.*, 2008) and glutamine synthetase (Alkhalfioui *et al.*, 2007b; Lemaire *et al.*, 2004; Marchand *et al.*, 2004, 2006; Yamazaki *et al.*, 2004). One of the target cysteines in adenosylhomocysteinase was identified (Hägglund *et al.*, 2008) and represents a conserved residue previously demonstrated to be essential for enzymatic activity and positioned near the active site in the ortholog from human placenta (Yuan *et al.*, 1996).

4. Carbohydrate metabolism

Enolases from plants were demonstrated to be redox sensitive *in vitro*, and suggested to be regulated by Trx (Anderson *et al.*, 1998). Indeed, enolase has been identified as a target of Trx-h in several proteomics surveys (Alkhalfioui *et al.*, 2007b, Hägglund *et al.*, 2008; Lemaire *et al.*, 2004; Wong *et al.*, 2004). Furthermore, one of the target cysteines identified by the ICAT labelling approach (Hägglund *et al.*, 2008) is suggested to form a regulatory disulfide bond (Anderson *et al.*, 1998). Other targets involved in carbohydrate metabolism include fructose bisphosphate aldolase (Marchand *et al.*, 2004, 2006; Yamazaki *et al.*, 2004), malate dehydrogenase (Alkhalfioui *et al.*, 2007b;

Balmer *et al.*, 2004; Marchand *et al.*, 2004; Yamazaki *et al.*, 2004) and succinyl-CoA ligase (Balmer *et al.*, 2004; Hägglund *et al.*, 2008).

V. SUMMARY AND PERSPECTIVES

Thanks to the outstanding research achievements in recent years, it is becoming increasingly apparent that plant NTS make use of a remarkable multitude of molecular mechanisms for control of cysteine oxidation. The recently discovered chloroplastic NTR-C represents a truly novel NTS with NTR and Trx functionalities fused into a single protein. Little is as yet known about the structure and catalytic mechanism of NTR-C, and this is unquestionably a highly prioritized target for future research efforts. Another topic which deserves attention, is the molecular basis for specific Trx–NTR interactions. Despite the large number of high-resolution structures of Trxs and NTRs from plants and other organisms, surprisingly little biochemical data is available regarding the molecular recognition of Trxs by NTR.

Proteomics studies have identified numerous proteins with different functions that are targeted by Trx *in vitro*. Furthermore, the recent application of quantitative proteomics to identify specific target disulfides has revealed site-specific targets that are candidates for further investigations. However, the biological relevance of the putative targets must be validated by other means. In this respect, recent developments of proteomics methods aimed at measuring *in vivo* redox states hold a great promise for the future. For example, an ICAT labelling strategy has been developed to identify oxidized cysteines in the proteome of *E. coli* mutants lacking Trx (Leichert *et al.*, 2008). It will indeed be a great challenge for the future to transfer this technology to plant systems, given the large number of NTR and Trx genes present in these organisms.

ACKNOWLEDGEMENTS

The Danish Technical Research Council (STVF, grant no. 26-03-0194), the Carlsberg Foundation and the Center for Advanced Food Studies (LMC) are acknowledged for financial support. K. G. K. holds a Ph.D. scholarship from the Technical University of Denmark.

REFERENCES

Akif, M., Suhre, K., Verma, C. and Mande, S. C. (2005). Conformational flexibility of *Mycobacterium tuberculosis* thioredoxin reductase: Crystal structure and normal-mode analysis. *Acta Crystallographica Section D* **61**, 1603–1611.

Alkhalfioui, F., Renard, M. and Montrichard, F. (2007a). Unique properties of NADP-thioredoxin reductase C in legumes. *Journal of Experimental Botany* **58**, 969–978.

Alkhalfioui, F., Renard, M., Vensel, W. H., Wong, J., Tanaka, C. K., Hurkman, W. J., Buchanan, B. B. and Montrichard, F. (2007b). Thioredoxin-linked proteins are reduced during germination of *Medicago truncatula* seeds. *Plant Physiology* **144**, 1559–1579.

Alkhalfioui, F., Renard, M., Frendo, P., Keichinger, C., Meyer, Y., Gelhaye, E., Hirasawa, M., Knaff, D. B., Ritzenthaler, C. and Montrichard, F. (2008). A novel type of thioredoxin dedicated to symbiosis in legumes. *Plant Physiology* **148**, 424–435.

Anderson, L. E., Li, A. D. and Stevens, F. J. (1998). The enolases of ice plant and *Arabidopsis* contain a potential disulphide and are redox sensitive. *Phytochemistry* **47**, 707–713.

Arnér, E. S. J. and Holmgren, A. (2000). Physiological functions of thioredoxin and thioredoxin reductase. *European Journal of Biochemistry* **267**, 6102–6109.

Balmer, Y., Vensel, W. H., Tanaka, C. K., Hurkman, W. J., Gelhaye, E., Rouhier, N., Jacquot, J. P., Manieri, W., Schürmann, P., Droux, M. and Buchanan, B. B. (2004). Thioredoxin links redox to the regulation of fundamental processes of plant mitochondria. *Proceedings of the National Academy of Sciences of the United States of America* **101**, 2642–2647.

Barranco-Medina, S., Krell, T., Bernier-Villamor, L., Sevilla, F., Lázaro, J. J. and Dietz, K. J. (2008). Hexameric oligomerization of mitochondrial peroxiredoxin PrxIIF and formation of an ultrahigh affinity complex with its electron donor thioredoxin Trx-o. *Journal of Experimental Botany* **59**, 3259–3269.

Behm, M. and Jacquot, J. P. (2000). Isolation and characterization of thioredoxin h from poplar xylem. *Plant Physiology and Biochemistry* **38**, 363–369.

Besse, I., Wong, J. H., Kobrehel, K. and Buchanan, B. B. (1996). Thiocalsin: A thioredoxin-linked, substrate-specific protease dependent on calcium. *Proceedings of the National Academy of Sciences of the United States of America* **93**, 3169–3175.

Bower, M. S., Matias, D. D., Fernandes-Carvalho, E., Mazzurco, M., Gu, T., Rothstein, S. J. and Goring, D. R. (1996). Two members of the thioredoxin-h family interact with the kinase domain of a Brassica S locus receptor kinase. *Plant Cell* **8**, 1641–1650.

Bréhélin, C., Mouaheb, N., Verdoucq, L., Lancelin, J. M. and Meyer, Y. (2000). Characterization of determinants for the specificity of *Arabidopsis* thioredoxins h in yeast complementation. *Journal of Biological Chemistry* **275**, 31641–31647.

Bréhélin, C., Laloi, C., Setterdahl, A. T., Knaff, D. B. and Meyer, Y. (2004). Cytosolic, mitochondrial thioredoxins and thioredoxin reductases in *Arabidopsis thaliana*. *Photosynthesis Research* **79**, 295–304.

Buchanan, B. B. and Balmer, Y. (2005). Redox regulation: A broadening horizon. *Annual Review of Plant Biology* **56**, 187–220.

Cabrillac, D., Cock, J. M., Dumas, C. and Gaude, T. (2001). The S-locus receptor kinase is inhibited by thioredoxins and activated by pollen coat proteins. *Nature* **410**, 220–223.

Cazalis, R., Pulido, P., Aussenac, T., Pérez-Ruiz, J. M. and Cejudo, F. J. (2006). Cloning and characterization of three thioredoxin h isoforms from wheat showing differential expression in seeds. *Journal of Experimental Botany* **257**, 2165–2172.

Chartron, J., Shiau, C., Stout, C. D. and Carroll, K. S. (2007). 3'-Phosphoadenosine-5'-phosphosulfate reductase in complex with thioredoxin: A structural snapshot in the catalytic cycle. *Biochemistry* **46**, 3942–3951.

Chivers, P. T. and Raines, R. T. (1997). General acid/base catalysis in the active site of *Escherichia coli* thioredoxin. *Biochemistry* **36**, 15810–15816.
Cho, M. J., Wong, J. H., Marx, C., Jiang, W., Lemaux, P. G. and Buchanan, B. B. (1999). Overexpression of thioredoxin h leads to enhanced activity of starch debranching enzyme (pullulanase) in barley grain. *Proceedings of the National Academy of Sciences of the United States of America* **96**, 14641–14646.
Coudevylle, N., Thureau, A., Hemmerlin, C., Gelhaye, E., Jacquot, J. P. and Cung, M. T. (2005). Solution structure of a natural CPPC active site variant, the reduced form of thioredoxin h1 from poplar. *Biochemistry* **44**, 2001–2008.
Dai, S., Saarinen, M., Ramaswamy, S., Meyer, Y., Jacquot, J. P. and Eklund, H. (1996). Crystal structure of *Arabidopsis thaliana* NADPH dependent thioredoxin reductase at 2.5 Å resolution. *Journal of Molecular Biology* **264**, 1044–1057.
Dixon, D. P., Davis, B. G. and Edwards, R. (2002). Functional divergence in the glutathione transferase superfamily in plants. Identification of two classes with putative functions in redox homeostasis in *Arabidopsis thaliana*. *Journal of Biological Chemistry* **277**, 30859–30869.
Eklund, H., Cambillau, C., Sjöberg, B. M., Holmgren, A., Jörnvall, H., Höög, J. O. and Brändén, C. I. (1984). Conformational and functional similarities between glutaredoxin and thioredoxins. *EMBO Journal* **3**, 1443–1449.
Gelhaye, E., Rouhier, N., Laurent, P., Sautière, P. E., Martin, F. and Jacquot, J. P. (2002). Isolation and characterization of an extended thioredoxin h from poplar. *Physiologia Plantarum* **114**, 165–171.
Gelhaye, E., Rouhier, N. and Jacquot, J. P. (2003). Evidence for a subgroup of thioredoxin h that requires GSH/Grx for its reduction. *FEBS Letters* **555**, 443–448.
Gelhaye, E., Rouhier, N., Gerard, J., Jolivet, Y., Gualberto, J., Navrot, N., Ohlsson, P. I., Wingsle, G., Hirasawa, M., Knaff, D. B., Wang, H. Dizengremel, P. *et al.* (2004). A specific form of thioredoxin h occurs in plant mitochondria and regulates the alternative oxidase. *Proceedings of the National Academy of Sciences of the United States of America* **101**, 14545–14550.
Gelhaye, E., Rouhier, N., Navrot, N. and Jacquot, J. P. (2005). The plant thioredoxin system. *Cellular and Molecular Life Sciences* **62**, 24–35.
Gustafsson, T. N., Sandalova, T., Lu, J., Holmgren, A. and Schneider, G. (2007). High-resolution stuctures of oxidized and reduced thioredoxin reductase from *Helicobacter pylori*. *Acta Crystallographica Section D* **63**, 833–843.
Hägglund, P., Bunkenborg, J., Maeda, K. and Svensson, B. (2008). Identification of thioredoxin disulfide targets using a quantitative proteomics approach based on isotope-coded affinity tags. *Journal of Proteome Research* **7**, 5270–5276.
Holmgren, A. (1979). Thioredoxin catalyzes the reduction of insulin disulfides by dithiothreitol and dihydrolipoamide. *Journal of Biological Chemistry* **254**, 9627–9632.
Holmgren, A. (1989). Thioredoxin and glutaredoxin systems. *Journal of Biological Chemistry* **264**, 13963–13966.
Holmgren, A., Söderberg, B. O., Eklund, H. and Brändén, C. I. (1975). Three-dimensional structure of *Escherichia coli* thioredoxin-S2 to 2.8 Å resolution. *Proceedings of the National Academy of Sciences of the United States of America* **72**, 2305–2309.
Ishiwatari, Y., Honda, C., Kawashima, I., Nakamura, S., Hirano, H., Mori, S., Fujiwara, T., Hayashi, H. and Chino, M. (1995). Thioredoxin h is one of the major proteins in rice phloem sap. *Planta* **195**, 456–463.

Ishiwatari, Y., Fujiwara, T., McFarland, K. C., Nemoto, K., Hayashi, H., Chino, M. and Lucas, W. J. (1998). Rice phloem thioredoxin h has the capacity to mediate its own cell-to-cell transport through plasmodesmata. *Planta* **205**, 12–22.

Jacquot, J. P., Rivera-Madrid, R., Marinho, P., Kollarova, M., Le Maréchal, P., Miginiac-Maslow, M. and Meyer, Y. (1994). *Arabidopsis thaliana* NADPH thioredoxin reductase. cDNA characterization and expression of the recombinant protein in *Escherichia coli*. *Journal of Molecular Biology* **235**, 1357–1363.

Jacquot, J. P., Eklund, H., Rouhier, N. and Schürmann, P. (2009). Structural and evolutionary aspects of thioredoxin reductases in photosynthetic plants. *Trends in Plant Science* **14**, 336–343.

Johansson, C., Lillig, C. H. and Holmgren, A. (2004). Human mitochondrial glutaredoxin reduces *S*-glutathionylated proteins with high affinity accepting electrons from either glutathione or thioredoxin reductase. *Journal of Biological Chemistry* **279**, 7537–7543.

Johnson, T. C., Wada, K., Buchanan, B. B. and Holmgren, A. (1987). Reduction of purothionin by the wheat seed thioredoxin system. *Plant Physiology* **85**, 446–451.

Juárez-Díaz, J. A., McClure, B., Vázquez-Santana, S., Guevara-García, A., León-Mejía, P., Márquez-Guzmán, J. and Cruz-García, F. (2006). A novel thioredoxin h is secreted in *Nicotiana alata* and reduces S-RNase *in vitro*. *Journal of Biological Chemistry* **281**, 3418–3424.

Juttner, J., Olde, D., Langridge, P. and Baumann, U. (2000). Cloning and expression of a distinct subclass of plant thioredoxins. *European Journal of Biochemistry* **267**, 7109–7117.

Kallis, G. B. and Holmgren, A. (1980). Differential reactivity of the functional sulfhydryl groups of cysteine32 and cysteine35 present in the reduced form of thioredoxin from *Escherichia coli*. *Journal of Biological Chemistry* **255**, 10261–10265.

Kanzok, S. M., Fechner, A., Bauer, H., Ulschmid, J. K., Muller, H. M., Botella-Munoz, J., Schneuwly, S., Schirmer, R. and Becker, K. (2001). Substitution of the thioredoxin system for glutathione reductase in *Drosophila melanogaster*. *Science* **291**, 643–646.

Katti, S. K., LeMaster, D. M. and Eklund, H. (1990). Crystal structure of thioredoxin from *Escherichia coli* at 1.68 Å resolution. *Journal of Molecular Biology* **212**, 167–184.

Kirkensgaard, K., Hägglund, P., Finnie, C., Svensson, B. and Henriksen, A. (2009). Crystal structure of *Hordeum vulgare* NADPH-dependent thioredoxin reductase 2. Unwinding the reaction mechanism. *Acta Crystallographica section D* **65**, 932–941.

Kobrehel, K., Yee, B. C. and Buchanan, B. B. (1991). Role of the NADP/thioredoxin system in the reduction of alpha-amylase and trypsin inhibitor proteins. *Journal of Biological Chemistry* **266**, 16135–16140.

Kobrehel, K., Wong, J. H., Balogh, A., Kiss, F., Yee, B. C. and Buchanan, B. B. (1992). Specific reduction of wheat storage proteins by thioredoxin h. *Plant Physiology* **99**, 919–924.

Koh, C. S., Navrot, N., Didierjean, C., Rouhier, N., Hirasawa, M., Knaff, D. B., Wingsle, G., Samian, R., Jacquot, J. P., Corbier, C. and Gelhaye, E. (2008). An atypical catalytic mechanism involving three cysteines of thioredoxin. *Journal of Biological Chemistry* **283**, 23062–23072.

Krause, G. and Holmgren, A. (1991). Substitution of the conserved tryptophan 31 in *Escherichia coli* thioredoxin by site-directed mutagenesis and structure–function analysis. *Journal of Biological Chemistry* **266**, 4056–4066.

Krimm, I., Lemaire, S., Ruelland, E., Miginiac-Maslow, M., Jacquot, J. P., Hirasawa, M., Knaff, D. B. and Lancelin, J. M. (1998). The single mutation Trp35->Ala in the 35–40 redox site of *Chlamydomonas reinhardtii* thioredoxin h affects its biochemical activity and the pH dependence of C36–C39 ^{1}H–^{13}C NMR. *European Journal of Biochemistry* **255**, 185–195.

Kuriyan, J., Krishna, T. S. R., Wong, L., Guenther, B., Pahler, A., Williams, C. H., Jr. and Model, P. (1991). Convergent evolution of similar function in two structurally divergent enzymes. *Nature* **352**, 172–174.

Laloi, C., Rayapuram, N., Chartier, Y., Grienenberger, J. M., Bonnard, G. and Meyer, Y. (2001). Identification and characterization of a mitochondrial thioredoxin system in plants. *Proceedings of the National Academy of Sciences of the United States of America* **98**, 14144–14149.

Leichert, L. I., Gehrke, F., Gudiseva, H. V., Blackwell, T., Ilbert, M., Walker, A. K., Strahler, J. R., Andrews, P. C. and Jakob, U. (2008). Quantifying changes in the thiol redox proteome upon oxidative stress *in vivo*. *Proceedings of the National Academy of Sciences of the United States of America* **105**, 8197–8202.

Lemaire, S. D., Guillon, B., Le Marechal, P., Keryer, E., Miginiac-Maslow, M. and Decottignies, P. (2004). New thioredoxin targets in the unicellular photosynthetic eukaryote *Chlamydomonas reinhardtii*. *Proceedings of the National Academy of Sciences of the United States of America* **101**, 7475–7480.

Lennon, B. W. and Williams, C. H. (1997). Reductive half-reaction of thioredoxin reductase from *Escherichia coli*. *Biochemistry* **36**, 9464–9477.

Lennon, B. W., Williams, C. H., Jr. and Ludwig, M. (1999). Crystal structure of reduced thioredoxin reductase from *Escherichia coli*: Structural flexibility in the isoalloxazine ring of the flavin adenine dinucleotide cofactor. *Protein Science* **8**, 2366–2379.

Lennon, B. W., Williams, C. H., Jr., and Ludwig, M. L. (2000). Twists in catalysis: Alternating conformations of *Escherichia coli* thioredoxin reductase. *Science* **289**, 1190–1194.

Li, Y., Hu, Y., Zhang, X., Xu, H., Lescop, E., Xia, B. and Jin, C. (2007). Conformational fluctuations coupled to the thiol-disulfide transfer between thioredoxin and arsenate reductase in *Bacillus subtilis*. *Journal of Biological Chemistry* **282**, 11078–11083.

Maeda, K., Finnie, C., Østergaard, O. and Svensson, B. (2003). Identification, cloning and characterization of two thioredoxin h isoforms, HvTrxh1 and HvTrxh2, from the barley seed proteome. *European Journal of Biochemistry* **270**, 2633–2643.

Maeda, K., Finnie, C. and Svensson, B. (2004). Cy5 maleimide labeling for sensitive detection of free thiols in native protein extracts: Identification of seed proteins targeted by barley thioredoxin h isoforms. *Biochemical Journal* **378**, 479–507.

Maeda, K., Finnie, C. and Svensson, B. (2005). Identification of thioredoxin h-reducible disulphides in proteomes by differential labelling of cysteines: Insight into recognition and regulation of proteins in barley seeds by thioredoxin h. *Proteomics* **5**, 1634–1644.

Maeda, K., Hägglund, P., Finnie, C., Svensson, B. and Henriksen, A. (2006). Structural basis for target protein recognition by the protein disulphide reductase thioredoxin. *Structure* **14**, 1701–1710.

Maeda, K., Hägglund, P., Finnie, C., Svensson, B. and Henriksen, A. (2008). Crystal structures of barley thioredoxin h isoforms HvTrxh1 and HvTrxh2 reveal features involved in protein recognition and possibly in discriminating the isoform specificity. *Protein Science* **17**, 1015–1024.

Manstein, D. J., Massey, V., Ghisla, S. and Pai, E. F. (1988). Stereochemistry and accessibility of prosthetic groups in flavoproteins. *Biochemistry* **27**, 2300–2305.

Marchand, C., Le Maréchal, P., Meyer, Y., Miginiac-Maslow, M., Issakidis-Bourguet, E. and Decottignies, P. (2004). New targets of *Arabidopsis* thioredoxins revealed by proteomic analysis. *Proteomics* **4**, 2696–2706.

Marchand, C., Le Marechal, P., Meyer, Y. and Decottignies, P. (2006). Comparative proteomic approaches for the isolation of proteins interacting with thioredoxin. *Proteomics* **6**, 6528–6537.

Marti, M. C., Olmos, E., Calvete, J. J., Diaz, I., Barranco-Medina, S., Whelan, J., Lazaro, J. J., Sevilla, F. and Jimenez, A. (2009). Mitochondrial and nuclear localization of a novel pea thioredoxin: Identification of its mitochondrial target proteins. *Plant Physiology* **150**, 646–657.

Martin, J. L. (1995). Thioredoxin—A fold for all reasons. *Structure* **3**, 245–250.

Marx, C., Wong, J. H. and Buchanan, B. B. (2003). Thioredoxin and germinating barley: Targets and protein redox changes. *Planta* **216**, 454–460.

Menchise, V., Corbier, C., Didierjean, C., Saviano, M., Benedetti, E., Jacquot, J. P. and Aubry, A. (2001). Crystal structure of the wild-type and D30A mutant thioredoxin h of *Chlamydomonas reinhardtii* and implications for the catalytic mechanism. *Biochemical Journal* **359**, 65–75.

Meyer, Y., Vignols, F. and Reichheld, J. P. (2002). Classification of plant thioredoxins by sequence similarity and intron position. *Methods in Enzymology* **347**, 394–402.

Mittard, V., Blackledge, M. J., Stein, M., Jacquot, J. P., Marion, D. and Lancelin, J. M. (1997). NMR solution structure of an oxidised thioredoxin h from the eukaryotic green alga *Chlamydomonas reinhardtii*. *European Journal of Biochemistry* **243**, 374–383.

Montrichard, F., Alkhalfioui, F., Yano, H., Vensel, W. H., Hurkman, W. J. and Buchanan, B. B. (2009). Thioredoxin targets in plants: The first 30 years. *Journal of Proteomics* **72**, 452–474.

Moon, J. C., Jang, H. H., Chae, H. B., Lee, J. R., Lee, S. Y., Jung, Y. J., Shin, M. R., Lim, H. S., Chung, W. S., Yun, D. J., Lee, K. O. and Lee, S. Y. (2006). The C-type *Arabidopsis* thioredoxin reductase ANTR-C acts as an electron donor to 2-Cys peroxiredoxins in chloroplasts. *Biochemical and Biophysical Research Communications* **348**, 478–484.

Mouaheb, N., Thomas, D., Verdoucq, L., Monfort, P. and Meyer, Y. (1998). *In vivo* functional discrimination between plant thioredoxins by heterologous expression in the yeast *Saccharomyces cerevisiae*. *Proceedings of the National Academy of Sciences of the United States of America* **95**, 3312–3317.

Muller, E. G. D. (1996). A glutathione reductase mutant of yeast accumulates high levels of oxidized glutathione and requires thioredoxin for growth. *Molecular Biology of the Cell* **7**, 1805–1813.

Nuruzzaman, M., Gupta, M., Zhang, C., Wang, L., Xie, W., Xiong, L., Zhang, Q. and Lian, X. (2008). Sequence and expression analysis of the thioredoxin protein gene family in rice. *Molecular Genetics and Genomics* **280**, 139–151.

Pai, E. F. (1991). Variations on a theme: the family of FAD-dependent NAD(P)H-(disulfide)-oxidoreductases. *Current Opinion in Structural Biology* **1**, 796–803.

Pérez-Ruiz, J. M., Spínola, M. C., Kirchsteiger, K., Moreno, J., Sahrawy, M. and Cejudo, F. J. (2006). Rice NTRC is a high-efficiency redox system for chloroplast protection against oxidative damage. *Plant Cell* **18**, 2356–2368.

Peterson, F. C., Lytle, B. L., Sampath, S., Vinarov, D., Tyler, E., Shahan, M., Markley, J. L. and Volkman, B. F. (2005). Solution structure of thioredoxin h1 from *Arabidopsis thaliana*. *Protein Science* **14**, 2195–2200.

Prinz, W. A., Åslund, F., Holmgren, A. and Beckwith, J. (1997). The role of the thioredoxin and glutaredoxin pathways in reducing protein disulfide bonds in the *Escherichia coli* cytoplasm. *Journal of Biological Chemistry* **272**, 15661–15667.

Pulido, P., Cazalis, R. and Cejudo, F. J. (2009). An antioxidant redox system in the nucleus of wheat seed cells suffering oxidative stress. *Plant Journal* **57**, 132–145.

Qin, J., Clore, G. M., Kennedy, W. M., Huth, J. R. and Gronenborn, A. M. (1995). Solution structure of human thioredoxin in a mixed disulfide intermediate complex with its target peptide from the transcription factor NF kappa B. *Structure* **3**, 289–297.

Qin, J., Clore, G. M., Kennedy, W. P., Kuszewski, J. and Gronenborn, A. M. (1996). The solution structure of human thioredoxin complexed with its target from Ref-1 reveals peptide chain reversal. *Structure* **4**, 613–620.

Reichheld, J. P., Mestres-Ortega, D., Laloi, C. and Meyer, Y. (2002). The multigenic family of thioredoxin h in *Arabidopsis thaliana*: Specific expression and stress response. *Plant Physiology and Biochemistry* **40**, 685–690.

Reichheld, J. P., Meyer, E., Khafif, M., Bonnard, G. and Meyer, Y. (2005). AtNTRB is the major mitochondrial thioredoxin reductase in *Arabidopsis thaliana*. *FEBS Letters* **579**, 337–342.

Reichheld, J. P., Khafif, M., Riondet, C., Droux, M., Bonnard, G. and Meyer, Y. (2007). Inactivation of thioredoxin reductases reveals a complex interplay between thioredoxin and glutathione pathways in *Arabidopsis* development. *Plant Cell* **19**, 1851–1865.

Ren, G., Stephan, D., Xu, Z., Zheng, Y., Tang, D., Harrison, R. S., Kurz, M., Jarrott, R., Shouldice, S. R., Hiniker, A., Martin, J. L. Heras, B. *et al.* (2009). Properties of the thioredoxin fold superfamily are modulated by a single amino acid residue. *Journal of Biological Chemistry* **284**, 10150–10159.

Risler, J. L., Delorme, M. O., Delacroix, H. and Henaut, A. (1988). Amino acid substitutions in structurally related proteins. A pattern recognition approach. Determination of a new and efficient scoring matrix. *Journal of Molecular Biology* **204**, 1019–1029.

Rivera-Madrid, R., Mestres, D., Marinho, P., Jacquot, J. P., Decottignies, P., Miginiac-Maslow, M. and Meyer, Y. (1995). Evidences for five divergent thioredoxin h sequences in *Arabidopsis thaliana*. *Proceedings of the National Academy of Sciences of the United States of America* **92**, 5620–5624.

Rouhier, N., Gelhaye, E. and Jacquot, J. P. (2002). Glutaredoxin-dependent peroxiredoxin from poplar: Protein–protein interaction and catalytic mechanism. *Journal of Biological Chemistry* **277**, 13609–13614.

Rouhier, N., Kauffmann, B., Tete-Favier, F., Palladino, P., Gans, P., Branlant, G., Jacquot, J. P. and Boschi-Muller, S. (2007). Functional and structural aspects of poplar cytosolic and plastidial type a methionine sulfoxide reductases. *Journal of Biological Chemistry* **282**, 3367–3378.

Rozhkova, A., Stirnimann, C. U., Frei, P., Grauschopf, U., Brunisholz, R., Grütter, M. G., Capitani, G. and Glockshuber, R. (2004). Structural basis and kinetics of inter- and intramolecular disulfide exchange in the redox catalyst DsbD. *EMBO Journal* **23**, 1709–1719.

Ruggiero, A., Masullo, M., Ruocco, M. R., Grimaldi, P., Lanzotti, M. A., Arcari, P., Zagari, A. and Vitagliano, L. (2009). Structure and stability of a thioredoxin reductase from *Sulfolobus solfataricus*: A thermostable protein with two functions. *Biochimica et Biophysica Acta* **1794**, 554–562.

Santandrea, G., Guo, Y., O'Connell, T. and Thompson, R. D. (2002). Post-phloem protein trafficking in the maize caryopsis: zmTRXh1, a thioredoxin specifically expressed in the pedicel parenchyma of *Zea mays* L., is found predominantly in the placentochalaza. *Plant Molecular Biology* **50**, 743–756.

Sarkar, N., Lemaire, S., Wu-Scharf, D., Issakidis-Bourguet, E. and Cerutti, H. (2005). Functional specialization of *Chlamydomonas reinhardtii* cytosolic thioredoxin h1 in the response to alkylation-induced DNA damage. *Eukaryotic Cell* **4**, 262–273.

Schürmann, P. and Buchanan, B. B. (2008). The ferredoxin/thioredoxin system of oxygenic photosynthesis. *Antioxidants & Redox Signaling* **10**, 1235–1274.

Serrato, A. J. and Cejudo, F. J. (2003). Type-h thioredoxins accumulate in the nucleus of developing wheat seed tissues suffering oxidative stress. *Planta* **217**, 392–399.

Serrato, A. J., Crespo, J. L., Florencio, F. J. and Cejudo, F. J. (2001). Characterization of two thioredoxins h with predominant localization in the nucleus of aleurone and scutellum cells of germinating wheat seeds. *Plant Molecular Biology* **46**, 361–371.

Serrato, A. J., Pérez-Ruiz, J. M. and Cejudo, F. J. (2002). Cloning of thioredoxin h reductase and characterization of the thioredoxin reductase-thioredoxin h system from wheat. *Biochemical Journal* **367**, 491–497.

Serrato, A. J., Pérez-Ruiz, J. M., Spínola, M. C. and Cejudo, F. J. (2004). A novel NADPH thioredoxin reductase, localized in the chloroplast, which deficiency causes hypersensitivity to abiotic stress in *Arabidopsis thaliana*. *Journal of Biological Chemistry* **279**, 43821–43827.

Shahpiri, A., Svensson, B. and Finnie, C. (2008). The NADPH-dependent thioredoxin reductase/thioredoxin system in germinating barley seeds: Gene expression, protein profiles and interactions between isoforms of thioredoxin h and thioredoxin reductase. *Plant Physiology* **146**, 789–799.

Shi, J. and Bhattacharyya, M. K. (1996). A novel plasma membrane-bound thioredoxin from soybean. *Plant Molecular Biology* **32**, 653–662.

Stenbaek, A., Hansson, A., Wulff, R. P., Hansson, M., Dietz, K. J. and Jensen, P. E. (2008). NADPH-dependent thioredoxin reductase and 2-Cys peroxiredoxins are needed for the protection of Mg-protoporphyrin monomethyl ester cyclase. *FEBS Letters* **582**, 2773–2778.

Suske, G., Wagner, W. and Follmann, H. (1979). NADPH thioredoxin reductase and a new thioredoxin from wheat. *Zeitschrift für Naturforschung* **34**, 214–221.

Sweat, T. A. and Wolpert, T. J. (2007). Thioredoxin h5 is required for victorin sensitivity mediated by a CC-NBS-LRR gene in *Arabidopsis*. *Plant Cell* **19**, 673–687.

Thompson, J. D., Higgins, D. G. and Gibson, T. J. (1994). CLUSTAL W: Improving the sensitivity of progressive multiple sequence alignment through sequence weighting, position-specific gap penalties and weight matrix choice. *Nucleic Acids Research* **22**, 4673–4680.

Veine, D. M., Mulrooney, S. B., Wang, P. F. and Williams, C. H., Jr. (1998). Formation and properties of mixed disulfides between thioredoxin reductase from *Escherichia coli* and thioredoxin: Evidence that cysteine-138 functions to initiate dithiol-disulfide interchange and to accept the reducing equivalent from reduced flavin. *Protein Science* **7**, 1441–1450.

Verdoucq, L., Vignols, F., Jacquot, J. P., Chartier, Y. and Meyer, Y. (1999). *In vivo* characterization of a thioredoxin h target protein defines a new peroxiredoxin family. *Journal of Biological Chemistry* **274**, 19714–19722.

Waksman, G., Krishna, T. S. R., Williams, C. H., Jr. and Kuriyan, J. (1994). Crystal structure of *Escherichia coli* thioredoxin reductase refined at 2 Å resolution. Implications for a large conformational change during catalysis. *Journal of Molecular Biology* **236**, 800–816.

Wang, P.-F., Veine, D. M., Ahn, S. H. and Williams, C. H., Jr. (1996). A stable mixed disulfide between thioredoxin reductase and its substrate, thioredoxin: Preparation and characterization. *Biochemistry* **35**, 4812–4819.

Weichsel, A., Gasdaska, J. R., Powis, G. and Montfort, W. R. (1996). Crystal structures of reduced, oxidized, and mutated human thioredoxins: Evidence for a regulatory homodimer. *Structure* **4**, 735–751.

Wieles, B., van Noort, J., Drijfhout, J. W., Offringa, R., Holmgren, A. and Ottenhoff, T. H. (1995a). Purification and functional analysis of the *Mycobacterium leprae* thioredoxin/thioredoxin reductase hybrid protein. *Journal of Biological Chemistry* **270**, 25604–25606.

Wieles, B., van Soolingen, D., Holmgren, A., Offringa, R., Ottenhoff, T. and Thole, J. (1995b). Unique gene organization of thioredoxin and thioredoxin reductase in *Mycobacterium leprae. Molecular Microbiology* **16**, 921–929.

Williams, C. H., Arscott, L. D., Muller, S., Lennon, B. W., Ludwig, M. L., Wang, P. F., Veine, K. B. and Schirmer, R. H. (2000). Thioredoxin reductase two modes of catalysis have evolved. *European Journal of Biochemistry* **267**, 6110–6117.

Wong, J. H., Kim, Y. B., Ren, P. H., Cai, N., Cho, M. J., Hedden, P., Lemaux, P. G. and Buchanan, B. B. (2002). Transgenic barley grain overexpressing thioredoxin shows evidence that the starchy endosperm communicates with the embryo and the aleurone. *Proceedings of the National Academy of Sciences of the United States of America* **99**, 16325–16330.

Wong, J. H., Balmer, Y., Cai, N., Tanaka, C. K., Vensel, W. H., Hurkman, W. J. and Buchanan, B. B. (2003). Unraveling thioredoxin-linked metabolic processes of cereal starchy endosperm using proteomics. *FEBS Letters* **547**, 151–156.

Wong, J. H., Cai, N., Balmer, Y., Tanaka, C. K., Vensel, W. H., Hurkman, W. J. and Buchanan, B. B. (2004). Thioredoxin targets of developing wheat seeds identified by complementary proteomic approaches. *Phytochemistry* **65**, 1629–1640.

Yamazaki, D., Motohashi, K., Kasama, T., Hara, Y. and Hisabori, T. (2004). Target proteins of the cytosolic thioredoxin in *Arabidopsis thaliana. Plant and Cell Physiology* **45**, 18–27.

Yano, H., Wong, J. H., Lee, Y. M., Cho, M. J. and Buchanan, B. B. (2001). A strategy for the identification of proteins targeted by thioredoxin. *Proceedings of the National Academy of Sciences of the United States of America* **98**, 4794–4799.

Yuan, C. S., Ault-Riche, D. B. and Borchardt, R. T. (1996). Chemical modification and site-directed mutagenesis of cysteine residues in human placental *S*-adenosylhomocysteine hydrolase. *Journal of Biological Chemistry* **271**, 28009–28016.

Zhang, Z., Bao, R., Zhang, Y., Yu, J., Zhou, C.-Z. and Chen, Y. (2009). Crystal structure of *Saccharomyces cerevisiae* cytoplasmic thioredoxin reductase Trr1 reveals the structure basis for species-specific recognition of thioredoxin. *Biochimica Biophysica Acta* **1794**, 124–128.

AUTHOR INDEX

Numbers in **bold** refer to pages on which full references are listed.

C

D

I

J

T

SUBJECT INDEX

D

H

I

R

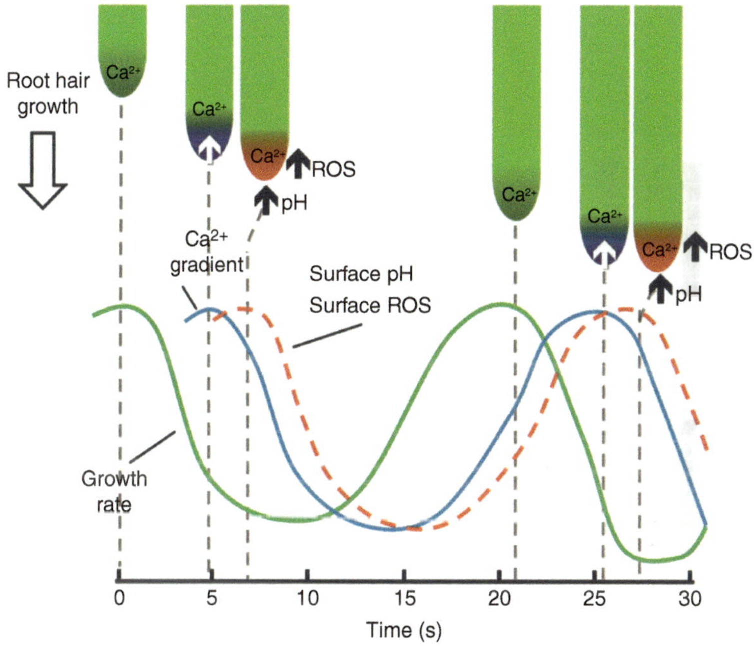

Kim *et al.*, Fig. 3. Oscillation model of tip growth of *Arabidopsis* root hairs. Oscillations of the Ca^{2+} gradient lagged oscillations in growth by 5 s. Dynamic increases in extracellular ROS and pH that oscillate with a similar frequency as growth but lag growth oscillations by 7–8 s. The maximal oscillatory increase in cytosolic Ca^{2+} is triggered by cell expansion associated with tip growth and plays a role in the subsequent restriction of growth. These three oscillation parameters, maximal oscillatory increases in cytosolic Ca^{2+}, extracellular ROS, and pH, may be linked as part of a system that limits growth after the initial burst of elongation has occurred. This figure was originally published in Monshausen *et al.* (2008) and is reproduced by permission from *Plant Physiology®/The Plant Cell.*

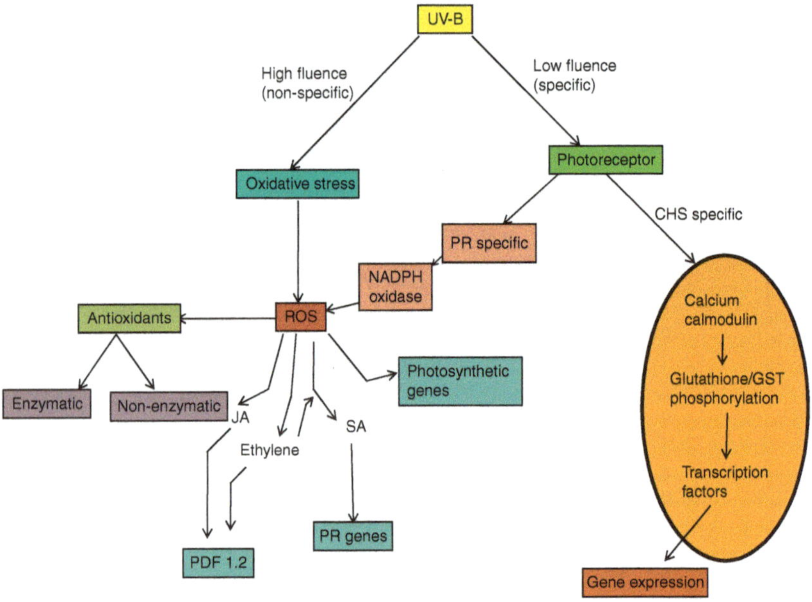

Agrawal *et al.*, Fig. 2. Signal transduction and multiple signaling pathways mediating responses to UV-B (redrawn with changes after Mackerness, 2000; Mackerness *et al.*, 1999). PR, pathogenesis related; JA, jasmonic acid; SA, salicylic acid; CHS, chalcone synthase; ROS, reactive oxygen species; PDF 1.2, plant defensin gene.

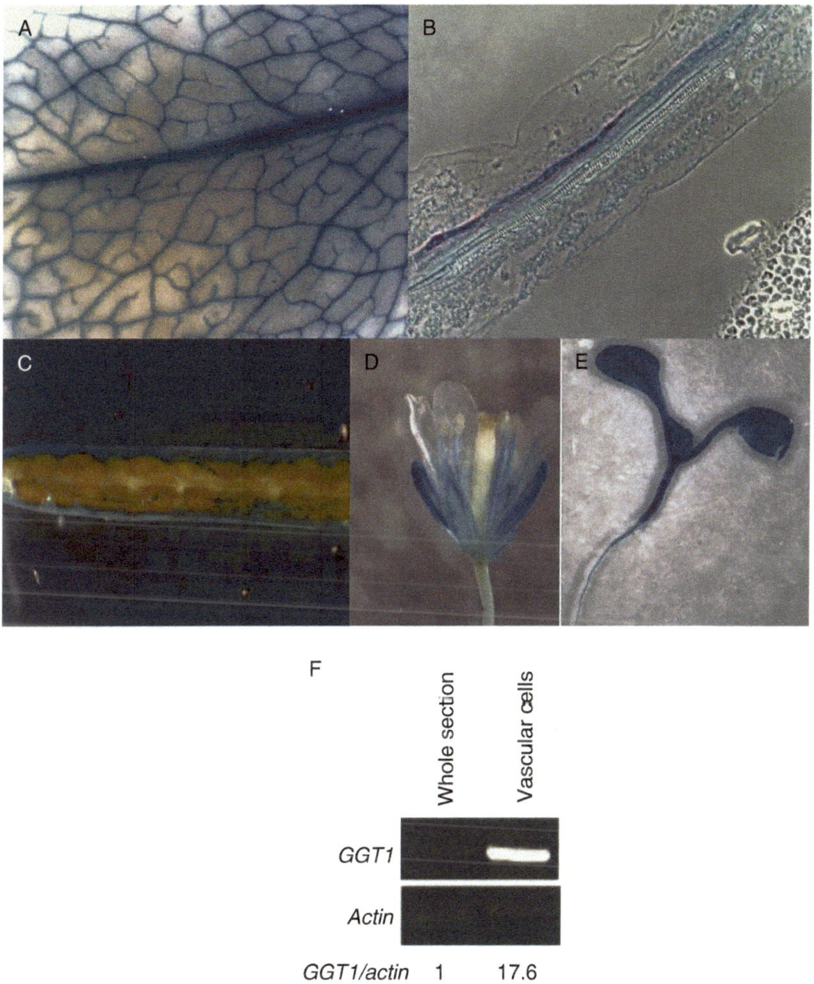

Ohkama-Ohtsu *et al.*, Fig. 2. GUS staining of *GGT1:: GUS* fusion. (A) GUS staining of a leaf. (B) Longitudinal section of a petiole. (C) Maturing silique showing staining in vascular tissue but no staining in developing seeds. (D) Flower with GUS staining in the vascular tissue of sepals, petals, and filaments of anthers. (E) One-week-old plant showing staining in the roots. (F) GGT1 transcript accumulation in vascular cells. Vascular cells in leaf cross sections were collected by laser microdissection. Total RNA extracted from vascular cells or whole sections was subjected to RT-PCR analysis. Values below the gel images are the levels of GGT1 mRNA relative to that of *ACTIN8* quantified using real-time PCR. These figures were originally published in Ohkama-Ohtsu *et al.* (2007a) and are reproduced with permission from Blackwell Publishing.

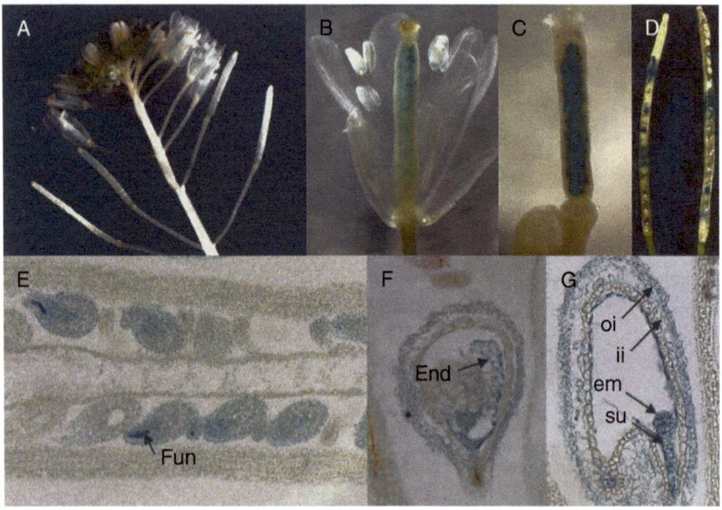

Ohkama-Ohtsu *et al.*, Fig. 3. GUS staining of *GGT2::GUS* fusion. (A) *Arabidopsis* shoot. (B) Flower stained before anthers open. (C) Pistil two days after pollination. (D) Mature siliques from a line that was still segregating for the *GGT2*::GUS insert. (E) Longitudinal section of a 5-day-old silique to illustrate staining in the funiculus (fun). (F) Longitudinal section of a seed three days after pollination (end, endosperm). (G) Longitudinal section of a seed 5 days after fertilization (oi, outer integument; ii, inner integument; em, embryo; su, suspensor). These figures were originally published in Ohkama-Ohtsu *et al.* (2007a) and are reproduced with permission from Blackwell Publishing.

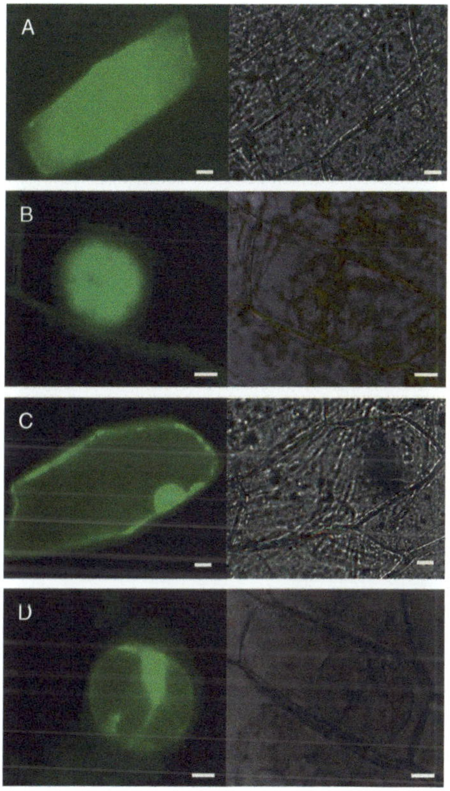

Ohkama-Ohtsu *et al.*, Fig. 4. Fluorescent micrographs of onion cells transiently expressing GGT4 ORF. N-terminal 100 amino acids of GGT4 fused to GFP (A and B), and unmodified GFP (C and D). In (B) and (D) cells were put in 10 mM mannitol solution. Images of the same cells taken under white light are on the right. Bars = 20 μm. These figures were originally published in Ohkama-Ohtsu *et al.* (2007b) and are reproduced with permission from Blackwell Publishing.

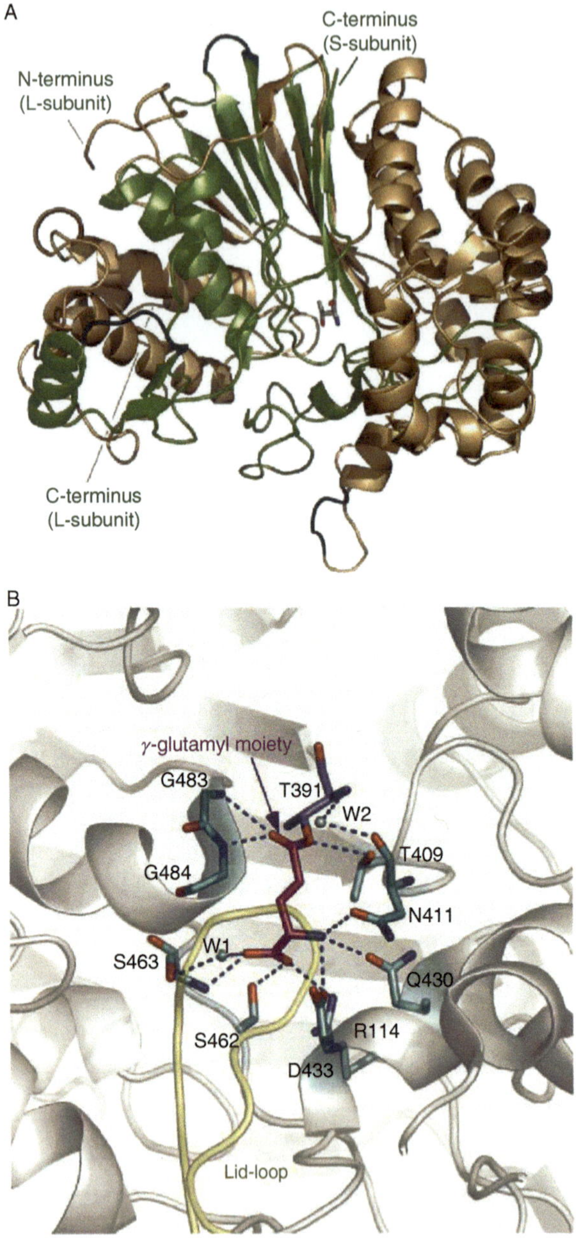

Ohkama-Ohtsu *et al.*, Fig. 5. (A) Ribbon drawing of *E. coli* GGT. Brown and green show the large subunit and the small subunit, respectively. The stick model at the center is the active residue Thr391, the N-terminus of the small subunit. The segments shown in black correspond to the regions in which large insertion/deletion of amino acid residues occur in the *Arabidopsis* GGTs. (B) Close-up view of the active site pocket of the γ-glutamyl-enzyme intermediate. The γ-glutamyl moiety is bound to Thr391 and the residues that interact with the moiety are colored. Hydrogen bonds are shown by broken lines. The lid loop (yellow) covers on the pocket. Except for the residues involved in the interaction with the γ-glutamyl moiety, individual atoms are omitted for clarity. PDB ID: 2DBW.

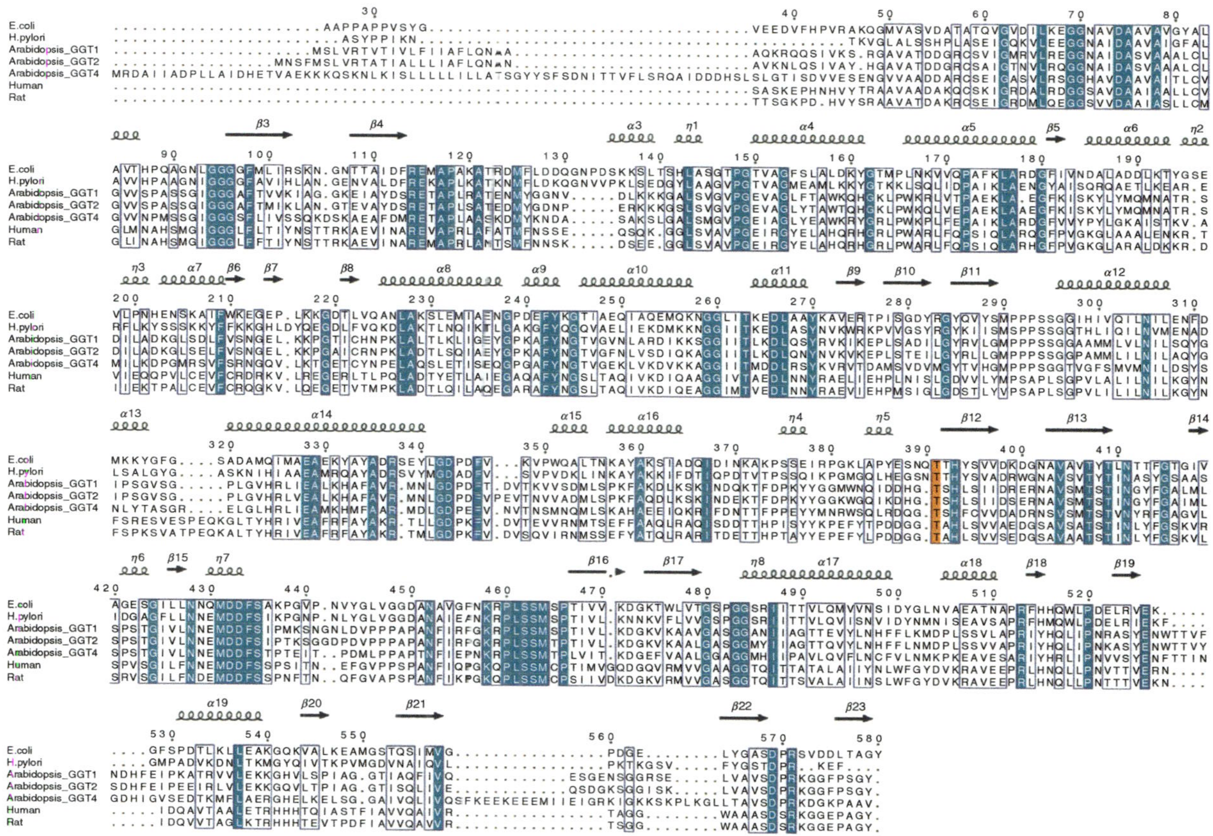

Ohkama-Ohtsu *et al.*, Fig. 6. Sequence alignment of bacterial, plant, and mammalian GGTs. The secondary structures observed in *E. coli* GGT are shown above the sequences. The residue numbers are for *E. coli* GGT. Thr391 shown in orange is the N-terminus of the small subunit of the mature GGT and the catalytic residue not only in the GGT reaction but also in the autoprocessing of the precursor protein.

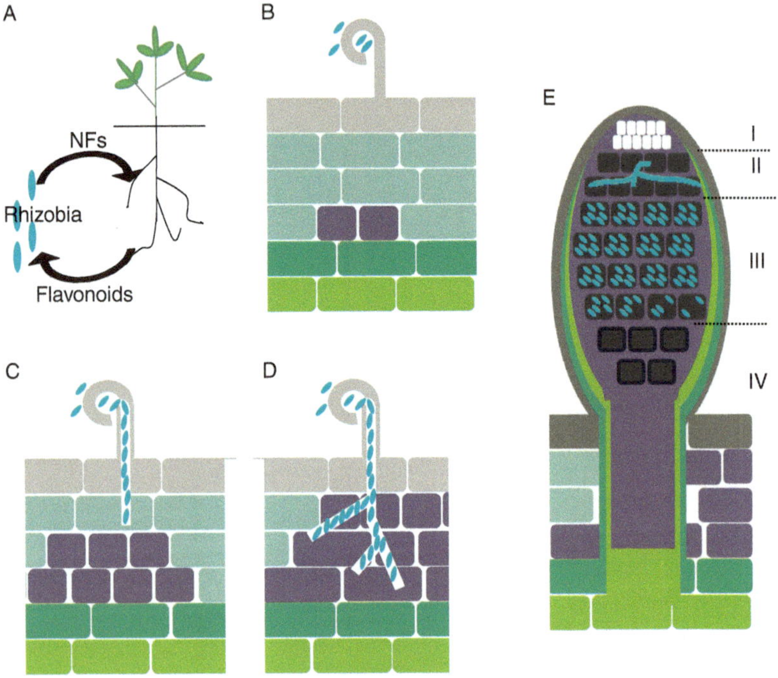

Marino *et al.*, Fig. 1. The different steps of nitrogen-fixing symbiosis. (A) Recognition of the partners, (B) root hair curling and nodule meristem formation, (C) infection thread penetration and meristem growth, (D) infection thread and meristem development, (E) indeterminate nodule with meristematic (I), infection (II), fixing (III), and senescent (IV) zones.

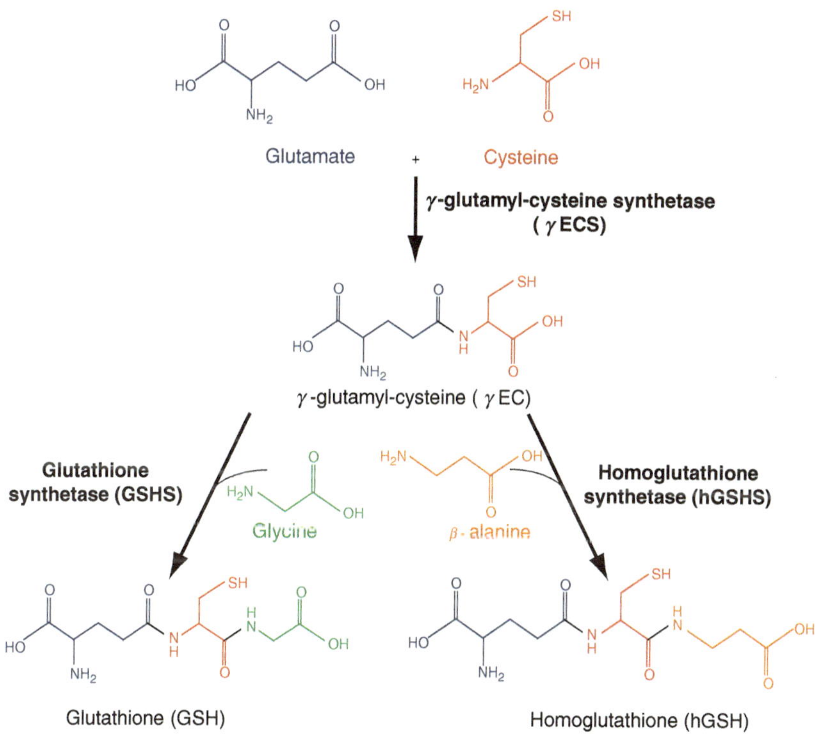

Marino *et al.*, Fig. 5. Biosynthesis pathway of GSH and hGSH. All enzymes are shown in bold.

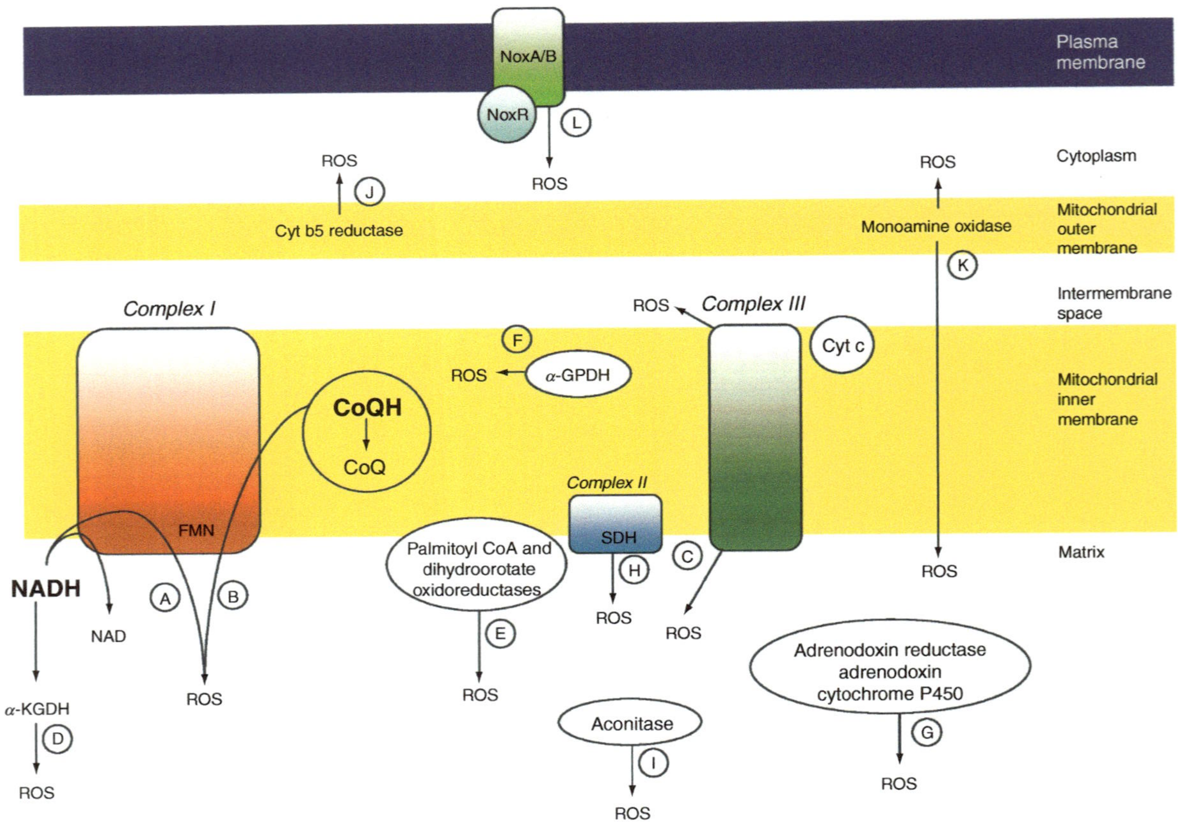

Morel *et al.*, Fig. 2. Reactive oxygen species (ROS) production sites. (A–I) ROS production sites in the mitochondrial inner membrane and matrix. (J, K) ROS production sites in the mitochondrial outer membrane. (L) ROS production by NADPH oxidases (NOX) at the plasma membrane. α-KGDH, α-ketoglutarate dehydrogenase; α-GPDH, α-glycerophosphate dehydrogenase; CoQ, Coenzyme Q; SDH, succinate dehydrogenase; cyt *c*, cytochrome *c*; FMN, flavin mononucleotide.

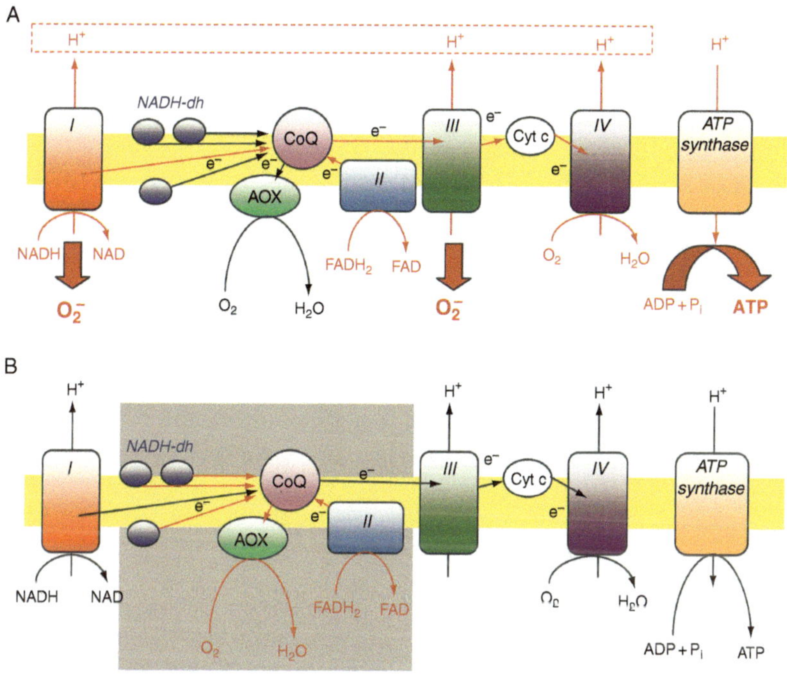

Morel *et al.*, Fig. 3. Regulation of the mitochondrial electron transport chain of *P. chrysosporium* grown in a ligninolytic condition compared to a control condition. (A) When the intracellular oxidative stress is low, all the classical mitochondrial complexes are involved in creating the proton gradient required to produce ATP. Complexes I and III are responsible for most of the superoxide production. To maintain a nontoxic amount of superoxide inside the cell, some detoxifying enzymes function in their degradation. (B) When there is a high oxidative stress, another level of regulation involves alternative mitochondrial pathways to bypass complexes I and III and limit ROS production. Red arrows indicate the preferential pathway in conditions (A) and (B).

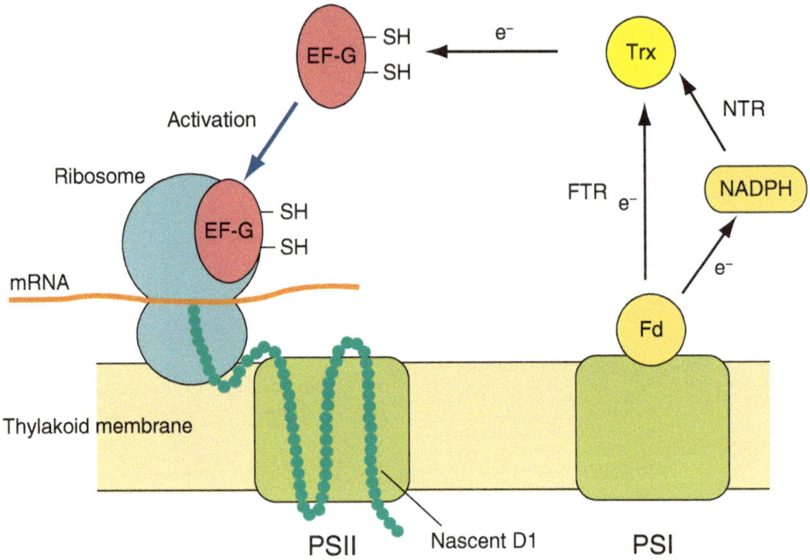

Nishiyama and Hisabori, Fig. 1. A model for the mechanism of the redox-regulation of the translational machinery in cyanobacteria. Reducing equivalents that are generated at PSI as a result of the photosynthetic electron transport are transmitted to EF-G in a Trx-mediated redox pathway. The resultant reduction of EF-G activates the light-dependent synthesis of proteins, such as the D1 protein. Thus, the photosynthetic machinery and the translational machinery are interconnected via redox regulation. Black arrows indicate the pathways of reducing equivalents. Fd, ferredoxin.

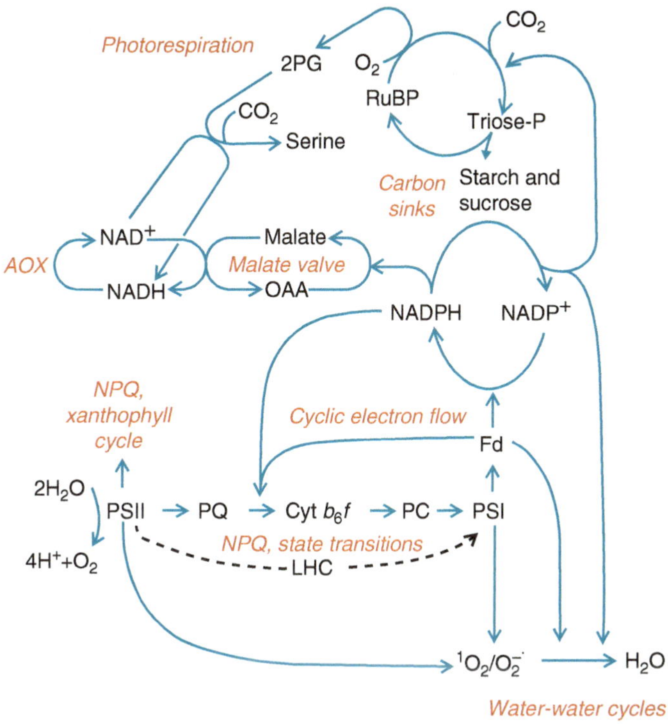

Hanke *et al.*, Fig. 2. Simplified picture showing redox balancing and buffering of photosynthetic electron flow (PET). Light excitation of photosystem II (PSII) enables the splitting of water to yield high-energy electrons and protons. These electrons are passed via plastoquinone (PQ) and the cytochrome b_6f complex (Cyt b_6f) to photosystem I (PSI), further generating the proton gradient used to drive ATP synthesis (not shown for reasons of clarity). Excited PSI donates electrons to ferredoxin (Fd), which can then be transferred to the NADPH that fuels triose phosphate (triose-P) generation in the Calvin cycle, eventually ending up in carbon sinks, such as starch and sucrose. Under conditions where high light, or other stress, causes PSII to generate oxidative species, the excess reductive pressure can be relieved by multiple pathways. Within PET, excess electrons can be released through NPQ. This phenomenon can include electron donation to the xanthophyll cycle and the movement of light-harvesting complexes (LHC) from PSII to PSI. Excitation of PSI then results in cyclic electron flow, either via ferredoxin (Fd) or NADPH. These soluble electron carriers are also used in reactions to remove excess oxidative species released from PSI and PSII by the Mehler reaction in water–water cycles. The ratio of $NADP^+$ to NADPH is exquisitely balanced in the chloroplast stroma by a number of compensatory pathways. NADPH can be oxidized in the Calvin cycle, where, depending on RuBisCO activity, it can fix either CO_2 or O_2 in photorespiration, or used to reduce oxaloacetate (OAA) to malate in the malate valve. Both photorespiration and the malate valve then result in export of the reduced products to other cellular compartments (not shown for clarity) to regenerate intermediates and remove excess reducing equivalents. A major contributor is the alternative oxidase of mitochondria (AOX), which can oxidize NADH without generating a proton gradient.

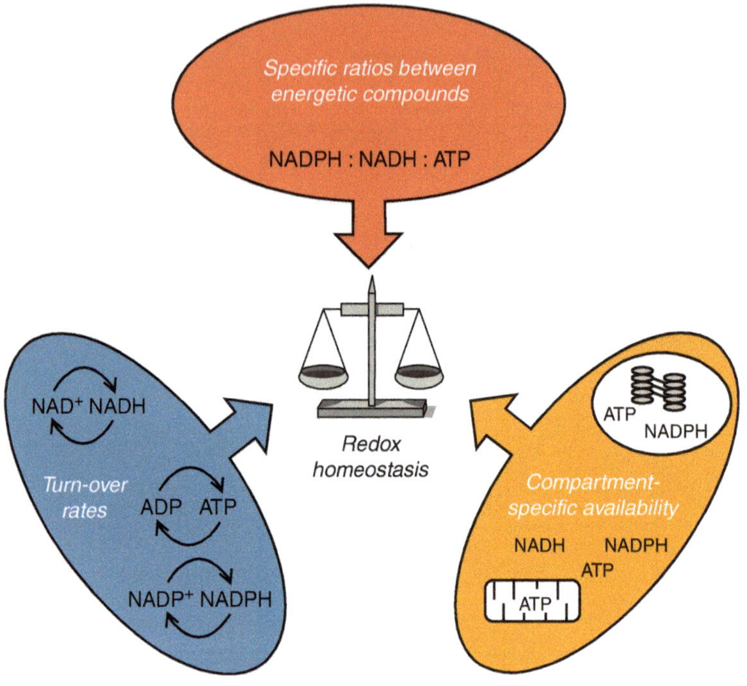

Hanke *et al.*, Fig. 3. Fine-tuning energy production and consumption, and the ratios between energy-carrying molecules. Energy generation and consumption in a plant cell needs to be tightly coupled with respect to species, turnover rate, ratio, and subcellular compartmentation of the energy carriers. The $NADP^+$/NADPH, NAD^+/NADH, and ADP/ATP ratios throughout the cell must be stably maintained to ensure acceptor molecules are available as sinks, and reductant and energy supply for metabolism is adequate. Energy-generating pathways produce these molecules in specific ratios, while demands for them in different cellular compartments can vary dramatically according to the metabolic demands of cell type, developmental stage, and environmental conditions. $NADP^+$/NADPH, NAD^+/NADH, and ADP/ATP ratios must be buffered to account for their rate of formation, consumption in specific compartments, and the ratio between the separate electron carriers and the adenylate pool.

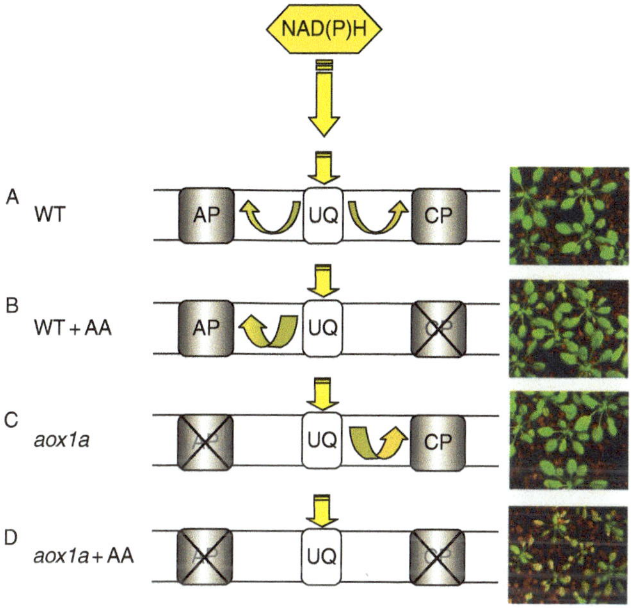

Hanke *et al.*, Fig. 4. Revealing a phenotype in mutants of the alternative oxidase (AOX1A) against a background of compensatory pathways. (A) Wild-type mitochondria possess two distinct pathways for the transfer of electrons from reduced ubiquinone to molecular oxygen: the cytochrome pathway (CP) via cytochrome oxidase (COX) that is linked to ATP production and the alternative pathway (AP) via alternative oxidase (AOX) without concomitant ATP production. (B) Under normal conditions, the loss of AOX1A by transgenic knockout (*aox1a*) does not yield an obvious phenotype, as it is compensated by the CP. (C) CP can be inhibited by antimycin a (AA), but on application to wild type this does not result in a stressed phenotype because increased AOX activity compensates for the inhibition. (D) After application of AA finally a phenotype for the *aox1a* mutant is visible due to elimination of the compensating CP resulting in profound stress, thus resolving the function of AOX in relieving oxidative stress.

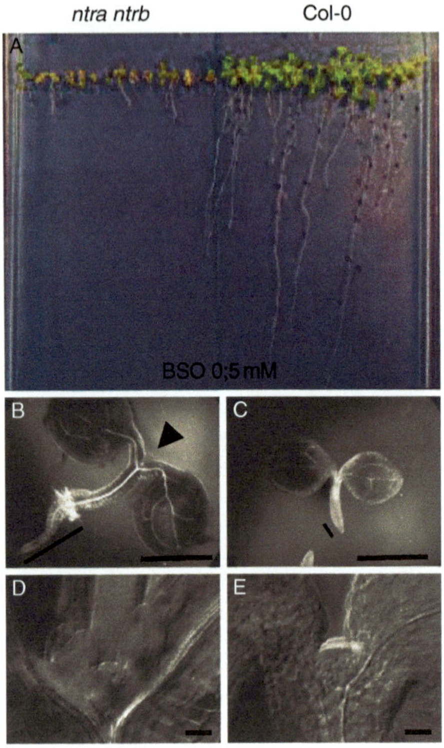

Reichheld *et al.*, Fig. 2. Hypersensitivity of the *ntra ntrb* mutant to glutathione biosynthesis inhibition. (A) Growth inhibition of the *ntra ntrb* mutant under the specific glutathione biosynthesis inhibitor BSO. Seeds were soaked on vertical MS/2 agar plates supplemented by 0.5 mM BSO and grown under a 16-h light/8-h dark regime. (B, C) Phenotypes of *rml1* (B) and *ntra ntrb rml1* (C) homozygous mutants at 8 days after germination on MS/2 medium. Leaf primordia are indicated by an arrow in (B). Approximate root length is indicated by black lines. Note that root hairs are emerging in *rml1* plants, in contrast with ntra ntrb rml1 plants. (D, E) Magnification of the shoot meristem of *rml1* (D) and *ntra ntrb rml1* (E) homozygous mutants at 8 days after germination on MS/2 medium. Note the absence of leaf primordia in *ntra ntrb rml1* (E) in contrast with *rml1* (D). Figure 2 is from Reichheld *et al.* (2007) with the permission of Plant Cell "Copyright American Society of Plant Biologists."

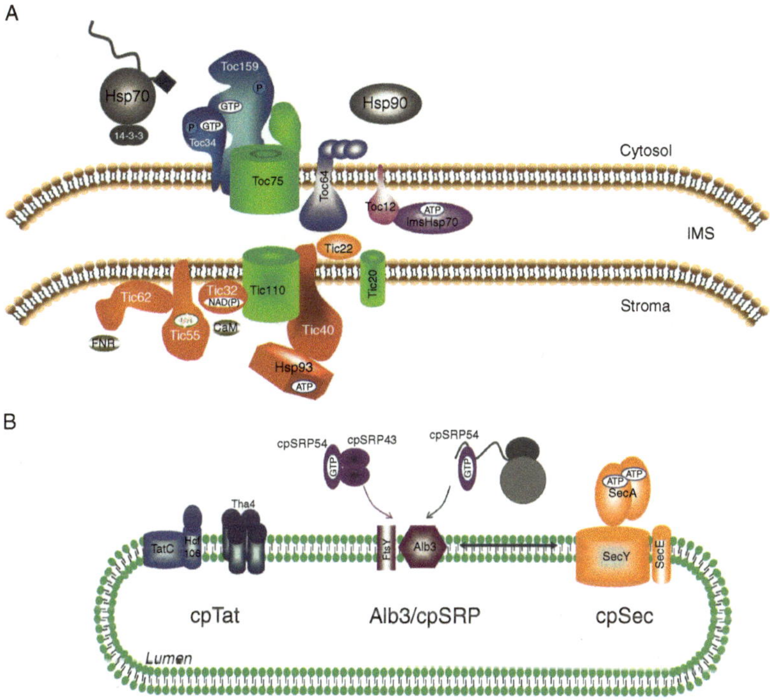

Balsera *et al.*, Fig. 1. Overview of protein machineries involved in protein import and routing in chloroplasts. (A) For translocation from the cytosol to the stroma, proteins that carry an N-terminal chloroplast transit peptide follow the general import pathway, involving the TOC and TIC complexes, respectively, at the outer and inner envelopes. (B) Proteins that are further transported to the thylakoid membrane or lumen follow different targeting pathways. Lumenal proteins are translocated through the thylakoids either by the cpSec or the cpTat machineries. Nuclear-encoded thylakoidal proteins are either inserted into the membrane posttranslationally by the cpSRP machinery or follow a mechanism of spontaneous insertion. Chloroplast-encoded thylakoidal proteins are cotranslationally inserted and folded into the membrane by the cpSRP system in cooperation with the cpSec machinery. The insertion of other thylakoidal proteins into the membrane is SRP-independent and seems to require only the cpSec machinery.

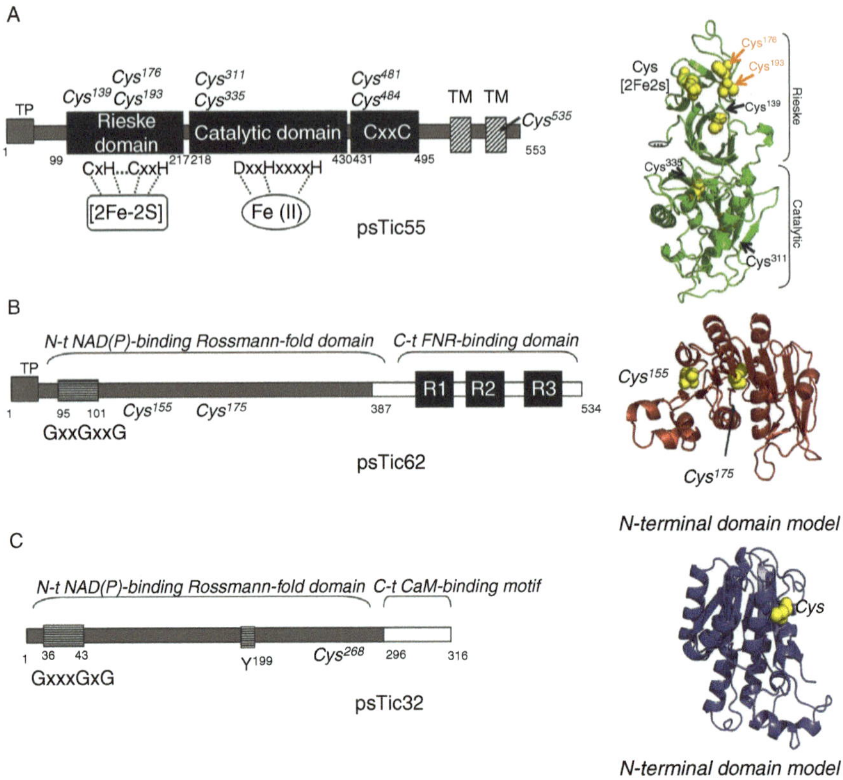

Balsera *et al.*, Fig. 6. Redox sensor proteins in the TIC complex. (A) Tic55 structural and functional module organization. TP and TM stand for transit peptide and transmembrane regions, respectively. Residues implicated in coordinating iron are indicated below the functional modules. The Tic55 sequence is composed of which Cys144 and Cys163 (not shown) with His146 and His166 participate in the coordination of the Rieske center. Two cysteines (Cys481-Cys484), which may form a disulfide bridge, are part of a soluble domain, the so-called CxxC domain, not found in other Rieske nonheme Fe proteins. The other cysteines are shown in the scheme. Tic55 has cyanobacterial homologs that present a similar modular organization (data not shown). Cys535 (in the second TM helix), Cys311 and Cys335 are the only conserved cysteine residues in plant Tic55 proteins. On the right, a homology model (using the protein structure with PDB code 1Z01 as template) is shown for the Rieske and catalytic domains. A circle with dots indicates where the protein sequence continues (CxxC domain and TM regions). Cysteines are represented in spacefill; (B) Tic62 bimodular organization and homology model prediction (Balsera *et al.*, 2007) of the N-terminal NAD(P)-binding domain. Tic62 sequence feature is a fully conserved GxxGxxG motif, involved in nucleotide coordination, and belongs to the SDR superfamily. The N-terminal catalytic domain is predicted to have a Rossmann-fold. Three repetitions (R1, R2 and R3) in pea Tic62, enriched in proline and serine residues, are involved in interaction with FNR in the stroma. Tic62 proteins have two conserved cysteines, shown in spacefill in the model on the right. Cys175 is in the protein core where the catalytic residues cluster and may participate in electron transfer reactions but likely not in a disulfide bridge formation; (C) Tic32 belongs also to the SDR superfamily, specifically to the tyrosine-dependent reductase family (tyrosine residue shown in the figure), and contains a fully conserved GxxxGxG motif. Tic32 contains a C-terminal extension involved in CaM interaction. A homology model has been obtained for the N-terminal reductase domain (using the protein structure with PDB code 2C07 as template). Just one conserved cysteine is found in Tic32 family of sequences.

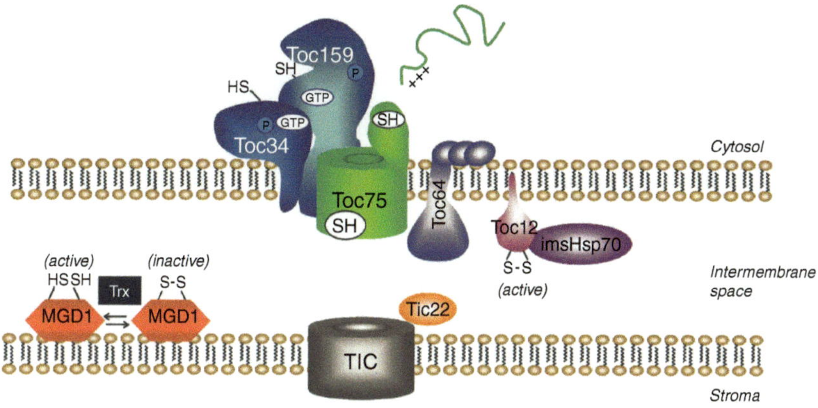

Balsera *et al.*, Fig. 7. Regulatory components in the outer membrane barrier and intermembrane space of chloroplasts. Different modes of regulation have been identified in the TOC complex: (i) modulation of the GTPase activities of the receptors; (ii) phosphorylation/dephosphorylation of the receptors; (iii) characteristic protein composition, according to the isoform of the multigene family present; (iv) intermolecular disulfide bridge formation involving cysteines of the TOC components. In the intermembrane space, oxidative folding machinery may exist for proteins that require the formation of disulfide bridges, like Toc12 (see text). Proteins, such as MGD1, contain disulfide bridges sensitive to thioredoxin, although no thioredoxin or thioredoxin-like protein has been described in this compartment.

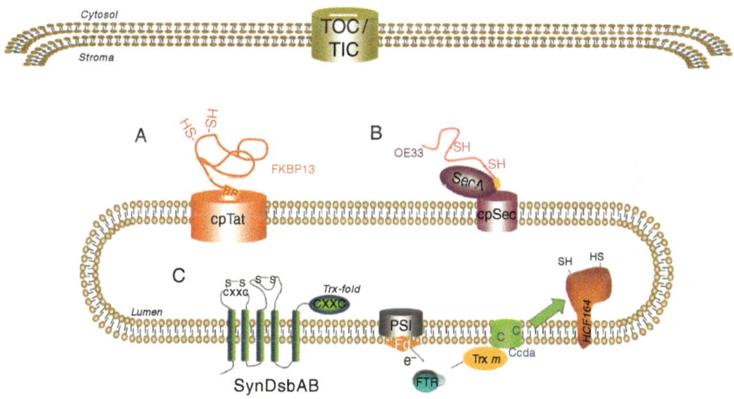

Balsera *et al.*, Fig. 10. Schematic view of transport of lumenal redox-active proteins through chloroplast thylakoids and enzymes possibly involved in redox regulation in the thylakoid lumen. Two lumenal redox-active proteins follow divergent pathways of translocation through the thylakoid membrane: (A) The immunophilin FKBP13 is transported folded via the cpTat pathway; (B) The extrinsic subunit of the oxygen evolving complex OE33 is kept unfolded during translocation by the cpSec machinery. Once in the lumen, both proteins require oxidation to achieve an active conformation; (C) Recent studies demonstrate the existence of a number of enzymes that may function to transport reducing equivalents from the stroma to the thylakoid lumen. HCF164, a protein bound to thylakoids, contains a lumenal thioredoxin-like domain, that has been proposed to accept reducing equivalents from stromal thioredoxin *m* via the polytopic membrane protein Ccda (cytochrome *c* defective A) (Motohashi and Hisabari, 2006). Further, SynDsbBA, a protein required for disulfide bond formation in the periplasmic space of bacteria, has been recently localized in thylakoids, adding a possible new participant in regulating thiol/disulfide formation of proteins within the lumen.

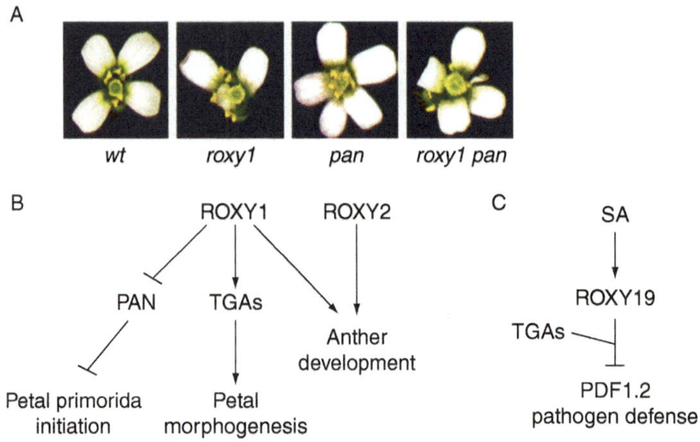

Li and Zachgo, Fig. 1. CC-type GRXs implicated in floral development and pathogen defense. (A) Comparison of petal phenotypes. In Arabidopsis wild-type (wt) flowers, four equally sized petals are formed. Petal number is reduced in the *roxy1* mutant and later petal morphogenesis is also disturbed. Floral organ numbers in the *pan* mutant are increased in whorls 1 and 2, such that a pentamerous pattern is formed. In *roxy1 pan* double mutants, similar to the *pan* single mutant, flowers develop a pentamerous phenotype in the outer whorls. Thus, *PAN* is epistatic to *ROXY1* with respect to its function of regulating petal primorida initiation. However, the pentamerous double mutant also displays an abnormal later petal morphogenesis, such as the formation of smaller or folded petals as observed in the *roxy1* mutant. (B) Regulatory interactions between ROXYs and TGAs in flower development. As revealed by the *pan* mutant phenotype, *PAN* represses petal primordia initiation. At early stages of petal development, ROXY1 controls petal primordia initiation probably by negatively regulating PAN activity. Additionally, acting independently from PAN, ROXY1 also affects later petal morphogenesis. ROXY1 likely positively modulates the activity of further TGA factors acting together in later petal differentiation. Furthermore, *ROXY1* and its closest homolog *ROXY2* act redundantly during anther development. (C) Regulatory interactions between ROXY19 and TGAs in pathogen defense. SA-inducible ROXY19 interacts with TGA factors and suppresses JA-responsive *PDF1.2* transcription. Arrows and hatchets denote positive and negative regulation, respectively.

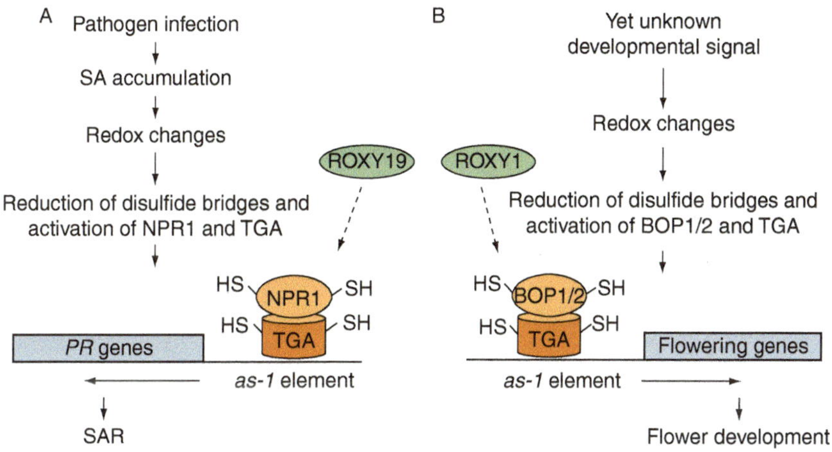

Li and Zachgo, Fig. 2. Signaling mechanisms involved in disease resistance and flower development employ similar protein molecules. (A) Upon pathogen attack, cellular redox changes cause reduction of inter- and intramolecular disulfide bonds of NPR1 and TGA proteins, respectively. Subsequently, active NPR1/TGA protein complexes can bind to *as-1* elements of target genes, thereby inducing expression of *PR* genes and establishment of SAR. ROXY19 interacts with and thus likely post-translationally modifies TGA factors, as indicated by the dashed arrow. (B) Similarly, redox changes caused by a yet unknown developmental signal might lead to the formation of active BOP/TGA protein complexes, which might in turn bind to *as-1* elements of flowering genes and activate transcription of flowering genes. Nuclear interactions of ROXY1 with TGA factors were shown to be crucial and indispensable for normal petal development (dashed arrow). Interestingly, ROXY19 can replace ROXY1 activity if appropriately expressed. Thus, ROXY19 can exert the same nuclear activity as ROXY1 in flower development.

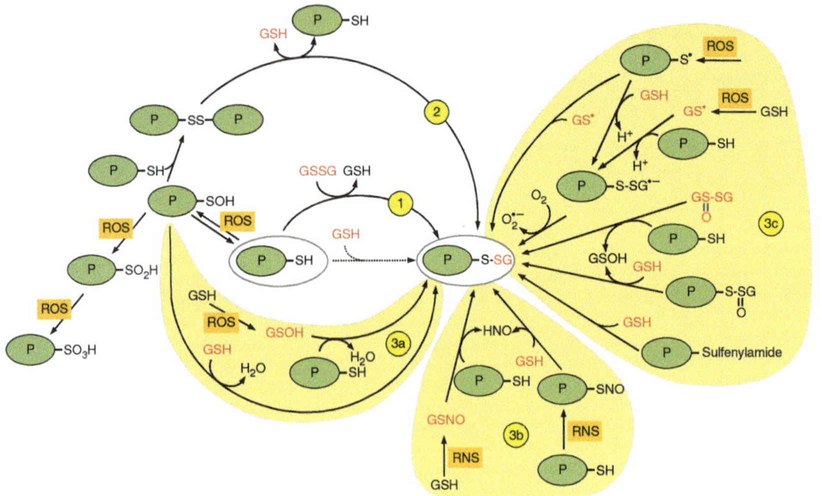

Gao *et al.*, Fig. 3. Potential mechanisms of protein glutathionylation. Several mechanisms can theoretically trigger protein glutathionylation. Glutathionylation can occur via a thiol/disulfide exchange between the protein and oxidized glutathione (GSSG) (1); by direct oxidation of both protein and reduced glutathione (GSH) (low probability, dotted line) or via reaction with intermolecular disulfide bridges (2). Glutathionylation can also involve activated thiol derivatives such as sulfenic acids (–SOH) (3a), *S*-nitrosylated thiols (–SNO) (3b) or thiyl radicals ($-S^{\cdot}$), thiosulfinates (–S(O)SG), and sulfenylamides (3c). See Section II. for further details.

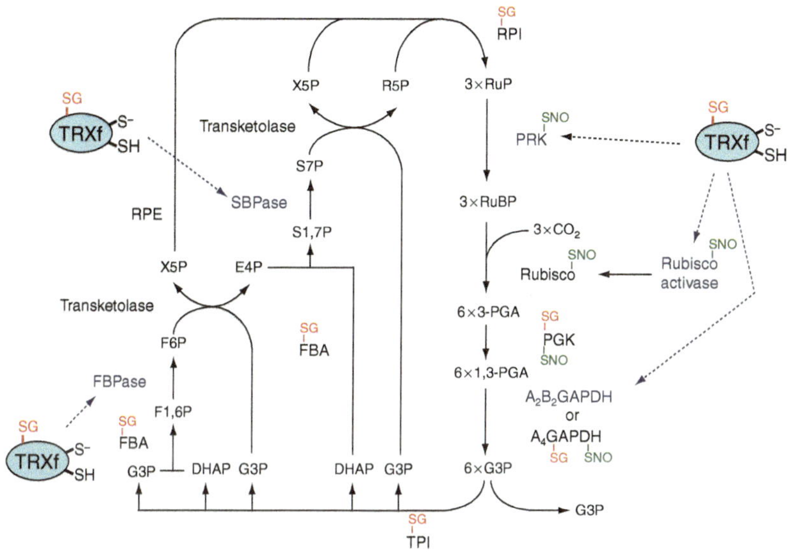

Gao *et al.*, Fig. 7. Redox regulation of Calvin cycle enzymes. Most Calvin cycle enzymes have been shown to undergo one or several post-translational redox modifications: oxidoreduction of disulfide bridges by thioredoxin f (dotted arrows pointed at established TRX targets indicated in blue), glutathionylation (–SG in red), and/or nitrosylation (–SNO in green). Enzymes (green): Rubisco, ribulose-1,5-bisphosphate carboxylase/oxygenase; PGK, phosphoglycerate kinase; GAPDH, glyceraldehyde-3-phosphate dehydrogenase; TPI, triose phosphate isomerase; FBA, fructose-1,6-bisphosphate aldolase; FBPase, fructose-1,6-bisphosphatase; SBPase, sedoheptulose-1,7-bisphosphatase; RPE, ribulose-5-phosphate epimerase; RPI, ribose-5-phosphate isomerase; PRK, phosphoribulokinase. Metabolites: RuBP, ribulose-1,5-bisphosphate; 3-PGA, 3-phosphoglycerate; 1,3-BPGA, 1,3-bisphosphoglycerate; G3P, glyceraldehyde-3-phosphate; DHAP, dihydroxyacetone phosphate; F1,6P, fructose-1,6-bisphosphate; F6P, fructose-6-phosphate; X5P, xylulose-5-phosphate; E4P, erythrose-4-phosphate; S1,7BP, sedoheptulose-1,7-bisphosphate; S7P, sedoheptulose-7-phosphate; R5P, ribose-5-phosphate; RuP, ribulose-5-phosphate.

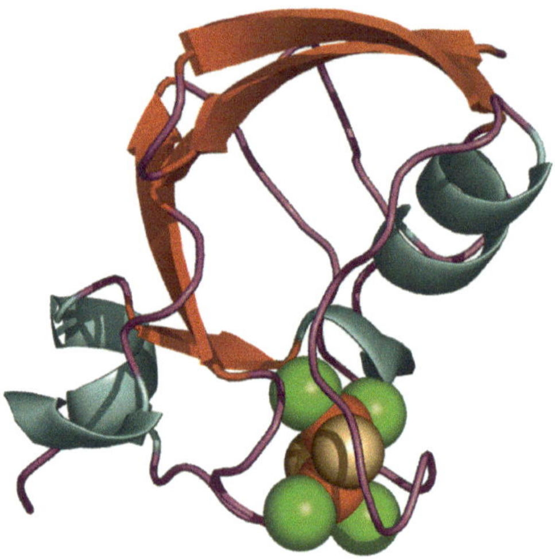

Selles *et al.*, Fig. 1. Crystal structure of the E92K mutant of [2Fe–2S] ferredoxin I from *Spinacia oleracea* (Binda *et al.*, 1998; pdb accession number 1A70). Sulfur atoms belonging to cysteine residues 38, 43, 46, and 76 are represented by green spheres, iron atoms by red, and labile sulfur atoms by yellow.

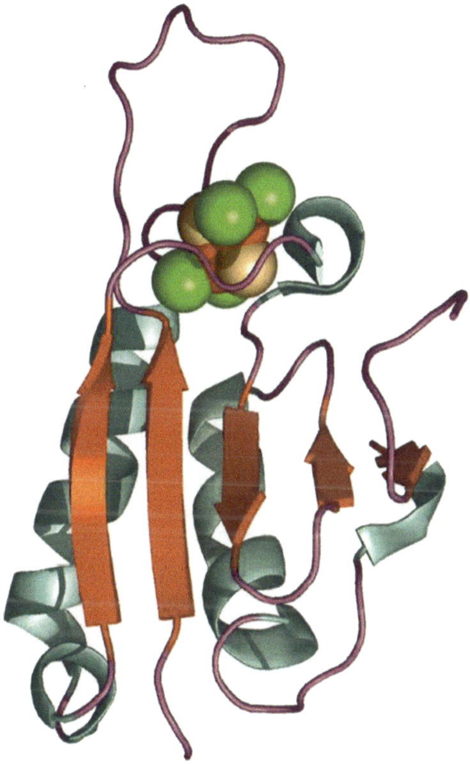

Selles *et al.*, Fig. 2. Crystal structure of a thioredoxin-like [2Fe–2S] ferredoxin from *Aquifex aeolicus* (Yeh *et al.*, 2000; pdb 1F37). Sulfur atoms belonging to cysteine residues 9, 22, 55, and 59 are represented by green spheres, iron atoms by red and labile sulfur atoms by yellow.

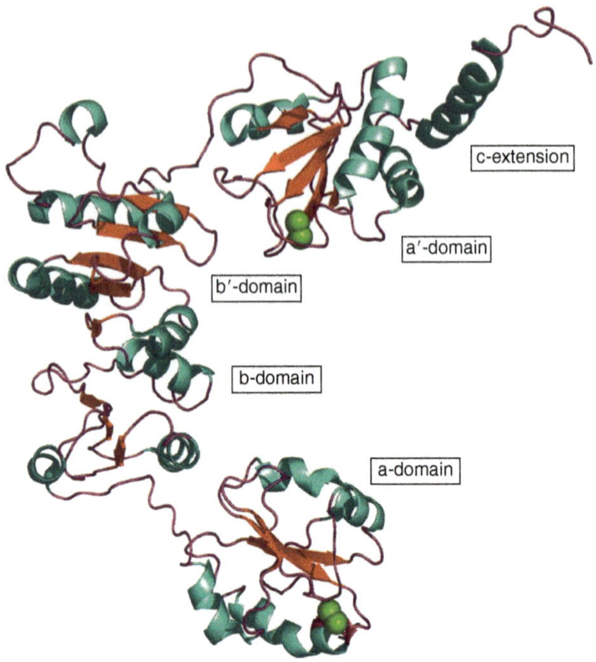

Selles *et al.*, Fig. 4. Crystal structure of protein disulfide isomerase *Sc*PDI1p from Baker's yeast *Saccharomyces cerevisiae* (Tian *et al.*, 2006; pdb accession number 3BOA). Active Trx domains are named a and a′, inactive Trx domain b and b′ and acidic C-terminal extension c. Sulfur atoms belonging to cysteines 43, 46, 388, and 391 are represented by green spheres.

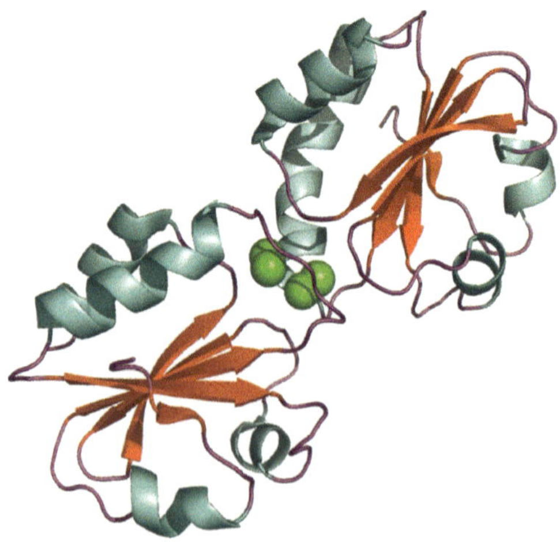

Selles *et al.*, Fig. 6. Crystal structure of the *E. coli* Trx A CACA mutant dimer (Collet *et al.*, 2005; pdb 1ZCP). Sulfur atoms belonging to protein cysteine residues 32 and 34 are shown as green spheres.

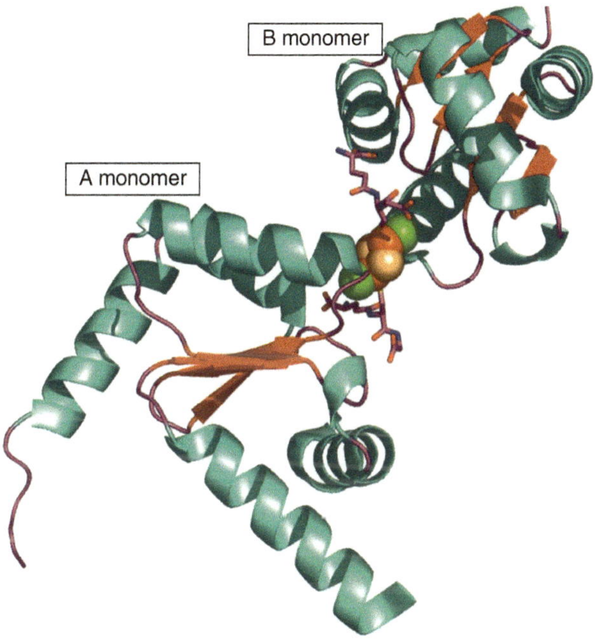

Selles *et al.*, Fig. 7. Crystal structure of the holo form of glutaredoxin C1 from *Populus Tremula X Tremuloides* (Rouhier *et al.*, 2007b; pdb 2E7P). Sulfur atoms belonging to protein cysteine residues 30 and 33 of each monomer (A and B) are shown as green spheres, iron atoms as red, labile sulfur atoms as yellow and glutathione in stick.

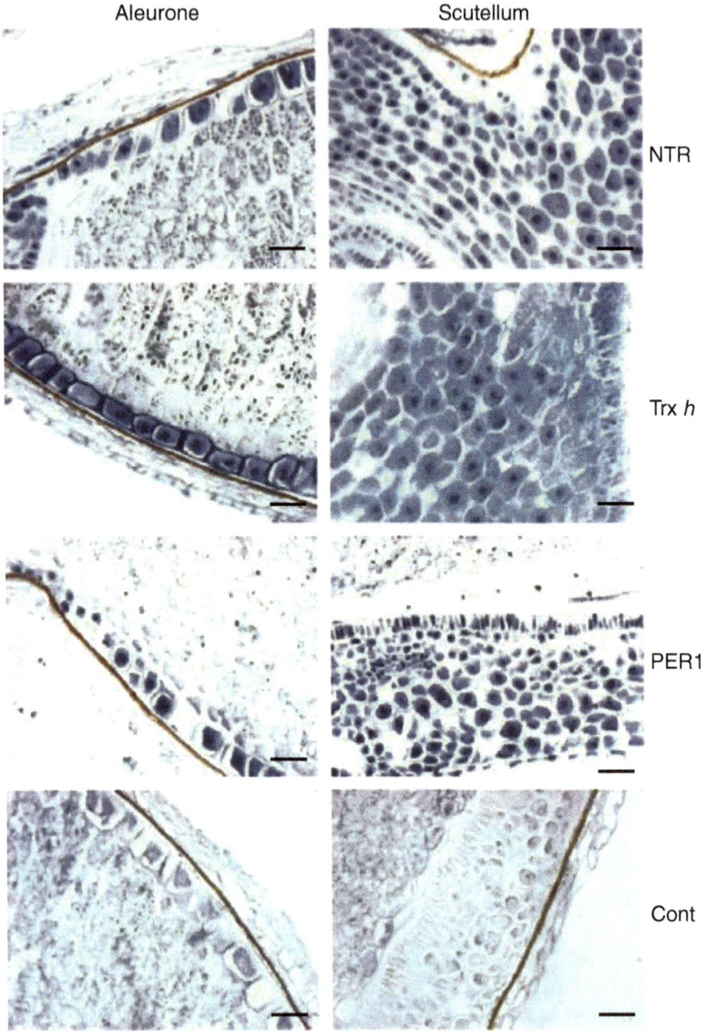

Pulido *et al.*, Fig. 2. Immunocytochemical analysis of 1-Cys Prx (PER1), NTR, and Trx *h* in aleurone and scutellum cells of developing wheat seeds. Sections (10 μm) of developing wheat seeds harvested at 25 days post anthesis were probed with anti-PER1, anti-NTR or anti-Trx *h* antibodies (as indicated) or pre-immune serum (Cont). Images show details of the aleurone layer and the scutellum cells. These three proteins show a similar pattern of localization in the two types of wheat seed cells that suffer oxidative stress, hence showing the presence of an antioxidant system in the nucleus of these cells. Bar, 50 μm.

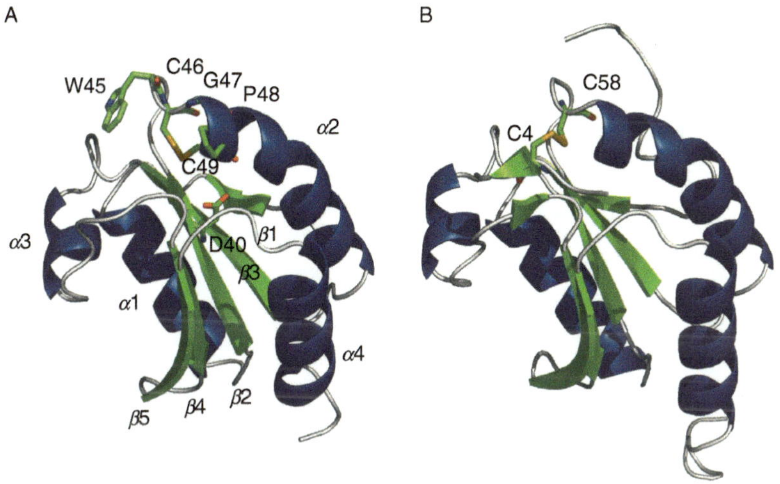

Hägglund *et al.*, Fig. 5. (A) The structure of *Hv*Trxh2 from barley (h-type Trx from subclass I) in the oxidized form (Maeda *et al.*, 2008). The two conserved redox-active cysteines in the $_{45}$WCGPC$_{49}$ motif are connected by a disulfide bond. D40 corresponds to the aspartic acid residue proposed to act as a general acid/base during catalysis. β-Strands and α-helices are labelled. (B) The structure of a $C_C \rightarrow S$ mutant of *Pt*Trxh4 from poplar (h-type Trx from subgroup III) in the oxidized form (Koh *et al.*, 2008). A disulfide bond is formed between C58 (C_N) and C4 (C_{NT}) conserved in the N-terminal part of Trx-h from subgroup III.

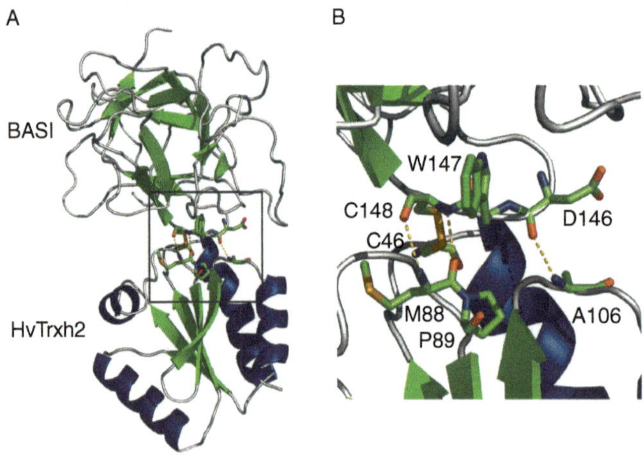

Hägglund *et al.*, Fig. 8. The crystal structure of *Hv*Trxh2-S-S-BASI (Maeda *et al.*, 2006). (i) Overall structure of *Hv*Trxh2-S-S-BASI in cartoon display with β-strands and α-helices in green and blue, respectively. (ii) Close up view of the interaction between *Hv*Trxh2 and BASI. The segment $_{146}$DWC$_{148}$ of BASI and C$_{46}$, $_{88}$MP$_{89}$ and A$_{106}$ of *Hv*Trxh2 are shown in stick representation. Carbon, oxygen, nitrogen and sulphur are shown in green, red, blue and yellow, respectively. Intermolecular hydrogen bonds are shown as dashed yellow lines.

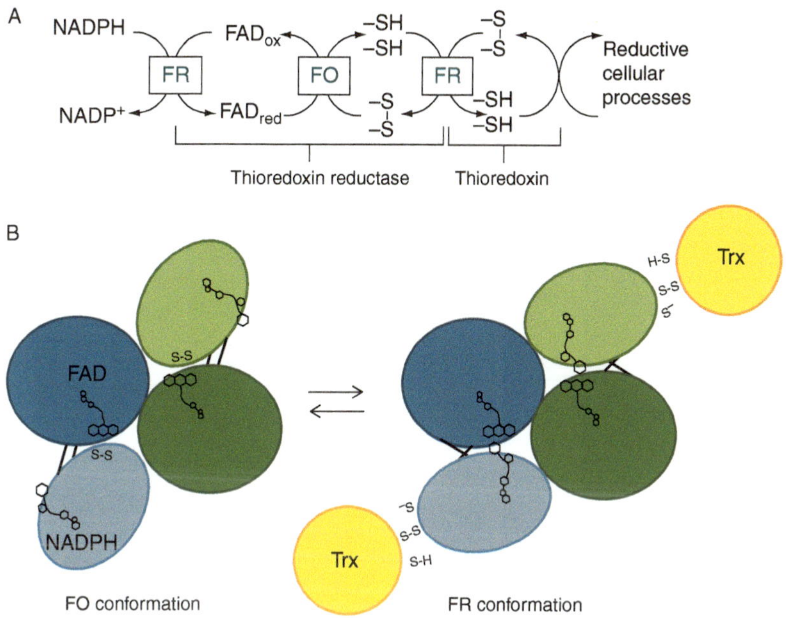

Hägglund *et al.*, Fig. 9. (A) The catalytic mechanism of NTR. Reducing equivalents are transferred from NADPH to a tightly bound FAD in NTR. From FAD, the electrons are transferred to a CXXC in the enzyme and subsequently to the oxidized target CXXC disulfide in Trx. In order to catalyze the entire reaction, NTR needs to shift between two conformations, the FO and FR conformation. The electron transfers linked to each conformation are indicated. (B) The two conformations as proposed by Waksman *et al.* (1994) are shown schematically. The two subunits of the NTR dimer are shown in blue and green, respectively. The darker coloured ovals symbolize the FAD domains, while the lighter coloured show the NADPH domains. Disulfides and dithiols are indicated as S–S and S–H, respectively. The black lines connecting the two domains symbolize the anti-parallel β-sheet around which a 66° rotation occurs to bring NTR from the FO to the FR conformation. The rotation position of the nicotinamide ring in proximity of the flavin ring system and the CXXC dithiol is brought to the surface of the protein, where it can interact with thioredoxin (shown in yellow). Adapted from Waksman *et al.* (1994).

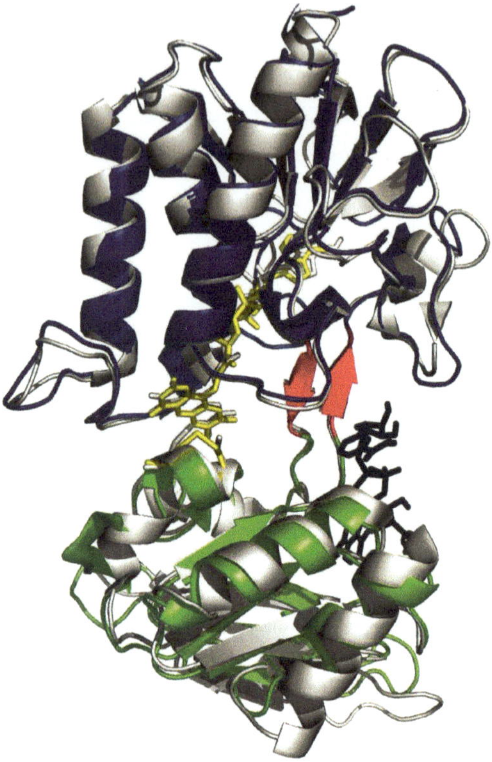

Hägglund *et al.*, Fig. 10. Superposition of one subunit of the functionally dimeric NTRs from *A. thaliana* (pdb: 1vdc, shown in blue/green) and *E. coli* (shown in white, pdb: 1tdf). The enzymes consist of two domains with similar Rossman-folds. One domain (blue) binds FAD (yellow sticks) and the other domain (green) binds NADPH. The two domains are connected by two anti-parallel β-strands (pink). The structures are in the FO conformation, having FAD in close proximity of the redox-active CXXC motif (yellow sticks) from the NADPH domain. The *E. coli* structure was crystallized in the presence of $NADP^+$ (black).

Kolibri 360 GmbH
Ferdinand-Jühlke-Straße 7
99095 Erfurt, Deutschland
produktsicherheit@kolibri360.de